Environmental Science

Environmental Science
Working with the Earth
ELEVENTH EDITION

G. TYLER MILLER, JR.

President, Earth Education and Research

THOMSON

BROOKS/COLE

Australia • Brazil • Canada • Mexico • Singapore • Spain • United Kingdom • United States

THOMSON

BROOKS/COLE

Publisher: *Jack Carey*
Developmental Editor: *Scott Spoolman*
Production Project Manager: *Andy Marinkovich*
Technology Project Manager: *Keli Amann*
Assistant Editor: *Carol Benedict*
Editorial Assistant: *Chris Ziemba*
Permissions Editor/Photo Researcher: *Abigail Reip*
Marketing Manager: *Ann Caven*
Marketing Assistant/Associate: *Brian Smith*
Advertising Project Manager: *Nathaniel Bergson-Michelson*
Print/Media Buyer: *Karen Hunt*

Production Management, Copyediting, and Composition: *Thompson Steele, Inc.*
Interior Illustration: *Precision Graphics; Sarah Woodward; Darwin and Vally Hennings; Tasa Graphic Arts, Inc.; Alexander Teshin Associates; John and Judith Walker; Rachel Ciemma; Victor Royer, Electronic Publishing Services, Inc.; J/B Woosley Associates; Kerry Wong; ScEYEnce*
Cover Image: *Creek in deciduous woodland (near Sedona, Arizona), James Nelson/Getty Images, Inc.*
Text Printer: *Courier Corporation/Kendallville*
Cover Printer: *Phoenix Color Corp.*
Title Page Photograph: *Salt Marsh in Peru, SuperStock*

For more information about our products, contact us at:
Thomson Learning Academic Resource Center
1-800-423-0563
For permission to use material from this text, contact us by:
Phone: 1-800-730-2214
Fax: 1-800-730-2215
Web: http://www.thomsonrights.com

Library of Congress Control Number: 2005926442

Student Edition (softcover): ISBN 0-534-42250-0
Student Edition (hardcover): ISBN 0-495-01475-3
Annotated Instuctor's Edition: ISBN: 0-534-42251-9

Thomson Higher Education
10 Davis Drive
Belmont, CA 94002-3098
USA

Asia (including India)
Thomson Learning
5 Shenton Way #01-01
UIC Building
Singapore 068808

Australia
Nelson Thomson Learning
102 Dodds Street
South Melbourne, Victoria 3205
Australia

Canada
Nelson Thomson Learning
1120 Birchmount Road
Toronto, Ontario M1K 5G4
Canada

Europe/Middle East/Africa
Thomson Learning
High Holborn House
50/51 Bedford Row
London WC1R 4LR
United Kingdom

Latin America
Thomson Learning
Seneca, 53
Colonia Polanco
11560 Mexico D.F.
Mexico

Spain
Paraninfo Thomson Learning
Calle/Magallanes, 25
28015 Madrid
Spain

Brief Contents

Detailed Contents   vii

Preface   xv

Learning Skills   1

HUMANS AND SUSTAINABILITY: AN OVERVIEW

1   Environmental Problems, Their Causes, and Sustainability   5

ECOLOGY AND SUSTAINABILITY

2   Science, Matter, and Energy   19

3   Ecosystems: What Are They and How Do They Work?   35

4   Evolution and Biodiversity   63

5   Climate and Biodiversity   78

6   Community Ecology, Population Ecology, and Sustainability   108

7   Applying Population Ecology: The Human Population   128

SUSTAINING BIODIVERSITY

8   Sustaining Biodiversity: The Ecosystem Approach   154

9   Sustaining Biodiversity: The Species Approach   183

SUSTAINING RESOURCES AND ENVIRONMENTAL QUALITY

10   Food, Soil, and Pest Management   206

11   Water and Water Pollution   236

12   Geology and Nonrenewable Minerals   269

13   Energy   285

14   Risk, Human Health, and Toxicology   327

15   Air Pollution   345

16   Climate Change and Ozone Loss   367

17   Solid and Hazardous Waste   388

SUSTAINING HUMAN SOCIETIES

18   Environmental Economics, Politics, and Worldviews   412

Science Supplements   S1

Glossary   G1

Index   I1

Detailed Contents

NASA

Learning Skills   1

HUMANS AND SUSTAINABILITY:
AN OVERVIEW

1   Environmental Problems,
    Their Causes, and
    Sustainability   5

    Case Study: Living in an Exponential
    Age   5

1-1   Living More Sustainably   6

1-2   Population Growth. Economic Growth
      and Economic Development   8

1-3   Resources   10

      Economics Case Study: The Tragedy
      of the Commons   10

1-4   Pollution   12

1-5   Environmental Problems: Causes and
      Connections   13

1-6   Cultural Changes and Sustainability   16

1-7   Is Our Present Course Sustainable?   17

ECOLOGY AND
SUSTAINABILITY

2   Science, Matter, and
    Energy   19

    Case Study: An Environmental Lesson
    from Easter Island   19

2-1   The Nature of Science   20

      Science Spotlight: What Is Harming
      the Robins?   21

2-2   Matter   22

2-3   Energy   29

2-4   Matter and Energy Change Laws and
      Sustainability   32

3   Ecosystems: What Are
    They and How Do They
    Work?   35

    Case Study: Have You Thanked the Insects
    Today?   35

3-1   The Nature of Ecology   36

      Science Spotlight: Which Species Rules
      the World?   36

3-2   The Earth's Life-Support Systems   38

Ray Pfortner/Peter Arnold, Inc.

Point-source air pollution from a pulp mill in New York State

3-3 Ecosystem Components 40

3-4 Energy Flow in Ecosystems 46

3-5 Soils 50

3-6 Matter Cycling in Ecosystems 53

3-7 How Do Ecologists Learn about Ecosystems? 60

4 Evolution and Biodiversity 63

Case Study: How Did We Become Such a Powerful Species So Quickly? 63

4-1 Origins of Life 64

4-2 Evolution and Adaptation 65

4-3 Ecological Niches and Adaptation 67

Science Spotlight: *Cockroaches: Nature's Ultimate Survivors* 69

4-4 Speciation, Extinction, and Biodiversity 70

4-5 What Is the Future of Evolution? 74

5 Climate and Biodiversity 78

Case Study: Blowing in the Wind: A Story of Connections 78

5-1 Climate: A Brief Introduction 79

5-2 Biomes: Climate and Life on Land 83

Coral reef in the Red Sea

5-3 Desert and Grassland Biomes 85

5-4 Forest and Mountain Biomes 89

5-5 Aquatic Environments: Types and Characteristics 95

5-6 Saltwater Life Zones 96
Case Study: Coral Reefs 100

5-7 Freshwater Life Zones 104

6 Community Ecology, Population Ecology, and Sustainability 108

Case Study: Why Should We Care about the American Alligator? 108

6-1 Community Structure and Species Diversity 109

6-2 Types of Species 111

Case Study: Why Are Amphibians Vanishing? 111

Case Study: Why Are Sharks Important Species? 113

6-3 Species Interactions 114

6-4 Ecological Succession: Communities in Transition 118

6-5 Population Dynamics and Carrying Capacity 120

6-6 Human Impacts on Ecosystems: Learning from Nature 123

Connections: *Ecological Surprises* 125

A white ukari in a Brazilian tropical forest

7   Applying Population Ecology: The Human Population   128

Case Study: Is the World Overpopulated?   128

7-1   Factors Affecting Human Population Size   129

Case Study: Fertility Rates in the United States   131

Economics and Politics Case Study: U.S. Immigration   134

7-2   Population Age Structure   134

7-3   Solutions: Influencing Population Size   137

7-4   Slowing Population Growth in India and China   139

Case Study: India   139

Case Study: China   140

7-5   Population Distribution: Urbanization and Urban Growth   140

Case Study: Urbanization in the United States   142

7-6   Urban Resource and Environmental Problems   144

Economics Case Study: The Urban Poor in Developing Countries   146

Connections: *How Can Reducing Crime Help the Environment?*   147

7-7   Transportation and Urban Development   148

Case Study: Motor Vehicles in the United States   148

7-8   Making Urban Areas More Livable and Sustainable   150

Comeback of the endangered American alligator

Crowded streets in China

Case Study: Curitiba, Brazil—One of the World's Most Sustainable Major Cities   152

SUSTAINING BIODIVERSITY

8   Sustaining Biodiversity: The Ecosystem Approach   154

Case Study: Reintroducing Wolves to Yellowstone   154

8-1   Human Impacts on Biodiversity   155

8-2   Public Lands in the United States   157

8-3   Managing and Sustaining Forests   159

8-4   Forest Resources and Management in the United States   165

Individuals Matter: *Butterfly in a Redwood Tree*   166

8-5   Tropical Deforestation   169

Individuals Matter: *Kenya's Green Belt Movement*   172

8-6   National Parks   172

Case Study: Stresses on U.S. National Parks   172

8-7   Nature Reserves   173

Science Case Study: Costa Rica—A Global Conservation Leader   174

Science and Politics Case Study: Wilderness Protection in the United States 176

8-8 Ecological Restoration 177

Science Case Study: Ecological Restoration of a Tropical Dry Forest in Costa Rica 178

8-9 Sustaining Aquatic Biodiversity 179

8-10 What Can We Do? 181

9 Sustaining Biodiversity: The Species Approach 183

Case Study: The Passenger Pigeon: Gone Forever 183

9-1 Species Extinction 184

9-2 Importance of Wild Species 188

9-3 Causes of Premature Extinction of Wild Species 189

Science Case Study: A Disturbing Message from the Birds 190

Science Case Study: Deliberate Introduction of the Kudzu Vine 194

Economics Case Study: The Rising Demand for Bushmeat in Africa 196

9-4 Protecting Wild Species: The Legal Approach 198

Case Study: What Has the Endangered Species Act Accomplished? 201

9-5 Protecting Wild Species: The Sanctuary Approach 202

9-6 Reconciliation Ecology 203

Science Spotlight: *Using Reconciliation Ecology to Protect Bluebirds* 204

Overgrazed rangeland (left) and lightly grazed rangeland (right)

U.S. Department of Agriculture, Natural Resources Conservation Service

SUSTAINING RESOURCES AND ENVIRONMENTAL QUALITY

10 Food, Soil, and Pest Management 206

Case Study: Would You Eat Winged Beans and Bug Cuisine? 206

10-1 Food Production 207

Science and Economics Case Study: Industrial Food Production in the United States 209

10-2 Soil Erosion and Degradation 211

Science Case Study: Soil Erosion in the United States 213

10-3 Sustainable Agriculture through Soil Conservation 216

10-4 Food Production, Nutrition, and Environmental Effects 217

10-5 Increasing Food Production 220

Science Case Study: Some Environmental Consequences of Meat Production 223

10-6 Protecting Food Resources: Pest Management 227

Individuals Matter: *Rachel Carson* 229

Science Spotlight: *How Successful Have Synthetic Pesticides Been in Reducing Crop Losses in the United States?* 230

Connections: *What Goes Around Can Come Around* 231

10-7 Solutions: Sustainable Agriculture 234

Endangered ring-tailed lemur in Madagascar

Women carrying firewood in India

11 Water and Water Pollution 236

Case Study: Water Conflicts in the Middle East 236

11-1 Water's Importance, Use, and Renewal 237

Science Case Study: Freshwater Resources in the United States 239

11-2 Supplying More Water 240

Politics and Ethics Case Study: Who Should Own and Manage Freshwater Resources? 241

Science Case Study: The Aral Sea Disaster 244

11-3 Reducing Water Waste 248

11-4 Too Much Water 251

Science and Poverty Case Study: Living on Floodplains in Bangladesh 252

11-5 Water Pollution: Types, Effects, and Sources 254

11-6 Pollution of Freshwater Streams, Lakes, and Aquifers 255

11-7 Ocean Pollution 259

Science Case Study: The Chesapeake Bay 261

11-8 Preventing and Reducing Surface Water Pollution 263

Science Case Study: Using Wetlands to Treat Sewage 265

11-9 Drinking Water Quality 266

12 Geology and Nonrenewable Minerals 269

Case Study: The General Mining Law of 1872 269

12-1 Geologic Processes 270

12-2 Internal and External Geologic Processes 271

12-3 Minerals, Rocks, and the Rock Cycle 274

12-4 Finding, Removing, and Processing Nonrenewable Minerals 276

12-5 Environmental Effects of Using Mineral Resources 278

12-6 Supplies of Mineral Resources 280

Science Case Study: Using Nanotechnology to Produce New Materials 282

13 Energy 285

Case Study: The Coming Energy-Efficiency and Renewable-Energy Revolution 285

13-1 Evaluating Energy Resources 286

13-2 Nonrenewable Fossil Fuels 290

Science, Economics, and Politics Case Study: How Much Oil Does the United States Have? 291

13-3 Nonrenewable Nuclear Energy 298

Science Case Study: The Chernobyl Nuclear Power Plant Accident 300

Science and Politics Case Study: High-Level Radioactive Wastes in the United States 303

13-4 Improving Energy Efficiency 306

Creek in Montana polluted by acidic mining wastes

Solar cells in a remote village in Niger, Africa

Connections: *Economics and Politics: The Real Cost of Gasoline in the United States*   308

13-5   Using Renewable Energy to Provide Heat and Electricity   312

13-6   Geothermal Energy   321

13-7   Hydrogen   322

Science Spotlight: *Producing Hydrogen from Green Algae Found in Pond Scum*   323

13-8   A Sustainable Energy Strategy   324

14   Risk, Human Health, and Toxicology   327

Case Study: The Big Killer   327

14-1   Risks and Hazards   328

14-2   Biological Hazards: Disease in Developed and Developing Countries   328

Science Case Study: Growing Germ Resistance to Antibiotics   329

Science Case Study: The Growing Global Threat from Tuberculosis   329

Science Case Study: HIV and AIDS   330

Science Case Study: Malaria   331

14-3   Chemical Hazards   334

14-4   Toxicology: Assessing Chemical Hazards   335

14-5   Risk Analysis   340

15   Air Pollution   345

Case Study: When Is a Lichen Like a Canary?   345

15-1   Structure and Science of the Atmosphere   346

15-2   Outdoor Air Pollution   347

15-3   Photochemical and Industrial Smog   348

Science Spotlight: *Air Pollution in the Past: The Bad Old Days*   350

15-4   Regional Outdoor Air Pollution from Acid Deposition   352

15-5   Indoor Air Pollution   357

Science Case Study: Exposure to Radioactive Radon Gas   358

15-6   Harmful Effects of Air Pollution   359

15-7   Preventing and Reducing Air Pollution   361

Economics Case Study: Using the Marketplace to Reduce Air Pollution?   362

16 Climate Change and Ozone Loss 367

Case Study: Studying a Volcano to Understand Climate Change 367

16-1 Past Climate Change and the Natural Greenhouse Effect 368

16-2 Climate Change and Human Activities 370

16-3 Factors Affecting the Earth's Temperature 374

16-4 Possible Effects of a Warmer World 375

16-5 Dealing with the Threat of Global Warming 378

16-6 Ozone Depletion in the Stratosphere 383

Science Case Study: Skin Cancer 384

16-7 Protecting the Ozone Layer 386

Individuals Matter: *Ray Turner and His Refrigerator* 386

17 Solid and Hazardous Waste 388

Case Study: Love Canal: There Is No "Away" 388

17-1 Wasting Resources 389

Case Study: Living in a High-Waste Society 390

17-2 Producing Less Waste 390

17-3 The Ecoindustrial Revolution and Selling Services Instead of Things 392

Individuals Matter: *Ray Anderson* 394

17-4 Reuse 394

17-5 Recycling 396

Science and Economics Case Study: Problems With Recycling Plastics 397

17-6 Burning and Burying Solid Waste 398

Global warming may submerge this low-lying island in the Maldives

17-7 Hazardous Waste 401

Science, Economics, and Ethics Case Study: A Black Day in Bhopal, India 403

17-8 Toxic Metals 406

Science Spotlight: *Lead* 406

Science Spotlight: *Mercury* 408

17-9 Achieving a Low-Waste Society 409

SUSTAINING HUMAN SOCIETIES

18 Environmental Economics, Politics, and Worldviews 412

Case Study: Biosphere 2: A Lesson in Humility 412

18-1 Economic Systems and Sustainability 413

18-2 Using Economics to Improve Environmental Quality 416

18-3 Reducing Poverty to Improve Environmental Quality and Human Well-Being 421

Solutions: *Global Outlook: Microloans for the Poor* 423

18-4 Politics and Environmental Policy 424

Case Study: Environmental Policy in the United States 425

Case Study: Environmental Action by Students in the United States 428

Ice cores used to monitor past climate history

The earth flag, a symbol of commitment to promoting environmental sustainability

18-5   Global Environmental Policy   430

18-6   Environmental Worldviews: Clashing Values and Cultures   431

18-7   Living More Sustainably   433

SCIENCE SUPPLEMENTS

1   Units of Measurement   S1

2   Ecological Footprints   S2

3   Major Events in U.S. Environmental History   S9

4   Balancing Chemical Equations   S15

5   Classifying and Naming Species   S17

6   Weather Basics   S19

7   Earthquakes, Tsunamis, and Volcanic Eruptions   S23

8   Brief History of the Age of Oil   S25

9   Environmental Science: Concepts and Connections   S26

Glossary   G1

Index   I1

Courtesy of Earth Flag Co.

For Instructors

This book is an interdisciplinary study of how nature works, how we interact with nature, and how we can live more sustainably. Throughout all editions of this and my other textbooks, I have sought to tell a story of how human societies can traverse a path to sustainability.

Sustainability as the Central Theme with Five Major Subthemes

In this edition, I hope to make the story of a path to sustainability even clearer to students. *Sustainability* is the central theme of this book as shown in the Brief Contents on p. v.

Five major subthemes—natural capital, natural capital degradation, solutions, trade-offs, and individuals matter—guide the way to sustainability. Figure 1-2 (p. 7) shows a path to sustainability based on these subthemes.

- *Natural capital. Sustainability* requires a focus on preserving *natural capital*—the natural resources and natural services that support all life and economies (Figure 1-3, p. 7). Logos based on this figure appear near many chapter titles to show the components of natural capital being discussed in those chapters. Some 90 diagrams illustrate natural capital. Examples are Figures 3-14 (p. 45), 3-25 (p. 56), 5-25 (p. 97), and Figure 8-7 (p. 160).

- *Natural capital degradation.* Certain human activities lead to *natural capital degradation*—the second subtheme. I describe how human activities can degrade natural capital, and some 124 diagrams illustrate this. Examples are Figures 5-33 (p. 103), 10-9 (p. 212), and 11-14 (p. 246).

- *Solutions.* The next step is a search for *solutions* to natural capital degradation and other environmental problems. I present solutions to environmental problems in a balanced manner and have students use critical thinking to evaluate proposed solutions. Some 95 diagrams and subsections present environmental solutions as strategies for prevention or control. Examples are Figures 8-21 (p. 171), 11-16 (p. 247), 15-17 (p. 363), and 17-23 (p. 409).

- *Trade-Offs.* The search for solutions involves *trade-offs*—the fourth subtheme—because any solution has advantages and disadvantages that must be evaluated. There are 35 Trade-Offs diagrams giving the advantages and disadvantages of various environmental technologies and solutions to environmental problems. Examples are Figures 7-21 and 7-22 (p. 149), 8-12 (p. 163), and 17-12 (p. 401). The caption in each of these diagrams has a Critical Thinking question asking readers to pick the single advantage and the single disadvantage they think are most important. Also see the discussion of the advantages and disadvantages of pesticides on pp. 228–331.

- *Individuals matter.* In the quest for finding and implementing solutions, I present many examples that illustrate the important contributions that individuals have made. I avoid gloom-and-doom and use uplifting stories to show how individuals working alone and together have changed the world. And I try to balance bad environmental news with good environmental news. Throughout the book *Individuals Matter* boxes describe what various scientists and concerned citizens have done to help us achieve sustainability. (pp. 172, 229, 386, 394). Also, 12 *What Can You Do?* boxes describe how readers can deal with various environmental problems. Examples are Figures 8-27 (p. 177), 11-21 (p. 251), and 13-48 (p. 326).

A Sound Science Basis

The path to sustainability is supported throughout this text by sound science—concepts and explanations widely accepted by scientists. Chapters 2–9 present the scientific principles of ecology and biodiversity (see Brief Contents, p. v) needed to understand how the earth works and to evaluate proposed solutions to the environmental problems. Science is then integrated throughout the book in text (pp. 220, 271, 315, 392), science spotlights (pp. 69, 204, 323, 408), science case studies (pp. 178, 244, 282), and figures (pp. 232, 289, 372, 385).

Environmental ScienceNow, a new feature of this edition, allows students to enhance their scientific understanding by viewing animations available on the website for this book. There are 92 *Environmental*

ScienceNow items, some of them related to figures (pp. 40, 353, 414) and others to text (pp. 118, 221, 392)

Science, Economics, Politics, and Ethics from a Global Perspective

Material from science, economics, political science, and environmental ethics is integrated throughout this text. Subsection labels, such as Science and Economics, Economics and Politics, and Science and Ethics (pp. 188, 226, 241, 288) show this integration. Chapter 18 ties much of this material together with its focus on environmental economics, politics, and ethics. It can be used anytime after Chapter 1.

This book assumes a global perspective on two levels. First, ecological principles reveal how all the world's life is connected and sustained within the biosphere (Chapter 3). Second, the book integrates information and images from around the world by presenting scientific, economic, political, and ethical issues. To emphasize this feature I have added the label *Global Outlook* to many of the book's figures (pp. 146, 351, and 395) and subsections (pp. 163, 259, 286–87). There are also numerous global maps (pp. 96, 116, 241).

Flexibility

There are hundreds of ways to organize the chapters in this book to fit the needs of different instructors. Since the publication of the first edition in 1975, my solution to this problem has been to design a highly flexible book that allows instructors to vary the order of chapters and sections within chapters.

I recommend that instructors start with Chapter 1 because it defines basic terms and gives an overview of sustainability, population, pollution, resources, and economic development issues that are treated throughout the book. This provides a springboard for instructors to use other chapters in almost any order. For example, Chapter 18 on Environmental Economics, Politics, and Worldviews could follow Chapter 1.

One often-used strategy is to follow Chapter 1 with Chapters 2–7, introducing basic science and ecological concepts. Instructors can then use the remaining chapters in any order desired.

In this edition, I have followed Chapter 7 with two chapters on the important topic of biodiversity. Thus, Chapters 2–9 include all of the basic biology (ecology and biodiversity). Many instructors have requested this chapter order, but the biodiversity chapters can easily be covered later.

Case Studies

Each chapter begins with a brief *Case Study* designed to capture interest and set the stage for the material that follows. In addition to these 18 case studies,

44 others appear throughout the book (see Detailed Contents, p. vii). These 62 case studies provide a more in-depth look at specific environmental problems and their possible solutions.

Critical Thinking

The introduction on *Learning Skills* describes critical thinking skills (pp. 3–4). And critical thinking exercises are used throughout the book in several ways:

- In all boxes (except *Individuals Matter*)
- In the captions of the book's 35 *Trade-Offs* figures
- As 68 *How Would You Vote?* exercises (see pp. 221, 298, and 398) —a new feature of this edition
- As end-of-chapter questions

Visual Learning

This book has 471 illustrations and photos (130 of them new to this edition) designed to present complex ideas in understandable ways relating to the real world.

To enhance learning and increase interest, I have put potentially boring tables and lists into a colorful visual format. This includes diagrams illustrating *natural capital* (p. 104), *natural capital degradation* (pp. 95, 103) *trade-offs* (pp. 231, 322), and *solutions* (pp. 307, 382).

I have also carefully selected 113 photographs—97 of them new to this edition—to illustrate key ideas. I have avoided the common practice of including numerous "filler" photographs that are not very effective or that show the obvious. When I cannot find a high-quality photograph, to illustrate an idea, I substitute a high-quality diagram.

Finally, to enhance visual learning, 92 *Environmental ScienceNow* interactive animations, referenced to the text and diagrams, are available online. This learning tool helps students assess their unique study needs through pretests, posttests, and personalized learning plans. It is FREE with every new copy of the book and can be purchased for used titles. Another feature of this learning tool is *How Do I Prepare?* It allows students to review basic math, chemistry, and other refresher skills.

The Most Significant Revision Ever

In preparing this new edition, I have consulted more than 10,000 research sources in the professional literature and about the same number of Internet sites. I have also benefited from the more than 250 experts and teachers (see list on pp. xix–xxi) who have provided detailed reviews of this and my other three books in this field. And I benefited from information sent to me by instructors and students using this and my other environmental science textbooks. This edition contains over 4,000 updates based on information and data published in 2002, 2003, 2004, and 2005.

This is the most significant revision since the first edition of *Environmental Science*. Major changes include the following:

- Reduction of the book's length by 2 chapters (from 20 to 18) and by 102 pages.

- Change of chapter order to put all of the biology (ecology and biodiversity) material together (see Brief Contents, p. v) based on feedback from many instructors. I plan to stick with this chapter order in future editions.

- Revision of the entire book with help from developmental editor Scott Spoolman and three in-depth reviews.

- Addition of *How Would You Vote?* feature, enabling students to vote online on 68 critical environmental issues and to get national and global tallies of the results (pp. 113, 159, 178). Most of these vote decisions are tied to trade-offs diagrams (pp. 221, 317, 401) or discussions of controversial issues and require students to use critical thinking skills to evaluate solutions and make decisions about environmental issues.

- Summary of key ideas at the beginning of each chapter (pp. 36, 155, 286) and at the beginning of each subsection (pp.10, 40, 175).

- 166 new figures (including 97 new photographs) and 120 improved figures. This means that 80% of the book's 471 figures are either new or improved.

- 35 *Trade-Offs* diagrams and 12 *What Can You Do?* diagrams.

- Important new topics include affluenza, future implications of genetic engineering, reconciliation ecology, ownership of water resources, nanotechnology, General Motors' fuel cell car of the future, bioterrorism, and expanded treatment of the stewardship environmental worldview.

In-Text Study Aids

Each chapter begins with a few general questions to reveal how it is organized and what students will be learning. When a new term is introduced and defined, it is printed in boldface type. A glossary of all key terms is located at the end of the book.

Each chapter ends with a set of *Critical Thinking* questions to encourage students to think critically and apply what they have learned to their lives. The website for the book also contains a set of *Review Questions* covering *all* of the material in the chapter, which can be used as a study guide for students. Some instructors download this from the website to give students a list of questions to answer as a course requirement.

New learning features for this edition, described above, include chapter opening summaries of key ideas, critical thinking questions in the captions of trade-offs diagrams, 92 *Environmental ScienceNow* animations keyed to the text, and 68 *How Would You Vote?* boxes.

Qualified users of this textbook have free access to the Companion Website for this book at

http://biology.brookscole.com/miller11.

At this website they will find the following material for each chapter:

- *Flash Cards*, which allow you to test your mastery of the Terms and Concepts to Remember for each chapter

- *Chapter Quizzes*, which provide a multiple-choice practice quiz

- *InfoTrac® College Edition articles* are listed by chapter and section online, as well as by relevant region of the country on InfoTrac® Map. These articles can also support *How Would You Vote?* exercises.

- *References*, which lists the major books and articles consulted in writing this chapter

- A brief *What Can You Do?* list addressing key environmental problems

- *Weblinks*, an extensive list of websites with news, research, and images related to individual sections of the chapter.

Qualified adopters of this textbook also have access to *WebTutor Advantage* on *WebCT* and *Blackboard*. The course pack contains our website content, *Environmental ScienceNow* media and tests, as well as other useful material. For more information, visit

http://webtutor.thomsonlearning.com

Teachers and students using *new* copies of this textbook also have free and unlimited access to *InfoTrac® College Edition*. This fully searchable online library gives users access to complete environmental articles from several hundred periodicals dating back over the past 24 years.

Other Student Learning Tools

- *Essential Study Skills for Science Students* by Daniel D. Chiras. This book includes chapters on developing good study habits, sharpening memory, getting the most out of lectures, labs, and reading assignments, improving test-taking abilities, and becoming a critical thinker. Instructors can have this book bundled FREE with the textbook.

- *Laboratory Manual* by C. Lee Rockett and Kenneth J. Van Dellen. This manual includes a variety of laboratory exercises, workbook exercises, and projects that require a minimum of sophisticated equipment.

Supplements for Instructors

■ *Multimedia Manager.* This CD-ROM, free to adopters, allows you to create custom lectures using over 2,000 pieces of high-resolution artwork, images, and animations from *Environmental ScienceNow* and the Web, assemble database files, and create Microsoft® PowerPoint lectures using text slides and figures from the textbook. This program's editing tools allow use of slides from other lectures, modification or removal of figure labels and leaders, insertion of your own slides, saving slides as JPEG images, and preparation of lectures for use on the Web.

■ *Transparency Masters and Acetates.* Includes 100 color acetates of line art from Miller's text and over 400 black-and-white master sheets of key diagrams for making overhead transparencies. Free to adopters.

■ *CNN™ Today Videos.* These informative news stories, available either on 3 VHS tapes or 3 CD-ROMs, contain 58 two- to three-minute video clips of current news stories on environmental issues from around the world.

■ *Instructor's Manual with Test Items.* Free to adopters.

■ *ExamView.* Allows an instructor to easily create and customize tests, see them on the screen exactly as they will print, and print them out.

Other Textbook Options

Instructors wanting a book with a different length and emphasis can use one of my three other books written for various types of environmental science courses: *Living in the Environment*, 14th edition (642 pages, Brooks/Cole, 2005), *Sustaining the Earth: An Integrated Approach*, 7th edition (327 pages, Brooks/Cole, 2005) and *Essentials of Ecology*, 3rd edition (272 pages, Brooks/Cole, 2005).

Help Me Improve This Book

Let me know how you think this book can be improved. If you find any errors, bias, or confusing explanations, please e-mail me about them at

http://mtg89@hotmail.com

Most errors can be corrected in subsequent printings of this edition rather than waiting for a new edition.

Acknowledgments

I wish to thank the many students and teachers who have responded so favorably to the 10 previous editions of *Environmental Science*, the 14 editions of *Living in the Environment*, the 7 editions of *Sustaining the Earth*, and the 3 editions of *Essentials of Ecology*, and who have corrected errors and offered many helpful suggestions for improvement. I am also deeply indebted to the more than 250 reviewers, who pointed out errors and suggested many important improvements in the various editions of these three books. Any errors and deficiencies left are mine.

I am particularly indebted to Scott Spoolman who served as a developmental editor for this new edition and who made many important suggestions for improving this book. I am also indebted to Michael L. Cain, John Pichtel, and Jennifer Rivers who did in-depth reviews and provided many helpful suggestions. I also thank Sue Holt for her helpful review of the environmental economics material.

The members of the talented production team, listed on the copyright page, have made vital contributions as well. My thanks also go to production editors Andy Marinkovich and Andrea Fincke, copy editor Jill Hobbs, Thompson Steele's page layout artist Bonnie Van Slyke, Brooks/Cole's hard-working sales staff, and Keli Amann, Joy Westberg, Carol Benedict, and the other members of the talented team who developed the multimedia, website, and advertising materials associated with this book.

I also thank C. Lee Rockett and Kenneth J. Van Dellen for developing the *Laboratory Manual* to accompany this book; Jane Heinze-Fry for her work on concept mapping and *Environmental Articles, Critical Thinking and the Environment: A Beginner's Guide*; Irene Kokkala for her excellent work on the *Instructor's Manual*, and the people who have translated this book into seven different languages for use throughout much of the world.

My deepest thanks go to Jack Carey, biology publisher at Brooks/Cole, for his encouragement, help, 39 years of friendship, and superb reviewing system. It helps immensely to work with the best and most experienced editor in college textbook publishing.

I dedicate this book to the earth and to Kathleen Paul Miller, my wife and research associate.

G. Tyler Miller, Jr.

Guest Essayists and Reviewers

Guest essays by the following authors are available online at the website for this book: **M. Kat Anderson,** ethnoecologist with the National Plant Center of the USDA's Natural Resource Conservation Center; **Lester R. Brown**, president, Earth Policy Institute; **Alberto Ruz Buenfil**, environmental activist, writer, and performer; **Robert D. Bullard**, professor of sociology and director of the Environmental Justice Resource Center at Clark Atlanta University; **Michael Cain,** ecologist and adjunct professor at Bowdoin College; **Herman E. Daly**, senior research scholar at the School of Public Affairs, University of Maryland; **Lois Marie Gibbs,** director, Center for Health, Environment, and Justice; **Garrett Hardin,** professor emeritus (now deceased) of human ecology, University of California, Santa Barbara; **John Harte**, professor of energy and resources, University of California, Berkeley; **Paul G. Hawken,** environmental author and business leader; **Jane Heinze-Fry**, environmental educator, **Paul F. Kamitsuja**, infectious disease expert and physician;

Amory B. Lovins, energy policy consultant and director of research, Rocky Mountain Institute; **Bobbi S. Low**, professor of resource ecology, University of Michigan; **Lester W. Milbrath,** director of the research program in environment and society, State University of New York, Buffalo; **Peter Montague**, director, Environmental Research Foundation, **Norman Myers,** tropical ecologist and consultant in environment and development; **David W. Orr,** professor of environmental studies, Oberlin College; **Noel Perrin,** adjunct professor of environmental studies, Dartmouth College, **David Pimentel,** professor of insect ecology and agricultural sciences, Cornell University; **John Pichtel**, Ball State University; **Andrew C. Revkin**, environmental author and environmental reporter for the New York Times; **Vandana Shiva**, physicist, educator, environmental consultant; **Nancy Wicks,** ecopioneer and director of Round Mountain Organics; **Donald Worster**, environmental historian and professor of American history, University of Kansas

Cumulative Reviewers

Barbara J. Abraham, Hampton College; Donald D. Adams, State University of New York at Plattsburgh; Larry G. Allen, California State University, Northridge; Susan Allen-Gil, Ithaca College; James R. Anderson, U.S. Geological Survey; Mark W. Anderson, University of Maine; Kenneth B. Armitage, University of Kansas; Samuel Arthur, Bowling Green State University; Gary J. Atchison, Iowa State University; Marvin W. Baker, Jr., University of Oklahoma; Virgil R. Baker, Arizona State University; Ian G. Barbour, Carleton College; Albert J. Beck, California State University, Chico; W. Behan, Northern Arizona University; Keith L. Bildstein, Winthrop College; Jeff Bland, University of Puget Sound; Roger G. Bland, Central Michigan University; Grady Blount II, Texas A&M University, Corpus Christi; Georg Borgstrom, Michigan State University; Arthur C. Borror, University of New Hampshire; John H. Bounds, Sam Houston State University; Leon F. Bouvier, Population Reference Bureau; Daniel J. Bovin, Université Laval; Michael F. Brewer, Resources for the Future, Inc.; Mark M. Brinson, East Carolina University; Dale Brown, University of Hartford; Patrick E. Brunelle, Contra Costa College; Terrence J. Burgess, Saddleback College North; David Byman, Pennsylvania State University, Worthington–Scranton; Michael L. Cain, Bowdoin College, Lynton K. Caldwell, Indiana University; Faith Thompson Campbell, Natural Resources Defense

Council, Inc.; Ray Canterbery, Florida State University; Ted J. Case, University of San Diego; Ann Causey, Auburn University; Richard A. Cellarius, Evergreen State University; William U. Chandler, Worldwatch Institute; F. Christman, University of North Carolina, Chapel Hill; Lu Anne Clark, Lansing Community College; Preston Cloud, University of California, Santa Barbara; Bernard C. Cohen, University of Pittsburgh; Richard A. Cooley, University of California, Santa Cruz; Dennis J. Corrigan; George Cox, San Diego State University; John D. Cunningham, Keene State College; Herman E. Daly, University of Maryland; Raymond F. Dasmann, University of California, Santa Cruz; Kingsley Davis, Hoover Institution; Edward E. DeMartini, University of California, Santa Barbara; Charles E. DePoe, Northeast Louisiana University; Thomas R. Detwyler, University of Wisconsin; Peter H. Diage, University of California, Riverside; Lon D. Drake, University of Iowa; David DuBose, Shasta College; Dietrich Earnhart, University of Kansas; T. Edmonson, University of Washington; Thomas Eisner, Cornell University; Michael Esler, Southern Illinois University; David E. Fairbrothers, Rutgers University; Paul P. Feeny, Cornell University; Richard S. Feldman, Marist College; Nancy Field, Bellevue Community College; Allan Fitzsimmons, University of Kentucky; Andrew J. Friedland, Dartmouth College; Kenneth O. Fulgham, Humboldt State University; Lowell L. Getz,

University of Illinois at Urbana–Champaign; Frederick F. Gilbert, Washington State University; Jay Glassman, Los Angeles Valley College; Harold Goetz, North Dakota State University; Jeffery J. Gordon, Bowling Green State University; Eville Gorham, University of Minnesota; Michael Gough, Resources for the Future; Ernest M. Gould, Jr., Harvard University; Peter Green, Golden West College; Katharine B. Gregg, West Virginia Wesleyan College; Paul K. Grogger, University of Colorado at Colorado Springs; L. Guernsey, Indiana State University; Ralph Guzman, University of California, Santa Cruz; Raymond Hames, University of Nebraska, Lincoln; Raymond E. Hampton, Central Michigan University; Ted L. Hanes, California State University, Fullerton; William S. Hardenbergh, Southern Illinois University at Carbondale; John P. Harley, Eastern Kentucky University; Neil A. Harriman, University of Wisconsin, Oshkosh; Grant A. Harris, Washington State University; Harry S. Hass, San Jose City College; Arthur N. Haupt, Population Reference Bureau; Denis A. Hayes, environmental consultant; Stephen Heard, University of Iowa; Gene Heinze-Fry, Department of Utilities, Commonwealth of Massachusetts; Jane Heinze-Fry, environmental educator; John G. Hewston, Humboldt State University; David L. Hicks, Whitworth College; Kenneth M. Hinkel, University of Cincinnati; Eric Hirst, Oak Ridge National Laboratory; Doug Hix, University of Hartford; S. Holling, University of British Columbia; Sue Holt, Cabrillo College; Donald Holtgrieve, California State University, Hayward; Michael H. Horn, California State University, Fullerton; Mark A. Hornberger, Bloomsberg University; Marilyn Houck, Pennsylvania State University; Richard D. Houk, Winthrop College; Robert J. Huggett, College of William and Mary; Donald Huisingh, North Carolina State University; Marlene K. Hutt, IBM; David R. Inglis, University of Massachusetts; Robert Janiskee, University of South Carolina; Hugo H. John, University of Connecticut; Brian A. Johnson, University of Pennsylvania, Bloomsburg; David I. Johnson, Michigan State University; Mark Jonasson, Crafton Hills College; Agnes Kadar, Nassau Community College; Thomas L. Keefe, Eastern Kentucky University; Nathan Keyfitz, Harvard University; David Kidd, University of New Mexico; Pamela S. Kimbrough; Jesse Klingebiel, Kent School; Edward J. Kormondy, University of Hawaii–Hilo/West Oahu College; John V. Krutilla, Resources for the Future, Inc.; Judith Kunofsky, Sierra Club; E. Kurtz; Theodore Kury, State University of New York at Buffalo; Steve Ladochy, University of Winnipeg; Mark B. Lapping, Kansas State University; Tom Leege, Idaho Department of Fish and Game; William S. Lindsay, Monterey Peninsula College;

E. S. Lindstrom, Pennsylvania State University; M. Lippiman, New York University Medical Center; Valerie A. Liston, University of Minnesota; Dennis Livingston, Rensselaer Polytechnic Institute; James P. Lodge, air pollution consultant; Raymond C. Loehr, University of Texas at Austin; Ruth Logan, Santa Monica City College; Robert D. Loring, DePauw University; Paul F. Love, Angelo State University; Thomas Lovering, University of California, Santa Barbara; Amory B. Lovins, Rocky Mountain Institute; Hunter Lovins, Rocky Mountain Institute; Gene A. Lucas, Drake University; Claudia Luke; David Lynn; Timothy F. Lyon, Ball State University; Stephen Malcolm, Western Michigan University; Melvin G. Marcus, Arizona State University; Gordon E. Matzke, Oregon State University; Parker Mauldin, Rockefeller Foundation; Marie McClune, The Agnes Irwin School (Rosemont, Pennsylvania); Theodore R. McDowell, California State University; Vincent E. McKelvey, U.S. Geological Survey; Robert T. McMaster, Smith College; John G. Merriam, Bowling Green State University; A. Steven Messenger, Northern Illinois University; John Meyers, Middlesex Community College; Raymond W. Miller, Utah State University; Arthur B. Millman, University of Massachusetts, Boston; Fred Montague, University of Utah; Rolf Monteen, California Polytechnic State University; Ralph Morris, Brock University, St. Catherine's, Ontario, Canada; Angela Morrow, Auburn University; William W. Murdoch, University of California, Santa Barbara; Norman Myers, environmental consultant; Brian C. Myres, Cypress College; A. Neale, Illinois State University; Duane Nellis, Kansas State University; Jan Newhouse, University of Hawaii, Manoa; Jim Norwine, Texas A&M University, Kingsville; John E. Oliver, Indiana State University; Carol Page, copyeditor; Eric Pallant, Allegheny College; Charles F. Park, Stanford University; Richard J. Pedersen, U.S. Department of Agriculture, Forest Service; David Pelliam, Bureau of Land Management, U.S. Department of Interior; Rodney Peterson, Colorado State University; Julie Phillips, De Anza College; John Pichtel, Ball State University; William S. Pierce, Case Western Reserve University; David Pimentel, Cornell University; Peter Pizor, Northwest Community College; Mark D. Plunkett, Bellevue Community College; Grace L. Powell, University of Akron; James H. Price, Oklahoma College; Marian E. Reeve, Merritt College; Carl H. Reidel, University of Vermont; Charles C. Reith, Tulane University; Roger Revelle, California State University, San Diego; L. Reynolds, University of Central Arkansas; Ronald R. Rhein, Kutztown University of Pennsylvania; Charles Rhyne, Jackson State University; Robert A. Richardson, University of Wisconsin; Benjamin F. Richason III, St. Cloud State

University; Jennifer Rivers, Northeastern University; Ronald Robberecht, University of Idaho; William Van B. Robertson, School of Medicine, Stanford University; C. Lee Rockett, Bowling Green State University; Terry D. Roelofs, Humboldt State University; Christopher Rose, California Polytechnic State University; Richard G. Rose, West Valley College; Stephen T. Ross, University of Southern Mississippi; Robert E. Roth, Ohio State University; Arthur N. Samel, Bowling Green State University; Floyd Sanford, Coe College; David Satterthwaite, I.E.E.D., London; Stephen W. Sawyer, University of Maryland; Arnold Schecter, State University of New York; Frank Schiavo, San Jose State University; William H. Schlesinger, Ecological Society of America; Stephen H. Schneider, National Center for Atmospheric Research; Clarence A. Schoenfeld, University of Wisconsin, Madison; Henry A. Schroeder, Dartmouth Medical School; Lauren A. Schroeder, Youngstown State University; Norman B. Schwartz, University of Delaware; George Sessions, Sierra College; David J. Severn, Clement Associates; Paul Shepard, Pitzer College and Claremont Graduate School; Michael P. Shields, Southern Illinois University at Carbondale; Kenneth Shiovitz; F. Siewert, Ball State University; E. K. Silbergold, Environmental Defense Fund; Joseph L. Simon, University of South Florida; William E. Sloey, University of Wisconsin, Oshkosh; Robert L. Smith, West Virginia University; Val Smith, University of Kansas; Howard M. Smolkin, U.S. Environmental Protection Agency; Patricia M. Sparks, Glassboro State College; John E. Stanley, University of Virginia; Mel Stanley, California State Polytechnic University, Pomona; Norman R. Stewart, University of Wisconsin, Milwaukee; Frank E. Studnicka, University of Wisconsin, Platteville; Chris Tarp, Contra Costa College; Roger E. Thibault, Bowling Green State University; William L. Thomas, California State University, Hayward; Shari Turney, copyeditor; John D. Usis, Youngstown State University; Tinco E. A. van Hylckama, Texas Tech University; Robert R. Van Kirk, Humboldt State University; Donald E. Van Meter, Ball State University; Gary Varner, Texas A&M University; John D. Vitek, Oklahoma State University; Harry A. Wagner, Victoria College; Lee B. Waian, Saddleback College; Warren C. Walker, Stephen F. Austin State University; Thomas D. Warner, South Dakota State University; Kenneth E. F. Watt, University of California, Davis; Alvin M. Weinberg, Institute of Energy Analysis, Oak Ridge Associated Universities; Brian Weiss; Margery Weitkamp, James Monroe High School (Granada Hills, California); Anthony Weston, State University of New York at Stony Brook; Raymond White, San Francisco City College; Douglas Wickum, University of Wisconsin, Stout; Charles G. Wilber, Colorado State University; Nancy Lee Wilkinson, San Francisco State University; John C. Williams, College of San Mateo; Ray Williams, Rio Hondo College; Roberta Williams, University of Nevada, Las Vegas; Samuel J. Williamson, New York University; Ted L. Willrich, Oregon State University; James Winsor, Pennsylvania State University; Fred Witzig, University of Minnesota at Duluth; George M. Woodwell, Woods Hole Research Center; Robert Yoerg, Belmont Hills Hospital; Hideo Yonenaka, San Francisco State University; Malcolm J. Zwolinski, University of Arizona.

Learning Skills

Students who can begin early in their lives to think of things as connected, even if they revise their views every year, have begun the life of learning.

MARK VAN DOREN

Why Is It Important to Study Environmental Science?

Environmental science may be the most important course you will ever take.

Welcome to *environmental science*—an *interdisciplinary* study of how nature works and how things in nature are interconnected. This book is an integrated and science-based study of environmental problems, connections, and solutions.

The following themes are interwoven throughout this book: *sustainability, natural capital, natural capital degradation, solutions* to environmental problems, *trade-offs* in finding acceptable solutions, the *importance of individual actions in implementing solutions* (individuals matter), and *sound science* (see Figure 1-2, p. 7).

Here are the key ideas in this book.

- We are in the process of degrading the earth's resources and services—called *natural capital*—that support all life and economies (see Figures 1-3, p. 7).

- There are four interconnected principles of sustainability based on understanding how nature has sustained itself for billions of years (see Figure 6-18, p. 125).

- We can learn to live more sustainability during this century by applying these four principles to our lifestyles and economies (see Figure 6-19, 126).

- There are always tradeoffs involved in making decisions about environmental problems and solutions.

- Evaluating our environmental choices requires critical thinking based on an ecological understanding of how the earth works.

Environmental issues affect every part of your life and are an important part of the news stories presented on television and in newspapers and magazines. The concepts, information, and issues discussed in this book and the course you are taking should be useful to you both now and in the future.

Understandably, I am biased. But *I strongly believe that environmental science is the single most important course in your education.* What could be more important than learning how the earth works, how we are affecting its life-support system, and how we can reduce our environmental impact?

We live in an incredibly exciting and challenging era. There is a growing awareness that during this century we need to make a new cultural transition in which we learn how to live more sustainably by not degrading our life-support system. I hope this book will stimulate you to become involved in promoting this change in the way we view and treat the earth that sustains other life, all economies, and us.

How Did I Become Involved with Environmental Problems?

I became involved in environmental science and education after hearing a scientific lecture in 1966.

In 1966, I heard Dean Cowie, a physicist with the U.S. Geological Survey, give a lecture on the problems of population growth and pollution. Afterward I went to him and said, "If even a fraction of what you have said is true, I will feel ethically obligated to give up my research on the corrosion of metals and devote the rest of my life to research and education on environmental problems and solutions. Frankly, I do not want to believe a word you have said, and I am going into the literature to try to show that your statements are either untrue or grossly distorted."

After six months of study, I was convinced of the seriousness of these and other environmental problems. Since then, I have been studying, teaching, and writing about them. This book summarizes what I have learned in more than three decades of trying to understand environmental principles, problems, connections, and solutions.

Improving Your Study and Learning Skills

Learning how to learn is life's most important skill.

Maximizing your ability to learn should be one of your most important lifetime educational goals. It involves continually trying to improve your study and learning skills.

This has a number of payoffs. You can learn more and do so more efficiently. You will also have more time to pursue other interests besides studying without feeling guilty about always being behind. Learning how to learn can also help you get better grades and live a more fruitful and rewarding life.

Here are some general study and learning skills.

Get organized. Becoming more efficient at studying gives you more time for other interests.

Make daily to-do lists in writing. Put items in order of importance, focus on the most important tasks, and assign a time to work on these items. Because life is full of uncertainties, you will be lucky to accomplish half of the items on your daily list. Shift your schedule to accomplish the most important items. Otherwise, you will fall behind and become increasingly frustrated.

Set up a study routine in a distraction-free environment. Develop a written daily study schedule and stick to it. Study in a quiet, well-lighted space. Work sitting at a desk or table—not lying down on a couch or bed. Take breaks every hour or so. During each break, take several deep breaths and move around to help you stay more alert and focused.

Avoid procrastination—putting work off until another time. Do not fall behind on your reading and other assignments. Set aside a particular time for studying each day and make it a part of your daily routine.

Do not eat dessert first. Otherwise, you may never get to the main meal (studying). When you have accomplished your study goals, reward yourself with play (dessert).

Make hills out of mountains. It is psychologically difficult to climb a mountain such as reading an entire book, reading a chapter in a book, writing a paper, or cramming to study for a test. Instead, break these large tasks (mountains) down into a series of small tasks (hills). Each day read a few pages of a book or chapter, write a few paragraphs of a paper, and review what you have studied and learned. As Henry Ford put it, "Nothing is particularly hard if you divide it into small jobs."

Look at the big picture first. Get an overview of an assigned reading by looking at the main headings or chapter outline. This textbook includes a list of the main questions that are the focus of each chapter.

Ask and answer questions as you read. For example, what is the main point of this section or paragraph? Each subsection in this book starts with a question that the material is designed to answer. This question is followed by a one-sentence summary of the key material in the subsection. I find this type of running summary to be more helpful than a harder-to-digest summary at the end of each chapter. My goal is to present the material in more manageable bites. You can also use the one-sentence summaries as a way to review what you have learned. Putting them all together gives you a summary of the chapter.

Focus on key terms. Use the glossary in your textbook to look up the meaning of terms or words you do not understand. Make flash cards for learning key terms and concepts, and review them frequently. This book shows all key terms in **boldfaced** type and lesser but still important terms in *italicized* type. Flash cards for testing your mastery of key terms for each chapter are available on the website for this book.

Interact with what you read. When I read, I mark key sentences and paragraphs with a highlighter or pen. I put an asterisk in the margin next to an idea I think is important and double asterisks next to an idea I think is especially important. I write comments in the margins, such as *Beautiful, Confusing, Misleading,* or *Wrong.* I fold down the top corner of pages with highlighted passages and the top and bottom corners of especially important pages. This way, I can flip through a chapter or book and quickly review the key ideas.

Review to reinforce learning. Before each class, review the material you learned in the previous class and read the assigned material. Review, fill in, and organize your notes as soon as possible after each class.

Become a better note taker. Do not try to take down everything your instructor says. Instead, jot down main points and key facts using your own shorthand system. Fill in and organize your notes after class.

Write out answers to questions to focus and reinforce learning. Answer questions at the end of each chapter or those assigned to you. Put your answers down in writing as if you were turning them in for a grade. Save your answers for review and preparation for tests.

Use the buddy system. Study with a friend or become a member of a study group to compare notes, review material, and prepare for tests. Explaining something to someone else is a great way to focus your thoughts and reinforce your learning. If available, attend review sessions offered by instructors or teaching assistants.

Learn your instructor's test style. Does your instructor emphasize multiple-choice, fill-in-the-blank, true-or-false, factual, thought, or essay questions? How much of the test will come from the textbook and how much from lecture material? Adapt your learning and studying methods to this style. You may disagree with it and feel that it does not adequately reflect what you know. But the reality is that your instructor is in charge and your grade (but not always your learning) usually depends heavily on going along with the instructor's system.

Become a better test taker. Avoid cramming. Eat well and get plenty of sleep before a test. Arrive on time or early. Calm yourself and increase your oxygen intake by taking several deep breaths. Do this about every 10–15 minutes. Look over the test and answer the questions you know well first. Then work on the harder ones. Use the process of elimination to narrow down the choices for multiple-choice questions. Paring them down to two choices gives you a 50% chance of guessing the right answer. For essay questions, organize your thoughts before you start writing. If you have no idea what a question means, make an educated guess—you might get some partial credit. Another strategy for getting some credit is to show your knowledge and reasoning by writing something like this: "If this question means so and so, then my answer is _____."

Develop an optimistic outlook. Try to be a "glass is half-full" rather than a "glass is half-empty" person. Pessimism, fear, anxiety, and excessive worrying (especially over things you cannot control) are destructive, they feed on themselves, and they lead to inaction. Try to keep your energizing feelings of optimism slightly ahead of your immobilizing feelings of pessimism. Then you will always be moving forward.

Take time to enjoy life. Every day take time to laugh and enjoy nature, beauty, and friendship. Becoming an effective and efficient learner is the best way to do this without falling behind and living under a cloud of guilt and anxiety.

Critical Thinking Skills: Detecting Baloney

Learning how to think critically is a skill you will need throughout your life.

Every day we are exposed to a sea of information, ideas, and opinions. How do we know what to believe and why? Do the claims seem reasonable or exaggerated?

Critical thinking involves developing skills to help you analyze and evaluate the validity of information and ideas you are exposed to and to make decisions. Critical thinking skills help you decide rationally what to believe or what to do. They involve examining information and conclusions or beliefs in terms of the evidence and chain of logical reasoning that supports them. Critical thinking helps you distinguish between facts and opinions, evaluate evidence and arguments, take and defend an informed position on issues, integrate information and see relationships, and apply your knowledge to dealing with new and different problems. Here are some basic skills for learning how to think more critically.

Question everything and everybody. Be skeptical, as any good scientist is. Do not believe everything you hear or read, including the content of this textbook. Evaluate all information you receive. Seek other sources and opinions. As Albert Einstein put it, "The important thing is not to stop questioning."

Do not believe everything you read on the Internet. The Internet is a wonderful and easily accessible source of information. It is also a useful way to find alternative information and opinions on almost any subject or issue—much of it not available in the mainstream media and scholarly articles. However, because the Internet is so open, anyone can write anything they want with no editorial control or peer evaluation—the method in which scientific or other experts in an area review and comment on an article before it is accepted for publication in a scholarly journal. As a result, evaluating information on the Internet is one of the best ways to put into practice the principles of critical thinking. Use and enjoy the Internet, but be skeptical and proceed with caution.

Identify and evaluate your personal biases and beliefs. Every person has biases and beliefs taught to us by our parents, teachers, friends, role models, and experience. What are your basic beliefs and biases? Where did they come from? What assumptions are they based on? How sure are you that your beliefs and assumptions are right and why? According to William James, "A great many people think they are thinking when they are merely rearranging their prejudices."

Be open-minded, flexible, and humble. Consider different points of view, suspend judgment until you gather more evidence, and change your mind when necessary. Recognize that many useful and acceptable solutions to a problem may exist and that very few issues are black or white. There are usually valid points on both (or many) sides of an issue. One way to evaluate divergent views is to get into another person's head. How do other people see or view the world? What are their basic assumptions and beliefs? Is their position logically consistent with their assumptions and beliefs? And always be humble about what you know. According to Will Durant, "Education is a progressive discovery of our own ignorance."

Evaluate how the information related to an issue was obtained. Are the statements made based on first-hand knowledge or research or on hearsay? Are unnamed sources used? Is the information based on reproducible and widely accepted scientific studies (*sound* or *consensus science*, p. 22) or on preliminary scientific results that may be valid but need further testing (*frontier science*, p. 22)? Is it based on a few isolated stories or experiences (*anecdotal information*) or on carefully controlled studies? Is it based on unsubstantiated and widely doubted scientific information or beliefs (*junk science* or *pseudoscience*)? You need to know how to detect junk science, as discussed on p. 22.

Question the evidence and conclusions presented. What are the conclusions or claims? What evidence is presented to support them? Does the evidence support them? Is there a need to gather more evidence to test the conclusions? Are there other, more reasonable conclusions?

Try to identify and assess the assumptions and beliefs of those presenting evidence and drawing conclusions. What is their expertise in this area? Do they have any unstated assumptions, beliefs, biases, or values? Do they have a personal agenda? Can they benefit financially or politically from acceptance of their evidence and conclusions? Would investigators with different basic assumptions or beliefs take the same data and come to different conclusions?

Do the arguments used involve logical fallacies or debating tricks? Here are six examples. *First,* attack the presenter of an argument rather than the argument itself. *Second,* appeal to emotion rather than facts and logic. *Third,* claim that if one piece of evidence or one conclusion is false, then all other pieces of evidence and conclusions are false. *Fourth,* say that a conclusion is false because it has not been scientifically proven (scientists can never prove anything absolutely, but they can establish degrees of reliability, as discussed on p. 22). *Fifth,* inject irrelevant or misleading information to divert attention from important points. *Sixth,* present only either/or alternatives when a number of options may exist.

Become a seeker of wisdom, not a vessel of information. Develop a written list of principles, concepts, and rules to serve as guidelines in evaluating evidence and claims and making decisions. Continually evaluate and modify this list based on your experiences. Many people believe that the main goal of education is to learn as much as you can by concentrating on gathering more and more information—much of it useless or misleading. I believe that the primary goal is to know as little as possible. That is, you need to learn how to sift through mountains of facts and ideas to find the few *nuggets of wisdom* that are the most useful in understanding the world and in making decisions. This takes a firm commitment to learning how to think logically and critically, and continually flushing less valuable and thought-clogging information from your mind. This book is full of facts and numbers, but they are useful only to the extent that they lead to an understanding of key and useful ideas, scientific laws, concepts, principles, and connections. The major goals of the study of environmental science are to find out how nature works and sustains itself (*environmental wisdom*) and to use *principles of environmental wisdom* to help make human societies and economies more sustainable and thus more beneficial and enjoyable. As Sandra Carey put it, "Never mis-

take knowledge for wisdom. One helps you make a living; the other helps you make a life."

Trade-Offs

There are no simple answers to the environmental problems we face.

There are always *trade-offs* involved in making and implementing environmental decisions. My challenge is to give a fair and balanced presentation of different viewpoints, advantages and disadvantages of various technologies and proposed solutions to environmental problems, and good and bad news about environmental problems without injecting personal bias.

Studying a subject as important as environmental science and ending up with no conclusions, opinions, and beliefs means that both teacher and student have failed. However, such conclusions should be based on using critical thinking to evaluate different ideas and understand the trade-offs involved.

Help Me Improve This Book

I welcome your help in improving this book.

Researching and writing a book that covers and connects ideas in such a wide variety of disciplines is a challenging and exciting task. Almost every day I learn about some new connection in nature.

In a book this complex, some errors are bound to arise—typographical mistakes that slip through or statements that you might question based on your knowledge and research. My goal is to provide an interesting, accurate, balanced, and challenging book that furthers your understanding of this vital subject. I have also attempted to balance the arguments offered on various sides of key environmental issues.

I invite you to contact me and point out any remaining bias, correct any errors you find, and suggest ways to improve this book. Over decades of teaching, some of my best teachers have been students taking my classes and reading my textbooks. Please e-mail your suggestions to me at

mtg89@hotmail.com

Now start your journey into this fascinating and important study of how the earth works and how we can leave the planet in at least as good a shape as we found it. Have fun.

Study nature, love nature, stay close to nature. It will never fail you.
FRANK LLOYD WRIGHT

1

Environmental Problems, Their Causes, and Sustainability

CASE STUDY
Living in an Exponential Age

Two ancient kings enjoyed playing chess, with the winner claiming a prize from the loser. After one match, the winning king asked the losing king to pay him by placing one grain of wheat on the first square of the chessboard, two grains on the second square, four on the third, and so on, with the number doubling on each square until all 64 were filled.

The losing king, thinking he was getting off easy, agreed with delight. It was the biggest mistake he ever made. He bankrupted his kingdom because the number of grains of wheat he had promised was probably more than all the wheat that has ever been harvested!

This fictional story illustrates the concept of **exponential growth,** in which a quantity increases at a constant rate per unit of time, such as 2% per year. Exponential growth is deceptive. It starts off slowly, but after only a few doublings, it grows to enormous numbers because each doubling is more than the total of all earlier growth.

Here is another example. Fold a piece of paper in half to double its thickness. If you could continue doubling the thickness of the paper 42 times, the stack would reach from the earth to the moon, 386,400 kilometers (240,000 miles) away. If you could double it

50 times, the folded paper would almost reach the sun, 149 million kilometers (93 million miles) away!

Between 1950 and 2005, world population increased from 2.5 billion to 6.5 billion. Unless death rates rise sharply, somewhere between 8 billion and 10 billion people will live on the earth by the end of this century (Figure 1-1).

We live in a world of haves and have-nots. Despite an 8-fold increase in economic growth between 1950 and 2005, *almost one of every two workers in the world tries to survive on an income of less than U.S.$2 per day.* Such poverty affects environmental quality because to survive many of the poor must deplete and degrade local forests, grasslands, soil, and wildlife.

Biologists estimate that human activities are causing premature extinction of the earth's species at an exponential rate of 0.1% to 1% per year—an irreversible loss of the earth's great variety of life forms, or *biodiversity.* In various parts of the world, forests, grasslands, wetlands, coral reefs, and topsoil from croplands continue to disappear or become degraded as the human ecological footprint spreads exponentially across the globe.

There is growing concern that exponential growth in human activities such as burning fossil fuels and clearing forests will change the earth's climate during this century. This could ruin some areas for farming, shift water supplies, alter and reduce biodiversity, and disrupt economies in various parts of the world.

Great news. We have solutions to these problems that we could implement within a few decades, as you will learn in this book.

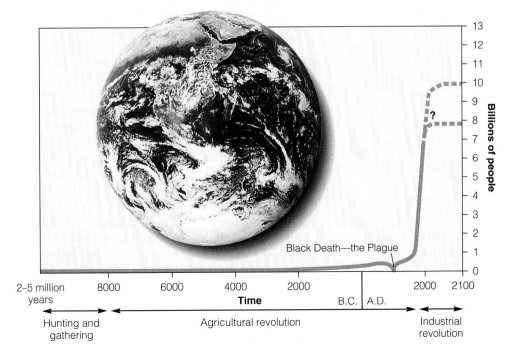

Black Death—the Plague

| 2–5 million years | 8000 | 6000 | 4000 | 2000 | B.C. | A.D. | 2000 | 2100 |

Time

Hunting and gathering | Agricultural revolution | Industrial revolution

Figure 1-1 The *J*-shaped curve of past exponential world population growth, with projections to 2100. Exponential growth starts off slowly, but as time passes the curve becomes increasingly steep. Unless death rates rise, the current world population of 6.5 billion people is projected to reach 8–10 billion people sometime this century. (This figure is not to scale.) (Data from the World Bank and United Nations; photo courtesy of NASA)

Alone in space, alone in its life-supporting systems, powered by inconceivable energies, mediating them to us through the most delicate adjustments, wayward, unlikely, unpredictable, but nourishing, enlivening, and enriching in the largest degree—is this not a precious home for all of us? Is it not worth our love?

BARBARA WARD AND RENÉ DUBOS

This chapter presents an overview of environmental problems, their causes, and ways we can live more sustainably. It discusses these questions:

- What are the six major themes of this book?

- What keeps us alive? What is an environmentally sustainable society?

- How fast is the human population increasing?

- What is the difference between economic growth and economic development?

- What are the earth's main types of resources? How can they be depleted or degraded?

- What are the principal types of pollution, and what can we do about pollution?

- What are the basic causes of today's environmental problems, and how are these causes connected?

- What are the harmful environmental effects of poverty and affluence?

- What major cultural changes have taken place since humans arrived?

- Is our current course sustainable? What is environmentally sustainable development?

KEY IDEAS

- Our lives and economies depend on energy from the sun and the earth's natural resources and natural services (natural capital) that nature provides for us at no cost.

- Today we are living unsustainably by depleting and degrading the earth's natural capital that supports us and our economies.

- The major causes of our environmental problems are population growth, wasteful resource use, poverty, failure to appreciate the value of the earth's natural capital, and ignorance about how the earth works.

- We have 50–100 years to make a transition to more sustainable human societies that mimic how the earth has sustained itself for billions of years.

1-1 LIVING MORE SUSTAINABLY

Environmental Science and Environmentalism

Environmental science is a study of how the earth works, how we interact with the earth, and how to deal with environmental problems.

Environment is everything that affects a living organism (any unique form of life). **Ecology** is a biological science that studies the relationships between living organisms and their environment.

This textbook is an introduction to **environmental science,** an interdisciplinary study that uses information from the physical sciences (such as biology, chemistry, and geology) and social sciences (such as economics, politics, and ethics) to learn how the earth works, how humans interact with the earth, and how to deal with the environmental problems we face. **Environmentalism** is a social movement dedicated to protecting the earth's life support systems for us and other species.

A Path to Sustainability: Major Themes of This Book

The central theme of this book is sustainability. It is built on the subthemes of natural capital, natural capital degradation, solutions, trade-offs, and individuals matter and is supported by sound science.

Sustainability is the ability of the earth's various systems, including human cultural systems and economies, to survive and adapt to changing environmental conditions. Figure 1-2 shows the steps along a path to sustainability, based on the subthemes for this book.

The first step is to sustain the earth's **natural capital**—the natural resources and natural services that keep us and other species alive and support our economies (Figure 1-3).

The first step toward sustainability is to understand the components and importance of natural capital and the natural or biological income it provides (Figure 1-2, Step 1). To economists, *capital* is wealth used to sustain a business and to generate more wealth. Invested financial capital can provide us with *financial income.* For example, suppose you invest $100,000 of capital and get a 10% return on your money. In one year you will get $10,000 in income from interest and increase your wealth to $110,000.

By analogy, the renewable resources that make up part of the earth's natural capital (Figure 1-3, left) can provide us with renewable *biological income* indefinitely as long as we do not use these resources faster than they are renewed by nature. For example, natural

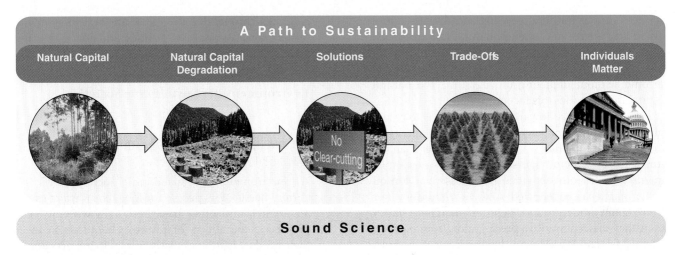

Figure 1-2 *A path to sustainability.* Five subthemes are used throughout this book to illustrate how we can make the transition to more environmentally sustainable societies and economies. Sound science is used to help us understand and implement this transition to sustainability.

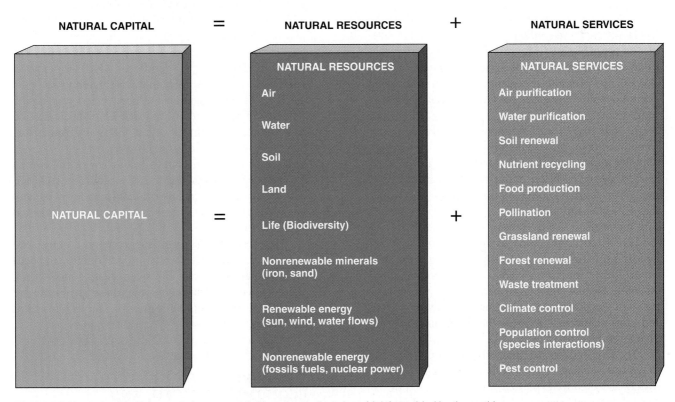

Figure 1-3 Natural capital: the natural resources (left) and natural services (right) provided by the earth's *natural capital* support and sustain the earth's life and economies. For example, *nutrients,* or chemicals such as carbon and nitrogen that plants and animals need as *resources,* are recycled through the air, water, soil, and organisms by the natural process of *nutrient cycling.* And the interactions and competition of different types of plants and animals (species) for *resources* (nutrients) keeps any single species from taking over through the natural service of *population control.* Colored wedges will be shown at the beginning of most chapters in this book to show the *natural resources* (blue wedges) and *natural services* (orange wedges) discussed in these chapters.

services such as nutrient recycling and climate control (including precipitation) renew natural resources such as topsoil and underground deposits of water (aquifers). Sustainability means living off such biological income without depleting or degrading the natural capital that provides this income.

The second step toward sustainability is to recognize that many human activities *degrade natural capital* by using normally renewable resources faster than nature can renew them (Figure 1-2, Step 2). For example, in many areas our activities erode essential topsoil faster than it can be renewed. And we withdraw water from some underground deposits (aquifers) faster than they are renewed by precipitation from nature's climate system. We also cut down (clear-cut) or burn diverse natural forests to grow crops, graze cattle, and supply us with wood and paper.

This leads us to search for *solutions* to these and other environmental problems (Figure 1-2, Step 3). For example, a solution might be to stop clear-cutting diverse mature forests.

The search for solutions often involves conflicts, and resolving these conflicts requires us to make *trade-offs,* or compromises (Figure 1-2, Step 4). To provide wood and paper and crops such as coffee, for example, we can promote the planting of tree and coffee plantations in areas that have already been cleared or degraded.

In the search for solutions, *individuals matter.* Sometimes one individual comes up with an idea for bringing about a solution. In other cases, individuals work together to bring about the political or social changes necessary for solving the problem (Figure 1-2, Step 5). Some individuals have found ways to eliminate the need to use trees to produce paper by using residues from crops (such as rice) and by planting rapidly growing plants and using their fiber to make paper. Other individuals are working politically to persuade elected officials to ban the clear-cutting of ancient forests while encouraging the planting of tree and crop plantations in areas that have already been cleared or degraded.

The five steps to sustainability must be supported by **sound science**—the concepts and ideas that are widely accepted by experts in a particular field of the natural or social sciences. For example, sound science tells us that we need to protect and sustain the many natural services provided by diverse mature forests. It also guides us in the design and management of tree and coffee plantations.

Moving toward sustainability also involves integrating information and ideas from the physical sciences (such as chemistry, biology, and geology) with those from the social sciences (such as economics, politics, and ethics). To show this integration, most sub-

headings throughout this book are tagged with the terms science, economics, politics, and ethics, or some combination of these words.

Environmentally Sustainable Societies

An environmentally sustainable society meets the basic resource needs of its people indefinitely without degrading or depleting the natural capital that supplies these resources.

An **environmentally sustainable society** meets the current needs of its people for food, clean water, clean air, shelter, and other basic resources without compromising the ability of future generations to meet their needs. *Living sustainably* means living off natural income replenished by soils, plants, air, and water and not depleting or degrading the earth's endowment of natural capital that supplies this biological income.

Imagine you win $1 million in a lottery. If you invest this money and earn 10% interest per year, you will have a sustainable annual income of $100,000 without depleting your capital. If you spend $200,000 per year, your $1 million will be gone early in the seventh year. Even if you spend only $110,000 per year, you will be bankrupt early in the eighteenth year.

The lesson here is an old one: *Protect your capital and live off the income it provides.* Deplete, waste, or squander your capital, and you will move from a sustainable to an unsustainable lifestyle.

The same lesson applies to our use of the earth's natural capital. According to many environmental scientists, we are living unsustainably by wasting, depleting, and degrading the earth's natural capital at an accelerating rate.

Some people disagree. They contend that the seriousness of human population, resource, and environmental problems has been exaggerated. They also believe we can overcome these problems by human ingenuity, economic growth, and technological advances.

1-2 POPULATION GROWTH, ECONOMIC GROWTH, AND ECONOMIC DEVELOPMENT

Human Population Growth

The rate at which world's population is growing has slowed but is still increasing rapidly.

Currently, the world's population is growing exponentially at a rate of about 1.2% per year—down from 2.2% in 1963. This may not seem like a very fast rate, but it added about 78 million people (6.5 billion ×

0.012 = 78 million) to the world's population in 2005, an average increase of 214,000 people per day, or 8,900 per hour. At this rate it takes only about 3 days to add the 652,000 Americans killed in battle in all U.S. wars and only 1.7 years to add the 129 million people killed in all wars fought in the past 200 years!

How much is 78 million? Suppose you spend 1 second saying hello to each of the 78 million new people added this year for 24 hours each day—no sleeping, eating, or anything else allowed. How long would your handshaking marathon take? Answer: about 2.5 years. By then you would have 195 million more people to shake hands with. Exponential growth is astonishing!

Economics: Economic Growth and Economic Development

Economic growth provides people with more goods and services, and economic development uses economic growth to improve living standards.

Economic growth is an increase in the capacity of a country to provide people with goods and services. Accomplishing this increase requires population growth (more producers and consumers), more production and consumption per person, or both.

Economic growth is usually measured by the percentage change in a country's **gross domestic product (GDP):** the annual market value of all goods and services produced by all firms and organizations, foreign and domestic, operating within a country. Changes in a country's economic growth per person are measured by **per capita GDP:** the GDP divided by the total population at midyear.

Economic development is the improvement of human living standards by economic growth. The United Nations (UN) classifies the world's countries as economically developed or developing based primarily on their degree of industrialization and their per capita GDP.

The **developed countries** (with 1.2 billion people) include the United States, Canada, Japan, Australia, New Zealand, and the countries of Europe. Most are highly industrialized and have high average per capita GDP. All other nations (with 5.3 billion people) are classified as **developing countries,** most of them in Africa, Asia, and Latin America. Some are *middle-income, moderately developed countries* and others are *low-income countries.*

Figure 1-4 compares some key characteristics of developed and developing countries. About 97% of the projected increase in the world's population is expected to take place in developing countries, as shown in Figure 1-5.

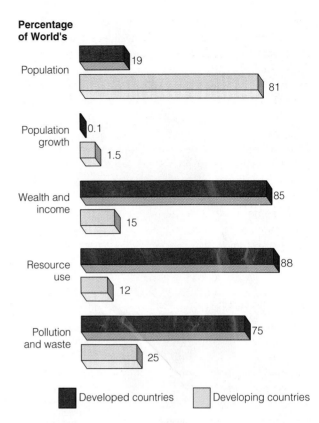

Figure 1-4 Global outlook: comparison of developed and developing countries, 2005. (Data from the United Nations and the World Bank)

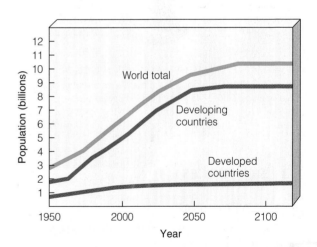

Figure 1-5 Global outlook: past and projected population size for developed countries, developing countries, and the world, 1950–2100. Developing countries are expected to account for 97% of the 2.4 billion people projected to be added to the world's population between 2005 and 2050. (Data from the United Nations)

Figure 1-6 (p. 10) summarizes some of the benefits (*good news*) and harm (*bad news*) caused mostly by economic development. It shows the effects of the wide and growing gap between the world's haves and have-nots.

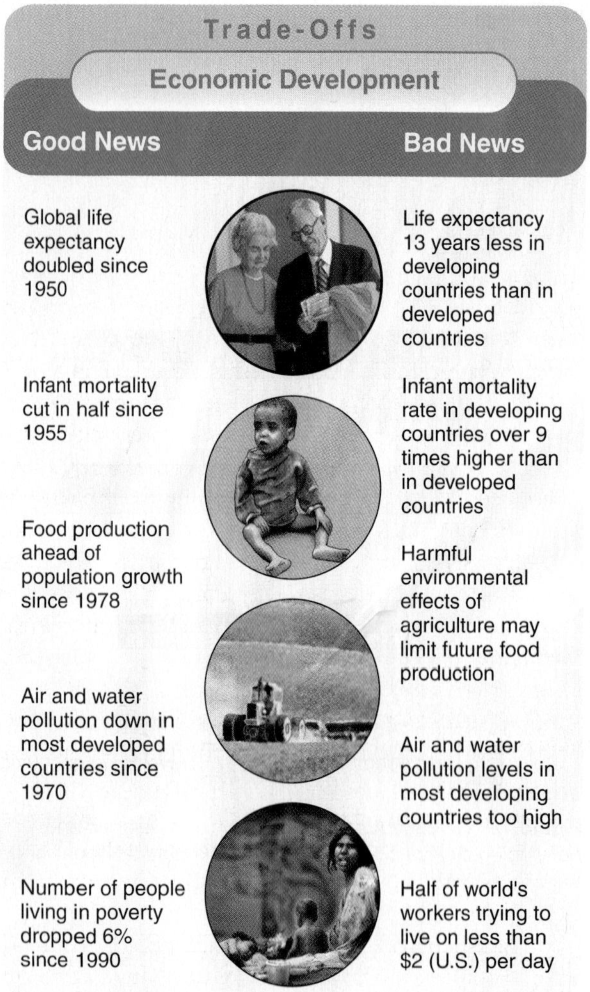

Trade-Offs

Economic Development

Good News	Bad News
Global life expectancy doubled since 1950	Life expectancy 13 years less in developing countries than in developed countries
Infant mortality cut in half since 1955	Infant mortality rate in developing countries over 9 times higher than in developed countries
Food production ahead of population growth since 1978	Harmful environmental effects of agriculture may limit future food production
Air and water pollution down in most developed countries since 1970	Air and water pollution levels in most developing countries too high
Number of people living in poverty dropped 6% since 1990	Half of world's workers trying to live on less than $2 (U.S.) per day

Figure 1-6 Trade-offs: good and bad news about economic development. *Critical thinking: pick the single pieces of good news and bad news that you believe are the most important.* (Data from the United Nations and World Health Organization)

1-3 RESOURCES

What Is a Resource?

We obtain *resources* from the environment to meet our needs and wants.

From a human standpoint, a **resource** is anything obtained from the environment to meet our needs and wants. Examples include food, water, shelter, manufactured goods, transportation, communication, and recreation. On our short human time scale, we classify the material resources we get from the environment as *perpetual* (such as sunlight, winds, and flowing water), *renewable* (such as fresh air and water, soils, forest products, and food crops), or *nonrenewable* (such as fossil fuels, metals, and sand).

Some resources, such as solar energy, fresh air, wind, fresh surface water, fertile soil, and wild edible plants, are directly available for use. Other resources, such as petroleum (oil), iron, groundwater (water found underground), and modern crops, are not directly available. They become useful to us only with some effort and technological ingenuity. For example, petroleum was a mysterious fluid until we learned how to find, extract, and convert (refine) it into gasoline, heating oil, and other products that we could sell at affordable prices. In such cases, resources are obtained by an interaction between natural capital and human capital.

Science: Perpetual and Renewable Resources

Resources renewed by natural processes are sustainable if we do not use them faster than they are replenished.

Solar energy is called a **perpetual resource** because on a human time scale it is renewed continuously. It also includes indirect forms of solar energy such as winds and flowing water. This solar capital is expected to last at least 6 billion years as the sun completes its life cycle.

On a human time scale, a **renewable resource** can be replenished fairly rapidly (hours to several decades) through natural processes as long as it is not used up faster than it is replaced. Examples include forests, grasslands, wild animals, fresh water, fresh air, and fertile soil.

Renewable resources can be depleted or degraded. The highest rate at which a renewable resource can be used *indefinitely* without reducing its available supply is called its **sustainable yield.**

When we exceed a resource's natural replacement rate, the available supply begins to shrink, a process known as **environmental degradation.** Examples of such degradation include urbanization of productive land, excessive topsoil erosion, pollution, deforestation (temporary or permanent removal of large expanses of forest for agriculture or other uses), groundwater depletion, overgrazing of grasslands by livestock, and reduction in the earth's forms of wildlife (biodiversity) by elimination of habitats and species.

Economics Case Study: The Tragedy of the Commons

Renewable resources that are freely available to everyone can be degraded.

One cause of environmental degradation is the overuse of **common-property** or **free-access resources.** No one owns these resources, and they are available to users at little or no charge. Examples include clean air, the open ocean and its fish, migratory birds, wildlife species, gases of the lower atmosphere, and space.

In 1968, biologist Garrett Hardin (1915–2003) called the degradation of renewable free-access resources the **tragedy of the commons.** It happens because each user reasons, "If I do not use this resource, someone else will. The little bit I use or pollute is not enough to matter, and such resources are renewable."

With only a few users, this logic works. Eventually, however, the cumulative effect of many people trying to exploit a free-access resource exhausts or ruins it. Then no one can benefit from it—and that is the tragedy.

One solution is to *use free-access resources at rates well below their estimated sustainable yields* by reducing population, regulating access to the resources, or both. Some communities have established rules and traditions to regulate access to common-property resources such as ocean fisheries, grazing lands, and forests. Governments have also enacted laws and international treaties to regulate access to commonly owned resources such as forests, national parks, rangelands, and fisheries in coastal waters.

Another solution is to *convert free-access resources to private ownership.* The reasoning is that if you own something, you are more likely to protect your investment.

That sounds good, but private ownership is not always the answer. Private owners do not always protect natural resources they own when this goal conflicts with protecting their financial capital or increasing their profits. For example, some private forest owners can make more money by clear-cutting timber, selling the degraded land, and investing their profits in other timberlands or businesses. Also, this approach is not practical for global common resources—such as the atmosphere, the open ocean, most wildlife species, and migratory birds—that cannot be divided up and converted to private property.

Science: Our Ecological Footprints

Supplying each person with resources and absorbing the wastes from such resource use creates a large ecological footprint or environmental impact.

The **per capita ecological footprint** is the amount of biologically productive land and water needed to supply each person with the resources he or she uses and to absorb the wastes from such resource use (Figure 1-7, left). It is an estimate of the average environmental impact of individuals in different countries and areas. The numbers shown in Figure 1-7 are estimates and should not be taken literally. But they can be used to show relative differences in resource use and waste production by countries and geographical areas. See Figures 2 and 3 in Science Supplement 2 at the end of this book for maps of our ecological footprints for the world and the United States.

Humanity's ecological footprint exceeds the earth's ecological capacity to replenish its renewable resources and absorb waste by about 21% (Figure 1-7, right). If these estimates are correct, we are drawing down renewable resources 21% faster than the earth can renew them. In other words, it will take the resources of 1.21 planet earths to support indefinitely our current production and consumption of renewable resources!

The ecological footprint of most people in developed countries is large because of their huge consumption of renewable resources. You can estimate

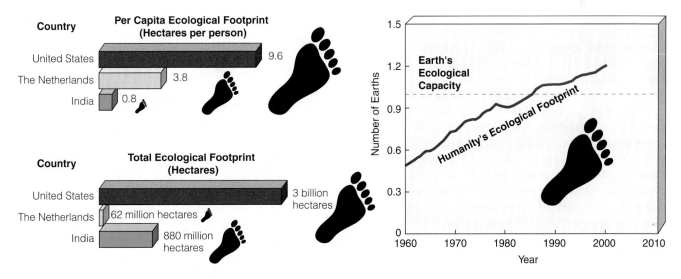

Figure 1-7 Natural capital use and degradation: relative *per capita and total ecological footprints* of the United States, the Netherlands, and India (left). The *per capita ecological footprint* is a measure of the biologically productive areas of the earth required to produce the resources required per person and absorb or break down the wastes produced by such resource use. By 2001, humanity's ecological footprint was about 21% higher than the earth's ecological capacity (right). (Data from World Wide Fund for Nature, UN Environmental Programme, and Global Footprint Network)

your own ecological footprint by visiting the website **www.redefiningprogress.org/**. Also see the Guest Essay by Michael Cain on the website for this chapter.

According to the developers of the ecological footprint concept, William Rees and Mathis Wackernagel, it would take the land area of about *four more planet earths* for the rest of the world to reach U.S. levels of consumption with existing technology. See the Guest Essay on the website for this chapter by Norman Myers about the growing ecological footprint of China.

Science: Nonrenewable Resources

Nonrenewable resources can be economically depleted to the point where it costs too much to obtain what is left.

Nonrenewable resources exist in a fixed quantity or stock in the earth's crust. On a time scale of millions to billions of years, geological processes can renew such resources. But on the much shorter human time scale of hundreds to thousands of years, these resources can be depleted much faster than they are formed.

These exhaustible resources include *energy resources* (such as coal, oil, and natural gas that cannot be recycled), *metallic mineral resources* (such as iron, copper, and aluminum that can be recycled), and *nonmetallic mineral resources* (such as salt, clay, sand, and phosphates, which usually are difficult or too costly to recycle).

Although we never completely exhaust a nonrenewable mineral resource, it becomes *economically depleted* when the costs of extracting and using what is left exceed its economic value. At that point, we have several choices: try to find more, recycle or reuse existing supplies (except for nonrenewable energy resources, which cannot be recycled or reused), waste less, use less, try to develop a substitute, or wait millions of years for more to be produced.

Some nonrenewable metallic mineral resources, such as copper and aluminum, can be recycled or reused to extend supplies. **Recycling** involves collecting waste materials, processing them into new materials, and selling these new products. For example, discarded aluminum cans can be crushed and melted to make new aluminum cans or other aluminum items that consumers can buy. **Reuse** means using a resource over and over in the same form. For example, glass bottles can be collected, washed, and refilled many times.

Recycling nonrenewable metallic resources takes much less energy, water, and other resources and produces much less pollution and environmental degradation than exploiting virgin metallic resources. Reusing such resources takes even less energy and other resources and produces less pollution and environmental degradation than recycling.

1-4 POLLUTION

Science: Sources and Harmful Effects of Pollutants

Pollutants are chemicals found at high enough levels in the environment to cause harm to people or other organisms.

Pollution is any addition to air, water, soil, or food that threatens the health, survival, or activities of humans or other living organisms. Pollutants can enter the environment naturally (for example, from volcanic eruptions) or through human activities (for example, from burning coal). Most pollution from human activities occurs in or near urban and industrial areas, where pollution sources such as cars and factories are concentrated. Industrialized agriculture also is a major source of pollution. Some pollutants contaminate the areas where they are produced; others are carried by wind or flowing water to other areas.

The pollutants we produce come from two types of sources. **Point sources** of pollutants are single, identifiable sources. Examples are the smokestack of a coal-burning power or industrial plant (Figure 1-8), the drainpipe of a factory, or the exhaust pipe of an automobile. **Nonpoint sources** of pollutants are dispersed and often difficult to identify. Examples are pesticides sprayed into the air or blown by the wind into the atmosphere, and runoff of fertilizers and pesticides from farmlands and suburban lawns and gardens into streams and lakes. It is much easier and cheaper to identify and control pollution from point sources than from widely dispersed nonpoint sources.

Pollutants can have three types of unwanted effects. *First*, they can disrupt or degrade life-support systems for humans and other species. *Second*, they can damage wildlife, human health, and property. *Third*, they can create nuisances such as noise and unpleasant smells, tastes, and sights.

Science: Solutions to Pollution

We can try to prevent production of pollutants or clean them up after they have been produced.

We use two basic approaches to deal with pollution. One is **pollution prevention,** or **input pollution control,** which reduces or eliminates the production of pollutants. The other is **pollution cleanup,** or **output pollution control,** which involves cleaning up or diluting pollutants after they have been produced.

Environmental scientists have identified three problems with relying primarily on pollution cleanup. *First,* it is only a temporary bandage as long as population and consumption levels grow without corresponding improvements in pollution control technology. For example, adding catalytic converters to car exhaust systems has reduced some forms of air pollution. At the same time, increases in the number of

Figure 1-8 Natural capital degradation: point-source air pollution from a pulp mill in New York State.

cars and in the total distance each travels have reduced the effectiveness of this cleanup approach.

Second, cleanup often removes a pollutant from one part of the environment only to cause pollution in another. For example, we can collect garbage, but the garbage is then *burned* (perhaps causing air pollution and leaving toxic ash that must be put somewhere), *dumped* into streams, lakes, and oceans (perhaps causing water pollution), or *buried* (perhaps causing soil and groundwater pollution).

Third, once pollutants become dispersed into the environment at harmful levels, it usually costs too much to reduce them to acceptable levels.

Pollution prevention (front-of-the-pipe) and pollution cleanup (end-of-the-pipe) solutions are both needed. Environmental scientists and some economists urge us to put more emphasis on prevention because it works better and is cheaper than cleanup.

1-5 ENVIRONMENTAL PROBLEMS: CAUSES AND CONNECTIONS

Key Environmental Problems and Their Basic Causes

The major causes of environmental problems are population growth, wasteful resource use, poverty, poor environmental accounting, and ecological ignorance.

We face a number of interconnected environmental and resource problems (Figure 1-9). The first step in dealing with these problems is to identify their underlying causes, listed in Figure 1-10 (p. 14).

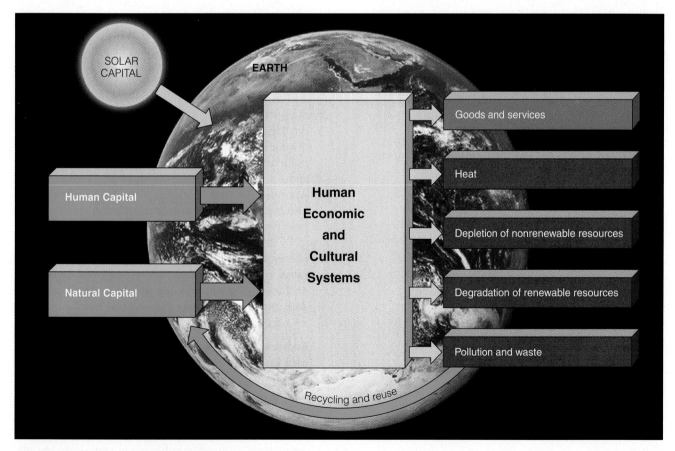

Figure 1-9 Natural capital use, depletion, and degradation: human and natural capital produce an amazing array of goods and services for most of the world's people. But this rapid flow of material resources through the world's economic systems depletes nonrenewable resources, degrades renewable resources, and adds heat, pollution, and waste to the environment.

Causes of Environmental Problems

Population growth

Unsustainable resource use

Poverty

Not including the environmental costs of economic goods and services in their market prices

Trying to manage and simplify nature with too little knowledge about how it works

Figure 1-10 Natural capital depletion and degradation: five basic causes of the environmental problems we face. *Critical thinking: can you think of any other basic causes?*

Economics: Poverty and Environmental Problems

Poverty is a major threat to human health and the environment.

Many of the world's poor do not have access to the basic necessities for a healthy, productive, and decent life (see Figure 1-11). Their daily lives are focused on getting enough food, water, and fuel to survive. Desperate for land to grow enough food, many of the world's poor people deplete and degrade forests, soil, grasslands, and wildlife for short-term survival. They do not have the luxury of worrying about long-term environmental quality or sustainability.

Poverty also affects population growth. Poor people often have many children as a form of economic security. Their children help them gather fuel (mostly wood and dung), haul drinking water, tend crops and livestock, work, and beg in the streets. The children also help their parents survive in their old age before they die, typically in their fifties in the poorest countries. The poor do not have retirement plans, social security, or government-sponsored health plans.

Many of the world's desperately poor die prematurely from four preventable health problems. The first problem is *malnutrition,* caused by a lack of protein and other nutrients needed for good health (Figure 1-12). The second problem is increased susceptibility to normally nonfatal infectious diseases, such as diarrhea and measles, caused by their weakened condition from malnutrition. A third factor is lack of access to clean drinking water. A fourth factor is severe respiratory disease and premature death from inhaling indoor air pollutants produced by burning wood or coal for heat and cooking in open fires or in poorly vented stoves. According to the World Health Organization, these four factors cause premature death for at least 7 million poor people each year or about 800 people per hour.

This premature death of about 19,200 humans per day is equivalent to 48 fully loaded 400-passenger jumbo jet planes crashing every day with no survivors! Two-thirds of those dying are children younger than age 5. The daily news rarely covers this ongoing human tragedy.

Economics and Ethics: Resource Consumption and Environmental Problems

Many consumers in developed countries have become addicted to buying more and more stuff in their search for fulfillment and happiness.

Affluenza ("af-loo-EN-zuh") is a term used to describe the unsustainable addiction to overconsumption and materialism exhibited in the lifestyles of affluent consumers in the United States and other developed coun-

Lack of access to	Number of people (% of world's population)
Adequate sanitation	2.4 billion (37%)
Enough fuel for heating and cooking	2 billion (31%)
Electricity	1.6 billion (25%)
Clean drinking water	1.1 billion (17%)
Adequate health care	1.1 billion (17%)
Enough food for good health	1.1 billion (17%)

Figure 1-11 Natural capital degradation: some harmful results of poverty. *Critical thinking: which two of these effects do you believe are the most harmful?* (Data from United Nations, World Bank, and World Health Organization)

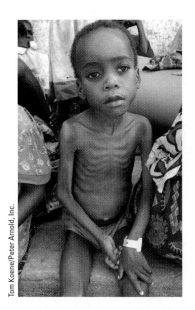

Figure 1-12 Global outlook: one in every three children under age 5, such as this child in Lunda, Angola, suffers from severe malnutrition caused by a lack of calories and protein. According to the World Health Organization, each day at least 13,700 children under age 5 die prematurely from malnutrition and infectious diseases, most from drinking contaminated water and being weakened by malnutrition.

Tom Koene/Peter Arnold, Inc.

tries. It is based on the assumption that buying more things can, should, and does buy happiness.

Most people infected with this virulent and contagious *shop-'til-you-drop virus* have some telltale symptoms. They feel overworked, have high levels of debt and bankruptcy, suffer from increasing stress and anxiety, have declining health, and feel unfulfilled in their quest to accumulate ever more stuff. As humorist Will Rogers said, "Too many people spend money they haven't earned to buy things they don't want, to impress people they don't like."

Globalization and global advertising are now spreading the virus throughout much of the world. Affluenza has an enormous environmental impact. It takes about 27 tractor-trailer loads of resources per year to support one American, or 7.9 billion truckloads per year to support the entire U.S. population. Stretched end-to-end, these trucks would more than reach the sun!

What can we do about affluenza? The first step in the rehabilitation of shopping addicts is to admit they have a problem. Next, they begin to kick their addiction by going on a "stuff" diet—that is, by living more simply. For example, before buying anything a person with affluenza should ask: Do I really need this or merely want it? Can I buy it secondhand (reuse)? Can I borrow it from a friend or relative? Another strategy: Do not hang out with other addicts. Shopaholics should avoid malls as much as they can.

After a lifetime of studying the growth and decline of the world's human civilizations, historian Arnold Toynbee summarized the true measure of a civilization's growth as the *law of progressive simplification:* "True growth occurs as civilizations transfer an increasing proportion of energy and attention from the material side of life to the nonmaterial side and thereby develop their culture, capacity for compassion, sense of community, and strength of democracy."

Economics: Positive Effects of Affluence on Environmental Quality

Affluent countries have more money for improving environmental quality.

Affluence need not lead to environmental degradation. Instead, it can prompt people to become more concerned about environmental quality, and it provides money for developing technologies to reduce pollution, environmental degradation, and resource waste. This explains why most of the important environmental progress made since 1970 has occurred in developed countries.

In the United States (and other developed countries), the air is cleaner, drinking water is purer, most rivers and lakes are cleaner, and the food supply is more abundant and safer than they were in 1970. Also, the country's total forested area is larger than it was in 1900, and most energy and material resources are used more efficiently. Similar advances have been made in most other affluent countries. Affluence financed these improvements in environmental quality.

Connections between Environmental Problems and Their Causes

Environmental quality is affected by interactions between population size, resource consumption, and technology.

Once we have identified environmental problems and their root causes, the next step is to understand how they are connected to one another. The three-factor model in Figure 1-13 (p. 16) is a starting point.

According to this simple model, the environmental impact (**I**) of a population on a given area depends on three key factors: the number of people (**P**), the average resource use per person (affluence, **A**), and the beneficial and harmful environmental effects of the technologies (**T**) used to provide and consume each unit of resource and control or prevent the resulting pollution and environmental degradation.

In developing countries, population size and the resulting degradation of renewable resources (as the poor struggle to stay alive) tend to be the key factors in total environmental impact (Figure 1-13, top). In such countries per capita resource use is low.

In developed countries, high rates of per capita resource use (affluenza) and the resulting high levels of pollution and environmental degradation per person usually are the key factors determining overall environmental impact (Figure 1-13, bottom) and a country's ecological footprint per person (Figure 1-7). For example, the average U.S. citizen consumes about 30 times as much as the average citizen of India and 100 times as much as the average person in the world's poorest countries. *Poor parents in a developing country would need 60–200 children to have*

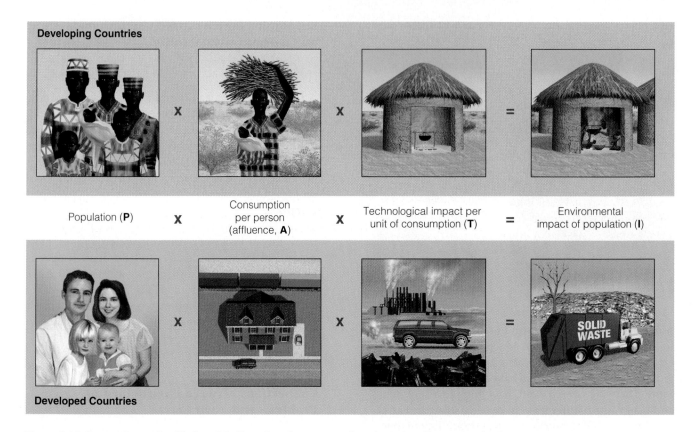

Developing Countries

Population (**P**)   **X**   Consumption per person (affluence, **A**)   **X**   Technological impact per unit of consumption (**T**)   **=**   Environmental impact of population (**I**)

Developed Countries

Figure 1-13 Connections: simplified model of how three factors—number of people, affluence, and technology—affect the environmental impact of the population in developing countries (top) and developed countries (bottom).

the same lifetime resource consumption as 2 children in a typical U.S. family.

Some forms of technology, such as polluting factories and motor vehicles and energy-wasting devices, increase environmental impact by raising the T factor in the equation. Other technologies, such as pollution control and prevention, solar cells, and energy-saving devices, lower environmental impact by decreasing the T factor. In other words, some forms of technology are *environmentally harmful* and some are *environmentally beneficial*.

1-6 CULTURAL CHANGES AND SUSTAINABILITY

Major Human Cultural Changes

Since our hunter–gatherer days, three major cultural changes have increased our impact on the environment.

Evidence from fossils and studies of ancient cultures suggests that the current form of our species, *Homo sapiens sapiens*, has walked the earth for only 60,000 years (some recent evidence suggests 90,000–195,000 years)—less than an eye-blink in this marvelous planet's 3.7 billion years of life.

Until about 12,000 years ago, we were mostly hunter–gatherers who typically moved as needed to find enough food for survival. Since then, three major cultural changes have occurred: the *agricultural revolution* (which began 10,000–12,000 years ago), the *industrial–medical revolution* (which began about 275 years ago), and the *information–globalization revolution* (which began about 50 years ago).

These major cultural changes have significantly increased our impact on the environment. They have given us much more energy and new technologies with which to alter and control more of the planet to meet our basic needs and increasing wants. They have also allowed expansion of the human population, mostly because of increased food supplies and longer life spans. In addition, they have greatly increased the resource use, pollution, and environmental degradation that threaten the long-term sustainability of human cultures.

Eras of Environmental History in the United States

The environmental history of the United States consists of the tribal, frontier, early conservation, and modern environmental eras.

The environmental history of the United States can be divided into four eras. During the *tribal era*, 5–10 million tribal people (now called Native Americans) occupied North America for at least 10,000 years before European settlers began arriving in the early 1600s. Many Native American cultures had a deep respect for the land and its animals and did not believe in land ownership.

The *frontier era* (1607–1890) began when European colonists started settling North America. They inhabited a continent offering seemingly inexhaustible forest and wildlife resources and rich soils. As a result, these early colonists developed a **frontier environmental worldview:** They viewed most of the continent as having vast resources and as a wilderness to be conquered by settlers clearing and planting land as they spread across the continent.

Next came the *early conservation era* (1832–1870), during which some people became alarmed at the scope of resource depletion and degradation in the United States during the latter part of the frontier era. They urged that part of the unspoiled wilderness on public lands owned jointly by all people (but managed by the government) be protected as a legacy to future generations.

This period was followed by an era—lasting from 1870 to the present—characterized by an increased role *of the federal government and private citizens in resource conservation, public health, and environmental protection.* Science Supplement 3 at the end of this book summarizes some of the major events during these periods.

1-7 IS OUR PRESENT COURSE SUSTAINABLE?

Are Things Getting Better or Worse?

There is good and bad environmental news.

Experts disagree about how serious our population and environmental problems are and what we should do about them. Some suggest that human ingenuity and technological advances will allow us to clean up pollution to acceptable levels, find substitutes for any scarce resources, and keep expanding the earth's ability to support more humans. They accuse most scientists and environmentalists of exaggerating the seriousness of the problems we face and of failing to appreciate the progress we have made in improving quality of life and protecting the environment.

Many leading environmental scientists disagree with this view. They cite evidence that we are degrading and disrupting many of the earth's life-support systems at an accelerating rate. They are greatly encouraged by the progress made in increasing average life expectancy, reducing infant mortality, increasing food supplies, and reducing many forms of pollution—especially in developed countries. At the same time, they point out that we need to use the earth in a way that is more sustainable for both present and future human generations and other species that support us and for other forms of life.

The most useful answer to the question of whether things are getting better or worse is *both.* Some things are getting better and some are getting worse.

Our challenge is not to get trapped into confusion and inaction by listening primarily to either of two groups of people. *Technological optimists* tend to overstate the situation by telling us to be happy and not to worry, because technological innovations and conventional economic growth and development will lead to a wonderworld for everyone. In contrast, *environmental pessimists* overstate the problems to the point where our environmental situation seems hopeless. According to the noted conservationist Aldo Leopold, "I have no hope for a conservation based on fear."

X *HOW WOULD YOU VOTE?** * Is the society you live in on an unsustainable path? Cast your vote online at http:// biology.brookscole.com/miller11.

Economics: Environmentally Sustainable Economic Development

Environmentally sustainable economic development rewards environmentally beneficial and sustainable activities and discourages environmentally harmful and unsustainable activities.

During this century, many analysts challenge us to put much greater emphasis on **environmentally sustainable economic development.** Figure 1-14 (p. 18) lists some of the shifts involved in implementing an *environmental,* or *sustainability, revolution* during this century based on this concept.

This type of development uses economic rewards (mostly government subsidies and tax breaks) to *encourage* environmentally beneficial and more sustainable forms of economic growth and uses economic penalties (mostly government taxes and regulations) to *discourage* environmentally harmful and unsustainable forms of economic growth.

This book tries to present a balanced view of good and bad environmental news and the trade-offs involved in dealing with the environmental problems we face. Try not to be overwhelmed or immobilized by the bad environmental news, because there is also some *great environmental news.* We have made immense progress in improving the human condition and dealing with many environmental problems. We are learning a great deal about how nature works and sustains itself. And we have numerous scientific, technological, and economic solutions available to deal with the environmental problems we face.

---

*To cast your vote, go the website for the book and then to the appropriate chapter (in this case, Chapter 1). In most cases, you will be able to compare how you voted with others using this book throughout the United States and the rest of the world.

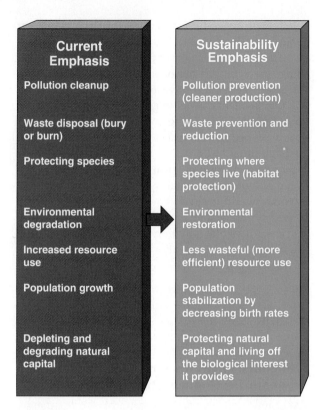

Figure 1-14 Solutions: some shifts involved in the *environmental* or *sustainability revolution.*

The challenge is to make creative use of our economic and political systems to implement such solutions. One key is to recognize that most economic and political change comes about as a result of individual actions and individuals acting together to bring about change by grassroots action from the bottom up. Social scientists suggest it takes only 5–10% of the population of a country or of the world to bring about major social change. Anthropologist Margaret Mead summarized our potential for change: "Never doubt that a small group of thoughtful, committed citizens can change the world. Indeed, it is the only thing that ever has." This means we have to accept our ethical responsibility for *stewardship* of the earth's natural capital by leaving the earth in as good, if not better, shape than we found it.

What's the use of a house if you don't have a decent planet to put it on?
HENRY DAVID THOREAU

CRITICAL THINKING

1. List **(a)** three forms of economic growth you believe are environmentally unsustainable and **(b)** three forms you believe are environmentally sustainable.

2. Give three examples of how you cause environmental degradation as a result of the tragedy of the commons.

3. Explain why you agree or disagree with the following propositions:

  a. Stabilizing population is not desirable because without more consumers, economic growth would stop.
  b. The world will never run out of resources because we can use technology to find substitutes and to help us reduce resource waste.

4. See if the affluenza bug infects you by indicating whether you agree or disagree with the following statements.
  a. I am willing to work at a job I despise so I can buy lots of stuff.
  b. When I am feeling down, I like to go shopping to make myself feel better.
  c. I would rather be shopping right now.
  d. I owe more than $1,000 on my credit cards.
  e. I usually make only the minimum payment on my monthly credit card bills.
  f. I am running out of room to store my stuff.
If you agree with two of these statements, you are infected with affluenza. If you agree with more than two, you have a serious case of affluenza. Compare your answers with those of your classmates and discuss the effects of the results on the environment and your feelings of happiness.

5. When you read that at least 19,200 people die prematurely each day (13 per minute) from preventable malnutrition and infectious disease, do you **(a)** doubt whether it is true, **(b)** not want to think about it, **(c)** feel hopeless, **(d)** feel sad, **(e)** feel guilty, or **(f)** want to do something about this problem?

6. How do you feel when you read that **(1)** the average American consumes 30 times more resources than the average citizen of India, and **(2)** human activities are projected to make the earth's climate warmer: **(a)** skeptical about those statements' accuracy, **(b)** indifferent, **(c)** sad, **(d)** helpless, **(e)** guilty, **(f)** concerned, or **(g)** outraged? Which of these feelings help perpetuate such problems, and which can help alleviate these problems?

7. Make a list of the resources you truly need. Then make another list of the resources you use each day only because you want them. Finally, make a third list of resources you want and hope to use in the future. Compare your lists with those compiled by other members of your class, and relate the overall result to the tragedy of the commons.

LEARNING ONLINE

The website for this book includes review questions for the entire chapter, flash cards for key terms and concepts, a multiple-choice practice quiz, interesting Internet sites, references, and a guide for accessing thousands of InfoTrac® College Edition articles.
  Visit

http://biology.brookscole.com/miller11

Then choose Chapter 1, and select a learning resource. For access to animations, additional quizzes, chapter outlines and summaries, register and log in to

Environmental Science ⊛ Now™

at **esnow.brookscole.com/miller11** using the access code card in the front of your book.

2

Science, Matter, and Energy

CASE STUDY

An Environmental Lesson from Easter Island

Let me tell you a sad story that has an important lesson for us.

Easter Island (Rapa Nui) is a small, isolated island in the great expanse of the South Pacific. Polynesians used double-hulled sea-going canoes to colonize this island about 2,900 years ago. They brought along their pigs, chickens, dogs, stowaway rats, taro roots, yams, bananas, and sugarcane.

The settlers found an island paradise with fertile soil that supported dense and diverse forests and lush grasses. The Polynesians developed a civilization based on two species of the island's trees, giant palms and basswoods (called hauhau). They used the towering palm trees for shelter, as tools, and to build large canoes used for catching fish such as porpoises. They felled the hauhau trees and burned them to cook and keep warm in the island's cool winters. In addition, the Polynesians made rope from these trees' fibers. Forests were also cleared to plant crops.

Life was good on Easter Island. The islanders had many children, with the population peaking at somewhere between 6,000 and 20,000 by 1400. But then residents began using the island's tree and soil resources faster than they could be renewed. As these resources became inadequate to support the growing population, the leaders of the island's different clans began appealing to the gods by carving at least 300 huge divine images from large stones (Figure 2-1). They directed the people to cut large trees to make huge platforms for the stone sculptures. They probably placed logs underneath to roll the platforms and statues or had 50 to 500 people use thick ropes to drag

Jeremy Woodhouse/WWI/Peter Arnold, Inc.

Figure 2-1 Natural capital degradation: these massive stone figures on Easter Island—some of them taller than the average five-story building—are the remains of the technology created by an ancient civilization of Polynesians. Their civilization collapsed because the people used up the trees (especially large palm trees) that were the basis of their livelihood. At least 300 of these huge stone statues once stood along the coast.

the platforms and statues across wooden rails to various locations on the island's coast.

In doing so, they used up the island's precious trees faster than they were regenerated—an example of the tragedy of the commons. By 1600, only a few small trees were left. Without large trees, the islanders could not build their traditional seagoing canoes for hunting large porpoises and catching fish in deeper offshore waters, and no one could escape the island by boat.

Without the once-great forests to absorb and slowly release water, springs and streams dried up, exposed soils eroded, crop yields plummeted, and famine struck. There was no firewood for cooking or keeping warm. The hungry islanders ate all of the island's seabirds and landbirds. Then they began raising and eating rats, descendants of hitchhikers on the first canoes to reach Easter Island.

Both the population and the civilization collapsed as rival clans fought one another for dwindling food supplies. Eventually, the islanders began to hunt and eat one another.

Dutch explorers reached the island on Easter Day, 1722, perhaps 1,000 years after the first Polynesians had landed. They found about 2,000 hungry Polynesians, living in caves on grassland dotted with shrubs.

Like Easter Island at its peak, the earth is an isolated island in the vastness of space with no other suitable planet to migrate to. As on Easter Island, our population and resource consumption are growing exponentially and our resources are finite.

Will the humans on Earth Island re-create the tragedy of Easter Island on a grander scale, or will we learn how to live more sustainably on this planet that is our only home? Scientific knowledge about how the earth works and sustains itself is a key to learning how to live more sustainably, as discussed in this chapter.

Science is an adventure of the human spirit. It is essentially an artistic enterprise, stimulated largely by curiosity, served largely by disciplined imagination, and based largely on faith in the reasonableness, order, and beauty of the universe.

WARREN WEAVER

To help us develop more sustainable societies, we need to know about the nature of science and the matter and energy that make up the earth's living and nonliving resources—the subjects of this chapter. It discusses these questions:

- What is science, and what do scientists do?

- What are the basic forms of matter, and what makes matter useful as a resource?

- What are the major forms of energy, and what makes energy useful as a resource?

- What scientific law governs changes of matter from one physical or chemical form to another?

- What three main types of nuclear changes can matter undergo?

- What are two scientific laws governing changes of energy from one form to another?

- How are the scientific laws governing changes of matter and energy from one form to another related to resource use and environmental degradation?

KEY IDEAS

- Scientists collect data and develop theories, models, and laws about how nature works.

- Scientists can establish that a particular theory, model, or law has a very high probability of being true (sound science).

- Matter consists of elements and compounds, which are in turn made up of atoms, ions, or molecules.

- When a physical or chemical change occurs, no atoms are created or destroyed (law of conservation of matter).

- We cannot create or destroy energy when it is converted from one form to another in a physical or chemical change (first law of thermodynamics).

- Whenever energy is changed from one form to another, we always end up with less usable energy than we started with (second law of thermodynamics).

- The earth's matter and energy laws that we cannot violate are lessons from nature about how to achieve more sustainable societies.

2-1 THE NATURE OF SCIENCE

What Is Science, and What Do Scientists Do?

Scientists collect data, form hypotheses, and develop theories, models, and laws about how nature works.

Science is an attempt to discover order in nature and use that knowledge to make predictions about what is likely to happen in nature. It is based on the assumption that events in the natural world follow orderly cause and effect patterns that can be understood through careful observation and experimentation. Figure 2-2 summarizes the scientific process. Trace the pathways in this figure.

The first thing scientists do is ask a question or identify a problem to be investigated. Then they collect **scientific data,** or facts, by making observations and measurements. They often conduct **experiments** to study some phenomenon under known conditions. The resulting scientific data or facts must be confirmed by repeated observations and measurements, ideally made by several different investigators.

The primary goal of science is not the data or facts themselves. Instead, science seeks new ideas, principles, or models that connect and explain certain scientific data and predict what is likely to happen in nature. Scientists working on a particular problem suggest a variety of possible explanations, or **scientific hypotheses,** for what they (or other scientists) observe in nature. A scientific hypothesis is an unconfirmed explanation of an observed phenomenon that can be tested by further research. To test a hypothesis, scientists may develop a **model,** an approximate representation or simulation of a system being studied. It may be an actual working model, a mental model, a pictorial model, a computer model, or a mathematical model.

Two important features of the scientific process are *reproducibility* and *peer review* of results by other scientists. Peers, or scientists working in the same field, check for reproducibility by repeating and analyzing one another's work to see if the data can be reproduced and whether proposed hypotheses are reasonable. *Peer review* happens when scientists publish details of the methods they used, the results of their experiments, and the reasoning behind their hypotheses for other scientists to examine and criticize.

If repeated experiments or tests using models support an hypothesis, it becomes a **scientific theory:** a highly reliable and widely accepted scientific hypothesis or a related group of scientific hypotheses.

Scientific theories are not to be taken lightly. They are not guesses, speculations, or suggestions. Instead, they are useful explanations of processes or natural phe-

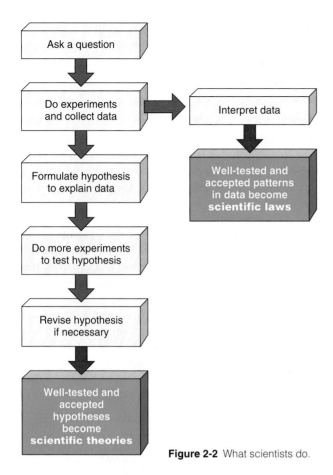

Figure 2-2 What scientists do.

nomena that have a high degree of certainty because they are supported by extensive evidence.

Nonscientists often use the word *theory* incorrectly when they actually mean *scientific hypothesis*, a tentative explanation that needs further evaluation. The statement, "Oh, that's just a theory," made in everyday conversation, implies a lack of knowledge and careful testing—the opposite of the scientific meaning of the word.

Another important result of science is a **scientific,** or **natural, law:** a description of what we find happening in nature over and over in the same way. For example, after making thousands of observations and measurements over many decades, scientists discovered the second law of thermodynamics. Simply stated, it says that heat always flows spontaneously from hot to cold—something you learned the first time you touched a hot object. Scientific laws describe repeated, consistent findings in nature, whereas scientific theories are widely accepted explanations of data and laws.

A scientific law is no better than the accuracy of the observations or measurements upon which it is based. But if the data are accurate, a scientific law cannot be broken.

Scientific Method

There are many scientific methods.

We often hear about *the* scientific method. In reality, many **scientific methods** exist: They are ways scientists gather data and formulate and test scientific hypotheses, models, theories, and laws.

Many *variables* or *factors* influence most processes or parts of nature scientists seek to understand. Ideally, scientists conduct a *controlled experiment* to isolate and study the effect of a single variable. To do such a *single-variable analysis,* scientists set up two groups. In the *experimental group*, the chosen variable is changed in a known way. In the *control group*, the chosen variable is not changed. If the experiment is designed properly, any difference between the two groups should result from a variable that was changed in the experimental group (see Science Spotlight, below).

A basic problem is that many of the phenomena environmental scientists investigate involve a huge number of interacting variables. This limitation is sometimes overcome by using *multivariable analysis*—

SCIENCE SPOTLIGHT

What Is Harming the Robins?

Suppose a scientist observes an abnormality in the growth of robin embryos in a certain area. The area has been sprayed with a pesticide, and the scientist suspects this chemical may be causing the abnormalities.

To test this hypothesis, the scientist carries out a controlled experiment. She maintains two groups of robin embryos of the same age in the laboratory. Each group is raised with the same conditions of light, temperature, food supply, and so on, except that the embryos in the experimental group are exposed to a known amount of the pesticide.

The embryos in both groups are then examined over an identical period of time. If the scientist finds a significantly larger number of the abnormalities in the experimental group than in the control group, the results support the idea that the pesticide is the culprit.

To be sure no errors occur during the procedure, the original researcher should repeat the experiment several times. Ideally, one or more other scientists should repeat the experiment independently.

Critical Thinking

Can you find flaws in this experiment that might lead you to question the scientist's conclusions? (*Hint:* What other factors in nature—but not in the laboratory—and in the embryos themselves could explain the results?)

that is, by running mathematical models on high-speed computers to analyze the interactions of many variables without having to carry out traditional controlled experiments.

Scientists use critical thinking skills (p. 3) and logical thinking to develop and evaluate scientific ideas. But scientists also try to come up with new and creative ideas to explain some of the things they observe in nature. According to physicist Albert Einstein, however, "There is no completely logical way to a new scientific idea." Intuition, imagination, and creativity are as important in science as they are in poetry, art, music, and other great adventures of the human spirit, as reflected by Warren Weaver's quotation found at the opening of this chapter.

How Valid Are the Results of Science?

Scientists try to establish that a particular theory or law has a very high probability of being true.

Scientists try to do two major things. *First,* they disprove things. *Second,* they establish that a particular theory or law has a very high probability or degree of certainty of being true. Ultimately, like scholars in any field, scientists cannot prove their theories and laws are *absolutely* true.

Although it may be extremely low, some degree of uncertainty is always present for any scientific theory, model, or law. Most scientists rarely say something like, "Cigarettes cause lung cancer." Rather, they might say, "Overwhelming evidence from thousands of studies indicates that there is a significant relationship between cigarette smoking and lung cancer."

Frontier Science and Sound Science

Scientific results fall into two categories: those that have not been confirmed (frontier science) and those that have been well tested and widely accepted (sound science).

News reports often focus on two things: so-called scientific breakthroughs, and disputes between scientists over the validity of preliminary and untested data and hypotheses. These preliminary results, called **frontier science,** are often controversial because they have not been widely tested and accepted. At the frontier stage, it is normal and healthy for reputable scientists to disagree about the meaning and accuracy of data and the validity of various hypotheses.

By contrast, **sound science** (also known as **consensus science**) consists of data, theories, and laws that are widely accepted by scientists who are considered experts in the field. The results of sound science are based on a self-correcting process of open peer review. To find out what scientists generally agree on, you can seek out reports by scientific bodies such as the U.S. National Academy of Sciences and the British Royal Society, which attempt to summarize consensus among experts in key areas of science.

Junk Science

Junk science is untested ideas presented as sound science.

Junk science consists of scientific results or hypotheses that are presented as sound science but have not undergone the rigors of peer review. Two problems arise in uncovering junk science. First, some scientists, politicians, and other analysts label as junk science any science that does not support or further their particular agenda. Second, reporters and journalists sometimes mislead their audiences by presenting sound or consensus science along with a quote from a scientist in the field who disagrees with the consensus view or from someone who is not an expert in the field. Such attempts to give a false sense of balance or fairness can mislead the public into distrusting well-established sound science. See the Guest Essay on environmental reporting by Andrew C. Revkin on the website for this chapter.

Here are some critical thinking questions you can use to uncover junk science.

- How reliable are the sources making a particular claim? Do they have a hidden agenda? Are they experts in this field? What is their source of funding?

- Do the conclusions follow logically from the observations?

- Has the claim been verified by impartial peer review?

- How does the claim compare with the consensus view of experts in this field?

2-2 MATTER

Elements and Compounds

Matter exists in chemical forms as elements and compounds.

Matter is anything that has mass (the amount of material in an object) and takes up space. Scientists classify matter as existing in various levels of organization (Figure 2-3). Chapter 3 gives details about the five levels of matter that make up the realm of ecology.

Matter is found in two chemical forms. One is **elements:** the distinctive building blocks of matter that make up every material substance. The other consists of **compounds:** two or more different elements held together in fixed proportions by attractive forces called *chemical bonds.*

To simplify things, chemists represent each element by a one- or two-letter symbol. Examples used in this book are hydrogen (H), carbon (C), oxygen (O), nitrogen (N), phosphorus (P), sulfur (S), chlorine (Cl),

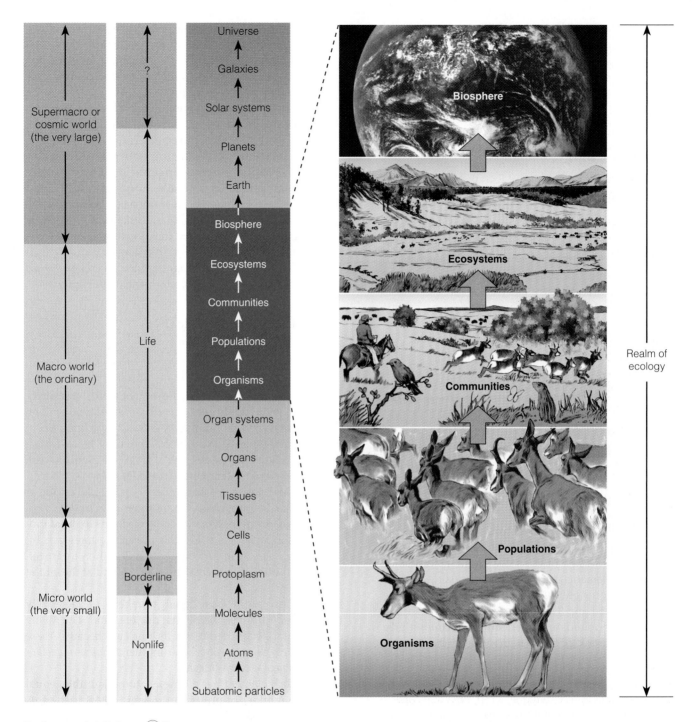

Environmental Science ⊕ Now™

Active Figure 2-3 Natural capital: levels of organization of matter in nature. Ecology focuses on five of these levels. *See an animation based on this figure and take a short quiz on the concept.*

fluorine (F), bromine (Br), sodium (Na), calcium (Ca), lead (Pb), mercury (Hg), arsenic (As), and uranium (U). *Good news.* The elements in the list above are the only ones you need to know to understand the material in this book.

From a chemical standpoint, how much are you worth? Not much. If we add up the market price per kilogram for each element in someone weighing 70 kilograms (154 pounds), the total value comes to about $120. Not very uplifting, is it?

Of course, you are worth much more because your body is not just a bunch of chemicals enclosed in a bag of skin. You are an incredibly complex system in which air, water, soil nutrients, energy-storing chemicals, and food chemicals interact in millions of ways to keep you alive and healthy. Feel better now?

Atoms, Ions, and Compounds

Atoms, ions, and compounds are the building blocks of matter.

If you had a supermicroscope capable of looking at individual elements and compounds, you could see they are made up of three types of building blocks. The first type is an **atom:** the smallest unit of matter that has the characteristics of a particular element.

The second type is an **ion:** an electrically charged atom or combination of atoms. Examples encountered in this book include *positive* hydrogen ions (H^+), sodium ions (Na^+), calcium ions (Ca^{2+}), and ammonium ions (NH_4^+) and *negative* chloride ions (Cl^-), nitrate ions (NO_2^-), sulfate ions (SO_4^{2-}), and phosphate ions (PO_4^{3-}).

A third building block is a *compound:* a substance containing atoms or ions of more than one element that are held together by chemical bonds. Chemists use a **chemical formula** to show the number of atoms (or ions) of each type in a compound. This shorthand contains the symbol for each element present and uses subscripts to represent the number of atoms or ions of each element in the compound's basic structural unit. Examples of compounds and their formulas encountered in this book are water (H_2O, read as "H-two-O"), oxygen (O_2), ozone (O_3), nitrogen (N_2), nitrous oxide (N_2O), nitric oxide (NO), hydrogen sulfide (H_2S), carbon monoxide (CO), carbon dioxide (CO_2), nitrogen dioxide (NO_2), sulfur dioxide (SO_2), ammonia (NH_3), sulfuric acid (H_2SO_4), nitric acid (HNO_3), methane (CH_4), and glucose ($C_6H_{12}O_6$).

Table sugar, vitamins, plastics, aspirin, penicillin, and most of the chemicals in your body are **organic compounds** that contain at least two carbon atoms combined with each other and with atoms of one or more other elements, such as hydrogen, oxygen, nitrogen, sulfur, phosphorus, chlorine, and fluorine. One exception, methane (CH_4), has only one carbon atom. All other compounds are called **inorganic compounds.**

The millions of known organic (carbon-based) compounds include the following:

- *Hydrocarbons:* compounds of carbon and hydrogen atoms. An example is methane (CH_4), the main component of natural gas, and the simplest organic compound.

- *Chlorinated hydrocarbons:* compounds of carbon, hydrogen, and chlorine atoms. An example is the insecticide DDT ($C_{14}H_9Cl_5$).

- *Simple carbohydrates* (simple sugars): certain types of compounds of carbon, hydrogen, and oxygen atoms. An example is glucose ($C_6H_{12}O_6$), which most plants and animals break down in their cells to obtain energy.

Larger and more complex organic compounds, called *polymers,* consist of a number of basic structural or molecular units (*monomers*) linked by chemical bonds, somewhat like cars linked in a freight train.

There are three major types of organic polymers: *complex carbohydrates,* consisting of two or more monomers of simple sugars (such as glucose) linked together; *proteins,* formed by linking together monomers of amino acids; and *nucleic acids* (such as DNA and RNA), made by linking sequences of monomers called nucleotides.

Genes consist of specific sequences of nucleotides found within a DNA molecule. These coded units of information about specific traits are passed from parents to offspring during reproduction. **Chromosomes** are combinations of genes that make up a single DNA molecule, together with a number of proteins. Each chromosome typically contains thousands of genes. Genetic information coded in your chromosomal DNA is what makes you different from an oak leaf, an alligator, or a flea, and from your parents. The relationships of genetic material to cells are depicted in Figure 2-4.

What Are Atoms Made Of?

Each atom has a tiny nucleus containing protons, and in most cases neutrons, and one or more electrons whizzing around somewhere outside the nucleus.

If you increased the magnification of your supermicroscope, you would find that each different type of atom contains a certain number of *subatomic particles.* There are three types of these atomic building blocks: positively charged **protons (p),** uncharged **neutrons (n),** and negatively charged **electrons (e).**

Each atom consists of an extremely small center, or **nucleus.** It contains one or more protons, and in most cases neutrons, and one or more electrons in rapid motion somewhere outside the nucleus. Atoms are incredibly small. In fact, more than 3 million hydrogen atoms could sit side by side on the period at the end of this sentence.

Each atom has equal numbers of positively charged protons inside its nucleus and negatively charged electrons outside its nucleus. Because these electrical charges cancel one another, *the atom as a whole has no net electrical charge.*

Each element has a unique **atomic number,** equal to the number of protons in the nucleus of each of its atoms. The simplest element, hydrogen (H), has only 1 proton in its nucleus, so its atomic number is 1. Carbon (C), with 6 protons, has an atomic number of 6, whereas uranium (U), a much larger atom, has 92 protons and an atomic number of 92.

Because electrons have so little mass compared with the masses of protons or neutrons, *most of an atom's mass is concentrated in its nucleus.* The mass of an atom is described in terms of its **mass number:** the total number of neutrons and protons in its nucleus. For example, a hydrogen atom with 1 proton and no neutrons in its nucleus has a mass number of 1, and a uranium atom with 92 protons and 143 neutrons in its nucleus has a mass number of 235 (92 + 143 = 235).

A human body contains trillions of cells, each with an identical set of genes.

There is a nucleus inside each human cell (except red blood cells).

Each cell nucleus has an identical set of chromosomes, which are found in pairs.

A specific pair of chromosomes contains one chromosome from each parent.

Each chromosome contains a long DNA molecule in the form of a coiled double helix.

Genes are segments of DNA on chromosomes that contain instructions to make proteins—the building blocks of life.

The genes in each cell are coded by sequences of nucleotides in their DNA molecules.

Figure 2-4 Relationships among cells, nuclei, chromosomes, DNA, and genes.

All atoms of an element have the same number of protons in their nuclei. They may have different numbers of uncharged neutrons in their nuclei, however, and therefore different mass numbers. Forms of an element having the same atomic number but different mass numbers are called **isotopes** of that element. Scientists identify isotopes by attaching their mass numbers to the name or symbol of the element. For example, hydrogen has three isotopes: hydrogen-1 (H-1, with one proton and no neutrons in its nucleus), hydrogen-2 (H-2, common name *deuterium*, with one proton and one neutron in its nucleus), and hydrogen-3

(H-3, common name *tritium*, with one proton and two neutrons in its nucleus).

Environmental Science ⊛ Now™
Examine atoms—their parts, how they work, and how they bond together to form molecules—at Environmental ScienceNow.

Matter Quality

Matter can be classified as having high or low quality depending on how useful it is to us as a resource.

Matter quality is a measure of how useful a form of matter is to humans as a resource, based on its availability and concentration, as shown in Figure 2-5.

Figure 2-5 Examples of differences in matter quality. *High-quality matter* (left column) is fairly easy to extract and is concentrated; *low-quality matter* (right column) is more difficult to extract and is more widely dispersed than high-quality matter.

High-quality matter is concentrated, is typically found near the earth's surface, and has great potential for use as a matter resource. **Low-quality matter** is dilute, is often located deep underground or dispersed in the ocean or the atmosphere, and has little potential for use as a material resource.

An aluminum can is a more concentrated, higher-quality form of aluminum than aluminum ore containing the same amount of aluminum. It takes less energy, water, and money to recycle an aluminum can than to make a new can from aluminum ore.

Material efficiency, or **resource productivity,** is the total amount of material needed to produce each unit of goods or services. Business expert Paul Hawken and physicist Amory Lovins contend that resource productivity in developed countries could be improved by 75–90% within two decades using existing technologies.

Physical and Chemical Changes

Matter can change from one physical form to another or change its chemical composition.

When a sample of matter undergoes a **physical change,** its chemical composition does not change. For example, a piece of aluminum foil cut into small pieces is still aluminum foil. When solid water (ice) melts or liquid water boils, none of the H_2O molecules involved changes; instead, the molecules are organized in different spatial (physical) patterns.

In a **chemical change,** or **chemical reaction,** the chemical compositions of elements or compounds change. Chemists use shorthand chemical equations to represent what happens in a chemical reaction. For example, when coal burns completely, the solid carbon (C) in the coal combines with oxygen gas (O_2) from the atmosphere to form the gaseous compound carbon dioxide (CO_2).

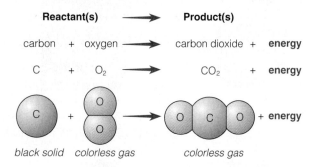

Energy is given off in this reaction, which explains why coal is a useful fuel. The reaction also shows how the complete burning of coal (or any of the carbon-containing compounds in wood, natural gas, oil, and gasoline) produces carbon dioxide. This gas helps warm the lower atmosphere (troposphere).

The Law of Conservation of Matter

When a physical or chemical change occurs, no atoms are created or destroyed.

We may change elements and compounds from one physical or chemical form to another, but we can never create or destroy any of the atoms involved in any physical or chemical change. All we can do is rearrange the elements and compounds into different spatial patterns (physical changes) or combinations (chemical changes). This finding, based on many thousands of measurements, is known as the **law of conservation of matter.** In describing chemical reactions, chemists use a shorthand system to account for all of the atoms, which they then use to balance chemical equations. See Science Supplement 4 at the end of this book for information on how to balance chemical equations.

The law of conservation of matter means there is no "away" in "to throw away." Everything we think we have thrown away remains here with us in some form. We can collect dust and soot from the smokestacks of industrial plants, but these solid wastes must then be put somewhere. We can remove substances from polluted water at a sewage treatment plant, but then we must burn them (producing some air pollution), bury them (possibly contaminating underground water supplies), or clean them up and apply the gooey sludge to the land as fertilizer (dangerous if the sludge contains toxic metals such as lead and mercury).

Types of Pollutants

The law of conservation of matter says that we will always produce some pollutants, but we can produce much less and clean up some of what we do produce.

We can make the environment cleaner and convert some potentially harmful chemicals into less harmful physical or chemical forms. But *the law of conservation of matter means we will always face the problem of what to do with some quantity of wastes and pollutants.*

Three factors determine the severity of a pollutant's harmful effects: its *chemical nature*, its *concentration*, and its *persistence*.

Concentration is sometimes expressed in terms of *parts per million (ppm)*, where 1 ppm corresponds to 1 part pollutant per 1 million parts of the gas, liquid, or solid mixture in which the pollutant is found. Smaller concentration units are parts per billion (ppb) and parts per trillion (ppt). Although we can reduce the concentration of a pollutant by dumping it into the air or into a large volume of water, there are limits to the effectiveness of this dilution approach. For example, the water flowing in a river can dilute or disperse some of the wastes dumped into it. If we dump in too much waste, however, this natural cleansing process does not work.

Persistence is a measure of how long the pollutant stays in the air, water, soil, or body. Pollutants can be classified into four categories based on their persistence:

- **Degradable,** or **nonpersistent, pollutants** are broken down completely or reduced to acceptable levels by natural physical, chemical, and biological processes.

- **Biodegradable pollutants** are complex chemical pollutants that living organisms (usually specialized bacteria) break down into simpler chemicals. Human sewage in a river, for example, is biodegraded fairly quickly by bacteria if the sewage is not added faster than it can be broken down.

- **Slowly degradable,** or **persistent, pollutants** take decades or longer to degrade. Examples include the insecticide DDT and most plastics.

- **Nondegradable pollutants** are chemicals that natural processes cannot break down. Examples include the toxic elements lead, mercury, and arsenic. Ideally, we should try not to use these chemicals. If we do, we should figure out ways to keep them from getting into the environment.

We can make the environment cleaner and convert some potentially harmful chemicals into less harmful physical or chemical forms. Nevertheless, the law of conservation of matter means we will always face the problem of what to do with some quantity of wastes and pollutants.

Nuclear Changes: Radioactive Decay, Fission, and Fusion

Nuclei of some atoms can spontaneously lose particles or give off high-energy radiation, split apart, or fuse together.

In addition to physical and chemical changes, matter can undergo a third type of change known as a *nuclear change.* There are three types of nuclear change: natural radioactive decay, nuclear fission, and nuclear fusion.

Natural radioactive decay is a nuclear change in which unstable isotopes spontaneously emit fast-moving chunks of matter (alpha particles or beta particles), high-energy radiation (gamma rays), or both at a fixed rate. The unstable isotopes are called **radioactive isotopes** or **radioisotopes.** Radioactive decay of these isotopes into other isotopes continues until it produces a stable isotope that is not radioactive.

Each type of radioisotope spontaneously decays at a characteristic rate into a different isotope. This rate of decay can be expressed in terms of *half-life:* the time needed for *one-half* of the nuclei in a given quantity of a radioisotope to decay and emit their radiation to form a different isotope.

An isotope's half-life cannot be changed by temperature, pressure, chemical reactions, or any other known factor. As a consequence, it can be used to esti-mate how long a sample of a radioisotope must be stored before it decays to a safe level. A rule of thumb is that such decay takes about 10 half-lives. Thus people must be protected from radioactive waste containing iodine-131 (which concentrates in the thyroid gland and has a half-life of 8 days) for 80 days (10 × 8 days). Plutonium-239, which is produced in nuclear reactors and used as the explosive in some nuclear weapons, can cause lung cancer when its particles are inhaled in even minute amounts. Its half-life is 24,000 years. Thus it must be stored safely for 240,000 years (10 × 24,000 years)—about four times longer than our species (*Homo sapiens sapiens*) has existed.

Exposure to alpha particles, beta particles, or gamma rays can alter DNA molecules in cells and in some cases lead to genetic defects in the next generation of offspring or several generations later. Such exposure can also damage body tissues and cause burns, miscarriages, eye cataracts, and certain cancers.

Environmental Science ⊕ Now™
Learn more about half-lives and how radioactive particles can be used by doctors to help us at Environmental ScienceNow.

Nuclear fission is a nuclear change in which the nuclei of certain isotopes with large mass numbers (such as uranium-235) are split apart into lighter nuclei when struck by neutrons; each fission releases two or three more neutrons plus energy (Figure 2-6, p. 28). Each of these neutrons, in turn, can trigger an additional fission reaction. For multiple fissions to take place, enough fissionable nuclei must be present to provide the **critical mass** needed for efficient capture of these neutrons.

Multiple fissions within a critical mass produce a **chain reaction,** which releases an enormous amount of energy (see Figure 2-6). This is somewhat like a room in which the floor is covered with spring-loaded mousetraps, each topped by a Ping-Pong ball. Open the door, throw in a single Ping-Pong ball, and watch the action in this simulated chain reaction of snapping mousetraps and balls flying around in every direction.

In an atomic bomb, an enormous amount of energy is released in a fraction of a second in an uncontrolled nuclear fission chain reaction. In the reactor of a nuclear power plant, the rate at which the nuclear fission chain reaction takes place is controlled. The heat released produces high-pressure steam to spin turbines, thereby generating electricity.

Nuclear fusion is a nuclear change in which two isotopes of light elements, such as hydrogen, are forced together at extremely high temperatures until they fuse to form a heavier nucleus. A tremendous amount of energy is released in this process. In fact, fusion of hydrogen nuclei to form helium nuclei is the source of energy in the sun and other stars.

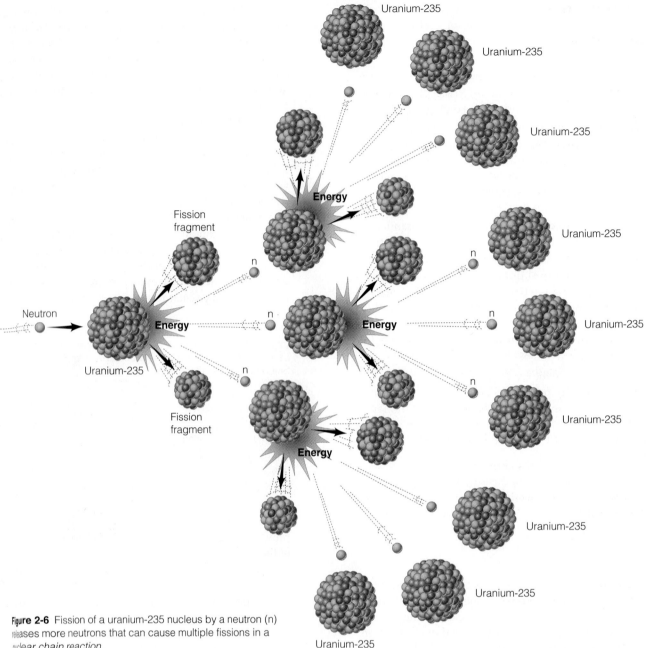

Figure 2-6 Fission of a uranium-235 nucleus by a neutron (n) releases more neutrons that can cause multiple fissions in a *nuclear chain reaction.*

After World War II, the principle of *uncontrolled nuclear fusion* was used to develop extremely powerful hydrogen, or thermonuclear, weapons. These weapons use the D–T fusion reaction, in which a hydrogen-2, or deuterium (D), nucleus and a hydrogen-3 (tritium, T) nucleus are fused to form a helium-4 nucleus, a neutron, and energy, as shown in Figure 2-7.

Scientists have also tried to develop *controlled nuclear fusion,* in which the D–T reaction produces heat that can be converted into electricity. After more than 50 years of research, this process remains in the laboratory stage. Even if it proves technologically and economically feasible, many energy experts do not expect it to become a practical source of energy until at least 2030.

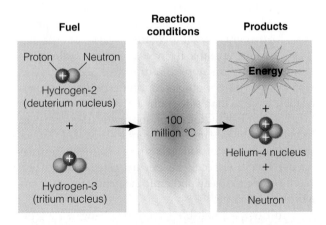

Figure 2-7 The deuterium–tritium (D–T) *nuclear fusion* reaction takes place at extremely high temperatures.

2-3 ENERGY

What Is Energy?

Energy is the work needed to move matter and the heat that flows from hot to cooler samples of matter.

Energy is the ability to do work and transfer heat. Using energy to do work means moving or lifting something such as this book, propelling a car or plane, warming your room, cooking your food, and using electricity to move electrons and light your room. The sun provides our planet with light and heat and provides plants with energy to produce the chemicals they need for growth. Animals get the energy they need from the chemical energy stored in the plant and animal tissue they eat. When you eat, your body transforms the energy stored in food into energy to do the work you need to stay alive, move, and think.

Energy exists in a number of different forms: *electrical energy* from the flow of electrons, *mechanical energy* used to move or lift matter, *light* or radiant energy produced by sunlight and electric light bulbs, *heat* when energy flows from a hot to a colder body, *chemical energy* stored in the chemical bonds holding matter together, and *nuclear* energy stored in the nuclei of atoms. These and other types of energy can be classified into two types based on whether they are *moving energy* (called kinetic energy) or *stored energy* (called potential energy).

Matter has **kinetic energy** because of its mass and its speed or velocity. Examples of this energy in motion are wind (a moving mass of air), flowing streams, heat flowing from a body at a high temperature to one at a lower temperature, and electricity (flowing electrons).

Another type of moving energy is **heat:** the total kinetic energy of all moving atoms, ions, or molecules within a given substance, excluding the overall motion of the whole object. When two objects at different temperatures contact one another, kinetic energy in the form of heat flows from the hotter object to the cooler object.

In **electromagnetic radiation,** another type of moving or kinetic energy, energy travels in the form of a *wave* as a result of the changes in electric and magnetic fields. Many different forms of electromagnetic radiation exist, each having a different *wavelength* (distance between successive peaks or troughs in the wave) and *energy content*, as shown in Figure 2-8. Such radiation travels through space at the speed of light, which is about 300,000 kilometers per second (186,000 miles per second). Visible light makes up most of the spectrum of electromagnetic radiation emitted by the sun (Figure 2-9, p. 30).

Environmental Science ⊛ Now™
Find out how color, wavelengths, and energy intensities of visible light are related at Environmental ScienceNow.

Besides moving energy, the other type of energy is **potential energy,** which is stored and potentially available for use. Examples of stored or potential energy include a rock held in your hand, an unlit match, still water behind a dam, the chemical energy stored in gasoline molecules, and the nuclear energy stored in the nuclei of atoms.

Potential energy can be changed to kinetic energy. Drop this book on your foot, and the book's potential energy when you held it changes into kinetic energy.

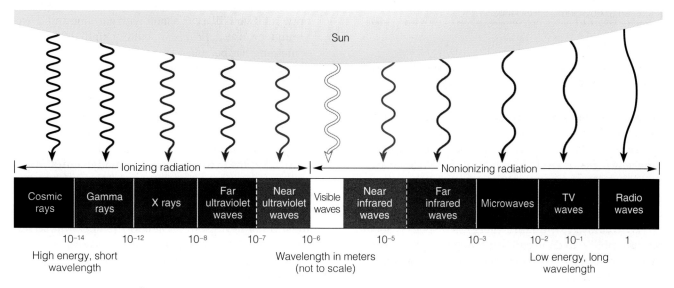

Environmental Science ⊛ Now™
Active Figure 2-8 The *electromagnetic spectrum:* the range of electromagnetic waves, which differ in wavelength (distance between successive peaks or troughs) and energy content. *See an animation based on this figure and take a short quiz on the concept.*

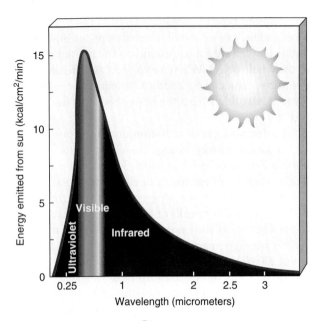

Energy emitted from sun (kcal/cm²/min)

Visible
Ultraviolet
Infrared

Wavelength (micrometers)

Active Figure 2-9 Solar capital: the spectrum of electromagnetic radiation released by the sun consists mostly of visible light. *See an animation based on this figure and take a short quiz on the concept.*

When a car engine burns gasoline, the potential energy stored in the chemical bonds of its molecules changes into heat, light, and mechanical (kinetic) energy that propel the car. Potential energy stored in a flashlight's batteries becomes kinetic energy in the form of light when the flashlight is turned on. When your body uses potential energy stored in various molecules to do work, it becomes kinetic energy.

Witness how kinetic and potential energy might be used by a Martian at Environmental ScienceNow.

Energy Quality

Energy can be classified as having high or low quality depending on how useful it is to us as a resource.

Energy quality is a measure of an energy source's ability to do useful work, as described in Figure 2-10.

High-quality energy is concentrated and can perform much useful work. Examples include electricity, the chemical energy stored in coal and gasoline, concentrated sunlight, and the nuclei of uranium-235, used as fuel in nuclear power plants.

By contrast, **low-quality energy** is dispersed and has little ability to do useful work. An example is heat dispersed in the moving molecules of a large amount of matter (such as the atmosphere or a large body of water) so that its temperature is low.

For example, the total amount of heat stored in the Atlantic Ocean is greater than the amount of high-

quality chemical energy stored in all the oil deposits of Saudi Arabia. Yet because the ocean's heat is so widely dispersed, it cannot be used to move things or to heat things to high temperatures.

The First Law of Thermodynamics

In a physical or chemical change, we can change energy from one form to another but we can never create or destroy any of the energy involved.

Scientists have observed energy being changed from one form to another in millions of physical and chemical changes. But they have never been able to detect the creation or destruction of any energy (except in nuclear changes). The results of their experiments have been summarized in the **law of conservation of energy,** also known as the **first law of thermodynamics:** *In all physical and chemical changes, energy is neither created nor destroyed, although it may be converted from one form to another.*

This scientific law tells us that when one form of energy is converted to another form in any physical or chemical change, *energy input always equals energy output.* No matter how hard we try or how clever we are, we cannot get more energy out of a system than we put in; in other words, *we cannot get something for nothing in terms of energy quantity.* This is one of Mother Nature's basic rules.

The Second Law of Thermodynamics

Whenever energy changes from one form to another, we always end up with less usable energy than we had initially.

Because the first law of thermodynamics states that energy can be neither created nor destroyed, you may be tempted to think there will always be enough energy. Yet if you fill a car's tank with gasoline and drive around or use a flashlight battery until it is dead, something has been lost. But what is it? The answer is *energy quality,* the amount of energy available that can perform useful work.

Countless experiments have shown that when energy changes from one form to another, a decrease in energy quality or ability to do useful work always occurs. The results of these experiments have been summarized in the **second law of thermodynamics:** *When energy changes from one form to another, some of the useful energy is always degraded to lower-quality, more dispersed, less useful energy.* This degraded energy usually takes the form of heat given off at a low temperature to the surroundings (environment). There it is dispersed by the random motion of air or water molecules and becomes even less useful as a resource.

In other words, *we cannot even break even in terms of energy quality because energy always goes from a more useful to a less useful form when it changes from one form to*

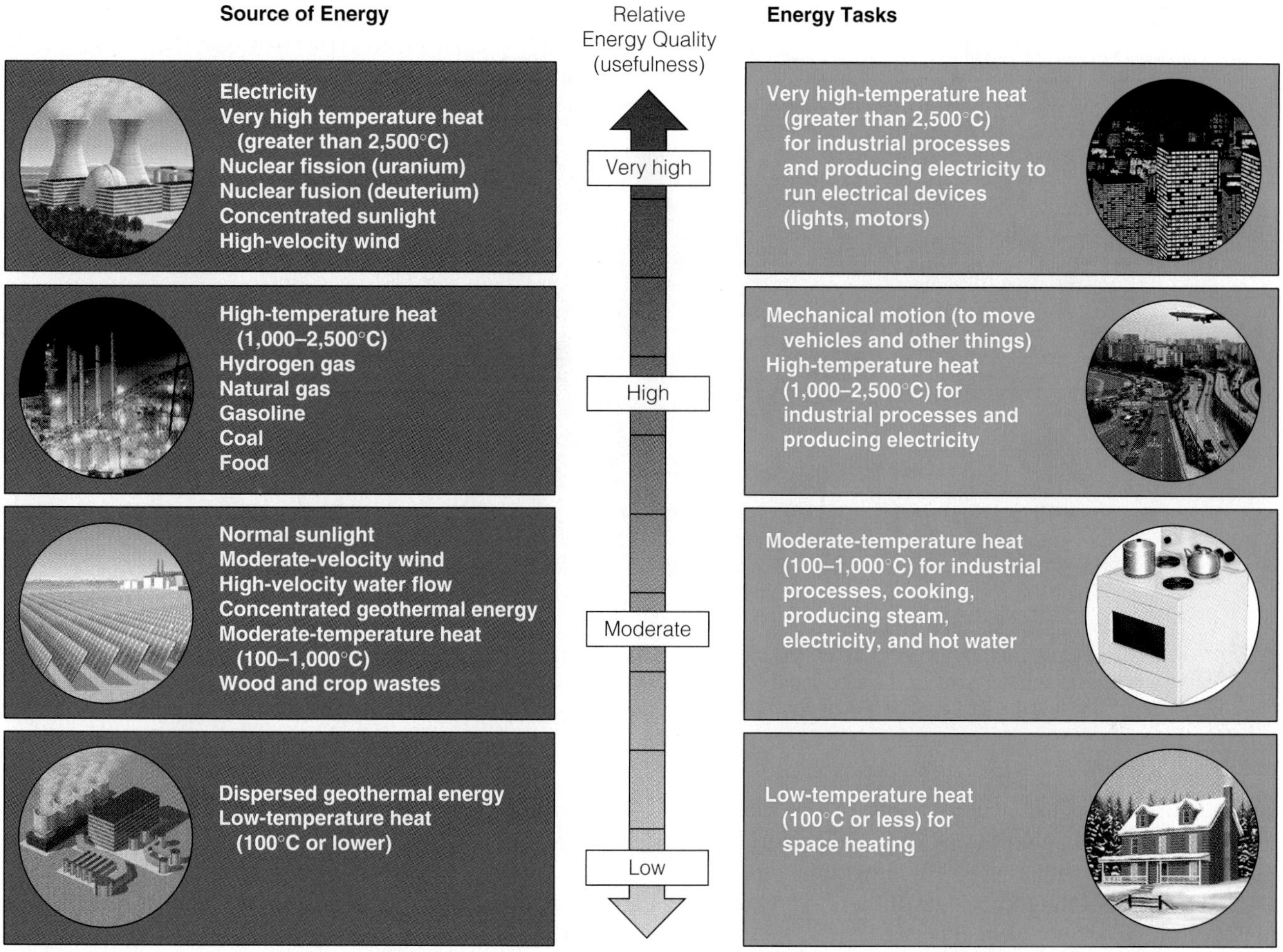

Source of Energy	Relative Energy Quality (usefulness)	Energy Tasks

Electricity
Very high temperature heat
  (greater than 2,500°C)
Nuclear fission (uranium)
Nuclear fusion (deuterium)
Concentrated sunlight
High-velocity wind

Very high

Very high-temperature heat
  (greater than 2,500°C)
  for industrial processes
  and producing electricity to
  run electrical devices
  (lights, motors)

High-temperature heat
  (1,000–2,500°C)
Hydrogen gas
Natural gas
Gasoline
Coal
Food

High

Mechanical motion (to move
  vehicles and other things)
High-temperature heat
  (1,000–2,500°C) for
  industrial processes and
  producing electricity

Normal sunlight
Moderate-velocity wind
High-velocity water flow
Concentrated geothermal energy
Moderate-temperature heat
  (100–1,000°C)
Wood and crop wastes

Moderate

Moderate-temperature heat
  (100–1,000°C) for industrial
  processes, cooking,
  producing steam,
  electricity, and hot water

Dispersed geothermal energy
Low-temperature heat
  (100°C or lower)

Low

Low-temperature heat
  (100°C or less) for
  space heating

Figure 2-10 Natural capital: categories of the quality of different sources of energy. *High-quality energy* is concentrated and has great ability to perform useful work. *Low-quality energy* is dispersed and has little ability to do useful work. To avoid unnecessary energy waste, you should match the quality of an energy source with the quality of energy needed to perform a task.

another. No one has ever found a violation of this fundamental scientific law. It is another one of Mother Nature's basic rules.

Consider three examples of the second law of thermodynamics in action. *First,* when you drive a car, only 20–25% of the high-quality chemical energy available in its gasoline fuel is converted into mechanical energy (to propel the vehicle) and electrical energy (to run its electrical systems). The remaining 75–80% is degraded to low-quality heat that is released into the environment and eventually lost into space. Thus, most of the money you spend for gasoline is not used to transport you anywhere.

Second, when electrical energy in the form of moving electrons flows through filament wires in an incandescent light bulb, it changes into about 5% useful light and 95% low-quality heat that flows into the environment. In other words, the *light bulb* is really a *heat bulb. Good news.* Scientists have developed compact fluorescent bulbs that are four times more efficient

than incandescent bulbs, and even more efficient versions are on the way. Do you use compact fluorescent bulbs?

Third, in living systems, solar energy is converted into chemical energy (food molecules) and then into mechanical energy (moving, thinking, and living). During each conversion, high-quality energy is degraded and flows into the environment as low-quality heat. Trace the flows and energy conversions in Figure 2-11 (p. 32) to see how.

The second law of thermodynamics also means *we can never recycle or reuse high-quality energy to perform useful work.* Once the concentrated energy in a serving of food, a liter of gasoline, a lump of coal, or a chunk of uranium is released, it is degraded to low-quality heat that is dispersed into the environment.

Energy efficiency, or **energy productivity,** is a measure of how much useful work is accomplished by a particular input of energy into a system. *Good news.* There is plenty of room for improving energy

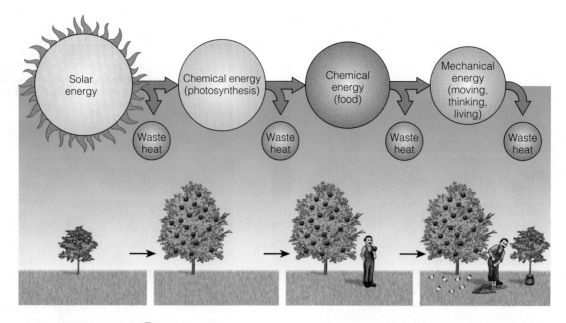

Environmental Science⊛Now™

Active Figure 2-11 The second law of thermodynamics in action in living systems. Each time energy changes from one form to another, some of the initial input of high-quality energy is degraded, usually to low-quality heat that is dispersed into the environment. *See an animation based on this figure and take a short quiz on the concept.*

efficiency. Scientists estimate that only 16% of the energy used in the United States ends up performing useful work. The remaining 84% is either unavoidably wasted because of the second law of thermodynamics (41%) or unnecessarily wasted (43%).

Thermodynamics teaches us an important lesson: The cheapest and quickest way to get more energy is to stop wasting almost half of the energy we use. We can do so by not driving gas-guzzling motor vehicles and by not living in poorly insulated and leaky houses. What are you doing to reduce your unnecessary waste of energy?

Environmental Science⊛Now™
See examples of how the first and second laws of thermodynamics apply in our world at Environmental ScienceNow.

2-4 MATTER AND ENERGY CHANGE LAWS AND SUSTAINABILITY

Unsustainable High-Throughput Economies

Most nations increase their economic growth by converting the world's resources to goods and services in ways that add large amounts of waste, pollution, and low-quality heat to the environment.

As a result of the law of conservation of matter and the second law of thermodynamics, individual resource use automatically adds some waste heat and waste matter to the environment. Most of today's advanced industrialized countries feature **high-throughput (high-**

waste) economies that attempt to boost economic growth by increasing the one-way flow of matter and energy resources through their economic systems (Figure 2-12). These resources flow through their economies into planetary *sinks* (air, water, soil, organisms), where pollutants and wastes can accumulate to harmful levels.

What happens if more people continue to use and waste more energy and matter resources at an increasing rate? In other words, what happens if most of the world's people become infected with the affluenza virus?

The law of conservation of matter and the two laws of thermodynamics discussed in this chapter tell us that eventually this resource consumption will exceed the capacity of the environment to dilute and degrade waste matter and absorb waste heat. However, they do not tell us how close we are to reaching such limits.

Matter-Recycling-and-Reuse Economies

Recycling and reusing more of the earth's matter resources slow down depletion of nonrenewable matter resources and reduce our environmental impact.

There is a way to slow down the resource use and reduce our environmental impact in a high-throughput economy. We can convert this linear economy into a circular **matter-recycling-and-reuse economy**, which mimics nature by recycling and reusing most of our matter outputs instead of dumping them into the environment.

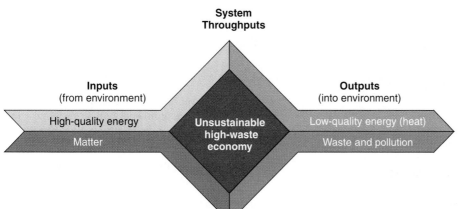

System
Throughputs

Inputs
(from environment)

High-quality energy

Matter

Unsustainable
high-waste
economy

Outputs
(into environment)

Low-quality energy (heat)

Waste and pollution

Environmental Science ⊕ Now™
Active Figure 2-12 Natural capital de-gredation: the *high-throughput economies* of most developed countries rely on continually increasing the rates of energy and matter flow. This practice produces valuable goods and services but also converts high-quality matter and energy resources into waste, pollution, and low-quality heat. *See an animation based on this figure and take a short quiz on the concept.*

Although changing to a matter-recycling-and-reuse economy will buy some time, it does not allow ever more people to use ever more resources indefinitely, even if all of them were somehow perfectly recycled and reused. The two laws of thermodynamics tell us that recycling and reusing matter resources always requires using high-quality energy (which cannot be recycled) and adds waste heat to the environment.

Sustainable Low-Throughput Economies

We can live more sustainably by reducing the throughput of matter and energy in our economies, not wasting matter and energy resources, recycling and reusing most of the matter resources we use, and stabilizing the size of our population.

Is there a better way out of our current dilemma? You bet. The three scientific laws governing matter and energy changes suggest that the best long-term solution to our environmental and resource problems is to shift from an economy based on increasing matter and energy flow (throughput) to a more sustainable **low-throughput (low-waste) economy,** as summarized in Figure 2-13. It means building on the concept of recycling and reusing as much matter as possible by also reducing the throughput of matter and energy through

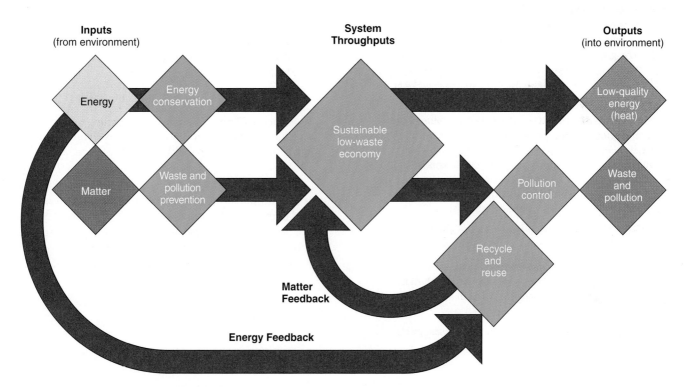

Environmental Science ⊕ Now™
Active Figure 2-13 Solutions: lessons from nature. A *low-throughput economy,* based on energy flow and matter recycling, works with nature to reduce the throughput of matter and energy resources (items shown in green). This is done by (1) reusing and recycling most nonrenewable matter resources, (2) using renewable resources no faster than they are replenished, (3) using matter and energy resources efficiently, (4) reducing unnecessary consumption, (5) emphasizing pollution prevention and waste reduction, and (6) controlling population growth. *See an animation based on this figure and take a short quiz on the concept.*

an economy. This can be done by wasting less matter and energy, living more simply to decrease per capita resource use, and slowing population growth to reduce the number of resource users. In other words, we learn to live more sustainably by heeding the *lessons from nature* revealed by the law of conservation of mass and the two laws of thermodynamics.

Since all forms of life depend on energy flow and matter recycling, their processes are governed by the three basic scientific laws of matter and thermodynamics. The next four chapters apply these laws to living systems and look at some biological principles that can teach us how to live more sustainably by learning from and working with nature.

Environmental Science ⊛ Now™

Compare how energy is used in high- and low-throughput economies at Environmental ScienceNow.

The second law of thermodynamics holds, I think, the supreme position among laws of nature. . . . If your theory is found to be against the second law of thermodynamics, I can give you no hope.

ARTHUR S. EDDINGTON

CRITICAL THINKING

1. Respond to the following statements:
   a. Scientists have not absolutely proven that anyone has ever died from smoking cigarettes.
   b. The greenhouse theory—that certain gases (such as water vapor and carbon dioxide) warm the atmosphere—is not a reliable idea because it is just a scientific theory.

2. Find an advertisement or an article describing some aspect of science in which **(a)** the concept of scientific proof is misused, **(b)** the term "theory" is used when it should have been "hypothesis," **(c)** a consensus or sound scientific finding is dismissed or downplayed because it is "only a theory," and **(d)** an example of sound science labeled as junk science for political purposes.

3. Use the library or Internet to find an example of junk science. Why is it junk science?

4. How does a scientific law (such as the law of conservation of matter) differ from a societal law (such maximum speed limits for vehicles)? Can each be broken?

5. A tree grows and increases its mass. Explain why this phenomenon is not a violation of the law of conservation of matter.

6. If there is no "away," why is the world not filled with waste matter?

7. Someone wants you to invest money in an automobile engine that will produce more energy than the energy in the fuel (such as gasoline or electricity) used to run the motor. What is your response? Explain.

8. Use the second law of thermodynamics to explain why a barrel of oil can be used only once as a fuel.

9. a. Imagine you have the power to revoke the law of conservation of matter for one day. What are the three most important things you would do with this power?
   b. Imagine you have the power to violate the first law of thermodynamics for one day. What are the three most important things you would do with this power?
   c. Imagine you have the power to violate the second law of thermodynamics for one day. What are the three most important things you would do with this power?

LEARNING ONLINE

The website for this book includes review questions for the entire chapter, flash cards for key terms and concepts, a multiple-choice practice quiz, interesting Internet sites, references, and a guide for accessing thousands of InfoTrac® College Edition articles.
   Visit

http://biology.brookscole.com/miller11

Then choose Chapter 2, and select a learning resource. For access to animations, additional quizzes, chapter outlines and summaries, register and log in to

Environmental Science ⊛ Now™

at **esnow.brookscole.com/miller11** using the access code card in the front of your book.

3

Ecosystems: What Are They and How Do They Work?

Air | Water | Soil | Energy | Biodiversity | Nutrient Recycling

CASE STUDY

Have You Thanked the Insects Today?

Insects have a bad reputation. We classify many insect species as *pests* because they compete with us for food, spread human diseases such as malaria, and invade our lawns, gardens, and houses. Some people have "bugitis"—they fear all insects and think the only good bug is a dead bug. This view fails to recognize the vital roles insects play in helping sustain life on earth.

Many of the earth's plant species (including trees) depend on insects to pollinate their flowers (Figure 3-1, left). Without pollinating insects, we would have very few fruits and vegetables to enjoy.

Insects that eat other insects—such as the praying mantis (Figure 3-1, right)—help control the populations of at least half the species of insects we call pests. This free pest control service is an important part of the earth's natural capital.

Insects have been around for at least 400 million years and are extremely successful forms of life. Some reproduce at an astounding rate. For example, a single housefly and her offspring can theoretically produce about 5.6 trillion flies in only one year.

Insects can rapidly develop new genetic traits, such as resistance to pesticides. They also have an exceptional ability to evolve into new species when faced with new environmental conditions, and they are very resistant to extinction.

The environmental lesson is clear: Although insects can thrive without newcomers such as humans, we and most other land organisms would perish without them.

Learning about insects' roles in nature requires us to understand how insects and other organisms living in a *biological community* such as a forest or pond interact with one another and with the nonliving environment. *Ecology* is the science that studies such relationships and interactions in nature.

Figure 3-1 Natural capital: the monarch butterfly feeding on pollen in a flower (left) and other insects pollinate flowering plants that serve as food for many plant eaters. The praying mantis eating a house cricket (right) and many other insect species help control the populations of at least half of the insect species we classify as pests.

The earth's thin film of living matter is sustained by grand-scale cycles of chemical elements.

G. EVELYN HUTCHINSON

This chapter describes the major components of ecosystems and the processes that sustain them. It discusses these questions:

- What is ecology?
- What basic processes keep us and other organisms alive?
- What are the major components of an ecosystem?
- What happens to energy in an ecosystem?
- What are soils, and how are they formed?
- What happens to matter in an ecosystem?
- How do scientists study ecosystems?

KEY IDEAS

- Life on earth is sustained by the flow of energy from the sun, the cycling of matter or key nutrients through the biosphere, and gravity that keeps the molecules in the atmosphere from flying off into space.

- Some organisms produce food; others consume food.

- The biodiversity found in the earth's genes, species, ecosystems, and ecosystem processes is a vital renewable resource.

- Soil supplies most of the nutrients needed for plant growth, helps purify water, and stores carbon that helps control atmospheric levels of carbon dioxide.

- Human activities are altering the flows of energy and the cycling of key nutrients through ecosystems.

3-1 THE NATURE OF ECOLOGY

What Is Ecology?

Ecology is a study of connections in nature.

Ecology (from the Greek words *oikos*, meaning "house" or "place to live," and *logos*, meaning "study of") is the study of how organisms interact with one another and with their nonliving environment. In effect, it examines *connections in nature*—the house for the earth's life. Ecologists focus on trying to understand the interactions among organisms, populations, communities, ecosystems, and the biosphere (Figure 2-3, p. 23).

An **organism** is any form of life. The **cell** is the basic unit of life in organisms. Organisms may consist of a single cell (bacteria, for instance) or many cells. Look in the mirror. You will see the result—about 10 trillion cells divided into about 200 different types.

Organisms are classified into **species,** groups of organisms that resemble one another in terms of their appearance, behavior, chemistry, and genetic makeup. Organisms that reproduce sexually by combining cells from both parents are classified as members of the same species if, under natural conditions, they can potentially breed with one another and produce live, fertile offspring. Scientists use a special system to classify and name species (see Science Supplement 5 at the end of this book).

How many species are on the earth? We do not know. Estimates range from 3.6 million to 100 million—most of them microorganisms too small to be seen with the naked eye. The best guess is that we share the planet with 10–14 million other species. So far biologists have identified and named about 1.4 million species, most of them insects (Figure 3-2 and Science Spotlight, below).

Science Spotlight: Which Species Rule the World?

Multitudes of tiny microbes such as bacteria, protozoa, fungi, and yeast help keep us alive.

They are everywhere and there are trillions of them. Billions are found inside your body, on your body, in a handful of soil, and in a cup of river water.

These mostly invisible rulers of the earth are *microbes,* a catchall term for many thousands of species of bacteria, protozoa, fungi, and yeasts—most too small to be seen with the naked eye.

Microbes do not get the respect they deserve. Most of us view them as threats to our health in the form of infectious bacteria or "germs," fungi that cause athlete's foot and other skin diseases, and protozoa that cause diseases such as malaria. In reality, these harmful microbes are in the minority.

You are alive because of multitudes of microbes toiling away mostly out of sight. Microbes convert nitrogen gas in the atmosphere into forms that plants can take up from the soil as nutrients. They also help produce foods such as bread, cheese, yogurt, vinegar, tofu, soy sauce, beer, and wine. Bacteria and fungi in the soil decompose organic wastes into nutrients that can be taken up by plants that we and most other animals eat. Without these wee creatures, we would be up to our eyeballs in waste matter.

Microbes, especially bacteria, help purify the water you drink by breaking down wastes. Bacteria in your intestinal tract break down the food you eat.

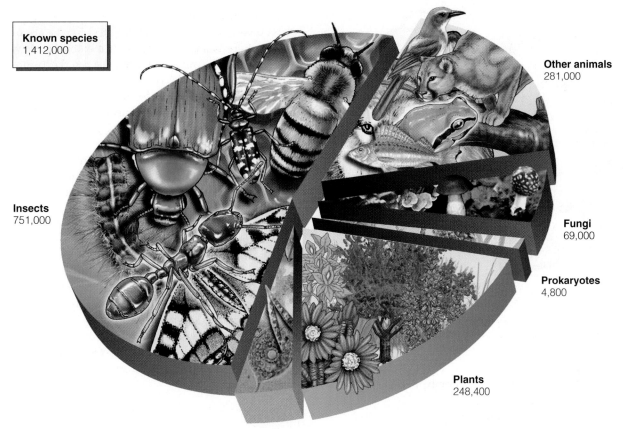

Other animals
281,000

Insects
751,000

Fungi
69,000

Prokaryotes
4,800

Plants
248,400

Protists
57,700

Figure 3-2 Natural capital: breakdown of
the earth's 1.4 million known species.

Some microbes in your nose prevent harmful bacteria
from reaching your lungs. Others are the source of dis-
ease-fighting antibiotics, including penicillin, ery-
thromycin, and streptomycin. Genetic engineers are
developing microbes that can extract metals from ores,
break down various pollutants, and help clean up
toxic waste sites.

Some microbes help control diseases that affect
plants and populations of insect species that attack our
food crops. Relying more on these microbes for pest
control can reduce the use of potentially harmful chemical pesticides.

Populations, Communities, and Ecosystems

Populations of different species living and interacting
in an area form a community, and a community
interacting with its physical environment of matter
and energy is an ecosystem.

A **population** is a group of interacting individuals of
the same species occupying a specific area (Figure 3-3).
Examples include sunfish in a pond, white oak trees in
a forest, and people in a country. In most natural populations, individuals vary slightly in their genetic makeup, which is why they do not all look or act alike. This

Figure 3-3 Natural capital: population of monarch butterflies.
The geographic distribution of this butterfly coincides with
that of the milkweed plant, on which monarch larvae and
caterpillars feed.

Figure 3-4 Natural capital: the *genetic diversity* among individuals of one species of Caribbean snail is reflected in the variations in shell color and banding patterns.

variation is called a population's **genetic diversity** (Figure 3-4).

The place where a population (or an individual organism) normally lives is its **habitat.** It may be as large as an ocean or as small as the intestine of a termite.

A **community,** or **biological community,** consists of all the populations of different species of plants, animals, and microorganisms living and interacting in an area.

An **ecosystem** is a community of different species interacting with one another and with their physical environment of matter and energy. Ecosystems can range in size from a puddle of water to a stream, a patch of woods, an entire forest, or a desert. Ecosystems can be natural or artificial (human created). Examples of artificial ecosystems include crop fields, farm ponds, and reservoirs. All of the earth's ecosystems together make up the **biosphere.**

Environmental Science ⊕ Now™
Learn more about how earth's life is organized on five levels in the study of ecology at Environmental ScienceNow.

3-2 THE EARTH'S LIFE-SUPPORT SYSTEMS

The Earth's Life-Support Systems: Four Spheres

The earth is made up of interconnected spherical layers that contain air, water, soil, minerals, and life.

We can think of the earth as consisting of several spherical layers (Figure 3-5). The **atmosphere** is a thin envelope or membrane of air around the planet. Its inner layer, the **troposphere,** extends only about 17 kilometers (11 miles) above sea level. It contains the majority of the planet's air, mostly nitrogen (78%) and oxygen (21%).

The next layer, stretching 17–48 kilometers (11–30 miles) above the earth's surface, is the **stratosphere.** Its lower portion contains enough ozone (O_3) to filter out most of the sun's harmful ultraviolet radiation. This allows life to exist on land and in the surface layers of bodies of water.

The **hydrosphere** consists of the earth's water. It is found as *liquid water* (on the planet's surface and underground), *ice* (polar ice, icebergs, and ice in frozen soil layers called *permafrost*), and *water vapor* in the atmosphere. The earth consists of an intensely hot *core,* a *mantle* consisting mostly of rock, and a thin *crust.* The **lithosphere** is the earth's crust and upper mantle.

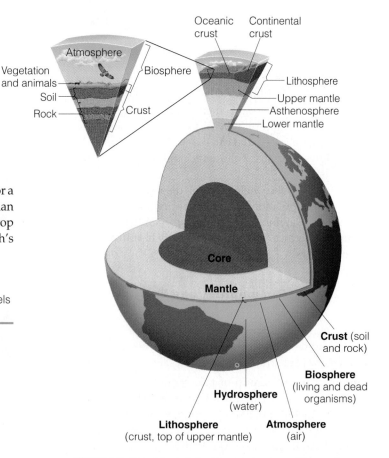

Figure 3-5 Natural capital: general structure of the earth.

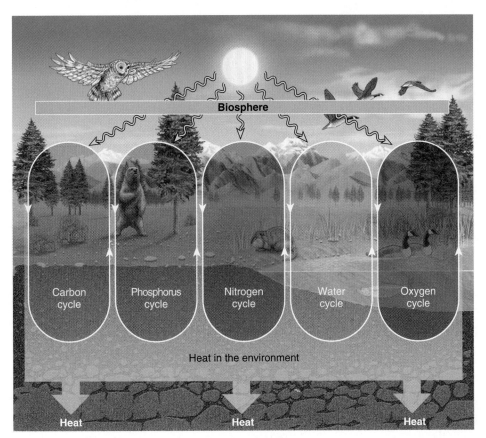

Active Figure 3-6 Natural capital: life on the earth depends on the *one-way flow of energy* (wavy arrows) from the sun through the biosphere, the *cycling of crucial elements* (solid lines around ovals), and *gravity*, which keeps atmospheric gases from escaping into space and enables chemicals to move through the matter cycles. This simplified model depicts only a few of the many cycling elements. *See an animation based on this figure and take a short quiz on the concept.*

All parts of the biosphere are interconnected, just as the parts of your body are connected. Any change in one of the biosphere's components or processes can have a ripple effect on other parts. If the earth were an apple, the biosphere would be no thicker than the apple's skin. *The goal of ecology is to understand the interactions in this thin, life-supporting global skin or membrane of air, water, soil, and organisms.*

What Sustains Life on Earth?

Solar energy, the cycling of matter, and gravity sustain the earth's life.

Life on the earth depends on three interconnected factors (Figure 3-6):

■ The *one-way flow of high-quality energy* from the sun through materials and living things in their feeding interactions, into the environment as low-quality energy (mostly heat dispersed into air or water molecules at a low temperature), and eventually back into space as heat. No round-trips are allowed because energy cannot be recycled.

■ The *cycling of matter* (the atoms, ions, or compounds needed for survival by living organisms)

through parts of the biosphere. Because the earth is closed to significant inputs of matter from space, its essentially fixed supply of nutrients must be continually recycled to support life. All nutrient trips in ecosystems are round-trips.

■ *Gravity,* which allows the planet to hold on to its atmosphere and enables the movement of chemicals between the air, water, soil, and organisms in the matter cycles.

What Happens to Solar Energy Reaching the Earth?

Solar energy flowing through the biosphere warms the atmosphere, evaporates and recycles water, generates winds, and supports plant growth.

About one-billionth of the sun's output of energy reaches the earth—a tiny sphere in the vastness of space—in the form of electromagnetic waves, mostly as visible light (Figure 2-9, p. 30). Much of this energy is either reflected away or absorbed by chemicals in the planet's atmosphere (Figure 3-7, p. 40).

Most solar radiation making it through the atmosphere is degraded into longer-wavelength infrared radiation. This infrared radiation encounters the

Environmental Science ⊕ Now™
Active Figure 3-7 Solar capital: flow of energy to
and from the earth. *See an animation based on this
figure and take a short quiz on the concept.*

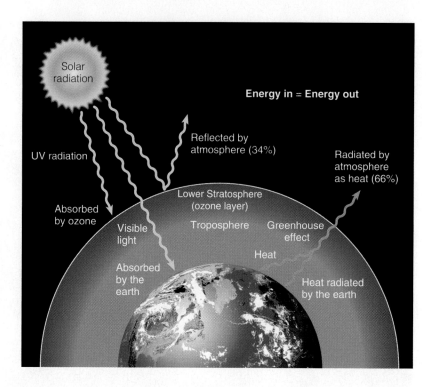

so-called *greenhouse gases* (such as water vapor, carbon dioxide, methane, nitrous oxide, and ozone) in
the troposphere. As it interacts with these gaseous
molecules, it increases their kinetic energy, helping
warm the troposphere and the earth's surface. Without this **natural greenhouse effect,** the earth would
be too cold for life as we know it to exist.

Environmental Science ⊕ Now™
Learn more about the flow of energy—from sun to earth and
within the earth's systems—at Environmental ScienceNow.

Why Is the Earth So Favorable for Life?

The earth's temperature range, distance from the sun,
and size result in conditions that are favorable for life
as we know it.

Life on our planet depends on the liquid water that
dominates the earth's surface. Temperature is crucial
because most life on earth needs average temperatures
between the freezing and boiling points of water.

The earth's orbit is the right distance from the sun
to provide these conditions. If the earth were much
closer to the sun, it would be too hot—like Venus—for
water vapor to condense to form rain. If it were much
farther away, the earth's surface would be so cold—
like Mars—that its water would exist only as ice. The
earth also spins; if it did not, the side facing the sun
would be hot and the other side too cold for water-
based life to exist.

The earth is also the right size: It has enough grav-
itational mass to keep its iron and nickel core molten

and to keep the light gaseous molecules (such as N_2,
O_2, CO_2, and H_2O) in its atmosphere from flying off
into space.

On a time scale of millions of years, the earth
is enormously resilient and adaptive. During the 3.7
billion years since life arose here, the planet's average
surface temperature has remained within the narrow
range of 10–20 °C (50–68 °F), even with a 30–40% increase in the sun's energy output. In short, this remarkable planet that we call home is uniquely suited
for life as we know it.

3-3 ECOSYSTEM COMPONENTS

Biomes and Aquatic Life Zones

Life exists on land systems called biomes and in
freshwater and ocean aquatic life zones.

Viewed from outer space, the earth resembles an enormous jigsaw puzzle consisting of large masses of land
and vast expanses of ocean.

Biologists have classified the terrestrial (land) portion of the biosphere into **biomes** ("BY-ohms"). Each of
these large regions—such as forests, deserts, and grasslands—is characterized by a distinct climate and specific species (especially vegetation) adapted to it (see
Figure 1 in Science Supplement 2 at the end of this
book). Figure 3-8 shows different major biomes along
the 39th parallel spanning the United States.

Scientists divide the watery parts of the biosphere
into **aquatic life zones,** each containing numerous eco-

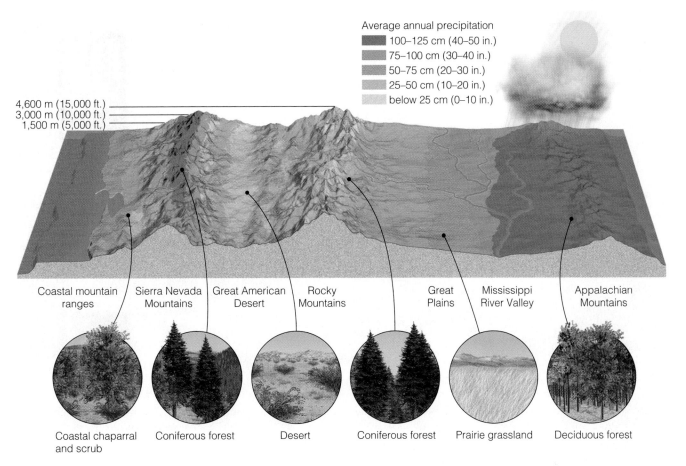

Average annual precipitation

■ 100–125 cm (40–50 in.)
■ 75–100 cm (30–40 in.)
■ 50–75 cm (20–30 in.)
■ 25–50 cm (10–20 in.)
□ below 25 cm (0–10 in.)

4,600 m (15,000 ft.)
3,000 m (10,000 ft.)
1,500 m (5,000 ft.)

| Coastal mountain ranges | Sierra Nevada Mountains | Great American Desert | Rocky Mountains | Great Plains | Mississippi River Valley | Appalachian Mountains |

| Coastal chaparral and scrub | Coniferous forest | Desert | Coniferous forest | Prairie grassland | Deciduous forest |

Figure 3-8 Natural capital: major biomes found along the 39th parallel across the United States. The differences reflect changes in climate, mainly differences in average annual precipitation and temperature.

systems. Examples include *freshwater life zones* (such as lakes and streams) and *ocean* or *marine life zones* (such as coral reefs, coastal estuaries, and the deep ocean).

Nonliving and Living Components of Ecosystems

Ecosystems consist of nonliving (abiotic) and living (biotic) components.

Two types of components make up the biosphere and its ecosystems: **abiotic** or nonliving components such as water, air, nutrients, and solar energy and **biotic** or living biological components such as plants, animals, and microbes.

Figures 3-9 and 3-10 (p. 42) are greatly simplified diagrams of some of the biotic and abiotic components in a freshwater aquatic ecosystem and a terrestrial ecosystem. Look carefully at these components and how they are connected to one another through the consumption habits of organisms.

Different species thrive under different physical conditions. Some need bright sunlight; others thrive in shade. Some need a hot environment; others prefer a

cool or cold one. Some do best under wet conditions; others succeed under dry conditions.

Each population in an ecosystem has a **range of tolerance** to variations in its physical and chemical environment, as shown in Figure 3-11 (p. 43). Individuals within a population may also have slightly different tolerance ranges for temperature or other factors because of small differences in genetic makeup, health, and age. For example, a trout population may do best within a narrow band of temperatures (*optimum level or range*), but a few individuals can survive above and below that band. Of course, if the water becomes too hot or too cold, none of the trout can survive.

These observations are summarized in the **law of tolerance:** *The existence, abundance, and distribution of a species in an ecosystem are determined by whether the levels of one or more physical or chemical factors fall within the range tolerated by that species.* A species may have a wide range of tolerance to some factors and a narrow range of tolerance to others. Most organisms are least tolerant during juvenile or reproductive stages of their life cycles. Highly tolerant species can live in a variety of habitats with widely different conditions.

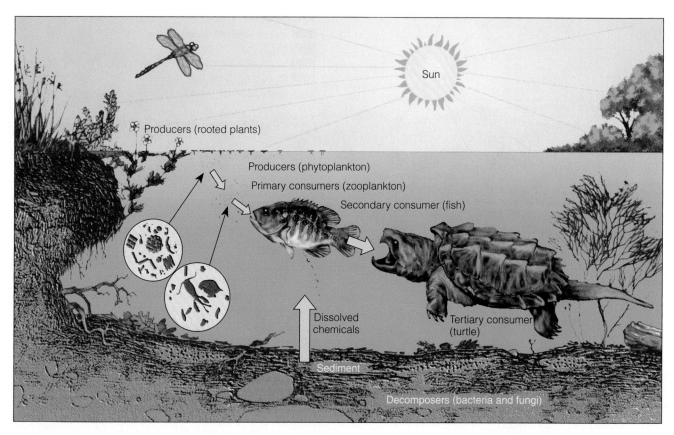

Figure 3-9 Natural capital: major components of a freshwater ecosystem.

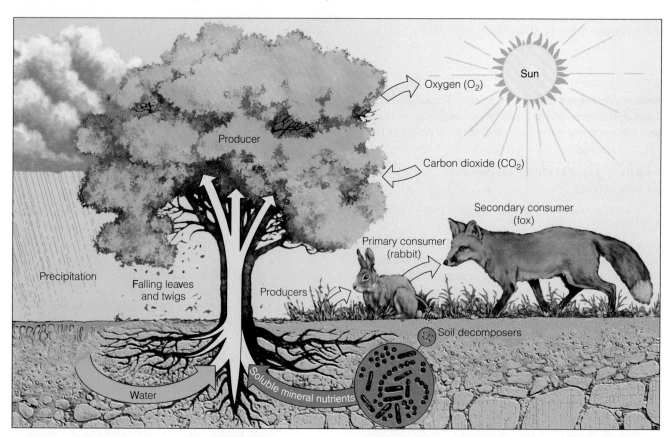

Environmental Science ⊛ Now™ **Active Figure 3-10** Natural capital: major components of an ecosystem in a field. *See an animation based on this figure and take a short quiz on the concept.*

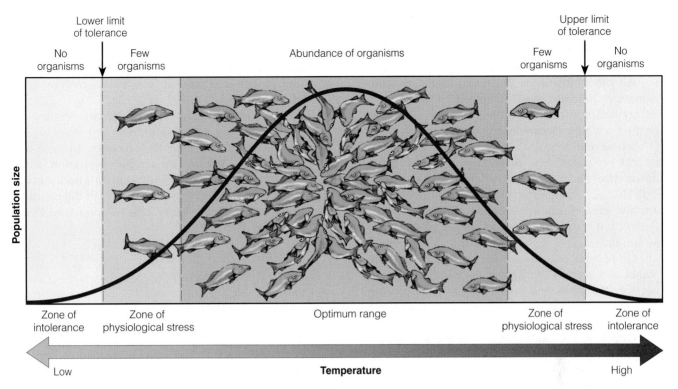

Figure 3-11 *Range of tolerance* for a population of organisms, such as fish, to an abiotic environmental factor—in this case, temperature.

Factors That Limit Population Growth

Availability of matter and energy resources can limit the number of organisms in a population.

A variety of factors can affect the number of organisms in a population. Sometimes one factor, known as a **limiting factor,** is more important in regulating population growth than other factors. This ecological principle, related to the law of tolerance, is called the **limiting factor principle:** *Too much or too little of any abiotic factor can limit or prevent growth of a population, even if all other factors are at or near the optimal range of tolerance.*

On land, precipitation often is the limiting factor. Lack of water in a desert limits plant growth. Soil nutrients also can act as a limiting factor on land. Suppose a farmer plants corn in phosphorus-poor soil. Even if water, nitrogen, potassium, and other nutrients are at optimal levels, the corn will stop growing when it uses up the available phosphorus.

Similarly, too much of an abiotic factor can be limiting. For example, providing excess water or fertilizer can kill plants—both common mistakes made by many beginning gardeners.

Important limiting factors for aquatic life zones include temperature, sunlight, nutrient availability, and **dissolved oxygen (DO) content**—the amount of oxygen gas dissolved in a given volume of water at a particular temperature and pressure. Another limiting factor in aquatic life zones is **salinity**—the amounts of various inorganic minerals or salts dissolved in a given volume of water.

Biological Components of Ecosystems: Producers, Consumers, and Decomposers

Some organisms in ecosystems produce food, while others consume food.

The earth's organisms either produce or consume food. **Producers,** sometimes called **autotrophs** (self-feeders), make their own food from compounds obtained from their environment.

On land, most producers are green plants. In freshwater and marine life zones, algae and plants are the major producers near shorelines. In open water, the dominant producers are *phytoplankton*—mostly microscopic organisms that float or drift in the water.

Most producers capture sunlight to make complex compounds (such as glucose, $C_6H_{12}O_6$) by **photosynthesis.** Although hundreds of chemical changes take place during photosynthesis, the overall reaction can be summarized as follows:

carbon dioxide + water + solar energy $\longrightarrow$ glucose + oxygen

$6\,CO_2$ + $6\,H_2O$ + solar energy $\longrightarrow$ $C_6H_{12}O_6$ + $6\,O_2$

(See Science Supplement 4 at the end of this book for information on how to balance chemical equations such as this one.)

A few producers, mostly specialized bacteria, can convert simple compounds from their environment into more complex nutrient compounds without sunlight, through a process called **chemosynthesis.**

All other organisms in an ecosystem are **consumers,** or **heterotrophs** ("other-feeders") that obtain energy and nutrients by feeding on other organisms or their remains. **Decomposers** (mostly certain types of bacteria and fungi) are specialized consumers that recycle organic matter in ecosystems. They break down (*biodegrade*) dead organic material or **detritus** ("di-TRI-tus," meaning "debris") to get nutrients. This activity releases simpler inorganic compounds into the soil and water, where producers can take them up as nutrients.

Omnivores play dual roles by feeding on both plants and animals. Examples are pigs, rats, foxes, bears, cockroaches, and humans. Were you an herbivore, a carnivore, or an omnivore when you had lunch today?

Detritivores include detritus feeders and decomposers that feed on detritus. Hordes of these waste eaters and degraders can transform a fallen tree trunk

into a powder and finally into simple inorganic molecules that plants can absorb as nutrients (Figure 3-12). *In natural ecosystems, there is little or no waste.* One organism's wastes serve as resources for other organisms, as the nutrients that make life possible are recycled again and again.

Producers, consumers, and decomposers use the chemical energy stored in glucose and other organic compounds to fuel their life processes. In most cells this energy is released by **aerobic respiration,** which uses oxygen to convert organic nutrients back into carbon dioxide and water. The net effect of the hundreds of steps in this complex process is represented by the following reaction:

$$\text{glucose} + \text{oxygen} \longrightarrow \text{carbon dioxide} + \text{water} + \text{energy}$$

$$C_6H_{12}O_6 + 6\,O_2 \longrightarrow 6\,CO_2 + 6\,H_2O + \text{energy}$$

Although the detailed steps differ, the net chemical change for aerobic respiration is the opposite of that for photosynthesis.

The survival of any individual organism depends on the *flow of matter and energy* through its body. By com-

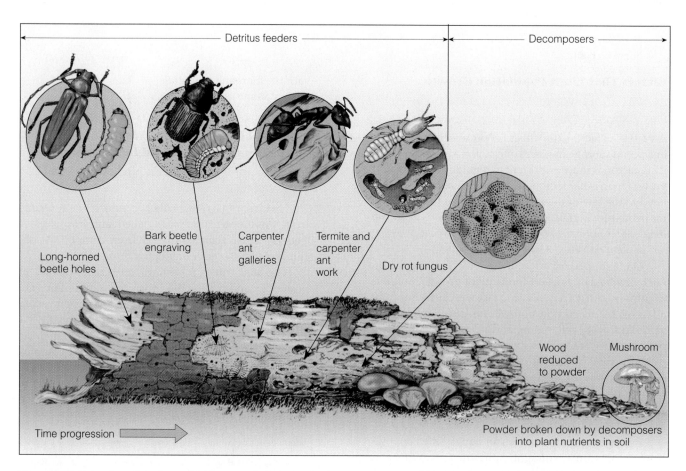

Figure 3-12 Natural capital: Some detritivores, called *detritus feeders*, directly consume tiny fragments of this log. Other detritivores, called *decomposers* (mostly fungi and bacteria), digest complex organic chemicals in fragments of the log, converting them into simpler inorganic nutrients that can be used again by producers.

parison, an ecosystem as a whole survives primarily through a combination of *matter recycling* (rather than one-way flow) and *one-way energy flow* (Figure 3-13).

Decomposers complete the cycle of matter by breaking down detritus into inorganic nutrients that can be reused by producers. These waste eaters and nutrient recyclers provide us with this crucial ecological service and never send us a bill. Without decomposers, our planet would be knee-deep in plant litter, dead animal bodies, animal wastes, and garbage, and most life as we know could not exist.

Environmental Science ⊕ Now™

Explore the components of ecosystems, how they interact, the roles of bugs and plants, and what a fox will eat, at Environmental ScienceNow.

Biodiversity: A Crucial Resource

A vital renewable resource is the biodiversity found in the earth's variety of genes, species, ecosystems, and ecosystem processes.

Biological diversity, or **biodiversity,** is one of the earth's most important renewable resources. It includes four components, as shown in Figure 3-14.

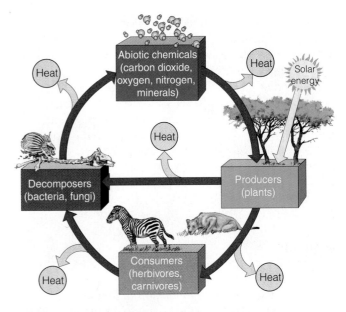

Environmental Science ⊕ Now™

Active Figure 3-13 Natural capital: the main structural components of an ecosystem (energy, chemicals, and organisms). Matter recycling and the flow of energy—first from the sun, then through organisms, and finally into the environment as low-quality heat—links these components. *See an animation based on this figure and take a short quiz on the concept.*

Functional Diversity
The biological and chemical processes such as energy flow and matter recycling needed for the survival of species, communities, and ecosystems.

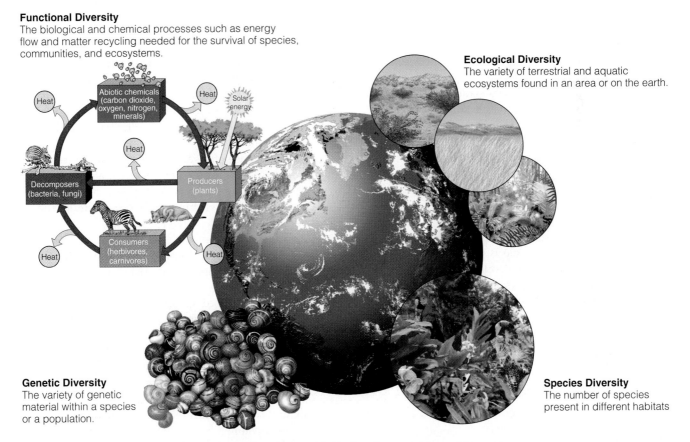

Ecological Diversity
The variety of terrestrial and aquatic ecosystems found in an area or on the earth.

Genetic Diversity
The variety of genetic material within a species or a population.

Species Diversity
The number of species present in different habitats

Figure 3-14 Natural capital: the major components of the earth's *biodiversity*—one of the earth's most important renewable resources.

Figure 3-15 Natural capital: two species found in tropical forests are part of the earth's biodiversity. On the left is an endangered white ukari in a Brazilian tropical forest. On the right is the world's largest flower, the flesh flower (Rafflesia), growing in a tropical rainforest of West Sumatra, Indonesia. The flower of this leafless plant can be as large as 1 meter (4.3 feet) in diameter and weigh 7 kilograms (15 pounds). The plant gives off a smell like rotting meat, presumably to attract flies and beetles that pollinate the flower. After blossoming once a year for a few weeks, the flower dissolves into a slimy black mass.

Figure 3-15 shows two tropical forest species that make up part of the earth's species diversity.

The earth's biodiversity is the biological wealth or capital that helps keep us alive. It supplies us with food, wood, fibers, energy, raw materials, industrial chemicals, and medicines—all of which pour hundreds of billions of dollars into the world economy each year. It also helps preserve the quality of the air and water, maintain the fertility of soils, dispose of wastes, and control populations of pests that attack crops and forests.

Biodiversity is a renewable resource as long we live off the biological income it provides instead of nibbling away at the natural capital that supplies this income. *Understanding, protecting, and sustaining biodiversity is a major goal of ecology and of this book.*

3-4 ENERGY FLOW IN ECOSYSTEMS

Food Chains and Food Webs

Food chains and webs show how eaters, the eaten, and the decomposed are connected to one another in an ecosystem.

All organisms, whether dead or alive, are potential sources of food for other organisms. A caterpillar eats a leaf, a robin eats the caterpillar, and a hawk eats the robin. Decomposers consume the leaf, caterpillar, robin, and hawk after they die. As a result, *there is little waste in natural ecosystems.*

A sequence of organisms, each of which serves as a source of food for the next, is called a **food chain.** It determines how energy and nutrients move from one organism to another through the ecosystem, as shown in Figure 3-16.

Ecologists assign each organism in an ecosystem to a *feeding level,* or **trophic level** (from the Greek word *trophos,* meaning "nourishment"), depending on whether it is a producer or a consumer and on what it eats or decomposes. Producers belong to the first trophic level, primary consumers to the second trophic level, secondary consumers to the third, and so on. Detritivores (detritus feeders and decomposers) process detritus from all trophic levels.

Of course, real ecosystems are more complex. Most consumers feed on more than one type of organism, and most organisms are eaten by more than one type of consumer. Because most species participate in several different food chains, the organisms in most

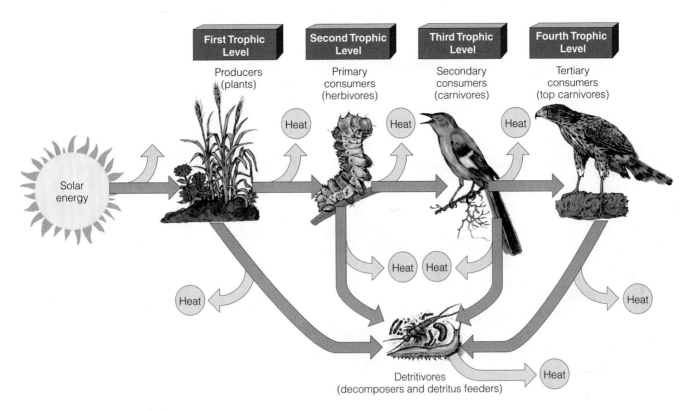

First Trophic Level	Second Trophic Level	Third Trophic Level	Fourth Trophic Level
Producers (plants)	Primary consumers (herbivores)	Secondary consumers (carnivores)	Tertiary consumers (top carnivores)

Detritivores
(decomposers and detritus feeders)

Environmental Science ⊕ Now™

Active Figure 3-16 Natural capital: a *food chain*. The arrows show how chemical energy in food flows through various *trophic levels* in energy transfers; most of the energy is degraded to heat, in accordance with the second law of thermodynamics. *Critical thinking: food chains rarely have more than four trophic levels. Can you figure out why? See an animation based on this figure and take a short quiz on the concept.*

ecosystems form a complex network of interconnected food chains called a **food web,** as shown in Figure 3-17 (p. 48). Trophic levels can be assigned in food webs just as in food chains.

Energy Flow in an Ecosystem

There is a decrease in the amount of energy available to each succeeding organism in a food chain or web.

Each trophic level in a food chain or web contains a certain amount of **biomass,** the dry weight of all organic matter contained in its organisms. The chemical energy stored in biomass is transferred from one trophic level to another.

The percentage of usable energy transferred as biomass from one trophic level to the next is called **ecological efficiency.** It ranges from 2% to 40% (that is, a loss of 60–98%) depending on the types of species and the ecosystem involved, but 10% is typical.

Assuming 10% ecological efficiency (90% loss) at each trophic transfer, if green plants in an area manage to capture 10,000 units of energy from the sun, then only about 1,000 units of energy will be available to support herbivores and only about 100 units to support carnivores.

The more trophic levels or steps in a food chain or web, the greater the cumulative loss of usable energy as energy flows through the various trophic levels. The **pyramid of energy flow** in Figure 3-18 (p. 49) illustrates this energy loss for a simple food chain, assuming a 90% energy loss with each transfer. How does this diagram help explain why there are so few tigers in the world?

Energy flow pyramids explain why the earth can support more people if they eat at lower trophic levels by consuming grains, vegetables, and fruits directly rather than passing such crops through another trophic level and eating grain eaters such as cattle.

The large loss in energy between successive trophic levels also explains why food chains and webs rarely have more than four or five trophic levels. In most cases, too little energy is left after four or five transfers to support organisms feeding at these high trophic levels. As a result, there are relatively few top carnivores such as eagles, hawks, tigers, and white sharks. This

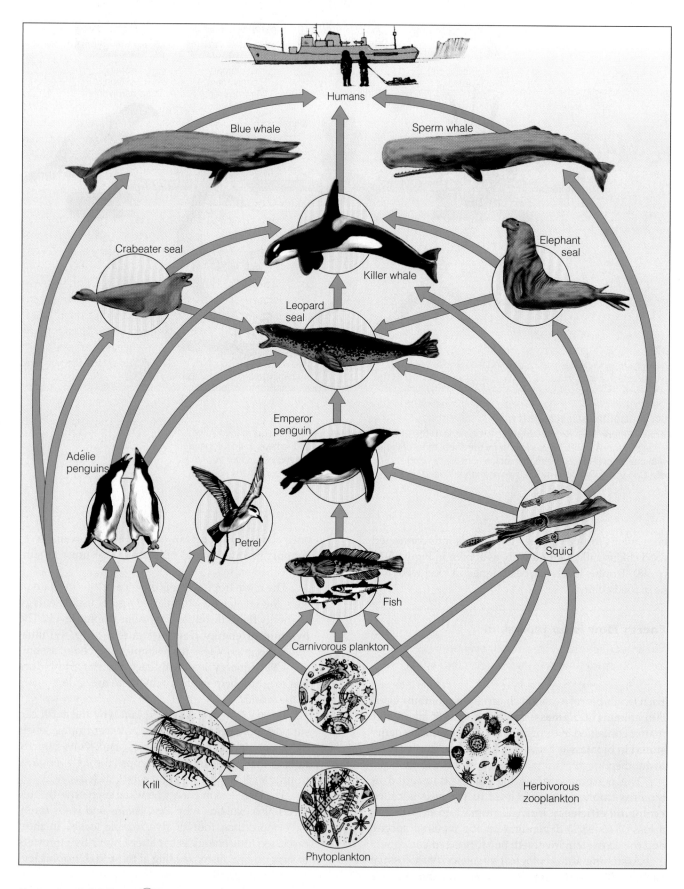

Environmental Science ⊕ Now™ **Active Figure 3-17** **Natural capital:** a greatly simplified *food web* in the Antarctic. Many more participants in the web, including an array of decomposer organisms, are not depicted here. *See an animation based on this figure and take a short quiz on the concept.*

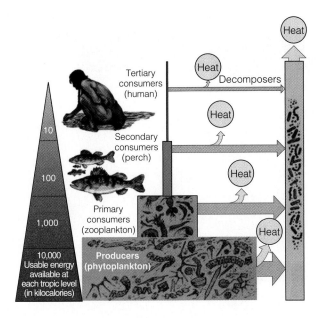

Environmental Science ⊕ Now™

Active Figure Figure 3-18 Natural capital: generalized *pyramid of energy flow* showing the decrease in usable energy available at each succeeding trophic level in a food chain or web. In nature, ecological efficiency varies from 2% to 40%, with 10% efficiency being common. This model assumes a 10% ecological efficiency (90% loss in usable energy to the environment, in the form of low-quality heat) with each transfer from one trophic level to another. *See an animation based on this figure and take a short quiz on the concept.*

phenomenon also explains why such species are usually the first to suffer when the ecosystems that support them are disrupted, and why these species are so vulnerable to extinction. Do you think humans are on this list?

Productivity of Producers

Different ecosystems use solar energy to produce and use biomass at different rates.

The *rate* at which an ecosystem's producers convert solar energy into chemical energy as biomass is the ecosystem's **gross primary productivity (GPP).** To stay alive, grow, and reproduce, an ecosystem's producers must use some of the biomass they produce for their own respiration. **Net primary productivity (NPP)** is the *rate* at which producers use photosynthesis to store energy *minus* the *rate* at which they use some of this stored energy through aerobic respiration, as shown in Figure 3-19. In other words, NPP = GPP − R, where R is energy used in respiration. NPP measures how fast producers can provide the food needed by other organisms (consumers) in an ecosystem.

Various ecosystems and life zones differ in their NPP, as illustrated in Figure 3-20 (p. 50). According to

this graph, what are nature's three most productive and three least productive systems? Despite its low NPP, so much open ocean is available that it produces more of the earth's NPP per year than any other ecosystem or life zone.

As we have seen, producers are the source of all food in an ecosystem. Only the biomass represented by NPP is available as food for consumers, and they use only a portion of this amount. Thus, *the planet's NPP ultimately limits the number of consumers (including humans) that can survive on the earth.* This is an important lesson from nature.

Peter Vitousek, Stuart Rojstaczer, and other ecologists estimate that humans now use, waste, or destroy about 27% of the earth's total potential NPP and 10–55% of the NPP of the planet's terrestrial ecosystems. They contend that this is the main reason why we are crowding out or eliminating the habitats and food supplies of other species. What might happen to us and to other consumer species as the human population grows over the next 40–50 years and per capita consumption of resources such as food, timber, and grassland rises sharply?

Environmental Science ⊕ Now™

Examine how energy flows among organisms at different trophic levels and across food webs in rain forests, prairies, and other ecosystems at Environmental ScienceNow.

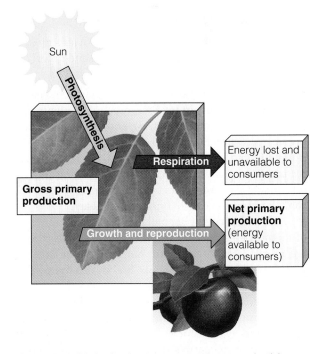

Figure 3-19 Distinction between gross primary productivity and net primary productivity. A plant uses some of its GPP to survive through respiration. The remaining energy is available to consumers.

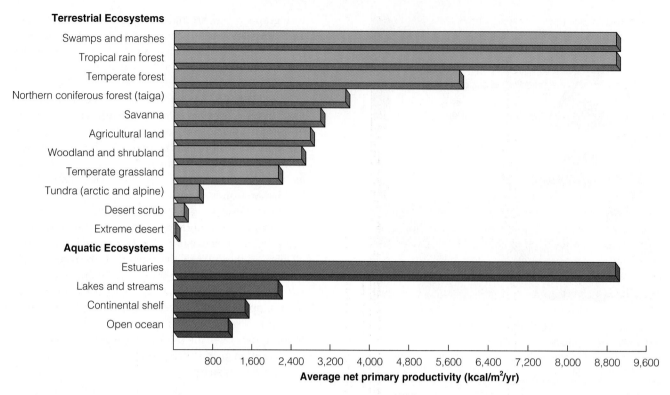

Figure 3-20 Natural capital: estimated *net primary productivity* per unit of area in major life zones and ecosystems, expressed as kilocalories of energy produced per square meter per year (kcal/m²/yr). (Data from R. H. Whittaker, *Communities and Ecosystems,* 2nd ed., New York: Macmillan, 1975)

3-5 SOILS

What Is Soil and Why Is It Important?

Soil is a slowly renewed resource that provides most of the nutrients needed for plant growth and helps purify water.

Soil is a thin covering over most land that comprises a complex mixture of eroded rock, mineral nutrients, decaying organic matter, water, air, and billions of living organisms. Study Figure 3-21, which shows a profile of different-aged soils. Although soil is a renewable resource, it is renewed very slowly. Depending mostly on climate, the formation of just 1 centimeter (0.4 inch) can take from 15 years to hundreds of years.

Soil is the base of life on land. It provides the bulk of the nutrients needed for plant growth. Indeed, you are mostly composed of soil nutrients imported into your body by the food you eat. Soil is also the earth's primary filter, cleansing water as it passes through. It helps decompose and recycle biodegradable wastes and is a major component of the earth's water recycling and water storage processes. In addition, it helps control the earth's climate by removing carbon dioxide from the atmosphere and storing it as carbon compounds.

Since the beginning of agriculture, human activities have led to rapid soil erosion, which can convert this renewable resource into a nonrenewable resource. Entire civilizations have collapsed because they mismanaged the topsoil that supported their populations.

Layers in Mature Soils

Most soils developed over a long time consist of several layers containing different materials.

Mature soils, which have developed over a long time, are arranged in a series of horizontal layers called **soil horizons,** each with a distinct texture and composition that varies with different types of soils. A cross-sectional view of the horizons in a soil is called a **soil profile.** Most mature soils have at least three of the possible horizons shown in Figure 3-21. Think of them as floors in the building of life underneath your feet.

The top layer is the *surface litter layer,* or *O horizon.* It consists mostly of freshly fallen undecomposed or partially decomposed leaves, twigs, crop wastes, animal waste, fungi, and other organic materials. Normally, it is brown or black.

The *topsoil layer,* or *A horizon,* is a porous mixture of partially decomposed organic matter, called **humus,** and some inorganic mineral particles. It is usually darker and looser than deeper layers. A fertile soil that produces high crop yields has a thick topsoil layer with lots of humus. This composition helps topsoil hold water and nutrients taken up by plant roots.

The roots of most plants and the majority of a soil's organic matter are concentrated in a soil's two upper layers. As long as vegetation anchors these layers, soil stores water and releases it in a nourishing trickle.

The two top layers of most well-developed soils teem with bacteria, fungi, earthworms, and small insects that interact in complex food webs. Bacteria and other decomposer microorganisms found by the billions in every handful of topsoil break down some of its complex organic compounds into simpler inorganic compounds that are soluble in water. Soil moisture carrying these dissolved nutrients is drawn up by the roots of plants and transported through stems and into leaves as part of the earth's chemical cycling processes.

The color of its topsoil suggests how useful a soil is for growing crops. Dark brown or black topsoil is rich in both nitrogen and organic matter. Gray, bright yellow, and red topsoils are low in organic matter and need nitrogen enrichment to support most crops.

The spaces, or pores, between the solid organic and inorganic particles in the upper and lower soil layers contain varying amounts of air (mostly nitrogen and oxygen gas) and water. Plant roots need the oxygen for cellular respiration.

Some precipitation that reaches the soil percolates through the soil layers and occupies many of the soil's open spaces or pores. This downward movement of water through soil is called **infiltration.** As the water seeps down, it dissolves various minerals and organic matter in upper layers and carries them to lower layers in a process called **leaching.**

Most of the world's crops are grown on soils exposed when grasslands and deciduous (leaf-shedding)

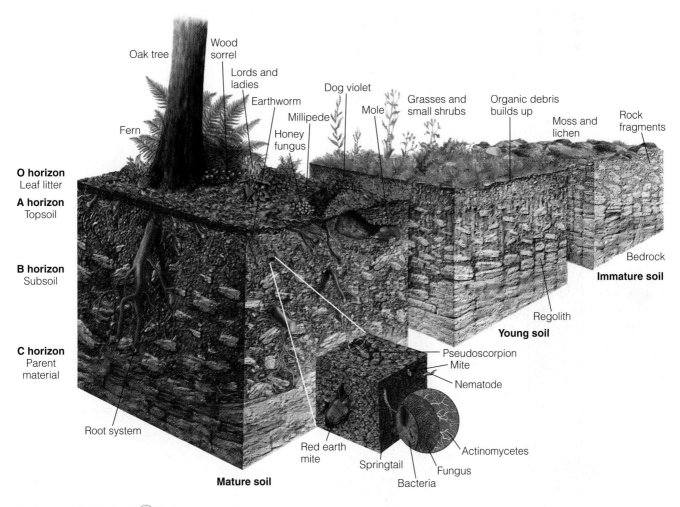

Environmental Science Now™

Active Figure 3-21 Natural capital: soil formation and generalized soil profile. Horizons, or layers, vary in number, composition, and thickness, depending on the type of soil. *See an animation based on this figure and take a short quiz on the concept.* (Used by permission of Macmillan Publishing Company from Derek Elsom, *Earth,* New York: Macmillan, 1992. Copyright © 1992 by Marshall Editions Developments Limited)

forests are cleared. Figure 3-22 profiles five important soil types.

Soils vary in their content of *clay* (very fine particles), *silt* (fine particles), *sand* (medium-size particles), and *gravel* (coarse to very coarse particles). Take a small amount of topsoil, moisten it, and rub it between your fingers and thumb. A gritty feel means it contains a lot of sand. A sticky feel means a high clay content, and you should be able to roll the soil into a clump. Silt-laden soil feels smooth, like flour. A *loam topsoil,* which is best suited to plant growth, has a texture between these extremes—a crumbly, spongy feeling—with many of its particles clumping loosely together.

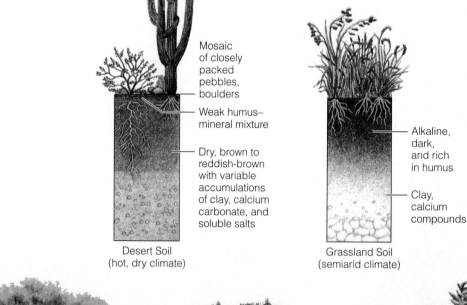

Desert Soil
(hot, dry climate)

- Mosaic of closely packed pebbles, boulders
- Weak humus–mineral mixture
- Dry, brown to reddish-brown with variable accumulations of clay, calcium carbonate, and soluble salts

Grassland Soil
(semiarid climate)

- Alkaline, dark, and rich in humus
- Clay, calcium compounds

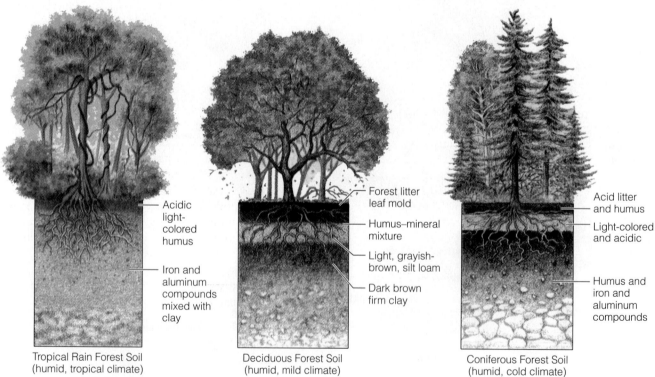

Tropical Rain Forest Soil
(humid, tropical climate)

- Acidic light-colored humus
- Iron and aluminum compounds mixed with clay

Deciduous Forest Soil
(humid, mild climate)

- Forest litter leaf mold
- Humus–mineral mixture
- Light, grayish-brown, silt loam
- Dark brown firm clay

Coniferous Forest Soil
(humid, cold climate)

- Acid litter and humus
- Light-colored and acidic
- Humus and iron and aluminum compounds

Environmental Science ⊕ Now™

Active Figure 3-22 Natural capital: soil profiles of the principal soil types typically found in five types of terrestrial ecosystems. *See an animation based on this figure and take a short quiz on the concept.*

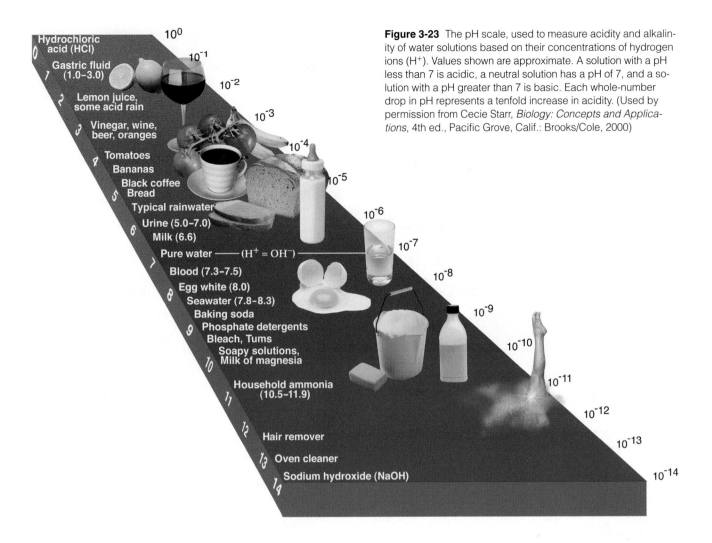

Figure 3-23 The pH scale, used to measure acidity and alkalinity of water solutions based on their concentrations of hydrogen ions (H^+). Values shown are approximate. A solution with a pH less than 7 is acidic, a neutral solution has a pH of 7, and a solution with a pH greater than 7 is basic. Each whole-number drop in pH represents a tenfold increase in acidity. (Used by permission from Cecie Starr, *Biology: Concepts and Applications*, 4th ed., Pacific Grove, Calif.: Brooks/Cole, 2000)

The acidity or alkalinity of a soil, as measured by its pH (Figure 3-23), influences the uptake of nutrients by plants. Chemicals can be added to raise or lower the acidity of soils.

Environmental Science ⊛ Now™
Compare soil profiles from grassland, desert, and three types of forests at Environmental ScienceNow.

3-6 MATTER CYCLING IN ECOSYSTEMS

Nutrient Cycles: Global Recycling

Global cycles recycle nutrients through the earth's air, land, water, and living organisms and, in the process, connect past, present, and future forms of life.

All organisms are interconnected by vast global recycling systems known as **nutrient cycles,** or **biogeochemical cycles** (literally, life–earth–chemical cycles). In these cycles, nutrient atoms, ions, and molecules that organisms need to live, grow, and reproduce are continuously cycled between air, water, soil, rock, and living organisms. These cycles, driven directly or indirectly by incoming solar energy and gravity, include the carbon, oxygen, nitrogen, phosphorus, and hydrologic (water) cycles (Figure 3-6, p. 39).

The earth's chemical cycles also connect past, present, and future forms of life. Some of the carbon atoms in your skin may once have been part of a leaf, a dinosaur's skin, or a layer of limestone rock. Your grandmother, Plato, or a hunter–gatherer who lived 25,000 years ago may have inhaled some of the oxygen molecules you just inhaled.

The Water Cycle

A vast global cycle collects, purifies, distributes, and recycles the earth's fixed supply of water.

The **hydrologic cycle,** or **water cycle,** which collects, purifies, and distributes the earth's fixed supply of water, as shown in Figure 3-24 (p. 54).

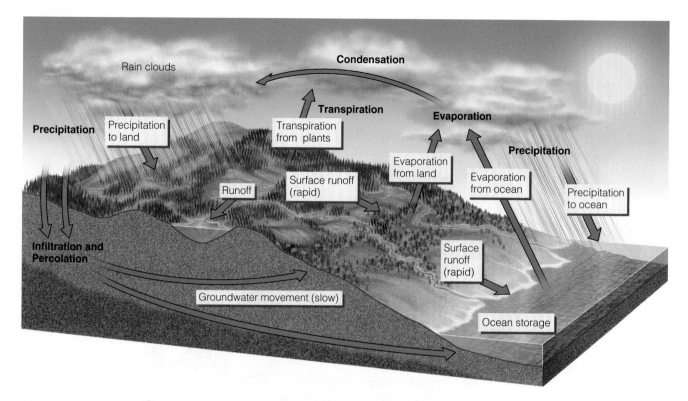

Rain clouds

Condensation

Precipitation

Precipitation to land

Transpiration

Transpiration from plants

Evaporation

Evaporation from land

Precipitation

Runoff

Surface runoff (rapid)

Evaporation from ocean

Precipitation to ocean

Infiltration and Percolation

Surface runoff (rapid)

Groundwater movement (slow)

Ocean storage

Environmental Science ⊕ Now™

Active Figure 3-24 Natural capital: simplified model of the *hydrologic cycle.* *See an animation based on this figure and take a short quiz on the concept.*

Solar energy evaporates water found on the earth's surface into the atmosphere. About 84% of water vapor in the atmosphere comes from the oceans; the rest comes from land.

Some of the fresh water returning to the earth's surface as precipitation in this cycle becomes locked in glaciers. And some infiltrates and percolates through soil and permeable rock formations to groundwater storage areas called *aquifers.* But most precipitation falling on terrestrial ecosystems becomes *surface runoff.* This water flows into streams and lakes, which eventually carry water back to the oceans, from which it can evaporate to repeat the cycle.

Besides replenishing streams and lakes, surface runoff causes soil erosion, which moves soil and rock fragments from one place to another. Water is the primary sculptor of the earth's landscape. Because water dissolves many nutrient compounds, it is a major medium for transporting nutrients within and between ecosystems.

Throughout the hydrologic cycle, many natural processes purify water. Evaporation and subsequent precipitation act as a natural distillation process that removes impurities dissolved in water. Water flowing above ground through streams and lakes and below ground in aquifers is naturally filtered and purified by

chemical and biological processes. Thus, *the hydrologic cycle can be viewed as a cycle of natural renewal of water quality.*

Effects of Human Activities on the Water Cycle

We alter the water cycle by withdrawing large amounts of fresh water, clearing vegetation and eroding soils, and polluting surface and underground water.

During the past 100 years, we have been intervening in the earth's current water cycle in three major ways. *First,* we withdraw large quantities of fresh water from streams, lakes, and underground sources.

Second, we clear vegetation from land for agriculture, mining, road and building construction, and other activities and sometimes cover the land with buildings, concrete, or asphalt. This increases runoff, reduces infiltration that recharges groundwater supplies, increases the risk of flooding, and accelerates soil erosion and landslides. We also increase flooding by destroying wetlands, which act like sponges to absorb and hold overflows of water.

Third, we add nutrients (such as phosphates and nitrates found in fertilizers) and other pollutants

The Carbon Cycle

Carbon recycles through the earth's air, water, soil, and living organisms.

Carbon, the basic building block of the carbohydrates, fats, proteins, DNA, and other organic compounds necessary for life, circulates through the biosphere in the carbon cycle shown in Figure 3-25 (p. 56).

The **carbon cycle** is based on carbon dioxide (CO_2) gas, which makes up 0.038% of the volume of the troposphere and is also dissolved in water. Carbon dioxide is a key component of nature's thermostat. If the carbon cycle removes too much CO_2 from the atmosphere, the atmosphere will cool; if it generates too much CO_2, the atmosphere will get warmer. Thus, even slight changes in this cycle can affect climate and ultimately the types of life that can exist on earth.

Terrestrial producers remove CO_2 from the atmosphere, and aquatic producers remove it from the water. They then use photosynthesis to convert CO_2 into complex carbohydrates such as glucose ($C_6H_{12}O_6$).

The cells in oxygen-consuming producers, consumers, and decomposers then carry out aerobic respiration. This process breaks down glucose and other complex organic compounds and converts the carbon back to CO_2 in the atmosphere or water for reuse by producers. This linkage between *photosynthesis* in producers and *aerobic respiration* in producers, consumers, and decomposers circulates carbon in the biosphere. Oxygen and hydrogen—the other elements in carbohydrates—cycle almost in step with carbon.

Some carbon atoms take a long time to recycle. Over millions of years, buried deposits of dead plant matter and bacteria are compressed between layers of sediment, where they form carbon-containing *fossil fuels* such as coal and oil (Figure 3-25). This carbon is not released to the atmosphere as CO_2 for recycling until these fuels are extracted and burned, or until long-term geological processes expose these deposits to air. In only a few hundred years, we have extracted and burned fossil fuels that took millions of years to form. This is why fossil fuels are nonrenewable resources on a human time scale.

Effects of Human Activities on the Carbon Cycle

Burning fossil fuels and clearing photosynthesizing vegetation faster than it is replaced can increase the earth's average temperature by adding excess carbon dioxide to the atmosphere.

Since 1800, and especially since 1950, we have been intervening in the earth's carbon cycle in two ways that add carbon dioxide to the atmosphere. *First,* we clear trees and other plants that absorb CO_2 through photosynthesis faster than they can grow back. *Second,* we add large amounts of CO_2 by burning fossil fuels (Figure 3-26, p. 56) and wood.

Computer models of the earth's climate systems suggest that increased concentrations of atmospheric CO_2 and other gases could enhance the planet's *natural greenhouse effect,* which helps warm the lower atmosphere (troposphere) and the earth's surface (Figure 3-7). The resulting *global warming* could disrupt global food production and wildlife habitats, alter temperature and precipitation patterns, and raise the average sea level in various parts of the world.

The Nitrogen Cycle: Bacteria in Action

Different types of bacteria help recycle nitrogen through the earth's air, water, soil, and living organisms.

Nitrogen is the atmosphere's most abundant element, with nitrogen gas (N_2) making up 78% of the volume of the troposphere. The N_2 in the atmosphere is a stable molecule that does not readily react with other elements, so it cannot be absorbed and used directly as a nutrient by multicellular plants or animals.

Fortunately, atmospheric electrical discharges in the form of lightning and certain types of bacteria in aquatic systems, in the soil, and in the roots of some plants can convert N_2 into compounds useful as nutrients for plants and animals as part of the **nitrogen cycle,** depicted in Figure 3-27 (p. 58).

The nitrogen cycle consists of several major steps. In *nitrogen fixation,* specialized bacteria in soil and aquatic environments convert (or fix) gaseous nitrogen (N_2) to ammonia (NH_3) that can be used by plants.

Ammonia not taken up by plants may undergo *nitrification.* In this two-step process, specialized soil bacteria convert most of the ammonia in soil first to *nitrite ions* (NO_2^-), which are toxic to plants, and then to *nitrate ions* (NO_3^-), which are easily taken up by the roots of plants. Animals, in turn, get their nitrogen by eating plants or plant-eating animals.

Plants and animals return nitrogen-rich organic compounds to the environment as wastes, cast-off particles, and dead bodies. In *ammonification,* vast armies of specialized decomposer bacteria convert this detritus into simpler nitrogen-containing inorganic compounds such as ammonia (NH_3) and water-soluble salts containing ammonium ions (NH_4^+).

Active Figure 3-25 Natural capital: simplified model of the global *carbon cycle.* Carbon moves through both marine ecosystems (left side) and terrestrial ecosystems (right side). Carbon reservoirs are shown as boxes; processes that change one form of carbon to another are shown in unboxed print. *See an animation based on this figure and take a short quiz on the concept.* (Cecie Starr, *Biology: Concepts and Applications,* 4th ed., Pacific Grove, Calif.: Brooks/Cole, © 2000)

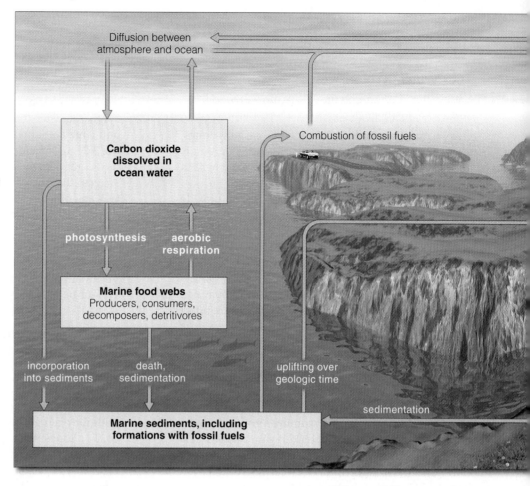

Diffusion between atmosphere and ocean

Carbon dioxide dissolved in ocean water

Combustion of fossil fuels

photosynthesis aerobic respiration

Marine food webs
Producers, consumers, decomposers, detritivores

incorporation into sediments

death, sedimentation

uplifting over geologic time

sedimentation

Marine sediments, including formations with fossil fuels

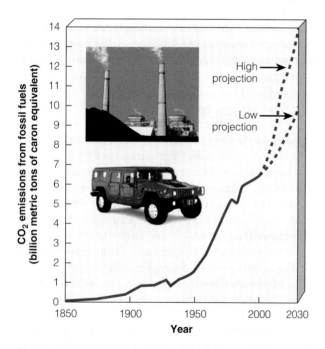

High projection

Low projection

Figure 3-26 Natural capital degradation: human interference in the global carbon cycle from carbon dioxide emissions when fossil fuels are burned, 1850–2004, and projections to 2030 (dashed lines). (Data from UN Environment Programme, British Petroleum, International Energy Agency, and U.S. Department of Energy)

In *denitrification,* nitrogen leaves the soil as specialized bacteria in waterlogged soil and in the bottom sediments of lakes, oceans, swamps, and bogs convert NH_3 and NH_4^+ back into nitrite and nitrate ions, and then into nitrogen gas (N_2) and nitrous oxide gas (N_2O). These gases are released to the atmosphere to begin the nitrogen cycle again.

Effects of Human Activities on the Nitrogen Cycle

We add large amounts of nitrogen-containing compounds to the earth's air and water and remove nitrogen from the soil.

We intervene in the nitrogen cycle in several ways. *First,* we add large amounts of nitric oxide (NO) into the atmosphere when N_2 and O_2 combine as we burn any fuel at high temperatures. In the atmosphere, this gas can be converted to nitrogen dioxide gas (NO_2) and nitric acid (HNO_3), which can return to the earth's surface as damaging *acid deposition,* commonly called *acid rain.*

Second, we add nitrous oxide (N_2O) to the atmosphere through the action of anaerobic bacteria on livestock wastes and commercial inorganic fertilizers

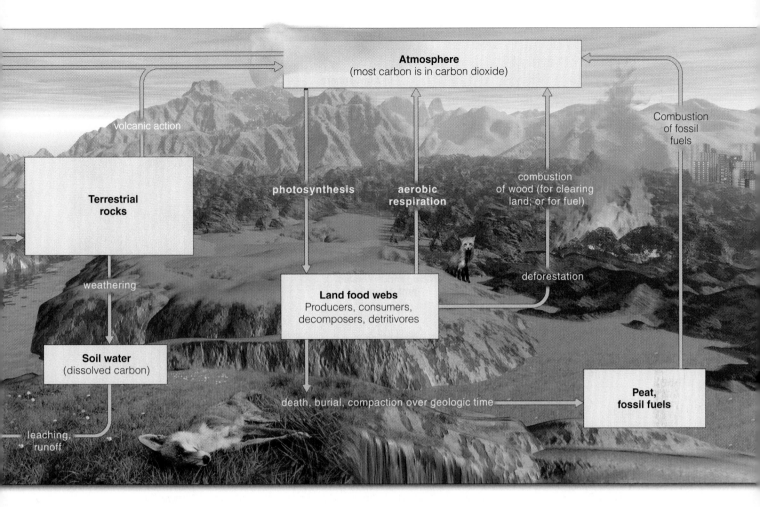

Atmosphere
(most carbon is in carbon dioxide)

volcanic action

Terrestrial
rocks

photosynthesis

aerobic
respiration

combustion
of wood (for clearing
land; or for fuel)

Combustion
of fossil
fuels

weathering

Land food webs
Producers, consumers,
decomposers, detritivores

deforestation

Soil water
(dissolved carbon)

death, burial, compaction over geologic time

Peat,
fossil fuels

leaching,
runoff

applied to the soil. This gas can warm the atmosphere and deplete ozone in the stratosphere.

Third, nitrate (NO_3^-) in inorganic fertilizers can leach through the soil and contaminate groundwater. This contaminated groundwater is harmful to drink, especially for infants and small children.

Fourth, we release large quantities of nitrogen stored in soils and plants as gaseous compounds into the troposphere through destruction of forests, grasslands, and wetlands.

Fifth, we upset aquatic ecosystems by adding excess nitrates in agricultural runoff and discharges from municipal sewage systems.

Sixth, we remove nitrogen from topsoil when we harvest nitrogen-rich crops, irrigate crops, and burn or clear grasslands and forests before planting crops.

Since 1950, human activities have more than doubled the annual release of nitrogen from the terrestrial portion of the earth into the rest of the environment (Figure 3-28, p. 58). This excessive input of nitrogen into the air and water represents a serious local, regional, and global environmental problem that has attracted relatively little attention compared to global environmental problems such as global warming and depletion of ozone in the stratosphere. Princeton University

physicist Robert Socolow calls for countries around the world to work out some type of nitrogen management agreement to help prevent this problem from reaching crisis levels.

The Phosphorus Cycle

Phosphorus cycles fairly slowly through the earth's water, soil, and living organisms.

Phosphorus circulates through water, the earth's crust, and living organisms in the **phosphorus cycle,** depicted in Figure 3-29 (p. 59). Bacteria play less important roles here than in the nitrogen cycle. Very little phosphorus circulates in the atmosphere because soil conditions do not allow bacteria to convert chemical forms of phosphorus to gaseous forms of phosphates. The phosphorus cycle is slow, and on a short human time scale much phosphorus flows one way from the land to the oceans.

Phosphorous typically is found as phosphate salts containing phosphate ions (PO_4^{3-}) in terrestrial rock formations and ocean bottom sediments. As water runs over phosphorus-containing rocks, it slowly erodes away inorganic compounds that contain phosphate ions.

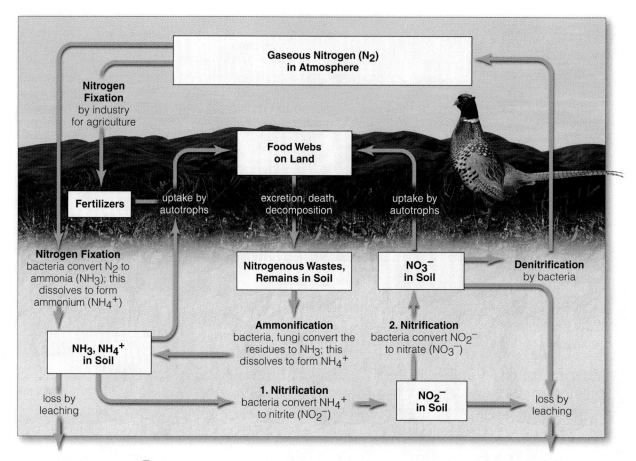

Active Figure 3-27 Natural capital: simplified model of the *nitrogen cycle* in a terrestrial ecosystem. Nitrogen reservoirs are shown as boxes; processes changing one form of nitrogen to another are shown in boxed print. *See an animation based on this figure and take a short quiz on the concept.* (Adapted from Cecie Starr and Ralph Taggart, *Biology: The Unity and Diversity of Life,* 9th ed., Belmont, Calif.: Wadsworth, © 2001)

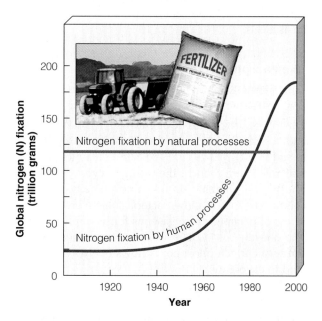

Figure 3-28 Natural capital degradation: human interference in the global nitrogen cycle. Human activities such as production of fertilizers now fix more nitrogen than all natural sources combined. (UN Environment Programme, UN Food and Agriculture Organization, and U.S. Department of Agriculture)

Phosphate can be lost from the cycle for long periods when it washes from the land into streams and rivers and is carried to the ocean. There it can be deposited as sediment and remain trapped for millions of years. Someday the geological processes of uplift may expose these seafloor deposits, from which phosphate can be eroded to start the cyclical process again.

Because most soils contain little phosphate, it is often the *limiting factor* for plant growth on land unless phosphorus (as phosphate salts mined from the earth) is applied to the soil as a fertilizer. Phosphorus also limits the growth of producer populations in many freshwater streams and lakes because phosphate salts are only slightly soluble in water.

Effects of Human Activities on the Phosphorus Cycle

We remove large amounts of phosphate from the earth to make fertilizer, reduce phosphorus in tropical soils by clearing forests, and add excess phosphates to aquatic systems.

We intervene in the earth's phosphorus cycle in three ways. *First,* we mine large quantities of phosphate

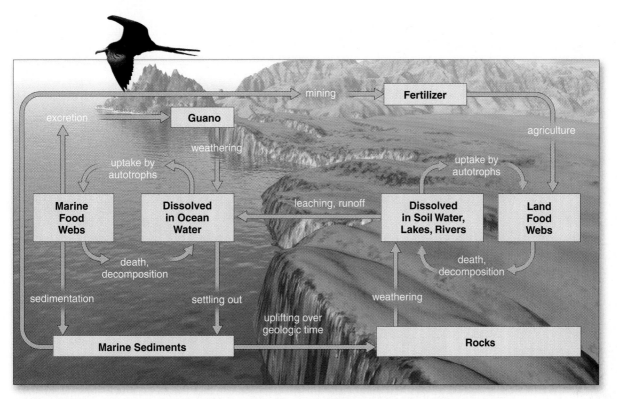

Figure 3-29 Natural capital: simplified model of the *phosphorus cycle*. Phosphorus reservoirs are shown as boxes; processes that change one form of phosphorus to another are shown in unboxed print. (From Cecie Starr and Ralph Taggart, *Biology: The Unity and Diversity of Life*, 9th ed., Belmont, Calif.: Wadsworth © 2001)

rock to make commercial inorganic fertilizers and detergents. *Second*, we reduce the available phosphate in tropical soils when we cut down tropical forests. *Third*, we disrupt aquatic systems with phosphates from runoff of animal wastes and fertilizers and discharges from sewage treatment systems.

Since 1900, human activities have increased the natural rate of phosphorus release into the environment by about 3.7-fold.

The Sulfur Cycle

Sulfur cycles through the earth's air, water, soil, and living organisms.

Sulfur circulates through the biosphere in the **sulfur cycle,** shown in Figure 3-30 (p. 60). Much of the earth's sulfur is stored underground in rocks and minerals, including sulfate (SO_4^-) salts buried deep under ocean sediments.

Sulfur also enters the atmosphere from several natural sources. Hydrogen sulfide (H_2S)—a colorless, highly poisonous gas with a rotten-egg smell—is released from active volcanoes and from organic matter in flooded swamps, bogs, and tidal flats broken down by anaerobic decomposers. Sulfur dioxide (SO_2), a colorless, suffocating gas, also comes from volcanoes.

Particles of sulfate (SO_4^{2-}) salts, such as ammonium sulfate, enter the atmosphere from sea spray, dust storms, and forest fires. Plant roots absorb sulfate

ions and incorporate the sulfur as an essential component of many proteins.

Certain marine algae produce large amounts of volatile dimethyl sulfide, or DMS (CH_3SCH_3). Tiny droplets of DMS serve as nuclei for the condensation of water into droplets found in clouds. In this way, changes in DMS emissions can affect cloud cover and climate. In the atmosphere, DMS is converted to sulfur dioxide.

In the atmosphere, sulfur dioxide (SO_2) from natural sources and human activities is converted to sulfur trioxide gas (SO_3) and to tiny droplets of sulfuric acid (H_2SO_4). In addition, it reacts with other atmospheric chemicals such as ammonia to produce tiny particles of sulfate salts. These droplets and particles fall to the earth as components of *acid deposition,* which along with other air pollutants can harm trees and aquatic life.

Effects of Human Activities on the Sulfur Cycle

We add sulfur dioxide to the atmosphere by burning coal and oil, refining oil, and producing some metals from ores.

We add sulfur dioxide to the atmosphere in three ways. *First*, we burn sulfur-containing coal and oil to produce electric power. *Second*, we refine sulfur-containing petroleum to make gasoline, heating oil, and other useful products. *Third*, we convert sulfur-containing

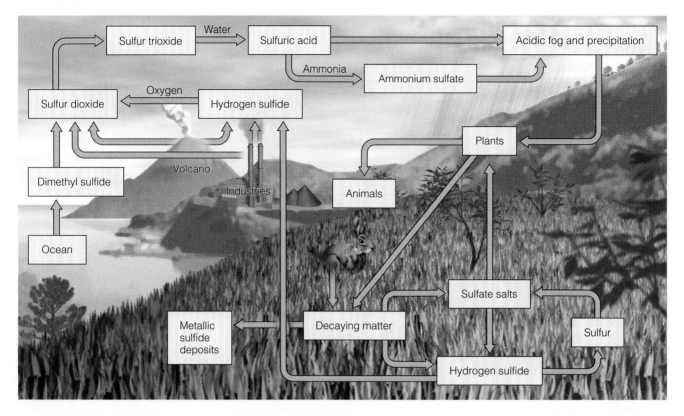

Environmental Science Now™

Active Figure 3-30 **Natural capital:** simplified model of the *sulfur cycle. See an animation based on this* figure and take a short quiz on the concept.

metallic mineral ores into free metals such as copper, lead, and zinc—an activity that releases large amounts of sulfur dioxide into the environment.

Environmental Science Now™

Learn more about the water, carbon, nitrogen, phosphorus, and sulfur cycles using interactive animations at Environmental ScienceNow.

3-7 HOW DO ECOLOGISTS LEARN ABOUT ECOSYSTEMS?

Field Research, Remote Sensing, and GIS

Ecologists go into ecosystems and hang out in treetops to learn what organisms live there and how they interact.

Field research, sometimes called muddy-boots biology, involves going into nature and observing and measuring the structure of ecosystems and what happens in them. Most of what we know about the structure and functioning of ecosystems has come from such research.

Ecologists trek through forests, deserts, and grasslands and wade or boat through wetlands, lakes, and streams collecting and observing species. Sometimes they carry out controlled experiments by isolating and

changing a variable in one part of an area and comparing the results with nearby unchanged areas.

Tropical ecologists erect tall construction cranes that stretch into the canopies of tropical forests to identify and observe the rich diversity of species living or feeding in these treetop habitats.

Increasingly, ecologists are using new technologies to collect field data. In *remote sensing* from aircraft and satellites and *geographic information systems* (GISs), data gathered from broad geographic regions are stored in spatial databases (Figure 3-31). Computers and GIS software can analyze and manipulate these data and combine them with ground and other data. The result: computerized maps of forest cover, water resources, air pollution emissions, coastal changes, relationships between cancers and sources of pollution, and changes in global sea temperatures.

Studying Ecosystems in the Laboratory

Ecologists use aquarium tanks, greenhouses, and controlled indoor and outdoor chambers to study ecosystems.

During the past 50 years, ecologists have increasingly supplemented field research by using *laboratory research* to set up, observe, and make measurements of

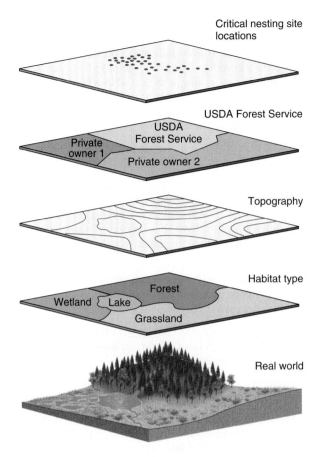

Critical nesting site
locations

USDA Forest Service

USDA
Forest Service

Private
owner 1

Private owner 2

Topography

Habitat type

Forest

Wetland   Lake

Grassland

Real world

Figure 3-31 *Geographic information systems* provide the computer technology for organizing, storing, and analyzing complex data collected over broad geographic areas. They enable scientists to overlay many layers of data (such as soils, topography, distribution of endangered populations, and land protection status).

model ecosystems and populations under laboratory conditions. Such simplified systems have been created in containers such as culture tubes, bottles, aquarium tanks, and greenhouses and in indoor and outdoor chambers where temperature, light, CO_2, humidity, and other variables can be controlled carefully.

Such systems make it easier for scientists to carry out controlled experiments. In addition, the laboratory experiments often are quicker and cheaper than similar experiments in the field.

But there is a catch. We must consider whether what scientists observe and measure in a simplified, controlled system under laboratory conditions takes place in the same way in the more complex and dynamic conditions found in nature. Thus the results of laboratory research must be coupled with and supported by field research.

Systems Analysis

Ecologists develop mathematical and other models to simulate the behavior of ecosystems.

Since the late 1960s, ecologists have explored the use of *systems analysis* to develop mathematical and other models that simulate ecosystems. Computer simulations can help us understand large and very complex systems (such as rivers, oceans, forests, grasslands, cities, and climate) that cannot be adequately studied and modeled in field and laboratory research. Figure 3-32 outlines the major stages of systems analysis.

Researchers can change values of the variables in their computer models to project possible changes in environmental conditions, anticipate environmental surprises, and analyze the effectiveness of various alternative solutions to environmental problems.

Of course, the simulations and projections made using ecosystem models are no better than the data and assumptions used to develop the models. Clearly, careful field and laboratory ecological research must be used to provide the baseline data and determine the causal relationships between key variables needed to develop and test ecosystem models.

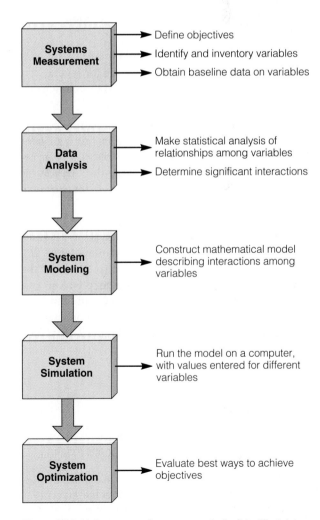

Systems Measurement
- Define objectives
- Identify and inventory variables
- Obtain baseline data on variables

Data Analysis
- Make statistical analysis of relationships among variables
- Determine significant interactions

System Modeling
- Construct mathematical model describing interactions among variables

System Simulation
- Run the model on a computer, with values entered for different variables

System Optimization
- Evaluate best ways to achieve objectives

Figure 3-32 Major stages of systems analysis. (Modified data from Charles Southwick)

Importance of Baseline Ecological Data

We need baseline data on the world's ecosystems so we can see how they are changing and develop effective strategies for preventing or slowing their degradation.

Before we can understand what is happening in nature and how best to prevent harmful environmental changes, we need to know about current conditions. In other words, we need *baseline data* about the condition of the earth's ecosystems.

By analogy, your doctor would like to have baseline data on your blood pressure, weight, and functioning of your organs and other systems as revealed by blood and other basic tests. If something happens to your health, the doctor can run new tests and compare the results with the baseline data to determine what has changed in an effort to come up with an effective treatment.

Bad news. According to a 2002 ecological study published by the Heinz Foundation and a 2005 Millennium Ecosystem Assessment, scientists have less than half of the basic ecological data they need to evaluate the status of ecosystems in the United States. Even fewer data are available for most other parts of the world.

Two Principles of Sustainability

Ecosystems have sustained themselves for several billion years by using solar energy and recycling the chemical nutrients their organisms need.

As described in this chapter, almost all natural ecosystems and the biosphere itself achieve *long-term* sustainability in two ways. *First*, they use *renewable solar energy* as their energy source. *Second*, they *recycle the chemical nutrients* their organisms need for survival, growth, and reproduction.

These two sustainability principles arise from the structure and function of natural ecosystems (Figures 3-6 and 3-13), the law of conservation of matter (p. 26), and the two laws of thermodynamics (p. 31). Thus the results of basic research in both the physical and biological sciences provide the same guidelines or lessons from nature on how we can live more sustainably on the earth, as summarized in Figure 2-13 (p. 33).

All things come from earth, and to earth they all return.
MENANDER (342–290 B.C.)

CRITICAL THINKING

1. **a.** A bumper sticker asks, "Have you thanked a green plant today?" Give two reasons for appreciating a green plant.
   b. Trace the sources of the materials that make up the bumper sticker, and decide whether the sticker itself is a sound application of the slogan.
   c. Explain how decomposers help keep you alive.

2. **a.** How would you set up a self-sustaining aquarium for tropical fish?
   b. Suppose you have a balanced aquarium sealed with a clear glass top. Can life continue in the aquarium indefinitely as long as the sun shines regularly on it?
   c. A friend cleans out your aquarium and removes all the soil and plants, leaving only the fish and water. What will happen? Explain.

3. Make a list of the food you ate for lunch or dinner today. Trace each type of food back to a particular producer species.

4. Use the second law of thermodynamics (p. 31) to explain why a sharp decrease in usable energy occurs as energy flows through a food chain or web. Does an energy loss at each step violate the first law of thermodynamics (p. 31)? Explain.

5. Use the second law of thermodynamics to explain why many poor people in developing countries live on a mostly vegetarian diet.

6. Why do farmers not need to apply carbon to grow their crops but often need to add fertilizer containing nitrogen and phosphorus?

7. Why are CO_2 levels in the atmosphere higher during the day than at night?

8. What would happen to an ecosystem if all its decomposers and detritus feeders were eliminated? If all its producers were eliminated?

9. Visit a nearby aquatic life zone or terrestrial ecosystem and try to identify its major producers, consumers, detritivores, and decomposers.

LEARNING ONLINE

The website for this book includes review questions for the entire chapter, flash cards for key terms and concepts, a multiple-choice practice quiz, interesting Internet sites, references, and a guide for accessing thousands of InfoTrac® College Edition articles.
   Visit

http://biology.brookscole.com/miller11

Then choose Chapter 3, and select a learning resource. For access to animations, additional quizzes, chapter outlines and summaries, register and log in to

Environmental Science ⊕ Now™

at **esnow.brookscole.com/miller11** using the access code card in the front of your book.

> ### Active Graphing
> Visit http://esnow.brookscole.com/miller11 to explore the graphing exercise for this chapter.

4 Evolution and Biodiversity

CASE STUDY

How Did We Become Such a Powerful Species So Quickly?

Like many other species, humans have survived and thrived because we have certain traits that allow us to adapt to and modify the environment to increase our survival chances. What are these adaptive traits?

First, consider the traits we do *not* have. We lack exceptional strength, speed, and agility. We do not have weapons such as claws or fangs, and we lack a protective shell or body armor.

Our senses are unremarkable. We see only visible light—a tiny fraction of the spectrum of electromagnetic radiation that bathes the earth. We cannot see infrared radiation, as a rattlesnake can, or the ultraviolet light that guides some insects to their favorite flowers.

We cannot see as far as an eagle or in the night like some owls and other nocturnal creatures. We cannot hear the high-pitched sounds that help bats maneuver

in the dark. Our ears cannot pick up low-pitched sounds like the songs of whales as they glide through the world's oceans. We cannot smell as keenly as a dog or a wolf. By such measures, our physical and sensory powers are pitiful.

Yet humans have survived and flourished within less than a twitch of the 3.7 billion years that life has existed on the earth. Analysts attribute our success to three adaptations: *strong opposable thumbs* that allow us to grip and use tools better than the few other animals that have thumbs, an ability to *walk* upright, and a *complex brain* (Figure 4-1). These adaptations have helped us develop technologies that extend our limited senses, weapons, and protective devices.

In only a short time, we have developed many powerful technologies to take over much of the earth's life-support systems and net primary productivity to meet our basic needs and rapidly growing wants. We named ourselves *Homo sapiens sapiens*—the doubly wise species. If we keep degrading the life-support system for us and other species, some say we should be called *Homo ignoramus*. During this century we will probably learn which of these names is more appropriate.

The *good news* is that we can change our ways. We can learn to work in concert with nature by understanding and copying the ways nature has sustained itself for several billion years despite major changes in environmental conditions. This means heeding Aldo Leopold's call for us to become earth citizens, not earth rulers.

To do so, we need to understand the basics of biological evolution, including how it affects the earth's biodiversity as species come and go and new ones arise to take their place. Here is the essence of this chapter. Each species here today represents a long chain of genetic changes in populations in response to changing environmental conditions. And each species plays a unique ecological role (called its *niche*) in the earth's communities and ecosystems

Figure 4-1 Humans have thrived as a species mostly because of our complex strong opposable thumbs (left), our ability to walk upright (right), and our complex brains.

There is a grandeur to this view of life . . . that, whilst this planet has gone cycling on . . . endless forms most beautiful and most wonderful have been, and are being, evolved.

CHARLES DARWIN

This chapter describes how scientists believe life on earth arose and developed into the diversity of species we find today. It discusses these questions:

- How do scientists account for the development of life on the earth?

- What is biological evolution by natural selection, and how has it led to the current diversity of organisms on the earth?

- What is an ecological niche, and how does it help a population adapt to changing environmental conditions?

- How do extinction of species and formation of new species affect biodiversity?

- What is the future of evolution, and what role should we play in this future?

KEY IDEAS

- Chemical, geological, and fossil evidence indicate that life on earth developed as a result of about 1 billion years of chemical evolution and 3.7 billion years of biological evolution.

- Populations can evolve when genes change or mutate, giving some individuals genetic traits that enhance their ability to survive and to produce offspring with these traits (natural selection).

- Each species in an ecosystem fills a unique ecological role called its ecological niche.

- As environmental conditions change, the balance between the formation of new species (speciation) and the disapearance or extinction of existing ones determines the earth's biodiversity.

- Human activities decrease the earth's biodiversity when they cause the premature extinction of species and destroy or degrade habitats that traditionally served as centers for the development of new species.

- Genetic engineering enables us to combine genes from different organisms—a process that both holds great promise and raises a number of legal, ethical, and environmental issues.

4-1 ORIGINS OF LIFE

Development of Life on the Primitive Earth

Scientific evidence indicates that the earth's life is the result of about 1 billion years of chemical evolution to form the first cells, followed by about 3.7 billion years of biological evolution to form the species we find on the earth today.

How did life on the earth evolve to its present incredible diversity of species? We do not know the full answers to these questions. But considerable evidence suggests that life on the earth developed in two phases over the past 4.6–4.7 billion years.

The first phase involved *chemical evolution* of the organic molecules, biopolymers, and systems of chemical reactions needed to form the first cells. It took about 1 billion years. Evidence for this phase comes from chemical analysis and measurements of radioactive elements in primitive rocks and fossils. Chemists have also conducted laboratory experiments showing how simple inorganic compounds in the earth's early atmosphere might have reacted to produce amino acids, simple sugars, and other organic molecules used as building blocks for the proteins, complex carbohydrates, RNA, and DNA needed for life.

Environmental Science Now™
Learn more about one of the most famous of these experiments exploring early life at Environmental ScienceNow.

Fossil and other evidence indicates that chemical evolution was followed by *biological evolution* from single-celled bacteria to multicellular protists, plants, fungi, and animals.

How Do We Know Which Organisms Lived in the Past?

Our knowledge about past life comes from fossils, chemical analysis, cores drilled out of buried ice, and DNA analysis.

Most of what we know of the earth's life history comes from **fossils**: mineralized or petrified replicas of skeletons, bones, teeth, shells, leaves, and seeds, or impressions of such items. Fossils provide physical evidence of ancient organisms and reveal what their internal structures looked like (Figure 4-2).

Despite its importance, the fossil record is uneven and incomplete. Some forms of life left no fossils, some fossils have decomposed, and others are yet to be found. The fossils we have found so far are believed to represent only 1% of all species that have ever lived.

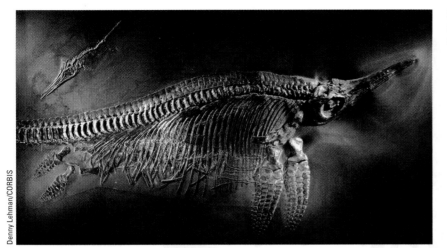

Figure 4-2 Fossilized skeleton of an ichthyo-saur, a marine reptile that lived more than 200 million years ago.

Evidence about the earth's early history also comes from chemical analysis and measurements of the half-lives of radioactive elements in primitive rocks and fossils. Scientists also drill cores from glacial ice and examine the kinds of life found at different layers. They also compare the DNA of past and current organisms.

4-2 EVOLUTION AND ADAPTATION

Biological Evolution

Biological evolution is the change in a population's genetic makeup over time.

According to scientific evidence, populations of organisms adapt to changes in environmental conditions through **biological evolution,** known more simply as **evolution.** Evolution involves the change in a population's genetic makeup through successive generations. *Populations—not individuals—evolve by becoming genetically different.*

According to the **theory of evolution,** all species descended from earlier, ancestral species. In other words, life comes from life. This widely accepted scientific theory explains how life has changed over the past 3.7 billion years (Figure 4-3, p. 66) and why it is so diverse today. Religious and other groups may offer other explanations, but evolution is the accepted scientific explanation.

Environmental Science ⊛ Now™
Get a detailed look at early evolution—the roots of the tree of life—at Environmental ScienceNow.

Biologists use the term **microevolution** to describe small genetic changes that occur in a population. They use the term **macroevolution** to describe long-term, large-scale evolutionary changes through which new species arise from ancestral species and other species are lost through extinction.

How Does Microevolution Work?

A population's gene pool changes over time when beneficial changes or mutations in its DNA molecules are passed on to offspring.

A population's **gene pool** consists of all of the genes (Figure 2-4, p. 25) in its individuals. *Microevolution* is a change in a population's gene pool over time.

The first step in microevolution is the development of *genetic variability* in a population. Genetic variability in a population originates through **mutations:** random changes in the structure or number of DNA molecules in a cell. Mutations can occur in two ways. *First,* DNA may be exposed to external agents such as radioactivity, X rays, and natural and human-made chemicals (called *mutagens*). *Second,* random mistakes sometimes occur in coded genetic instructions when DNA molecules are copied each time a cell divides and whenever an organism reproduces.

Mutations can occur in any cells, but only those in reproductive cells are passed on to offspring. Although some mutations are harmless, most are lethal. *Occasionally,* a mutation results in beneficial genetic traits may give that individual and its offspring better chances for survival and reproduction under existing environmental conditions or when such conditions change.

Natural Selection

Some members of a population may have genetic traits that enhance their ability to survive and produce offspring with these traits.

Natural selection occurs when some individuals of a population have genetically based traits that increase their chances of survival and their ability to produce offspring with the same traits.

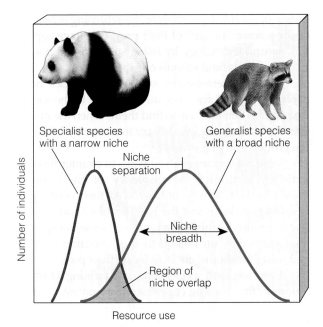

Specialist species with a narrow niche

Generalist species with a broad niche

Niche separation

Niche breadth

Region of niche overlap

Number of individuals

Resource use

Figure 4-4 Overlap of the niches of two different species: a specialist and a generalist. In the overlap area, the two species compete for one or more of the same resources. As a result, each species can occupy only a part of its fundamental niche; the part it occupies is its realized niche. Generalist species have a broad niche (right), and specialist species have a narrow niche (left).

Generalist and Specialist Species

Some species have broad ecological roles; others have narrower or more specialized roles.

Scientists use the niches of species to broadly classify them as *generalists* or *specialists*. **Generalist species** have broad niches (Figure 4-4, right curve). They can live in many different places, eat a variety of foods,

and tolerate a wide range of environmental conditions. Flies, cockroaches (Science Spotlight, p. 69), mice, rats, white-tailed deer, raccoons, coyotes, copperheads, channel catfish, and humans are generalist species.

Specialist species occupy narrow niches (Figure 4-4, left curve). They may be able to live in only one type of habitat, use one or a few types of food, or tolerate a narrow range of climatic and other environmental conditions. This makes specialists more prone to extinction when environmental conditions change.

For example, *tiger salamanders* breed only in fishless ponds where their larvae will not be eaten. Threatened *red-cockaded woodpeckers* carve nest holes almost exclusively in longleaf pines that are at least 75 years old. China's highly endangered *giant pandas* feed almost exclusively on various types of bamboo. Some shorebirds also occupy specialized niches, feeding on crustaceans, insects, and other organisms on sandy beaches and their adjoining coastal wetlands (Figure 4-5).

Is it better to be a generalist or a specialist? It depends. When environmental conditions are fairly constant, as in a tropical rain forest, specialists have an advantage because they have fewer competitors. Under rapidly changing environmental conditions, the generalist usually is better off than the specialist.

Natural selection can lead to an increase in specialized species when several species must compete intensely for scarce resources. Over time one species may evolve into a variety of species with different adaptations that reduce competition and allow them to share limited resources.

This *evolutionary divergence* of a single species into a variety of similar species with specialized niches can be illustrated by considering the honeycreepers that live on the island of Hawaii. Starting from a single ancestor species, numerous honeycreeper species evolved with

Black skimmer seizes small fish at water surface

Brown pelican dives for fish, which it locates from the air

Scaup and other diving ducks feed on mollusks, crustaceans, and aquatic vegetation

Avocet sweeps bill through mud and surface water in search of small crustaceans, insects, and seeds

Flamingo feeds on minute organisms in mud

Louisiana heron wades into water to seize small fish

Oystercatcher feeds on clams, mussels, and other shellfish into which it pries its narrow beak

Cockroaches: Nature's Ultimate Survivors

Cockroaches (Figure 4A), the bugs many people love to hate, have been around for 350 million years. One of evolution's great success stories, they have thrived because they are *generalists*.

The earth's 3,500 cockroach species can eat almost anything, including algae, dead insects, fingernail clippings, salts in tennis shoes, electrical cords, glue, paper, and soap. They can also live and breed almost anywhere except in polar regions.

Some cockroach species can go for months without food, survive for a month on a drop of water from a dishrag, and withstand massive doses of radiation. One species can survive being frozen for 48 hours.

Cockroaches usually can evade their predators (and a human foot in hot pursuit) because most species have antennae that can detect minute movements of air, vibration sensors in their knee joints, and rapid response times (faster than you can blink). Some even have wings.

Figure 4A As generalists, cockroaches are among the earth's most adaptable and prolific species.

They also have high reproductive rates. In only a year, a single Asian cockroach (especially prevalent in Florida) and its offspring can add about 10 million new cockroaches to the world. Their high reproductive rate helps them quickly develop genetic resistance to almost any poison we throw at them.

Most cockroaches sample food before it enters their mouths, so they learn to shun foul-tasting poisons. They also clean up after themselves by eating their own dead and, if food is scarce enough, their living.

Only about 25 species of cockroach live in homes. Unfortunately, such species can carry viruses and bacteria that cause diseases such as hepatitis, polio, typhoid fever, plague, and salmonella. They can also cause people to have allergic reactions ranging from watery eyes to severe wheezing. About 60% of Americans suffering from asthma are allergic to live or dead cockroaches.

Critical Thinking

If you could, would you exterminate all cockroach species? What might be some ecological consequences of this action?

---

different types of beaks specialized to feed on food sources such as specific types of insects, nectar from particular types of flowers, and certain types of seeds and fruit (Figure 4-6, p. 70).

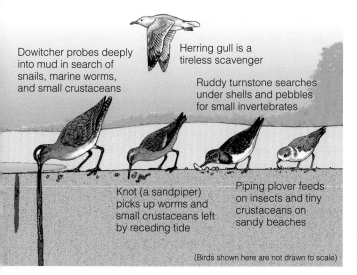

Dowitcher probes deeply into mud in search of snails, marine worms, and small crustaceans

Herring gull is a tireless scavenger

Ruddy turnstone searches under shells and pebbles for small invertebrates

Knot (a sandpiper) picks up worms and small crustaceans left by receding tide

Piping plover feeds on insects and tiny crustaceans on sandy beaches

(Birds shown here are not drawn to scale)

Limits on Adaptation

A population's ability to adapt to new environmental conditions is limited by its gene pool and the speed with which it can reproduce.

Will adaptations to new environmental conditions in the not too distant future allow our skin to become more resistant to the harmful effects of ultraviolet radiation, our lungs to cope with air pollutants, and our liver to better detoxify pollutants? The answer is *no* because of two limits to adaptations in nature.

Figure 4-5 Natural capital: specialized feeding niches of various bird species in a coastal wetland. Such resource partitioning reduces competition and allows sharing of limited resources.

Fruit and seed eaters

Greater Koa-finch

Kona Grosbeak

Akiapolaau

Maui Parrotbill

Insect and nectar eaters

Kuai Akialaoa

Amakihi

Crested Honeycreeper

Apapane

Unkown finch ancestor

Figure 4-6 Natural capital: evolutionary divergence of honey-creepers into specialized ecological niches. Each species has a beak specialized to take advantage of certain types of food resources.

First, a change in environmental conditions can lead to adaptation only for genetic traits already present in a population's gene pool. You must have genetic dice to play the genetic dice game.

Second, even if a beneficial heritable trait is present in a population, the population's ability to adapt may be limited by its reproductive capacity. Populations of genetically diverse species that reproduce quickly—such as weeds, mosquitoes, rats, bacteria, or cockroaches—often adapt to a change in environmental conditions in a short time. In contrast, species that cannot produce large numbers of offspring rapidly—such as elephants, tigers, sharks, and humans—take a long time (typically thousands or even millions of years) to adapt through natural selection. You have to be able to throw the genetic dice fast.

Here is some *bad news* for most members of a population. Even when a favorable genetic trait is present in a population, most of the population would have to die or become sterile so that individuals with the trait could predominate and pass the trait on. As a result, most players get kicked out of the genetic dice game before they have a chance to win. Most members of the human population would have to die prematurely for hundreds of thousands of generations for a new ge-

netic trait to predominate—hardly a desirable solution to the environmental problems we face.

Two Commonly Misunderstood Aspects of Evolution

Evolution is about leaving the most descendants, and there is no master plan leading to genetic perfection.

There are two common misconceptions about biological evolution. One is that "survival of the fittest" means "survival of the strongest." To biologists, *fitness* is a measure of reproductive success, not strength. The fittest individuals are those that leave the most descendants.

The other misconception is that evolution involves some grand plan of nature in which species become more perfectly adapted. From a scientific standpoint, no plan or goal of genetic perfection has been identified in the evolutionary process.

Environmental Science ⊛ Now™
Learn more about two special types of natural selection, one stabilizing and the other disruptive, at Environmental ScienceNow.

4-4 SPECIATION, EXTINCTION, AND BIODIVERSITY

Speciation: How New Species Develop

A new species arises when members of a population are isolated from other members for so long that changes in their genetic makeup prevent them from producing fertile offspring if they get together again.

Under certain circumstances, natural selection can lead to an entirely new species. In this process, called **speciation,** two species arise from one. For sexually reproducing species, a new species is formed when some members of a population can no longer breed with other members to produce fertile offspring.

The most common mechanism of speciation (especially among animals) takes place in two phases: geographic isolation and reproductive isolation. In **geographic isolation,** different groups of the same population of a species become physically isolated from one another for long periods. For example, part of a population may migrate in search of food and then begin living in another area with different environmental conditions, as shown in Figure 4-7. Populations can become separated by a physical barrier (such as a mountain range, stream, lake, or road), by a change such as a volcanic eruption or earthquake, or by a few individuals being carried to a new area by wind or water.

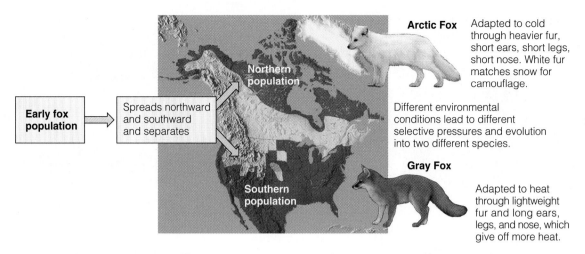

Figure 4-7 *Geographic isolation* can lead to reproductive isolation, divergence, and speciation.

In **reproductive isolation,** mutation and natural selection operate independently in the gene pools of geographically isolated populations. If this process continues long enough, members of the geographically and reproductively isolated populations of sexually reproducing species may become so different in genetic makeup that they can never produce live, fertile offspring. Then one species has become two, and speciation has occurred.

For some rapidly reproducing organisms, this type of speciation may occur within hundreds of years. For most species, it takes from tens of thousands to millions of years.

Given this time scale, it is difficult to observe and document the appearance of a new species. Thus many controversial hypotheses attempt to explain the details of speciation.

Extinction: Lights Out

A species becomes extinct when its populations cannot adapt to changing environmental conditions.

Another process affecting the number and types of species on the earth is **extinction,** in which an entire species ceases to exist. When environmental conditions change drastically enough, a species must evolve (become better adapted), move to a more favorable area if possible, or become extinct.

For most of the earth's geological history, species have faced incredible challenges to their existence. Continents have broken apart and moved over millions of years (Figure 4-8, p. 72). The earth's land area has repeatedly shrunk when continents have been flooded, has expanded when the world's oceans have shrunk, and has sometimes been covered with ice.

The earth's life has also had to cope with volcanic eruptions, meteorites and asteroids crashing onto the

planet, and releases of large amounts of methane trapped beneath the ocean floor. Some of these events created dust clouds that shut down or sharply reduced photosynthesis long enough to eliminate huge numbers of producers and, soon thereafter, the consumers that fed on them.

In some places, populations of existing species have been reduced or eliminated by newly arrived migrant species or species that are accidentally or deliberately introduced into new areas. More recently, humans have taken over or degraded many of the earth's resources and habitats. Today's biodiversity represents the species that have survived and thrived despite environmental upheavals.

Background Extinction, Mass Extinction, and Mass Depletion

All species eventually become extinct, but drastic changes in environmental conditions can eliminate large groups of species.

Extinction is the ultimate fate of all species, just as death awaits all individual organisms. Biologists estimate that 99.9% of all the species that ever existed are now extinct. The human species will not escape this ultimate fate.

As local environmental conditions change, a certain number of species disappear at a low rate, called **background extinction.** Based on the fossil record and analysis of ice cores, biologists estimate that the average annual background extinction rate is one to five species for each million species on the earth.

In contrast, **mass extinction** is a significant rise in extinction rates above the background level. In such a catastrophic, widespread (often global) event, large groups of existing species (perhaps 25–70%) are wiped out. Fossil and geological evidence indicates that the earth's species have experienced five mass extinctions

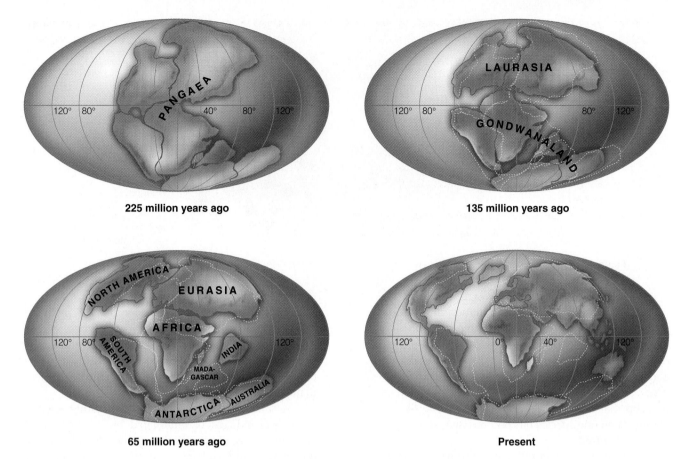

Figure 4-8 Geological processes and biological evolution: over millions of years, the earth's continents have moved very slowly on several gigantic plates. This process plays a role in the extinction of species as land areas split apart and in the rise of new species where once isolated areas of land combine. Rock and fossil evidence indicates that 200–250 million years ago all of the earth's present-day continents were locked together in a supercontinent called Pangaea (top left). About 180 million years ago, Pangaea began splitting apart as the earth's huge plates separated and eventually resulted in today's locations of the continents (bottom right).

225 million years ago

135 million years ago

65 million years ago

Present

(20–60 million years apart) during the past 500 million years (Figure 4-9).

In periods of **mass depletion,** extinction rates are much higher than normal but not high enough to classify as a mass extinction. In both types of events, large numbers of species have become extinct.

A mass extinction or mass depletion crisis for some species is an opportunity for other species. The existence of millions of species today means that speciation, on average, has kept ahead of extinction, especially during the last 250 million years (Figure 4-10, p. 74).

Effects of Human Activities on the Earth's Biodiversity

Human activities are decreasing the earth's biodiversity.

Speciation minus extinction equals *biodiversity*, the planet's genetic raw material for future evolution in response to changing environmental conditions. Although extinction is a natural process, humans have become a major force in the premature extinction of species.

According to biologists Stuart Primm and Edward O. Wilson, during the 20th century, extinction rates increased by 100–1,000 times the natural background extinction rate. As human population and resource consumption increase over the next 50–100 years, we are expected to take over a larger share of the earth's surface and net primary productivity (NPP) (Figure 3-20, p. 50). According to Wilson and Primm, this may cause the premature extinction of at least one-fifth of the earth's current species by 2030 and half of those species by the end of this century. If not checked, this trend could bring about a new *mass depletion* and possibly a new *mass extinction.* Wilson says that if we make an "all-out effort to save the biologically richest parts of the world, the amount of loss can be cut at least by half."

On our short time scale, such major losses cannot be recouped by formation of new species; it took millions of years after each of the earth's past mass extinctions and depletions for life to recover to the previous level of biodiversity. We are also destroying or degrading ecosystems such as tropical forests, coral reefs, and wetlands that are centers for future speciation. (See the Guest Essay on this topic by Norman Myers on the website for this chapter.) Genetic engineering cannot stop this loss of biodiversity because genetic engineers rely on natural biodiversity for their genetic raw material.

We can summarize the 3.7-billion-year biological history of the earth in one sentence: *Organisms convert solar energy to food, chemicals cycle, and species filling different biological roles (niches) have evolved in response to changing environmental conditions.*

Each species here today represents the outcome of a long chain of evolution, and each plays a unique ecological role in the earth's communities and ecosystems. These species, communities, and ecosystems are essential for future evolution, as populations of species continue to adapt to changes in environmental conditions by changing their genetic makeup.

Environmental Science ⊕ Now™

Learn more about different types of speciation and ways in which they can occur at Environmental ScienceNow.

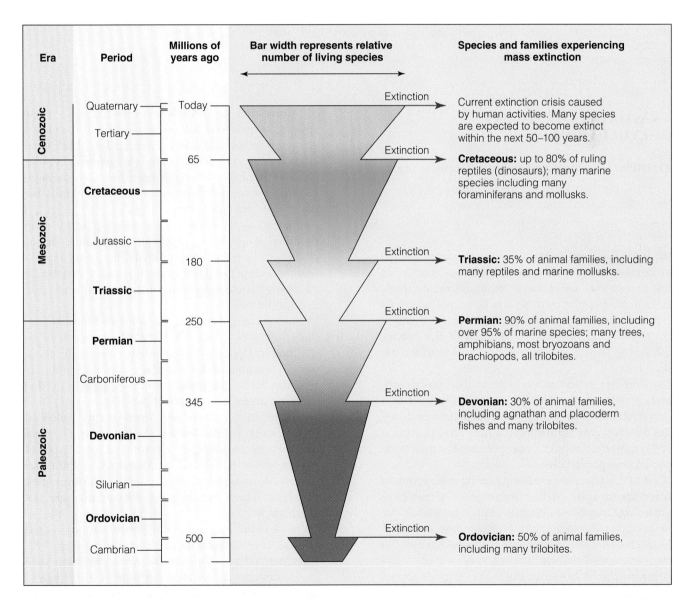

Figure 4-9 Fossils and radioactive dating indicate that five major *mass extinctions* (indicated by arrows) have taken place over the past 500 million years. Mass extinctions leave many organism roles (niches) unoccupied and create new ones. As a result, each mass extinction has been followed by periods of recovery (represented by the wedge shapes) called *adaptive radiations*. During these periods, which last 10 million years or longer, new species evolve to fill new or vacated ecological roles. Many scientists say that we are now in the midst of a sixth mass extinction, caused primarily by human activities.

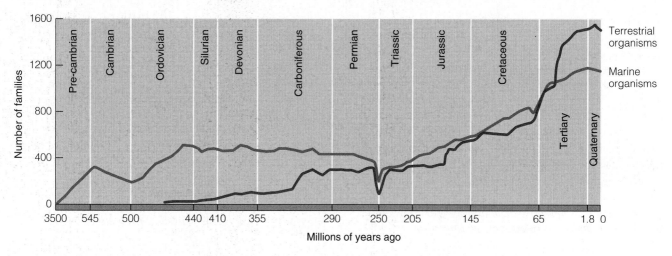

Figure 4-10 Natural capital: changes in the earth's biodiversity over geological time. The biological diversity of life on land and in the oceans has increased dramatically over the last 3.5 billion years, especially during the past 250 million years. During the last 1.8 million years this increase has leveled off. *Critical thinking: will the human species be a major factor in decreasing the earth's biodiversity over this century?*

4-5 WHAT IS THE FUTURE OF EVOLUTION?

Artificial Selection and Genetic Engineering

We have learned how to selectively breed members of populations and to use genetic engineering to produce plants and animals with certain genetic traits.

We have used **artificial selection** to change the genetic characteristics of populations of a species. In this process, we select one or more desirable genetic traits in the population of a plant or animal, such as a type of wheat, fruit, or dog. Then we use *selective breeding* to end up with populations of the species containing large numbers of individuals with the desired traits.

Artificial selection has yielded food crops with higher yields, cows that give more milk, trees that grow faster, and many different types of dogs and cats. But traditional crossbreeding is a slow process. Also, it can combine traits only from species that are close to one another genetically.

Today's scientists are using genetic engineering to speed up our ability to manipulate genes. **Genetic engineering,** or **gene splicing,** is a set of techniques for isolating, modifying, multiplying, and recombining genes from different organisms. It enables scientists to transfer genes between different species that would never interbreed in nature. For example, genes from a fish species can be put into a tomato or strawberry.

The resulting organisms are called **genetically modified organisms (GMOs)** or **transgenic organisms.** Figure 4-11 outlines the steps involved in developing a genetically modified plant.

Compared to traditional crossbreeding, gene splicing takes about half as much time to develop a new crop or animal variety and costs less. Traditional crossbreeding involves mixing the genes of similar types of organisms through breeding. Genetic engineering, by contrast, allows us to transfer traits between different types of organisms.

Scientists have used gene splicing to develop modified crop plants, genetically engineered drugs, pest-resistant plants, and animals that grow rapidly (Figure 4-12, p. 76). They have also created genetically engineered bacteria to clean up spills of oil and other toxic pollutants.

Genetic engineers have also learned how to produce a *clone,* or genetically identical version, of an individual in a population. Scientists have made clones of domestic animals such as sheep and cows and may someday be able to clone humans—a possibility that excites some people and horrifies others.

Bioengineers have developed many beneficial GMOs: chickens that lay low-cholesterol eggs, tomatoes with genes that can help prevent some types of cancer, and bananas and potatoes that contain oral vaccines to treat certain viral diseases in developing countries where needles and refrigeration are not available.

Researchers envision using genetically engineered animals to act as biofactories for producing drugs, vaccines, antibodies, hormones, industrial chemicals such as plastics and detergents, and human body organs. This new field is called *biopharming.* For example, cows may be able to produce insulin for treating diabetes, perhaps more cheaply than making the insulin in laboratories. Have you considered this field as a career choice?

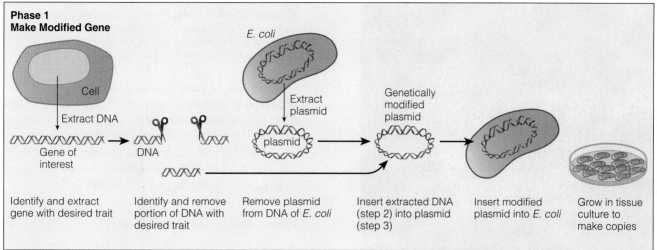

Phase 1
Make Modified Gene

Cell

Extract DNA

Gene of interest

DNA

E. coli

Extract plasmid

plasmid

Genetically modified plasmid

Insert modified plasmid into *E. coli*

Identify and extract gene with desired trait

Identify and remove portion of DNA with desired trait

Remove plasmid from DNA of *E. coli*

Insert extracted DNA (step 2) into plasmid (step 3)

Insert modified plasmid into *E. coli*

Grow in tissue culture to make copies

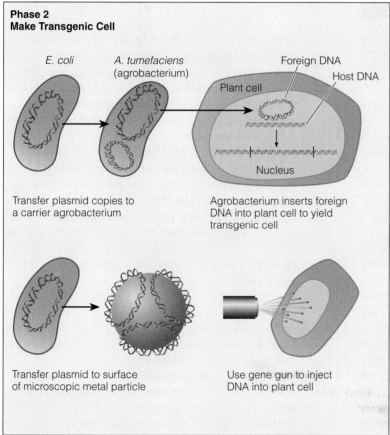

Phase 2
Make Transgenic Cell

E. coli

A. tumefaciens (agrobacterium)

Foreign DNA

Host DNA

Plant cell

Nucleus

Transfer plasmid copies to a carrier agrobacterium

Agrobacterium inserts foreign DNA into plant cell to yield transgenic cell

Transfer plasmid to surface of microscopic metal particle

Use gene gun to inject DNA into plant cell

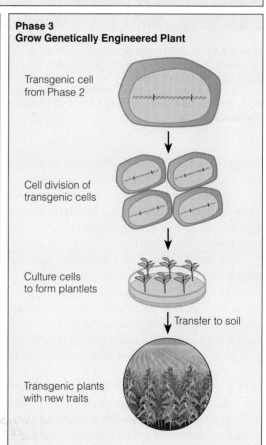

Phase 3
Grow Genetically Engineered Plant

Transgenic cell from Phase 2

Cell division of transgenic cells

Culture cells to form plantlets

Transfer to soil

Transgenic plants with new traits

Figure 4-11 *Genetic engineering.* Steps in genetically modifying a plant.

Ethics: Concerns about the Genetic Revolution

Genetic engineering has great promise for improving the human condition, but it is an unpredictable process and raises a number of privacy, ethical, legal, and environmental issues.

The hype about genetic engineering suggests that its results are controllable and predictable. In reality, most current forms of genetic engineering are messy and unpredictable. Genetic engineers can insert a gene into the nucleus of a cell, but with current technology they cannot know whether the cell will incorporate the new gene into its DNA. They also do not know where the

Figure 4-12 An example of genetic engineering. The six-month-old mouse on the left is normal; the same-age mouse on the right has a human growth hormone gene inserted in its cells. Mice with the human growth hormone gene grow two to three times faster and reach a size twice that of mice without the gene.

new gene will be located in the DNA molecule's structure and how this will affect the organism.

Genetic engineering is a trial-and-error process with many failures and unexpected results. Indeed, the average success rate of genetic engineering experiments is only about 1%.

The application of our increasing genetic knowledge holds great promise, but it raises some serious ethical and privacy issues. For example, some people have genes that make them more likely to develop certain genetic diseases or disorders. We now have the power to detect these genetic deficiencies, even before birth.

This possibility raises some important issues. If gene therapy is developed for correcting these deficiencies, who will get it? Will it be reserved mostly for the rich? Will it lead to more abortions of genetically defective fetuses? Will health insurers refuse to insure people with certain genetic defects that could lead to health problems? Will employers refuse to hire them?

Some people dream of a day when our genetic engineering prowess could eliminate death and aging altogether. As a person's cells, organs, or other parts wear out or become damaged, they would be replaced with new ones. These replacement parts might be grown in genetic engineering laboratories or biopharms. Or people might choose to have a clone available for spare parts.

Is it moral to take this path? Who will decide? Who will regulate this activity? Will genetically designed humans and clones have the same legal rights as people?

X *HOW WOULD YOU VOTE?* Should we legalize the production of human clones if a reasonably safe technology for doing so becomes available? Cast your vote online at http://biology.brookscole.com/miller11.

How might such genetic developments affect resource use, pollution, and environmental degradation? If everyone could live with good health as long as they wanted for a price, sellers of body makeovers would encourage customers to line up. Each of these affluent, long-lived people could have an enormous ecological footprint for perhaps centuries.

Ethics: What Are Our Options?

Arguments persist about how much we should regulate genetic engineering research and development.

In the 1990s, a backlash developed against the increasing use of genetically modified food plants and animals. Some protesters argue passionately against this new technology for a variety of mostly ethical reasons. Others advocate slowing down the technological rush and taking a closer look at the short- and long-term advantages and disadvantages of this and other rapidly emerging genetic technologies.

At the very least, they say, all genetically modified crops and animal products and foods containing such components should be clearly labeled as such. Such labels would give consumers a more informed choice, much like the food labels that list ingredients and nutritional information. Makers of genetically modified products strongly oppose such labeling because they fear it would hurt sales.

Supporters of genetic engineering wonder why there is so much concern. After all, we have been genetically modifying plants and animals for centuries. Now we have a faster, better, and perhaps cheaper way to do it, so why not use it?

Proponents of more careful control of genetic engineering counter that most new technologies have had unintended harmful consequences. For example, pesticides have helped protect crops from insect pests and disease. Wonderful. At the same time, their overuse has accelerated genetic evolution in many species, which have become resistant to many of the most widely used pesticides. The pesticides have also unintentionally wiped out many natural predator insects that helped keep crop pest populations under control.

The ecological lesson: Whenever we intervene in nature we must pause and ask, "What happens next?" That explains why many analysts urge caution before rushing into genetic engineering and other forms of biotechnology without more careful evaluation of their unintended consequences and more stringent regulation of this new technology.

All we have yet discovered is but a trifle in comparison with what lies hid in the great treasury of nature.

ANTOINE VAN LEEUWENHOCK

CRITICAL THINKING

1. An important adaptation of humans is a strong opposable thumb, which allows us to grip and manipulate things with our hands. Fold each of your thumbs into the palm of its hand and then tape them securely in that position for an entire day. After the demonstration, make a list of the things you could not do without the use of your thumbs.

2. a. How would you respond to someone who tells you that he or she does not believe in biological evolution because it is "just a theory"?

b. How would you respond to a statement that we should not worry about air pollution because natural selection will enable humans to develop lungs that can detoxify pollutants?

3. How would you respond to someone who says that because extinction is a natural process, we should not worry about the loss of biodiversity?

4. Describe the major differences between the ecological niches of humans and cockroaches. Are these two species in competition? If so, how do they manage to coexist?

5. Explain why you are for or against each of the following: **(a)** requiring labels indicating the use of genetically modified components in any food item, **(b)** using genetic engineering to develop "superior" human beings, and **(c)** using genetic engineering to eliminate aging and death.

6. Congratulations! You are in charge of the future evolution of life on the earth. What are the three most important things you would do?

LEARNING ONLINE

The website for this book includes review questions for the entire chapter, flash cards for key terms and concepts, a multiple-choice practice quiz, interesting Internet sites, references, and a guide for accessing thousands of InfoTrac® College Edition articles.

Visit

http://biology.brookscole.com/miller11

Then choose Chapter 4, and select a learning resource. For access to animations, additional quizzes, chapter outlines and summaries, register and log in to

Environmental Science ⊛ Now™

at **esnow.brookscole.com/miller11** using the access code card in the front of your book.

Active Graphing

Visit http://esnow.brookscole.com/miller11 to explore the graphing exercise for this chapter.

CASE STUDY
Blowing in the Wind: A Story of Connections

Wind, a vital part of the planet's circulatory system, connects most life on the earth. Without wind, the tropics would be unbearably hot and most of the rest of the planet would freeze.

Winds also transport nutrients from one place to another. Dust rich in phosphates and iron blows across the Atlantic from the Sahara Desert in Africa (Figure 5-1). This movement helps build up agricultural soils in the Bahamas and supplies nutrients for plants in the rain forest's upper canopy in Brazil. Iron-rich dust blowing from China's Gobi Desert falls into the Pacific Ocean between Hawaii and Alaska. This input of iron stimulates the growth of phytoplankton, the minute producers that support ocean food webs. This is the *good news.*

Now for the *bad news:* Wind also transports harmful viruses, bacteria, fungi, and particles of long-lived pesticides and toxic metals. Particles of reddish-brown soil and pesticides banned in the United States are blown from Africa's deserts and eroding farmlands into the sky over Florida. This makes it difficult for the state to meet federal air pollution standards during summer months.

More *bad news.* Some types of fungi in this dust may play a role in degrading or killing coral reefs in the Florida Keys and the Caribbean. Scientists are currently studying possible links between contaminated African dust and a sharp rise in rates of asthma in the Caribbean since 1973.

Particles of iron-rich dust from Africa that enhance the productivity of algae have been linked to outbreaks of toxic algal blooms—referred to as *red tides*—in Florida's coastal waters. People who eat shellfish contaminated by a toxin produced in red tides can become paralyzed or even die. Europe and the Middle East also receive contaminated African dust.

Pollution and dust from rapidly industrializing China and central Asia blow across the Pacific Ocean and degrade air quality over parts of the western United States. In 2001, climate scientists reported that a huge dust storm of soil particles blown from northern China had blanketed areas from Canada to Arizona with a layer of dust. Asian pollution contributes as much as 10% of West Coast smog—a threat that is expected to increase.

There is also *mixed news.* Particles from volcanic eruptions ride the winds, circle the globe, and change the earth's climate for a while. Emissions from the 1991 eruption of Mount Pinatubo in the Philippines cooled the earth slightly for 3 years, temporarily masking signs of global warming. Like the blowing desert dust, volcanic ash adds valuable trace minerals to the soil where it settles.

The familiar lesson: *There is no away* because *everything is connected.* Wind acts as part of the planet's circulatory system for heat, moisture, plant nutrients, and long-lived pollutants. Movement of soil particles from one place to another by wind and water is a natural phenomenon. When we disturb the soil and leave it unprotected, we accelerate this process.

Wind also acts as an important factor in climate through its influence on global air circulation patterns. Climate, in turn, is crucial for determining what kinds of plant and animal life are found in the major biomes of the biosphere, as discussed in this chapter.

NOAA. USGS/MND EROS Data Center

Figure 5-1 Some of the dust shown here blowing from Africa's Sahara Desert can end up as soil nutrients in Amazonian rain forests and toxic air pollutants in Florida and the Caribbean.

To do science is to search for repeated patterns, not simply to accumulate facts, and to do the science of geographical ecology is to search for patterns of plant and animal life that can be put on a map.

ROBERT H. MACARTHUR

This chapter provides an introduction to the earth's climate and how it affects the types of life found in different parts of the earth. It discusses addresses these questions:

- What key factors determine the earth's climate?

- How does climate determine where the earth's major biomes are found?

- What are the major types of desert and grassland biomes, and how do human activities affect them?

- What are the major types of forest and mountain biomes, and how do human activities affect them?

- What are the major types of saltwater life zones, and how do human activities affect them?

- What are the major types of freshwater life zones, and how do human activities affect them?

KEY IDEAS

- Climate is an area's average weather conditions—especially temperature and precipitation—over a long time.

- The climate of a region is determined by incoming solar radiation, global patterns of air and water movement, gases in the atmosphere, and major features of the earth's surface.

- The location of terrestrial biomes such as deserts, grassland, and forests is determined mostly by climate and human activities that remove or alter vegetation.

- Saltwater and freshwater aquatic life zones cover almost three-fourths of the earth's surface.

- Human activities are disrupting and degrading many of the ecological and economic services provided by the earth's terrestrial biomes and aquatic systems.

5-1 CLIMATE: A BRIEF INTRODUCTION

Factors That Affect Climate

Climate is the average weather conditions—especially temperature and precipitation—of an area over a long time. It is affected by global air circulation.

Weather is an area's short-term temperature, precipitation, humidity, wind speed, cloud cover, and other physical conditions of the lower atmosphere over a short period of time. Science Supplement 5 at the end of this book introduces you to weather basics.

Climate is a region's general pattern of atmospheric or weather conditions over a long time. *Average temperature* and *average precipitation* are the two main factors determining climate. Figure 5-2 (p. 80) depicts the earth's major climate zones.

Many factors contribute to the global and local climate. The main ones are the amount of solar radiation reaching the area, the earth's daily rotation and annual path around the sun, air circulation over the earth's surface, the global distribution of landmasses and seas, the circulation of ocean currents, and the elevation of landmasses.

Three major factors determine how air circulates over the earth's surface. The first factor is the *uneven heating of the earth's surface.* Air is heated much more at its fattest part, the equator, where the sun's rays strike directly, than at the poles, where sunlight strikes at a slanted angle and spreads out over a much greater area. You can observe this effect by shining a flashlight in a darkened room on the middle of a spherical object such as a basketball and moving the light up and down. The differences in the amount of incoming solar energy help explain why tropical regions near the equator are hot, polar regions are cold, and temperate regions in between generally have intermediate average temperatures.

A second factor is the *rotation of the earth on its axis*—an imaginary line connecting the north and south poles. As the earth rotates around its north–south axis, its equator rotates faster than its polar regions. As a result, the air masses rising above the earth and moving north and south are deflected to the west or east over different parts of the planet's surface (Figure 5-3, p. 80). The direction of air movement in these cells sets up belts of *prevailing winds*—major surface winds that blow almost continuously and distribute air and moisture over the earth's surface.

The third factor affecting global air circulation is *properties of air, water, and land.* Heat from the sun evaporates ocean water and transfers heat from the oceans to the atmosphere, especially near the hot equator. Also, the earth's landmasses give up heat faster than its oceans. These properties create giant cyclical convection cells that circulate air, heat, and moisture both vertically and from place to place in the troposphere. The earth's air circulation patterns, prevailing winds, and mixture of continents and oceans result in six giant convection cells—three north of the equator and three south of the equator—in which warm, moist air rises and cool, dry air sinks. This leads an irregular distribution of climates and patterns of vegetation, as shown in Figure 5-4 (p. 81).

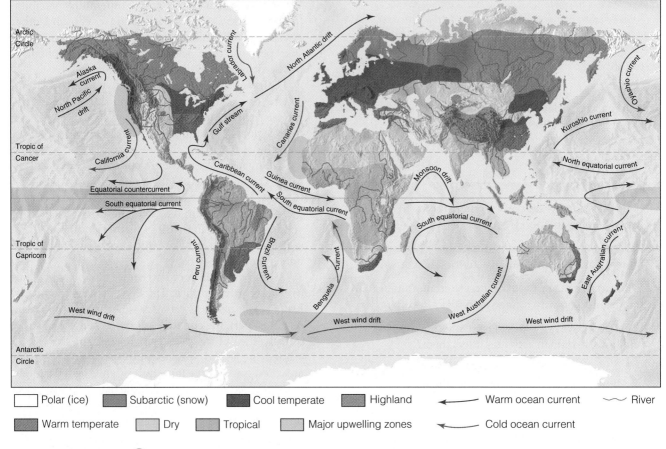

☐ Polar (ice)	■ Subarctic (snow)	■ Cool temperate	■ Highland	← Warm ocean current	~ River	
■ Warm temperate	☐ Dry	■ Tropical	■ Major upwelling zones	← Cold ocean current		

Active Figure 5-2 Natural capital: generalized map of the earth's current climate zones, showing the major contributing ocean currents and drifts. *See an animation based on this figure and take a short quiz on the concept.*

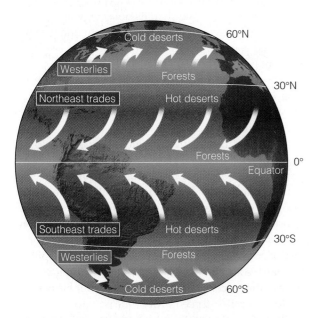

Figure 5-3 Natural capital: the earth's rotation deflects the movement of the air over different parts of the earth, creating global patterns of prevailing winds.

Watch the formation of six giant convection cells and learn more about how they affect climates at Environmental ScienceNow.

Effects of Ocean Currents and Winds on Regional Climates

Ocean currents and winds influence climate by redistributing heat received from the sun from one place to another.

Ocean currents also have a major effect on the climate of different regions. The oceans absorb heat from the air circulation patterns just described, with the bulk of this heat being absorbed near the warm tropical areas. This heat and differences in water density create warm and cold ocean currents (Figure 5-2). These currents tend to flow clockwise between the continents in the northern hemisphere and counterclockwise in the southern hemisphere (Figure 5-2). Driven by winds and the earth's rotation, they redistribute heat received from the sun from one place to another, thereby influ-

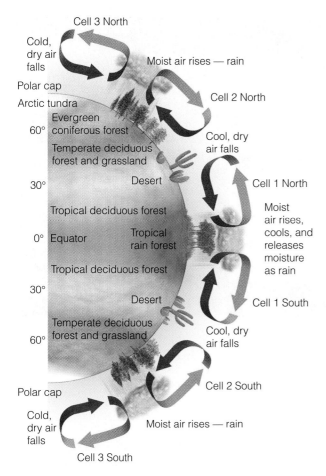

Cell 3 North

Cold, dry air falls

Moist air rises — rain

Polar cap

Arctic tundra

Cell 2 North

60° Evergreen coniferous forest

Temperate deciduous forest and grassland

Cool, dry air falls

30° Desert

Cell 1 North

Tropical deciduous forest

Moist air rises, cools, and releases moisture as rain

0° Equator

Tropical rain forest

Tropical deciduous forest

30° Desert

Cell 1 South

Temperate deciduous forest and grassland

Cool, dry air falls

60° Cell 2 South

Polar cap

Cold, dry air falls

Moist air rises — rain

Cell 3 South

Figure 5-4 Natural capital: global air circulation and biomes. Heat and moisture are distributed over the earth's surface by vertical currents, which form six giant convection cells at different latitudes. The direction of air flow and the ascent and descent of air masses in these convection cells determine the earth's general climatic zones. The uneven distribution of heat and moisture over the planet's surface leads to the forests, grasslands, and deserts that make up the earth's biomes.

encing climate and vegetation, especially near coastal areas. They also help mix ocean waters and distribute the nutrients and dissolved oxygen needed by aquatic organisms.

The warm Gulf Stream (Figure 5-2), for example, transports 25 times more water than all of the world's rivers combined. Without the warming effect of the heat carried northward from the equator, the climate of northwestern Europe would be subarctic to arctic. If this ocean current suddenly stopped flowing, deserts would appear in the tropics and thick ice sheets would cover northern Europe, Siberia, and Canada.

Environmental Science ⊕ Now™

Learn more about how oceans affect air movements, near where you live and all over the world, at Environmental ScienceNow.

Effects of Gases in the Atmosphere: The Natural Greenhouse Effect

Water vapor, carbon dioxide, and other gases influence climate by warming the lower troposphere and the earth's surface.

Small amounts of certain gases in the atmosphere play a key role in determining the earth's average temperatures and thus its climates. These gases include water vapor (H_2O), carbon dioxide (CO_2), methane (CH_4), and nitrous oxide (N_2O).

These **greenhouse gases** allow mostly visible light and some infrared radiation and ultraviolet (UV) radiation from the sun to pass through the troposphere. The earth's surface absorbs much of this solar energy and transforms it into longer-wavelength infrared radiation (heat), which then rises into the troposphere.

Some of this heat escapes into space, but some is absorbed by molecules of greenhouse gases and emitted into the troposphere as even longer-wavelength infrared radiation. Some of this released energy radiates into space, and some warms the troposphere and the earth's surface. This natural warming effect of the troposphere is called the **greenhouse effect** (Figure 5-5, p. 82). Without its current greenhouse gases (especially water vapor, which is found in the largest concentration), the earth would be a cold and mostly lifeless planet.

Human activities such as burning fossil fuels, clearing forests, and growing crops release carbon dioxide, methane, and nitrous oxide into the atmosphere. Considerable evidence and climate models indicate that these large inputs of these greenhouse gases into the troposphere can enhance the earth's natural greenhouse effect and lead to *global warming*. This could alter precipitation patterns, shift areas where we can grow crops, raise average sea levels, and change the areas where some types of plants and animals can live.

Environmental Science ⊕ Now™

Witness the greenhouse effect and see how human activity has affected it at Environmental ScienceNow.

Topography of the Earth's Surface and Local Climate: Land Matters

Interactions between land and oceans, as well as disruptions of air flows by mountains and cities, affect local climates.

Heat is absorbed and released more slowly by water than by land. This difference creates land and sea breezes. As a result, the world's oceans and large lakes moderate the climate of nearby lands.

Various topographic features of the earth's surface create local and regional climatic conditions that differ from the general climate of a region. For example,

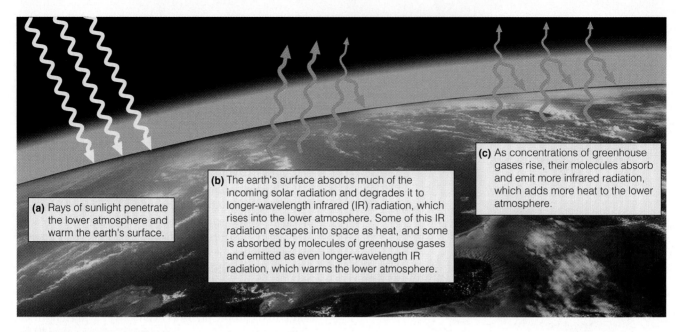

(a) Rays of sunlight penetrate the lower atmosphere and warm the earth's surface.

(b) The earth's surface absorbs much of the incoming solar radiation and degrades it to longer-wavelength infrared (IR) radiation, which rises into the lower atmosphere. Some of this IR radiation escapes into space as heat, and some is absorbed by molecules of greenhouse gases and emitted as even longer-wavelength IR radiation, which warms the lower atmosphere.

(c) As concentrations of greenhouse gases rise, their molecules absorb and emit more infrared radiation, which adds more heat to the lower atmosphere.

Environmental Science ⊕ Now™

Active Figure 5-5 Natural capital: the *natural greenhouse effect*. Without the atmospheric warming provided by this natural effect, the earth would be a cold and mostly lifeless planet. When concentrations of greenhouse gases in the atmosphere rise, the average temperature of the troposphere rises. *See an animation based on this figure and take a short quiz on the concept.* (Modified by permission from Cecie Starr, *Biology: Concepts and Applications,* 4th ed., Pacific Grove, Calif.: Brooks/Cole, 2000.)

mountains interrupt the flow of prevailing surface winds and the movement of storms. When moist air blowing inland from an ocean reaches a mountain range, it cools as it is forced to rise and expand. The air then loses most of its moisture as rain and snow on the windward (wind-facing) slopes.

As the drier air mass flows down the leeward (away from the wind) slopes, it draws moisture out of the plants and soil below. The lower precipitation and the resulting semiarid or arid conditions on the lee-

ward side of high mountains create the **rain shadow effect** (Figure 5-6), sometimes leading to the formation of deserts. For example, in the Sierra Nevada mountains in North America, about five times more precipitation falls on the side of the range facing the wind as on the side facing away from the wind.

Cities also create distinct microclimates. Bricks, concrete, asphalt, and other building materials absorb and hold heat, and buildings block wind flow. Motor vehicles and the climate control systems of buildings

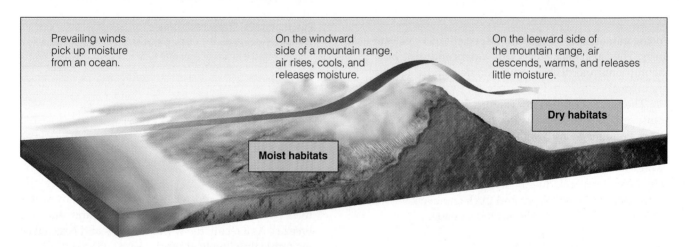

Prevailing winds pick up moisture from an ocean.

On the windward side of a mountain range, air rises, cools, and releases moisture.

On the leeward side of the mountain range, air descends, warms, and releases little moisture.

Dry habitats

Moist habitats

Figure 5-6 Natural capital: the *rain shadow effect* is a reduction of rainfall on the side of mountains facing away from prevailing surface winds. Warm, moist air in prevailing onshore winds loses most of its moisture as rain and snow on the windward (wind-facing) slopes of a mountain range. This leads to semiarid and arid conditions on the leeward side of the mountain range and the land beyond. California's Mojave Desert and Asia's Gobi Desert are both produced by this effect.

release large quantities of heat and pollutants. As a result, cities tend to have more haze and smog, higher temperatures, and lower wind speeds than the surrounding countryside.

5-2 BIOMES: CLIMATE AND LIFE ON LAND

Why Do Different Organisms Live in Different Places?

Different climates lead to different communities of organisms, especially vegetation.

Why is one area of the earth's land surface a desert, another a grassland, and yet another a forest? Why do different types of deserts, grasslands, and forests exist? The general answer to these questions is differences in *climate* (Figure 5-2), resulting mostly from differences in average temperature and precipitation caused by global air circulation (Figure 5-4). Different climates support different communities of organisms.

Figure 5-7 shows how scientists have divided the world into 12 major *biomes*. These terrestrial regions have characteristic types of natural communities adapted to the climate of each region. What kind of biome do you live in?

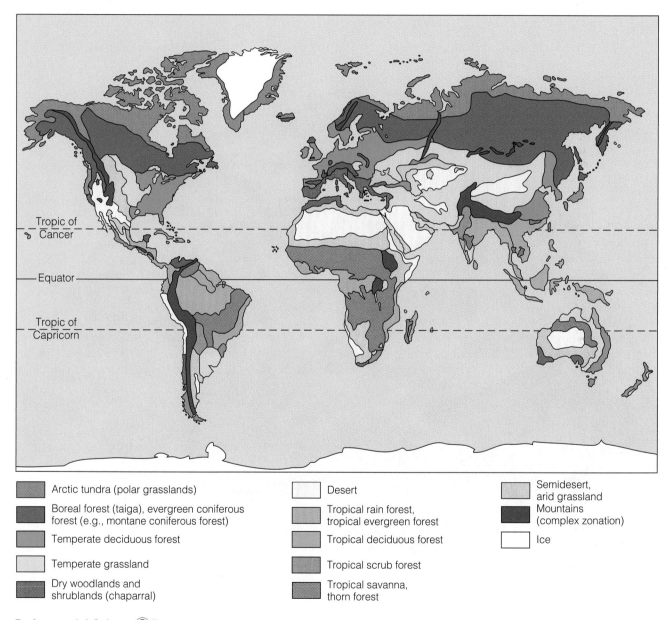

Arctic tundra (polar grasslands)	Desert	Semidesert, arid grassland
Boreal forest (taiga), evergreen coniferous forest (e.g., montane coniferous forest)	Tropical rain forest, tropical evergreen forest	Mountains (complex zonation)
Temperate deciduous forest	Tropical deciduous forest	Ice
Temperate grassland	Tropical scrub forest	
Dry woodlands and shrublands (chaparral)	Tropical savanna, thorn forest	

Environmental Science ⊕ Now™

Active Figure 5-7 Natural capital: the earth's major *biomes*—the main types of natural vegetation in various undisturbed land areas—result primarily from differences in climate. Each biome contains many ecosystems whose communities have adapted to differences in climate, soil, and other environmental factors. Humans have removed or altered much of this natural vegetation in some areas for farming, livestock grazing, lumber and fuelwood, mining, and construction. *See an animation based on this figure and take a short quiz on the concept.*

By comparing Figure 5-7 with Figure 5-2 and Figure 1 in Science Supplement 2 at the end of this book, you can see how the world's major biomes vary with climate. Figure 3-8 (p. 41) shows major biomes in the United States as one moves through different climates along the 39th parallel.

Average annual precipitation and temperature (as well as soil type; see Figure 3-22, p. 52) are the most important factors in producing tropical, temperate, or polar deserts, grasslands, and forests (Figure 5-8).

On maps such as the one in Figure 5-7, biomes are depicted as having sharp boundaries and being covered with the same general type of vegetation. In reality, *biomes are not uniform.* They consist of a *mosaic of patches,* each having a somewhat different biological community but sharing similarities unique to the biome. These patches occur mostly because the resources that plants and animals need are not uniformly distributed and because human activities remove and alter natural vegetation.

Figure 5-9 shows how climate and vegetation vary with **latitude** (distance from the equator) and **altitude** (elevation above sea level). If you climb a tall mountain from its base to its summit, you can observe changes in plant life similar to those you would encounter in traveling from the equator to the earth's poles.

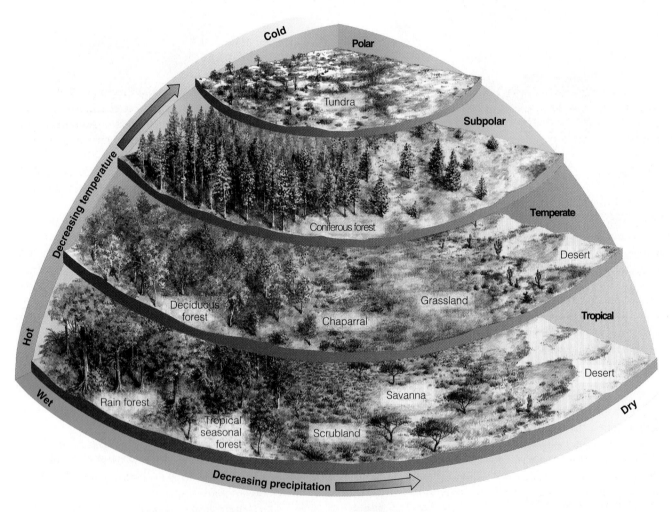

Figure 5-8 Natural capital: average precipitation and average temperature, acting together as limiting factors over a period of 30 or more years, determine the type of desert, grassland, or forest biome in a particular area. Although the actual situation is much more complex, this simplified diagram explains how climate determines the types and amounts of natural vegetation found in an area left undisturbed by human activities. (Used by permission of Macmillan Publishing Company, from Derek Elsom, *The Earth,* New York: Macmillan, 1992. Copyright © 1992 by Marshall Editions Developments Limited.)

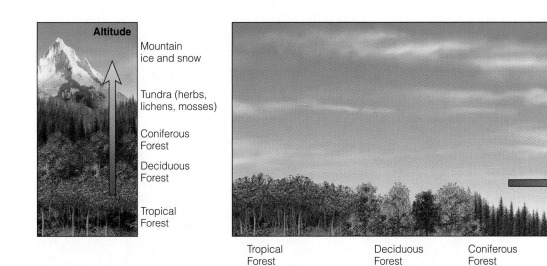

Figure 5-9 Natural capital: generalized effects of altitude (left) and latitude (right) on climate and biomes. Parallel changes in vegetation type occur when we travel from the equator to the poles or from lowlands to mountaintops.

5-3 DESERT AND GRASSLAND BIOMES

Types of Deserts

Deserts have little precipitation and little vegetation and are found in tropical, temperate, and polar regions.

A **desert** is an area where evaporation exceeds precipitation. Annual precipitation is low and often scattered unevenly throughout the year. Deserts have sparse, widely spaced, mostly low vegetation.

During the day, the baking sun warms the ground in the desert. At night, most of the heat stored in the ground radiates quickly into the atmosphere. Desert soils have little vegetation and moisture to help store the heat and the skies above deserts are usually clear. This explains why in a desert you may roast during the day but shiver at night.

A combination of low rainfall and different average temperatures creates tropical, temperate, and cold deserts (Figure 5-8). *Tropical deserts* are hot and dry most of the year. They have few plants and a hard, windblown surface strewn with rocks and some sand. They are the deserts we often see in the movies.

In *temperate deserts,* such as the Mojave in southern California, daytime temperatures are high in summer

and low in winter and more precipitation falls than in tropical deserts. The sparse vegetation consists mostly of widely dispersed, drought-resistant shrubs and cacti or other succulents, and animals are adapted to the lack of water and temperature variations, as shown in Figure 5-10 (p. 86). In *cold deserts,* such as the Gobi Desert in China, winters are cold, summers are warm or hot, and precipitation is low.

Figure 5-11 (p. 87) shows major human impacts on deserts. Deserts take a long time to recover from disturbances because of their slow plant growth, low species diversity, slow nutrient cycling (because of sparse bacterial activity in their soils), and lack of water. Desert vegetation destroyed by livestock overgrazing and off-road vehicles may take decades to grow back.

Types of Grasslands

Grasslands have enough precipitation to support grasses but not enough to support large stands of trees.

Grasslands, or **prairies,** are regions with enough average annual precipitation to support grass and, in some areas, a few trees. Most grasslands are found in the interiors of continents (Figure 5-7).

Grasslands persist because of a combination of seasonal drought, grazing by large herbivores, and occasional fires—all of which keep large numbers of shrubs and trees from growing. The three main types of grasslands—tropical, temperate, and polar (tundra)—result from combinations of low average

Figure 5-10 Natural capital:
some components and interactions in a *temperate desert
ecosystem.* When these organisms die, decomposers break
down their organic matter into
minerals that plants use. Colored
arrows indicate transfers of
matter and energy between
producers; primary consumers
(herbivores); secondary, or
higher-level, consumers (carnivores); and decomposers. Organisms are not drawn to scale.
The photo shows a temperate
desert south of Fallon, Nevada.

Red-tailed hawk

Gambel's
quail

Yucca

Agave

Jack
rabbit

Collared
lizard

Prickly
pear
cactus

Roadrunner

Darkling
beetle

Bacteria

Fungi

Diamondback rattlesnake

Kangaroo rat

Producer
to primary
consumer

Primary
to secondary
consumer

Secondary to
higher-level
consumer

All producers and
consumers to
decomposers

Bob Rowan/Progressive Image/CORBIS

Large desert cities

Soil destruction by off-road vehicles and urban development

Soil salinization from irrigation

Depletion of underground water supplies

Land disturbance and pollution from mineral extraction

Storage of toxic and radioactive wastes

Large arrays of solar cells and solar collectors used to produce electricity

Figure 5-11 Natural capital degradation: major human impacts on the world's deserts. *Critical thinking: what are the direct and indirect effects of your lifestyle on deserts?*

precipitation and various average temperatures (Figure 5-8).

One type of tropical grassland, called a *savanna*, usually has warm temperatures year-round, two prolonged dry seasons, and abundant rain during the rest of the year. African tropical savannas boast enormous herds of *grazing* (grass- and herb-eating) and *browsing* (twig- and leaf-nibbling) hoofed animals, including wildebeests, gazelles, zebras, giraffes, and antelopes. These and other large herbivores have evolved specialized eating habits that minimize competition between species for the vegetation found on the savanna. For example, giraffes eat leaves and shoots from the tops of trees, elephants eat leaves and branches farther down, Thomson's gazelles and wildebeests prefer short grass, and zebras graze on longer grass and stems.

In *temperate grasslands,* winters are bitterly cold, summers are hot and dry, and annual precipitation is fairly sparse and falls unevenly through the year. Drought, occasional fires, and intense grazing inhibit the growth of trees and bushes, except along rivers.

Because the aboveground parts of most of the grasses die and decompose each year, organic matter accumulates to produce a deep, fertile soil (Figure 3-22, top right, p. 52). This soil is held in place by a thick net-

work of intertwined roots of drought-tolerant grasses unless the topsoil is plowed up and allowed to blow away by prolonged exposure to high winds found in these biomes.

Types of temperate grasslands include the *tall-grass prairies* (Figure 5-12, p. 88) and *short-grass prairies* of the midwestern and western United States and Canada. Here winds blow almost continuously, and evaporation is rapid, often leading to fires in the summer and fall.

Many of the world's natural temperate grasslands have disappeared because they are great places to grow crops (Figure 5-13, p. 88) and graze cattle. They are often flat, are easy to plow, and have fertile, deep soils. Plowing breaks up the soil and leaves it vulnerable to erosion by wind and water, and overgrazing can transform grasslands into semidesert or desert.

Environmental Science ⊛ Now™

Learn more about how plants and animals in a temperate prairie are connected across a food web at Environmental ScienceNow.

Polar grasslands, or *arctic tundra,* occur just south of the arctic polar ice cap (Figure 5-7). During most of the year, these treeless plains are bitterly cold, swept by frigid winds, and covered with ice and snow (Figure 5-14, left, p. 89). Winters are long and dark, and the scant precipitation falls mostly as snow.

This biome is carpeted with a thick, spongy mat of low-growing plants, primarily grasses, mosses, and dwarf woody shrubs. Trees or tall plants cannot survive in the cold and windy tundra because they would lose too much of their heat. Most of the annual growth of the tundra's plants occurs during the 5- to 8-week summer, when the sun shines almost around the clock (Figure 5-14, right, p. 89).

One outcome of the extreme cold is the formation of **permafrost,** soil in which the water it contains stays frozen more than 2 years in a row. During the brief summer the permafrost layer keeps melted snow and ice from soaking into the ground. As a consequence, the waterlogged tundra forms a large number of shallow lakes, marshes, bogs, ponds, and other seasonal wetlands. Hordes of mosquitoes, black flies, and other insects thrive in these shallow surface pools. They serve as food for large colonies of migratory birds (especially waterfowl) that return from the south to nest and breed in the bogs and ponds.

Another type of tundra, called *alpine tundra,* occurs above the limit of tree growth but below the permanent snow line on high mountains (Figure 5-9, left). The vegetation is similar to that found in arctic tundra but receives more sunlight than arctic vegetation, and alpine tundra has no permafrost layer.

Figure 5-15 (p. 89) lists the major human impacts on grasslands.

Active Figure 5-12 Natural capital: some components and interactions in a *temperate tall-grass prairie ecosystem* in North America. When these organisms die, decomposers break down their organic matter into minerals that plants can use. Colored arrows indicate transfers of matter and energy between producers, primary consumers (herbivores), secondary consumers (carnivores), and decomposers. Organisms are not drawn to scale. *See an animation based on this figure and take a short quiz on the concept.*

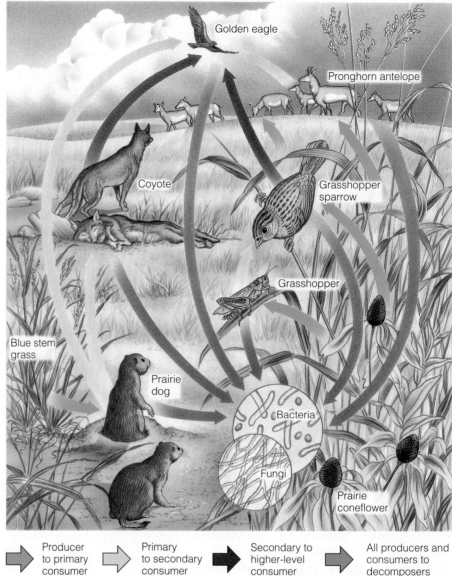

⇒ Producer to primary consumer	⇒ Primary to secondary consumer	⇒ Secondary to higher-level consumer	⇒ All producers and consumers to decomposers

National Archives/EPA Documerica

Figure 5-13 Natural capital degradation: replacement of a biologically diverse temperate grassland with a monoculture crop in California. When humans remove the tangled root network of natural grasses, the fertile topsoil becomes subject to severe wind erosion unless it is covered with some type of vegetation.

Figure 5-14 Natural capital: Arctic tundra (polar grassland) near Nome, Alaska, in winter (left) and in Alaska's Denali National Park in summer (right).

Natural Capital Degradation

Grasslands

Conversion of savanna and temperate grassland to cropland

Release of CO_2 to atmosphere from burning and conversion of grassland to cropland

Overgrazing of tropical and temperate grasslands by livestock

Damage to fragile arctic tundra by oil production, air and water pollution, and off-road vehicles

Figure 5-15 Natural capital degradation: major human impacts on the world's grasslands. *Critical thinking: what are the direct and indirect effects of your lifestyle on grasslands?*

5-4 FOREST AND MOUNTAIN BIOMES

Types of Forests

Forests have enough precipitation to support stands of trees and are found in tropical, temperate, and polar regions.

Undisturbed areas with moderate to high average annual precipitation tend to be covered with **forest,** which contains various species of trees and smaller forms of vegetation. The three main types of forest—*tropical, temperate,* and *boreal* (polar)—result from combinations of the precipitation level and various average temperatures (Figure 5-8).

Tropical rain forests (Figure 5-16, p. 90) are found near the equator (Figure 5-7), where hot, moisture-laden air rises and dumps its moisture (Figure 5-4). These forests have a warm annual mean temperature (which varies little, either daily or seasonally), high humidity, and heavy rainfall almost daily.

Tropical rain forests are dominated by *broadleaf evergreen plants,* which keep most of their leaves year-round. The tops of the trees form a dense canopy that blocks most light from reaching the forest floor. Many of the plants that live at the ground level have enormous leaves to capture what little sunlight filters through to the dimly lit forest floor.

Tropical rain forests are teeming with life and boast incredible biological diversity. These life forms occupy a variety of specialized niches in distinct layers—in the plants' case, based mostly on their need for sunlight, as shown in Figure 5-17. Stratification of specialized plant and animal niches in a tropical rain forest enables the coexistence of a great variety of species. Although tropical rain forests cover only 2% of the earth's land surface, at least half of the earth's terrestrial species reside there.

Dropped leaves, fallen trees, and dead animals decompose quickly because of the warm, moist conditions and hordes of decomposers. This rapid recycling of scarce soil nutrients explains why little litter is found on the ground. Instead of being stored in the soil, most minerals released by decomposition are taken up quickly by plants and stored by trees, vines, and other plants. As a result, tropical rain forest soils contain very few plant nutrients (Figure 3-22, bottom left, p. 52). Rain forests are not good places to clear and grow crops or graze cattle on a sustainable basis.

Environmental Science⊕Now™
Learn more about how plants and animals in a rain forest are connected across a food web at Environmental ScienceNow.

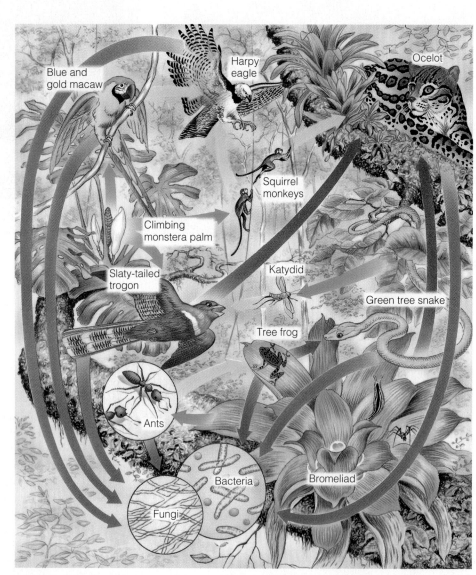

Environmental Science⊕Now™
Active Figure 5-16 Natural capital: some components and interactions in a *tropical rain forest ecosystem.* When these organisms die, decomposers break down their organic matter into minerals that plants use. Colored arrows indicate transfers of matter and energy between producers; primary consumers (herbivores); secondary, or higher-level, consumers (carnivores); and decomposers. Organisms are not drawn to scale. *See an animation based on this figure and take a short quiz on the concept.*

 Producer to primary consumer

 Primary to secondary consumer

 Secondary to higher-level consumer

 All producers and consumers to decomposers

Height (meters)

- 45
- 40 — Harpy eagle
- 35 — Toco toucan
- 30
- 25
- 20 — Wooly opossum
- 15
- 10
- 5 — Black-crowned antpitta — Brazilian tapir
- 0

Emergent layer

Canopy

Understory

Shrub layer

Ground layer

Figure 5-17 Natural capital: stratification of specialized plant and animal niches in a *tropical rain forest*. Filling such specialized niches enables species to avoid or minimize competition for resources and results in the coexistence of a great variety of species.

Temperate deciduous forests (Figure 5-18, p. 92) grow in areas with moderate average temperatures that change significantly with the season. These areas have long, warm summers, cold but not too severe winters, and abundant precipitation, often spread fairly evenly throughout the year.

This biome is dominated by a few species of *broadleaf deciduous trees* such as oak, hickory, maple, poplar, and beech. They survive cold winters by becoming dormant through the winter (photo in Figure 5-18, right). Each spring they grow new leaves whose colors change in the fall into an array of reds and golds before the leaves drop (photo in Figure 5-18, left). Because of a slow rate of decomposition, these forests accumulate a thick layer of slowly decaying leaf litter that is a storehouse of nutrients

Evergreen coniferous forests, also called *boreal forests* and *taigas* ("TIE-guhs"), are found just south of the arc-

tic tundra in northern regions across North America, Asia, and Europe (Figure 5-7). In this subarctic climate, winters are long, dry, and extremely cold; in the northernmost taiga, sunlight is available only 6–8 hours per day. Summers are short, with mild to warm temperatures, and the sun typically shines 19 hours per day.

Most boreal forests are dominated by a few species of *coniferous* (cone-bearing) *evergreen trees* such as spruce, fir, cedar, hemlock, and pine that keep some of their narrow-pointed leaves (needles) year-round (Figure 5-19, p. 93). The small, needle-shaped, waxy-coated leaves of these trees can withstand the intense cold and drought of winter when snow blankets the ground. The trees are ready to take advantage of the brief summers in these areas without taking time to grow new needles. Plant diversity is low in these forests because few species can survive the winters when soil moisture is frozen.

Figure 5-18 Natural capital: some components and interactions in a *temperate deciduous forest ecosystem.* When these organisms die, decomposers break down their organic matter into minerals that plants use. Colored arrows indicate transfers of matter and energy between producers; primary consumers (herbivores); secondary, or higher-level, consumers (carnivores); and decomposers. Organisms are not drawn to scale. The photos show a temperate deciduous forest in Rhode Island during fall (left) and winter (right).

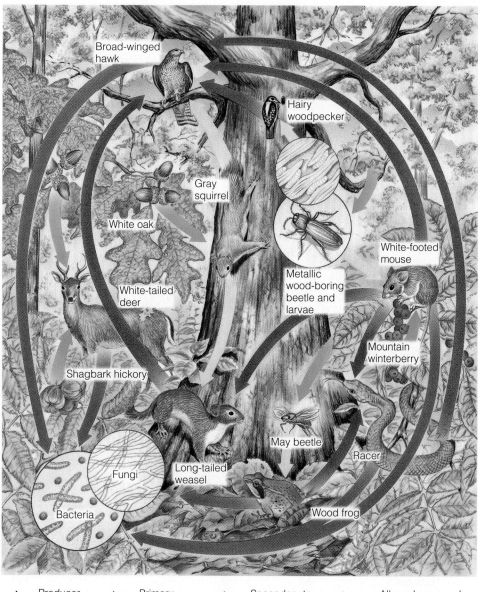

Broad-winged hawk

Hairy woodpecker

Gray squirrel

White oak

White-tailed deer

White-footed mouse

Metallic wood-boring beetle and larvae

Mountain winterberry

Shagbark hickory

Fungi

May beetle

Racer

Bacteria

Long-tailed weasel

Wood frog

Producer to primary consumer

Primary to secondary consumer

Secondary to higher-level consumer

All producers and consumers to decomposers

Paul W. Johnson/Biological Photo Service

Paul W. Johnson/Biological Photo Service

Blue jay

Balsam fir

Moose

Wolf

Bebb
willow

Snowshoe
hare

Starflower

Bacteria

Great
horned
owl

Marten

White
spruce

Pine sawyer
beetle
and larvae

Fungi

Bunchberry

 Producer
to primary
consumer

 Primary
to secondary
consumer

 Secondary to
higher-level
consumer

 All producers and
consumers to
decomposers

Figure 5-19 Natural capital: some components and interactions in an *evergreen coniferous* (*boreal* or *taiga*) *forest ecosystem*. When these organisms die, decomposers break down their organic matter into minerals that plants use. Colored arrows indicate transfers of matter and energy between producers; primary consumers (herbivores); secondary, or higher-level, consumers (carnivores); and decomposers. Organisms are not drawn to scale. The photo shows an evergreen coniferous forest in Alaska's Tongass National Forest.

Tom Bean/CORBIS

Beneath the stands of trees is a deep layer of partially decomposed conifer needles and leaf litter. Decomposition occurs slowly because of the low temperatures, waxy coating of conifer needles, and high soil acidity. The decomposing conifer needles make the thin, nutrient-poor soil acidic and prevent most other plants (except certain shrubs) from growing on the forest floor.

These biomes contain a variety of wildlife. During the brief summer the soil becomes waterlogged, forming acidic bogs, or *muskegs,* in low-lying areas of these forests. Warblers and other insect-eating birds feed on hordes of flies, mosquitoes, and caterpillars that breed in these habitats.

Coastal coniferous forests or *temperate rain forests* are found in scattered coastal temperate areas with ample rainfall or moisture from dense ocean fogs. Dense stands of large conifers such as Sitka spruce, Douglas fir, and redwoods dominate undisturbed areas of biomes along the coast of North America, from Canada to northern California.

Figure 5-20 lists major human impacts on the world's forests.

Some of the world's most spectacular environments are *mountains* (Figure 5-21), which cover about one-fifth of the earth's land surface. Mountains are places where dramatic changes in altitude, climate, soil, and vegetation take place over a very short distance (Figure 5-9, left). Because of the steep slopes,

Mark Hamblin/WWI/Peter Arnold, Inc.

Figure 5-21 **Natural capital degradation:** mountains such these in Washington state's Mount Ranier National Park play important ecological roles.

Natural Capital Degradation

Forests

Clearing and degradation of tropical forests for agriculture, livestock grazing, and timber harvesting

Clearing of temperate deciduous forests in Europe, Asia, and North America for timber, agriculture, and urban development

Clearing of evergreen coniferous forests in North America, Finland, Sweden, Canada, Siberia, and Russia

Conversion of diverse forests to less biodiverse tree plantations

Damage to soils from off-road vehicles

Figure 5-20 **Natural capital degradation:** major human impacts on the world's forests. *Critical thinking: what are the direct and indirect effects of your lifestyle on forests.*

mountain soils are especially prone to erosion when the vegetation holding them in place is removed by natural disturbances (such as landslides and avalanches) or human activities (such as timber cutting and agriculture). Many freestanding mountains are *islands of biodiversity* surrounded by a sea of lower-elevation landscapes transformed by human activities.

Mountains play a number of important ecological roles. They contain the majority of the world's forests, which are habitats for much of our planet's terrestrial biodiversity. They often are habitats for endemic species found nowhere else on earth, and they serve as sanctuaries for animal species driven from lowland areas.

They also help regulate the earth's climate. Mountaintops covered with ice and snow reflect solar radiation back into space. Mountains affect sea levels as a result of decreases or increases in glacial ice—most of which is locked up in Antarctica (the most mountainous of all continents).

Landless poor migrating uphill to survive by growing crops

Timber extraction

Mineral resource extraction

Hydroelectric dams and reservoirs

Increasing tourism (such as hiking and skiing)

Air pollution from industrial and urban centers

Increased ultraviolet radiation from ozone depletion

Soil damage from off-road vehicles

Figure 5-22 Natural capital degradation: major human impacts on the world's mountains. *Critical thinking: what are the direct and indirect effects of your lifestyle on mountains?*

Finally, mountains play a critical role in the hydrologic cycle by gradually releasing melting ice, snow, and water stored in the soils and vegetation of mountainsides to small streams.

Despite their ecological, economic, and cultural importance, the fate of mountain ecosystems has not been a high priority of governments or many environmental organizations. Mountain ecosystems are coming under increasing pressure from several human activities (Figure 5-22).

5-5 AQUATIC ENVIRONMENTS: TYPES AND CHARACTERISTICS

Types of Aquatic Life Zones

Saltwater and freshwater aquatic life zones cover almost three-fourths of the earth's surface.

The aquatic equivalents of biomes are called *aquatic life zones*. The major types of organisms found in aquatic environments are determined by the water's *salinity*—the amounts of various salts such as sodium chloride (NaCl) dissolved in a given volume of water. As a result, aquatic life zones are classified into two major types: *saltwater* or *marine* (particularly estuaries, coastlines, coral reefs, coastal marshes, mangrove swamps, and oceans) and *freshwater* (particularly lakes and ponds, streams and rivers, and inland wetlands). Figure 5-23 (p. 96) shows the distribution of the world's major oceans, lakes, rivers, coral reefs, and mangroves.

What Kinds of Organisms Live in Aquatic Life Zones?

Aquatic systems contain floating, drifting, swimming, bottom-dwelling, and decomposer organisms.

Saltwater and freshwater life zones contain several major types of organisms. One group consists of weakly swimming, free-floating plankton. They consist mostly of *phytoplankton* (plant plankton) and *zooplankton* (animal plankton).

A second group of organisms consists of **nekton,** strongly swimming consumers such as fish, turtles, and whales. A third group, **benthos,** dwells on the bottom. Examples include barnacles and oysters that anchor themselves to one spot, worms that burrow into the sand or mud, and lobsters and crabs that walk about on the seafloor. A fourth group consists of **decomposers** (mostly bacteria) that break down the organic compounds in the dead bodies and wastes of aquatic organisms into simple nutrient compounds for use by producers.

Factors That Limit Life at Different Depths in Aquatic Life Zones

Life in aquatic systems is found in surface, middle, and bottom layers.

Most aquatic life zones can be divided into three layers: *surface, middle,* and *bottom.* Important environmental factors determining the types and numbers of organisms found in these layers are *temperature, access to sunlight for photosynthesis, dissolved oxygen content,* and *availability of nutrients* such as carbon (as dissolved CO_2 gas), nitrogen (as NO_3^-), and phosphorus (mostly as PO_4^{3-}) for producers.

In deep aquatic systems, photosynthesis is largely confined to the upper layer, or euphotic zone, through which sunlight can penetrate. The depth of the euphotic zone in oceans and deep lakes can be reduced when excessive algal growth (algal blooms) clouds the water.

In shallow waters in streams, ponds, and oceans, ample supplies of nutrients for primary producers are usually available. By contrast, nitrates, phosphates, iron, and other nutrients are often in short supply in the open ocean and limit net primary productivity (NPP; see Figure 3-20, p. 50). NPP is much higher in some parts of the open ocean, where upward flows of

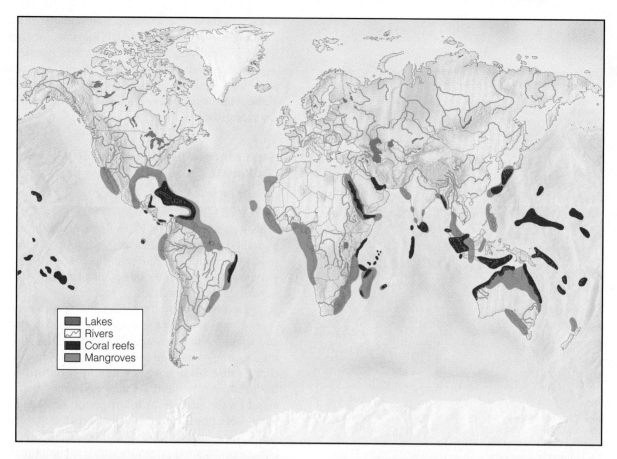

Figure 5-23 Natural capital: distribution of the world's major saltwater oceans, coral reefs, mangroves, and freshwater lakes and rivers. The photo on the left shows a coral reef in the Red Sea. The photo on the right shows a mangrove forest in Daintree National Park in Queensland, Australia.

Lakes
Rivers
Coral reefs
Mangroves

water bring nutrients from the ocean bottom to the surface for use by producers.

Most creatures living on the bottom of the deep ocean and deep lakes depend on animal and plant plankton that die and drift into deep waters. Because this food is limited, deep-dwelling species tend to reproduce slowly. As a consequence, they are especially vulnerable to depletion from overfishing.

5-6 SALTWATER LIFE ZONES

Why Should We Care about the Oceans?

Although oceans occupy most of the earth's surface and provide many ecological and economic services, we know less about them than we know about the moon.

A more accurate name for the earth would be *Ocean*—after all, saltwater oceans cover about 71% of the planet's surface (Figure 5-24). They contain about 250,000 known species of marine plants and animals and provide many important ecological and economic services (Figure 5-25).

As landlubbers, humans have a distorted and limited view of our watery home planet. We know more about the surface of the moon than about the oceans that cover most of the earth. According to aquatic scientists, the greatly increased scientific investigation of poorly understood marine and freshwater aquatic systems could yield immense ecological and economic benefits.

The Coastal Zone: Where Most of the Action Is

The coastal zone accounts for less than 10% of the world's ocean area but contains 90% of all marine species.

Oceans have two major life zones: the *coastal zone* and the *open sea* (Figure 5-26, p. 98). The **coastal zone** is the warm, nutrient-rich, shallow water that extends from the high-tide mark on land to the gently sloping, shallow edge of the *continental shelf* (the submerged part of the continents). This zone has numerous interactions with the land, so human activities readily affect it.

Although it makes up less than 10% of the world's ocean area, the coastal zone contains 90% of all marine species and is the site of most large commercial marine fisheries. Most ecosystems found in the coastal zone have a very high NPP per unit of area, thanks to the zone's ample supplies of sunlight and plant nutrients that flow from land and are distributed by wind and ocean currents.

Ocean hemisphere   Land–ocean hemisphere

Figure 5-24 Natural capital: the ocean planet. The salty oceans cover 71% of the earth's surface. About 97% of the earth's water is in the interconnected oceans, which cover 90% of the planet's mostly ocean hemisphere (left) and 50% of its land–ocean hemisphere (right). Freshwater systems cover less than 1% of the earth's surface.

Natural Capital
Marine Ecosystems

Ecological Services	Economic Services
Climate moderation	Food
CO₂ absorption	Animal and pet feed (fish meal)
Nutrient cycling	
Waste treatment and dilution	Pharmaceuticals
Reduced storm impact (mangrove, barrier islands, coastal wetlands)	Harbors and transportation routes
	Coastal habitats for humans
Habitats and nursery areas for marine and terrestrial species	Recreation
	Employment
	Offshore oil and natural gas
Genetic resources and biodiversity	
	Minerals
Scientific information	Building materials

Figure 5-25 Natural capital: major ecological and economic services provided by marine systems.

Environmental Science ⊕ Now™

Learn about ocean provinces where all ocean life exists at Environmental ScienceNow.

Estuaries, Coastal Wetlands, and Mangrove Swamps: Centers of Productivity

Several highly productive coastal ecosystems face increasing stress from human activities.

One highly productive area in the coastal zone is an **estuary,** a partially enclosed area of coastal water where seawater mixes with fresh water and nutrients from rivers, streams, and runoff from land (Figure 5-27, p. 98). Estuaries and their associated **coastal wetlands** (land areas covered with water all or part of the year) include river mouths, inlets, bays, sounds, mangrove forest swamps in tropical waters, and salt marshes in temperate zones (Figure 5-28, p. 99).

The constant water movement stirs up the nutrient-rich silt, making it available to producers. For

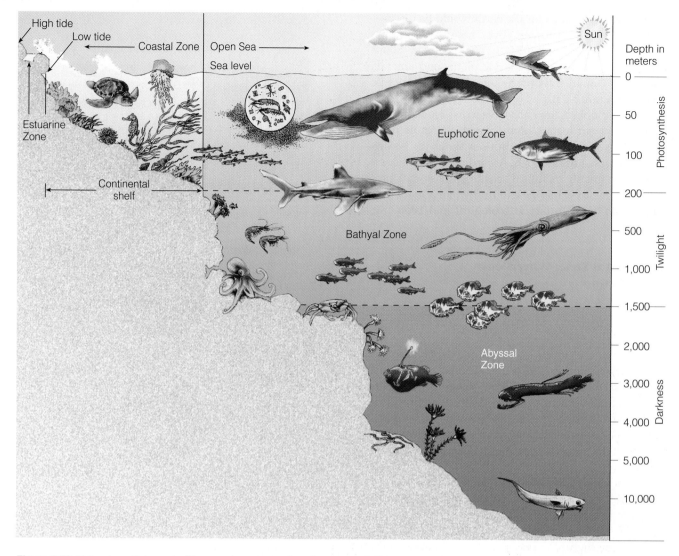

High tide
Low tide
Coastal Zone
Open Sea
Sea level
Sun
Depth in meters
Estuarine Zone
Continental shelf
Euphotic Zone
Bathyal Zone
Abyssal Zone
Photosynthesis
Twilight
Darkness

0
50
100
200
500
1,000
1,500
2,000
3,000
4,000
5,000
10,000

Figure 5-26 Natural capital: major life zones in an ocean (not drawn to scale). Actual depths of zones may vary.

Figure 5-27 Natural capital degradation: view of an *estuary* taken from space. The photo shows the sediment plume at the mouth of Madagascar's Betsiboka River as it flows through the estuary and into the Mozambique Channel. Because of its topography, heavy rainfall, and the clearing of forests for agriculture, Madagascar is the world's most eroded country.

NASA

Herring gulls

Peregrine falcon

Snowy egret

Cordgrass

Short-billed dowitcher

Marsh periwinkle

Phytoplankton

Smelt

Zooplankton and small crustaceans

Soft-shelled clam

Clamworm

Bacteria

Producer to primary consumer	Primary to secondary consumer	Secondary to higher-level consumer	All consumer and producers to decomposers

SuperStock

Figure 5-28 Natural capital: components and interactions in a *salt marsh ecosystem* in a temperate area such as the United States. When these organisms die, decomposers break down their organic matter into minerals used by plants. Colored arrows indicate transfers of matter and energy between consumers (herbivores); secondary, or higher-level, consumers (carnivores); and decomposers. Organisms are not drawn to scale. The photo shows a salt marsh in Peru.

this reason, estuaries and their associated coastal wetlands are some of the earth's most productive ecosystems and provide other important ecological and economic services.

These systems filter out toxic pollutants, excess plant nutrients, sediments, and other pollutants. They reduce storm damage by absorbing the energy of waves and storing excess water produced by storms. They also provide food, habitats, and nursery sites for numerous aquatic species. *Bad news:* We are degrading or destroying some of the ecological services that these important ecosystems provide at no cost.

Rocky and Sandy Shores: Living with the Tides

Organisms in coastal areas experiencing daily low and high tides have evolved many ways to survive under harsh and changing conditions.

The area of shoreline that appears between low and high tides is called the **intertidal zone.** Organisms living in this zone must be able to avoid being swept away or crushed by waves, and evade or cope with being immersed during high tides and left high and dry (and much hotter) at low tides. They must also survive changing levels of salinity when heavy rains dilute salt water. To deal with these stresses, most intertidal organisms hold on to something, dig in, or hide in protective shells.

On some coasts, steep *rocky shores* are pounded by waves. The numerous pools and other niches in the rocks in their intertidal zones contain a great variety of species that occupy different niches (Figure 5-29, top).

Other coasts have gently sloping *barrier beaches,* or *sandy shores,* with niches for different marine organisms (Figure 5-29, bottom). Most of them keep hidden from view and survive by burrowing, digging, and tunneling in the sand. These sandy beaches and their adjoining coastal wetlands are also home to numerous birds that occupy specialized niches by feeding on different types of crustaceans, insects, and other organisms (Figure 4-5, p. 69).

Barrier islands are low and narrow sandy islands that form offshore from a coastline. These beautiful but very limited pieces of real estate are prime targets for real estate development. Living on these islands can be risky. Sooner or later many of the structures humans build on low-lying barrier islands, such as Atlantic City, New Jersey, and Miami Beach, Florida, are damaged or destroyed by flooding, severe beach erosion, or major storms (including hurricanes; see Figure 3 in Science Supplement 6 at the end of this book).

Undisturbed barrier beaches have one or more rows of natural sand dunes in which the sand is held in place by the roots of grasses (Figure 5-30, p. 102). These dunes serve the first line of defense against the ravages of the sea. Such real estate is so scarce and valuable that coastal developers frequently remove the protective dunes or build behind the first set of dunes and cover them with buildings and roads. Large storms can then flood and even sweep away seaside buildings and severely erode the sandy beaches. Some people inaccurately call these human-influenced events "natural disasters."

Case Study: Coral Reefs

Coral reefs are biologically diverse and productive ecosystems that are increasingly stressed by human activities.

Coral reefs form in clear, warm coastal waters of the tropics and subtropics (Figure 5-23). These stunningly beautiful natural wonders are among the world's oldest, most diverse, and productive ecosystems, and they provide homes for one-fourth of all marine species (Figure 5-31, p. 102).

Coral reefs are formed by massive colonies of tiny animals called *polyps* (close relatives of jellyfish). They slowly build reefs by secreting a protective crust of limestone (calcium carbonate) around their soft bodies. When the polyps die, their empty crusts remain behind as a platform for more reef growth. The resulting elaborate network of crevices, ledges, and holes serves as calcium carbonate "condominiums" for a variety of marine animals.

Coral reefs represent a mutually beneficial relationship between the polyps and the single-celled algae called *zooxanthellae* ("zoh-ZAN-thel-ee") that live in the tissues of the polyps. The algae provide the polyps with color, food, and oxygen through photosynthesis. The polyps, in turn, provide the algae with a well-protected home and some of their nutrients.

Although coral reefs occupy only about 0.1% of the world's ocean area, they provide numerous free ecological and economic services. They help moderate atmospheric temperatures by removing CO_2 from the atmosphere, act as natural barriers that help protect 15% of the world's coastlines from erosion by battering waves and storms, and support at least one-fourth of all identified marine species and almost two-thirds of marine fish species. Economically, they produce about one-tenth of the global fish catch and one-fourth of the catch in developing countries, and provide jobs and building materials for some of the world's poorest countries. They also support fishing and tourism industries worth billions of dollars each year.

Coral reefs are vulnerable to damage because they grow slowly and are disrupted easily. They also thrive only in clear and fairly shallow water of constant high salinity. This water must have a temperature of 18–30°C (64–86°F).

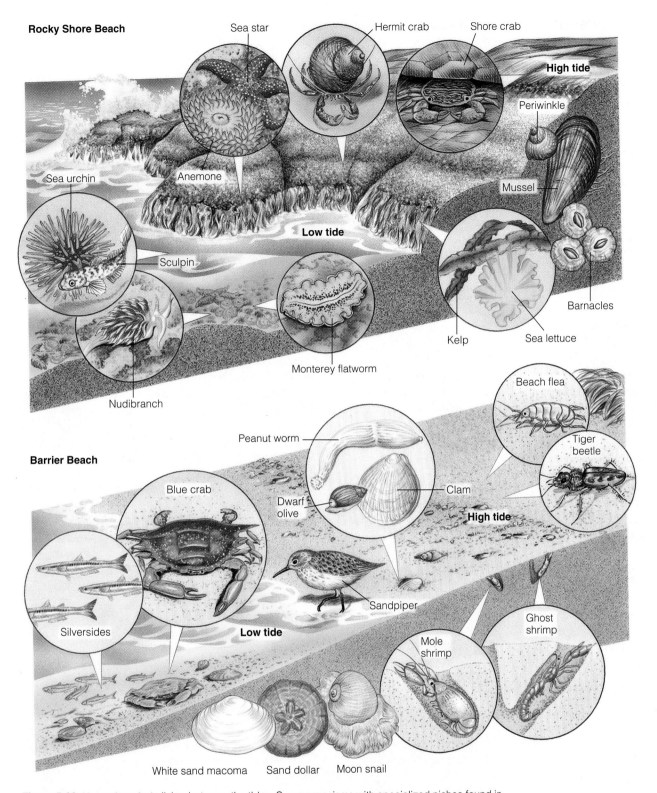

Rocky Shore Beach

Sea star

Hermit crab

Shore crab

High tide

Periwinkle

Sea urchin

Anemone

Mussel

Sculpin

Barnacles

Low tide

Kelp

Sea lettuce

Monterey flatworm

Nudibranch

Barrier Beach

Beach flea

Tiger beetle

Peanut worm

Blue crab

Dwarf olive

Clam

High tide

Silversides

Sandpiper

Low tide

Ghost shrimp

Mole shrimp

White sand macoma

Sand dollar

Moon snail

Figure 5-29 Natural capital: living between the tides. Some organisms with specialized niches found in various zones on rocky shore beaches (top) and barrier or sandy beaches (bottom). Organisms are not drawn to scale.

The biodiversity of coral reefs can be reduced by natural disturbances such as severe storms, freshwater floods, and invasions of predatory fish. Throughout their very long geologic history, coral reefs have been able to adapt to such natural environmental changes. Today the biggest threats to the biodiversity of the world's coral reefs come from human activities listed in Figure 5-32 (p. 103).

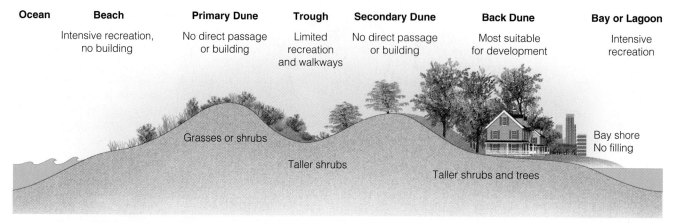

Ocean	Beach	Primary Dune	Trough	Secondary Dune	Back Dune	Bay or Lagoon
	Intensive recreation, no building	No direct passage or building	Limited recreation and walkways	No direct passage or building	Most suitable for development	Intensive recreation

Bay shore
No filling

Grasses or shrubs

Taller shrubs

Taller shrubs and trees

Figure 5-30 Natural capital: primary and secondary dunes on a gently sloping sandy barrier beach or a barrier island help protect land from erosion by the sea. The roots of various grasses that colonize the dunes help hold the sand in place. Ideally, construction would occur only behind the second strip of dunes, and walkways to the beach would be built over the dunes to keep them intact. This helps preserve barrier beaches and protect buildings from damage by wind, high tides, beach erosion, and storm surges. Such protection is rare, however, because the short-term economic value of ocean-front land is considered much higher than its long-term ecological and economic values. Rising sea levels from global warming may put many barrier islands and beaches under water by the end of this century.

Figure 5-31 Natural capital: components and interactions in a *coral reef ecosystem*. When these organisms die, decomposers break down their organic matter into minerals used by plants. Colored arrows indicate transfers of matter and energy between producers; primary consumers (herbivores); secondary, or higher-level, consumers (carnivores); and decomposers. Organisms are not drawn to scale. See the photo of a coral reef in Figure 5-23 (left).

Producer to primary consumer

Primary to secondary consumer

Secondary to higher-level consumer

All consumer and producers to decomposers

Natural Capital Degradation

Coral Reefs

Ocean warming

Soil erosion

Algae growth from fertilizer runoff

Mangrove destruction

Coral reef bleaching

Rising sea levels

Increased UV exposure from ozone depletion

Using cyanide and dynamite to harvest coral reef fish

Coral removal for building material, aquariums, and jewelery

Damage from anchors, ships, and tourist divers

Figure 5-32 Natural capital degradation: major threats to coral reefs. *Critical thinking: which two of these threats do you believe are the most serious?*

Biological Zones in the Open Sea: Light Rules

The open ocean consists of a brightly lit surface layer, a dimly lit middle layer, and a dark bottom zone.

The sharp increase in water depth at the edge of the continental shelf separates the coastal zone from the vast volume of the ocean called the **open sea.** Primarily on the basis of the penetration of sunlight, it is divided into three vertical zones (see Figure 5-26).

The *euphotic zone* is the brightly lit upper zone where floating drifting phytoplankton carry out photosynthesis. Nutrient levels are low (except around upwellings), and levels of dissolved oxygen are high. Large, fast-swimming predatory fish such as swordfish, sharks, and bluefin tuna populate this zone.

The *bathyal zone* is the dimly lit middle zone that does not contain photosynthesizing producers because of a lack of sunlight. Various types of zooplankton and smaller fish, many of which migrate to feed on the surface at night, populate this zone.

The lowest zone, called the *abyssal zone*, is dark and very cold and has little dissolved oxygen. Nevertheless, the ocean floor contains enough nutrients to support 98% of the 250,000 identified species living in the ocean.

Average primary productivity and NPP per unit of area are quite low in the open sea except at an occasional equatorial upwelling, where currents bring up nutrients from the ocean bottom. However, because the open sea covers so much of the earth's surface, it makes the largest contribution to the earth's overall NPP.

Effects of Human Activities on Marine Systems: Red Alert

People are destroying or degrading many of the coastal areas' most vital resources.

In their desire to live near the coast, people are destroying or degrading the resources that make coastal areas so enjoyable and economically and ecologically valuable. Currently, 45% of the world's people and more than half of the U.S. population live along or near coasts. Some 13 of the world's 19 megacities with populations of 10 million or more people are located in coastal zones. By 2030, at least 6.3 billion people are expected to live in or near coastal areas. Figure 5-33 lists major human impacts on marine systems.

Natural Capital Degradation

Marine Ecosystems

Half of coastal wetlands lost to agriculture and urban development

Over one-third of mangrove forests lost since 1980 to agriculture, development, and aquaculture shrimp farms

About 10% of world's beaches eroding because of coastal development and rising sea level

Ocean bottom habitats degraded by dredging and trawler fishing boats

At least 20% of coral reefs severely damaged and 30–50% more threatened

Figure 5-33 Natural capital degradation: major human impacts on the world's marine systems. *Critical thinking: how does your lifestyle directly or indirectly contribute to this degradation?*

5-7 FRESHWATER LIFE ZONES

Freshwater Life Zones

Freshwater ecosystems provide important ecological and economic services even though they cover less than 1% of the earth's surface.

Freshwater life zones occur where water with a dissolved salt concentration of less than 1% by volume accumulates on or flows through the surfaces of terrestrial biomes. Examples include *standing* (lentic) bodies of fresh water such as lakes, ponds, and inland wetlands and *flowing* (lotic) systems such as streams and rivers. Although freshwater systems cover less than 1% of the earth's surface, they provide a number of important ecological and economic services (Figure 5-34).

Lakes: Water-Filled Depressions

Lakes consist of sunlit surface layers near and away from the shore; at deeper levels, they contain a dark layer and a bottom zone.

Lakes are large natural bodies of standing fresh water formed when precipitation, runoff, or groundwater seepage fills depressions in the earth's surface. Causes of such depressions include glaciation (the Great Lakes of North America), crustal displacement (Lake Nyasa in East Africa), and volcanic activity (Crater Lake in Oregon). Lakes are supplied with water from rainfall, melting snow, and streams that drain the surrounding watershed.

Lakes normally consist of four distinct zones that are defined by their depth and distance from shore (Figure 5-35). The top layer, called the *littoral* ("LIT-tore-el") *zone,* consists of the shallow sunlit waters near the shore to the depth at which rooted plants stop growing. It has a high biological diversity.

Next is the *limnetic* ("lim-NET-ic") *zone:* the open, sunlit water surface layer away from the shore that extends to the depth penetrated by sunlight. The main photosynthetic body of the lake, it produces the food and oxygen that support most of the lake's consumers.

Next comes the *profundal* ("pro-FUN-dahl") *zone:* the deep, open water where it is too dark for photosynthesis. Without sunlight and plants, oxygen levels are low here. Fish adapted to the lake's cooler and darker water are found in this zone.

Finally, the bottom of the lake contains the *benthic* ("BEN-thic") *zone.* Mostly decomposers, detritus feeders, and fish that swim from one zone to the other inhabit it. The benthic zone is nourished mainly by detritus that falls from the littoral and limnetic zones and by sediment washing into the lake.

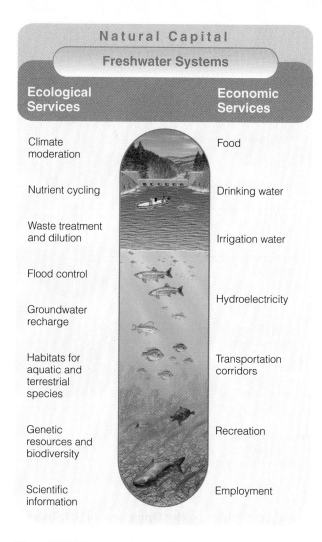

Figure 5-34 Natural capital: major ecological and economic services provided by freshwater systems.

Environmental Science ⊕ Now™
Learn more about the zones of a lake, how its water turns over between seasons, and how lakes differ below their surfaces at Environmental ScienceNow.

Freshwater Streams and Rivers

Water flowing from mountains to the sea creates different aquatic conditions and habitats.

Precipitation that does not sink into the ground or evaporate is **surface water.** It becomes **runoff** when it flows into streams. The land area that delivers runoff, sediment, and dissolved substances to a stream is called a **watershed,** or **drainage basin.** Small streams join to form rivers, and rivers flow downhill to the ocean (Figure 5-36, p. 106).

In many areas, streams begin in mountainous or hilly areas that collect and release water falling to the

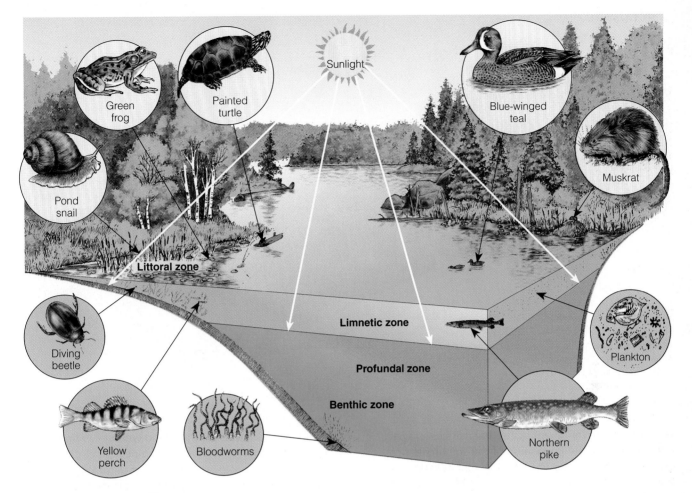

Sunlight

Green frog

Painted turtle

Blue-winged teal

Muskrat

Pond snail

Littoral zone

Limnetic zone

Profundal zone

Plankton

Diving beetle

Benthic zone

Yellow perch

Bloodworms

Northern pike

Environmental Science⊕Now™

Active Figure 5-35 Natural capital: distinct zones of life in a fairly deep temperate zone lake. *See an animation based on this figure and take a short quiz on the concept.*

earth's surface as rain or snow that melts during warm seasons. The downward flow of surface water and groundwater from mountain highlands to the sea typically takes place in three aquatic life zones characterized by different environmental conditions: the *source zone,* the *transition zone,* and the *floodplain zone* (Figure 5-36).

As streams flow downhill, they shape the land through which they pass. Over millions of years the friction of moving water may level mountains and cut deep canyons, and the rock and soil removed by the water are deposited as sediment in low-lying areas.

Streams receive many of their nutrients from bordering land ecosystems. Such nutrient inputs come from falling leaves, animal feces, insects, and other forms of biomass washed into streams during heavy rainstorms or by melting snow. To protect a stream or river system from excessive inputs of nutrients and pollutants, we must protect its watershed.

Freshwater Inland Wetlands: Vital Sponges

Inland wetlands absorb and store excess water from storms and provide a variety of wildlife habitats.

Inland wetlands are lands covered with fresh water all or part of the time (excluding lakes, reservoirs, and streams) that are located away from coastal areas. They include *marshes* (dominated by grasses and reeds with few trees), *swamps* (dominated by trees and shrubs), and *prairie potholes* (depressions carved out by glaciers). Other examples are *floodplains* (which receive excess water during heavy rains and floods) and the wet *arctic tundra* in summer. Some wetlands are huge; others are small.

Some wetlands are covered with water year-round. In contrast, *seasonal wetlands* remain underwater or soggy for only a short time each year. They include prairie potholes, floodplain wetlands, and bottomland hardwood swamps. Some stay dry for years

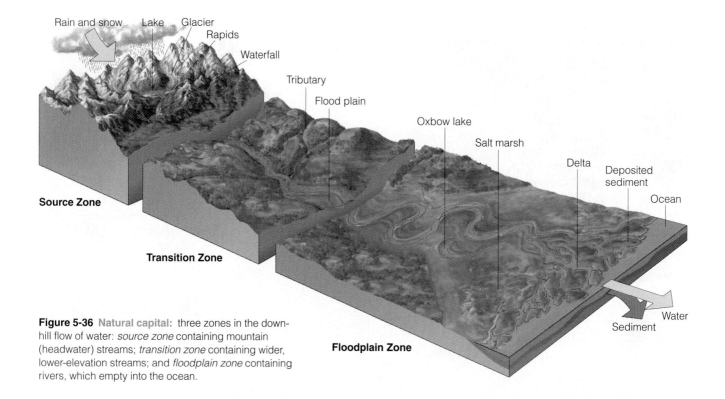

Figure 5-36 Natural capital: three zones in the down-hill flow of water: *source zone* containing mountain (headwater) streams; *transition zone* containing wider, lower-elevation streams; and *floodplain zone* containing rivers, which empty into the ocean.

before water covers them again. In such cases, scientists must use the composition of the soil or the presence of certain plants (such as cattails, bulrushes, or red maples) to determine that a particular area is really a wetland. Inland wetlands provide a number of free ecological and economic services. For example, they filter toxic wastes and pollutants, absorb and store excess water from storms, and provide habitats for many species.

Natural Capital Degradation: Effects of Human Activities on Freshwater Systems

We have built dams, levees, and dikes that reduce the flow of water and alter wildlife habitats in rivers; established cities and farmlands that pollute streams and rivers; and filled in inland wetlands to grow food and build cities.

Human activities affect freshwater systems in four major ways. *First,* dams, diversions, or canals fragment almost 60% of the world's 237 large rivers. They alter and destroy wildlife habitats along rivers and in coastal deltas and estuaries by reducing water flow.

Second, flood control levees and dikes built along rivers alter and destroy aquatic habitats. *Third,* cities and farmlands add pollutants and excess plant nutrients to nearby streams and rivers. *Fourth,* many inland wetlands have been drained or filled to grow crops or have been covered with concrete, asphalt, and build-

ings. More than half of the inland wetlands estimated to have existed in the continental United States during the 1600s no longer exist. This loss of natural capital has been an important factor in increased flood and drought damage in the United States. Many other countries have suffered similar losses. For example, 80% of all wetlands in Germany and France have been destroyed.

Scientists have made a good start in understanding important aspects of the ecology of the world's terrestrial and aquatic systems. One of the major lessons from their research: In nature, *everything is connected.* According to scientists, we urgently need more research on how the world's terrestrial and aquatic systems work and are connected to one another and to the atmosphere. With such information we will have a clearer picture of how our activities affect the earth's biodiversity and what we can do to live more sustainably.

When we try to pick out anything by itself, we find it hitched to everything else in the universe.
JOHN MUIR

CRITICAL THINKING

1. List a limiting factor for each of the following ecosystems: **(a)** a desert, **(b)** arctic tundra, **(c)** alpine tundra, **(d)** the floor of a tropical rain forest, **(e)** a temperate deciduous forest, **(f)** the surface layer of the open sea, and **(g)** the bottom of a deep lake.

2. Why do deserts and arctic tundra support a much smaller biomass of animals than do tropical forests?

3. Why do most animals in a tropical rain forest live in its trees?

4. Which biomes are best suited for **(a)** raising crops and **(b)** grazing livestock?

5. Some biologists have suggested placing large herds of bison on public lands in the North American plains as a way of restoring remaining tracts of tall-grass prairie. Ranchers with permits to graze cattle and sheep on federally managed lands have strongly opposed this idea. Do you agree or disagree with the idea of returning large numbers of bison to the plains of North America? Explain.

6. What type of biome do you live in? How have human activities over the past 50 years affected the characteristic vegetation and animal life normally found in the biome you live in? How is your lifestyle affecting this biome?

7. Which factors in your lifestyle contribute to the destruction and degradation of coastal and inland wetlands?

8. You are a defense attorney arguing in court for sparing an undeveloped old-growth tropical rain forest and a coral reef from severe degradation or destruction by development. Write your closing statement for the defense of each of these ecosystems. If the judge decides you can save only one of the ecosystems, which one would you choose, and why?

9. Congratulations! You are in charge of the world. What are the three most important features of your plan to help sustain **(a)** the earth's terrestrial biodiversity and **(b)** the earth's aquatic biodiversity?

LEARNING ONLINE

The website for this book includes review questions for the entire chapter, flash cards for key terms and concepts, a multiple-choice practice quiz, interesting Internet sites, references, and a guide for accessing thousands of InfoTrac® College Edition articles.

Visit

http://biology.brookscole.com/miller11

Then choose Chapter 5, and select a learning resource. For access to animations, additional quizzes, chapter outlines and summaries, register and log in to

Environmental Science ⊛ Now™

at **esnow.brookscole.com/miller11** using the access code card in the front of your book.

6 Community Ecology, Population Ecology, and Sustainability

Population Control

Biodiversity

CASE STUDY

Why Should We Care about the American Alligator?

The American alligator (Figure 6-1), North America's largest reptile, has no natural predators except for humans. This species, which has survived for nearly 200 million years, has been able to adapt to numerous changes in the earth's environmental conditions.

But matters changed when hunters began killing large numbers of these animals for their exotic meat and their supple belly skin, used to make shoes, belts, and pocketbooks.

Other people hunted alligators for sport or out of hatred. Between 1950 and 1960, hunters wiped out 90% of the alligators in Louisiana. By the 1960s, the alligator population in the Florida Everglades was also near extinction.

People who say "So what?" are overlooking the alligator's important ecological role—its *niche*—in subtropical wetland communities. Alligators dig deep depressions, or gator holes. These holes hold fresh water during dry spells, serve as refuges for aquatic life, and supply fresh water and food for many animals.

Large alligator nesting mounds provide nesting and feeding sites for species of herons and egrets. Alligators eat large numbers of gar (a predatory fish). This helps maintain populations of game fish such as bass and bream.

As alligators move from gator holes to nesting mounds, they help keep areas of open water free of invading vegetation. Without these free ecosystem ser-

A. & J. Visage/Peter Arnold, Inc.

Figure 6-1 Natural capital: the American alligator plays important ecological roles in its marsh and swamp habitats in the southeastern United States. Since being classified as an endangered species in 1967, it has recovered enough to have its status changed from endangered to threatened species—an outstanding success story in wildlife conservation.

vices, freshwater ponds and coastal wetlands found in the alligator's habitat, shrubs and trees would fill in, and dozens of species would disappear.

Some ecologists classify the American alligator as a *keystone species* because of its important ecological roles in helping maintain the structure, function, and sustainability of the communities where it is found.

In 1967, the U.S. government placed the American alligator on the endangered species list. Protected from hunters, the population had made a strong comeback in many areas by 1975—too strong, according to those who find alligators in their backyards and swimming pools, as well as to duck hunters, whose retriever dogs are sometimes eaten by alligators.

In 1977, the U.S. Fish and Wildlife Service reclassified the American alligator as a *threatened* species in Florida, Louisiana, and Texas, where 90% of the animals live. In 1987, this reclassification was extended to seven more states.

The recent increase in demand for alligator meat and hides has created a booming business for alligator farms, especially in Florida. Such farms reduce the need for illegal hunting of wild alligators.

To biologists, the comeback of the American alligator is an important success story in wildlife conservation. Its tale illustrates the unique role (niche) filled by each species in a community or ecosystem and highlights how interactions between species can affect ecosystem structure and function. Understanding the roles and interactions of species and the changes populations, communities, and ecosystems undergo is the subject of this chapter.

Animal and vegetable life is too complicated a problem for human intelligence to solve, and we can never know how wide a circle of disturbance we produce in the harmonies of nature when we throw the smallest pebble into the ocean of organic life.

GEORGE PERKINS MARSH

This chapter looks at the roles and interactions of species in a community, the ways in which communities and populations respond to changes in environmental conditions, and actions we can take to help sustain populations, communities, and ecosystems.

- What determines the number of species in a community?

- How can we classify species according to their roles in a community?

- How do species interact with one another?

- How do communities respond to changes in environmental conditions?

- How do populations respond to changes in environmental conditions?

- How do species differ in their reproductive patterns?

- What are the major impacts of human activities on populations, communities, and ecosystems?

- What lessons can we learn from ecology about living more sustainably?

KEY IDEAS

- Biological communities differ in their physical structure, species diversity, and the ecological roles their species play.

- Species interactions have profound effects on communities and population size.

- Communities change their species composition and structure in response to changing environmental conditions.

- No population can continue grow indefinitely because of limitations on resources and competition between species for those resources.

- Nature has survived for billions of years by relying on solar energy, recycling nutrients, using biodiversity to sustain itself and adapt to new environmental conditions, and controlling population growth.

- We can live more sustainably by mimicking the four major ways nature has used to adapt and sustain itself.

6-1 COMMUNITY STRUCTURE AND SPECIES DIVERSITY

Community Structure: Appearance Matters

Biological communities differ in their structure and physical appearance.

One way that ecologists distinguish between biological communities is by describing their overall *physical appearance:* the relative sizes, stratification, and distribution of the populations and species in each community, as shown in Figure 6-2 (p. 110) for various terrestrial communities. There are also differences in the physical structures and zones of communities in aquatic life zones such as oceans, rocky shores and sandy beaches, lakes, river systems, and inland wetlands.

The physical structure within a particular type of community or ecosystem can also vary. Most large terrestrial communities and ecosystems consist of a mosaic of different-sized vegetation patches. Life is patchy.

Likewise, community structure varies around its *edges* where one type of community makes a transition to a different type of community. For example, the edge area between a forest and an open field may be sunnier, warmer, and drier than the forest interior and support different species than the forest and field interiors.

Increasing the edge area through habitat fragmentation makes many species more vulnerable to stresses such as predators and fire. It also creates barriers that can prevent some species from colonizing new areas and finding food and mates.

Species Diversity and Niche Structure in Communities

Biological communities differ in the types and numbers of species they contain and the ecological roles those species play.

A second characteristic of a community is its **species diversity:** the number of different species it contains (*species richness*) combined with the abundance of individuals within each of those species (*species evenness*). For example, two communities, each with a total of 20 different species and 200 individuals, have the same species diversity.

But these communities could differ in their species richness and species evenness. For example, community A might have 10 individuals in each of its 20 species. Community B might have 10 species, each with 2 individuals, and 10 other species, each with 18 individuals. Which community has the highest species evenness?

Another characteristic is a community's *niche structure:* how many ecological niches occur, how they resemble or differ from one another, and how the various

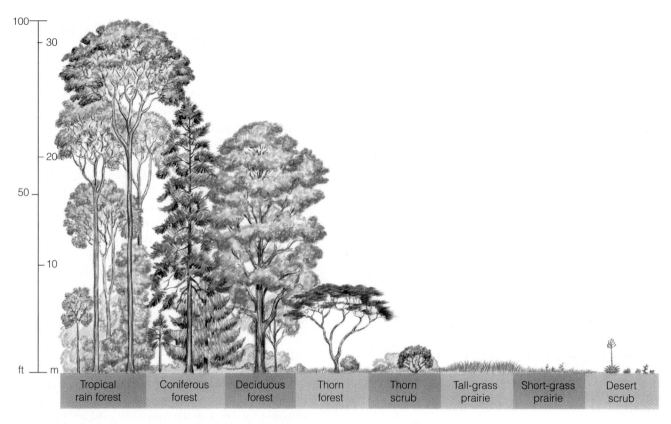

100 —
30
20
50 —
10
ft —— m

| Tropical rain forest | Coniferous forest | Deciduous forest | Thorn forest | Thorn scrub | Tall-grass prairie | Short-grass prairie | Desert scrub |

Figure 6-2 Natural capital: generalized types, relative sizes, and stratification of plant species in various terrestrial communities.

species interact. For most terrestrial plants and animals, species diversity is highest in the tropics and declines as we move from the equator toward the poles.

The most species-rich environments are tropical rain forests, coral reefs, the deep sea, and large tropical lakes. A community such as a tropical rain forest or a coral reef with a large number of different species (high species richness) generally has only a few members of each species (low species evenness).

Environmental Science ⊕ Now™
Learn about how latitude affects species diversity and about the differences between big and small islands at Environmental ScienceNow.

Are Complex Communities More Sustainable Than Simple Ones?

Having many different species can provide some ecological stability or sustainability for communities, but we do not know whether this applies to all communities or what the minimum number of species needed to ensure stability is.

In the 1960s, most ecologists believed the greater the species diversity and the accompanying web of feeding and biotic interactions, the greater an ecosystem's

stability. According to this hypothesis, a complex community with a diversity of species and feeding paths has more ways to respond to most environmental stresses because it does not have "all its eggs in one basket." This is a useful hypothesis, but recent research has found exceptions to this intuitively appealing idea.

Because no community can function without some producers and decomposers, there is a minimum threshold of species diversity below which communities and ecosystems cannot function. Beyond this, it is difficult to know whether simple communities are less stable than complex and biodiverse ones or to identify the threshold of species diversity needed to maintain community stability. Research by ecologist David Tilman and others suggests that communities with more species tend to have a higher net primary productivity (NPP) and can be more resilient than simpler ones.

Many studies support the idea that some level of biodiversity provides insurance against catastrophe. But how much biodiversity is needed in various communities remains uncertain. Recent research suggests that the average annual NPP of an ecosystem reaches a peak with 10–40 producer species. Many ecosystems contain even more producer species, but it is difficult to distinguish between the essential species and the nonessential species. We need much more research into this important area of ecology.

6-2 TYPES OF SPECIES

Types of Species in Communities

Communities can contain native, nonnative, indicator, keystone, and foundation species that play different ecological roles.

Ecologists often use labels—such as *native, nonnative, indicator, keystone,* or *foundation*—to describe the major niches filled by various species in communities. Any given species may play more than one of these five ecological roles in a particular community.

Native species are those species that normally live and thrive in a particular community. Other species that migrate into or are deliberately or accidentally introduced into an community are called **nonnative species, invasive species,** or **alien species.**

Many people tend to think of nonnative species as villains. In fact, most introduced and domesticated species of crops and animals such as chickens, cattle, and fish from around the world are beneficial to us.

Sometimes, however, a nonnative species can crowd out native species and cause unintended and unexpected consequences. In 1957, for example, Brazil imported wild African bees to help increase honey production. Instead, the bees displaced domestic honeybees and reduced the honey supply.

Since then, these nonnative bee species—popularly known as "killer bees"—have moved northward into Central America and parts of the southwestern United States. They are still heading north but should be stopped eventually by the harsh winters in the central United States unless they can adapt genetically to cold weather.

The wild African bees are not the fearsome killers portrayed in some horror movies, but they are aggressive and unpredictable. They have killed thousands of domesticated animals and an estimated 1,000 people in the western hemisphere. Most of their human victims died because they were allergic to bee stings or because they fell down or became trapped and could not flee.

Indicator Species: Biological Smoke Alarms

Some species can alert us to harmful changes taking place in biological communities.

Species that serve as early warnings of damage to a community or an ecosystem are called **indicator species.** For example, the presence or absence of trout species in water at temperatures within their range of tolerance (Figure 3-11, p. 43) is an indicator of water quality because trout need clean water with high levels of dissolved oxygen.

Birds are excellent biological indicators because they are found almost everywhere and are affected quickly by environmental changes such as loss or fragmentation of their habitats and introduction of chemical pesticides. The population of many bird species is declining.

Butterflies are good indicator species because their association with various plant species makes them vulnerable to habitat loss and fragmentation. Some amphibians are also believed to be indicator species, as discussed next.

Case Study: Why Are Amphibians Vanishing?

The disappearance of many amphibian species may indicate a decline in environmental quality in many parts of the world.

Amphibians (frogs, toads, and salamanders) live part of their lives in water and part on land, and some are classified as indicator species. Frogs, for example, are especially vulnerable to environmental disruption at various points in their life cycle, shown in Figure 6-3 (p. 112). As tadpoles, they live in water and eat plants; as adults, they live mostly on land and eat insects that can expose them to pesticides. Frogs' eggs have no protective shells to block ultraviolet (UV) radiation or pollution. As adults, they take in water and air through their thin, permeable skins, which can readily absorb pollutants from water, air, or soil.

Since 1980, populations of hundreds of the world's estimated 5,280 amphibian species have been vanishing or declining in almost every part of the world, even in protected wildlife reserves and parks. According to the World Conservation Union, one-fourth of all known amphibian species are extinct, endangered, or vulnerable to extinction.

No single cause has been identified to explain the amphibian declines. However, scientists have identified a number of factors that can affect frogs and other amphibians at various points in their life cycles:

- *Habitat loss and fragmentation* (especially from draining and filling of inland wetlands, deforestation, and development)

- *Prolonged drought* (which dries up breeding pools so few tadpoles survive)

- *Pollution* (particularly from exposure to pesticides, which can make frogs more vulnerable to bacterial, viral, and fungal diseases and cause sexual abnormalities)

- *Increases in ultraviolet radiation* caused by reductions in stratospheric ozone (which can harm young embryos of amphibians in shallow ponds)

Figure 6-3 Typical *life cycle of a frog.* Populations of various frog species can decline because of the effects of harmful factors at different points in their life cycle. Such factors include habitat loss, drought, pollution, increased ultraviolet radiation, parasitism, disease, overhunting for food (frog legs), and nonnative predators and competitors.

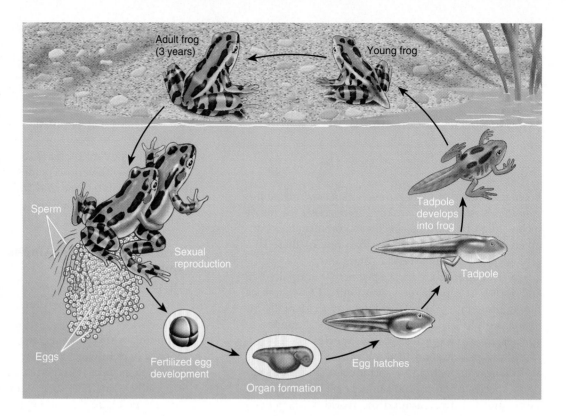

Adult frog (3 years)

Young frog

Sperm

Sexual reproduction

Tadpole develops into frog

Tadpole

Eggs

Fertilized egg development

Organ formation

Egg hatches

- *Parasites*
- *Overhunting* (especially in Asia and France, where frog legs are a delicacy)
- *Viral and fungal diseases*
- *Natural immigration or deliberate introduction of nonnative predators and competitors* (such as fish) and *disease organisms*

Why should we care if various amphibian species become extinct? Scientists give three reasons. *First,* this trend suggests that environmental health is deteriorating in parts of the world because amphibians are sensitive biological indicators of changes in environmental conditions such as habitat loss and degradation, pollution, UV exposure, and climate change.

Second, adult amphibians play important ecological roles in biological communities. For example, amphibians eat more insects (including mosquitoes) than do birds. In some habitats, extinction of certain amphibian species could lead to extinction of other species, such as reptiles, birds, aquatic insects, fish, mammals, and other amphibians that feed on them or their larvae.

Third, amphibians represent a genetic storehouse of pharmaceutical products waiting to be discovered. Compounds in secretions from amphibian skin have been isolated and used as painkillers and antibiotics and as treatment for burns and heart disease.

The plight of some amphibian indicator species is a warning signal. They may not need us, but we and other species need them.

Keystone Species: Major Players

Keystone species help determine the types and numbers of various other species in a community.

A keystone is the wedge-shaped stone placed at the top of a stone archway. Remove this stone and the arch collapses. In some communities, **keystone species** serve a similar role. These species have a much larger effect on the types and abundances of other species in a community than their numbers would suggest. Eliminating a keystone species may dramatically alter the structure and function of a community.

Keystone species play critical ecological roles. One is *pollination* of flowering plant species by bees, hummingbirds, bats, and other species. In addition, *top predator* keystone species feed on and help regulate the populations of other species. Examples are the wolf, leopard, lion, alligator (Case Study, p. 108), and great white shark (Case Study, right).

Have you thanked a *dung beetle* today? Perhaps you should: These keystone species rapidly remove, bury, and recycle dung. Without them, we would be up to our eyeballs in animal wastes, and many plants would be starved for nutrients. These beetles also churn and aerate the soil, making it more suitable for plant life.

The loss of a keystone species can lead to population crashes and extinctions of other species that depend on it for certain ecological services. According to biologist Edward O. Wilson, "The loss of a keystone species is like a drill accidentally striking a power line. It causes lights to go out all over."

Case Study: Why Are Sharks Important Species?

Some shark species eat and remove sick and injured ocean animals. Some can help us learn how to fight cancer and immune system disorders.

The world's 370 shark species vary widely in size. The smallest is the dwarf dog shark, about the size of a large goldfish. The largest is the whale shark, the world's largest fish. It can grow to 15 meters (50 feet) long and weigh as much as two full-grown African elephants.

Various shark species, feeding at the top of food webs, cull injured and sick animals from the ocean, thus playing an important ecological role. Without their services, the oceans would be teeming with dead and dying fish.

Many people—influenced by movies (such as *Jaws*), popular novels, and widespread media coverage of a fairly small number of shark attacks per year—think of sharks as people-eating monsters. In reality, the three largest species—the whale shark, basking shark, and megamouth shark—are gentle giants. They swim through the water with their mouths open, filtering out and swallowing huge quantities of *plankton* (small free-floating sea creatures).

Every year, members of a few species—mostly great white, bull, tiger, gray reef, lemon, hammerhead, shortfin mako, and blue sharks—injure 60–100 people worldwide. Between 1990 and 2003, sharks killed a total of 8 people off U.S. coasts and 88 people worldwide—an average of 7 people per year. Most attacks involve great white sharks, which feed on sea lions and other marine mammals and sometimes mistake divers and surfers for their usual prey. Whose fault is this?

Media coverage of shark attacks greatly distorts the danger from sharks. You are 30 times more likely to be killed by lightning than by a shark each year—and your chance of being killed by lightning is already extremely small.

For every shark that injures a person, we kill at least 1 million sharks, or a total of about 100 million sharks each year. Sharks are caught mostly for their fins and then thrown back alive into the water to bleed to death or drown because they can no longer swim.

Shark fins are widely used in Asia as a soup ingredient and as a pharmaceutical cure-all. In high-end restaurants in China, a bowl of shark fin soup can cost as much as $100. As affluence—and demand for this delicacy—increases among the Chinese middle class, a single large fin from a whale shark may become worth more than $10,000.

According to a 2001 study by Wild Aid, shark fins sold in restaurants throughout Asia and in Chinese communities in cities such as New York, San Francisco, and London contain dangerously high levels of toxic mercury. Consumption of high levels of mercury is especially dangerous for pregnant women, fetuses, and infants feeding on breast milk.

Sharks are also killed for their livers, meat (chiefly mako and thresher), hides (a source of exotic, high-quality leather), and jaws (especially great whites, whose jaws can sell for as much as $10,000). Some are killed simply because we fear them. Sharks (especially blue, mako, and oceanic whitetip) also die when they are trapped in nets or lines deployed to catch swordfish, tuna, shrimp, and other commercially important species.

Despite our unkind treatment of them, sharks save human lives. They are teaching us how to fight cancer, which sharks almost never get. Scientists are also studying their highly effective immune system, which allows wounds to heal without becoming infected.

Sharks are especially vulnerable to overfishing because they grow slowly, mature late, and have only a few young each generation. Today, they are among the most vulnerable and least protected animals on earth.

In 2003, experts at the National Aquarium in Baltimore, Maryland, estimated that populations of some shark species have decreased by 90% since 1992. Eight of the world's shark species are considered critically endangered or endangered and 82 species are threatened with extinction.

In response to a public outcry over depletion of some species, the United States and several other countries have banned hunting sharks for their fins in their territorial waters. Unfortunately, such bans are difficult to enforce.

With more than 400 million years of evolution behind them, sharks have had a long time to get things right. Preserving their genetic development begins with the knowledge that sharks may not need us, but we and other species need them.

X *HOW WOULD YOU VOTE?* Do we have an ethical obligation to protect shark species from premature extinction and treat them humanely? Cast your vote online http://biology .brookscole.com/miller11.

Foundation Species: Other Major Players

Foundation species create and enhance habitats that can benefit other species in a community.

Some ecologists think the keystone species should be expanded to include **foundation species,** which play a major role in shaping communities by creating and enhancing their habitats in ways that benefit other species. For example, elephants push over, break, or uproot trees, creating forest openings in the savanna grasslands and woodlands of Africa. Their work promotes the growth of grasses and other forage plants that benefit smaller grazing species such as antelope. It also accelerates nutrient cycling rates. Some bat and bird foundation species can regenerate deforested areas and spread fruit plants by depositing plant seeds in their droppings.

6-3 SPECIES INTERACTIONS

How Do Species Interact?

Species can interact and increase their ability to survive through competition, predation, parasitism, mutualism, and commensalism.

When different species in an ecosystem have activities or resource needs in common, they may interact with one another. Members of these species may be harmed by, benefit from, or be unaffected by the interaction. Ecologists identify five basic types of interactions between species: *interspecific competition, predation, parasitism, mutualism,* and *commensalism.*

The most common interaction between species is *competition* for shared or scarce resources such as space and food. Ecologists call such competition between species **interspecific competition.**

As long as commonly used resources are abundant, different species can share them. Each species can then come closer to occupying the *fundamental niche* it would occupy if it faced no competition from other species.

Of course, most species face competition from other species for one or more limited resources (such as food, sunlight, water, soil nutrients, space, nesting sites, and good places to hide). As a result, the fundamental niches of the competing species may overlap. With significant niche overlap, one of the competing species must migrate, if possible, to another area; shift its feeding habits or behavior through natural selection and evolution; suffer a sharp population decline; or become extinct in that area.

Humans compete with other species for space, food, and other resources. As we convert more of the earth's land and aquatic resources and NPP to our uses, we deprive many other species of resources they need to survive.

Reducing or Avoiding Competition: Sharing Resources

Some species develop adaptations through natural selection that allow them to reduce or avoid competition for resources with other species.

Over a time scale long enough for natural selection to occur, some species competing for the same resources develop adaptations that reduce or avoid competition. One way this happens is through **resource partitioning.** It occurs when species competing for similar

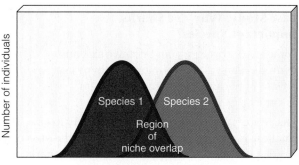

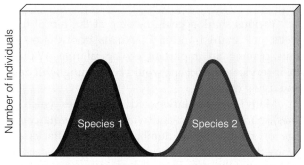

Figure 6-4 Natural capital: *resource partitioning* and *niche specialization* as a result of competition between two species. The top diagram shows the overlapping niches of two competing species. The bottom diagram shows that through natural selection the niches of the two species become separated and more specialized (narrower) so that they avoid competing for the same resources.

scarce resources evolve more specialized traits that allow them to use shared resources at different times, in different ways, or in different places.

Through natural selection, the fairly broad niches of two competing species (Figure 6-4, top) can become more specialized (Figure 6-4, bottom) so that the species can share limited resources. When lions and leopards live in the same area, for example, lions take mostly larger animals as prey, and leopards take smaller ones. Hawks and owls feed on similar prey, but hawks hunt during the day and owls hunt at night.

Figure 6-5 shows resource partitioning by some insect-eating bird species. Figure 4-5 (p. 69) shows the specialized feeding niches of bird species in a coastal wetland.

Predators and Prey: Eating and Being Eaten

Species called predators feed on all or parts of other species called prey.

In **predation,** members of one species (the *predator*) feed directly on all or part of a living organism of another species (the *prey*). Together, the two kinds of organisms, such as lions (the predator or hunter) and zebras (the prey or hunted), form a *predator–prey* *relationship.* Such relationships are depicted in Figures 3-10 (p. 42) and 3-16 (p. 47).

At the individual level, members of the prey species are clearly harmed. At the population level, predation plays a role in evolution by natural selection. It can benefit the prey species because predators such as some shark species (Case Study, p. 113) kill the sick, weak, aged, and least fit members of a population. The remaining prey gain better access to food supplies and avoid excessive population growth. Predation also helps successful genetic traits to become more dominant in the prey population through natural selection, which can enhance that species' reproductive success and long-term survival.

Some people tend to view predators with contempt. When a hawk tries to capture and feed on a rabbit, some root for the rabbit. Yet the hawk, like all predators, is merely trying to get enough food to feed itself and its young. In doing so, it plays an important ecological role in controlling rabbit populations.

How Do Predators Increase Their Chances of Getting a Meal?

Some predators are fast enough to catch their prey, some hide and lie in wait, and some inject chemicals to paralyze their prey.

Predators have a variety of methods that help them capture prey. *Herbivores* can simply walk, swim, or fly up to the plants they feed on.

Carnivores feeding on mobile prey have two main options: *pursuit* and *ambush.* Some, such as the cheetah, catch prey by running fast; others, such as the American bald eagle, fly and have keen eyesight; still others, such as wolves and African lions, cooperate in capturing their prey by hunting in packs.

Other predators use *camouflage*—a change in shape or color—to hide in plain sight and ambush their prey. For example, praying mantises (Figure 3-1, right, p. 35) sit in flowers of a similar color and ambush visiting insects. White ermines (a type of weasel) and snowy owls hunt in snow-covered areas. People camouflage themselves to hunt wild game and use camouflaged traps to ambush wild game.

Some predators use *chemical warfare* to attack their prey. For example, spiders and poisonous snakes use venom to paralyze their prey and to deter their predators.

How Do Prey Defend Themselves Against or Avoid Predators?

Some prey escape their predators or have protective shells or thorns, some camouflage themselves, and some use chemicals to repel or poison predators.

Prey species have evolved many ways to avoid predators, including the ability to run, swim, or fly fast, and a

Figure 6-5 Sharing the wealth: *resource partitioning* of five species of insect-eating warblers in the spruce forests of Maine. Each species minimizes competition with the others for food by spending at least half its feeding time in a distinct portion (shaded areas) of the spruce trees, and by consuming somewhat different insect species. (After R. H. MacArthur, "Population Ecology of Some Warblers in Northeastern Coniferous Forests," *Ecology* 36 (1958): 533–536.)

highly developed sense of sight or smell that alerts them to the presence of predators. Other avoidance adaptations include protective shells (as on armadillos, which roll themselves up into an armor-plated ball, and turtles), thick bark (giant sequoia), spines (porcupines), and thorns (cacti and rosebushes). Many lizards have brightly colored tails that break off when they are attacked, often giving them enough time to escape.

Other prey species use the camouflage of certain shapes or colors or the ability to change color (chameleons and cuttlefish). Some insect species have evolved shapes that look like twigs (Figure 6-6a), bark, thorns, or even bird droppings on leaves. A leaf insect may be almost invisible against its background (Figure 6-6b), and an arctic hare in its white winter fur appears invisible against the snow.

Chemical warfare is another popular strategy. Some prey species discourage predators with chemicals that are *poisonous* (oleander plants), *irritating* (stinging nettles and bombardier beetles, Figure 6-6c), *foul smelling* (skunks, skunk cabbages, and stinkbugs), or *bad tasting* (buttercups and monarch butterflies, Figure 6-6d). When attacked, some species of squid and octopus emit clouds of black ink to confuse predators and allow them to escape.

Many bad-tasting, bad-smelling, toxic, or stinging prey species have evolved *warning coloration*, brightly colored advertising that enables experienced predators to recognize and avoid them. They flash a warning, "Eating me is risky." Examples include brilliantly colored poisonous frogs (Figure 6-6e); red-, yellow-, and black-striped coral snakes; and foul-tasting monarch butterflies (Figure 6-6d) and grasshoppers.

Based on coloration, biologist Edward O. Wilson gives us two rules for evaluating possible danger from an unknown animal species we encounter in nature. *First,* if it is small and strikingly beautiful, it is probably poisonous. *Second,* if it is strikingly beautiful and easy to catch, it is probably deadly.

Some butterfly species, such as the non-poisonous viceroy (Figure 6-6f), gain some protection by looking and acting like the monarch, a protective device known as *mimicry*. Other prey species use *behavioral strategies* to avoid predation. Some attempt to scare off predators by puffing up (blowfish), spreading

(a) Span worm

(b) Wandering leaf insect

(c) Bombardier beetle

(d) Foul-tasting monarch butterfly

(e) Poison dart frog

(f) Viceroy butterfly mimics monarch butterfly

(g) Hind wings of Io moth resemble eyes of a much larger animal.

(h) When touched, snake caterpillar changes shape to look like head of snake.

Figure 6-6 Natural capital: some ways in which prey species avoid their predators: **(a, b)** *camouflage,* **(c–e)** *chemical warfare,* **(d, e)** *warning coloration,* **(f)** *mimicry,* **(g)** *deceptive looks,* and **(h)** *deceptive behavior.*

their wings (peacocks), or mimicking a predator (Figure 6-6h). Some moths have wings that look like the eyes of much larger animals (Figure 6-6g). Other prey species gain some protection by living in large groups (schools of fish, herds of antelope, flocks of birds).

Parasites: Sponging Off of Others

Although parasites can harm their host organisms, they may also promote community biodiversity.

Parasitism occurs when one species (the *parasite*) feeds on part of another organism (the *host*), usually by living on or in the host. Parasitism can be viewed as a special form of predation. But unlike a conventional predator, a parasite usually is much smaller than its host (prey) and rarely kills its host. Also, most parasites remain closely associated with, draw nourishment from, and may gradually weaken their host over time.

Tapeworms, microorganisms that cause disease (pathogens), and other parasites live *inside* their hosts. Other parasites attach themselves to the *outside* of their hosts. Examples include ticks, fleas, mosquitoes, mistletoe plants, and sea lampreys that use their sucker-like mouths to attach themselves to their fish hosts and feed on their blood. Some parasites move from one host to another, as fleas and ticks do; others, such as tapeworms, spend their adult lives with a single host.

From the host's point of view, parasites are harmful. Nevertheless, parasites play important ecological roles. Collectively, the matrix of parasitic relationships in a community acts like glue to help hold the species in a community together. Parasites also promote biodiversity by helping keep some species from becoming so plentiful that they eliminate other species.

Mutualism: Win–Win Relationships

Pollination and fungi that help plant roots take up nutrients are examples of species interactions that benefit both species.

In **mutualism**, two species interact in a way that benefits both. Such benefits include having pollen and seeds dispersed for reproduction, being supplied with food, or receiving protection. For example, honeybees, caterpillars and other insects may feed on a male flower's nectar, picking up pollen in the process, and then pollinate female flowers when they feed on them.

Figure 6-7 shows three examples of mutualistic relationships that combine *nutrition* and *protection*. One involves birds that ride on the backs of large animals like African buffalo, elephants, and rhinoceroses (Figure 6-7a). The birds remove and eat parasites from the animal's body and often make noises warning the animal when predators approach.

A second example involves clownfish species, which live within sea anemones, whose tentacles sting and paralyze most fish that touch them (Figure 6-7b). The clownfish, which are not harmed by the tentacles, gain protection from predators and feed on the

(a) Oxpeckers and black rhinoceros

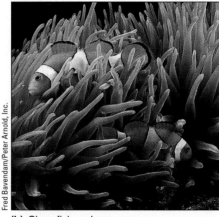

(b) Clownfish and sea anemone

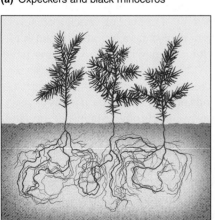

(c) Mycorrhizae fungi on juniper seedlings in normal soil

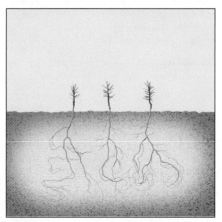

(d) Lack of mycorrhizae fungi on juniper seedlings in sterilized soil

Figure 6-7 Natural capital: examples of *mutualism.* (a) Oxpeckers (or tickbirds) feed on parasitic ticks that infest large, thick-skinned animals such as the endangered black rhinoceros. (b) A clownfish gains protection and food by living among deadly stinging sea anemones and helps protect the anemones from some of their predators. (c) Beneficial effects of mycorrhizal fungi attached to roots of juniper seedlings on plant growth compared to (d) growth of such seedlings in sterilized soil without mycorrhizal fungi.

detritus left from the meals of the anemones. The sea anemones benefit because the clownfish protect them from some of their predators.

A third example is the highly specialized fungi that combine with plant roots to form mycorrhizae (from the Greek words for fungus and roots). The fungi get nutrition from the plant's roots. In turn, the fungi benefit the plant by using their myriad networks of hairlike extensions to improve the plant's ability to extract nutrients and water from the soil (Figure 6-7c).

In *gut inhabitant mutualism,* vast armies of bacteria in the digestive systems of animals break down (digest) their food. The bacteria receive a sheltered habitat and food from their host. In turn, they help break down (digest) their host's food. Hundreds of millions of bacteria in your gut help digest the food you eat. Thank these little critters for helping keep you alive.

It is tempting to think of mutualism as an example of cooperation between species. In reality, each species benefits by exploiting the other.

Commensalism: Using without Harming

Some species interact in a way that helps one species but has little, if any, effect on the other.

Commensalism is a species interaction that benefits one species but has little, if any, effect on the other species. One example is *redwood sorrel,* a small herb. It benefits from growing in the shade of tall redwood trees, with no known harmful effects on the redwood trees.

Another example involves *epiphytes* (such as some types of orchids and bromeliads), plants that attach themselves to the trunks or branches of large trees in tropical and subtropical forests (Figure 6-8). These *air plants* benefit by having a solid base on which to grow. They also live in an elevated spot that gives them better access to sunlight, water from the humid air and rain, and nutrients falling from the tree's upper leaves and limbs. Their presence apparently does not harm the tree.

Environmental Science ⊕ Now™
Review the ways species can interact and see the results of an experiment on species interaction at Environmental ScienceNow.

6-4 ECOLOGICAL SUCCESSION: COMMUNITIES IN TRANSITION

Ecological Succession: How Communities Change over Time

New environmental conditions can cause changes in community structure that lead to one group of species being replaced by other groups.

Figure 6-8 Natural capital: *commensalism.* This bromeliad—an epiphyte or air plant in Brazil's Atlantic tropical rainforest—roots on the trunk of a tree rather than the soil without penetrating or harming the tree. In this interaction, the epiphyte gains access to water, other nutrient debris, and sunlight; the tree apparently remains unharmed.

All communities change their structure and composition over time in response to changing environmental conditions. The gradual change in species composition of a given area is called **ecological succession.**

Ecologists recognize two types of ecological succession, depending on the conditions present at the beginning of the process. **Primary succession** involves the gradual establishment of biotic communities on nearly lifeless ground, where there is no soil in a terrestrial community (Figure 6-9) or no bottom sediment in an aquatic community. Examples include bare rock exposed by a retreating glacier or severe soil erosion, newly cooled lava, an abandoned highway or parking lot, or a newly created shallow pond or reservoir.

Primary succession usually takes a long time— typically thousands or even tens of thousands of years. Before a community can become established on land, there must be soil. Depending mostly on the climate, it takes natural processes several hundred to several thousand years to produce fertile soil.

With the other, more common type of ecological succession, called **secondary succession,** a series of communities with different species can develop in places containing soil or bottom sediment. This development begins in an area where the natural community of organisms has been disturbed, removed, or destroyed, but the soil or bottom sediment remains. Candidates for secondary succession include abandoned farmlands (Figure 6-10, p. 120), burned or cut forests, heavily polluted streams, and land that has been dammed or flooded. Because some soil or sediment is present, new vegetation usually can begin to germinate within a few weeks. Seeds can be present in soils, or they can be carried from nearby plants by wind or by birds and other animals.

During primary or secondary succession, disturbances such as natural or human-caused fires or deforestation can convert a particular stage of succession to an earlier stage. Such disturbances create new conditions that encourage some species and discourage or eliminate others.

Environmental Science ⊛ Now™
Explore the differences between primary and secondary succession at Environmental ScienceNow.

Can We Predict the Path of Succession and Is Nature in Balance?

Scientists cannot project the course of a given succession or view it as preordained progress toward a stable climax community that is in balance with its environment.

According to the classic view, succession proceeds in an orderly sequence along an expected path until a certain stable type of *climax community* occupies an area. Such a community is dominated by a few long-lived plant species and is in balance with its environment. This equilibrium model of succession is what ecologists once meant when they talked about the *balance of nature*.

Over the last several decades, many ecologists have changed their views about balance and equilibrium in nature. Under the old balance-of-nature view, a large terrestrial community undergoing succession eventually became covered with an expected type of climax vegetation. In reality, a close look at almost any community reveals that it consists of an ever-changing mosaic of vegetation patches at different stages of succession.

The modern view is that we cannot project the course of a given succession or view it as preordained progress toward an ideally adapted climax community. Rather, succession reflects the ongoing struggle by different species for enough light, nutrients, food, and space. This competition allows them to survive and gain reproductive advantages over other species.

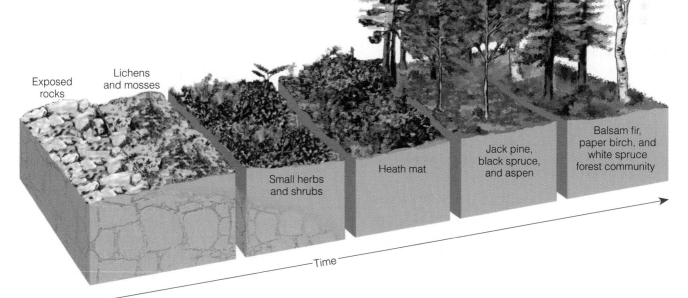

Exposed rocks

Lichens and mosses

Small herbs and shrubs

Heath mat

Jack pine, black spruce, and aspen

Balsam fir, paper birch, and white spruce forest community

Time

Environmental Science ⊛ Now™
Active Figure 6-9 **Natural capital:** starting from scratch. *Primary ecological succession* over several hundred years of plant communities on bare rock exposed by a retreating glacier on Isle Royal in northern Lake Superior. *See an animation based on this figure and take a short quiz on the concept.*

Active Figure 6-10 Natural capital: natural restoration of disturbed land. *Secondary ecological succession* of plant communities on an abandoned farm field in North Carolina. It took 150–200 years after the farmland was abandoned for the area to become covered with a mature oak and hickory forest. A new disturbance such as deforestation or fire would create conditions favoring pioneer species. In the absence of new disturbances, secondary succession would recur over time, but not necessarily in the same sequence shown here. *See an animation based on this figure and take a short quiz on the concept.*

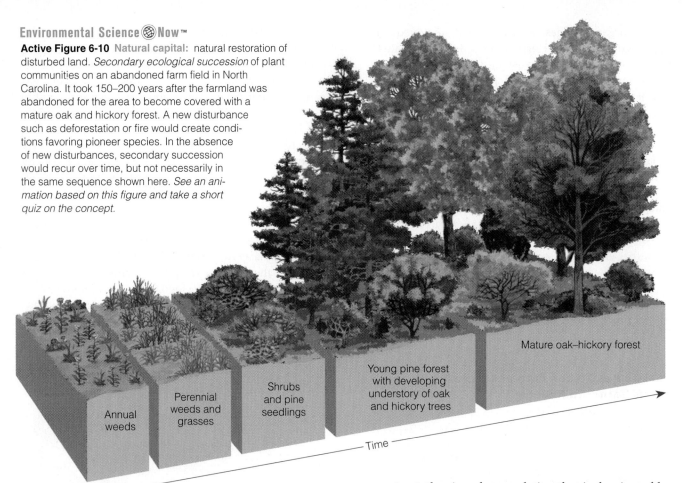

Annual weeds

Perennial weeds and grasses

Shrubs and pine seedlings

Young pine forest with developing understory of oak and hickory trees

Mature oak–hickory forest

Time

6-5 POPULATION DYNAMICS AND CARRYING CAPACITY

Changes in Population Size: Entrances and Exits

Populations increase through births and immigration and decrease through deaths and emigration.

Four variables—*births, deaths, immigration,* and *emigration*—govern changes in population size. A population increases by birth and immigration and decreases by death and emigration:

Population change = (Births + Immigration) − (Deaths + Emigration)

A population's *age structure* can have a strong effect on how rapidly its size increases or decreases. Age structures are usually described in terms of organisms that are not mature enough to reproduce (the *prereproductive age*), those that are capable of reproduction (the *reproductive stage*), and those that are too old to reproduce (the *post-reproductive stage*).

The size of a population that includes a large proportion of young organisms in their reproductive stage or that will soon enter this stage is likely to increase. In contrast, the size of a population that is dominated by individuals past their reproductive stage is likely to decrease. The size of a population with a fairly even distribution between the three reproductive stages will likely remain stable because the reproduction by younger individuals will be roughly balanced by the deaths of older individuals.

Limits on Population Growth

No population can increase its size indefinitely because resources such as light, water, and nutrients are limited and competitors or predators are usually plentiful.

Populations vary in their capacity for growth, also known as the **biotic potential** of a population. The **intrinsic rate of increase** (*r*) is the rate at which a population would grow if it had unlimited resources. Most populations grow at a rate slower than this maximum.

Some species have an astounding biotic potential. Without any controls on their population growth, the descendants of a single female housefly could total about 5.6 trillion flies within about 13 months. If this rapid exponential growth continued, within a few years flies would cover the earth's entire surface!

Fortunately, this is not a realistic scenario because *no population can increase its size indefinitely.* In the real world, a rapidly growing population reaches some size limit imposed by one or more limiting factors, such as light, water, space, or nutrients, or by too many competitors or predators. *There are always limits to population growth in nature.* This is an important lesson from nature.

Environmental resistance consists of all factors that act to limit the growth of a population. Together, biotic potential and environmental resistance determine the **carrying capacity (*K*):** the number of individuals of a given species that can be sustained indefinitely in a given space (area or volume). The growth rate of a population decreases as its size nears the carrying capacity of its environment because resources such as food and water begin to dwindle.

Exponential and Logistic Population Growth: J-Curves and S-Curves

With ample resources, a population can grow rapidly. As resources become limited, its growth rate slows and levels off.

A population with few, if any, resource limitations grows exponentially at a fixed rate such as 1% or 2%. *Exponential growth* starts slowly but then accelerates as the population increases because the base size of the population is increasing. Plotting the number of individuals against time yields a J-shaped growth curve (Figure 6-11, bottom half of curve).

Logistic growth involves exponential population growth followed by a steady decrease in population growth with time until the population size levels off (Figure 6-11, top half of curve). This slowdown occurs as the population encounters environmental resistance and approaches the carrying capacity of its environment. After leveling off, a population with this type of growth typically fluctuates slightly above and below the carrying capacity.

A plot of the number of individuals against time yields a sigmoid, or S-shaped, logistic growth curve (the whole curve in Figure 6-11). Figure 6-12 depicts such a curve for sheep on the island of Tasmania, south of Australia, in the early 19th century.

Environmental Science ⊛ Now™
Learn how to estimate a population of butterflies and see a mouse population growing exponentially at Environmental ScienceNow.

Exceeding Carrying Capacity: Move, Switch Eating Habits, or Decline in Size

When a population exceeds its resource supplies, many of its members will die unless they can switch to new resources or move to an area with more resources.

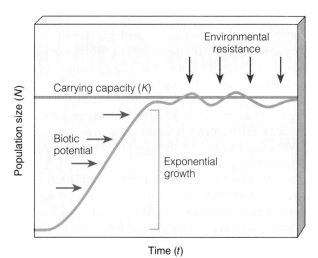

Environmental Science ⊛ Now™
Active Figure 6-11 Natural capital: no population can grow forever. *Exponential growth* (lower part of the curve) occurs when resources are not limiting and a population can grow at its *intrinsic rate of increase* (*r*) or biotic potential. Such exponential growth is converted to *logistic growth*, in which the growth rate decreases as the population becomes larger and faces environmental resistance. Over time, the population size stabilizes at or near the *carrying capacity* (*K*) of its environment, which results in a sigmoid (S-shaped) population growth curve. Depending on resource availability, the size of a population often fluctuates around its carrying capacity. *See an animation based on this figure and take a short quiz on the concept.*

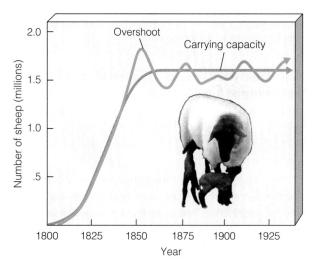

Figure 6-12 Boom and bust: *logistic growth* of a sheep population on the island of Tasmania between 1800 and 1925. After sheep were introduced in 1800, their population grew exponentially thanks to an ample food supply. By 1855, they had overshot the land's carrying capacity. Their numbers then stabilized and fluctuated around a carrying capacity of about 1.6 million sheep.

Some species do not make a smooth transition from exponential growth to logistic growth. Such populations use up their resource supplies and temporarily *overshoot*, or exceed, the carrying capacity of their environment. This occurs because of a *reproductive time lag:* the period needed for the birth rate to fall and the death rate to rise in response to resource overconsumption. Sometimes it takes a while to get the message out.

In such cases, the population suffers a *dieback,* or *crash,* unless the excess individuals can switch to new resources or move to an area with more resources. Such a crash occurred when reindeer were introduced onto a small island off the southwest coast of Alaska (Figure 6-13).

Humans are not exempt from population overshoot and dieback. Ireland experienced a population crash after a fungus destroyed the potato crop in 1845. About 1 million people died, and 3 million people migrated to other countries. Polynesians on Easter Island experienced a population crash after using up most of the island's trees that helped keep them alive (p. 19).

Technological, social, and other cultural changes have extended the earth's carrying capacity for the human species. We have increased food production and used large amounts of energy and matter resources to make normally uninhabitable areas become habitable. How long will we be able to keep doing so on a planet with a finite size, finite resources, and a human population whose size and per capita resource use are growing exponentially? We do not know the answer to this important question.

Reproductive Patterns: Opportunists and Competitors

Some species have a large number of small offspring and give them little parental care; other species have a few larger offspring and take care of them until they can reproduce.

Species use different reproductive patterns to help ensure their survival. Species with a capacity for a high rate of population increase (*r*) are called **r-selected species** (Figure 6-14 and Figure 6-15, left). Such species reproduce early and put most of their energy into reproduction. Examples include algae, bacteria, rodents, annual plants (such as dandelions), and most insects.

These species have many—usually small—offspring and give them little or no parental care or protection. They overcome the massive loss of offspring by producing so many that a few will survive to reproduce many more offspring and thus begin the cycle again.

Such species tend to be *opportunists.* They reproduce and disperse rapidly when conditions are favorable or when a disturbance opens up a new habitat or niche for invasion, as in the early stages of ecological succession.

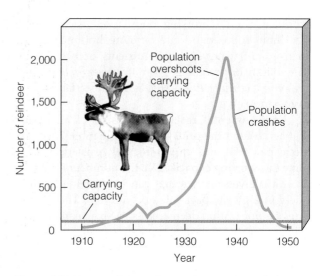

Figure 6-13 Exponential growth, overshoot, and population crash of reindeer introduced to a small island off the southwest coast of Alaska. When 26 reindeer (24 of them female) were introduced in 1910, lichens, mosses, and other food sources were plentiful. By 1935, the herd size had soared to 2,000, overshooting the island's carrying capacity. This led to a population crash, with the herd size plummeting to only 8 reindeer by 1950.

Environmental changes caused by disturbances can allow opportunist species to gain a foothold. However, once established, their populations may crash because of unfavorable changes in environmental conditions or invasion by more competitive species. This helps explain why most r-selected or opportunist species go through irregular and unstable boom-and-bust cycles in their population size.

At the other extreme are *competitor* or **K-selected species** (Figure 16-14 and Figure 16-15, right). They

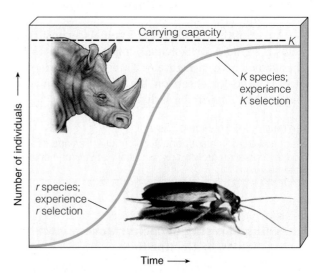

Figure 6-14 Positions of *r-selected* and *K-selected* species on the sigmoid (S-shaped) population growth curve.

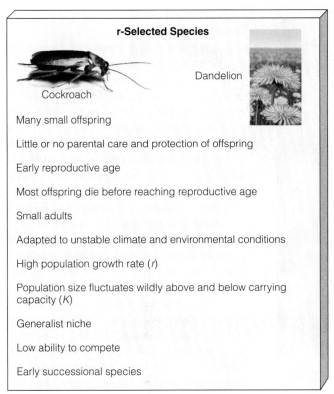

r-Selected Species

Cockroach   Dandelion

Many small offspring

Little or no parental care and protection of offspring

Early reproductive age

Most offspring die before reaching reproductive age

Small adults

Adapted to unstable climate and environmental conditions

High population growth rate (r)

Population size fluctuates wildly above and below carrying capacity (K)

Generalist niche

Low ability to compete

Early successional species

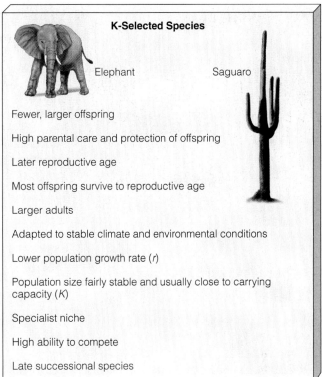

K-Selected Species

Elephant   Saguaro

Fewer, larger offspring

High parental care and protection of offspring

Later reproductive age

Most offspring survive to reproductive age

Larger adults

Adapted to stable climate and environmental conditions

Lower population growth rate (r)

Population size fairly stable and usually close to carrying capacity (K)

Specialist niche

High ability to compete

Late successional species

Figure 6-15 Natural capital: generalized characteristics of *r-selected* (opportunist) species and *K-selected* (competitor) species. Many species have characteristics between these two extremes.

tend to reproduce late in life and have a small number of offspring with fairly long life spans. Typically the offspring of such species develop inside their mothers (where they are safe), are born fairly large, mature slowly, and are cared for and protected by one or both parents until they reach reproductive age. This reproductive pattern results in a few big and strong individuals that can compete for resources and reproduce a few young to begin the cycle again.

Such species are called K-selected species because they tend to do well in competitive conditions when their population size is near the carrying capacity (K) of their environment. Their populations typically follow a logistic growth curve (Figure 6-11).

Most large mammals (such as elephants, whales, and humans), birds of prey, and large and long-lived plants (such as the saguaro cactus, and most tropical rain forest trees) are K-selected species. Many K-selected species—especially those with long generation times and low reproductive rates like elephants, rhinoceroses, and sharks—are prone to extinction.

Most organisms have reproductive patterns between the extremes of r-selected species and K-selected species, or they change from one extreme to the other under certain environmental conditions. In agriculture we raise both r-selected species (crops) and K-selected species (livestock).

The reproductive pattern of a species may give it a temporary advantage, but *the availability of a suitable habitat for individuals of a population in a particular area determines its ultimate population size.* No matter how fast a species can reproduce, there can be no more dandelions than there is dandelion habitat and no more zebras than there is zebra habitat in a particular area.

Environmental Science ⊕ Now™
Explore survivorship curves, showing how certain species differ in their typical lifespans, at Environmental ScienceNow.

6-6 HUMAN IMPACTS ON ECOSYSTEMS: LEARNING FROM NATURE

Human Modification of Natural Ecosystems: Our Big Footprints

We have used technology to alter nature to meet our growing needs and wants in eight major ways that threaten the survival of many other species and ultimately could reduce the quality of life for our own species.

Property	Natural Systems	Human-Dominated Systems
Complexity	Biologically diverse	Biologically simplified
Energy source	Renewable solar energy	Mostly nonrenewable fossil fuel energy
Waste production	Little, if any	High
Nutrients	Recycled	Often lost or wasted
Net primary productivity	Shared among many species	Used, destroyed, or degraded to support human activities

Figure 6-16 Some typical characteristics of natural and human-dominated systems.

To survive and provide resources for growing numbers of people, we have modified, cultivated, built on, or degraded a greatly increasing number and area of the earth's natural systems. Excluding Antarctica, our activities have, to some degree, directly affected about 83% of the earth's land surface. See Figures 2 and 3 in Science Supplement 2 at the end of this book. Figure 6-16 compares some of the characteristics of natural and human-dominated systems.

We have used technology to alter much of the rest of nature to meet our growing needs and wants in eight major ways (Figure 6-17). First, we have *reduced biodiversity by destroying, fragmenting, degrading, and simplifying wildlife habitats.* We have cleared forests, dug up grasslands, and filled in wetlands to grow food or to construct buildings, highways, and parking lots. This represents a loss of overall biodiversity and a degradation of the earth's natural capital.

Second, we have *used, wasted, or destroyed an increasing percentage of the earth's net primary productivity that supports all consumer species (including humans).* This type of alteration is the main reason we are crowding out or eliminating the habitats and food supplies of a growing number of species.

Third, we have unintentionally *strengthened some populations of pest species and disease-causing bacteria.* Our overuse of pesticides and antibiotics has speeded up natural selection among rapidly reproducing pest and bacterial populations and led to genetic resistance to these chemicals.

Fourth, we have *eliminated some predators.* Some ranchers want to eradicate bison or prairie dogs that compete with their sheep or cattle for grass. They also want to eliminate wolves, coyotes, eagles, and other predators that occasionally kill sheep. A few big-game hunters have pushed for elimination of predators that prey on game species. This simplifies and disrupts natural ecosystems and food webs.

Fifth, we have *deliberately or accidentally introduced new or nonnative species into communities.* Most of these species, such as food crops and domesticated livestock, are beneficial to us but a few are harmful to us and other species.

Sixth, we have *used some renewable resources faster than they can be regenerated.* Ranchers and nomadic herders sometimes allow livestock to overgraze grasslands until erosion converts these communities to less productive semideserts or deserts. Farmers sometimes deplete soil of its nutrients by excessive crop growing. Some fish species are overharvested. Illegal hunting or poaching endangers wildlife species with economically valuable parts such as elephant tusks, rhinoceros horns, and tiger skins. In some areas, fresh water is being pumped out of underground aquifers faster than it is being replenished.

Seventh, some human activities *interfere with the normal chemical cycling and energy flows in ecosystems.* Soil nutrients can erode from monoculture crop fields, tree plantations, construction sites, and other simplified communities, and then overload and disrupt other communities such as lakes and coastal waters. Our inputs of carbon dioxide into the carbon cycle have been increasing sharply (Figure 3-26, p. 56)—mostly from burning fossil fuels and from clearing and burning forests and grasslands. This and other inputs of greenhouse gases from human activities can trigger global climate change by altering energy flow through the troposphere. The human input of nitrogen into the nitrogen cycle exceeds the earth's natural input (Figure 3-28, p. 58). We are also altering energy flows through the biosphere by releasing chemicals into the atmosphere that can increase the amount of harmful ultraviolet energy reaching the troposphere by reducing ozone levels in the troposphere.

Eighth, while most natural systems run on sunlight, *human-dominated ecosystems have become increasingly dependent on nonrenewable energy from fossil fuels.* Fossil fuel systems typically produce pollution, add

Reduction of biodiversity

Increasing use of the earth's net primary productivity

Increasing genetic resistance of pest species and disease-causing bacteria

Elimination of many natural predators

Deliberate or accidental introduction of potentially harmful species into communities

Using some renewable resources faster than they can be replenished

Interfering with the earth's chemical cycling and energy flow processes

Relying mostly on polluting fossil fuels

Environmental Science ⊕ Now™

Active Figure 6-17 Natural capital degradation: major ways humans have altered the rest of nature to meet our growing population, needs, and wants. *See an animation based on this figure and take a short quiz on the concept.*

the greenhouse gas carbon dioxide to the atmosphere, and waste a great deal of energy.

To survive, we must exploit and modify parts of nature. However, we are beginning to understand that any human intrusion into nature has multiple effects, most of them unintended and unpredictable (see Connections, at right).

Environmental Science ⊕ Now™

Examine how resources have been depleted or degraded around the world at Environmental ScienceNow.

We face two major challenges. *First*, we need to maintain a balance between simplified, human-altered communities and the more complex natural communities on which we and other species depend. *Second*, we need to slow down the rates at which we are simplify-

ing, homogenizing, and degrading nature for our purposes. Otherwise, what is at risk is not the resilient earth but rather the quality of life for our own species and the existence of other species we drive to premature extinction. We cannot save the earth; it can get along very nicely without us, just as it has done for 3.7 billion years. However, by learning how the earth works and by working with its natural processes, we can sustain the quality of life for the human species and avoid the projected premature extinction of as many as half of the world's species during this century as a result of the eight factors just discussed.

Ecological Surprises

CONNECTIONS

Malaria once infected nine out of ten people in North Borneo, now known as Sabah. In 1955, the World Health Organization (WHO) began spraying the island with dieldrin (a DDT relative) to kill malaria-carrying mosquitoes. The program was so successful that the dreaded disease was nearly eliminated.

Then unexpected things began to happen. The dieldrin also killed other insects, including flies and cockroaches living in houses. The islanders applauded. Next, small insect-eating lizards that also lived in the houses died after gorging themselves on dieldrin-contaminated insects.

Cats began dying after feeding on the lizards. In the absence of cats, rats flourished and overran the villages. When the people became threatened by sylvatic plague carried by rat fleas, the WHO parachuted healthy cats onto the island to help control the rats. Operation Cat Drop worked.

But then the villagers' roofs began to fall in. The dieldrin had killed wasps and other insects that fed on a type of caterpillar that either avoided or was not affected by the insecticide. With most of its predators eliminated, the caterpillar population exploded, munching its way through its favorite food: the leaves used in thatched roofs.

Ultimately, this episode ended happily: Both malaria and the unexpected effects of the spraying program were brought under control. Nevertheless, this chain of unintended and unforeseen events emphasizes the unpredictability of interfering with a community. It reminds us that when we intervene in nature, we need to ask, "Now what will happen?"

Critical Thinking

Do you believe the beneficial effects of spraying pesticides on Sabah outweighed the resulting unexpected and harmful effects? Explain.

Four Principles of Sustainability: Copy Nature

We can develop more sustainable economies and societies by mimicking the four major ways that nature has adapted and sustained itself for several billion years.

How can we live more sustainably? According to ecologists, we can find out how nature has survived and adapted for several billion years and copy its strategy. Figure 6-18 summarizes the four major ways in which life on earth has survived and adapted for several billions of years. Figure 6-19 gives an expanded description of these principles (left side) and summarizes how we can live more sustainably by mimicking these fundamental but amazingly simple lessons from nature in designing our societies, products, and economies (right side). *Figures 6-18 and 6-19 summarize the major message of this book. Study them carefully.*

Biologists have used these lessons from their ecological study of nature to formulate four guidelines for developing more sustainable societies and lifestyles:

- *Our lives, lifestyles, and economies are totally dependent on the sun and the earth.* We need the earth, but the earth does not need us. As a species, we are very expendable.

- *Everything is connected to, and interdependent with, everything else.* The primary goal of ecology is to discover what connections in nature are the strongest, most important, and most vulnerable to disruption for us and other species.

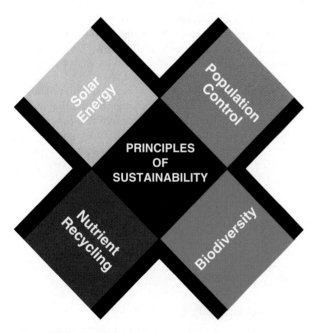

Figure 6-18 Sustaining natural capital: four interconnected principles of sustainability derived from learning how nature sustains itself. This diagram also appears on the bottom half of this book's back cover.

Solutions
Principles of Sustainability

How Nature Works		Lessons for Us
Runs on renewable solar energy.	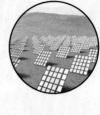	Rely mostly on renewable solar energy.
Recycles nutrients and wastes. There is little waste in nature.		Prevent and reduce pollution and recycle and reuse resources.
Uses biodiversity to maintain itself and adapt to new environmental conditions.		Preserve biodiversity by protecting ecosystem services and habitats and preventing premature extinction
Controls a species' population size and resource use by interactions with its environment and other species.		Reduce human births and wasteful resource use to prevent environmental overload and depletion and degradation of resources.

Figure 6-19 Solutions: implications of the four principles of sustainability (left), derived from observing nature, for the long-term sustainability of human societies (right). These four operating principles of nature are connected to one another. Failure of any single principle can lead to temporary or long-term unsustainability, and disruption of ecosystems and human economies and societies.

- *We can never do just one thing.* Any human intrusion into nature has unexpected and mostly unintended side effects (Connections, p. 124). When we alter nature, we need to ask, "Now what will happen?"

- *We cannot indefinitely sustain a civilization that depletes and degrades the earth's natural capital, but we can sustain one that lives off the biological income provided by the earth's natural capital.*

Increasingly, environmental scientists and ecologists are urging that we base our efforts to prevent damage to the earth's life-support system on the **precautionary principle:** When evidence indicates that an activity can seriously harm human health or the environment, we should take precautionary measures to prevent or minimize such harm, even if some of the cause-and-effect relationships have not been fully established scientifically. This principle is based on the commonsense idea behind many adages such as "Better safe than sorry," "Look before you leap," "First, do no harm," and "Slow down for speed bumps."

As an analogy, we know that eating too much of certain types of foods and not getting enough exercise can greatly increase our chances of a heart attack, diabetes, and other disorders. The exact connections between these health problems, chemicals in various foods, exercise, and genetics are still under study and often debated. People with such conditions could use this uncertainty and unpredictability as an excuse to continue overeating and not exercising. In reality, the wise course is to eat better and exercise more to help *prevent* potentially serious health problems.

Recently, the precautionary principle has served as the basis of several international environmental treaties. For example, a global treaty developed by 122 countries in 2000 seeks to ban or phase out 12 *persistent organic pollutants (POPs)*.

Some analysts point out that we should be selective in applying the precautionary principle. Although we need to project possible unintended effects carefully, we can never predict all of the unintended effects of our actions and technologies. We must always be willing to take some risks. Otherwise, we would stifle creativity and innovation and severely limit the development of new technologies and products.

In Chapter 7, we will apply the principles of population dynamics and sustainability discussed in this chapter to the growth of the human population. In Chapters 8 and 9, we apply those principles to understand the earth's terrestrial and aquatic biodiversity and to consider how to help sustain them.

We cannot command nature except by obeying her.
SIR FRANCIS BACON

CRITICAL THINKING

1. Some homeowners in Florida believe they should have the right to kill any alligator found on their property. Others argue against this policy, noting that alligators are a threatened species, and that housing developments have invaded the habitats of alligators, not the other way around. Others go further and believe the American alligator species has an inherent right to exist. What is your opinion on this issue? Explain.

2. How would you respond to someone who claims it is not important to protect areas of temperate and polar biomes because most of the world's biodiversity is found in the tropics?

3. How would you determine whether a particular species found in a given area is a keystone species?

4. Many conservationists consider the practice of catching sharks, removing their fins, and then throwing the live shark back into the water to die cruel and unethical. Others support this practice because they say sharks do not experience pain, have no inherent right to exist as a species, and bite or kill a few people each year. Do you believe we have an ethical obligation to treat sharks humanely? Explain.

5. How would you reply to someone who argues that (a) we should not worry about our effects on natural systems because succession will heal the wounds of human activities and restore the balance of nature, (b) efforts to preserve natural systems are not worthwhile because nature is largely unpredictable, and (c) because there is no balance in nature and no stability in species diversity we should cut down diverse old-growth forests and replace them with tree farms?

6. Why are pest species likely to be extreme r-selected species? Why are many endangered species likely to be extreme K-selected species?

7. A bumper sticker reads, "Nature always bats last and owns the stadium." What does this mean in ecological terms? What is its lesson for the human species?

8. Identify aspects of your lifestyle that follow or violate each of the four sustainability principles shown in Figures 6-18 and 6-19 (p. 126). Would you be willing to change aspects of your lifestyle that violate these sustainability principles? List the three most important ways you could do so.

LEARNING ONLINE

The website for this book includes review questions for the entire chapter, flash cards for key terms and concepts, a multiple-choice practice quiz, interesting Internet sites, references, and a guide for accessing thousands of InfoTrac® College Edition articles.
Visit

http://biology.brookscole.com/miller11

Then choose Chapter 6, and select a learning resource. For access to animations, additional quizzes, chapter outlines and summaries, register and log in to

Environmental Science ⊛ Now™

at **esnow.brookscole.com/miller11** using the access code card in the front of your book.

Active Graphing

Visit http://esnow.brookscole.com/miller11 to explore the graphing exercise for this chapter.

---

7 Applying Population Ecology: The Human Population

Population Control / Land

CASE STUDY
Is the World Overpopulated?

The world's human population is projected to increase from 6.5 to 8–9 billion or more between 2005 and 2050 (Figure 1-1, p. 5), with growth occurring especially rapidly in developing countries such as China (Figure 7-1). This raises an important question: *Can the world provide an adequate standard of living for 2.4 billion more people without causing widespread environmental damage?*

Is the earth overpopulated? If so, what measures should be taken to slow population growth? Some argue that the planet is already too crowded, especially in developed countries such as the United States where high resource consumption rates magnify the environmental impact of each person (Figure 1-7, p. 11). Others would encourage population growth as a means to stimulate economic growth.

Those who do not believe the earth is overpopulated point out that the average life span of the world's 6.5 billion people is longer today than at any time in the past and is projected to get longer. According to them, the world can support billions more people. They also see more people as the world's most valuable resource for solving environmental and other problems and for stimulating economic growth by increasing the number of consumers.

Some view any form of population regulation as a violation of their religious beliefs. Others see it as an intrusion into their privacy and personal freedom to

Figure 7-1 Crowded streets in China. Together, China and India have 37% of the world's population and the resource use per person in these countries is projected to grow rapidly as they become more modernized.

have as many children as they want. Some developing countries and some members of minorities in developed countries regard population control as a form of genocide to keep their numbers and political power from growing.

Proponents of slowing and eventually stopping population growth have a different view. They point out that we fail to provide the basic necessities for about one of every five people on the earth today. If we cannot or will not provide basic support for about 1.4 billion people today, they ask, how will we be able to do so for the projected 2.4 billion more people by 2050?

Proponents of slowing population growth warn of two serious consequences if we do not sharply lower birth rates. *First,* the death rate may increase because of declining health and environmental conditions in some areas—something that is already happening in parts of Africa. *Second,* resource use and environmental harm may intensify as more consumers increase their already large ecological footprint in developed countries and in some developing countries, such as China and India, which are undergoing rapid economic growth.

Population increase and its accompanying rise in consumption can increase environmental stresses such as *infectious disease, biodiversity losses, loss of tropical forests, fisheries depletion, increasing water scarcity, pollution of the seas,* and *climate change.*

Proponents of this view recognize that population growth is not the only cause of these problems (Figure 1-10, p. 14). But they argue that adding several hundred million more people in developed countries and several billion more in developing countries can only intensify existing environmental and social problems.

These analysts believe people should have the freedom to produce as many children as they want, but only if it does not reduce the quality of other people's lives now and in the future, either by impairing the earth's ability to sustain life or by causing social disruption. According to their view, limiting the freedom of individuals to do anything they want, in an effort to protect the freedom of other individuals, is the basis of most laws in modern societies.

HOW WOULD YOU VOTE? Should the population of the country where you live should be stabilized as soon as possible? Cast your vote online at http://biology.brookscole.com/miller11.

The problems to be faced are vast and complex, but come down to this: 6.5 billion people are breeding exponentially. The process of fulfilling their wants and needs is stripping earth of its biotic capacity to produce life; a climactic burst of consumption by a single species is overwhelming the skies, earth, waters, and fauna.

PAUL HAWKEN

This chapter looks at the factors that affect the growth of the human population and the distribution of people between urban and rural areas. It addresses the following questions:

- How is population size affected by birth, death, fertility, and migration rates?

- How is population size affected by the percentage of males and females at each age level?

- How can we slow population growth?

- What success have India and China had in slowing population growth?

- How is the world's population distributed between rural and urban areas, and what factors determine how urban areas develop?

- What are the major resource and environmental problems faced by urban areas?

- How do transportation systems shape urban areas and growth, and what are the advantages and disadvantages of various forms of transportation?

- How can cities be made more sustainable and more desirable places to live?

KEY IDEAS

- The average number of children that a woman bears has dropped sharply since 1950, but it is not low enough to stabilize the world's population in the near future.

- The number of people in young, middle, and older age groups determines how fast populations grow or decline.

- The best way to slow population growth is by investing in family planning, reducing poverty, and elevating the status of women.

- Cities are rarely self-sustaining. Supplying them with resources and absorbing and diluting their wastes can threaten biodiversity.

- An ecocity allows people to walk, bike, or take mass transit for most of their travel. It recycles and reuses most of its wastes, grows much of its own food, and protects biodiversity by preserving surrounding land.

7-1 FACTORS AFFECTING HUMAN POPULATION SIZE

Human Population History: An Overview

We have postponed reaching the limits of disease, food, water, and energy supplies on human population growth mostly by taking over much of the earth and sharply reducing death rates.

For most of history, the human population grew slowly (Figure 1-1, left part of curve, p. 5). But about 200 years ago, our population growth took off (Figure 1-1, right part of curve, and Figure 1-5, p. 9).

Three major reasons explain this development. *First*, thanks to our highly complex brain and tool-using abilities, humans developed the ability to expand into diverse new habitats and different climate zones.

Second, the emergence of early and modern agriculture allowed more people to be fed per unit of land area. This increase in the carrying capacity of land allowed our population to grow.

Third, we have managed to put off reaching the limits of disease, food, water, and energy supplies on overall population growth. We have used better sanitation and developed antibiotics and vaccines to help control infectious disease agents, and we have tapped into concentrated sources of energy (fossil fuels). These changes allowed births to exceed deaths.

Can the world's life-support systems sustain the 8–11 billion people projected to be around when this century ends? At what average level of resource consumption will they live?

Birth Rates and Death Rates: Entrances and Exits

The human population in a particular area increases because of births and immigration and decreases through deaths and emigration.

Human populations grow or decline through the interplay of three factors: *births, deaths,* and *migration.* **Population change** is calculated by subtracting the number of people leaving a population (through death and emigration) from the number entering it (through birth and immigration) during a specified period of time (usually one year):

$$\text{Population change} = (\text{Births} + \text{Immigration}) - (\text{Deaths} + \text{Emigration})$$

When births plus immigration exceeds deaths plus emigration, population increases; when the reverse is true, population declines.

Instead of using the total numbers of births and deaths per year, demographers use the **birth rate,** or **crude birth rate** (the number of live births per 1,000 people in a population in a given year), and the **death rate,** or **crude death rate** (the number of deaths per 1,000

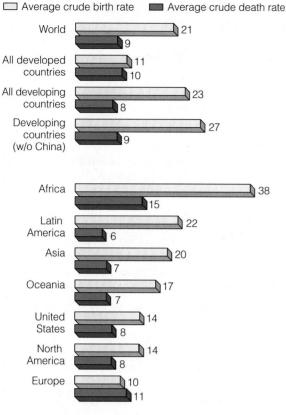

□ Average crude birth rate ■ Average crude death rate

World — 21 / 9
All developed countries — 11 / 10
All developing countries — 23 / 8
Developing countries (w/o China) — 27 / 9

Africa — 38 / 15
Latin America — 22 / 6
Asia — 20 / 7
Oceania — 17 / 7
United States — 14 / 8
North America — 14 / 8
Europe — 10 / 11

Figure 7-2 Average crude birth and death rates for various groupings of countries in 2005. (Data from Population Reference Bureau)

people in a population in a given year) in their analyses. Figure 7-2 shows the estimated crude birth and death rates for various parts of the world in 2005.

Population Growth Today: Slowing but Still Growing

The rate at which the world's population is increasing has slowed, but the population continues to grow fairly rapidly.

Birth rates and death rates are coming down worldwide, but death rates have fallen more sharply than birth rates. As a result, more births are occurring than deaths: Every time your heart beats, 2.4 more babies are added to the world's population. At this rate, we are sharing the earth and its resources with 214,000 more people per day—97% of them in developing countries.

The rate of the world's annual population change usually is expressed as a percentage:

$$\text{Annual rate of natural population change (\%)} = \frac{\text{Birth rate} - \text{Death rate}}{1,000 \text{ persons}} \times 100$$

$$= \frac{\text{Birth rate} - \text{Death rate}}{10}$$

Exponential population growth has not disappeared but rather has declined to a slower rate. The rate of the world's annual population growth (natural increase) dropped by almost half between 1963 and 2005, from 2.2% to 1.2%. This is *good news*. Nevertheless, during the same period the population base doubled from 3.2 billion to almost 6.5 billion. The drop in the rate of population increase is somewhat like learning that a truck heading straight at you has slowed from 100 kilometers per hour (kph) to 45 kph while its weight has more than doubled.

Although an exponential growth rate of 1.2% may seem small, consider this: During 2005, it added about 78 million people to the world's population, compared to 69 million in 1963 when the world's population growth reached its peak. This increase is roughly equal to adding another New York City every month, a Germany every year, and a United States every 3.7 years.

There is a big difference between exponential population growth rates in developed and developing countries. In 2005, the population of developed countries was growing at a rate of 0.1% per year. That of the developing countries was 1.5% per year—15 times faster.

As a result of these trends, the population of the developed countries, currently 1.2 billion, is expected to change little in the next 50 years. In contrast, the population of the developing countries is projected to rise steadily from 5.3 billion in 2005 to 8 billion in 2050 and then to level off by the end of this century (Figure 1-5, p. 9). The six nations expected to produce most of this growth are, in order, India, China, Pakistan, Nigeria, Bangladesh, and Indonesia.

China—1.3 billion in 2005, or one of every five people in the world—and India—1.1 billion—are the world's most populous countries. Together they have 37% of the world's population. The United States—296 million people in 2005—has the world's third largest population but only 4.6% of its people.

Environmental Science ⊛ Now™
Can you guess what regions of the world will likely have the greatest population increases by 2025? Find out at Environmental ScienceNow.

Doubling Time

Doubling time is how long it takes for a population growing at a specified rate to double its size.

One measure of population growth is **doubling time:** the time (usually in years) it takes for a population growing at a specified rate to double its size. A quick way to calculate doubling time is to use the **rule of 70:** 70/percentage growth rate = doubling time in years.

For example, in 2005 the world's population grew by 1.2%. If that rate continues, the earth's population will double in about 58 years (70/1.2).

The population of Nigeria is increasing by 2.4% per year. How long will it take for its population to double?

Declining Fertility Rates: Fewer Babies per Woman

The average number of children that a woman bears has dropped sharply since 1950, but is not low enough to stabilize the world's population in the near future.

Fertility is the number of births that occur to an individual woman or in a population. Two types of fertility rates affect a country's population size and growth rate.

The first type, **replacement-level fertility,** is the number of children a couple must bear to replace themselves. It is slightly higher than two children per couple (2.1 in developed countries and as high as 2.5 in some developing countries), mostly because some female children die before reaching their reproductive years.

Does reaching replacement-level fertility bring an immediate halt in population growth? No, because so many future parents are alive. If each of today's couples had an average of 2.1 children and their children also had 2.1 children, the world's population would continue to grow for 50 years or more (assuming death rates do not rise).

The second type of fertility rate, the **total fertility rate (TFR),** is the average number of children a woman typically has during her reproductive years. In 2005, the average global TFR was 2.7 children per woman: 1.6 in developed countries (down from 2.5 in 1950) and 3.0 in developing countries (down from 6.5 in 1950). Although the decline in TFR in developing countries is impressive, this level of fertility remains far above the replacement level of 2.1.

How many of us are likely to be here in 2050? Answer: 7.2–10.6 billion people, depending on the world's projected average TFR (Figure 7-3). The medium projection is 8.9 billion people. About 97% of this growth is projected to take place in developing countries, where acute poverty (living on less than $1 per day) is a way of life for about 1.4 billion people.

Case Study: Fertility Rates in the United States

Population growth in the United States has slowed but is not close to leveling off.

The population of the United States grew from 76 million in 1900 to 296 million in 2005, despite oscillations in the country's TFR (Figure 7-4). In 1957, the peak of the baby boom after World War II, the TFR reached 3.7 chil-

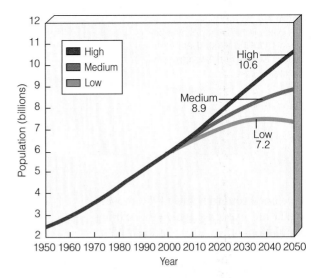

Figure 7-3 Global outlook: United Nations world population projections to 2050, assuming the world's total fertility rate is 2.6 (high), 2.1 (medium), or 1.5 (low) children per woman. The most likely projection is the medium one—8.9 billion people by 2050. (Data from United Nations, *World Population Prospects: The 2000 Revision*, 2001)

dren per woman. Since then, it has generally declined, remaining at or below replacement level since 1972.

The drop in the TFR has led to a decline in the rate of population growth in the United States. But the country's population is still growing faster than that of any other developed country and is not close to leveling off.

Nearly 2.9 million people were added to the U.S. population in 2005. About 59% of this growth occurred because births outnumbered deaths; the rest came from legal and illegal immigration.

According to U.S. Census Bureau, the U.S. population is likely to increase from 296 million in 2005 to 457 million by 2050 and then to 571 million by 2100. In contrast, population growth has slowed in other major

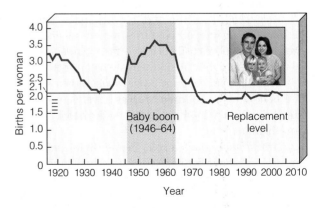

Figure 7-4 Total fertility rates for the United States, 1917–2005. (Data from Population Reference Bureau and U.S. Census Bureau)

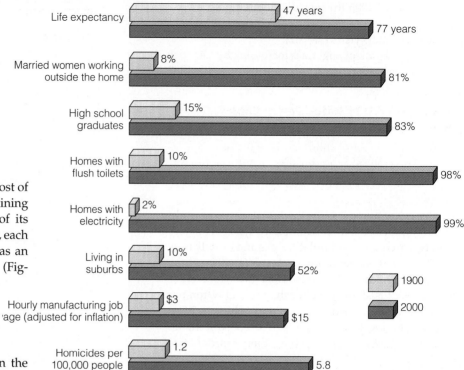

Figure 7-5 Some major changes that took place in the United States between 1900 and 2000. (Data from U.S. Census Bureau and Department of Commerce)

Life expectancy — 47 years / 77 years

Married women working outside the home — 8% / 81%

High school graduates — 15% / 83%

Homes with flush toilets — 10% / 98%

Homes with electricity — 2% / 99%

Living in suburbs — 10% / 52%

Hourly manufacturing job wage (adjusted for inflation) — $3 / $15

Homicides per 100,000 people — 1.2 / 5.8

1900
2000

developed countries since 1950, most of which are expected to have declining populations after 2010. Because of its high per capita rate of resource use, each addition to the U.S. population has an enormous environmental impact (Figure 1-7, p. 11, and Figure 3 in Science Supplement 2 at the end of this book). In addition to the almost fourfold increase in population growth, some amazing changes in lifestyles took place in the United States during the 20th century (Figure 7-5).

Factors Affecting Birth Rates and Fertility Rates

The number of children women have is affected by the cost of raising and educating children, educational and employment opportunities for women, infant deaths, marriage age, and availability of contraceptives and abortions.

Many factors affect a country's average birth rate and TFR. One is the *importance of children as a part of the labor force.* Proportions of children working tend to be higher in developing countries—especially in rural areas, where children begin working to help raise crops at an early age.

Another economic factor is the *cost of raising and educating children.* Birth and fertility rates tend to be lower in developed countries, where raising children is much more costly because they do not enter the labor force until they are in their late teens or twenties.

The *availability of private and public pension systems* affects how many children couples have. Pensions reduce parents' need to have many children to help support them in old age.

Urbanization plays a role. People living in urban areas usually have better access to family planning services and tend to have fewer children than those living in rural areas, where children are often needed to perform essential tasks.

Another important factor is *the educational and employment opportunities available for women.* TFRs tend to be low when women have access to education and paid employment outside the home. In developing countries, women with no education generally have two more children than women with a secondary school education.

Another factor is the *infant mortality rate.* In areas with low infant mortality rates, people tend to have a smaller number of children because fewer children die at an early age.

Average age at marriage (or, more precisely, the average age at which women have their first child) also plays a role. Women normally have fewer children when their average age at marriage is 25 or older.

Birth rates and TFRs are also affected by the *availability of legal abortions.* Each year about 190 million women become pregnant. The United Nations and the World Bank estimate that 46 million of these women get abortions—26 million legal and 20 million illegal (and often unsafe).

The *availability of reliable birth control methods* (Figure 7-6) allows women to control the number and spacing of the children they have. *Religious beliefs, traditions, and cultural norms* also play a role. In some countries, these factors favor large families and strongly oppose abortion and some forms of birth control.

Factors Affecting Death Rates

Death rates have declined because of increased food supplies, better nutrition, advances in medicine, improved sanitation and personal hygiene, and safer water supplies.

The rapid growth of the world's population over the past 100 years is not the result of a rise in the crude birth rate. Instead, it largely reflects a decline in crude death rates, especially in developing countries.

More people started living longer and fewer infants died because of increased food supplies and distribution, better nutrition, medical advances such as immunizations and antibiotics, improved sanitation, and safer water supplies (which curtailed the spread of many infectious diseases).

Two useful indicators of the overall health of people in a country or region are **life expectancy** (the average number of years a newborn infant can expect to live) and the **infant mortality rate** (the number of babies out of every 1,000 born who die before their first birthday).

Great news. The global life expectancy at birth increased from 48 years to 67 years (76 years in developed countries and 65 years in developing countries) between 1955 and 2005; it is projected to reach 74 by 2050. Between 1900 and 2005, life expectancy in the United States increased from 47 to 78 years and is projected to reach 82 years by 2050.

Bad news. In the world's poorest countries, life expectancy is 49 years or less. In many African countries, life expectancy is expected to fall further because of more deaths from AIDS.

Infant mortality is viewed as the best single measure of a society's quality of life because it reflects a country's general level of nutrition and health care. A high infant mortality rate usually indicates insufficient food (undernutrition), poor nutrition (malnutrition), and a high incidence of infectious disease (usually from contaminated drinking water and weakened disease resistance from undernutrition and malnutrition).

Good news. Between 1965 and 2005, the world's infant mortality rate dropped from 20 (per 1,000 live births) to 6.5 in developed countries and from 118 to 61 in developing countries.

Bad news. At least 7.6 million infants (most in developing countries) die of preventable causes during their first year of life—an average of 21,000 mostly unnecessary infant deaths per day. This is equivalent to 55 jumbo jets, each loaded with 400 infants younger than age 1, crashing each day with no survivors!

The U.S. infant mortality rate declined from 165 in 1900 to 6.6 in 2005. This sharp decline was a major factor in the marked increase in U.S. average life expectancy during this period.

Some 46 countries (most in Europe) had lower infant mortality rates than the United States in 2005. Three factors helped keep the U.S. infant mortality rate relatively high: *inadequate health care for poor women during pregnancy and for their babies after birth, drug addiction among pregnant women,* and *a high birth rate among teenagers.* Although the United States has the highest teenage birth rate of any developed country, the rate dropped by 40% between 1991 and 2002.

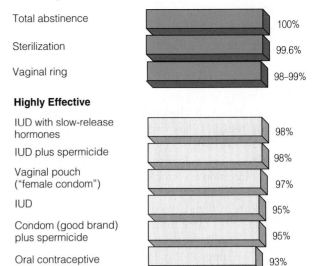

Extremely Effective

Total abstinence — 100%
Sterilization — 99.6%
Vaginal ring — 98–99%

Highly Effective

IUD with slow-release hormones — 98%
IUD plus spermicide — 98%
Vaginal pouch ("female condom") — 97%
IUD — 95%
Condom (good brand) plus spermicide — 95%
Oral contraceptive — 93%

Effective

Cervical cap — 89%
Condom (good brand) — 86%
Diaphragm plus spermicide — 84%
Rhythm method (Billings, Sympto-Thermal) — 84%
Vaginal sponge impregnated with spermicide — 83%
Spermicide (foam) — 82%

Moderately Effective

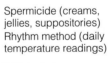

Spermicide (creams, jellies, suppositories) — 75%
Rhythm method (daily temperature readings) — 74%
Withdrawal — 74%
Condom (cheap brand) — 70%

Unreliable

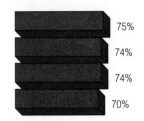

Douche — 40%
Chance (no method) — 10%

Figure 7-6 Typical effectiveness rates of birth control methods in the United States. Percentages are based on the number of undesired pregnancies per 100 couples using a specific method as their sole form of birth control for a year. For example, an effectiveness rating of 93% for oral contraceptives means that for every 100 women using the pill regularly for one year, 7 will get pregnant. Effectiveness rates tend to be lower in developing countries, primarily because of lack of education. Globally, 39% of the world's people using contraception rely on sterilization (32% of females and 7% of males), followed by intrauterine devices (IUDs; 22%), the pill (14%), and male condoms (7%). Preferences in the United States are female sterilization (26%), the pill (25%), male condoms (19%), and male sterilization (10%). (Data from Alan Guttmacher Institute, Henry J. Kaiser Family Foundation, and the United Nations Population Division)

Economics and Politics Case Study: U.S. Immigration

Immigration has played, and continues to play, a major role in the growth and cultural diversity of the U.S. population.

Since 1820, the United States has admitted almost twice as many immigrants and refugees as all other countries combined! The number of legal immigrants (including refugees) has varied during different periods because of changes in immigration laws and rates of economic growth (Figure 7-7). Currently, legal and illegal immigration account for about 41% of the country's annual population growth.

Between 1820 and 1960, most legal immigrants to the United States came from Europe. Since 1960, most have come from Latin America (53%) and Asia (25%), followed by Europe (14%). Latinos (67% of them from Mexico) made up 14% of the U.S. population in 2005. By 2050, Latinos are projected to account for one of every four people in the United States.

In 1995, the U.S. Commission on Immigration Reform recommended reducing the number of legal immigrants from about 900,000 to 700,000 per year for a transition period and then to 550,000 per year. Some analysts want to limit legal immigration to 20% of the country's annual population growth. They would accept new entrants only if they can support themselves, arguing that providing immigrants with public services makes the United States a magnet for the world's poor.

There is also support for efforts to sharply reduce illegal immigration. Some remain concerned that a crackdown on the country's 8–10 million illegal immigrants might lead to discrimination against legal immigrants.

Proponents of reducing immigration argue that it would allow the United States to stabilize its population sooner and help reduce the country's enormous environmental impact. The public strongly supports this position. In a January 2002 Gallup poll, 58% of people polled believed that immigration rates should be reduced (up from 45% in January 2001). A 1993 Hispanic Research Group survey found that 89% of Hispanic Americans supported an immediate moratorium on immigration.

Others oppose reducing current levels of legal immigration. They argue that it would diminish the United States' historical role as a place of opportunity for the world's poor and oppressed. In addition, immigrants pay taxes, take many menial and low-paying jobs that most other Americans shun, open businesses, and create jobs. Moreover, according to the U.S. Census Bureau, after 2020 higher immigration levels will be needed to supply enough workers as baby boomers retire.

X *HOW WOULD YOU VOTE?* Should immigration into the United States (or the country where you live) be reduced? Cast your vote online at http://biology.brookscole.com /miller11.

7-2 POPULATION AGE STRUCTURE

Age-Structure Diagrams

The number of people in young, middle, and older age groups determines how fast populations grow or decline.

As mentioned earlier, even if the replacement-level fertility rate of 2.1 were magically achieved globally tomorrow, the world's population would keep growing for at least another 50 years (assuming no large increase in the death rate). This results mostly from the **age structure**: the distribution of males and females in each age group in the world's population.

Population experts (demographers) construct a population age-structure diagram by plotting the percentages or numbers of males and females in the total population in each of three age categories: *prereproductive* (ages 0–14), *reproductive* (ages 15–44), and *postreproductive* (ages 45 and older). Figure 7-8 presents generalized age-structure diagrams for countries with rapid, slow, zero, and negative population growth rates. Which of these figures best represents the country where you live?

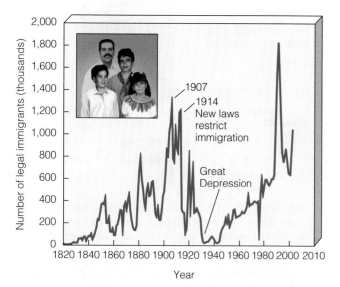

Figure 7-7 Legal immigration to the United States, 1820–2002. The large increase in immigration since 1989 resulted mostly from the Immigration Reform and Control Act of 1986, which granted legal status to illegal immigrants who could show they had been living in the country for several years. In 2005, almost one in eight people in the United States was born in another country. (Data from U.S. Immigration and Naturalization Service)

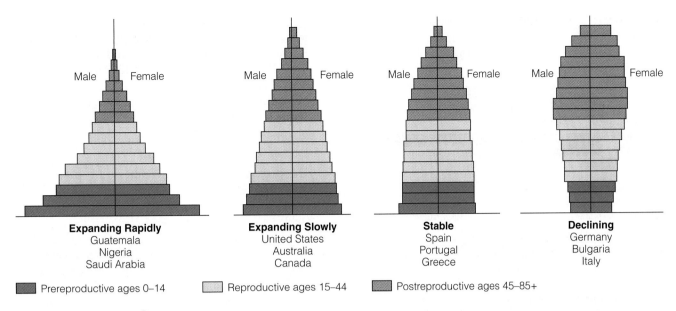

| Prereproductive ages 0–14 | Reproductive ages 15–44 | Postreproductive ages 45–85+ |

Environmental Science ⊕ Now™

Active Figure 7-8 Generalized population age-structure diagrams for countries with rapid population growth (1.5–3%), slow population growth (0.3–1.4%), a stable population (0–0.2%), and a declining population. Populations with a large proportion of its people in the prereproductive ages of 0–14 (at left) have a large potential for rapid population growth. *See an animation based on this figure and take a short quiz on the concept.* (Data from Population Reference Bureau)

Effects of Age Structure on Population Growth

The number of people younger than age 15 is the major factor determining a country's future population growth.

Any country with many people younger than age 15 (represented by a wide base in Figure 7-8, far left) has a powerful built-in momentum to increase its population size unless death rates rise sharply. The number of births rises even if women have only one or two children because a large number of girls will soon be moving into their reproductive years.

What is perhaps the world's most important population statistic? *Twenty-nine percent of the people on the planet were younger than 15 years old in 2005.* These 1.9 billion young people are poised to move into their prime reproductive years. In developing countries, the number is even higher: 32%, compared with 17% in developed countries. We live in a *demographically divided world*, as shown by population data for the United States, Brazil, and Nigeria (Figure 7-9, p. 136).

Using Age-Structure Diagrams to Make Population and Economic Projections

Changes in the distribution of a country's age groups have long-lasting economic and social impacts.

Between 1946 and 1964, the United States had a *baby boom* that added 79 million people to its population. Over time, this group looks like a bulge moving up

through the country's age structure, as shown in Figure 7-10 (p. 136).

Baby boomers now represent nearly half of all adult Americans. As a result, they dominate the population's demand for goods and services. They are also playing increasingly important roles in deciding who gets elected and what laws are passed. Baby boomers who created the youth market in their teens and twenties are now creating the 50-something market and will soon move on to create a 60-something market. In 2011, the first baby boomers will turn 65, and the number of Americans older than age 65 will consequently grow sharply through 2029.

According to some analysts, the retirement of baby boomers is likely to create a shortage of workers in the United States unless immigrant workers replace some of them. Retired baby boomers are likely to use their political clout to force the smaller number of people in the baby-bust generation that followed them (Figure 7-4) to pay higher income, health-care, and Social Security taxes.

In other respects, the baby-bust generation should have an easier time than the baby-boom generation. Fewer people will be competing for educational opportunities, jobs, and services. Also, labor shortages may drive up their wages, at least for jobs requiring education or technical training beyond high school.

As these projections illustrate, any booms or busts in the age structure of a population create social and economic changes that ripple through a society for decades.

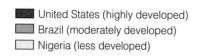

■ United States (highly developed)
■ Brazil (moderately developed)
□ Nigeria (less developed)

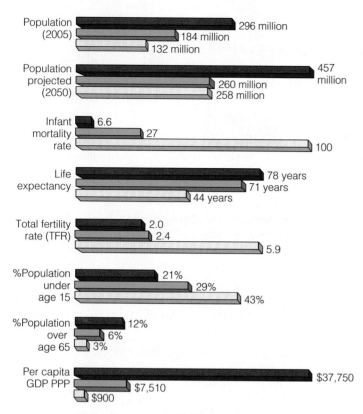

Population
(2005): 296 million / 184 million / 132 million

Population
projected
(2050): 457 million / 260 million / 258 million

Infant
mortality
rate: 6.6 / 27 / 100

Life
expectancy: 78 years / 71 years / 44 years

Total fertility
rate (TFR): 2.0 / 2.4 / 5.9

%Population
under
age 15: 21% / 29% / 43%

%Population
over
age 65: 12% / 6% / 3%

Per capita
GDP PPP: $37,750 / $7,510 / $900

Figure 7-9 Global outlook: comparison of key demographic indicators in highly developed (United States), moderately developed (Brazil), and less developed (Nigeria) countries in 2005. (Data from Population Reference Bureau)

Environmental Science⊕Now™
Examine how the baby boom affects the U.S. age structure over several decades at Environmental ScienceNow.

Economics: Rapid Population Decline from Reduced Fertility

Rapid population decline can lead to long-lasting economic and social problems.

The populations of most countries are projected to grow throughout most of this century. By 2005, however, 40 countries had populations that were either stable (annual growth rates at or below 0.3%) or declining. All, except Japan, are in Europe. Thus about 14% of humanity (896 million people) lives in countries with stable or declining populations. By 2050, according to the UN, the population size of most developed countries (but not the United States) will have stabilized.

As the age structure of the world's population changes and the percentage of people age 60 or older increases, more countries will begin experiencing population declines. If population decline is gradual, its harmful effects usually can be managed.

However, rapid population decline can lead to severe economic and social problems. A country that experiences a "baby bust" or a "birth dearth" has a sharp rise in the proportion of older people. They consume an increasingly larger share of medical care, social security funds, and other costly public services funded by an ever smaller number of working taxpayers. Such countries can also face labor shortages unless they rely more heavily on automation or immigration of foreign workers.

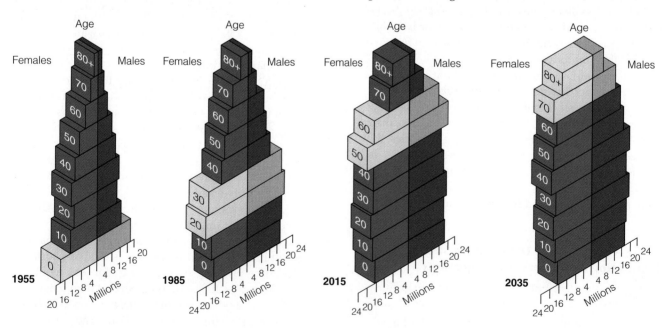

Environmental Science⊕Now™
Active Figure 7-10 Tracking the baby-boom generation in the United States. *See an animation based on this figure and take a short quiz on the concept.* (Data from Population Reference Bureau and U.S. Census Bureau)

Population Decline from a Rising Death Rate: The AIDS Tragedy

A large number of deaths from AIDS disrupts a country's social and economic structure by removing large numbers of young adults from its age structure.

Between 2000 and 2050, AIDS is projected to cause the premature deaths of 278 million people in 53 countries—including 38 countries in Africa. These premature deaths are almost equal to the entire current population of the United States. Read this paragraph again, and think hard about the enormity of this tragedy.

Unlike hunger and malnutrition, which kill mostly infants and children, AIDS kills many young adults. This change in the young-adult age structure of a country has a number of harmful effects. *First*, it produces a sharp drop in average life expectancy. In 8 African countries, where 16–39% of the adult population is infected with HIV, life expectancy could drop to 34–40 years.

Second, it leads to a loss of a country's most productive young adult workers and trained personnel such as scientists, farmers, engineers, teachers, and government, business, and health-care workers. As a result, the number of productive adults available to support the young and the elderly and to grow food and provide essential services declines sharply.

Analysts have called for the international community—especially developed countries—to develop and fund a massive program to help countries ravaged by AIDS in Africa and elsewhere. This program would have two major goals. *First*, it would reduce the spread of HIV through a combination of improved education and health care. *Second*, it would provide financial assistance for education and health care as well as volunteer teachers and health-care and social workers to help compensate for the missing young-adult generation.

7-3 SOLUTIONS: INFLUENCING POPULATION SIZE

Economics: The Demographic Transition

As counties become economically developed, their birth and death rates tend to decline.

Demographers have closely examined the birth and death rates of western European countries that became industrialized during the 19th century. From these data they have developed a hypothesis of population change known as the **demographic transition**: As countries become industrialized, first their death rates and then their birth rates decline. This transition takes place in four distinct stages (Figure 7-11).

Some economists believe that today's developing countries will make the demographic transition over the next few decades. Other population analysts fear that the still-rapid population growth in many developing countries will outstrip economic growth and overwhelm some local life-support systems. As a consequence, many of these countries could become caught in a *demographic trap* at stage 2, the transition stage. This is now happening as death rates rise in a number of developing countries, especially in Africa. Indeed, countries in Africa being ravaged by the HIV/AIDS epidemic are falling back to stage 1.

Analysts also point out that some of the conditions that allowed developed countries to develop in the past are not available to many of today's developing

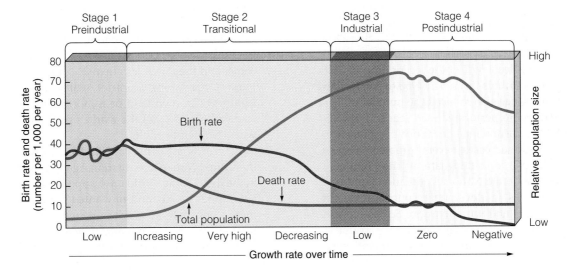

Environmental Science ⊛ Now™

Active Figure 7-11 Generalized model of the *demographic transition*. There is uncertainty over whether this model will apply to some of today's developing countries. *See an animation based on this figure and take a short quiz on the concept.*

countries. One problem is a shortage of skilled workers needed to produce the high-tech products necessary to compete in today's global economy. Another is a lack of capital and other resources that allow rapid economic development.

Two other problems hinder economic development in many developing countries. The first obstacle is a sharp rise in their debt to developed countries. Such countries devote much of their income to paying the interest on their debts. This leaves too little money for improving social, health, and environmental conditions.

A second problem is that developing countries now receive less economic assistance from developed countries. Indeed, since the mid-1980s, developing countries have paid developed countries $40–50 billion per year (mostly in debt interest) more than they have received from these countries.

Environmental Science ⊕ Now™
Explore the effects of economic development on birth and death rates and population growth at Environmental ScienceNow.

Family Planning: Planning for Babies Works

Family planning has been a major factor in reducing the number of births and abortions throughout most of the world.

Family planning provides educational and clinical services that help couples choose how many children to have and when to have them. Such programs vary from culture to culture, but most provide information on birth spacing, birth control, and health care for pregnant women and infants.

Family planning has helped increase the proportion of married women in developing countries who use modern forms of contraception from 10% of married women of reproductive age in the 1960s to 51% of these women in 2005. In addition, family planning is responsible for at least 55% of the recent drop in TFRs in developing countries, from 6 in 1960 to 3.0 in 2005. Family planning has also reduced the number of legal and illegal abortions performed each year and lowered the risk of maternal and fetal death from pregnancy.

Despite such successes, several problems remain. *First*, according to John Bongaarts of the Population Council and the United Nations Population Fund, 42% of all pregnancies in developing countries are unplanned and 26% end with abortion. *Second*, an estimated 150 million women in developing countries want to limit the number and determine the spacing of their children, but they lack access to contraceptive services. According to the United Nations, extending family planning services to these women and to others who will soon be entering their reproductive years

could prevent an estimated 5.8 million births per year and more than 5 million abortions per year!

Some analysts call for expanding family planning programs to include teenagers and sexually active unmarried women, who are excluded from many existing programs. For teenagers, many advocate much greater emphasis on abstinence.

Another suggestion is to develop programs that educate men about the importance of having fewer children and taking more responsibility for raising them. Proponents also call for greatly increased research on developing more effective and more acceptable birth control methods for men.

Finally, a number of analysts urge pro-choice and pro-life groups to join forces in greatly reducing unplanned births and abortions, especially among teenagers.

Empowering Women: Ensuring Education, Jobs, and Human Rights

Women tend to have fewer children if they are educated, hold a paying job outside the home, and do not have their human rights suppressed.

Three key factors lead women to have fewer and healthier children: education, paying jobs outside the home, and living in societies where their rights are not suppressed.

Women make up roughly half of the world's population. They do almost all of the world's domestic work and child care for little or no pay. In addition, they provide more unpaid health care than all of the world's organized health services combined.

Women also do 60–80% of the work associated with growing food, gathering fuelwood, and hauling water in rural areas of Africa, Latin America, and Asia. As one Brazilian woman put it, "For poor women the only holiday is when you are asleep."

Globally, women account for two-thirds of all hours worked but receive only 10% of the world's income, and they own less than 2% of the world's land. In most developing countries, women do not have the legal right to own land or to borrow money. Women also make up 70% of the world's poor and 60% of the world's illiterate adults.

According to Thorya Obaid, executive director of the United Nations Population Agency, "Many women in the developing world are trapped in poverty by illiteracy, poor health, and unwanted high fertility. All of these contribute to environmental degradation and tighten the grip of poverty. If we are serious about sustainable development, we must break this vicious cycle."

That means giving women everywhere full legal rights and the opportunity to become educated and earn income outside the home. Achieving these goals would slow population growth, promote human rights and freedom, reduce poverty, and slow environmental degradation—a win–win result.

Empowering women by seeking gender equality will require some major social changes. Although it will be difficult to achieve in male-dominated societies, it can be done.

Good news. An increasing number of women in developing countries are taking charge of their lives and reproductive behavior—they are not waiting for the slow processes of education and cultural change. Such bottom-up change by individual women will play an important role in stabilizing population and providing women with equal rights.

Solutions: Reducing Population Growth

Experience suggests that the best way to slow population growth is a combination of investing in family planning, reducing poverty, and elevating the status of women.

In 1994, the United Nations held its third Conference on Population and Development in Cairo, Egypt. One of the conference's goals was to encourage action to stabilize the world's population at 7.8 billion by 2050 instead of the projected 8.9 billion.

The major goals of the resulting population plan, endorsed by 180 governments, are to do the following by 2015:

- Provide universal access to family planning services and reproductive health care

- Improve health care for infants, children, and pregnant women

- Develop and implement national population polices

- Improve the status of women and expand educational and job opportunities for young women

- Provide more education, especially for girls and women

- Increase the involvement of men in child-rearing responsibilities and family planning

- Sharply reduce poverty

- Sharply reduce unsustainable patterns of production and consumption

The experiences of Japan, Thailand, South Korea, Taiwan, Iran, and China indicate that a country can achieve or come close to replacement-level fertility within a decade or two. Such experiences also suggest that the best way to slow population growth is through the combination of *investing in family planning*, *reducing poverty*, and *elevating the status of women*.

7-4 SLOWING POPULATION GROWTH IN INDIA AND CHINA

Case Study: India

For more than five decades, India has tried to control its population growth with only modest success.

The world's first national family planning program began in India in 1952, when its population was nearly 400 million. In 2005, after 53 years of population control efforts, India was the world's second most populous country, with a population of 1.1 billion.

In 1952, India added 5 million people to its population. In 2005, it added 18 million. Figure 7-12 compares demographic data for India and China.

India faces a number of already serious poverty, malnutrition, and environmental problems that could worsen as its population continues to grow rapidly. By global standards, one of every four people in India is poor. Nearly half of the country's labor force is unemployed or can find only occasional work.

India currently is self-sufficient in food grain production. Nevertheless, 40% of its population and more than half of its children suffer from malnutrition, mostly because of poverty.

Furthermore, India faces critical resource and environmental problems. With 17% of the world's people, it has just 2.3% of the world's land resources and 2% of the world's forests. About half of the country's cropland is degraded as a result of soil erosion, waterlogging,

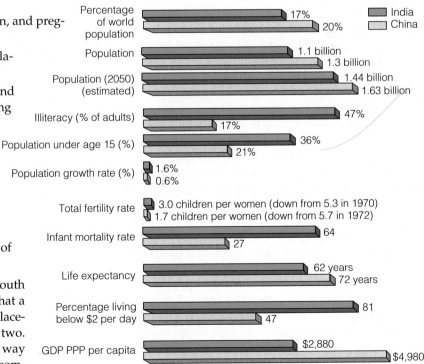

Figure 7-12 Global outlook: basic demographic data for India and China in 2005. (Data from United Nations and Population Reference Bureau)

salinization, overgrazing, and deforestation. In addition, more than two-thirds of its water is seriously polluted and sanitation services often are inadequate.

India's huge and growing middle class is larger than the entire U.S. population. As these members of the middle class increase their resource use per person, India's ecological footprint will expand and increase the pressure on the country's and the earth's natural capital.

Without its long-standing family planning program, India's population and environmental problems would be growing even faster. Still, to its supporters the results of the program have proved disappointing for several reasons: poor planning, bureaucratic inefficiency, the low status of women (despite constitutional guarantees of equality), extreme poverty, and lack of administrative and financial support.

The government has provided information about the advantages of small families for years. Even so, Indian women have an average of 3.0 children. Most poor couples still believe they need many children to do work and care for them in old age. The country's strong cultural preference for male children also means some couples keep having children until they produce one or more boys. The result: Even though 90% of Indian couples know of at least one modern birth control method, only 43% actually use one.

Case Study: China

Since 1970, China has used a government-enforced program to cut its birth rate in half and sharply reduce its fertility rate.

Since 1970, China has made impressive efforts to feed its people and bring its population growth under control. Between 1972 and 2005, the country cut its crude birth rate in half and trimmed its TFR from 5.7 to 1.6 children per woman (see Figure 7-12).

To achieve its sharp drop in fertility, China has established the world's most extensive, intrusive, and strict population control program. Couples are strongly urged to postpone marriage and to have no more than one child. Married couples who pledge to have no more than one child receive extra food, larger pensions, better housing, free medical care, salary bonuses, free school tuition for their child, and preferential treatment in employment when their child enters the job market. Couples who break their pledge lose such benefits.

The government also provides married couples with ready access to free sterilization, contraceptives, and abortion. As a consequence, 86% of married women in China use modern contraception. In the 1960s, government officials realized that the only alternative to strict population control was mass starvation.

In China (as in India), there is a strong preference for male children because of the lack of a real social security system. A folk saying goes, "Rear a son, and protect yourself in old age." Many pregnant women use ultrasound to determine the gender of their fetus and often get an abortion if it is female. The result: a growing *gender imbalance* in China's population, with a projected 30–40 million surplus of men expected by 2020.

China has 20% of the world's population, but only 7% of the world's fresh water and cropland, 4% of its forests, and 2% of its oil. Soil erosion in China is serious and apparently getting worse.

Population experts expect China's population to peak around 2040, then begin a slow decline. This projection has led some members of China's parliament to call for amending the country's one-child policy so that some urban couples can have a second child. The goal would be to provide more workers to help support China's aging population.

China's economy is growing at one of the world's highest rates as the country undergoes rapid industrialization. Its burgeoning middle class will consume more resources per person, increase China's ecological footprint, and increase the strain on the country's and the earth's natural capital.

What lesson can other countries learn from China? They should try to curb population growth before they must choose between mass starvation and coercive measures that severely restrict human freedom.

7-5 POPULATION DISTRIBUTION: URBANIZATION AND URBAN GROWTH

Factors Affecting Urban Growth: Moving to the City

Urban populations are growing rapidly throughout the world, and many cities in developing countries have become centers of poverty.

Almost half of the world's people live in densely populated urban areas. Rural people are *pulled* to urban areas in search of jobs, food, housing, a better life, entertainment, and freedom from religious, racial, and political conflicts. Some are also *pushed* from rural areas into urban areas by factors such as poverty, lack of land to grow food, declining agricultural jobs, famine, and war.

Five major trends are important in understanding the problems and challenges of urban growth. First, *the proportion of the global population living in urban areas is increasing.* Between 1850 and 2005, the percentage of people living in urban areas increased from 2% to 48%. According to UN projections, 60% of the world's people will live in urban areas by 2030. Thus, between

2005 and 2030, the world's urban population is projected to increase from 3.1 billion to 5 billion. Almost all of this growth will occur in already overcrowded cities in developing countries (Figure 7-13).

Second, *the number of large cities is mushrooming.* In 2005, more than 400 cities had 1 million or more people, and this number is projected to increase to 564 cities by 2015. Today 18 *megacities* or *meagalopolises* (up from 8 in 1985) are home to 10 million or more people each—most of them in developing countries (Figure 7-13). As they grow and sprawl outward, separate urban areas may merge to form a *megalopolis.* For example, the remaining open space between Boston, Massachusetts, and Washington, D.C., is rapidly urbanizing and coalescing. The result is an almost 800-kilometer-long (500-mile-long) urban area that is sometimes called *Bowash* (Figure 7-14).

Third, *the urban population is increasing rapidly in developing countries.* Between 2005 and 2030, the percentage of people living in urban areas in developing countries is expected to increase from 41% to 56%.

Fourth, *urban growth is much slower in developed countries* (with 76% urbanization) *than in developing countries.* Developed countries are projected to reach 84% urbanization by 2030.

Fifth, *poverty is becoming increasingly urbanized as more poor people migrate from rural to urban areas, mostly in developing countries.* The United Nations estimates

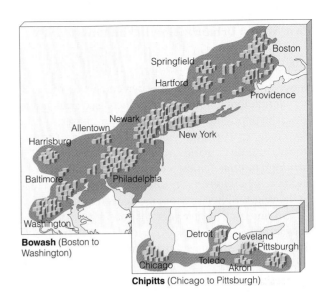

Bowash (Boston to Washington)

Chipitts (Chicago to Pittsburgh)

Figure 7-14 Two megalopolises: *Bowash*, consisting of urban sprawl and coalescence between Boston and Washington, D.C., and *Chipitts*, extending from Chicago to Pittsburgh.

that at least 1 billion people live in crowded *slums* (tenements and rooming houses where 3–6 people live in a single room) of central cities and in *squatter settlements* and *shantytowns* (where people build shacks from scavenged building materials) that surround the outskirts of most cities in developing countries.

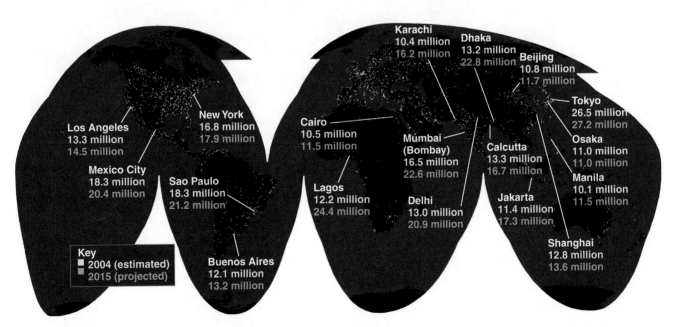

Figure 7-13 Global outlook: major urban areas throughout the world based on satellite images of the earth at night that show city lights. Currently, the 48% of the world's people living in urban areas occupy about 2% of the earth's land area. Note that most of the world's urban areas are found along the coasts of continents, and most of Africa and much of the interior of South America, Asia, and Australia are dark at night. This figure also shows the populations of the world's 18 *megacities* with 10 or more million people in 2004, and their projected populations in 2015. All but four are located in developing countries. (Data from National Geophysics Data Center, National Oceanic and Atmospheric Administration, and United Nations)

Case Study: Urbanization in the United States

Eight of every ten Americans live in urban areas, about half of them in sprawling suburbs.

Between 1800 and 2005, the percentage of the U.S. population living in urban areas increased from 5% to 79%. The population has shifted in four phases.

First, *people migrated from rural areas to large central cities.* Currently, three-fourths of Americans live in 271 *metropolitan areas* (cities with at least 50,000 people), and nearly half live in consolidated metropolitan areas containing 1 million or more residents (Figure 7-15).

Second, *many people migrated from large central cities to suburbs and smaller cities.* Currently, about 51% of Americans live in the suburbs and 30% live in central cities.

Third, *many people migrated from the North and East to the South and West.* Since 1980, about 80% of the U.S. population increase has occurred in the South and West, particularly near the coasts. California in the West, with 34.5 million people, is the most populous state, followed by Texas in the Southwest with 21.3 million residents. This shift is expected to continue.

Fourth, *some people have migrated from urban and suburban areas back to rural areas* since the 1970s, and especially since 1990. The result is rapid growth of *exurbs*, vast sprawling areas that are not related to central cities and have no center.

During these shifts in the last century, the quality of life for most Americans improved significantly (Figure 7-5). Since 1920, many of the worst urban envi-

Figure 7-15 Major urban areas in the United States based on satellite images of the earth at night that show city lights (top). About 8 out of 10 Americans live in urban areas that occupy about 1.7% of the land area of the lower 48 states. Areas with names in white are the fastest-growing metropolitan areas. Nearly half (48%) of Americans live in *consolidated metropolitan areas* with 1 million or more people that are projected to grow and merge into huge urban areas shown as shaded areas in the bottom map. (Data from National Geophysical Data Center/National Oceanic and Atmospheric Administration, U.S. Census Bureau)

ronmental problems in the United States have been reduced significantly. Most people have better working and housing conditions, and air and water quality have improved. Better sanitation, public water supplies, and medical care have slashed death rates and the prevalence of sickness from malnutrition and infectious diseases. Concentrating most of the population in urban areas has also helped protect the country's biodiversity by reducing the destruction and degradation of wildlife habitat.

However, a number of U.S. cities—especially older ones—have *deteriorating services* and *aging infrastructures* (streets, schools, bridges, housing, and sewers). Many face *budget crunches* from rising costs as some businesses and people move to the suburbs or exurbs areas and reduce revenues from property taxes. There is also *rising poverty* in the centers of many older cities, where unemployment rates are typically 50% or higher.

Urban Sprawl

Where land is ample and affordable, urban areas tend to sprawl outward, swallowing up the surrounding countryside.

A major problem in the United States and some other countries with lots of room for expansion is **urban sprawl** (Figure 7-16). Growth of low-density development on the edges of cities and towns gobbles up the surrounding countryside—frequently prime farmland or forests—and increases dependence on cars. The result is a far-flung hodgepodge of housing developments, shopping malls, parking lots, and office complexes—loosely connected by multilane highways and freeways. Urban sprawl is the product of increased prosperity, ample and affordable land, automobiles, cheap gasoline, and poor urban planning.

Figure 7-17 (p. 144) shows some of the undesirable consequences of urban sprawl. Sprawl has increased travel time in automobiles, decreased energy efficiency, increased urban flooding problems, and destroyed prime cropland, forests, open space, and wetlands. It has also led to the economic death of many central cities.

To pay for heavily mortgaged houses and cars, adults in a typical suburban family spend most of their nonworking hours driving to and from work or running errands over a vast suburban landscape. Many have little energy and time left for their children or themselves, or getting to know their neighbors. On the other hand, many people prefer living in sprawling exurbs that are not dependent on a central city for jobs, shopping, or entertainment.

Environmental Science ⊛ Now™
Examine how the San Francisco Bay area grew in population between 1900 and 1990 at Environmental ScienceNow.

1952

1967

1972

1995

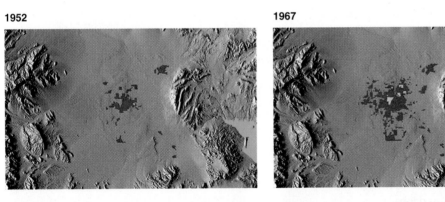

Images provided courtesy of the U.S. Geological Survey

Figure 7-16 Natural capital degradation: *urban sprawl* in and around Las Vegas, Nevada, 1952–1995. Between 1970 and 2004, the population of water-short Clark County (which includes Las Vegas) more than quadrupled from 463,000 to 2 million. This growth is expected to continue.

Natural Capital Degradation

Urban Sprawl

Land and Biodiversity	Human Health and Aesthetics	Water	Energy, Air, and Climate	Economic Effects
Loss of cropland	Contaminated drinking water and air	Increased runoff	Increased energy use and waste	Higher taxes
Loss of forests and grasslands	Weight gain	Increased surface water and groundwater pollution	Increased air pollution	Decline of downtown business districts
Loss of wetlands	Noise pollution	Increased use of surface water and groundwater	Increased greenhouse gas emissions	Increased unemployment in central city
Loss and fragmentation of wildlife habitats	Sky illumination at night	Decreased storage of surface water and groundwater	Enhanced global warming	Loss of tax base in central city
Increased wildlife roadkill	Traffic congestion	Increased flooding	Warmer microclimate (urban heat island effect)	
Increased soil erosion		Decreased natural sewage treatment		

Figure 7-17 Natural capital degradation: some undesirable impacts of urban sprawl or car-dependent development. Do you live in an area suffering from urban sprawl?

7-6 URBAN RESOURCE AND ENVIRONMENTAL PROBLEMS

Advantages of Urbanization

Urban areas can offer more job opportunities and better education and health, and can help protect biodiversity by concentrating people.

Urbanization has many benefits. From an *economic standpoint*, cities are centers of economic development, education, technological developments, and jobs. They serve as centers of industry, commerce, and transportation. However, this is changing as suburban and exurban areas not dependent on central cities grow.

In terms of *health*, urban residents in many parts of the world live longer and have lower infant mortality rates and fertility rates than do rural populations. In addition, urban dwellers generally have better access to medical care, family planning, education, and social services than do their rural counterparts.

Urban areas also enjoy some environmental advantages. For example, recycling is more economically feasible because concentrations of recyclable materials and per capita expenditures on environmental protection are higher in urban areas. Also, concentrating people in urban areas helps preserve biodiversity by reducing the stress on wildlife habitats.

Disadvantages of Urbanization

Cities are rarely self-sustaining. They threaten biodiversity, lack trees, grow little of their food, concentrate pollutants and noise, spread infectious diseases, and are centers of poverty, crime, and terrorism.

Although urban populations occupy only about 2% of the earth's land area, they consume three-fourths of its resources. Because of this high rate of consumption and their high waste output (Figure 7-18), most of the world's cities are not self-sustaining systems.

Urbanization can help preserve biodiversity in some areas. On the other hand, large areas of land must be disturbed and degraded to provide urban dwellers with food, water, energy, minerals, and other resources. This activity decreases and degrades the earth's biodiversity. As cities expand and sprawl out-

144   CHAPTER 7   Applying Population Ecology: The Human Population

ward (Figure 7-16), they destroy rural cropland, fertile soil, forests, wetlands, and wildlife habitats. At the same time, most provide little of the food they use. From an environmental standpoint, urban areas are somewhat like gigantic vacuum cleaners, sucking up much of the world's matter, energy, and living resources and spewing out pollution, wastes, and heat. As a consequence, they have large ecological footprints that extend far beyond their boundaries. If you live in a city, you can calculate its ecological footprint by going to the website **www.redefiningprogress.org/**. Also, see the Guest Essay on this topic by Michael Cain on this chapter's website.

In urban areas, most trees, shrubs, or other plants are destroyed to make way for buildings, roads, and parking lots. As a result, most cities do not benefit from vegetation that might otherwise absorb air pollutants, give off oxygen, help to cool the air through transpiration, provide shade, reduce soil erosion, muffle noise, provide wildlife habitats, and give aesthetic pleasure. As one observer remarked, "Most cities are places where they cut down most of the trees and then name the streets after them."

As cities grow and their water demands increase, expensive reservoirs and canals must be built and deeper wells drilled. This activity can deprive rural and wild areas of surface water and deplete groundwater faster than it is replenished.

Flooding also tends to be greater in central cities and their suburbs, sometimes because they are often built on floodplain areas or along low-lying coastal areas subject to natural flooding. Covering land with buildings, asphalt, and concrete can also cause precipitation to run off quickly and overload storm drains. In addition, urban development and sprawl often destroys or degrades wetlands that act as natural sponges to help absorb excess water. Many of the world's largest cities face another threat: They are located in coastal areas (Figure 7-13) that could be flooded sometime in this century if sea levels rise as projected due to global warming.

Because of their high population densities and high resource consumption, urban dwellers produce most of the world's air pollution, water pollution, and solid and hazardous wastes. Pollutant levels in urban areas are generally higher than in rural areas because pollution is produced in a smaller area and cannot be dispersed and diluted as readily as pollution produced in rural areas. In addition, high population densities in urban areas can increase the spread of *infectious diseases*

Figure 7-18 Natural capital degradation: urban areas are rarely sustainable systems. The typical city depends on large nonurban areas for huge inputs of matter and energy resources and for large outputs of waste matter and heat. According to an analysis by Mathis Wackernagel and William Rees, an area 58 times as large as that of London is needed to supply its residents with resources. They estimate that meeting the needs of all the world's people at the same rate of resource use as that of London would take at least three more earths.

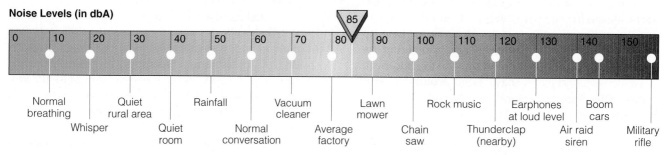

Permanent damage begins after 8-hour exposure

Noise Levels (in dbA)

| 0 | 10 | 20 | 30 | 40 | 50 | 60 | 70 | 80 | 85 | 90 | 100 | 110 | 120 | 130 | 140 | 150 |

Normal breathing

Whisper

Quiet rural area

Quiet room

Rainfall

Normal conversation

Vacuum cleaner

Average factory

Lawn mower

Chain saw

Rock music

Thunderclap (nearby)

Earphones at loud level

Air raid siren

Boom cars

Military rifle

Figure 7-19 *Noise levels* (in decibel-A [dbA] sound pressure units) of some common sounds. You are being exposed to a sound level high enough to cause permanent hearing damage if you need to raise your voice to be heard above the racket, if a noise causes your ears to ring, or if nearby speech seems muffled. Prolonged exposure to lower noise levels and occasional loud sounds may not damage your hearing but can greatly increase internal stress. Noise pollution can be reduced by modifying noisy activities and devices, shielding noisy devices or processes, shielding workers or other receivers from the noise, moving noisy operations or things away from people, and using antinoise (a technology that cancels out one noise with another).

United Nations

Figure 7-20 **Global outlook:** extreme poverty forces hundreds of millions of people to live in slums such as this one in Rio de Janeiro, Brazil, where adequate water supplies, sewage disposal, and other services don't exist.

(especially if adequate drinking water and sewage systems are not available), physical injuries (mostly from industrial and traffic accidents), and excessive noise (Figure 7-19).

Cities generally are warmer, rainier, foggier, and cloudier than suburbs and nearby rural areas. The enormous amounts of heat generated by cars, factories, furnaces, lights, air conditioners, and heat-absorbing dark roofs and roads in cities create an *urban heat island* that is surrounded by cooler suburban and rural areas. As cities grow and merge (Figure 7-14), their heat islands may merge and keep polluted air from being diluted and cleansed.

The artificial light created by cities also hinders astronomers from conducting their research and affects various plant and animal species. Species vulnerable to *light pollution* include endangered sea turtles that lay their eggs on beaches at night and migrating birds that are lured off course by the lights of high-rise buildings and fatally collide with them.

Urban areas can intensify poverty and social problems. Crime rates also tend to be higher in these areas than in rural areas (Connections, right). And, of course, urban areas are more likely and desirable targets for terrorist acts.

Economics Case Study: The Urban Poor in Developing Countries

Most of the urban poor in developing countries live in crowded, unhealthy, and dangerous conditions, but many are better off than the rural poor.

Many of the world's poor live in *squatter settlements* and *shantytowns* on the outskirts of most cities in developing countries (Figure 7-20), some perched precariously on steep hillsides subject to landslides. In these illegal settlements, people take over unoccupied land and build shacks from corrugated metal, plastic sheets, scrap wood, packing crates, and other scavenged building materials. Still others live or sleep on the streets, having nowhere else to go.

Squatters living near the edge of survival in these areas usually lack clean water supplies, sewers, electricity, and roads, and they often are subject to severe air and water pollution and hazardous wastes from nearby factories. Their locations may also be especially prone to landslides, flooding, earthquakes, or volcanic eruptions.

Most cities cannot afford to provide squatter settlements and shantytowns with basic services and protections, and their officials fear that improving services will attract even more of the rural poor. In fact, many city governments regularly bulldoze squatter shacks and send police to drive the illegal settlers out. The people then move back in or develop another shantytown somewhere else.

X *HOW WOULD YOU VOTE?* Should squatters around cities of developing countries be given title to land they do not own? Cast your vote online at http://biology.brookscole.com/miller11.

Despite joblessness, squalor, overcrowding, and environmental and health hazards, most of these poor urban residents are better off than their rural counterparts. Thanks to the greater availability of family planning programs, they tend to have fewer children and better access to schools. Many squatter settlements provide a sense of community and a vital safety net of neighbors, friends, and relatives for the poor.

Mexico City is an example of an urban area in crisis. About 18.3 million people—roughly one of every six Mexicans—live there (Figure 7-13). It is the world's second most populous city, and each year at least 200,000 new residents arrive.

Mexico City suffers from severe air pollution, close to 50% unemployment, deafening noise, overcrowding, traffic congestion, inadequate public transportation, and a soaring crime rate. More than one-third of its residents live in slums called *barrios* or in squatter settlements that lack running water and electricity.

At least 3 million people have no sewer facilities. As a consequence, huge amounts of human waste are deposited in gutters, vacant lots, and open sewers every day, attracting armies of rats and swarms of flies. When the winds pick up dried excrement, a *fecal snow* blankets parts of the city. Open garbage dumps also contribute dust and bacteria to the atmosphere. This bacteria-laden fallout leads to widespread salmonella and hepatitis infections, especially among children.

How Can Reducing Crime Help the Environment?

CONNECTIONS

Most people do not realize that reducing crime can also help improve environmental quality. For example, events such as robbery, assault, and shootings can have several harmful environmental effects.

Crime can drive people out of cities, which are our most energy-efficient living arrangements. Every brick in an abandoned urban building represents an energy waste equivalent to burning a 100-watt light bulb for 12 hours. Each new suburb means replacing farmland or reservoirs of natural biodiversity with dispersed, energy- and resource-wasting roads, houses, and shopping centers.

Crime can also make people less willing to walk, bicycle, and use energy-efficient public transit systems. It forces many people to use more energy to deter burglars. For example, trees and bushes planted near a house help save energy by reducing solar heat gain in the summer and providing windbreaks in the winter. To thwart break-ins, many homeowners clear away those trees and bushes. Many homeowners also use more energy by leaving lights, TVs, and radios on to deter burglars.

Finally, the threat of crime causes overpackaging of many items to deter shoplifting or poisoning of food or drug items.

Critical Thinking

Can you think of any environmental benefits of certain types of crimes?

Mexico City has one of the world's worst photochemical smog problems because of a combination of too many cars and polluting industries, a sunny climate, and topographical bad luck. The city sits in a high-elevation bowl-shaped valley surrounded on three sides by mountains—conditions that trap air pollutants at ground level. Breathing its air is said to be roughly equivalent to smoking three packs of cigarettes per day.

The city's air and water pollution cause an estimated 100,000 premature deaths per year. Writer Carlos Fuentes has nicknamed this megacity "Makesicko City."

Some progress has been made. The percentage of days each year in which air pollution standards are violated has fallen from 50% to 20%. At the same time, the city has an inadequate mass transportation system and retains weak, poorly enforced air pollution standards for industries and motor vehicles.

7-7 TRANSPORTATION AND URBAN DEVELOPMENT

Land Availability, Transportation Systems, and Urban Development

Land availability determines whether a city must grow vertically or spread out horizontally and whether it relies mostly on mass transportation or the automobile.

If a city cannot spread outward, it must grow vertically—upward and downward (below ground)—so that it occupies a small land area with a high population density. Most people living in *compact cities* like Hong Kong and Tokyo walk, ride bicycles, or use energy-efficient mass transit.

A combination of cheap gasoline, plentiful land, and a network of highways produce *dispersed cities.* These are found in countries such the United States, Canada, and Australia, where ample land often is available for outward expansion. Sprawling cities depend on the automobile for most travel, which has a number of undesirable effects (Figure 7-17). Nevertheless, motor vehicles are increasing in both compact and dispersed cities.

Case Study: Motor Vehicles in the United States

Passenger vehicles account for almost all urban transportation in the United States. Each year Americans drive as far as everyone else in the world combined.

America showcases the advantages and disadvantages of living in a society dominated by motor vehicles. With 4.6% of the world's people, the United States has almost one-third of the world's motor vehicles, with one-third of them being gas guzzling sport utility vehicles (SUVs), pickup trucks, and vans.

Mostly because of urban sprawl and convenience, passenger vehicles are used for 98% of all urban transportation and 91% of travel to work in the United States. About 75% of Americans drive to work alone, 5% commute to work on public transit, and 0.5% bicycle to work. Each year Americans drive about the same distance driven by all other drivers in the world and in the process use about 43% of the world's gasoline! According to the American Public Transit System, if Americans increased their use of mass transit from the current rate of 5% to 10%, it would reduce U.S. dependence on oil by 40%.

Many governments in rapidly industrializing countries such as China want to develop an automobile-centered transportation system like that in the United States. Suppose China succeeds in having one or two cars in every garage and consumes oil at the

U.S. rate. According to environmental leader Lester R. Brown, China would need slightly more oil each year than the world now produces and would have to pave an area equal to half of the land it now uses to produce food.

Advantages and Disadvantages of Motor Vehicles

Motor vehicles provide personal benefits and help fuel economies, but they also kill many people, pollute the air, promote urban sprawl, and lead to time- and gas-wasting traffic jams.

On a personal level, motor vehicles provide mobility and offer a convenient and comfortable way to get from one place to another. They also are symbols of power, sex, social status, and success for many people. For some, they provide temporary escape from an increasingly hectic world.

From an economic standpoint, much of the world's economy is built on producing motor vehicles and supplying roads, services, and repairs for them. In the United States, for example, $1 of every $4 spent and one of every six nonagricultural jobs is connected to the automobile.

Despite their important benefits, motor vehicles have many harmful effects on people and the environment. They have killed almost 18 million people since 1885, when Karl Benz built the first automobile. Throughout the world, they kill approximately 1.2 million people each year—an average of 3,300 deaths per day—and injure another 15 million people.

In the United States, motor vehicle accidents kill more than 43,000 people per year and injure another 5 million, at least 300,000 of them severely. *Car accidents have killed more Americans than all wars in the country's history.*

Motor vehicles are the world's largest source of air pollutants, which cause 30,000–60,000 premature deaths per year in the United States, according to the Environmental Protection Agency. They are also the fastest-growing source of climate-changing carbon dioxide emissions—now producing almost one-fourth of them.

Motor vehicles have helped create urban sprawl. At least a third of urban land worldwide and half in the United States is devoted to roads, parking lots, gasoline stations, and other automobile-related uses. This fact prompted urban expert Lewis Mumford to suggest that the U.S. national flower should be the concrete cloverleaf.

Another problem is congestion. If current trends continue, U.S. motorists will spend an average of two years of their lives in traffic jams. Building more roads may not be the answer. Many analysts agree with economist Robert Samuelson that "cars expand to fill available concrete."

Solutions: Reducing Automobile Use

Although it would be politically unpopular, we could reduce our reliance on automobiles by making users pay for their harmful effects.

Some environmentalists and economists suggest that one way to reduce the harmful effects of automobile use is to make drivers pay directly for most of the damage they cause—a *user-pays* approach based on honest environmental accounting. One way to phase in such full-cost pricing would be to charge a tax on gasoline that covers the estimated harmful costs of driving. Such taxes would amount to about $1.30–2.10 per liter ($5–8 per gallon) of gasoline in the United States and would spur the use of more energy-efficient motor vehicles and mass transit.

Proponents of this approach urge governments to use gasoline tax revenues to help finance mass transit systems, bike paths, and sidewalks. The government could reduce taxes on income and wages to offset the increased taxes on gasoline, thereby making this *tax shift* more politically acceptable. Another way to reduce automobile use and congestion would be to raise parking fees and charge tolls on roads, tunnels, and bridges—especially during peak traffic times.

Most analysts doubt that these approaches would be feasible in the United States, for three reasons. *First*, they face strong political opposition from two groups: the public, which is largely unaware of the huge hidden costs they are already paying, and powerful transportation-related industries such as oil and tire companies, road builders, carmakers, and many real estate developers. However, taxpayers might accept sharp increases in gasoline taxes if the extra costs were offset by decreases in taxes on wages and income.

Second, fast, efficient, reliable, and affordable mass transit options and bike paths are not widely available in most of the United States. In addition, the dispersed nature of most U.S. urban areas makes people dependent on cars.

Third, most people who can afford cars are virtually addicted to them.

Solutions: Alternatives to the Car

Alternatives include walking, bicycling, and taking subways, trolleys, trains, and buses.

Several alternatives to motor vehicles exist, each with its own advantages and disadvantages. Examples include *bicycles* (Figure 7-21), *mass transit rail systems in urban areas* (Figure 7-22), *bus systems in urban areas* (Figure 7-23, p. 150), and *rapid rail systems between urban areas* (Figure 7-24, p. 150).

Figure 7-21 Trade-offs: advantages and disadvantages of *bicycles. Critical thinking: pick the single advantage and disadvantage that you think are the most important.*

Figure 7-22 Trade-offs: advantages and disadvantages of *mass transit rail systems in urban areas. Critical thinking: pick the single advantage and disadvantage that you think are the most important.*

HOW WOULD YOU VOTE? Should urban areas in the country where you live emphasize developing a mass transportation system or developing a bus system? Cast your vote online at http://biology.brookscole.com/miller1.

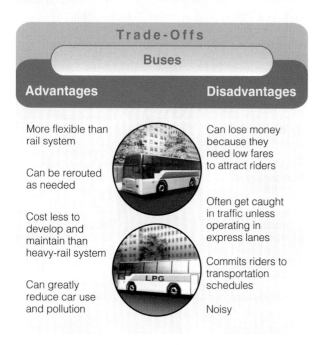

Trade-Offs

Buses

Advantages	Disadvantages
More flexible than rail system	Can lose money because they need low fares to attract riders
Can be rerouted as needed	
Cost less to develop and maintain than heavy-rail system	Often get caught in traffic unless operating in express lanes
	Commits riders to transportation schedules
Can greatly reduce car use and pollution	Noisy

Figure 7-23 Trade-offs: advantages and disadvantages of *bus systems in urban areas. Critical thinking: pick the single advantage and disadvantage that you think are the most important.*

HOW WOULD YOU VOTE? Should the United States (or the country where you live) develop rapid rail systems between urban areas that are less than 960 kilometers (600 miles) apart even though this will be quite costly? Cast your vote online at http://biology.brookscole.com/miller11.

7-8 MAKING URBAN AREAS MORE LIVABLE AND SUSTAINABLE

Solutions: Smart Growth

Smart growth can control growth patterns, discourage urban sprawl, reduce car dependence, and protect ecologically sensitive areas.

Smart growth is emerging as a means to encourage more environmentally sustainable development that requires less dependence on cars, controls and directs sprawl, and reduces wasteful resource use. It recognizes that urban growth will occur. At the same time, it uses zoning laws and other tools to channel growth into areas where it can cause less harm, discourage sprawl, protect ecologically sensitive and important lands and waterways, and develop more environmentally sustainable urban areas and neighborhoods that are more enjoyable places to live. Figure 7-25 lists popular smart growth tools. Are any of them being employed in your community?

Some communities are using the principles of *new urbanism* to develop entire villages and recreate mixed neighborhoods within existing cities. These principles include *walkability*, with most things being located within a 10-minute walk of home and work; *mixed use and diversity*, which seeks a mix of pedestrian-friendly shops, offices, apartments, and homes and people of different ages, classes, cultures, and races; *quality urban design* emphasizing beauty, aesthetics, and architectural diversity; *environmental sustainability* based on development with minimal environmental impact; and *smart transportation* in which high-quality trains connect neighborhoods, towns, and cities. The goal is to create places that uplift, enrich, and inspire the human spirit.

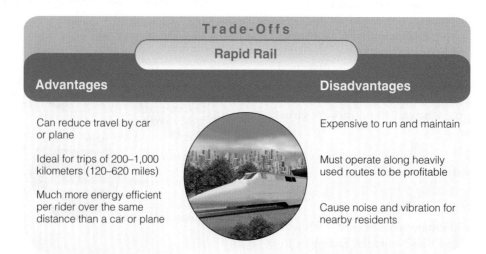

Trade-Offs

Rapid Rail

Advantages	Disadvantages
Can reduce travel by car or plane	Expensive to run and maintain
Ideal for trips of 200–1,000 kilometers (120–620 miles)	Must operate along heavily used routes to be profitable
Much more energy efficient per rider over the same distance than a car or plane	Cause noise and vibration for nearby residents

Figure 7-24 Trade-offs: advantages and disadvantages of *rapid rail systems between urban areas. Critical thinking: pick the single advantage and disadvantage that you think are the most important.*

Limits and Regulations

Limit building permits

Urban growth boundaries

Greenbelts around cities

Public review of new development

Protection

Preserve existing open space

Buy new open space

Buy development rights that prohibit certain types of development on land parcels

Zoning

Encourage mixed use

Concentrate development along mass transportation routes

Promote high-density cluster housing developments

Taxes

Tax land, not buildings

Tax land on value of actual use (such as forest and agriculture) instead of highest value as developed land

Tax Breaks

For owners agreeing legally to not allow certain types of development (conservation easements)

For cleaning up and developing abandoned urban sites (brownfields)

Planning

Ecological land-use planning

Environmental impact analysis

Integrated regional planning

State and national planning

Revitalization and New Growth

Revitalize existing towns and cities

Build well-planned new towns and villages within cities

Figure 7-25 Solutions: *smart growth* or *new urbanism tools* used to prevent and control urban growth and sprawl. *Critical thinking: which five of the tools do you believe are the most important ways to prevent or control urban sprawl?*

Solutions: Making Cities More Sustainable and Desirable Places to Live

An ecocity allows people to walk, bike, or take mass transit for most of their travel. It recycles and reuses most of its wastes, grows much of its own food, and protects biodiversity by preserving surrounding land.

According to most environmentalists and urban planners, our primary problem is not urbanization but rather our failure to make cities more sustainable and livable. They call for us to make new and existing urban areas more self-reliant, sustainable, and enjoyable places to live through good ecological design. See the Guest Essay on this topic by David Orr on the website for this chapter.

A more environmentally sustainable city, called an *ecocity* or *green city*, emphasizes the following goals:

- Preventing pollution and reducing waste
- Using energy and matter resources efficiently
- Recycling, reusing, and composting at least 60% of all municipal solid waste
- Using solar and other locally available, renewable energy resources
- Protecting and encouraging biodiversity by preserving surrounding land

An ecocity is a people-oriented city, not a car-oriented city. Its residents are able to walk, bike, or use low-polluting mass transit for most of their travel. Its buildings, vehicles, and appliances meet high energy-efficiency standards. Trees and plants adapted to the local climate and soils are planted throughout to provide shade and beauty, supply wildlife habitats, and reduce pollution, noise, and soil erosion. Small organic gardens and a variety of plants adapted to local climate conditions often replace monoculture grass lawns.

Abandoned lots, industrial sites, and polluted creeks and rivers are cleaned up and restored. Nearby forests, grasslands, wetlands, and farms are preserved. Much of an ecocity's food comes from nearby organic farms, solar greenhouses, community gardens, and small gardens on rooftops, in yards, and in window boxes. People designing and living in ecocities take seriously the advice Lewis Mumford gave more than three decades ago: "Forget the damned motor car and build cities for lovers and friends."

The ecocity is not a futuristic dream. Examples of cities that have attempted to become more environmentally sustainable and livable include Curitiba, Brazil (described in the Case Study that follows); Waitakere City, New Zealand; Leicester, England; Portland, Oregon; Davis, California; Olympia, Washington; and Chattanooga, Tennessee.

Case Study: Curitiba, Brazil—One of the World's Most Sustainable Major Cities

One of the world's most livable and sustainable major cities is Curitiba, Brazil, with more than 2.5 million people.

The "ecological capital" of Brazil, Curitiba decided in 1969 to focus on mass transit. Today the city has the world's best bus system. Each day it carries about 60% of Curitiba's more than 2.5 million people (up from 300,000 in 1950) throughout the city along express lanes dedicated to buses (Figure 7-26). Only high-rise apartment buildings are allowed near major bus routes, and each building must devote its bottom two floors to stores, a practice that reduces the need for residents to travel.

Bike paths run throughout most of the city. Cars are banned from 49 blocks of the city's downtown area, which features a network of pedestrian walkways connected to bus stations, parks, and bike paths. Because Curitiba relies less on automobiles, it uses less energy per person and has less air pollution, greenhouse gas emissions, and traffic congestion than most comparable cities.

Trees have been planted throughout the city. No tree in the city can be cut down without a permit, and two trees must be planted for each one harvested.

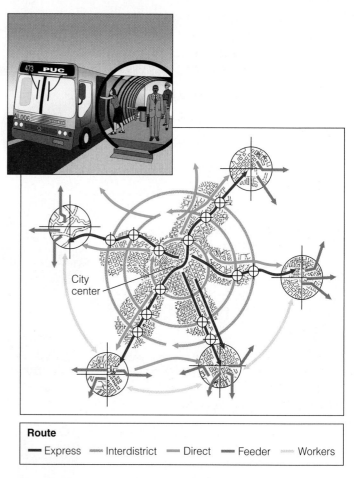

Route
— Express ····· Interdistrict — Direct ···· Feeder ···· Workers

Figure 7-26 Solutions: bus system in Curitiba, Brazil. This system moves large numbers of passengers around rapidly because each of the five major spokes has two express lanes used only by buses. Double- and triple-length bus sections are hooked together as needed, and boarding is facilitated by the use of extra-wide doors and raised tubes that allow passengers to pay before entering the bus (top left).

The city recycles roughly 70% of its paper and 60% of its metal, glass, and plastic, which is sorted by households for collection three times a week. Recovered materials are sold mostly to the city's more than 500 major industries, which must meet strict pollution standards. Most of these industries are located in an industrial park outside the city limits. A major bus line runs to the park, but many of the workers live nearby and can walk or bike to work.

The city uses old buses as roving classrooms to give its poor people basic skills needed for jobs. Other retired buses have become classrooms, health clinics, soup kitchens, and day-care centers. The day-care centers are open 11 hours a day and are free for low-income parents.

The poor receive free medical, dental, and child care, and 40 feeding centers are available for street children. The city has a *build-it-yourself* system that gives each poor family a plot of land, building materials, two trees, and an hour's consultation with an architect.

In Curitiba, virtually all households have electricity, drinking water, and trash collection. About 95% of its citizens can read and write, and 83% of adults have at least a high school education. All schoolchildren study ecology.

This global model of urban planning and sustainability is the brainchild of architect and former college teacher Jaime Lerner, who has served as Curitiba's mayor three times since 1969. Under his leadership, the municipal government dedicated itself to two goals. *First,* it sought solutions to problems that are simple, innovative, fast, cheap, and fun. *Second,* the government vowed to be honest, accountable, and open to public scrutiny.

An exciting challenge during this century will be to reshape existing cities and design new ones like Curitiba that are more livable and sustainable and have a lower environmental impact.

The city is not an ecological monstrosity. It is rather the place where both the problems and the opportunities of modern technological civilization are most potent and visible.

PETER SELF

CRITICAL THINKING

1. Why is it rational for a poor couple in a developing country such as India to have four or five children? What changes might induce such a couple to consider their behavior irrational?

2. Identify a major local, national, or global environmental problem, and describe the role of population growth in this problem.

3. Suppose that all women in the world today began bearing children at the replacement-level fertility rate of 2.1 children per woman. Explain why this would not immediately stop global population growth. Roughly how long would it take for population growth to stabilize (assuming death rates do not rise)?

4. Do you believe that the population is too high in **(a)** your own country and **(b)** the area where you live? Explain.

5. Should everyone have the right to have as many children as they want? Explain.

6. How environmentally sustainable is the area where you live? List five ways to make it more environmentally sustainable.

7. If you were in charge of Mexico City, what are the three most important things you would do?

8. Do you believe the United States or the country where you live should develop a comprehensive and integrated mass transit system over the next 20 years, including building an efficient rapid-rail network for travel within and between its major cities? How would you pay for such a system?

9. If you own a car or hope to own one, what conditions, if any, would encourage you to rely less on the automobile and to travel to school or work by bicycle, on foot, by mass transit, or by a carpool or vanpool?

10. Some analysts suggest phasing out federal, state, and government subsidies that encourage sprawl from building roads, single-family housing, and large malls and superstores. These would be replaced by subsidies that encourage walking and bicycle paths, multifamily housing, high-density residential development, and development with a mix of housing, shops, and offices (mixed-use development). Do you support this approach? Explain.

11. Congratulations! You are in charge of the world. List the three most important features of your **(a)** population policy and **(b)** urban policy.

LEARNING ONLINE

The website for this book includes review questions for the entire chapter, flash cards for key terms and concepts, a multiple-choice practice quiz, interesting Internet sites, references, and a guide for accessing thousands of InfoTrac® College Edition articles.
Visit

http://biology.brookscole.com/miller11

Then choose Chapter 7, and select a learning resource. For access to animations, additional quizzes, chapter outlines and summaries, register and log in to

Environmental Science ⊕ Now™

at **esnow.brookscole.com/miller11** using the access code card in the front of your book.

> ### Active Graphing
> Visit http://esnow.brookscole.com/miller11 to explore the graphing exercise for this chapter.

Sustaining Biodiversity: The Ecosystem Approach

Biodiversity · Land · Forest Renewal

CASE STUDY

Reintroducing Wolves to Yellowstone

At one time, the gray wolf, also known as the eastern timber wolf (Figure 8-1), roamed over most of North America. Then between 1850 and 1900, an estimated 2 million wolves were shot, trapped, and poisoned by ranchers, hunters, and government employees. The idea was to make the West and the Great Plains safe for livestock and for big-game animals prized by hunters.

It worked. When Congress passed the U.S. Endangered Species Act in 1973, only about 400–500 gray wolves remained in the lower 48 states, primarily in Minnesota and Michigan. In 1974, the U.S. Fish and Wildlife Service (USFWS) listed the gray wolf as endangered in all 48 lower states except Minnesota.

Ecologists recognize the important role this keystone predator species once played in parts of the West and the Great Plains. These wolves culled herds of bison, elk, caribou, and mule deer, and kept down coyote populations. They also provided uneaten meat for scavengers such as ravens, bald eagles, ermines, and foxes.

In recent years, herds of elk, moose, and antelope have expanded. Their larger numbers have devastated some vegetation, increased erosion, and threatened the niches of other wildlife species. Reintroducing a keystone species such as the gray wolf into a terrestrial ecosystem is one way to help sustain the biodiversity of the ecosystem and prevent further environmental degradation.

In 1987, the USFWS proposed reintroducing gray wolves into the Yellowstone ecosystem. The suggestion brought angry protests. Some ranchers feared the

Tom Kitchin/Tom Stack & Associates

Figure 8-1 Natural capital restoration: the *gray wolf*. Ranchers, hunters, miners, and loggers have vigorously opposed efforts to return this keystone species to its former habitat in the Yellowstone National Park—which is bigger than the U.S. states of Delaware and Rhode Island combined. Wolves were reintroduced beginning in 1995 and now number around 850.

wolves would attack their cattle and sheep; one enraged rancher said that the idea was "like reintroducing smallpox." Other objections came from hunters who feared the wolves would kill too many big-game animals, and from mining and logging companies that worried the government would halt their operations on wolf-populated federal lands.

Since 1995, federal wildlife officials have caught gray wolves in Canada and relocated them in Yellowstone National Park and northern Idaho. By 2005, about 850 gray wolves lived in or around these two areas.

With wolves around, elk are gathering less near streams and rivers. Their diminished presence has spurred the growth of aspen and willow trees that attract beavers. And leftovers of elk killed by wolves are an important food source for grizzly bears.

The wolves have also cut coyote populations in half. This has increased populations of smaller animals such as ground squirrels and foxes hunted by coyotes, providing more food for eagles and hawks. Between 1995 and 2002, the wolves also killed 792 sheep, 278 cattle, and 62 dogs in the Northern Rockies.

In 2004, the U.S. Fish and Wildlife Service (USFWS) proposed removing wolves from protection under the Endangered Species Act in Idaho and Montana. Private citizens in these states could then kill wolves that attack livestock or pets on private lands. Conservationists say this action is premature, warning that it could undermine one of the nation's most successful conservation efforts.

Population growth, economic development, and poverty are exerting increasing pressure on the world's forests, grasslands, parks, wilderness, oceans, rivers, and other storehouses of biodiversity, a topic explored in this chapter.

Forests precede civilizations, deserts follow them.

FRANCOIS-AUGUSTE-RENÉ DE CHATEAUBRIAND

This chapter discusses how we can help sustain the earth's terrestrial and aquatic biodiversity by protecting places where wild species live. It addresses the following questions:

- How have human activities affected the earth's biodiversity?

- What are the major types of public lands in the United States, and how are they used?

- How should forest resources be used, managed, and sustained globally and in the United States?

- How serious is tropical deforestation, and how can we help sustain tropical forests?

- What problems do parks face, and how should we manage them?

- How should we establish, design, protect, and manage terrestrial nature reserves?

- How can we protect and sustain aquatic biodiversity?

- What is ecological restoration, and why is it important?

- What can we do to help sustain the earth's biodiversity?

KEY IDEAS

- Human activities have depleted and degraded some of the earth's terrestrial and aquatic biodiversity. These threats are expected to increase.

- We should protect the earth's biodiversity because of the economic and ecological services it provides.

- Cutting down large areas of forests reduces biodiversity, eliminates the ecological services forests provide, and can contribute to regional and global climate change.

- We can use forests more sustainably by emphasizing the economic value of their ecological services, harvesting trees no faster than they are replenished, and protecting old-growth and vulnerable areas.

- We need to protect more of the earth's terrestrial and aquatic systems from unsustainable use for their resources by employing adaptive ecosystem management and protecting the most endangered biodiversity hot spots.

- We need to mount a global effort to rehabilitate and restore ecosystems we have damaged.

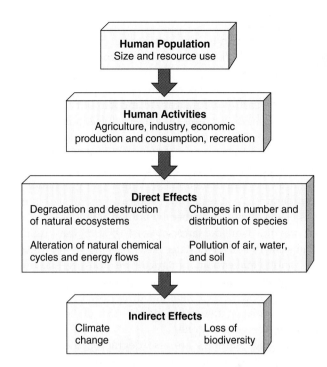

Figure 8-2 Natural capital degradation: major connections between human activities and the earth's biodiversity.

8-1 HUMAN IMPACTS ON BIODIVERSITY

Science and Economics: Effects of Human Activities on Global Biodiversity

We have depleted and degraded some of the earth's biodiversity, and these threats are expected to increase.

Figure 8-2 summarizes how many human activities decrease biodiversity. You can also get an idea of our impact on the earth's natural terrestrial systems in Science Supplement 2 at the end of this book by comparing the two-page satellite map of the earth's natural terrestrial systems (Figure 1), the two-page map of our large and growing ecological footprint on these natural systems (Figure 2), and the map of our ecological footprint in the United States (Figure 3). According to biodiversity expert Edward O. Wilson, "The natural world is everywhere disappearing before our eyes— cut to pieces, mowed down, plowed under, gobbled up, replaced by human artifacts."

Consider a few examples of how human activities have decreased and degraded the earth's biodiversity. According a 2002 study on the impact of the human ecological footprint on the earth's land, we have disturbed to some extent at least half and probably about 83% of the earth's land surface (excluding Antarctica and Greenland). Most of this damage has come from filling in wetlands and converting grasslands and forests to crop fields and urban areas. The 2005 UN Millennium Report, prepared by nearly 1,400 experts, found that human activities are causing massive damage to the earth's natural capital and laid out common-sense strategies for protecting the earth's biodiversity.

About 82% of temperate deciduous forests have been cleared, fragmented, and dominated because their soils and climate are very favorable for growing food and urban development. Temperate grasslands, temperate rain forests, and tropical dry forests have also been greatly disturbed by human activities.

In the United States, at least 95% of the virgin forests in the lower 48 states have been logged for lumber and to make room for agriculture, housing, and industry. In addition, 98% of the tallgrass prairie in the Midwest and Great Plains has disappeared, and 99% of California's native grassland and 85% of its original redwood forests are gone.

Human activities are also degrading the earth's *aquatic biodiversity*. About half of the world's wetlands (including half of U.S. wetlands) were lost during the last century. An estimated 27% of the world's diverse coral reefs have been severely damaged. By 2050, another 70% may be severely damaged or eliminated.

Three-fourths of the world's 200 commercially valuable marine fish species are either overfished or fished to their estimated sustainable yield. According to a 2003 report by the U.S. Commission on Ocean Policy, about 40% of U.S. commercial fish stocks are depleted or overfished.

Human activities also contribute to the *premature extinction of species*. Biologists estimate that the current global extinction rate of species is at least 100 times and probably 1,000–10,000 times what it was before humans existed. These threats to the world's biodiversity are projected to increase sharply during the next few decades.

Figure 8-3 outlines the goals, strategies, and tactics for preserving and restoring the terrestrial ecosystems and aquatic systems that provide habitats and resources for the world's species (as discussed in this chapter) and preventing the premature extinction of species (as discussed in Chapter 9).

Environmental Science ⊕ Now™
Examine the details on how human activities are threatening biodiversity around the world at Environmental ScienceNow.

Science, Economics, and Ethics: Why Should We Care about Biodiversity?

Biodiversity should be protected from degradation by human activities because it exists and because of its usefulness to us and other species.

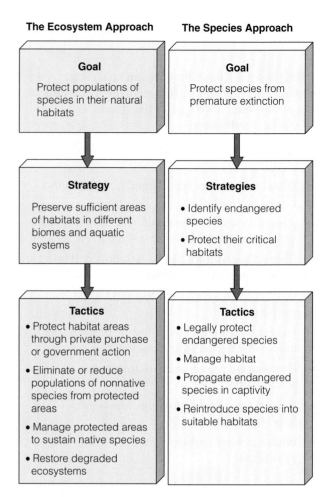

Figure 8-3 Solutions: goals, strategies, and tactics for protecting biodiversity.

Biodiversity researchers contend that we should act to preserve the earth's overall diversity because its genes, species, ecosystems, and ecological processes have two types of value. *First,* they have **intrinsic value** because these components of biodiversity exist, regardless of their use to us. *Second,* they have **instrumental value** because of their usefulness to us.

Two major types of instrumental values exist. One consists of *use values* that benefit us in the form of economic goods and services, ecological services, recreation, scientific information, and preserving options for such uses in the future. The other type consists of *nonuse values*. For example, there is *existence value*—knowing that a redwood forest, wilderness, or endangered species (Figure 8-4) exists, even if we will never see it or get direct use from it. *Aesthetic value* is another nonuse value—many people appreciate a tree, a forest, a wild species (Figure 8-5), or a vista because of its beauty. *Bequest value,* a third type of nonuse value, is based on the willingness of some people to pay to protect some forms of natural capital to ensure their availability for use by future generations.

age fotostock/SuperStock

Figure 8-5 Natural capital: many species of wildlife, such as this scarlet macaw in Brazil's Amazon rainforest, are a source of beauty and pleasure. These and other colorful species of parrots can become endangered when they are removed from the wild and sold (sometimes illegally) as pets.

Figure 8-4 Natural capital degradation: endangered orangutans in a tropical forest in Gunung Leuser National Park, Indonesia. In 1900 there were over 315,000 wild orangutans. Now there are less than 20,000.and they are disappearing at a rate of over 2,000 individuals a year.

© Ernest Manewal/SuperStock

8-2   PUBLIC LANDS IN THE UNITED STATES

Using and Preserving Natural Capital: U.S. Public Lands

More than one-third of the land in the United States consists of publicly owned national forests, resource lands, parks, wildlife refuges, and protected wilderness areas.

No nation has set aside as much of its land for public use, resource extraction, enjoyment, and wildlife as has the United States. The federal government manages roughly 35% of the country's land, which ultimately belongs to every American. About 73% of this federal public land is in Alaska and another 22% is in the western states (Figure 8-6, p. 158).

Some federal public lands are used for many purposes. For example, the *National Forest System* consists of 155 forests (Figure 5-19, bottom, p. 93) and 22 grass-

lands. These forests, which are managed by the U.S. Forest Service (USFS), are used for logging, mining, livestock grazing, farming, oil and gas extraction, recreation, hunting, fishing, and conservation of watershed, soil, and wildlife resources.

The Bureau of Land Management (BLM) manages *National Resource Lands*. These lands are used primarily for mining, oil and gas extraction, and livestock grazing.

The U.S. Fish and Wildlife Service (USFWS) manages the nation's 542 *National Wildlife Refuges*. Most refuges protect habitats and breeding areas for waterfowl and big game to provide a harvestable supply for hunters; a few protect endangered species from extinction. Permitted activities in most refuges include hunting, trapping, fishing, oil and gas development, mining, logging, grazing, some military activities, and farming.

Uses of some other public lands are more restricted. Consider the *National Park System* managed by

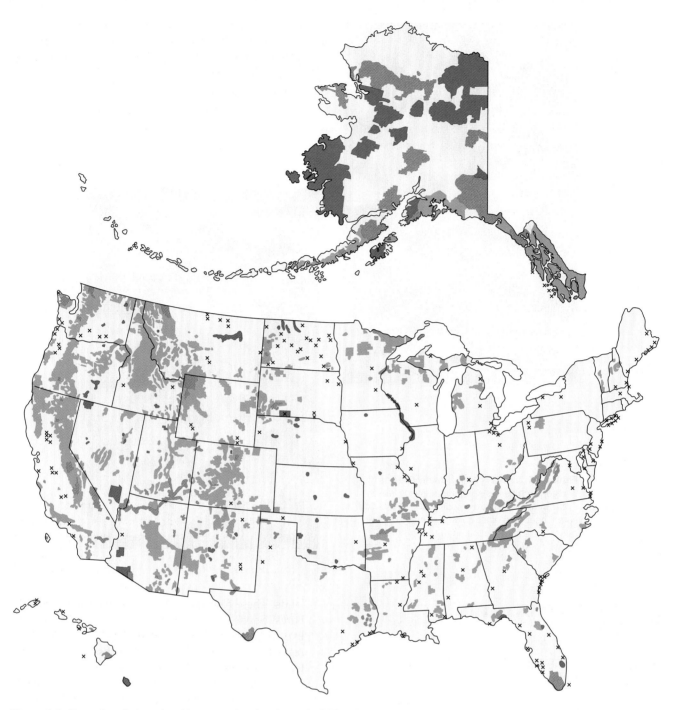

Figure 8-6 Natural capital: national forests, national parks, and wildlife refuges managed by the U.S. federal government. U.S. citizens jointly own these and other public lands. (Data from U.S. Geological Survey)

the National Park Service (NPS). It includes 58 major parks (mostly in the West, Figure 5-21, p. 94) and 331 national recreation areas, monuments, memorials, battlefields, historic sites, parkways, trails, rivers, seashores, and lakeshores. Only camping, hiking, sport fishing, and boating can take place in the national parks, whereas sport hunting, mining, and oil and gas drilling are allowed in national recreation areas.

The most restricted public lands are 660 roadless areas that make up the *National Wilderness Preservation System.* These areas lie within the national parks, national wildlife refuges, national forests, and national resource lands and are managed by the agencies in charge of those lands. Most of these areas are open only for recreational activities such as hiking, sport fishing, camping, and nonmotorized boating.

Science, Economics, and Politics: Managing U.S. Public Lands

Since the 1800s, controversy has swirled over how U.S. public lands should be used because of the valuable resources they contain.

Many federal public lands contain valuable oil, natural gas, coal, timber, and mineral resources. Since the 1800s, debates have focused on how the resources on these lands should be used and managed.

Most conservation biologists, environmental economists, and many free-market economists believe that four principles should govern use of public land:

- Protecting biodiversity, wildlife habitats, and the ecological functioning of public land ecosystems should be the primary goal.

- No one should receive subsidies or tax breaks for using or extracting resources on public lands.

- The American people deserve fair compensation for the use of their property.

- All users or extractors of resources on public lands should be fully responsible for any environmental damage they cause.

There is strong and effective opposition to these ideas. Economists, developers, and resource extractors tend to view public lands in terms of their usefulness in providing mineral, timber, and other resources and their ability to increase short-term economic growth. They have succeeded in blocking implementation of the four principles just listed. For example, in recent years, the government has given more than $1 billion per year in subsidies to privately owned mining, fossil fuel extraction, logging, and grazing interests that use U.S. public lands.

Some developers and resource extractors have sought to go further, mounting a campaign to get the U.S. Congress to pass laws that would do the following:

- Sell public lands or their resources to corporations or individuals, usually at less than market value

- Slash federal funding for regulatory administration of public lands

- Cut all old-growth forests in the national forests and replace them with tree plantations

- Open all national parks, national wildlife refuges, and wilderness areas to oil drilling, mining, off-road vehicles, and commercial development

- Do away with the National Park Service and launch a 20-year construction program of new concessions and theme parks run by private firms in the national parks

- Continue mining on public lands under the provisions of the 1872 Mining Law, which allows mining interests to pay no royalties to taxpayers for hard-rock minerals they remove

- Repeal the Endangered Species Act or modify it to allow economic factors to override protection of endangered and threatened species

- Redefine government-protected wetlands so that about half of them would no longer be protected

- Prevent individuals or groups from legally challenging these uses of public land for private financial gain

X *How Would You Vote?* Should much more of U.S. public lands (or government-owned lands in the country where you live) be opened up to the extraction of timber, mineral, and energy resources? Cast your vote online at http://biology.brookscole.com/miller11.

8-3 MANAGING AND SUSTAINING FORESTS

Science: Benefits and Types of Forests

Some forests have not been disturbed by human activities, others have grown back after being cut, and some consist of planted stands of a particular tree species.

Forests with at least 10% tree cover occupy about 30% of the earth's land surface (excluding Greenland and Antarctica). Figure 5-7 (p. 83) shows the distribution of the world's boreal, temperate, and tropical forests. These forests provide many important ecological and economic services (Figure 8-7, p. 160).

Forest managers and ecologists classify forests into three major types based on their age and structure. The first type is an **old-growth forest:** an uncut or regenerated forest that has not been seriously disturbed by human activities or natural disasters for at least several hundred years. Old-growth forests are storehouses of biodiversity because they provide ecological niches for a multitude of wildlife species (Figure 5-17, p. 91).

The second type is a **second-growth forest:** a stand of trees resulting from secondary ecological succession (Figure 6-10, p. 120). These forests develop after the trees in an area have been removed by *human activities* (such as clear-cutting for timber or conversion to cropland) or by *natural forces* (such as fire, hurricanes, or volcanic eruption).

A **tree plantation,** also called a **tree farm,** is a third type (Figure 8-8, 160). This managed tract contains uniformly aged trees of one or two species that are

Natural Capital

Forests

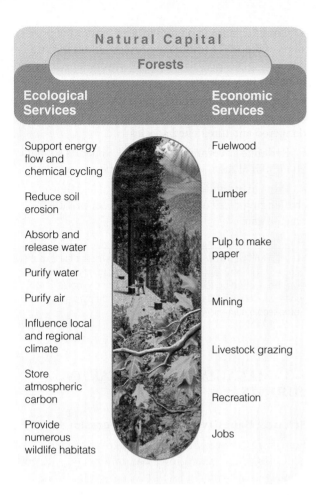

Ecological Services	Economic Services
Support energy flow and chemical cycling	Fuelwood
Reduce soil erosion	Lumber
Absorb and release water	Pulp to make paper
Purify water	
Purify air	Mining
Influence local and regional climate	Livestock grazing
Store atmospheric carbon	Recreation
Provide numerous wildlife habitats	Jobs

Figure 8-7 Natural capital: major ecological and economic services provided by forests.

Figure 8-8 Natural capital degradation: *tree plantation* in North Carolina. Some diverse old-growth and second-growth forests are cleared and replanted with a single tree (monoculture), often for harvest as Christmas trees (shown here), timber, or wood converted to pulp to make paper. This shift from a polyculture to a monoculture system decreases the area's biodiversity.

Gene Alexander/USDA

harvested by clear-cutting as soon as they become commercially valuable. The land is then replanted and clear-cut again in a regular cycle.

Science and Economics: Types of Forest Management

Some forests consist of one or two species of commercially important tree species that are cut down and replanted; others contain diverse tree species harvested individually or in small groups.

Two forest management systems exist. In **even-aged management,** trees in a given stand are maintained at about the same age and size. With this approach, which is sometimes called *industrial forestry*, a simplified *tree plantation* replaces a biologically diverse old-growth or second-growth forest. The plantation consists of one or two fast-growing and economically desirable species that can be harvested every six to ten years, depending on the species (Figure 8-9).

In **uneven-aged management,** a stand includes a variety of tree species with many ages and sizes, in an effort to foster natural regeneration. Here the goals are biological diversity, long-term sustainable production of high-quality timber, selective cutting of individual mature or intermediate-aged trees, and multiple uses of the forest for timber, wildlife, watershed protection, and recreation.

According to a 2001 study by the Worldwide Fund for Nature (WWF), intensive but sustainable management of as little as one-fifth of the world's forests—an area twice the size of India—could meet the world's current and future demand for commercial wood and fiber. This intensive use of the world's tree plantations and some of its secondary forests would leave the world's remaining old-growth forest untouched.

Science and Economics: Harvesting Trees

Trees can be harvested individually from diverse forests, or an entire forest stand can be cut down in one or several phases.

The first step in forest management is to build roads for access and timber removal. Even carefully designed logging roads have a number of harmful effects (Figure 8-10)—namely, increased erosion and sediment runoff into waterways, habitat fragmentation, and biodiversity loss. Logging roads also expose forests to invasion by nonnative pests, diseases, and wildlife species. And they open once-inaccessible forests to farmers, miners, ranchers, hunters, and off-road vehicle users. In addition, logging roads on public lands in the United States disqualify the land for protection as wilderness.

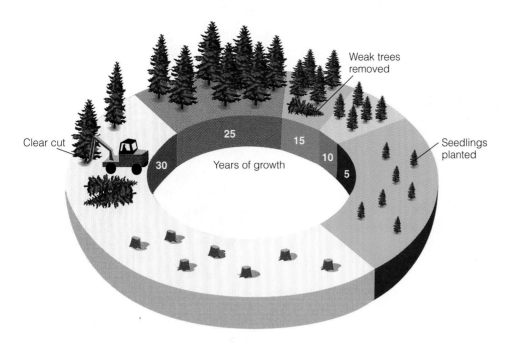

Figure 8-9 Natural capital degradation: short (25- to 30-year) rotation cycle of cutting and regrowth of a monoculture tree plantation in modern industrial forestry. In tropical countries, where trees can grow more rapidly year-round, the rotation cycle can be 6–10 years.

Weak trees removed

Clear cut

25

15

30   Years of growth   10

5

Seedlings planted

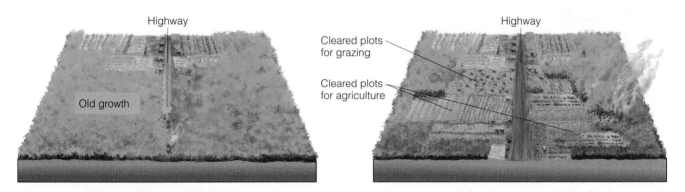

Highway

Old growth

Highway

Cleared plots for grazing

Cleared plots for agriculture

Figure 8-10 Natural capital degradation: Building roads into previously inaccessible forests paves the way to fragmentation, destruction, and degradation.

Once loggers reach a forest area, they use a variety of methods to harvest the trees (Figure 8-11, p. 162). With *selective cutting,* intermediate-aged or mature trees in an uneven-aged forest are cut singly or in small groups (Figure 8-11a).

Some tree species that grow best in full or moderate sunlight are cleared completely in one or several cuttings by *shelterwood cutting* (Figure 8-11b), *seed-tree cutting* (Figure 8-11c), or *clear-cutting* (in which all trees on a site are removed in a single cut; Figures 8-11d and 8-12, p. 163). Figure 8-13 (p. 163) lists the advantages and disadvantages of clear-cutting.

✗ *HOW WOULD YOU VOTE?* Should conventional clear-cutting of trees on publicly owned lands be banned except for tree species needing full or moderate sunlight for growth? Cast your vote online at http://biology.brookscole.com/miller11.

A clear-cutting variation that can allow a more sustainable timber yield without widespread destruction is *strip cutting* (Figure 8-11e). It involves clear-cutting a strip of trees along the contour of the land, with the corridor narrow enough to allow natural regeneration within a few years. After regeneration, loggers cut another strip above the first, and so on. This approach allows clear-cutting of a forest in narrow strips over several decades with minimal damage.

Science: Harmful Environmental Effects of Deforestation

Cutting down large areas of forests reduces biodiversity and the ecological services forests provide, and it can contribute to regional and global climate change.

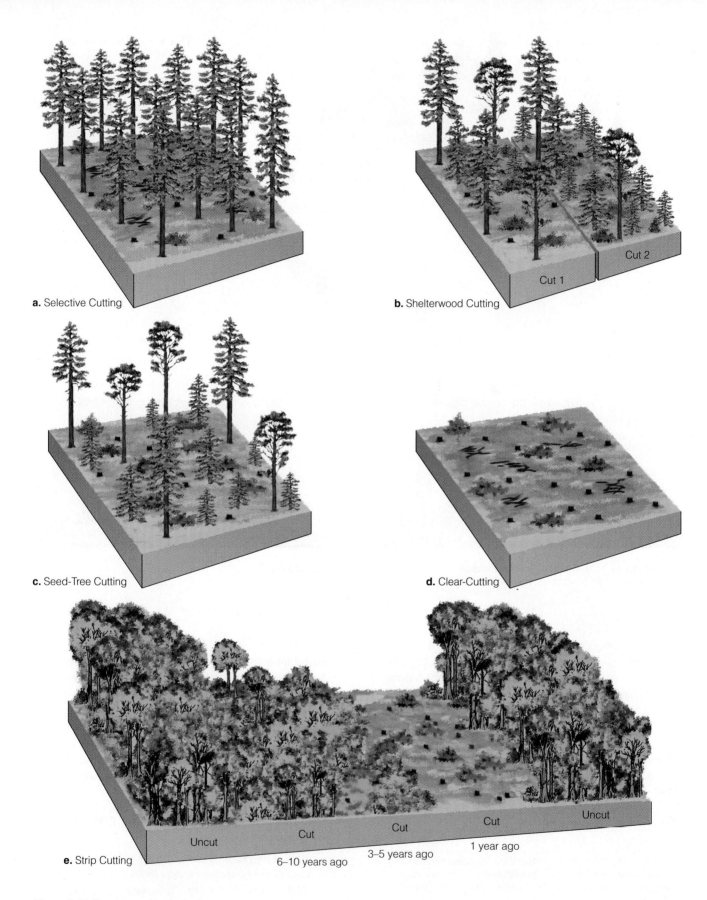

a. Selective Cutting

b. Shelterwood Cutting

Cut 1

Cut 2

c. Seed-Tree Cutting

d. Clear-Cutting

e. Strip Cutting

Uncut

Cut

Cut

Cut

Uncut

6–10 years ago

3–5 years ago

1 year ago

Figure 8-11 Tree-harvesting methods.

Deforestation is the temporary or permanent removal of large expanses of forest for agriculture or other uses. Harvesting timber (Figure 8-12) and fuelwood from forests provides many economic benefits (Figure 8-7, right). However, deforestation can have many harmful environmental effects (Figure 8-14, p. 164) that can reduce the ecological services provided by forests (Figure 8-7, left).

If deforestation occurs over a large enough area, it can cause a region's climate to become hotter and drier and prevent the regrowth of a forest. Deforestation can also contribute to projected global warming if trees are removed faster than they grow back. When forests are cleared for agriculture or other purposes and burned, the carbon stored as biomass in the trees is released into the atmosphere as the greenhouse gas carbon dioxide (CO_2).

Environmental Science ⬤ Now™
Learn more about how deforestation can affect the drainage of a watershed and disturb its ecosystem at Environmental ScienceNow.

Figure 8-12 Natural capital degradation: clear-cut logging in Washington state.

Trade-Offs

Clear-Cutting Forests

Advantages	Disadvantages
Higher timber yields	Reduces biodiversity
Maximum economic return in shortest time	Disrupts ecosystem processes
Can reforest with genetically improved fast-growing trees	Destroys and fragments some wildlife habitats
Short time to establish new stand of trees	Leaves moderate to large openings
Needs less skill and planning	Increases soil erosion
Best way to harvest tree plantations	Increases sediment water pollution and flooding when done on steep slopes
Good for tree species needing full or moderate sunlight for growth	Eliminates most recreational value for several decades

Figure 8-13 Trade-offs: advantages and disadvantages of clear-cutting forests. *Critical thinking: pick the single advantage and disadvantage that you think are the most important.*

Global Outlook: Extent of Deforestation

Human activities have reduced the earth's forest cover by 20–50%, and deforestation is continuing at a fairly rapid rate, except in most temperate forests in North America and Europe.

Forests are renewable resources as long as the rate of cutting and degradation does not exceed the rate of regrowth. There are two pieces of *bad news*.

First, surveys by the World Resources Institute (WRI) indicate that over the past 8,000 years human activities have reduced the earth's original forest cover by 20–50%.

Second, surveys by the UN Food and Agricultural Organization (FAO) and the World Resources Institute (WRI) indicate that the global rate of forest cover loss during the 1990s was between 0.2% and 0.5% per year, and at least another 0.1–0.3% of the world's forests were degraded. If these estimates are correct, the world's forests are being cleared and degraded at a

- Decreased soil fertility from erosion

- Runoff of eroded soil into aquatic systems

- Premature extinction of species with specialized niches

- Loss of habitat for migratory species such as birds and butterflies

- Regional climate change from extensive clearing

- Releases CO_2 into atmosphere from burning and tree decay

- Accelerates flooding

Figure 8-14 Natural capital degradation: harmful environmental effects of deforestation that can reduce the ecological services provided by forests. *Critical thinking: what are the direct and indirect effects of your lifestyle on deforestation?*

rate of 0.3–0.8% per year, with much higher rates in some areas. More than four-fifths of these losses took place in the tropics. According to the WRI, if current deforestation rates continue, about 40% of the world's remaining intact forests will have been logged or converted to other uses within 10–20 years, if not sooner.

There are also two encouraging trends. *First,* the total area occupied by many temperate forests in North America and Europe has increased slightly because of reforestation from secondary ecological succession on cleared forest areas and abandoned croplands.

Second, some of the cut areas of tropical forest display increased tree cover from regrowth and planting of tree plantations. But ecologists do not believe that tree plantations, which have much lower biodiversity, should be counted as forest any more than croplands should be counted as grassland. According to ecologist Michael L. Rosenzweig, "Forest plantations are just cornfields whose stalks have gotten very tall and turned to wood. They display nothing of the majesty of natural forests."

Solutions: Managing Forests More Sustainably

We can use forests more sustainably by emphasizing the economic value of their ecological services, harvesting trees no faster than they are replenished, and protecting old-growth and vulnerable areas.

Biodiversity researchers and a growing number of foresters have called for more sustainable forest management. Figure 8-15 lists ways to achieve this goal.

Currently, forests are valued mostly for their economic services (Figure 8-7, right). But suppose we took into account the estimated monetary value of the ecological services provided by forests (Figure 8-7, left). According to a 1997 appraisal by a team of ecologists, economists, and geographers—led by ecological economist Robert Costanza of the University of Vermont—the world's forests provide us with ecological services worth about $4.7 trillion per year—hundreds of times more than the economic value of forests. The researchers noted that their estimates could easily be too low by a factor of 10 to 1 million or more because their calculations included only estimates of the ecosystem services provided by forests themselves, not the natural capital that generates them.

Based on this accounting system, most of the world's old-growth and second-growth forests should not be clear-cut. Instead, their ecological services could be sustained indefinitely by selectively harvesting trees no faster than they are replenished and by using them primarily for recreation and as centers of biodiversity.

According to ecological economist Robert Costanza, "We have been cooking the books for a long time by leaving out the worth of nature." Biologist David Suzuki warns, "Our economic system has been con-

Solutions

Sustainable Forestry

- Conserve biodiversity along with water and soil resources

- Grow more timber on long rotations

- Rely more on selective cutting and strip cutting

- No clear-cutting, seed-tree, or shelterwood cutting on steeply sloped land

- No fragmentation of remaining large blocks of forest

- Sharply reduce road building into uncut forest areas

- Leave most standing dead trees and fallen timber for wildlife habitat and nutrient recycling

- Certify timber grown by sustainable methods

- Include ecological services of trees and forests in estimating economic value

Figure 8-15 Solutions: ways to manage forests more sustainably. *Critical thinking: which two of these solutions do you believe are the most important?*

structed under the premise that natural services are free. We can't afford that luxury any more."

Why have we not changed our accounting system to reflect these losses? One reason is that the economic savings provided by conserving forests (and other parts of nature) benefit everyone now and in the future, whereas the profits made by exploiting them are immediate and benefit a relatively small group of individuals. A second reason is that many current government subsidies and tax incentives support destruction and degradation of forests and other ecosystems for short-term economic gain. Also, most people are unaware of the value of the ecological services and income provided by forests and other parts of nature.

Another way to encourage sustainable use of forests is to establish and use methods to evaluate timber that has been grown sustainably, as discussed next.

Solutions: Certifying Sustainably Grown Timber

Organizations have developed standards for certifying that timber has been harvested sustainably and that wood products have been produced from sustainably harvested timber.

Collins Pine owns and manages a large area of productive timberland in northeastern California. Since 1940, the company has used selective cutting to help maintain the ecological, economic, and social sustainability of its timberland.

Since 1993, Scientific Certification Systems (SCS) has evaluated the company's timber production. SCS, which is part of the nonprofit Forest Stewardship Council (FSC), was formed to develop a list of environmentally sound practices for use in certifying timber and products made from such timber.

Each year, SCS evaluates Collins Pine's landholdings to ensure that cutting has not exceeded long-term forest regeneration, roads and harvesting systems have not caused unreasonable ecological damage, soils are not damaged, downed wood (boles) and standing dead trees (snags) are left to provide wildlife habitat, and the company is a good employer and a good steward of its land and water resources.

In 2001, the Worldwide Fund for Nature (WWF) called on the world's five largest companies that harvest and process timber and buy wood products to adopt the FSC's sustainable forest management principles. According to the WWF, doing so would halt virtually all logging of old-growth forests yet still meet the world's industrial wood and wood fiber needs using one-fifth of the world's forests.

In 2002, Mitsubishi, one of the world's largest forestry companies, announced that third parties would certify its forestry operations using standards developed by the Forest Stewardship Council (FSC). Home Depot, Lowes, Andersen, and other major sellers of wood products in the United States have also agreed to sell only wood certified as being sustainably grown by independent groups such as the FSC (to the degree that certified wood is available).

8-4 FOREST RESOURCES AND MANAGEMENT IN THE UNITED STATES

U.S. Forests: Encouraging News

U.S. forests cover more area than they did in 1920, more wood is grown than is cut, and the country has set aside large areas of protected forests.

Forests cover about 30% of the U.S. land area, providing habitats for more than 80% of the country's wildlife species and supplying about two-thirds of the nation's total surface water.

Good news. Forests (including tree plantations) in the United States cover more area than they did in 1920. Many of the old-growth forests that were cleared or partially cleared between 1620 and 1960 have grown back naturally through secondary ecological succession as fairly diverse second-growth (and in some cases third-growth) forests in every region of the United States, except much of the West. In 1995, environmental writer Bill McKibben cited forest regrowth in the United States—especially in the East—as "the great environmental story of the United States, and in some ways the whole world."

Every year more wood is grown in the United States than is cut, and each year the total area planted with trees increases. By 2000, protected forests accounted for 40% of the country's total forest area, mostly in the national forests (Figure 5-19, bottom, p. 93).

Bad news. Since the mid-1960s, an increasing area of the nation's remaining old-growth and fairly diverse second-growth forests has been clear-cut and replaced with biologically simplified tree plantations. According to biodiversity researchers, this practice reduces overall forest biodiversity and disrupts ecosystem processes such as energy flow and chemical cycling. Some environmentally concerned citizens have protested the cutting down of ancient trees and forests (see Individuals Matter, p. 166).

Science: Fires and U.S. Forests

Forest fires can burn away flammable underbrush and small trees, burn large trees and leap from treetop to treetop, or burn flammable materials found under the ground.

Butterfly in a Redwood Tree

"Butterfly" is the nickname given to Julia Hill. This young woman spent two years of her life on a small platform near the top of a giant redwood tree in California to protest the clear-cutting of a forest of these ancient trees, some of them more than 1,000 years old.

She and other protesters were illegally occupying these trees as a form of *nonviolent civil disobedience,* similar to that used decades ago by Mahatma Gandhi in his efforts to end the British occupation of India. Butterfly had never participated in any environmental protest or act of civil disobedience.

Initially, she went to the site to express her belief that it was wrong to cut down these ancient giants for short-term economic gain, even if you own them. She planned to stay for only a few days.

After seeing the destruction and climbing one of these magnificent trees, Butterfly ended up staying in the tree for two years to publicize what was happening and help save the surrounding trees. She became a symbol of the protest and during her stay used a cell phone to communicate with members of the mass media throughout the world to help develop public support for saving the trees.

Can you imagine spending two years of your life in a tree on a platform not much bigger than a king-sized bed, hovering 55 meters (180 feet) above the ground and enduring high winds, intense rainstorms, snow, and ice? And Butterfly was not living in a quiet pristine forest. All round her was noise from trucks, chainsaws, and helicopters trying to scare her into returning to the ground.

Although Butterfly lost her courageous battle to save the sur-rounding forest, she persuaded Pacific Lumber MAXXAM to save her tree (called Luna) and a 60-meter (200-foot) buffer zone around it. Not too long after she descended from her perch, someone used a chainsaw to seriously damage the tree. Cables and steel plates are now used to preserve it.

But maybe Butterfly and the earth did not lose. A book she wrote about her stand, and her subsequent travels to campuses all over the world, have inspired a number of young people to stand up for protecting biodiversity and other environmental causes.

Butterfly led others by following in the tradition of Gandhi, who said, "My life is my message." Would you spend a day or a week of your life protesting something that you believed to be wrong?

Three types of fires can affect forest ecosystems. *Surface fires* (Figure 8-16, left) usually burn only undergrowth and leaf litter on the forest floor. They may kill seedlings and small trees but spare most mature trees and allow most wild animals to escape.

Occasional surface fires have a number of ecological benefits. They burn away flammable ground material and help prevent more destructive fires. They also release valuable mineral nutrients (tied up in slowly decomposing litter and undergrowth), stimulate the

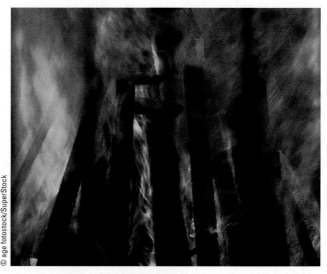

David J. Moorhead, The University of Georgia www.forestryimages.org

© age fotostock/SuperStock

Figure 8-16 Surface fires, such as this one in Tifton, Georgia (left), usually burn undergrowth and leaf litter on a forest floor. They can help prevent more destructive crown fires (right) by removing flammable ground material. Sometimes carefully controlled surface fires are deliberately set to prevent buildup of flammable ground material in forests.

germination of certain tree seeds (such as those of the giant sequoia and jack pine), and help control pathogens and insects. In addition, wildlife species such as deer, moose, elk, muskrat, woodcock, and quail depend on occasional surface fires to maintain their habitats and provide food in the form of vegetation that sprouts after fires.

Some extremely hot fires, called *crown fires* (Figure 8-16, right), may start on the ground but eventually burn whole trees and leap from treetop to treetop. They usually occur in forests that have not experienced surface fires for several decades, which allows dead wood, leaves, and other flammable ground litter to accumulate. These rapidly burning fires can destroy most vegetation, kill wildlife, increase soil erosion, and burn or damage human structures in their paths.

Sometimes surface fires go underground and burn partially decayed leaves or peat. Such *ground fires* are most common in northern peat bogs. They may smolder for days or weeks and are difficult to detect and extinguish.

Solutions: Reducing Forest Damage from Fire

To reduce fire damage, we can set controlled surface fires to prevent buildup of flammable material, allow fires on public lands to burn unless they threaten human structures and lives, and clear small areas around buildings in areas subject to fire.

In the United States, the Smokey Bear educational campaign undertaken by the Forest Service and the National Advertising Council has prevented countless forest fires. It has also saved many lives and prevented billions of dollars in losses of trees, wildlife, and human structures.

At the same time, this educational program has convinced much of the public that all forest fires are bad and should be prevented or put out. Ecologists warn that trying to prevent all forest fires increases the likelihood of destructive crown fires by allowing accumulation of highly flammable underbrush and smaller trees in some forests.

According to the U.S. Forest Service, severe fires could threaten 40% of all federal forest lands, mainly through fuel buildup from past rigorous fire protection programs (the Smokey Bear era), increased logging in the 1980s that left behind highly flammable logging debris (called *slash*), and greater public use of federal forest lands. In addition, an estimated 40 million people now live in remote forested areas or areas populated by dry chaparral vegetation with a high wildfire risk.

Ecologists and forest fire experts have proposed several strategies for reducing fire-related harm to forests and people. One approach is to set small, contained surface fires or clear out (thin) flammable small trees and underbrush in the highest-risk forest areas. Such *prescribed fires* require careful planning and monitoring to keep them from getting out of control.

Another strategy is to allow many fires in national parks, national forests, and wilderness areas to burn and thereby remove flammable underbrush and smaller trees as long as the fires do not threaten human structures and life. A third approach is to protect houses or other buildings in fire-prone areas by thinning a zone of about 46 meters (200 feet) around them and eliminating the use of flammable materials such as wooden roofs.

In 2003, the U.S. Congress passed the *Healthy Forests Restoration Act*. Under this law, timber companies are allowed to cut down economically valuable medium-size and large trees in most national forests in return for clearing away smaller, more fire-prone trees and underbrush. The law also exempts most thinning projects from environmental reviews and appeals currently required by forest protection laws.

According to biologists and many forest fire scientists, this law is likely to *increase* the chances of severe forest fires for two reasons. *First*, removing the most fire-resistant large trees—the ones that are valuable to timber companies—encourages dense growth of highly flammable young trees and underbrush. *Second*, removing the large and medium trees leaves behind highly flammable slash. Many of the worst fires in U.S. history—including some of those during the 1990s—burned through cleared forest areas containing slash.

Fire scientists agree that some national forests need thinning to reduce the chances of catastrophic fires, but suggest focusing on two goals. The first goal would be to reduce ground-level fuel and vegetation in dry forest types and leave widely spaced medium and large trees that are the most fire resistant and thus can help forest recovery after a fire. These trees also provide critical wildlife habitat, especially as standing dead trees (snags) and logs where many animals live. The second goal would emphasize clearing of flammable vegetation around individual homes and buildings and near communities that are especially vulnerable to wildfire.

Critics of the Healthy Forests Restoration Act say that these goals could be accomplished at a much lower cost to taxpayers by giving grants to communities that seem especially vulnerable to wildfires for thinning forests and protecting homes and buildings in their areas.

X *HOW WOULD YOU VOTE?* Do you support repealing or modifying the Healthy Forests Restoration Act? Cast your vote online at http://biology.brookscole.com/miller11.

Science, Economics, and Politics:
U.S. National Forests

Debate continues regarding whether U.S. national forests should be managed primarily for timber, their ecological services, recreation, or a mix of these uses.

For decades, controversy has swirled over the use of resources in the national forests. Timber companies push to cut as much of this timber in these forests as possible at low prices. Biodiversity experts and environmentalists call for sharply reducing or eliminating tree harvesting in national forests and using more sustainable forest management practices (Figure 8-15) for timber cutting in these forests. They believe that national forests should be managed primarily to provide recreation and to sustain biodiversity, water resources, and other ecological services.

The Forest Service's timber-cutting program loses money because the revenues from timber sales do not cover the expenses of road building, timber sale preparation, administration, and other overhead costs. Because of such government subsidies, timber sales from U.S. federal lands have lost money for taxpayers in 97 of the last 100 years!

Figure 8-17 lists the advantages and disadvantages of logging in national forests. According to a 2000 study by the accounting firm Econorthwest, recreation, hunting, and fishing in national forests add ten times more money to the national economy and provide seven times more jobs than does extraction of timber and other resources.

✗ *HOW WOULD YOU VOTE?* Should logging in U.S. national forests be banned? Cast your vote online at http://biology .brookscole.com/miller11.

One way to reduce the pressure to harvest trees for paper production in national and private forests is to make paper by using fiber that does not come from trees. *Tree-free fibers* for making paper come from two sources: *agricultural residues* left over from crops (such as wheat, rice, and sugar) and *fast-growing crops* (such as kenaf and industrial hemp).

China uses tree-free pulp from rice straw and other agricultural wastes to make almost two-thirds of its paper. Most of the small amount of tree-free paper produced in the United States is made from the fibers of a rapidly growing woody annual plant called *kenaf* (pronounced "kuh-NAHF"; Figure 8-18).

Economics: American Forests in a Globalized Economy

Efficient production of timber from tree plantations in temperate and tropical countries in the southern hemisphere is decreasing the need for timber production in the United States.

Trade-Offs

Logging in U.S. National Forests

Advantages	Disadvantages
Helps meet country's timber needs	Provides only 4% of timber needs
Cut areas grow back	Ample private forest land to meet timber needs
Keeps lumber and paper prices down	Has little effect on timber and paper prices
Provides jobs in nearby communities	Damages nearby rivers and fisheries
Promotes economic growth in nearby communities	Recreation in national forests provides more local jobs and income for local communities than logging
	Decreases recreational opportunities

Figure 8-17 Trade-offs: advantages and disadvantages of allowing logging in U.S. national forests. *Critical thinking: pick the single advantage and disadvantage that you think are the most important.*

In today's global economy, the United States faces a surplus of domestic timber and a declining role in the global wood products industry. Timber and pulpwood for making paper can be produced faster and more efficiently in the cultivated tree plantations found in the temperate and tropical regions of the southern hemisphere.

As a result, the United States will likely become a minor player in the global production of wood-based products such as lumber and pulp. Imports of timber and paper from other parts of the world will likely supply an increasing share of U.S. demand for wood products.

This trend could help preserve the nation's biodiversity by decreasing the pressure to clear-cut old-growth and second-growth forests on public and private lands. It is being enhanced by the increased replacement of solid wood for structural purposes with engineered wood products made by gluing small pieces of wood together.

Conversely, this shift in timber production to other countries could decrease biodiversity and watershed protection in the United States. Most forested land in the United States is privately owned, and pri-

U.S. Department of Agriculture

Figure 8-18 Solutions: pressure to cut trees to make paper could be greatly reduced by planting and harvesting a fast-growing plant known as kenaf.

vate owners who can no longer make a profit by selling off some of their timber might be tempted to sell their land to housing developers.

The lower income from harvesting trees also means that both the government and private owners will have less money and fewer incentives for managing forests, forest restoration, and forest thinning projects to reduce damage from fire and insects. As a result, stewardship of public and private forests in the United States may decline unless the country overhauls its forest policy to meet this new reality and the challenges it presents.

8-5 TROPICAL DEFORESTATION

Global Outlook: Clearing and Degradation of Tropical Forests

Large areas of ecologically and economically important tropical forests are being cleared and degraded at a fast rate.

Tropical forests cover about 6% of the earth's land area—roughly the area of the lower 48 U.S. states. Climatic and biological data suggest that mature tropical

forests once covered at least twice as much area as they do today, with most of the destruction occurring since 1950. Satellite scans and ground-level surveys used to estimate forest destruction indicate that large areas of tropical forests are being cut rapidly in parts of South America (especially Brazil), Africa, and Asia.

Studies indicate that at least half of the world's species of terrestrial plants and animals live in tropical rain forests (Figures 3-15, p. 46), 8-4, and 8-5. Because of their specialized niches, these species are highly vulnerable to extinction when their tropical forest habitats are cleared or degraded. Brazil has about 40% of the world's remaining tropical rain forest and an estimated 30% of the world's terrestrial plant and animal species in the vast Amazon basin, which is about two-thirds the size of the continental United States.

In 1970, deforestation affected only 1% of the area of the Amazon basin. By 2005, an estimated 16–47% had been deforested or degraded and converted mostly to tropical grassland (savanna). Between 2002 and 2004, the rate of deforestation increased sharply—mostly to make way for cattle ranching and large plantations for growing crops such as soybeans used for cattle feed. In 2004, loggers and farmers cleared and burned (Figure 8-19, p. 170) an area of tropical forest equivalent to a loss of 11 football fields a minute.

According to Pedro Dias and a number of other tropical forest researchers, without immediate and aggressive action to reduce current forest destruction and degradation practices, at least a third and perhaps as much as 60% of Brazil's original Amazon rain forest will disappear and turn into tropical grassland sometime during this century.

You probably have not heard about the loss of most of Brazil's Atlantic coastal rain forest. This less famous forest once covered about 12% of Brazil's land area. Now 93% of it has been cleared and most of what is left is recovering from previous cutting episodes. The result: a major loss of biodiversity because an area in this forest a little larger than two typical suburban house lots in the United States has 450 tree species! The entire United States has only about 865 native tree species.

Observers disagree about the how rapidly tropical forests are being deforested and degraded for three reasons. *First*, it is difficult to interpret satellite images. *Second*, some countries hide or exaggerate deforestation rates for political and economic reasons. *Third*, governments and international agencies define forest, deforestation, and forest degradation in different ways.

Because of these factors, estimates of global tropical forest loss vary from 50,000 square kilometers (19,300 square miles) to 170,000 square kilometers (65,600 square miles) per year. This rate is high enough to lose or degrade half of the world's remaining tropical forests in 35–117 years.

http://biology.brookscole.com/miller11 **169**

Figure 8-19 Natural capital degradation: each year large areas of tropical forest in Brazil's Amazon basin are burned to make way for cattle ranches and plantation crops. According to a 2003 study by NASA, the Amazon is slowly getting drier due to this practice. If this trend continues, it will prevent the restoration of forest by secondary ecological succession and convert a large area to tropical grasslands.

Herbert Giradet/Peter Arnold, Inc.

Economics and Politics: Causes of Tropical Deforestation and Degradation

The primary causes of tropical deforestation and degradation are population growth, poverty, environmentally harmful government subsidies, debts owed to developed countries, and failure to value their ecological services.

Tropical deforestation results from a number of inter-connected primary and secondary causes (Figure 8-20). Population growth and poverty combine to drive sub-sistence farmers and the landless poor to tropical forests, where they try to grow enough food to survive. Government subsidies can accelerate deforestation by making timber or other tropical forest resources such as land for cattle grazing cheap, relative to the economic value of the ecological services the forests provide.

Governments in Indonesia, Mexico, and Brazil also encourage the poor to

colonize tropical forests by giving them title to land that they clear. This practice can help reduce poverty but may lead to environmental degradation unless the new settlers are taught how to use such forests more sustainably, which is rarely done. In addition, interna-tional lending agencies encourage developing coun-tries to borrow huge sums of money from developed

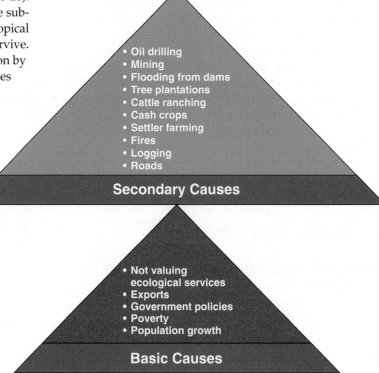

- Oil drilling
- Mining
- Flooding from dams
- Tree plantations
- Cattle ranching
- Cash crops
- Settler farming
- Fires
- Logging
- Roads

Secondary Causes

- Not valuing ecological services
- Exports
- Government policies
- Poverty
- Population growth

Basic Causes

Figure 8-20 Natural capital degradation: major interconnected causes of the destruction and degradation of tropical forests. The importance of specific secondary causes varies in different parts of the world.

countries to finance projects such as roads, mines, logging operations, oil drilling, and dams in tropical forests. Most countries also fail to value the ecological services of their forests (Figure 8-7, left).

The depletion and degradation of a tropical forest begin when a road is cut deep into the forest interior for logging and settlement (Figure 8-10). Loggers then use selective cutting to remove the best timber. Many other trees are toppled in the process because of their shallow roots and the network of vines connecting trees in the forest's canopy. Although timber exports to developed countries contribute significantly to tropical forest depletion and degradation, domestic use accounts for more than 80% of the trees cut in developing countries.

After the best timber has been removed, timber companies often sell the land to ranchers. Within a few years, they typically overgraze it and sell it to settlers who have migrated to the forest hoping to grow enough food to survive. Then they move their land-degrading ranching operations to another forest area. According to a 2004 report by the Center for International Forestry Research, the rapid spread of cattle ranching poses the biggest threat to the Amazon's tropical forests.

The settlers cut most of the remaining trees, burn the debris after it has dried, and plant crops in the cleared patches. They can also endanger some wild species by hunting them for meat. After a few years of crop growing and rain erosion, the nutrient-poor tropical soil is depleted of nutrients. Then the settlers move on to newly cleared land.

In some areas—especially Africa and Latin America—large sections of tropical forest are cleared for raising cash crops such as sugarcane, bananas, pineapples, strawberries, soybeans, and coffee—mostly for export to developed countries. Tropical forests are also cleared for mining and oil drilling and to build dams on rivers that flood large areas of the forest.

Healthy rain forests do not burn. But increased burning (Figure 8-19), logging, settlements, grazing, and farming along roads built in these forests results in patchy fragments of forest (Figure 8-10, right). When these areas dry out, they are readily ignited by lightning and burned by farmers and ranchers. In addition to destroying and degrading biodiversity, their combustion releases large amounts of carbon dioxide into the atmosphere.

Solutions: Reducing Tropical Deforestation and Degradation

There are a number of ways to slow and reduce the deforestation and degradation of tropical forests.

Analysts have suggested various ways to protect tropical forests and use them more sustainably (Figure 8-21). One method is to help new settlers in tropical

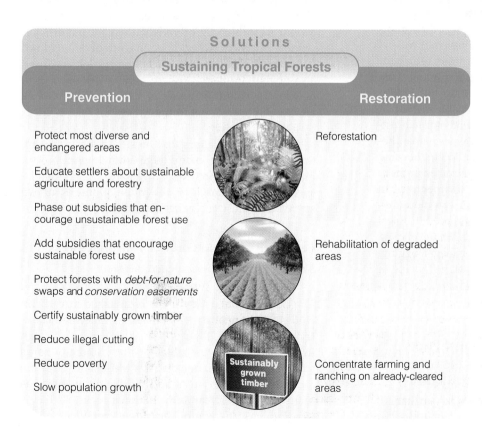

Solutions

Sustaining Tropical Forests

Prevention	Restoration
Protect most diverse and endangered areas	Reforestation
Educate settlers about sustainable agriculture and forestry	
Phase out subsidies that encourage unsustainable forest use	
Add subsidies that encourage sustainable forest use	Rehabilitation of degraded areas
Protect forests with *debt-for-nature* swaps and *conservation easements*	
Certify sustainably grown timber	
Reduce illegal cutting	
Reduce poverty	Concentrate farming and ranching on already-cleared areas
Slow population growth	

Figure 8-21 Solutions: ways to protect tropical forests and use them more sustainably. *Critical thinking: which three of these solutions do you believe are the most important?*

Kenya's Green Belt Movement

Wangari Maathai (Figure 8-A) founded the Green Belt Movement in Kenya in 1977. The goals of this highly regarded women's self-help group are to establish tree nurseries, raise seedlings, and plant and protect a tree for each of Kenya's 34 million people. By 2004, the 50,000 members of this grassroots group had established 6,000 village nurseries and planted and protected more than 30 million trees.

The success of this project has sparked the creation of similar programs in more than 30 other African countries. According to this inspiring leader:

© RADU SIGHETI/Reuters/Corbis

I don't really know why I care so much. I just have something inside me that tells me that there is a problem and I have got to do something about it. And I'm sure it's the *same voice that is speaking to everyone on this planet, at least everybody who seems to be concerned about the fate of the world, the fate of this planet.*

Figure 8-A Wangari Maathai, the first Kenyan woman to earn a Ph.D. (in anatomy) and to head an academic department (veterinary medicine) at the University of Nairobi, organized the internationally acclaimed Green Belt Movement in 1977. For her work in protecting the environment, she has received many honors, including the Goldman Prize, the Right Livelihood Award, the UN Africa Prize for Leadership, and the 2004 Nobel Peace Prize. After years of being harassed, beaten, and jailed for opposing government policies, she was elected to Kenya's parliament as a member of the Green Party in 2002. In 2003, she was appointed Assistant Minister for Environment and Natural Resources.

---

forests learn how to practice small-scale sustainable agriculture and forestry. Another method is to sustainably harvest some of the renewable resources such as fruits and nuts in rain forests.

We can also use *debt-for-nature swaps* to make it financially attractive for countries to protect their tropical forests. In such a swap, participating countries act as custodians of protected forest reserves in return for foreign aid or debt relief. Another important tool is using an international system for evaluating and certifying tropical timber produced by sustainable methods.

Loggers can also use gentler methods for harvesting trees. For example, cutting canopy vines (lianas) before felling a tree can reduce damage to neighboring trees by 20–40%, and using the least obstructed paths to remove the logs can halve the damage to other trees. In addition, governments and individuals can mount efforts to reforest and rehabilitate degraded tropical forests and watersheds (see Individuals Matter, above). Another suggestion is to clamp down on illegal logging.

8-6 NATIONAL PARKS

Global Outlook: Threats to National Parks

Countries have established more than 1,100 national parks, but most are threatened by human activities.

Today, more than 1,100 national parks larger than 10 square kilometers (4 square miles) each are located in more than 120 countries.

Unfortunately, according to a 1999 study by the World Bank and the Worldwide Fund for Nature, only 1% of the parks in developing countries receive protection. Local people invade most of them in search of wood, cropland, game animals, and other natural products for their daily survival. Loggers, miners, and wildlife poachers (who kill animals to obtain and sell items such as rhino horns, elephant tusks, and furs) also operate in many of these parks. Park services in developing countries typically have too little money and too few personnel to fight these invasions, either by force or by education.

Another problem is that most national parks are too small to sustain many large-animal species. Also, many parks suffer from invasions by nonnative species that can reduce the populations of native species and cause ecological disruption.

Case Study: Stresses on U.S. National Parks

National parks in the United States face many threats.

The U.S. national park system, established in 1912, includes 58 national parks (sometimes called the country's *crown jewels;* Figure 5-21, p. 94), most of them in the West (Figure 8-6). State, county, and city parks supplement these national parks. Most state parks are located near urban areas and receive about twice as many visitors per year as the national parks.

Popularity is one of the biggest problems of national and state parks in the United States. During the summer, users entering the most popular U.S. national and state parks often face hour-long backups and experience noise, congestion, eroded trails, and stress instead of peaceful solitude.

Many visitors expect parks to have grocery stores, laundries, bars, golf courses, video arcades, and other facilities found in urban areas. U.S. Park Service rangers spend an increasing amount of their time on law enforcement and crowd control instead of conservation, management, and education. Many overworked and underpaid rangers are leaving for better-paying jobs. In some parks, noisy dirt bikes, dune buggies, snowmobiles, and other off-road vehicles (ORVs) degrade the aesthetic experience for many visitors, destroy or damage fragile vegetation, and disturb wildlife.

Many parks suffer damage from the migration or deliberate introduction of nonnative species. European wild boars (imported to North Carolina in 1912 for hunting) threaten vegetation in part of the Great Smoky Mountains National Park. Nonnative mountain goats in Washington's Olympic National Park trample native vegetation and accelerate soil erosion.

Nearby human activities that threaten wildlife and recreational values in many national parks include mining, logging, livestock grazing, coal-burning power plants, water diversion, and urban development. Polluted air, drifting hundreds of kilometers, kills ancient trees in California's Sequoia National Park and often blots out the awesome views at Arizona's Grand Canyon. According to the National Park Service, air pollution affects scenic views in national parks more than 90% of the time.

Figure 8-22 lists suggestions that various analysts have made for sustaining and expanding the national park system in the United States.

8-7 NATURE RESERVES

Science, Economics, and Politics: Protecting Land from Human Exploitation

Ecologists believe that we should protect more land to help sustain the earth's biodiversity, but powerful economic and political interests oppose doing this.

Most ecologists and conservation biologists believe the best way to preserve biodiversity is to create a worldwide network of protected areas. Currently, 12% of the earth's land area is protected strictly or partially in nature reserves, parks, wildlife refuges, wilderness, and other areas. In other words, we have reserved 88% of the earth's land for us, and most of the remaining area consists of ice, tundra, or desert where we do not want to live because it is too cold or too hot.

The 12% figure is actually misleading because no more than 5% of these areas are truly protected. Thus

Solutions
National Parks

- Integrate plans for managing parks and nearby federal lands

- Add new parkland near threatened parks

- Buy private land inside parks

- Locate visitor parking outside parks and use shuttle buses for entering and touring heavily used parks

- Increase funds for park maintenance and repairs

- Survey wildlife in parks

- Raise entry fees for visitors and use funds for park management and maintenance

- Limit the number of visitors to crowded park areas

- Increase the number and pay of park rangers

- Encourage volunteers to give visitor lectures and tours

- Seek private donations for park maintenance and repairs

Figure 8-22 Solutions: suggestions for sustaining and expanding the national park system in the United States. *Critical thinking: which two of these solutions do you believe are the most important?* (Data from Wilderness Society and National Parks and Conservation Association)

we have strictly protected only 7% of the earth's terrestrial areas from potentially harmful human activities. See the map of our ecological footprints (Figures 2 and 3) in Science Supplement 2 at the end of this book.

Conservation biologists call for full protection of at least 20% of the earth's land area in a global system of biodiversity reserves that includes multiple examples of all the earth's biomes. Achieving this goal will require action and funding by national governments, private groups, and cooperative ventures involving governments, businesses, and private conservation groups.

Private groups play an important role in establishing wildlife refuges and other reserves to protect biological diversity. For example, since its founding by a group of professional ecologists in 1951, the *Nature Conservancy*—with more than 1 million members worldwide—has created the world's largest system of private natural areas and wildlife sanctuaries in 30 countries.

Most developers and resource extractors oppose protecting even the current 12% of the earth's remaining undisturbed ecosystems. They contend that these areas might contain valuable resources that would add to economic growth.

Ecologists and conservation biologists disagree. They view protected areas as islands of biodiversity that help sustain all life and economies, and serve as centers of future evolution. See Norman Myer's Guest Essay on this topic on the website for this chapter.

Whenever possible, conservation biologists call for using the *buffer zone concept* to design and manage nature reserves. This means protecting an inner core of a reserve by establishing two buffer zones in which local people can extract resources sustainably, in ways that do not harm the inner core. Doing so can enlist local people as partners in protecting a reserve from unsustainable uses. The United Nations has used this principle in creating its global network of 425 biosphere reserves in 95 countries (Figure 8-23).

Science Case Study: Costa Rica—A Global Conservation Leader

Costa Rica has devoted a larger proportion of land to conserving its significant biodiversity than any other country.

Tropical forests once completely covered Costa Rica, a Central American country that is smaller in area than West Virginia and about one-tenth the size of France.

Biosphere Reserve

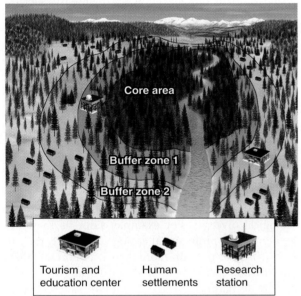

Figure 8-23 Solutions: a model *biosphere reserve*. Each reserve contains a protected inner core surrounded by two buffer zones that local and indigenous people can use for sustainable logging, food growing, cattle grazing, hunting, fishing, and eco-tourism.

Figure 8-24 Solutions: Costa Rica has consolidated its parks and reserves into eight *megareserves* designed to sustain about 80% of the country's rich biodiversity.

Between 1963 and 1983, politically powerful ranching families cleared much of the country's forests to graze cattle. They exported most of the beef produced to the United States and western Europe.

Despite such widespread forest loss, tiny Costa Rica is a superpower of biodiversity, with an estimated 500,000 plant and animal species. A single park in Costa Rica is home to more bird species than all of North America.

In the mid-1970s, the country established a system of reserves and national parks that by 2004 included about a quarter of its land—6% of it in reserves for indigenous peoples. Costa Rica now devotes a larger proportion of land to biodiversity conservation than does any other country!

The country's parks and reserves are consolidated into eight *megareserves* designed to sustain about 80% of Costa Rica's biodiversity (Figure 8-24). Each reserve contains a protected inner core surrounded by two buffer zones that local and indigenous people can use for sustainable logging, food growing, cattle grazing, hunting, fishing, and eco-tourism. Costa Rica's biodiversity conservation strategy has paid off. Today, the country's largest source of income is its $1-billion-a-year tourism business—almost two-thirds of it from eco-tourists.

To reduce deforestation, the government has eliminated subsidies for converting forest to cattle grazing land. It also pays landowners to maintain or restore tree coverage. The goal is to make sustaining forests profitable. The strategy has worked: Costa Rica has gone from having one of the world's highest deforestation rates to one of the lowest.

Solution: Adaptive Ecosystem Management

People with competing interests can work together to develop adaptable plans for managing and sustaining nature reserves.

Managing and sustaining a nature reserve is difficult. Reserves are constantly changing in response to environmental changes. In addition, their size, shape, and biological makeup often are determined by political, legal, and economic factors that depend on land ownership and conflicting public demands rather than by ecological principles and considerations.

One way to deal with these uncertainties and conflicts is through *adaptive ecosystem management.* This strategy relies on four principles.

First, integrate ecological, economic, and social principles to help maintain and restore the sustainability and biological diversity of reserves while supporting sustainable economies and communities.

Second, seek ways to get government agencies, private conservation organizations, scientists, business interests, and private landowners to reach a consensus on how to achieve common conservation objectives.

Third, view all decisions and strategies as scientific and social experiments, and use failures as opportunities for learning and improvement.

Fourth, emphasize continual information gathering, monitoring, reassessment, flexibility, adaptation, and innovation in the face of uncertainty and usually unpredictable change.

Figure 8-25 summarizes the adaptive ecosystem management process.

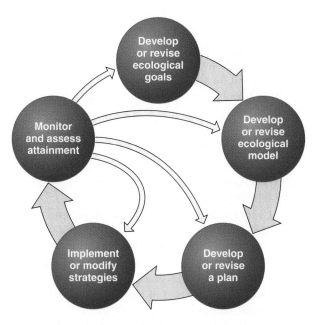

Figure 8-25 Solutions: the adaptive ecosystem management process.

Science and Stewardship: Biodiveresity Hot Spots

We can prevent or slow down losses of biodiversity by concentrating our efforts on protecting hot spots where there is an immediate threat to an area's biodiversity.

In reality, few countries are physically, politically, or financially able to set aside and protect large biodiversity reserves. To protect as much of the earth's remaining biodiversity as possible, conservation biologists use an *emergency action* strategy that identifies and quickly protects *biodiversity hot spots.* These "ecological arks" are areas especially rich in plant and animal species that are found nowhere else and are in great danger of extinction or serious ecological disruption.

Figure 8-26 (p. 176) shows 25 such hot spots. They contain almost two-thirds of the earth's terrestrial biodiversity and are the only homes for more than one-third of the planet's known terrestrial plant and animal species. Says Norman Myers, "I can think of no other biodiversity initiative that could achieve so much at a comparatively small cost, as the hot spots strategy."

Environmental Science ⊛ Now™
Learn more about hot spots around the world, what is at stake there, and how they are threatened at Environmental ScienceNow.

Natural Capital: Wilderness

Wilderness is land legally set aside in a large enough area to prevent or minimize harm from human activities.

One way to protect undeveloped lands from human exploitation is by legally setting them aside as wilderness. According to the U.S. Wilderness Act of 1964, *wilderness* consists of areas "of undeveloped land affected primarily by the forces of nature, where man is a visitor who does not remain." U.S. president Theodore Roosevelt summarized what we should do with wilderness: "Leave it as it is. You cannot improve it."

The U.S. Wilderness Society estimates that a wilderness area should contain at least 4,000 square kilometers (1,500 square miles). Otherwise, it can be affected by air, water, and noise pollution from nearby human activities.

Wild places are areas where people can experience the beauty of nature and observe natural biological diversity. They can also enhance the mental and physical health of visitors by allowing them to get away from noise, stress, development, and large numbers of people. Wilderness preservationist John Muir advised us:

Climb the mountains and get their good tidings. Nature's peace will flow into you as the sunshine into the

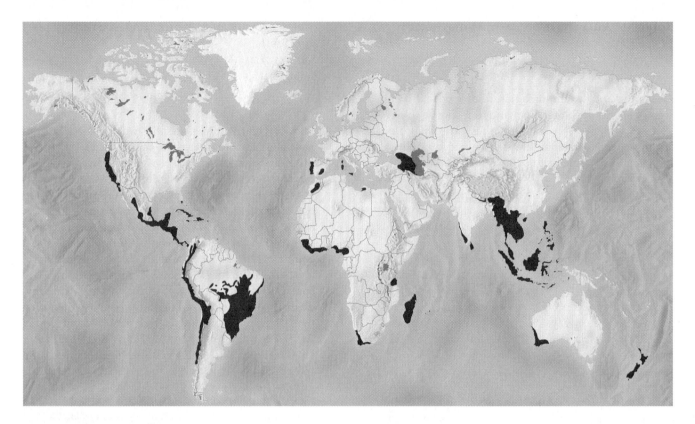

Environmental Science ⊕ Now™

Active Figure 8-26 Endangered natural capital: 25 *hot spots* identified by ecologists as important but endangered centers of biodiversity that contain a large number of endemic plant and animal species found nowhere else. Recent research has added nine additional hot spots to this list. *See an animation based on this figure and take a short quiz on the concept.* (Data from Center for Applied Biodiversity Science at Conservation International)

trees. The winds will blow their freshness into you, and the storms their energy, while cares will drop off like autumn leaves.

Even those who never use the wilderness areas may want to know they are there, a feeling expressed by novelist Wallace Stegner:

Save a piece of country . . . and it does not matter in the slightest that only a few people every year will go into it. This is precisely its value. . . . We simply need that wild country available to us, even if we never do more than drive to its edge and look in. For it can be a means of reassuring ourselves of our sanity as creatures, a part of the geography of hope.

Some critics oppose protecting wilderness for its scenic and recreational value for a small number of people. They believe this policy to be an outmoded concept that keeps some areas of the planet from being economically useful to humans.

Most biologists disagree. To them, the most important reasons for protecting wilderness and other areas from exploitation and degradation are to *preserve their biodiversity* as a vital part of the earth's natural capital

and to *protect them as centers for evolution* in response to mostly unpredictable changes in environmental conditions. In other words, wilderness is a biodiversity and wildness bank and an eco-insurance policy.

Some analysts also believe wilderness should be preserved because the wild species it contains have a right to exist (or struggle to exist) and play their roles in the earth's ongoing saga of biological evolution and ecological processes, without human interference.

Science and Politics Case Study: Wilderness Protection in the United States

Only a small percentage of the land area of the United States has been protected as wilderness.

In the United States, conservationists have been trying to save wild areas from development since 1900. Overall, they have fought a losing battle. Not until 1964 did Congress pass the Wilderness Act. It allowed the government to protect undeveloped tracts of public land from development as part of the National Wilderness Preservation System.

The area of protected wilderness in the United States increased tenfold between 1970 and 2000. Even so, only 4.6% of U.S. land is protected as wilderness—almost three-fourths of it in Alaska. Only 1.8% of the land area of the lower 48 states is protected, most of it in the West. In other words, Americans have reserved 98% of the continental United States to be used as they see fit and have protected only 2% as wilderness. According to a 1999 study by the World Conservation Union, the United States ranks 42nd among nations in terms of terrestrial area protected as wilderness, and Canada is in 36th place.

In addition, only 4 of the 413 wilderness areas in the lower 48 states are larger than 4,000 square kilometers (1,500 square miles). Also, the system includes only 81 of the country's 233 distinct ecosystems. Most wilderness areas in the lower 48 states are threatened habitat islands in a sea of development.

Almost 400,000 square kilometers (150,000 square miles) in scattered blocks of public lands could qualify for designation as wilderness—about 60% of it in the national forests. For two decades, these areas have been temporarily protected while they were evaluated for wilderness protection. Wilderness supporters would like to see all of them protected as part of the wilderness system. This is unlikely because of the political strength of industries that see these areas as resources for increased profits and short-term economic growth.

Figure 8-27 lists some ways you can help sustain the earth's terrestrial biodiversity.

What Can You Do?
Sustaining Terrestrial Biodiversity

- Plant trees and take care of them.

- Recycle paper and buy recycled paper products.

- Buy wood and wood products made from trees that have been grown sustainably.

- Help rehabilitate or restore a degraded area of forest or grassland near your home.

- When building a home, save all the trees and as much natural vegetation and soil as possible.

- Landscape your yard with a diversity of plants natural to the area instead of having a monoculture lawn.

- Live in town because suburban sprawl reduces biodiversity.

Figure 8-27 Individuals matter: ways to help sustain terrestrial biodiversity. *Critical thinking: which two of these actions do you believe are the most important? Which things in this list do you plan to do?*

8-8 ECOLOGICAL RESTORATION

Science: Rehabilitating and Restoring Damaged Ecosystems

Scientists have developed a number of techniques for rehabilitating and restoring degraded ecosystems and creating artificial ecosystems.

Almost every natural place on the earth has been affected or degraded to some degree by human activities. Much of the harm we have inflicted on nature is at least partially reversible through **ecological restoration**: the process of repairing damage caused by humans to the biodiversity and dynamics of natural ecosystems. Examples include replanting forests, restoring grasslands, restoring wetlands and stream banks, reclaiming urban industrial areas (brownfields), reintroducing native species (p. 154), removing invasive species, and freeing river flows by removing dams.

Farmer and philosopher Wendell Berry says we should try to answer three questions in deciding whether and how to modify or rehabilitate natural ecosystems. *First*, what is here? *Second*, what will nature permit us to do here? *Third*, what will nature help us do here? An important strategy is to mimic nature and natural processes and ideally let nature do most of the work, usually through secondary ecological succession.

By studying how natural ecosystems recover, scientists are learning how to speed up repair operations using a variety of approaches. They include the following measures:

- *Restoration:* trying to return a particular degraded habitat or ecosystem to a condition as similar as possible to its natural state. Unfortunately, lack of knowledge about the previous composition of a degraded area can make it impossible to restore an area to its earlier state.

- *Rehabilitation:* attempts to turn a degraded ecosystem back into a functional or useful ecosystem without trying to restore it to its original condition.

- *Replacement:* replacing a degraded ecosystem with another type of ecosystem. For example, a productive pasture or tree farm may replace a degraded forest.

- *Creating artificial ecosystems:* for example, the creation of artificial wetlands.

Researchers have suggested four basic science-based principles for carrying out ecological restoration.

- Mimic nature and natural processes and ideally let nature do most of the work, usually through secondary ecological succession (Figure 8-28, p. 178).

- Recreate important ecological niches that have been lost.

Figure 8-28 Natural capital restoration: in the mid-1980s, cattle had degraded the vegetation and soil on this stream bank along Arizona's San Pedro River (left). Within ten years, the area was restored through natural regeneration after banning grazing and off-road vehicles (right).

- Rely on pioneer species, keystone species (p. 154), foundation species, and natural ecological succession to facilitate the restoration process.

- Control or remove harmful nonnative species.

Some analysts worry that environmental restoration could encourage continuing environmental destruction and degradation by suggesting that any ecological harm we do can be undone. Some go further and say that we do not understand the incredible complexity of ecosystems well enough to restore or manage damaged natural ecosystems.

Restorationists agree that restoration should not be used as an excuse for environmental destruction. But they point out that so far we have been able to protect or preserve no more than about 7% of nature from the effects of human activities. Ecological restoration is badly needed for much of the world's ecosystems that we have already damaged. They also point out that if a restored ecosystem differs from the original system, it is better than nothing, and that increased experience will improve the effectiveness of ecological restoration.

✗ *How Would You Vote?* Should we mount a massive effort to restore ecosystems we have degraded even though it will be quite costly? Cast your vote online at http://biology .brookscole.com/miller11.

Science Case Study: Ecological Restoration of a Tropical Dry Forest in Costa Rica

A degraded tropical dry forest in Costa Rica is being restored in a cooperative project between tropical ecologists and local people.

Costa Rica is the site of one of the world's largest *ecological restoration* projects. In the lowlands of its Guanacaste National Park (Figure 8-24), a small tropical dry deciduous forest has been burned, degraded, and fragmented by large-scale conversion to cattle ranches and farms.

Today, it is being restored and relinked to the rain forest on adjacent mountain slopes. The goal is to eliminate damaging nonnative grass and cattle and reestablish a tropical dry forest ecosystem over the next 100–300 years.

Daniel Janzen, professor of biology at the University of Pennsylvania and a leader in the field of restoration ecology, has helped galvanize international support and has raised more than $10 million for this restoration project. He recognizes that ecological restoration and protection of the park will fail unless the people in the surrounding area believe they will benefit from such efforts. Janzen's vision to have the nearly 40,000 people who live near the park become an essential part of the restoration of the degraded forest, a concept he calls *biocultural restoration.*

By actively participating in the project, local residents reap educational, economic, and environmental benefits. Local farmers make money by sowing large areas with tree seeds and planting seedlings started in Janzen's lab. Local grade school, high school, and university students and citizens' groups study the park's ecology and visit it on field trips. The park's location near the Pan American Highway makes it an ideal area for eco-tourism, which stimulates the local economy.

The project also serves as a training ground in tropical forest restoration for scientists from all over the world. Research scientists working on the project give guest classroom lectures and lead some of the field trips.

In a few decades, today's children will be running the park and the local political system. If they understand the ecological importance of their local environment, they are more likely to protect and sustain its biological resources. Janzen believes that education, awareness, and involvement—not guards and fences—are the best ways to restore degraded ecosystems and protect largely intact ecosystems from unsustainable use.

8-9 SUSTAINING AQUATIC BIODIVERSITY

Science: What Do We Know about Aquatic Biodiversity?

We know fairly little about the biodiversity of the world's marine and freshwater systems that provides us with important economic and ecological services.

Although ocean water covers about 71% of the earth's surface, we have explored only about 5% of the earth's global ocean and know relatively little about its biodiversity and how it works. We also know relatively little about freshwater biodiversity. Scientific investigation of poorly understood marine and freshwater aquatic systems is a *research frontier* whose study could result in immense ecological and economic benefits.

Scientists have established three general patterns of marine biodiversity. *First,* the greatest marine biodiversity occurs in coral reefs, estuaries, and the deep-sea floor. *Second,* biodiversity is higher near coasts than in the open sea because of the greater variety of producers, habitats, and nursery areas in coastal areas. *Third,* biodiversity is higher in the bottom region of the ocean than in the surface region because of the greater variety of habitats and food sources on the ocean bottom.

We should care about aquatic biodiversity because it helps keep us alive and supports our economies. Marine systems provide a variety of important ecological and economic services (Figure 5-25, p. 97), as do freshwater systems (Figure 5-34, p. 104).

Natural Capital Degradation: Major Human Impacts on Aquatic Biodiversity

Humans have destroyed or degraded a large proportion of the world's coastal wetlands, coral reefs, mangroves, and ocean bottom, and overfished many marine and freshwater species.

The greatest threat to the biodiversity of the world's oceans is loss and degradation of habitats, as summarized in Figure 5-33, p. 103. For example, more than one-fourth of the world's diverse coral reefs have been severely damaged, mostly by human activities, and another 70% of these reefs may be severely damaged or eliminated by 2050.

Many sea-bottom habitats are also being degraded and destroyed by dredging operations and trawler boats, which, like giant submerged bulldozers, drag huge nets weighted down with heavy chains and steel plates over ocean bottoms to harvest bottom fish and shellfish. Each year, thousands of trawlers scrape and disturb an area of ocean bottom about 150 times larger than the area of forests clear-cut each year (Figure 8-29). In 2004, some 1,134 scientists signed a statement urging the United Nations to declare a moratorium on bottom trawling on the high seas.

Figure 8-29 Natural capital degradation: area of ocean bottom before (left) and after (right) a trawler net scraped it like a gigantic plow. These ocean floor communities can take decades or centuries to recover. According to marine scientist Elliot Norse, "Bottom trawling is probably the largest human-caused disturbance to the biosphere." Trawler fishers disagree, claiming that ocean bottom life recovers after trawling.

Studies indicate that about three-fourths of the world's 200 commercially valuable marine fish species (40% in U.S. waters) are either overfished or fished to their estimated sustainable yield. One result of the increasingly efficient global hunt for fish is that big fish in many populations of commercially valuable species are becoming scarce. Smaller fish are the next targets, as the fishing industry has begun working its way down marine food webs. This practice can reduce the breeding stock needed for recovery of depleted species, unravel food webs, and disrupt marine ecosystems. Today, we throw away almost one-third of the fish we catch because they are gathered unintentionally in our harvest of commercially valuable species.

According to marine biologists, at least 1,200 marine species have become extinct in the past few hundred years, and many thousands of additional marine species could disappear during this century. Indeed, *fish are threatened with extinction by human activities more than any other group of species.* And, freshwater animals are disappearing five times faster than land animals.

Science, Education, and Politics: Why Is It Difficult to Protect Marine Biodiversity?

Coastal development, the invisibility and vastness of the world's oceans, citizen unawareness, and lack of legal jurisdiction hinder protection of marine biodiversity.

Protecting marine biodiversity is difficult for several reasons. One problem is rapidly growing coastal development and the accompanying massive inputs of sediment and other wastes from land into coastal water. This practice harms shore-hugging species and threatens biologically diverse and highly productive coastal ecosystems such as coral reefs, marshes, and mangrove forest swamps.

Another problem is that much of the damage to the oceans and other bodies of water is not visible to most people. Also, many people incorrectly view the seas as an inexhaustible resource that can absorb an almost infinite amount of waste and pollution.

Finally, most of the world's ocean area lies outside the legal jurisdiction of any country. Thus, it is an open-access resource, subject to overexploitation because of the tragedy of the commons.

Solutions: Protecting and Sustaining Marine Biodiversity

Marine biodiversity can be sustained by protecting endangered species, establishing protected sanctuaries, managing coastal development, reducing water pollution, and preventing overfishing.

There are several ways to protect and sustain marine biodiversity. For example, we can *protect endangered*

and threatened aquatic species, as discussed in Chapter 9.

We can also *establish protected marine sanctuaries.* Since 1986, the World Conservation Union has helped establish a global system of *marine protected areas* (MPAs), mostly at the national level. Approximately 90 of the world's 350 biosphere reserves include coastal or marine habitats. In 2004, Australia's Great Barrier Reef—the world's largest living structure—became the world's biggest marine protected area when the government banned fishing and shipping on one-third of the reef.

Scientific studies show that within fully protected marine reserves, fish populations double, fish size grows by almost one-third, fish reproduction triples, and species diversity increases by almost one-fourth. Furthermore, this improvement happens within two to four years after strict protection begins and lasts for decades.

Unfortunately, less than 0.01% of the world's ocean area consists of fully protected marine reserves. In the United States, the total area of fully protected marine habitat is only 130 square kilometers (50 square miles). In other words, *we have failed to strictly protect 99.9% of the world's ocean area from human exploitation.* Also, many current marine sanctuaries are too small to protect most of the species within them and do not provide adequate protection from pollution that flows into coastal waters as a result of land use.

In 1997, a group of international marine scientists called for governments to increase fully protected marine reserves to at least 20% of the ocean's surface by 2020. A 2003 study by the Pew Fisheries Commission recommended establishing many more protected marine reserves in U.S. coastal waters and connecting them with protected corridors so that fish can move back and forth between such areas.

We can also establish *integrated coastal management* in which fishermen, scientists, conservationists, citizens, business interests, developers, and politicians collaborate to identify shared problems and goals. Then they attempt to develop workable and cost-effective solutions that preserve biodiversity and environmental quality while meeting economic and social needs, ideally using adaptive ecosystem management (Figure 8-25). Currently, more than 100 integrated coastal management programs are being developed throughout the world, including the Chesapeake Bay in the United States.

Another important strategy is to protect existing coastal and inland wetlands from being destroyed or degraded. We can also prevent overfishing. Figure 8-30 lists potential measures for managing global fisheries more sustainably and protecting marine biodiversity. Most of these approaches rely on some sort of government regulation. Finally, we can regulate and prevent aquatic pollution, as discussed in Chapter 11.

Fishery Regulations

Set catch limits well below the maximum sustainable yield

Improve monitoring and enforcement of regulations

Economic Approaches

Sharply reduce or eliminate fishing subsidies

Charge fees for harvesting fish and shellfish from publicly owned offshore waters

Certify sustainable fisheries

Protected Areas

Establish no-fishing areas

Establish more marine protected areas

Rely more on integrated coastal management

Consumer Information

Label sustainably harvested fish

Publicize overfished and threatened species

Bycatch

Use wide-meshed nets to allow escape of smaller fish

Use net escape devices for seabirds and sea turtles

Ban throwing edible and marketable fish back into the sea

Aquaculture

Restrict coastal locations for fish farms

Control pollution more strictly

Depend more on herbivorous fish species

Nonative Invasions

Kill organisms in ship ballast water

Filter organisms from ship ballast water

Dump ballast water far at sea and replace with deep-sea water

Figure 8-30 Solutions: ways to manage fisheries more sustainably and protect marine biodiversity. *Critical thinking: which five of these actions do you believe are the most important?*

8-10 WHAT CAN WE DO?

Global Outlook: The IUCN World Conservation Strategy

The IUCN has developed a global strategy for sustaining the world's biodiversity.

The IUCN, also known as the World Conservation Union, has developed a *World Conservation Strategy* for preserving and sustaining the world's biodiversity. It has the following three main objectives:

■ Maintain essential ecological services and life-support sysems that support human life and sustainable development

■ Preserve genetic diversity as a source of plants and domesticated animals needed as human sources of food

■ Ensure sustainable use of all species and ecosystems

Plans for accomplishing these goals include educating the public and policymakers about the importance of the earth's biological resources and services,

designing national conservation strategies, establishing a much larger global network of protected areas, greatly expanding biodiversity educational training and research, and establishing economic incentives to promote the conservation of species and ecosystems throughout the world.

Solutions: Establishing Priorities

Biodiversity expert Edward O. Wilson has proposed eight priorities for protecting most of the world's remaining ecosystems and species.

In 2002, Edward O. Wilson, considered to be one of the world's foremost experts on biodiversity, proposed the following priorities for protecting most of the world's remaining ecosystems and species:

■ *Take immediate action to preserve the world's biological hot spots* (Figure 8-26).

■ *Keep intact the world's remaining old-growth forests and cease all logging of such forests.*

■ *Complete the mapping of the world's terrestrial and aquatic biodiversity so we know what we have and can make conservation efforts more precise and cost-effective.*

- *Determine the world's marine hot spots and assign them the same priority for immediate action as for those on land.*

- *Concentrate on protecting and restoring everywhere the world's lakes and river systems, which are the most threatened ecosystems of all.*

- *Ensure that the full range of the earth's terrestrial and aquatic ecosystems is included in a global conservation strategy.*

- *Make conservation profitable.* This involves finding ways to raise the income of people who live in or near nature reserves so they can become partners in their protection and sustainable use.

- *Initiate ecological restoration products worldwide* to heal some of the damage we have done and increase the share of the earth's land and water allotted to the rest of nature.

According to Wilson, such a conservation strategy would cost about $30 billion per year—an amount that could be provided by a penny tax per cup of coffee.

This strategy for protecting the earth's precious biodiversity will not be implemented without bottom-up political pressure on elected officials from individual citizens and groups. It will also require cooperation among key people in government, the private sector, science, and engineering using adaptive ecosystem management (Figure 8-25).

We abuse land because we regard it as a commodity belonging to us. When we see land as a community to which we belong, we may begin to use it with love and respect.

ALDO LEOPOLD

CRITICAL THINKING

1. Explain why you agree or disagree with **(a)** the four principles that biologists and some economists have suggested for using public land in the United States (p. 159) and **(b)** the nine suggestions made by developers and resource extractors for managing and using U.S. public land (p. 159).

2. Explain why you agree or disagree with each of the proposals for providing more sustainable use of forests throughout the world, listed in Figure 8-15, p. 164. Which two of these proposals do you believe are the most important? Explain.

3. Should developed countries provide most of the money to preserve remaining tropical forests in developing countries? Explain.

4. In the early 1990s, Miguel Sanchez, a subsistence farmer in Costa Rica, was offered $600,000 by a hotel developer for a piece of land that he and his family had been using sustainably for many years. The land contained an old-growth rain forest and a black sand beach in an area under rapid development. Sanchez refused the offer. What would you have done if you were a poor subsistence farmer in Miguel Sanchez's position? Explain your decision.

5. Should there be a ban on the use of off-road motorized vehicles and snowmobiles on all public lands in the United States or in the country where you live? Explain.

6. Are you in favor of establishing more wilderness areas in the United States, especially in the lower 48 states (or in the country where you live)? Explain. What might be some drawbacks of this move?

7. If ecosystems are undergoing constant change, why should we **(a)** establish and protect nature reserves and **(b)** carry out ecological restoration?

8. Congratulations! You are in charge of the world. List the three most important features of your policies for using and managing **(a)** forests, **(b)** parks, and **(c)** aquatic biodiversity.

LEARNING ONLINE

The website for this book includes review questions for the entire chapter, flash cards for key terms and concepts, a multiple-choice practice quiz, interesting Internet sites, references, and a guide for accessing thousands of InfoTrac® College Edition articles.
    Visit

http://biology.brookscole.com/miller11

Then choose Chapter 8, and select a learning resource. For access to animations, additional quizzes, chapter outlines and summaries, register and log in to

Environmental Science ⊕ Now™

at **esnow.brookscole.com/miller11** using the access code card in the front of your book.

Active Graphing

Visit http://esnow.brookscole.com/miller11 to explore the graphing exercise for this chapter.

CASE STUDY

The Passenger Pigeon: Gone Forever

In 1813, bird expert John James Audubon saw a single flock of passenger pigeons that he estimated was 16 kilometers (10 miles) wide and hundreds of kilometers long, and contained perhaps a billion birds. The flock took three days to fly past him and was so dense that it darkened the skies.

By 1914, the passenger pigeon (Figure 9-1) had disappeared forever. How could a species that was once the most common bird in North America (and probably the world) become extinct in only a few decades? Humans wiped them out. The extinction of this species largely resulted from uncontrolled commercial hunting and loss of the bird's habitat and food supply as forests were cleared to make room for farms and cities.

Passenger pigeons were good to eat, their feathers made good filling for pillows, and their bones were widely used for fertilizer. They were easy to kill because they flew in gigantic flocks and nested in long, narrow colonies.

Commercial hunters would capture one pigeon alive, sew its eyes shut, and tie it to a perch called a stool. Soon a curious flock would land beside this "stool pigeon"—a term we now use to describe someone who turns in another person for breaking the law. Then the birds would be shot or ensnared by nets that might trap more than 1,000 birds at once.

Beginning in 1858, passenger pigeon hunting became a big business. Shotguns, traps, artillery, and even dynamite were used. People burned grass or sulfur below their roosts to suffocate the birds. Shooting galleries used live birds as targets. In 1878, one professional pigeon trapper made $60,000 by killing 3 million birds at their nesting grounds near Petoskey, Michigan!

By the early 1880s, only a few thousand birds remained. At that point, recovery of the species was doomed because the females laid only one egg per nest each year. On March 24, 1900, a young boy in Ohio shot the last known wild passenger pigeon. The last passenger pigeon on earth, a hen named Martha (after Martha Washington), died in the Cincinnati Zoo in 1914. Her stuffed body is now on view at the National Museum of Natural History in Washington, D.C.

Eventually all species become extinct or evolve into new species. Biologists estimate that every day 2–200 species become prematurely extinct primarily because of human activities. This rate of loss of biodiversity is expected to increase as the human population grows, consumes more resources, disturbs more of the earth's land and aquatic systems, and uses more of the earth's net primary productivity that supports all species.

Michael Sewell/Peter Arnold, Inc.

Figure 9-1 Lost natural capital: passenger pigeons have been extinct in the wild since 1900. The last known passenger pigeon died in the Cincinnati Zoo in 1914.

The last word in ignorance is the person who says of an animal or plant: "What good is it?" . . . If the land mechanism as a whole is good, then every part of it is good, whether we understand it or not. . . . Harmony with land is like harmony with a friend; you cannot cherish his right hand and chop off his left.

ALDO LEOPOLD

This chapter looks at the problem of premature extinction of species by human activities and ways to reduce this threat to the world's biodiversity. It addresses the following questions:

- How do biologists estimate extinction rates, and how do human activities affect these rates?

- Why should we care about protecting wild species?

- Which human activities endanger wildlife?

- How can we help prevent premature extinction of species?

- What is reconciliation ecology, and how can it help prevent premature extinction of species?

KEY IDEAS

- The current rate of extinction is at least 1,000 to 10,000 times the rate before humans arrived on earth, and it is expected to increase in the future.

- The greatest threat to a species is the loss and degradation of the place where it lives, followed by the accidental or deliberate introduction of harmful nonnative species.

- One of the world's most far-reaching and controversial environmental laws is the U.S. Endangered Species Act, which was passed in 1973.

- Reconciliation ecology involves finding ways to share the places that we dominate with other species.

9-1 SPECIES EXTINCTION

Science: Three Types of Species Extinction

Species can become extinct locally, ecologically, or globally.

Biologists distinguish among three levels of species extinction. *Local extinction* occurs when a species is no longer found in an area it once inhabited but is still found elsewhere in the world. Most local extinctions involve losses of one or more populations of species.

Ecological extinction occurs when so few members of a species are left that it can no longer play its eco-

logical roles in the biological communities where it is found.

In *biological extinction*, a species is no longer found anywhere on the earth (Figure 9-2 and Case Study, p. 183). Biological extinction is forever.

Science: Endangered and Threatened Species—Ecological Smoke Alarms

An endangered species could soon become extinct; a threatened species is likely to become extinct.

Biologists classify species heading toward biological extinction as either *endangered* or *threatened* (Figure 9-3, p. 186). An **endangered species** has so few individual survivors that the species could soon become extinct over all or most of its natural range. A **threatened species** (also known as a vulnerable species) is still abundant in its natural range but because of declining numbers is likely to become endangered in the near future.

Some species have characteristics that make them especially vulnerable to ecological and biological extinction (Figure 9-4, p. 188). As biodiversity expert Edward O. Wilson puts it, "the first animal species to go are the big, the slow, the tasty, and those with valuable parts such as tusks and skins."

One 2000 study found that human activities threaten several types of species with premature extinction (Figure 9-5, p. 188). Another 2000 survey by the Nature Conservancy and the Association for Biodiversity Information found that about one-third of the 21,000 plant and animal species in the United States are vulnerable to premature extinction.

Science: Estimating Extinction Rates

Scientists use measurements and models to estimate extinction rates.

Biologists estimate that more than 99.9% of all species that have ever existed are now extinct because of a combination of background extinctions, mass extinctions, and mass depletions taking place over thousands to millions of years. Biologists also talk of an *extinction spasm*, wherein large numbers of species are lost over a period of a few centuries or at most 1,000 years.

Biologists trying to catalog extinctions have three problems. *First*, the extinction of a species typically takes such a long time that it is not easy to document. *Second*, we have identified only 1.4–1.8 million of the world's estimated 5–100 million species. *Third*, we know little about most of the species we have identified.

The truth is that we do not know how many species become extinct each year mostly because of our activities. Scientists simply do the best they can with the tools they have to estimate past and projected future extinction rates.

Figure 9-2 Lost natural capital: some animal species that have become prematurely extinct largely because of human activities, mostly habitat destruction and overhunting.

| Passenger pigeon | Great auk | Dodo | Dusky seaside sparrow | Aepyornis (Madagascar) |

One approach is to study records documenting the rate at which mammals and birds have become extinct since humans arrived and comparing that rate with the fossil records of such extinctions prior to our arrival. To project future extinction rates, biologists may observe how the number of species present increases with the size of an area. This *species–area relationship* suggests that on average a 90% loss of habitat causes the extinction of 50% of the species living in that habitat.

Scientists also use models to estimate the risk that a particular species will become endangered or extinct within a certain period of time, based on factors such as trends in population size, changes in habitat availability, and interactions with other species. Estimates of extinction rates can vary because differing assumptions are made about the earth's total number of species, the proportion of these species that are found in tropical forests, the rate at which tropical forests are being cleared, and the reliability of the methods used to make these estimates.

Science: Effects of Human Activities on Extinction Rates

Biologists estimate that the current rate of extinction is at least 1,000 to 10,000 times the rate before humans arrived on earth.

In due time, all species become extinct. Nevertheless, considerable evidence indicates that humans are hastening the final exit for a growing number of species. Before we came on the scene, the estimated extinction rate was roughly one extinct species per million species on earth annually. This amounted to an annual extinction rate of about 0.0001% per year.

Using the methods just described, biologists conservatively estimate that the current rate of extinction is at least 1,000 to 10,000 times the rate before we ar-

rived. This amounts to an annual extinction rate of at least 0.1% to 1% per year.

How many species are we losing prematurely each year? The answer depends on how many species are on the earth. Assuming that the extinction rate is 0.1%, we lose 5,000 species per year if there are 5 million species on earth, 14,000 species if there are 14 million species (biologists' current best guess), and 100,000 species if there are 100 million species.

Most biologists would consider the premature loss of 1 million species over 100–200 years to be an extinction crisis or spasm that, if it continued, would lead to a mass depletion or even a mass extinction. At an extinction rate of 0.1% per year, the time it would take to lose 1 million species would be 200 years if there were a total of 5 million species, 71 years with a total of 14 million species, and 10 years with 100 million species.

According to researchers Edward O. Wilson and Stuart Primm, at a 1% extinction rate, at least one-fifth of the world's current animal and plant species could be gone by 2030 and half could vanish by the end of this century. In the words of biodiversity expert Norman Myers, "Within just a few human generations, we shall—in the absence of greatly expanded conservation efforts—impoverish the biosphere to an extent that will persist for at least 200,000 human generations or twenty times longer than the period since humans emerged as a species."

Most biologists consider extinction rates of 0.1%–1% to be conservative estimates for several reasons. *First,* both the rate of species loss and the extent of biodiversity loss are likely to increase during the next 50–100 years because of the projected growth of the world's human population and resource use per person. In other words, the size of our already large ecological footprint (Figure 1-7, p. 11, and Figure 2 in

Grizzly bear Kirkland's warbler White top pitcher plant Arabian oryx African elephant

Mojave desert tortoise Swallowtail butterfly Humpback chub Golden lion tamarin Siberian tiger

West Virginia spring salamander Giant panda Whooping crane Knowlton cactus Blue whale

Mountain gorilla Pine barrens tree frog Swamp pink Hawksbill sea turtle El Segunda blue butterfly

Figure 9-3 Endangered natural capital: species that are endangered or threatened with premature extinction largely because of human activities. Almost 30,000 of the world's species and 1,260 of those in the United States are officially listed as being in danger of becoming extinct. Most biologists believe the actual number of species at risk is much larger.

Florida manatee

Northern spotted owl

Gray wolf

Florida panther

Bannerman's turaco

Devil's hole pupfish

Snow leopard

Symphonia

Black-footed ferret

Utah prairie dog

Ghost bat

California condor

Black lace cactus

Black rhinoceros

Oahu tree snail

Science Supplement 2 at the end of this book) is likely to increase.

Second, current and projected extinction rates are much higher than the global average in parts of the world that are endangered centers of the world's biodiversity. Conservation biologists urge us to focus our efforts on slowing the much higher rates of extinction in such *hot spots* (Figure 8-26, p. 176) as the best and quickest way to protect much of the earth's biodiversity from being lost prematurely.

Third, we are eliminating, degrading, and simplifying many biologically diverse environments—such as tropical forests, tropical coral reefs, wetlands, and estuaries—that serve as potential colonization sites for the emergence of new species. Thus, in addition to increasing the rate of extinction, we may be limiting the long-term recovery of biodiversity by reducing the rate of speciation for some types of species. In other words, we are creating a *speciation crisis.* See the Guest Essay by Normal Myers on this topic on the website for this chapter.

Some people—but few biologists—say the current estimated extinction rates are too high and are based on inadequate data and models. Researchers agree that their estimates of extinction rates are based on inadequate data and sampling. They are continually striving to get better data and improve the models they use to estimate extinction rates.

Characteristic	Examples
Low reproductive rate (K-strategist)	Blue whale, giant panda, rhinoceros
Specialized niche	Blue whale, giant panda, Everglades kite
Narrow distribution	Many island species, elephant seal, desert pupfish
Feeds at high trophic level	Bengal tiger, bald eagle, grizzly bear
Fixed migratory patterns	Blue whale, whooping crane, sea turtles
Rare	Many island species, African violet, some orchids
Commercially valuable	Snow leopard, tiger, elephant, rhinoceros, rare plants and birds
Large territories	California condor, grizzly bear, Florida panther

Figure 9-4 Characteristics of species that are prone to ecological and biological extinction.

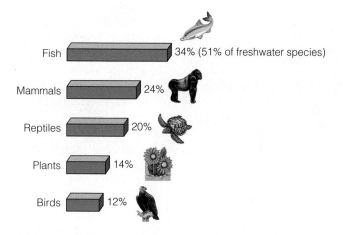

Fish — 34% (51% of freshwater species)

Mammals — 24%

Reptiles — 20%

Plants — 14%

Birds — 12%

Figure 9-5 **Endangered natural capital:** percentage of various types of species threatened with premature extinction because of human activities. (Data from World Conservation Union, Conservation International, and World Wildlife Fund)

At the same time, they point to clear evidence that human activities have increased the rate of species extinction and that this rate is likely to rise. According to these biologists, arguing over the numbers and waiting to get better data and models should not be used as excuses for inaction. They call for us to implement a *precautionary strategy* now to help prevent a significant decrease in the earth's biodiversity as a result of our activities.

To these biologists, we are not heeding Aldo Leopold's warning about preserving biodiversity as we tinker with the earth: "To keep every cog and wheel is the first precaution of intelligent tinkering."

Environmental Science⊕Now™
Learn more about how growing food, cutting trees, and other human activities affect biodiversity at Environmental ScienceNow.

9-2 IMPORTANCE OF WILD SPECIES

Science and Economics: Why Should We Preserve Wild Species?

We should not cause the premature extinction of species because of the economic and ecological services they provide.

So what is all the fuss about? If all species eventually become extinct, why should we worry about losing a few more because of our activities? Does it matter that the passenger pigeon, the remaining organutans (Figure 8-4, p. 157), or some unknown plant or insect in a tropical forest becomes prematurely extinct because of human activities?

New species eventually evolve to take the place of ones lost through extinction spasms, mass depletions, or mass extinctions (Figure 4-10, p. 74). So why should we care if we speed up the extinction rate over the next 50–100 years? *Because it will take at least 5 million years for natural speciation to rebuild the biodiversity we are likely to destroy during this century!*

Conservation biologists and ecologists say we should act now to prevent premature extinction of species because of their *instrumental value* based on their usefulness to us in the form of economic and ecological services (see top half of back cover). For example, some species provide economic value in the form of food crops, fuelwood and lumber, paper, and medicine (Figure 9-6).

Another instrumental value is the *genetic information* in species that allows them to adapt to changing environmental conditions and to form new species. Genetic engineers use this information to produce new types of crops (Figure 4-11, p. 75) and foods as well as edible vaccines for viral diseases such as hepatitis B. Carelessly eliminating many of the species making up the world's vast genetic library is like burning books before we read them. Wild species also provide a way for us to learn how nature works and sustains itself.

The earth's wild plants and animals also provide us with *recreational pleasure*. Each year, Americans

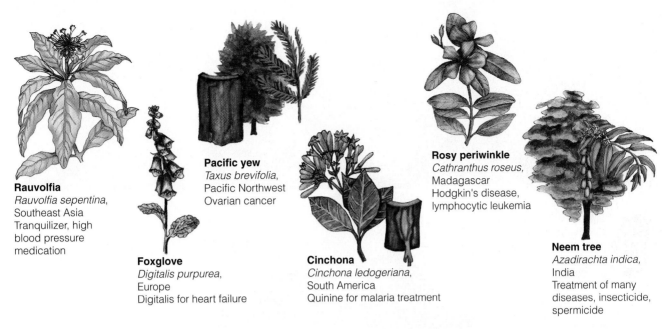

Rauvolfia
Rauvolfia sepentina,
Southeast Asia
Tranquilizer, high
blood pressure
medication

Foxglove
Digitalis purpurea,
Europe
Digitalis for heart failure

Pacific yew
Taxus brevifolia,
Pacific Northwest
Ovarian cancer

Cinchona
Cinchona ledogeriana,
South America
Quinine for malaria treatment

Rosy periwinkle
Cathranthus roseus,
Madagascar
Hodgkin's disease,
lymphocytic leukemia

Neem tree
Azadirachta indica,
India
Treatment of many
diseases, insecticide,
spermicide

Figure 9-6 Natural capital: *nature's pharmacy.* Parts of these and a number of other plants and animals (many of them found in tropical forests) are used to treat a variety of human ailments and diseases. Nine of the ten leading prescription drugs originally came from wild organisms. Approximately 2,100 of the 3,000 plants identified by the National Cancer Institute as sources of cancer-fighting chemicals come from tropical forests. Despite their economic and health potential, fewer than 1% of the estimated 125,000 flowering plant species in tropical forests (and a mere 1,100 of the world's 260,000 known plant species) have been examined for their medicinal properties. Once the active ingredients in the plants have been identified, they can usually be produced synthetically. Many of these tropical plant species are likely to become extinct before we can study them.

spend more than three times as many hours watching wildlife—doing nature photography and bird watching, for example—as they spend watching movies or professional sporting events.

Conservation biologist Michael Soulé estimates that one male lion living to age 7 generates $515,000 in eco-tourist dollars in Kenya but only $1,000 if it is killed for its skin. Similarly, over a lifetime of 60 years, a Kenyan elephant is worth about $1 million in eco-tourist revenue—many times more than its tusks are worth when they are sold illegally for their ivory.

Eco-tourism should not cause ecological damage but some of it does. The website for this chapter lists some guidelines for evaluating eco-tours.

Ethics: The Intrinsic Value of Wild Species

Some believe that each wild species has an inherent right to exist.

Some people believe that each wild species has *intrinsic* or *existence* value based on its inherent right to exist and play its ecological roles, regardless of its usefulness to humans. According to this view, we have an ethical responsibility to protect species from becoming prematurely extinct as a result of human activities, and to prevent the degradation of the world's ecosystems and its overall biodiversity.

Some people distinguish between the survival rights of plants and those of animals, mostly for practical reasons. Poet Alan Watts once said he was a vegetarian "because cows scream louder than carrots."

Other people distinguish among various types of species. For example, they might think little about getting rid of the world's mosquitoes, cockroaches, rats, or disease-causing bacteria.

Some conservation biologists caution us not to focus primarily on protecting relatively large organisms—the plants and animals we can see and are familiar with. They remind us that the true foundation of the earth's ecosystems and ecological processes are invisible bacteria and the algae, fungi, and other *microorganisms* that decompose the bodies of larger organisms and recycle the nutrients needed by all life.

9-3 CAUSES OF PREMATURE EXTINCTION OF WILD SPECIES

Science: Habitat Loss and Degradation— Remember HIPPO

The greatest threat to a species is the loss and degradation of the place where it lives.

Figure 9-7 (p. 190) shows the basic and secondary causes of the endangerment and premature extinction

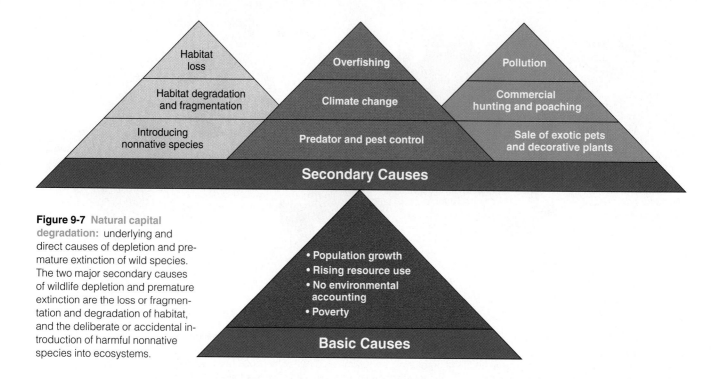

Figure 9-7 Natural capital degradation: underlying and direct causes of depletion and premature extinction of wild species. The two major secondary causes of wildlife depletion and premature extinction are the loss or fragmentation and degradation of habitat, and the deliberate or accidental introduction of harmful nonnative species into ecosystems.

The figure contains the following labels:

Secondary Causes:
- Habitat loss
- Habitat degradation and fragmentation
- Introducing nonnative species
- Overfishing
- Climate change
- Predator and pest control
- Pollution
- Commercial hunting and poaching
- Sale of exotic pets and decorative plants

Basic Causes:
- Population growth
- Rising resource use
- No environmental accounting
- Poverty

of wild species. Conservation biologists sometimes summarize the main secondary factors leading to premature extinction using the acronym **HIPPO**: **H**abitat destruction and fragmentation, **I**nvasive (alien) species, **P**opulation growth (too many people consuming too many resources), **P**ollution, and **O**verharvesting.

According to biodiversity researchers, the greatest threat to wild species is habitat loss (Figure 9-8), degradation, and fragmentation. Many species have a hard time surviving when we take over their ecological "house" and their food supplies and make them homeless.

Deforestation of tropical forests is the greatest eliminator of species, followed by the destruction of coral reefs and wetlands, plowing of grasslands, and pollution of streams, lakes, and oceans. Globally, temperate biomes have been affected more by habitat loss and degradation than have tropical biomes because of widespread development in temperate countries over the past 200 years. Development is currently shifting to many tropical biomes.

Island species—many of them *endemic species* found nowhere else on earth—are especially vulnerable to extinction when their habitats are destroyed, degraded, or fragmented (Figure 9-9, p. 192). Any habitat surrounded by a different one can be viewed as a *habitat island* for most of the species that live there. Most national parks and other protected areas are habitat islands, many of them encircled by potentially damaging logging, mining, energy extraction, and industrial activities. Freshwater lakes are also habitat islands that

are especially vulnerable to the introduction of nonnative species and pollution.

Science: Habitat Fragmentation

Species are more vulnerable to extinction when their habitats are divided into smaller, more isolated patches.

Habitat fragmentation occurs when a large, continuous area of habitat is reduced in area and divided into smaller, more scattered, and isolated patches or "habitat islands." This process divides populations of a species into smaller and more isolated groups that are more vulnerable to predators, competitive species, disease, and catastrophic events such as a storm or fire. Also, it creates barriers that limit the abilities of some species to disperse and colonize new areas, get enough to eat, and find mates.

Environmental Science Now™
See how serious the habitat fragmentation problem is for elephants, tigers, and rhinos at Environmental ScienceNow.

Science Case Study: A Disturbing Message from the Birds

Human activities are causing serious declines in the populations of many bird species.

Approximately 70% of the world's 9,800 known bird species are declining in numbers, and roughly one of every six bird species is threatened with extinction,

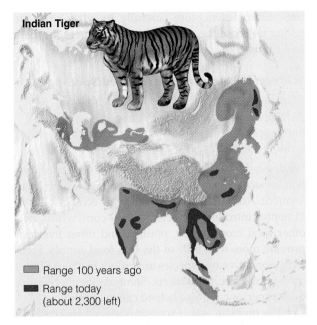

Indian Tiger

Range 100 years ago

Range today
(about 2,300 left)

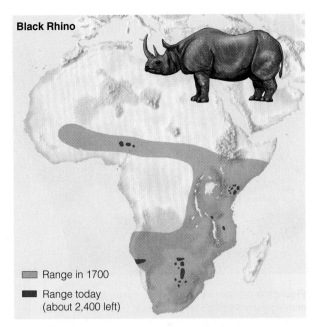

Black Rhino

Range in 1700

Range today
(about 2,400 left)

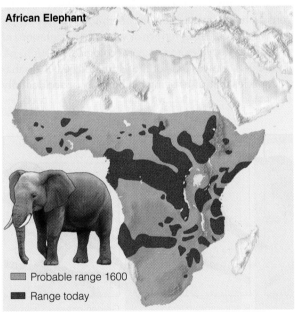

African Elephant

Probable range 1600

Range today

Asian or Indian Elephant

Former range

Range today
(34,000–54,000 left)

Environmental Science ⊕ Now™

Active Figure 9-8 Natural capital degradation: reductions in the ranges of four wildlife species, mostly as the result of habitat loss and hunting. What will happen to these and millions of other species when the world's human population doubles and per capita resource consumption rises sharply in the next few decades? *See an animation based on this figure and take a short quiz on the concept.* (Data from International Union for the Conservation of Nature and World Wildlife Fund)

mostly because of habitat loss and fragmentation. A 2002 National Audubon Society study found that one-fourth of all U.S. bird species are declining in numbers or are at risk of disappearing. Figure 9-10 (p. 192) shows the 10 most threatened U.S. songbird species.

Conservation biologists view this decline of bird species with alarm. Why? Birds are excellent *environmental indicators* because they live in every climate and

biome, respond quickly to environmental changes in their habitats, and are easy to track and count.

In addition, birds play important ecological roles. They help control populations of rodents and insects (which decimate many tree species), pollinate a variety of flowering plants, spread plants throughout their habitats by consuming and excreting plant seeds, and scavenge dead animals. Conservation biologists

In 1950, an estimated 100,000 tigers existed in the world. Despite international protection, fewer than 7,500 tigers now remain in the wild, about 4,000 of them in India. Bengal tigers are at risk because a tiger fur sells for $100,000 in Tokyo. With the body parts of a single tiger worth $5,000–20,000, it is not surprising that illegal hunting has skyrocketed, especially in India. Without emergency action to curtail poaching and preserve their habitat, few if any tigers may be left in the wild within 20 years.

As commercially valuable species become endangered, their black market demand soars. This increases their chances of premature extinction from poaching. For poachers, the money they can make far outweighs the small risk of being caught, fined, and imprisoned.

Economics Case Study: The Rising Demand for Bushmeat in Africa

Rapid population growth in parts of Africa has increased the number of people hunting wild animals for food or for sale of their meat to restaurants.

Indigenous people in much of West and Central Africa have sustainably hunted wildlife for *bushmeat,* a source of food, for centuries. But in the last two decades bushmeat hunting in some areas has skyrocketed as local people try to provide food for a rapidly growing population and to supply restaurants (Figure 9-15). The

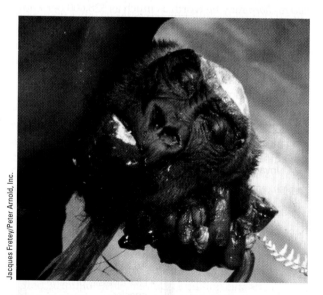

Jacques Fretey/Peter Arnold, Inc.

Figure 9-15 Natural capital degradation: *Bushmeat,* such as this severed head of a lowland gorilla in the Congo, is consumed as a source of protein by local people in parts of West Africa and sold in the national and international marketplace. You can find bushmeat on the menu in Cameroon and the Congo in West Africa as well as in Paris, France, and Brussels, Belgium. It is often supplied by illegal poaching. Wealthy patrons of some restaurants regard gorilla meat as a source of status and power.

bushmeat trade is also increasing in Southeast Asia, the Caribbean, and Central and South America.

So what is the big deal? After all, people have to eat. For most of our existence, humans have survived by hunting and gathering wild species.

The problem is that bushmeat hunting has caused the local extinction of many animals in parts of West Africa and has driven one species—Miss Waldron's red colobus monkey—to complete extinction. It is also a factor in greatly reducing gorilla, orangutan (Figure 8-4, p. 157), and chimpanzee populations.

This practice also threatens forest carnivores such as crowned eagles and leopards by depleting their main prey species. The forest itself is changed because of the decrease in seed-dispersing animals—those connections again.

Economics and Ethics: Killing Predators and Pest Control

Killing predators that bother us or cause economic losses threatens some species with premature extinction.

People often try to exterminate species that compete with them for food and game animals. For example, African farmers kill large numbers of elephants to keep them from trampling and eating food crops. Each year, U.S. government animal control agents shoot, poison, or trap thousands of coyotes, prairie dogs, wolves, bobcats, and other species that prey on livestock, on species prized by game hunters, on crops, or on fish raised in aquaculture ponds.

Since 1929, U.S. ranchers and government agencies have poisoned 99% of North America's prairie dogs (Figure 9-3) because horses and cattle sometimes step into the burrows and break their legs. This practice has also nearly wiped out the endangered black-footed ferret (Figure 9-3; about 600 left in the wild), which preyed on the prairie dog.

Economics and Ethics: Exotic Pets and Decorative Plants

Legal and illegal trade in wildlife species used as pets or for decorative purposes threatens some species with extinction.

The global legal and illegal trade in wild species for use as pets is a huge and very profitable business. However, for every live animal captured and sold in the pet market, an estimated 50 others are killed.

About 25 million U.S. households have exotic birds as pets, 85% of them imported. More than 60 bird species, mostly parrots, (Figure 8-5, p. 157), are endangered or threatened because of this wild bird trade. According to the U.S. Fish and Wildlife Service, collectors of exotic birds may pay $10,000 for a threatened hyacinth macaw smuggled out of Brazil. During its lifetime, a single macaw left in the wild might yield as much as $165,000 in tourist income. A 1992 study suggested that keeping a pet bird indoors for more than 10 years doubles a person's chances of getting lung cancer from inhaling tiny particles of bird dander.

Other wild species whose populations are depleted because of the pet trade include amphibians, reptiles, mammals, and tropical fish (taken mostly from the coral reefs of Indonesia and the Philippines). Divers catch tropical fish by using plastic squeeze bottles of cyanide to stun them. For each fish caught alive, many more die. In addition, the cyanide solution kills the coral animals that create the reef, which is a center for marine biodiversity.

Things do not have to be this way. Pilai Poonswad decided to do something about poachers taking hornbills—large, beautiful, and rare birds—from a rain forest in Thailand. She visited the poachers in their villages and showed them why the birds are worth more alive than dead. Today, some ex-poachers are earning much more money by taking eco-tourists into the forest to see these magnificent birds. Because of their vested financial interest in preserving the hornbills, they help protect them from poachers.

Some exotic plants, especially orchids and cacti, are endangered because they are gathered (often illegally) and sold to collectors to decorate houses, offices, and landscapes. A collector may pay $5,000 for a single rare orchid. A mature crested saguaro cactus can earn cactus rustlers as much as $15,000.

Clearly, collecting exotic pets and plants kills large numbers of them and endangers many of these species and others that depend on them. Are such collectors lovers or haters of the species they collect? Should we leave most exotic species in the wild?

Science: Climate Change and Pollution

Projected climate change and exposure to pollutants such as pesticides can threaten some species with premature extinction.

In the past, most natural climate changes have taken place over long periods of time—giving species more time to adapt or evolve into new species to cope with the change. Considerable evidence indicates that human activities such as greenhouse gas emissions and deforestation may bring about rapid climate change during this century. This could change the habitats of many species and accelerate the extinction of some species.

Pollution threatens populations and species in a number of ways. Unintended effects of pesticides threatens some species with extinction. According to the U.S. Fish and Wildlife Service, each year pesticides kill about one-fifth of the United States' beneficial honeybee colonies, more than 67 million birds, and 6–14 million fish. They also threaten one-fifth of the country's endangered and threatened species.

During the 1950s and 1960s, populations of fish-eating birds such as the osprey, cormorant, brown pelican, and bald eagle plummeted. A chemical derived from the pesticide DDT, when biologically magnified in food webs (Figure 9-16), made the birds' eggshells so fragile they could not reproduce successfully. Also hard hit were such predatory birds as the prairie falcon, sparrow hawk, and peregrine falcon, which help control rabbits, ground squirrels, and other crop eaters. *Good news:* Since the U.S. ban on DDT in 1972, most of these species have made a comeback.

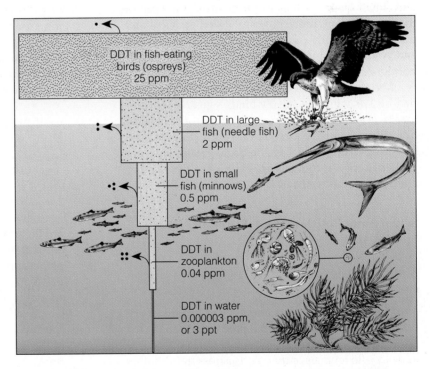

DDT in fish-eating birds (ospreys) 25 ppm

DDT in large fish (needle fish) 2 ppm

DDT in small fish (minnows) 0.5 ppm

DDT in zooplankton 0.04 ppm

DDT in water 0.000003 ppm, or 3 ppt

Figure 9-16 **Natural capital degradation:** *bioaccumulation* and *biomagnification.* DDT is a fat-soluble chemical that can accumulate in the fatty tissues of animals. In a food chain or web, the accumulated DDT can be biologically magnified in the bodies of animals at each higher trophic level. The concentration of DDT in the fatty tissues of organisms was biomagnified about 10 million times in this food chain in an estuary near Long Island Sound in New York. If each phytoplankton organism takes up from the water and retains one unit of DDT, a small fish eating thousands of zooplankton (which feed on the phytoplankton) will store thousands of units of DDT in its fatty tissue. Each large fish that eats 10 of the smaller fish will ingest and store tens of thousands of units, and each bird (or human) that eats several large fish will ingest hundreds of thousands of units. Dots represent DDT, and arrows show small losses of DDT through respiration and excretion.

9-4 PROTECTING WILD SPECIES: THE LEGAL APPROACH

Global Outlook and Politics: International Treaties

International treaties have helped reduce the international trade of endangered and threatened species, but enforcement is difficult.

Several international treaties and conventions help protect endangered or threatened wild species. One of the most far reaching is the 1975 *Convention on International Trade in Endangered Species (CITES)*. This treaty, now signed by 166 countries, lists some 900 species that cannot be commercially traded as live specimens or wildlife products because they are in danger of extinction. It also restricts international trade of 29,000 other species because they are at risk of becoming threatened.

CITES has helped reduce international trade in many threatened animals, including elephants, crocodiles, and chimpanzees. Unfortunately, the effects of this treaty are limited because enforcement varies from country to country, and convicted violators often pay only small fines. Also, member countries can exempt themselves from protecting any listed species, and much of the highly profitable illegal trade in wildlife and wildlife products goes on in countries that have not signed the treaty.

The *Convention on Biological Diversity (CBD)*, ratified by 190 countries, legally binds signatory governments to reversing the global decline of biological diversity. Its implementation has been slow because some key countries such as the United States have not ratified it. Also, it contains no severe penalties or other enforcement mechanisms.

Science and Politics: The U.S. Endangered Species Act

One of the world's most far-reaching and controversial environmental laws is the Endangered Species Act, which was passed in 1973.

The United States controls imports and exports of endangered wildlife and wildlife products through two laws. The *Lacey Act of 1900* prohibits transporting live or dead wild animals or their parts across state borders without a federal permit. The *Endangered Species Act of 1973 (ESA; amended in 1982, 1985, and 1988)* was designed to identify and legally protect endangered species in the United States and abroad. This act is probably the most far-reaching environmental law ever adopted by any nation, which has made it controversial. Canada and a number of other countries have similar laws.

The National Marine Fisheries Service (NMFS) is responsible for identifying and listing endangered and threatened ocean species. The U.S. Fish and Wildlife Services (USFWS) identifies and lists all other endangered and threatened species. Any decision by either agency to add or remove a species from the list must be based on biological factors alone, without consideration of economic or political factors. Economic factors can be used in deciding whether and how to protect endangered habitat, and in developing recovery plans for listed species.

The ESA also forbids federal agencies (except the Defense Department) to carry out, fund, or authorize projects that would jeopardize an endangered or threatened species or to destroy or modify the critical habitat it needs to survive. For offenses committed on private lands, fines as high as $100,000 and one year in prison can be imposed to ensure protection of the habitats of endangered species.

The ESA makes it illegal for Americans to sell or buy any product made from an endangered or threatened species. These species cannot be hunted, killed, collected, or injured in the United States. This protection has been extended to threatened and endangered foreign species.

In 2003, the George W. Bush administration proposed eliminating protection of foreign species, creating an uproar among conservationists. With this rule change, U.S. hunters, circuses, and the pet industry could pay individuals or governments to kill, capture, and import animals that are on the brink of extinction in other countries.

X *HOW WOULD YOU VOTE?* Should the U.S. Endangered Species Act no longer protect threatened and endangered species in other countries? Cast your vote online at http://www.biology.brookscole.com/miller11.

Between 1973 and 2005, the number of U.S. species on the official endangered and threatened list increased from 92 to about 1,260 species—60% of them plants and 40% animals. According to a 2000 study by the Nature Conservancy, one-third of the country's species are actually at risk of extinction, and 15% of all species are at high risk. This is far more than the 1,260 species on the ESA list. The study also found that many of the country's rarest and most imperiled species are concentrated in a few hot spots (Figure 9-17).

The ESA generally requires the secretary of the interior to designate and protect the *critical habitat* needed for the survival and recovery of each listed species. So far, critical habitats have been established for only one-third of the species on the ESA list, mostly because of political pressure and a lack of funds. Since 2001, the government has stopped listing new species and designating critical habitats for listed species unless required to do so by court order.

Getting a species listed is only half the battle. Next, the USFWS or the NMFS is supposed to prepare

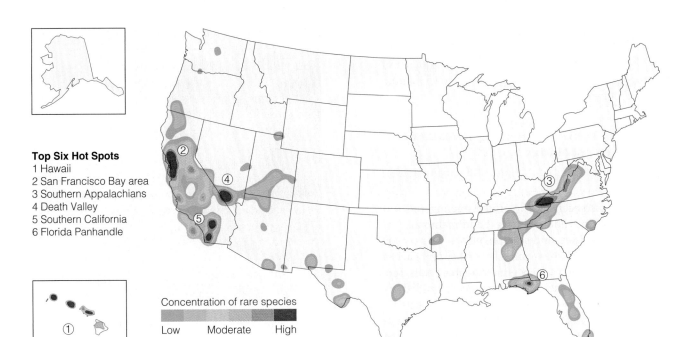

Concentration of rare species

Low   Moderate   High

Figure 9-17 Threatened natural capital: biodiversity hot spots in the United States. The shaded areas contain the largest concentrations of rare and potentially endangered species. (Data from State Natural Heritage Programs, the Nature Conservancy, and Association for Biodiversity Information)

a plan to help the species recover. By 2004, only one-fourth of the species on the endangered and threatened list had active plans. Most of the other plans existed only on paper, mostly because of political opposition and limited funds.

In 1982, Congress amended the ESA to allow the secretary of the interior to use *habitat conservation plans (HCP)*. They are designed to strike a compromise between the interests of private landowners and those of endangered and threatened species, with the goal of not reducing the recovery chances of a protected species.

With an HCP, landowners, developers, or loggers are allowed to destroy some critical habitat in exchange for taking steps to protect members of the species. Such measures might include setting aside a part of the species' habitat as a protected area, paying to relocate the species to another suitable habitat, or paying money to have the government buy suitable habitat elsewhere. Once the plan is approved it cannot be changed, even if new data show that the plan is inadequate to protect a species and help it recover.

Concern is growing that many of the plans may have been approved without enough scientific evaluation of their effects on a species' recovery. Also, many plans represent political compromises that do not protect the species or make inadequate provisions for its recovery.

The ESA also requires that all commercial shipments of wildlife and wildlife products enter or leave the country through one of nine designated ports. Few illegal shipments are confiscated (Figure 9-18) because

Steve Hillebrand/U.S. Fish and Wildlife Service

Figure 9-18 Natural capital degradation: confiscated products made from endangered species. Because of a scarcity of funds and inspectors, probably no more than one-tenth of the illegal wildlife trade in the United States is discovered. The situation is even worse in most other countries.

the 60 USFWS inspectors can examine only one-fourth of the approximately 90,000 shipments that enter and leave the United States each year. Even if caught, many violators are not prosecuted, and convicted violators often pay only a small fine.

Science and Politics: Protecting Threatened and Endangered Marine Species

We can use laws to reduce the premature extinction of marine species.

The ESA has also been used to protect endangered and threatened marine reptiles (turtles) and mammals (especially whales, seals, and sea lions). Each year plastic items dumped from ships and left as litter on beaches threaten the lives of millions of marine mammals, turtles, and seabirds that ingest, become entangled in, choke on, or are poisoned by such debris (Figure 9-19).

The world's eight major sea turtle species are endangered or threatened (Figure 9-20). Two major threats to these turtles are loss or degradation of beach habitat (where they come ashore to lay their eggs) and legal and illegal taking of their eggs. Other threats include unintentional capture and drowning by commercial fishing boats (especially shrimp trawlers) and increased use of the turtles as sources of food, medicinal ingredients, tortoiseshell (for jewelry), and leather from their flippers. In China, for example, some sea turtles sell for as much as $1,500.

Two major problems hinder our efforts to protect marine biodiversity by protecting endangered species. One is lack of knowledge about marine species. The other is the difficulty of monitoring and enforcing treaties to protect marine species, especially in the open ocean.

Figure 9-20 Threatened natural capital: endangered green sea turtle.

Figure 9-19 Threatened natural capital: before the discarded piece of plastic was removed, this Hawaiian monk seal was slowly starving to death. Each year plastic items dumped from ships and left as litter on beaches threaten the lives of millions of marine mammals, turtles, and seabirds.

Politics: Should the Endangered Species Act Be Weakened?

Some believe that the Endangered Species Act should be weakened or repealed because it has been a failure and hinders the economic development of private land.

Since 1992, Congress has been debating its reauthorization of the ESA. Proposals include those that would eliminate the act, weaken it, or strengthen it.

Opponents of the ESA contend that it puts the rights and welfare of endangered plants and animals above those of people, has not been effective in protecting endangered species, and has caused severe economic losses by hindering development on private land that contains endangered or threatened species. Since 1995, efforts to weaken the ESA have included the following suggested changes:

- Making protection of endangered species on private land voluntary.

- Having the government compensate landowners if it forces them to stop using part of their land to protect endangered species.

- Making it harder and more expensive to list newly endangered species by requiring government wildlife officials to navigate through a series of hearings and peer-review panels.

- Eliminating the need to designate critical habitats because developing and implementing a recovery plan is more important. Also, dealing with lawsuits for failure to develop critical habitats takes up most of the limited funds for carrying out the ESA.

- Allowing the secretary of the interior to permit a listed species to become extinct without trying to save it and to determine whether a species should be listed.

- Allowing the secretary of the interior to give any state, county, or landowner permanent exemption from the law, with no requirement for public notification or comment.

Other critics would go further and do away with this act. Because this step is politically unpopular with the American public, most efforts are designed to weaken the act and reduce its meager funding.

Science: Should the Endangered Species Act Be Strengthened?

According to most conservation biologists, the Endangered Species Act should be strengthened and modified to develop a new system to protect and sustain the country's biodiversity.

Most conservation biologists and wildlife scientists agree that the ESA has some deficiencies and needs to be simplified and streamlined. But they contend that the ESA has not been a failure (Case Study, below).

They also contest the charge that the ESA has caused severe economic losses. According to government records, since 1979 only 0.05% of the almost 200,000 projects evaluated by the USFWS have been blocked or canceled as a result of the ESA.

A study by the U.S. National Academy of Sciences recommended three major changes to make the ESA more scientifically sound and effective:

- Greatly increase the meager funding for implementing the act.

- Develop recovery plans more quickly.

- When a species is first listed, establish a core of its survival habitat as a temporary emergency measure that could support the species for 25–50 years.

Most biologists and wildlife conservationists believe the ESA should be changed to emphasize protecting and sustaining biological diversity and ecological functioning rather than focusing mostly on saving individual species. This new ecosystems approach would follow three principles:

- Find out what species and ecosystems the country has.

- Locate and protect the most endangered ecosystems (Figure 9-17) and species.

- Provide private landowners who agree to help protect specific endangered ecosystems with significant financial incentives (tax breaks and write-offs) and technical help.

CASE STUDY

What Has the Endangered Species Act Accomplished?

Critics of the ESA call it an expensive failure because only 37 species have been removed from the endangered list. Most biologists agree that the act needs strengthening and modification, but insist that it has not been a failure, for four reasons.

First, species are listed only when they face serious danger of extinction. This is like setting up a poorly funded hospital emergency room that takes only the most desperate cases, often with little hope for recovery, and saying it should be shut down because it has not saved enough patients.

Second, it takes decades for most species to become endangered or threatened. Not surprisingly, it also takes decades to bring a species in critical condition back to the point where it can be removed from the list. Expecting the ESA—which has been in existence only since 1973—to quickly repair the biological depletion of many decades is unrealistic.

Third, the conditions of almost 40% of the listed species are stable or improving. A hospital emergency room taking only the most desperate cases and then stabilizing or improving the condition of 40% of its patients would be considered an astounding success.

Fourth, the ESA budget was only $58 million in 2005—about what the Department of Defense spends in a little more than an hour or 20¢ per year per U.S. citizen. To its supporters, it seems amazing that the ESA has managed to stabilize or improve the conditions of almost 40% of the listed species on a shoestring budget.

Yes, the act can be improved and federal regulators have sometimes been too heavy handed in enforcing it. But instead of gutting or doing away with the ESA, biologists call for it to be strengthened and modified to help protect ecosystems and the nation's overall biodiversity.

Some critics say that only 20% of the endangered species are stable or improving. If correct, the ESA is still an incredible bargain.

Critical Thinking

Should the budget for the Endangered Species Act be drastically increased? Explain.

9-5 PROTECTING WILD SPECIES: THE SANCTUARY APPROACH

Science and Stewardshp: Wildlife Refuges and Other Protected Areas

The United States has set aside 542 federal refuges for wildlife, but many refuges are suffering from environmental degradation.

In 1903, President Theodore Roosevelt established the first U.S. federal wildlife refuge at Pelican Island, Florida. Since then, the National Wildlife Refuge System has grown to include 542 refuges. More than 35 million Americans visit these refuges each year to hunt, fish, hike, or watch birds and other wildlife.

More than three-fourths of the refuges serve as vital wetland sanctuaries for protecting migratory waterfowl. One-fifth of U.S. endangered and threatened species have habitats in the refuge system, and some refuges have been set aside for specific endangered species. These areas have helped Florida's key deer, the brown pelican, and the trumpeter swan to recover.

Conservation biologists call for setting aside more refuges for endangered plants. They also urge Congress and state legislatures to allow abandoned military lands that contain significant wildlife habitat to become national or state wildlife refuges. *Bad news:* According to a General Accounting Office study, activities considered harmful to wildlife occur in nearly 60% of the nation's wildlife refuges.

Science and Stewardshp: Gene Banks, Botanical Gardens, and Farms

Establishing gene banks and botanical gardens, and using farms to raise threatened species can help protect species from extinction, but these options lack funding and storage space.

Gene or *seed banks* preserve genetic information and endangered plant species by storing their seeds in refrigerated, low-humidity environments. More than 100 seed banks around the world collectively hold about 3 million samples.

Scientists urge the establishment of many more such banks, especially in developing countries. Unfortunately, some species cannot be preserved in gene banks. The banks are also expensive to operate and can be destroyed by accidents.

The world's 1,600 *botanical gardens* and *arboreta* contain living plants, representing almost one-third of the world's known plant species. However, they contain only about 3% of the world's rare and threatened plant species.

Botanical gardens also help educate an estimated 150 million visitors each year about the need for plant conservation. But these sanctuaries have too little storage capacity and too little funding to preserve most of the world's rare and threatened plants.

We can take pressure off some endangered or threatened species by raising individuals on *farms* for commercial sale. For example, farms in Florida raise alligators for their meat and hides. *Butterfly farms* flourish in Papua New Guinea, where many butterfly species are threatened by development activities.

Science and Stewardshp: Zoos and Aquariums

Zoos and aquariums can help protect endangered animal species, but efforts lack funding and storage space.

Zoos, aquariums, game parks, and animal research centers are being used to preserve some individuals of critically endangered animal species, with the long-term goal of reintroducing the species into protected wild habitats.

Two techniques for preserving endangered terrestrial species are egg pulling and captive breeding. *Egg pulling* involves collecting wild eggs laid by critically endangered bird species and then hatching them in zoos or research centers. In *captive breeding,* some or all of the wild individuals of a critically endangered species are captured for breeding in captivity, with the aim of reintroducing the offspring into the wild, such as with the peregrine falcon and the California condor (Figure 9-3).

Lack of space and money limits efforts to maintain populations of endangered species in zoos and research centers. The captive population of each species must number 100–500 individuals to avoid extinction through accident, disease, or loss of genetic diversity through inbreeding. Recent genetic research indicates that 10,000 or more individuals are needed for an endangered species to maintain its capacity for biological evolution.

Zoos and research centers contain only about 3% of the world's rare and threatened plant species. The major conservation role of these facilities will be to help educate the public about the ecological importance of the species they display and the need to protect their habitat.

Public aquariums that exhibit unusual and attractive fish and some marine animals such as seals and dolphins also help educate the public about the need to protect such species. In the United States, more than 35 million people visit aquariums each year. Public aquariums have not served as effective gene banks for endangered marine species, especially marine mammals that need large volumes of water.

Instead of seeing zoos and aquariums as sanctuaries, some critics claim that most of them imprison once-wild animals. They also contend that zoos and aquariums can foster the notion that we do not need to preserve large numbers of wild species in their natural habitats.

Other people criticize zoos and aquariums for putting on shows in which animals wear clothes, ride bicycles, or perform tricks. They see such exhibitions as fostering the idea that the animals exist primarily to entertain us and, in the process, raise money for their keepers.

Regardless of their benefits and drawbacks, zoos, aquariums, and botanical gardens are not biologically or economically feasible solutions for most of the world's current endangered species and the much larger number of species expected to be threatened over the next few decades.

9-6   RECONCILIATION ECOLOGY

Science and Stewardshp: Reconciliation Ecology

Reconciliation ecology involves finding ways to share the places that we dominate with other species.

In 2003, ecologist Michael L. Rosenzweig wrote a book entitled *Win-Win Ecology: How Earth's Species Can Survive in the Midst of Human Enterprise.* Rosenzweig strongly supports the eight-point program of Edward O. Wilson to help save the earth's natural habitats by establishing and protecting nature reserves (p. 181). He also supports the species protection strategies discussed in this chapter. But he contends that, in the long run, these approaches will fail for two reasons.

First, current reserves are devoted to saving only about 7% of the world's terrestrial area. To Rosenzweig, the real challenge is to help sustain wild species in the human-dominated portion of nature that makes up 97% of the planet's terrestrial ecological "cake," excluding polar and other uninhabitable areas.

Second, setting aside funds and refuges and passing laws to protect endangered and threatened species are essentially desperate attempts to save species that are in deep trouble. They can help a few species, but the real challenge is learning how to keep more species away from the brink of extinction.

Rosenzweig suggests that we develop a new form of conservation biology, called **reconciliation ecology**. This science focuses on inventing, establishing, and maintaining new habitats to conserve species diversity in places where people live, work, or play. In other words, we need to learn how to share the spaces we dominate with other species.

Science and Stewardshp: Implementing Reconciliation Ecology

Some people are finding creative ways to practice reconciliation ecology in their neighborhoods and cities.

Practicing reconciliation ecology begins by looking at the habitats we prefer. Given a choice, most people prefer a grassy and fairly open habitat with a few scattered trees and many people prefer to live near a stream, lake, river, or ocean. We also love flowers.

The problem is that most species do not like what we like or cannot survive in the habitats we prefer. No wonder so few of them live with us.

So what do we do? Reconciliation ecology goes far beyond efforts to attract birds to backyards. For example, providing a self-sustaining habitat for a butterfly species may require 20 or so neighbors to band together. Maintaining a habitat for an insect-eating bat species could help keep down mosquitoes and other pesky insects in a neighborhood.

Some monoculture grass yards could be replaced with diverse yards using plant species adapted to local climates that are selected to attract certain species. This would make neighborhoods more interesting, keep down insect pests, and require less use of noisy and polluting lawnmowers.

Communities could have contests and awards for people who design the most biodiverse and species-friendly yards and gardens. Signs could describe the type of ecosystem being mimicked and the species being protected as a way to educate and encourage experiments by other people. Some creative person might be able to design more biologically diverse golf courses and cemeteries. People have already worked together to help preserve bluebirds within human-dominated habitats (Science Spotlight, p. 204).

San Francisco's Golden Gate Park is a large oasis of gardens and trees in the midst of a major city. It is a good example of reconciliation ecology because it was designed and planted by humans who transformed it from a system of sand dunes.

The Department of Defense controls about 10 million hectares (25 million acres) of land in the United States. Perhaps some of this land could serve as laboratories for developing and testing reconciliation ecology ideas. Some college campuses and schools might also serve as reconciliation ecology laboratories. How about yours?

Clearly, protecting the species that make up part of the earth's biodiversity from premature extinction is a difficult, controversial, and challenging responsibility.

Using Reconciliation Ecology to Protect Bluebirds

SCIENCE SPOTLIGHT

Here is some *bad news:* Populations of bluebirds in much of the eastern United States are declining, for two reasons.

First, these birds nest in tree holes of a certain size. In the past, dead and dying trees provided plenty of these holes. Today, timber companies often cut down all of the trees, and many homeowners manicure their property by removing dead and dying trees.

Second, two aggressive, abundant, and nonnative bird species—starlings and house sparrows—like to nest in tree holes and take them away from bluebirds. To make matters worse, starlings eat blueberries that bluebirds need to survive during the winter.

Good news: People have come up with a creative way to help save the bluebird. They have designed nest boxes with holes large enough to accommodate bluebirds but too small for starlings. Because house sparrows like shallow boxes, they made the bluebird boxes deep enough to make them unattractive nesting sites for the sparrows.

In 1979, the North American Bluebird Society was founded to spread the word and encourage people to use the bluebird boxes on their properties and to keep house cats away from nesting bluebirds. Now bluebird numbers are building back up.

Restoration ecology works! Perhaps you might want to consider a career in this exciting new field.

Critical Thinking

Can you come up with a reconciliation project to help protect a threatened bird or other species in your neighborhood or school?

What Can You Do?

Protecting Species

- Do not but furs, ivory products, and other materials made from endangered or threatened animal species.
- Do not buy wood and paper products produced by cutting remaining old-growth forests in the tropics.
- Do not buy birds, snakes, turtles, tropical fish, and other animals that are taken from the wild.
- Do not buy orchids, cacti, and other plants that are taken from the wild.

Figure 9-21 Individuals matter: ways to help prevent the premature extinction of species. *Critical thinking: which two of these actions do you believe are the most important? Which of the actions do you plan to do?*

Figure 9-21 lists some things you can do to help prevent the premature extinction of species.

We know what to do. Perhaps we will act in time.

EDWARD O. WILSON

CRITICAL THINKING

1. How can **(a)** population growth, **(b)** poverty, and **(c)** affluence increase the premature extinction of wild species?

2. Discuss your gut-level reaction to the following statement: "Eventually, all species become extinct. Thus, it does not really matter that the passenger pigeon is extinct, and that the whooping crane and the world's remaining tiger species are endangered mostly because of human activities." Be honest about your reaction, and give arguments for your position.

3. Make a log of your own consumption of all products for a single day. Relate your level and types of consumption to the decline of wildlife species and the increased destruction and degradation of wildlife habitats in the United States (or the country where you live), in tropical forests, and in aquatic ecosystems.

4. Do you accept the ethical position that each *species* has the inherent right to survive without human interference, regardless of whether it serves any useful purpose for humans? Explain. Would you extend this right to the *Anopheles* mosquito, which transmits malaria, and to infectious bacteria? Explain.

5. Your lawn and house are invaded by fire ants, which can cause painful bites. What would you do?

6. Which of the following statements best describes your feelings toward wildlife:

(a) As long as it stays in its space, wildlife is okay.
(b) As long as I do not need its space, wildlife is okay.
(c) I have the right to use wildlife habitat to meet my own needs.
(d) When you have seen one redwood tree, fox, elephant, or some other form of wildlife, you have seen them all, so lock up a few of each species in a zoo or wildlife park and do not worry about protecting the rest.
(e) Wildlife should be protected.

7. List your three favorite species. Why are they your favorites? Are they cute and cuddly looking, like the giant panda and the koala? Do they have humanlike qualities, like apes or penguins that walk upright? Are they large, like elephants or blue whales? Are they beautiful, like tigers and monarch butterflies? Are any of them plants? Are any of them species such as bats, sharks, snakes, or spiders that many people fear? Are any of them microorganisms that help keep you alive? Reflect on what your choice of favorite species tells you about your attitudes toward most wildlife.

8. Environmental groups in a heavily forested state want to restrict logging in some areas to save the habitat of an endangered squirrel. Timber company officials argue that the well-being of one type of squirrel is not as important as the well-being of the many families who would be affected if the restriction causes the company to lay off hundreds of workers. If you had the power to decide this issue, what would you do and why? Can you come up with a compromise?

9. Congratulations! You are in charge of preventing the premature extinction of the world's existing species from human activities. What would be the three major components of your program to accomplish this goal?

LEARNING ONLINE

The website for this book includes review questions for the entire chapter, flash cards for key terms and concepts, a multiple-choice practice quiz, interesting Internet sites, references, and a guide for accessing thousands of InfoTrac® College Edition articles.

Visit

http://biology.brookscole.com/miller11

Then choose Chapter 9, and select a learning resource. For access to animations, additional quizzes, chapter outlines and summaries, register and log in to

Environmental Science ⊛ Now™

at **esnow.brookscole.com/miller11** using the access code card in the front of your book.

Active Graphing

Visit http://esnow.brookscole.com/miller11 to explore the graphing exercise for this chapter.

---

10 Food, Soil, and Pest Management

Soil

Pest Control

Biodiversity

Food Production

Soil Renewal

CASE STUDY

Would You Eat Winged Beans and Bug Cuisine?

Most of the world's people live mainly on a diet of wheat, rice, or corn—but these foods do not provide enough protein so that they can avoid malnutrition. Those who can afford it get protein by supplementing these grains with various types of beans and meat such as beef, chicken, pork, and fish.

The conventional approach to food production presents two problems. *First,* most of the poor cannot afford to eat meat. *Second,* modern agriculture has a greater harmful environmental impact than any other human activity.

Some analysts recommend greatly increased cultivation of less widely known plants to partially overcome these problems. One possibility is the *winged bean* (Figure 10-1), a fast-growing, protein-rich plant found in New Guinea and Southeast Asia. Its edible winged pods, spinach-like leaves, tendrils, and seeds contain as much protein as soybeans. The seeds can be ground into flour or used to make a caffeine-free beverage that tastes like coffee.

This plant has so many different edible parts that it has been called a *supermarket on a stalk.* It also needs little fertilizer because of nitrogen-fixing nodules in its roots—a trait that reduces the harmful environmental effects of cultivating it as a source of food.

Some edible insects—called *microlivestock*—are also important potential sources of protein, vitamins, and minerals in many parts of the world. There are about 1,500 edible insect species.

One example is *Mopani* (Figure 10-2)—emperor moth caterpillars—one of several insects eaten in South Africa. This food is so popular that the caterpillars

Figure 10-1 Natural capital: the winged bean, a fast-growing, protein-rich plant, could be grown to help reduce malnutrition and the harmful environmental effects from applying large amounts of inorganic fertilizer.

Sir Ghillean Prance/Visuals Unlimited

Anthony Bannister/CORBIS

Figure 10-2 Natural capital: Emperor moth caterpillars, called *Mopani*, and a number of other insects are used as a low-carbohydrate, low-fat, high-protein sources of food in many parts of the world.

(known as mopane worms) are being overharvested. Kalahari desert dwellers eat cockroaches, giant waterbugs are crushed and used in vegetable dip in Thailand, lightly toasted butterflies are a favorite food in Bali, black ant larvae are served in tacos in Mexico, and French-fried ants are sold and eaten like peanuts on the streets of Bogota, Colombia. Other examples are fried scorpions in peanut sauce, giant waterbugs tucked inside corn tamales, and a powder made from crickets and worms that is used to flavor foods.

A few cooks and chefs are having "critter tasting" parties. Numerous websites provide bug menus, bug nutritional information, and phone numbers for distributors of edible bugs.

Most of these insects are low-carbohydrate, low-fat sources of food and are 58–78% protein by weight—three to four times the level found in protein-rich foods such as beef, fish, chicken, or eggs. Consuming more of these plentiful insects would help reduce malnutrition and lessen the large environmental impact of producing conventional forms of meat.

Rapidly growing and reproducing edible bugs could be produced in small "bug farms" with little or no need for water, fertilizers, and pesticides. Indeed, people could have a small container for producing their favorite bugs in a garage, basement, or back yard.

Why don't we depend more on animals such as protein-rich insects and plants such as the winged bean? One problem is persuading farmers to assume the financial risk of cultivating new types of food crops. Another is convincing consumers to try new foods. Would you eat a bug soup?

A third problem is that seed companies and producers of conventional sources of meat are not pushing these alternative food sources because even a slight move toward adopting these sources of food would sharply reduce their profits.

There are two spiritual dangers in not owning a farm. One is the danger of supposing that breakfast comes from the grocery, and the other that heat comes from the furnace.

ALDO LEOPOLD

This chapter examines how food is produced and the resulting environmental effects, ways to control soil erosion, and methods for controlling populations of pests. It addresses the following questions:

- How is the world's food produced?

- How are green revolution and traditional methods used to raise crops?

- How are soils being degraded and eroded, and what can be done to reduce these losses?

- How much has food production increased, how serious is malnutrition, and what are the environmental effects of producing food?

- How can we increase production of crops, meat, and fish and shellfish?

- How do government policies affect food production?

- How can we design and shift to more sustainable or organic agricultural systems?

KEY IDEAS

- Food production from croplands, rangelands, and ocean fisheries has increased dramatically since 1950.

- Soil is eroding faster than it is forming on more than one-third of the world's cropland, and much of this land also suffers from salt buildup and waterlogging.

- One of every six people in developing countries is chronically undernourished or malnourished; one of every four people in the world is overweight.

- Modern agriculture has a greater harmful environmental impact than any human activity, and these effects may limit future food production.

- Three-fourths of the world's commercially valuable marine fish species are overfished or fished at their biological limit.

- We can produce food more sustainably by reducing resource throughput and working with nature.

10-1 FOOD PRODUCTION

Science: The Success of Food Production

Since 1950, food production from croplands, rangelands, and ocean fisheries has increased dramatically.

We depend on three systems for our food supply. *Croplands* mostly produce grains and provide about 77% of the world's food. *Rangelands* produce meat, mostly from grazing livestock, and supply about 16% of the world's food. *Oceanic fisheries* supply about 7% of the world's food.

Since 1950, there has been a staggering increase in global food production from all three systems. The growth occurred because of technological advances such as increased use of tractors and farm machinery and high-tech fishing boats and gear; inorganic chemical fertilizers; irrigation; pesticides; high-yield varieties of wheat, rice, and corn; densely populated feedlots and enclosed pens for raising cattle, pigs, and chickens; and aquaculture ponds and ocean cages for raising some types of fish and shellfish.

We face important challenges in increasing food production without causing serious environmental harm. To feed the 8.9 billion people projected to exist in 2050, we must produce and equitably distribute more food than has been produced since agriculture began about 10,000 years ago, and do so in an environmentally sustainable manner.

Can we achieve this goal? Some analysts say we can, mostly by using genetic engineering (Figure 4-11, p. 75). Others have doubts. They are concerned that environmental degradation, pollution, lack of water for irrigation, overgrazing by livestock, overfishing, and loss of vital ecological services may limit future food production.

We also face the challenge of sharply reducing poverty. One out of every five people does not have enough land to grow food or enough money to buy enough food—regardless of how much is available.

Science: Plants and Animals That Feed the World—The Big Three

Wheat, rice, and corn provide more than half of the calories in the food we consume.

Of the estimated 30,000 plant species with parts that people can eat only 14 plant and 8 terrestrial animal species supply an estimated 90% of our global intake of calories. Just three types of grain crops—*wheat, rice, and corn*—provide more than half the calories people consume.

Two-thirds of the world's people survive primarily on traditional grains (mainly rice, wheat, and corn), mostly because they cannot afford meat. As incomes rise, most people consume more meat and other products of domesticated livestock, which in turn means more grain consumption by those animals.

Fish and shellfish are an important source of food for about 1 billion people, mostly in Asia and in coastal areas of developing countries. On a global scale, however, fish and shellfish supply only 7% of the world's food and about 6% of the protein in the human diet.

Science: Industrial and Traditional Food Production

About 80% of the world's food supply is produced by industrialized agriculture and 20% by subsistence agriculture.

There are two major types of agricultural systems: *industrialized* and *traditional*. **Industrialized agriculture,** or **high-input agriculture,** uses large amounts of fossil fuel energy, water, commercial fertilizers, and pesticides to produce single crops (monocultures) or livestock animals for sale. Practiced on one-fourth of all cropland, mostly in developed countries (Figure 10-3), this form of agriculture has spread since the mid-1960s to some developing countries.

Plantation agriculture is a form of industrialized agriculture used primarily in tropical developing coun-

tries. It involves growing *cash crops* (such as bananas, coffee, soybeans, sugarcane, cocoa, and vegetables) on large monoculture plantations, mostly for sale in developed countries.

An increasing amount of livestock production in developed countries is industrialized. Large numbers of cattle are brought to densely populated *feedlots*, where they are fattened up for about 4 months before slaughter. Most pigs and chickens in developed countries spend their lives in densely populated pens and cages and eat mostly grain grown on cropland.

Traditional agriculture consists of two main types, which together are practiced by 2.7 billion people (42% of the world's people) in developing countries, and provide about one-fifth of the world's food supply. **Traditional subsistence agriculture** uses mostly human labor and draft animals to produce only enough crops or livestock for a farm family's survival. In **traditional intensive agriculture,** farmers increase their inputs of human and draft-animal labor, fertilizer, and

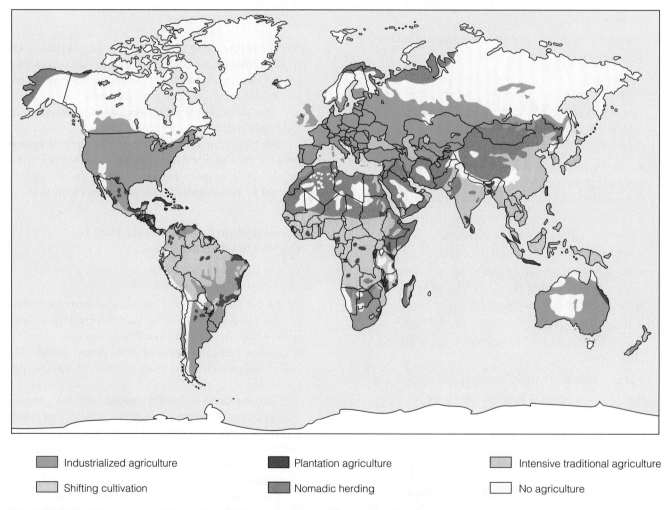

▨ Industrialized agriculture	■ Plantation agriculture	▨ Intensive traditional agriculture
▨ Shifting cultivation	▨ Nomadic herding	□ No agriculture

Figure 10-3 Global outlook: locations of the world's principal types of food production. Excluding Antarctica and Greenland, agricultural systems cover almost one-third of the earth's land surface.

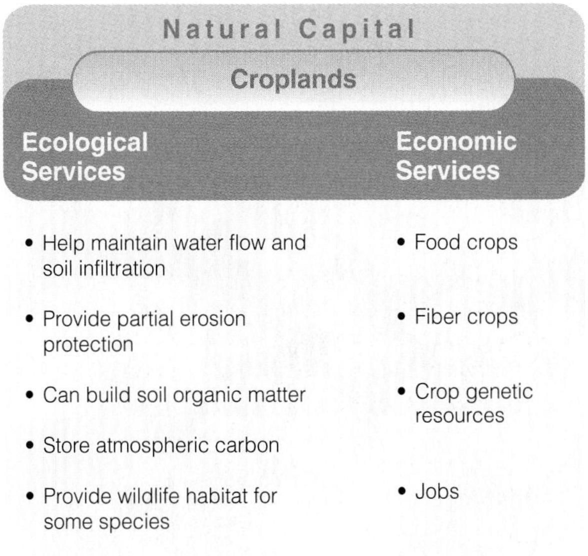

Natural Capital

Croplands

Ecological Services	Economic Services
• Help maintain water flow and soil infiltration	• Food crops
• Provide partial erosion protection	• Fiber crops
• Can build soil organic matter	• Crop genetic resources
• Store atmospheric carbon	
• Provide wildlife habitat for some species	• Jobs

Figure 10-4 Natural capital: ecological and economic services provided by croplands.

water to obtain a higher yield per area of cultivated land. They produce enough food to feed their families and to sell for income.

Croplands, like natural ecosystems, provide ecological and economic services (Figure 10-4). Indeed, agriculture is the world's largest industry, providing a living for one of every five people.

Science: Using Green Revolutions to Increase Food Production

Since 1950, most of the increase in global food production has come from using high-input agriculture to produce more crops on each unit of land.

Farmers can produce more food by farming more land or getting higher yields per unit of area from existing cropland. Since 1950, most of the increase in global food production has come from increased yields per unit of area of cropland in a process called the **green revolution.**

The green revolution involves three steps. *First,* develop and plant monocultures of selectively bred or genetically engineered high-yield varieties of key crops such as rice, wheat, and corn. *Second,* produce high yields by using large inputs of fertilizer, pesticides, and water. *Third,* increase the number of crops grown per year on a plot of land through *multiple cropping.*

This high-input approach dramatically increased crop yields in most developed countries between 1950 and 1970 in what is called the *first green revolution* (Figure 10-5, blue shading, p. 210).

A *second green revolution* has been taking place since 1967. Fast-growing dwarf varieties of rice (Figure 10-6, p. 210) and wheat (developed by Norman Bourlag, who later received a Nobel Peace Prize for his work), specially bred for tropical and subtropical climates, have been introduced into several developing countries (Figure 10-5, green). Producing more food on less land is also an important way to protect biodiversity by saving large areas of forests, grasslands, wetlands, and easily eroded mountain terrain from being used to grow food.

Yield increases depend not only on fertile soil and ample water but also on high inputs of fossil fuels to run machinery, produce and apply inorganic fertilizers and pesticides, and pump water for irrigation. In total, high-input, green revolution agriculture uses about 8% of the world's oil output.

Science and Economics Case Study: Industrial Food Production in the United States

The United States uses industrialized agriculture to produce about 17% of the world's grain in a very efficient manner.

In the United States, industrialized farming has evolved into *agribusiness,* as big companies and larger family-owned farms have taken control of almost three-fourths of U.S. food production.

In total annual sales, agriculture is bigger than the country's automotive, steel, and housing industries combined. It generates about 18% of the nation's gross domestic product and almost one-fifth of all jobs in the private sector, employing more people than any other industry. With only 0.3% of the world's farm labor force, U.S. farms produce about 17% of the world's grain and nearly half of its grain exports.

Since 1950, U.S. farmers have used green revolution techniques to more than double the yield of key crops such as wheat, corn, and soybeans without cultivating more land. Such increases in the yield per hectare of key crops have kept large areas of forests, grasslands, wetlands, and easily erodible land from being converted to farmland.

In addition, the country's agricultural system has become increasingly efficient. While the U.S. output of crops, meat, and dairy products has been increasing steadily since 1975, the major inputs of labor and resources—with the exception of pesticides—to produce each unit of that output have fallen steadily since 1950.

This industrialization of agriculture has been made possible by the availability of cheap energy, most of it from oil. Putting food on the table consumes about 17% of all commercial energy used in the United States each year (Figure 10-7, p. 211). The input of energy

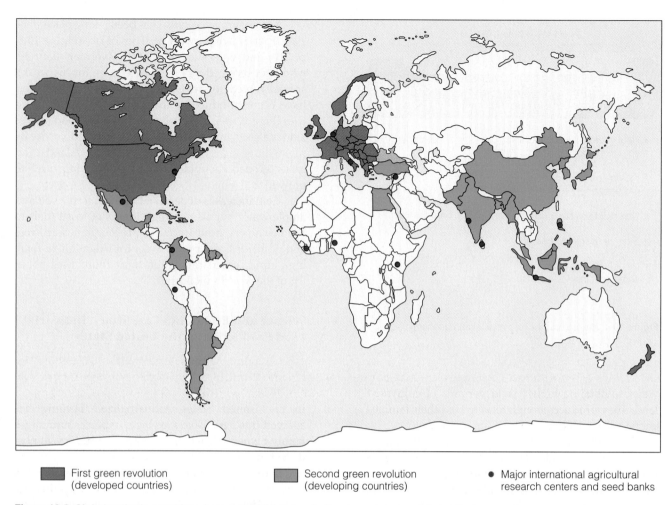

First green revolution (developed countries)

Second green revolution (developing countries)

• Major international agricultural research centers and seed banks

Figure 10-5 Global outlook: countries whose crop yields per unit of land area increased during the two green revolutions. The first (blue) took place in developed countries between 1950 and 1970; the second (green) has occurred since 1967 in developing countries with enough rainfall or irrigation capacity. Several agricultural research centers and gene or seed banks (red dots) play a key role in developing high-yield crop varieties.

Figure 10-6 Solutions: high-yield, semidwarf variety of rice called IR-8 (left), developed as part of the second green revolution. Crossbreeding two parent strains of rice produced it: PETA from Indonesia (center) and DGWG from China (right). The shorter and stiffer stalks of the new variety allow the plants to support larger heads of grain without toppling over and increase the benefit from applying more fertilizer.

International Rice Research Institute, Manila

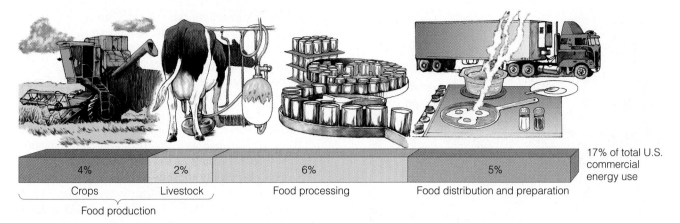

4%	2%	6%	5%	17% of total U.S. commercial energy use
Crops	Livestock	Food processing	Food distribution and preparation	

Food production

Figure 10-7 In the United States, industrialized agriculture uses about 17% of all commercial energy. Food travels an average 2,400 kilometers (1,500 miles) from farm to table. (Data from David Pimentel and Worldwatch Institute)

needed to produce a unit of food has fallen considerably, so that today most plant crops in the United States provide more food energy than the energy used to grow them.

Energy efficiency is much lower if we look at the whole U.S. food system. Considering the energy used to grow, store, process, package, transport, refrigerate, and cook all plant and animal food, *about 10 units of nonrenewable fossil fuel energy are needed to put 1 unit of food energy on the table.* By comparison, every unit of energy from human labor in traditional subsistence farming provides at least 1 unit of food energy and as many as 10 units of food energy using traditional intensive farming.

Global Outlook: Traditional Agriculture

Many traditional farmers in developing countries use low-input agriculture to produce a variety of crops on each plot of land.

Traditional farmers in developing countries today grow about one-fifth of the world's food on about three-fourths of its cultivated land. Many traditional farmers grow several crops on the same plot simultaneously, a practice known as **interplanting.** Such crop diversity reduces the chance of losing most or all of the year's food supply to pests, bad weather, and other misfortunes.

Interplanting strategies vary. **Polyvarietal cultivation** involves planting a plot with several genetic varieties of the same crop. In **intercropping,** two or more different crops are grown at the same time on a plot—for example, a carbohydrate-rich grain that uses soil nitrogen and a nitrogen-fixing plant (legume) that puts it back. In **agroforestry,** or **alley cropping,** crops and trees are grown together.

A fourth type of interplanting is **polyculture,** in which many different plants maturing at various times

are planted together. Low-input polyculture offers a number of advantages. There is less need for fertilizer and water because root systems at different depths in the soil capture nutrients and moisture efficiently. This practice provides more protection from wind and water erosion because the soil is covered with crops year-round. Insecticides are rarely needed because multiple habitats are created for natural predators of crop-eating insects. Also, there is little or no need for herbicides because weeds have trouble competing with the multitude of crop plants. The diversity of crops raised provides insurance against bad weather. Polyculture is a way of growing food by copying nature.

Recent ecological research has shown that, on average, low-input polyculture produces higher yields per hectare of land than high-input monoculture. For example, a 2001 study by ecologists Peter Reich and David Tilman found that carefully controlled polyculture plots with 16 different species of plants consistently outproduced plots with 9, 4, or only 1 type of plant species.

10-2 SOIL EROSION AND DEGRADATION

Science: Causes of Soil Erosion

Water, wind, and people cause soil erosion.

Most people in developed countries get their food from grocery stores, fast-food chains, and restaurants. But ultimately *all food comes from the earth or soil*—the base of life. This explains why preserving the world's topsoil (Figure 3-21, p. 51) is the key to producing enough food to feed the world's growing population.

Land degradation occurs when natural or human-induced processes decrease the ability of land to support crops, livestock, or wild species in the future. One

Figure 10-8 Natural capital degradation: erosion of vital top-soil from irrigated cropland in Arizona.

type of land degradation is **soil erosion:** the movement of soil components, especially surface litter and top-soil, from one place to another. The two main agents of erosion are *flowing water* and *wind,* with water causing most soil erosion (Figure 10-8).

Some soil erosion is natural; some is caused by human activities. In undisturbed vegetated ecosystems, the roots of plants help anchor the soil, and usually soil is not lost faster than it forms. Soil becomes more vulnerable to erosion through human activities that destroy plant cover, including farming, logging, construction, overgrazing by livestock, off-road vehicle use, and deliberate burning of vegetation.

More severe *gully erosion* (Figure 10-9) occurs when rivulets of fast-flowing water join together and

with each succeeding rain cut the channels wider and deeper until they become ditches or gullies. Gully erosion usually happens on steep slopes where all or most vegetation has been removed.

Soil erosion has two major harmful effects. One is *loss of soil fertility* through depletion of plant nutrients in topsoil. The other occurs when eroded soil ends up as sediment in nearby surface waters, where it can pollute water, kill fish and shellfish, and clog irrigation ditches, boat channels, reservoirs, and lakes.

Soil—especially topsoil—is classified as a renewable resource because natural processes regenerate it. If topsoil erodes faster than it forms on a piece of land, it eventually becomes a nonrenewable resource.

Science: Global Soil Erosion

Soil is eroding faster than it is forming on more than one-third of the world's cropland.

A 1992 joint survey by the United Nations (UN) Environment Programme and the World Resources Institute estimated that topsoil is eroding faster than it forms on about 38% of the world's cropland (Figure 10-10). According to a 2000 study by the Consultative Group on International Agricultural Research, soil erosion and degradation have reduced food production on 16% of the world's cropland. See the Guest Essay on soil erosion by David Pimentel on the website for this chapter.

Some analysts contend that erosion estimates are overstated because they underestimate the abilities of some local farmers to restore degraded land. The UN Food and Agriculture Organization (FAO) also points out that much of the eroded topsoil does not go far and

Figure 10-9 Natural capital degradation: severe gully erosion on cropland in Bolivia after trees that held the soil in place were cut.

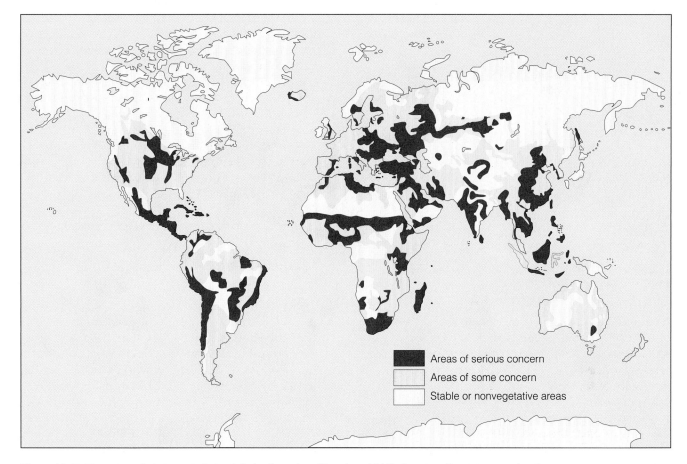

Figure 10-10 Natural capital degradation: global soil erosion. (Data from UN Environment Programme and the World Resources Institute)

is deposited farther down a slope, valley, or plain. In some places, the loss in crop yields in one area could be offset by increased yields elsewhere.

Science Case Study: Soil Erosion in the United States

Soil in the United States is eroding faster than it forms on most cropland, but since 1987, erosion has been cut by about two-thirds.

Bad news. According to the Natural Resources Conservation Service, soil on cultivated land in the United States is eroding about 16 times faster than it can form. Erosion rates are even higher in heavily farmed regions. The Great Plains, for example, has lost one-third or more of its topsoil in the 150 years since it was first plowed.

Good news. Of the world's major food-producing nations, only the United States is sharply reducing some of its soil losses through a combination of planting crops without disturbing the soil and government-sponsored soil conservation programs.

These efforts to slow soil erosion are an important step and, since 1985, have cut soil losses on U.S. cropland by about two-thirds. However, effective soil conservation is practiced today on only half of all U.S. agricultural land and on half of the country's most erodible cropland.

Science: Desertification

About one-third of the world's land has lost some of its productivity from a combination of drought and human activities that reduce or degrade topsoil.

In **desertification,** the productive potential of drylands (arid or semiarid land) falls by 10% or more because of a combination of natural climate change that causes prolonged drought and human activities that reduce or degrade topsoil. The process can be *moderate* (a 10–25% drop in productivity), *severe* (a 25–50% drop), or *very severe* (a drop of 50% or more, usually creating huge gullies and sand dunes). Only in extreme cases does desertification lead to what we call desert.

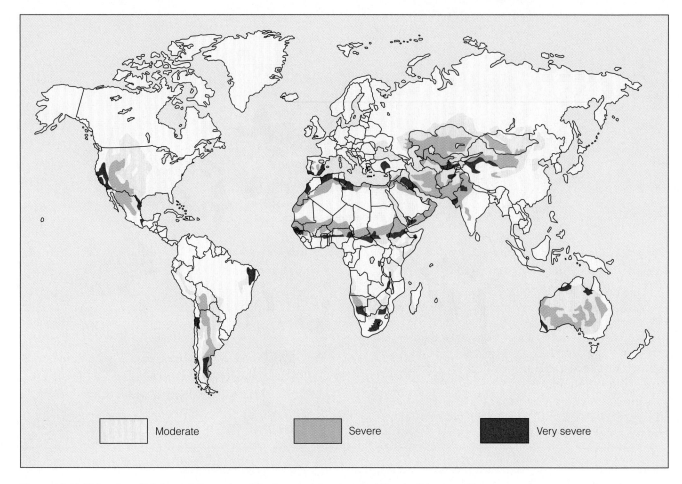

| | Moderate | | Severe | | Very severe |

Figure 10-11 Natural capital degradation: desertification of arid and semiarid lands. It is caused by a combination of prolonged drought and human activities that expose soil to erosion. (Data from UN Environment Programme and Harold E. Drengue)

Over thousands of years the earth's deserts have expanded and contracted, mostly because of natural climate changes. However, human activities can accelerate desertification in some parts of the world (Figure 10-11).

According to a 2004 report by the United Nations, an area the size of Brazil and 12 times the size of Texas has become desertified in the past 50 years. Each year since 1990 about 6 million hectares (15 million acres) of land has become degraded and less fertile from desertification. According to a 2003 UN conference on desertification, one-third of the world's land and 70% of all drylands are suffering from the effects of desertification. UN officials estimate that this loss of soil productivity threatens the livelihoods of at least 250 million people in 110 countries (70 in Africa). China is facing serious desertification, as its portion of the Gobi Desert expanded by an area half the size of Pennsylvania between 1994 and 1999.

Figure 10-12 summarizes the major causes and consequences of desertification. We cannot control when or where prolonged droughts may occur, but we can reduce overgrazing, deforestation, and destructive forms of planting, irrigation, and mining that leave behind barren soil. We can also restore land suffering from desertification by planting trees and grasses that anchor soil and hold water, establishing windbreaks, and growing trees and crops together (agroforestry).

Science: Salinization and Waterlogging of Soils

Repeated irrigation can reduce crop productivity by salt buildup in the soil and waterlogging of crop plants.

The one-fifth of the world's cropland that is irrigated produces almost 40% of the world's food. But irrigation has a downside. Most irrigation water is a dilute solution of various salts, picked up as the water flows over or through soil and rocks. Irrigation water not absorbed into the soil evaporates, leaving behind a thin crust of dissolved salts (such as sodium chloride) in the topsoil.

Repeated annual applications of irrigation water lead to the gradual accumulation of salts in the upper soil layers, a process called **salinization** (Figure 10-13). It stunts crop growth, lowers crop yields, and eventually kills plants and ruins the land.

Causes	Consequences
Overgrazing	Worsening drought
Deforestation	Famine
Erosion	Economic losses
Salinization	Lower living standards
Soil compaction	Environmental refugees
Natural climate change	

Figure 10-12 Natural capital degradation: causes and consequences of desertification.

According to a 1995 study, severe salinization has reduced yields on about one-fifth of the world's irrigated cropland, and another one-third has been moderately salinized. The most severe salinization occurs in Asia, especially in China, India, and Pakistan.

Salinization affects almost one-fourth of irrigated cropland in the United States. The proportion of damaged land is much higher in some heavily irrigated western states (Figure 10-14).

We know how to prevent and deal with soil salinization, as summarized in Figure 10-15. Unfortunately, some of these remedies are expensive.

Figure 10-14 Natural capital degradation: Because of high evaporation, poor drainage, and severe salinization, white alkaline salts have displaced crops that once grew on this heavily irrigated land in Colorado.

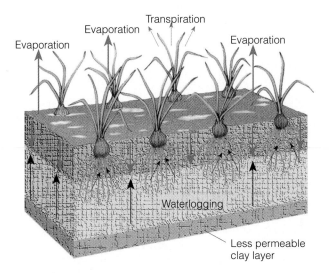

Salinization

1. Irrigation water contains small amounts of dissolved salts.

2. Evaporation and transpiration leave salts behind.

3. Salt builds up in soil.

Waterlogging

1. Precipitation and irrigation water percolate downward.

2. Water table rises.

Figure 10-13 Natural capital degradation: *Salinization* and *waterlogging* of soil on irrigated land without adequate drainage can decrease crop yields.

Solutions

Soil Salinization

Prevention	Cleanup
Reduce irrigation	Flushing soil (expensive and wastes water)
	Not growing crops for 2–5 years
Switch to salt-tolerant crops (such as barley, cotton, sugarbeet)	Installing underground drainage systems (expensive)

Figure 10-15 Solutions: methods for preventing and cleaning up soil salinization. *Critical thinking: which two of these solutions do you believe are the most important?*

Another problem with irrigation is **waterlogging** (Figure 10-13). Farmers often apply large amounts of irrigation water to leach salts deeper into the soil. Without adequate drainage, water may accumulate underground and gradually raise the water table. Saline water then envelops the deep roots of plants, lowering their productivity and killing them after prolonged exposure. At least one-tenth of the world's irrigated land suffers from waterlogging, and the problem is getting worse.

10-3 SUSTAINABLE AGRICULTURE THROUGH SOIL CONSERVATION

Science: Conservation Tillage

Modern farm machinery can plant crops without disturbing the soil.

Soil conservation involves using a variety of ways to reduce soil erosion and restore soil fertility, mostly by keeping the soil covered with vegetation.

Eliminating plowing and tilling is the key to reducing erosion and restoring healthy soil. Many U.S. farmers use **conservation-tillage farming,** which disturbs the soil as little as possible while planting crops.

With *minimum-tillage farming*, the soil is not disturbed over the winter. At planting time, special tillers break up and loosen the subsurface soil without turning over the topsoil, previous crop residues, or any cover vegetation. In *no-till farming*, special planting machines inject seeds, fertilizers, and weed killers (herbicides) into thin slits made in the unplowed soil and then smooth over the cut.

In 2004, farmers used conservation tillage on about 45% of U.S. cropland. The U.S. Department of Agriculture (USDA) estimates that using conservation tillage on 80% of U.S. cropland would reduce soil erosion by at least half. Conservation tillage also has great potential to reduce soil erosion and raise crop yields in the Middle East and in Africa.

Science: Other Methods for Reducing Soil Erosion

Farmers have developed a number of ways to grow crops that reduce soil erosion.

Figure 10-16 shows some of the methods farmers have used to reduce soil erosion. **Terracing** can reduce soil erosion on steep slopes by converting the land into a series of broad, nearly level terraces that run across the land's contours (Figure 10-16a). This practice retains water for crops at each level and reduces soil erosion by controlling runoff.

Contour farming involves plowing and planting crops in rows across the slope of the land rather than up and down (Figure 10-16b). Each row acts as a small dam to help hold soil and to slow water runoff.

Strip cropping (Figure 10-16b) involves planting alternating strips of a row crop (such as corn or cotton) and another crop that completely covers the soil (such as a grass or a grass–legume mixture). The cover crop traps soil that erodes from the row crop, catches and reduces water runoff, and helps prevent the spread of pests and plant diseases.

Another way to reduce erosion is to leave crop residues on the land after the crops are harvested. Farmers can also plant *cover crops* such as alfalfa, clover, or rye immediately after harvest to help protect and hold the soil.

Another method for slowing erosion is *alley cropping* or *agroforestry*, in which one or more crops are planted together in strips or alleys between trees and shrubs that can provide fruit or fuelwood (Figure 10-16c). The trees or shrubs provide shade (which reduces water loss by evaporation) and help retain and slowly release soil moisture. They also can provide fruit, fuelwood, and trimmings that can be used as mulch (green manure) for the crops and as fodder for livestock.

Some farmers establish **windbreaks,** or **shelterbelts,** of trees (Figure 10-16d) to reduce wind erosion, help retain soil moisture, supply wood for fuel, and provide habitats for birds, pest-eating and pollinating insects, and other animals.

Science: Organic Fertilizers

Soil conservation can reduce the loss of soil nutrients, and applying inorganic and organic fertilizers can help restore lost nutrients.

The best way to maintain soil fertility is through soil conservation. The next best option is to restore some of the plant nutrients that have been washed, blown, or leached out of soil or removed by repeated crop harvesting.

Fertilizers are used to partially restore lost plant nutrients. Farmers can use **organic fertilizer** from plant and animal materials or **commercial inorganic fertilizer** produced from various minerals.

Several types of *organic fertilizers* are available. One is **animal manure:** the dung and urine of cattle, horses, poultry, and other farm animals. It improves soil structure, adds organic nitrogen, and stimulates beneficial soil bacteria and fungi.

A second type of organic fertilizer called **green manure** consists of freshly cut or growing green vegetation that is plowed into the soil to increase the organic matter and humus available to the next crop. A third type is **compost,** produced when microorganisms in soil break down organic matter such as leaves, food wastes, paper, and wood in the presence of oxygen.

(a) Terracing

(b) Contour planting and strip cropping

(c) Alley cropping

(d) Windbreaks

Figure 10-16 Solutions: In addition to conservation tillage, soil conservation methods include **(a)** terracing, **(b)** contour planting and strip cropping, **(c)** alley cropping or agroforestry, and **(d)** windbreaks.

Crops such as corn, tobacco, and cotton can deplete nutrients (especially nitrogen) in the topsoil if they are planted on the same land several years in a row. **Crop rotation** provides one way to reduce these losses. Farmers plant areas or strips with nutrient-depleting crops one year. The next year, they plant the same areas with legumes whose root nodules add nitrogen to the soil. In addition to helping restore soil nutrients, this method reduces erosion by keeping the soil covered with vegetation.

Science: Inorganic Fertilizers

Inorganic fertilizers can help restore soil fertility if they are used with organic fertilizers and if their harmful environmental effects are controlled.

Many farmers (especially in developed countries) rely on *commercial inorganic fertilizers*. The active ingredients typically are inorganic compounds that contain *nitrogen, phosphorus,* and *potassium*. Other plant nutrients may be present in low or trace amounts. These fertilizers account for about one-fourth of the world's crop yield. However, without careful control these fertilizers can run off the land and pollute nearby bodies of water.

These fertilizers can replace depleted inorganic nutrients, but they do not replace organic matter. Thus, for healthy soil, both inorganic and organic fertilizers should be used.

10-4 FOOD PRODUCTION, NUTRITION, AND ENVIRONMENTAL EFFECTS

Science: Global Grain Production

After increasing significantly since 1950, global grain production has mostly leveled off since 1985, and per capita grain production has declined since 1978.

After almost tripling between 1950 and 1985, world grain production has essentially slowed down (Figure 10-17, left, p. 218). Also, after rising by about 36% between 1950 and 1978, per capita food production has declined (Figure 10-17, right). The sharpest drop in per capita food production has occurred in Africa since 1970.

Figure 10-17 Global outlook: total worldwide grain production of wheat, corn, and rice (left), and per capita grain production (right), 1950–2004. In order, the world's three largest grain-producing countries are China, the United States, and India. (Data from U.S. Department of Agriculture, Worldwatch Institute, UN Food and Agriculture Organization, and Earth Policy Institute)

Good news. We produce more than enough food to meet the basic nutritional needs of every person on the earth. *Bad news.* One of every six people in developing countries is not getting enough to eat because food is not distributed equally among the world's people. Such unequal distribution occurs because of differences in soil, climate, political and economic power, and average income per person.

Most agricultural experts agree that *the root cause of hunger and malnutrition is and will continue to be poverty,* which prevent poor people from growing or buying enough food regardless of how much is available. Other factors are war and corruption, which make it hard for poor people to have access to food that they or others produce.

Science: Chronic Hunger and Malnutrition

Some people cannot grow or buy enough food to meet their basic energy needs. Others do not get enough protein and other key nutrients.

To maintain good health and resist disease, we need fairly large amounts of *macronutrients* (such as protein, carbohydrates, and fats), as well as smaller amounts of *micronutrients* consisting of various vitamins (such as A, C, and E) and minerals (such as iron, iodine, and calcium).

People who cannot grow or buy enough food to meet their basic energy needs suffer from chronic **undernutrition.** Chronically undernourished children are likely to suffer from mental retardation and stunted growth. They also are susceptible to infectious diseases such as measles and diarrhea, which rarely kill children in developed countries.

Many of the world's poor can afford only a low-protein, high-carbohydrate diet consisting of grains such as wheat, rice, or corn. Many suffer from **malnu-**

trition resulting from deficiencies of protein, calories, and other key nutrients (Figure 1-12, p. 15).

Good news. According to the FAO, the average daily food intake in calories per person in the world and in developing countries rose sharply between 1961 and 2004, and it is projected to continue increasing through 2030. Also, the estimated number of chronically undernourished or malnourished people fell from 918 million in 1970 to 825 million in 2001—about 95% of them in developing countries.

Bad news. One of every six people in developing countries (including about one of every three children younger than age 5) is chronically undernourished or malnourished. The FAO estimates that each year at least 5.5 million people die prematurely from undernutrition, malnutrition, and increased susceptibility to normally nonfatal infectious diseases (such as measles and diarrhea) because of their weakened condition. Each day at least 15,100 people—80% of them children younger than age 5—die prematurely from these causes related to poverty.

Science: Not Getting Enough Vitamins and Minerals

One of every three persons has a deficiency of one or more vitamins and minerals, especially vitamin A, iron, and iodine.

According to the World Health Organization (WHO), one out of every three people suffers from a deficiency of one or more vitamins and minerals. The most widespread micronutrient deficiencies in developing countries involve *vitamin A, iron,* and *iodine.*

According to the WHO, 120–140 million children in developing countries are deficient in vitamin A. Globally, 250,000 children younger than age 6 go blind each year from a lack of vitamin A and as many as 80% of them die within a year.

Other nutritional deficiency diseases are caused by a lack of minerals. Too little *iron*—a component of the hemoglobin that transports oxygen in the blood—causes *anemia*. According to a 1999 survey by the WHO, one of every three people in the world—mostly women and children in tropical developing countries—suffers from iron deficiency. This condition causes fatigue, makes infection more likely, and increases a woman's chances of dying in childbirth and an infant's chances of dying of infection in its first year of life.

Elemental *iodine* is essential for proper functioning of the thyroid gland, which produces a hormone that controls the body's rate of metabolism. Iodine is found in seafood and crops grown in iodine-rich soils. Chronic lack of iodine can cause stunted growth, mental retardation, and goiter—an abnormal enlargement of the thyroid gland that can lead to deafness. According to the United Nations, 26 million children suffer brain damage each year from lack of iodine. Approximately 600 million people—mostly in South and Southeast Asia—suffer from goiter.

Solutions: Reducing Hunger and Malnutrition

There are several ways to reduce childhood deaths from nutrition-related causes.

Studies by the United Nations Children's Fund (UNICEF) indicate that one-half to two-thirds of childhood deaths from nutrition-related causes could be prevented at an average annual cost of $5–10 per child by the following measures:

- Immunizing children against childhood diseases such as measles

- Encouraging breast-feeding (except for mothers with AIDS)

- Preventing dehydration from diarrhea by giving infants a mixture of sugar and salt in a glass of water

- Preventing blindness by giving children a vitamin A capsule twice a year at a cost of about 75¢ per child, or fortifying common foods with vitamin A and other micronutrients at a cost of about 10¢ per child annually

- Providing family planning services to help mothers space births at least 2 years apart

- Increasing education for women, with emphasis on nutrition, drinking water sterilization, and child care

Science and Affluence: Eating Too Much Unhealthy Food

Overnutrition is a cause of poor health and preventable deaths.

Overnutrition occurs when food energy intake exceeds energy use and causes excess body fat. Too many calories, too little exercise, or both can cause overnutrition.

People who are underfed and underweight and those who are overfed and overweight face similar health problems: *lower life expectancy, greater susceptibility to disease and illness,* and *lower productivity and life quality.* We live in a world where 1 billion people have health problems because they do not get enough to eat and another 1.2 billion worry about health problems from eating too much. According to 2004 study by the International Obesity Task Force, one of every four people in the world is overweight and 5% are obese.

In developed countries, overnutrition causes preventable deaths, mostly from heart disease, cancer, stroke, and diabetes. According to the U.S. Centers for Disease Control and Prevention (CDC), almost two-thirds of Americans adults are overweight or obese—the highest overnutrition rate in any developed country. The $40 billion that Americans spend each year trying to lose weight is 1.7 times more than the $24 billion per year needed to eliminate undernutrition and malnutrition in the world.

A study of thousands of Chinese villagers indicates that the healthiest diet for humans is largely vegetarian, with only 10–15% of calories coming from fat. This stands in contrast to the typical meat-based diet, in which 30–40% of calories come from fat.

In 2004, the WHO urged governments to discourage food and beverage ads that exploit children; tax less-healthy foods; and limit the availability of high-fat and high-sugar foods in schools.

Science: Environmental Effects of Producing Food

Modern agriculture has a greater harmful environmental impact than any human activity, and these effects may limit future food production.

Modern agriculture has significant harmful effects on air, soil, water, and biodiversity, as Figure 10-18 (p. 220) shows. According to many analysts, agriculture has a greater harmful environmental impact than any human activity!

Some analysts believe these harmful environmental effects can be overcome and will not limit future food production. Other analysts disagree. For example, according to environmental expert Norman Myers, a combination of environmental factors may limit future food production. They include *soil erosion, salt buildup and waterlogging of soil on irrigated lands, water deficits and droughts,* and *loss of wild species* that provide the genetic resources for improved forms of foods.

According to a 2002 study by the UN Department for Economic and Social Affairs, nearly 30% of the world's cropland has been degraded to some degree

Natural Capital Degradation

Food Production

Biodiversity Loss	Soil	Water	Air Pollution	Human Health
Loss and degradation of habitat from clearing grasslands and forests and draining wetlands	Erosion	Water waste	Greenhouse gas emissions from fossil fuel use	Nitrates in drinking water
	Loss of fertility	Aquifer depletion		Pesticide residues in drinking water, food, and air
Fish kills from pesticide runoff	Salinization	Increased runoff and flooding from land cleared to grow crops	Other air pollutants from fossil fuel use	
	Waterlogging			Contamination of drinking and swimming water with disease organisms from livestock wastes
Killing of wild predators to protect livestock	Desertification	Sediment pollution from erosion	Greenhouse gas emissions of nitrous oxide from use of inorganic fertilizers	
		Fish kills from pesticide runoff		
Loss of genetic diversity from replacing thousands of wild crop strains with a few monoculture strains		Surface and groundwater pollution from pesticides and fertilizers	Belching of the greenhouse gas methane into the atmosphere by cattle	Bacterial contamination of meat
		Overfertilization of lakes and slow-moving rivers from runoff of nitrates and phosphates from fertilizers, livestock wastes, and food processing wastes	Pollution from pesticide sprays	

Figure 10-18 Natural capital degradation: major environmental effects of food production. According to UN studies, land degradation reduced cumulative food production worldwide by 13% on cropland and 4% on pastureland between 1950 and 2000.

by soil erosion, salt buildup, and chemical pollution, and 17% has been seriously degraded. Such environmental factors may limit food production in India and China, the world's two most populous countries.

10-5 INCREASING FOOD PRODUCTION

Science: Traditional Crossbreeding and Genetic Engineering

We can increase crop yields by using crossbreeding to mix the genes of similar types of organism and by using genetic engineering to mix the genes of different organisms.

For centuries, farmers and scientists have used *crossbreeding* to develop genetically improved varieties of crop strains. Such selective breeding has yielded amazing results. Ancient ears of corn were about the size of your little finger and wild tomatoes were once the size of a grape.

Traditional crossbreeding is a slow process, typically taking 15 years or more to produce a commercially valuable new variety, and it can combine traits only from species that are close to one another genetically. It also provides varieties that remain useful for only 5–10 years before pests and diseases reduce their effectiveness.

Today, scientists are creating a *third green revolution*—actually a *gene revolution*—by using *genetic engineering* to develop genetically improved strains of crops and livestock animals. This technology involves splicing a gene from one species and transplanting it into the DNA of another species (Figure 4-11, p. 75). Compared to traditional crossbreeding, gene splicing takes about half as long to develop a new crop, cuts costs, and allows the insertion of genes from almost any other organism into crop cells.

Ready or not, the world is entering the *age of genetic engineering*. More than two-thirds of the food products on U.S. supermarket shelves now contain some form of genetically engineered crops, and the proportion is increasing rapidly.

Despite its promise, considerable controversy has arisen over the use of *genetically modified food* (GMF) and other forms of genetic engineering. Its producers and investors see this kind of food as a potentially sustainable way to solve world hunger problems. Some critics consider it potentially dangerous "Frankenfood." Figure 10-19 summarizes the projected advantages and disadvantages of this new technology.

Critics recognize the potential benefits of genetically modified crops. At the same time, they warn that we know too little about the potential harm to human health and ecosystems from the widespread use of such crops. Also, genetically modified organisms cannot be recalled if they cause some unintended harmful genetic and ecological effects—as some scientists expect.

Most scientists and economists who have evaluated the genetic engineering of crops believe that its enormous potential benefits outweigh the much smaller risks. Critics call for more controlled field experiments, more research and long-term safety testing to better understand the risks, and stricter regulation of this rapidly growing technology. A 2004 study by

the Ecological Society of America recommended more caution in releasing genetically engineered organisms into the environment. Although a 2004 National Academy of Sciences study found no evidence that genetically engineered crops have harmed human health more than crops created by conventional crossbreeding, it called for federal regulators to look more closely at the potential health effects of genetically modified plants before approving them for use as commercial crops.

X *HOW WOULD YOU VOTE?* Do the advantages of genetically engineered foods outweigh their disadvantages? Cast your vote online at http://biology.brookscole.com/miller11.

Many analysts and consumer advocates believe governments should require mandatory labeling of GMFs. Consumers would then have more information to help them make informed choices about the foods they buy. Such labeling is required in Japan, Europe, South Korea, Canada, Australia, and New Zealand and was favored by 81% of Americans polled in 1999.

Industry representatives and the USDA oppose such labeling, claiming that GMFs are not substantially different from foods developed by conventional crossbreeding methods. Also, they fear—probably correctly—that labeling such foods would hurt sales by arousing suspicion.

Environmental Science ⊕ Now™
Learn more about how genes are modified and used to create plants with new traits at Environmental ScienceNow.

X *HOW WOULD YOU VOTE?* Should all genetically engineered foods be so labeled? Cast your vote online at http://biology.brookscole.com/miller11.

Science: Problems with Expanding the Green Revolution

Lack of resources, such as water and fertile soil, and environmental factors may limit our ability to continue increasing crop yields.

Many analysts believe we can produce all the food we need in the future by spreading the use of existing high-yield green revolution crops and genetically engineered crops to more of the world.

Other analysts disagree. They point to several factors that have limited the success of the green and gene revolutions to date and may continue to do so. Without huge amounts of fertilizer and water, most green revolution crop varieties produce yields that are no higher (and are sometimes lower) than those from traditional strains. Also, green revolution and genetically engineered crop strains and their high inputs of water, fertilizer, and pesticides cost too much for most subsistence farmers in developing countries.

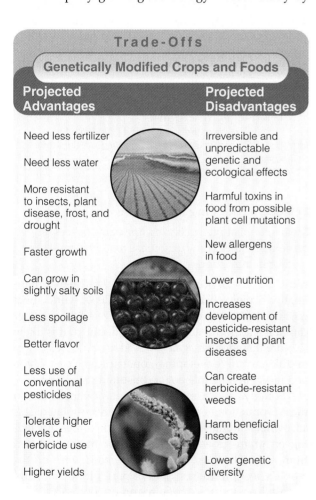

Trade-Offs

Genetically Modified Crops and Foods

Projected Advantages	Projected Disadvantages
Need less fertilizer	Irreversible and unpredictable genetic and ecological effects
Need less water	
More resistant to insects, plant disease, frost, and drought	Harmful toxins in food from possible plant cell mutations
Faster growth	New allergens in food
Can grow in slightly salty soils	Lower nutrition
Less spoilage	Increases development of pesticide-resistant insects and plant diseases
Better flavor	
Less use of conventional pesticides	Can create herbicide-resistant weeds
Tolerate higher levels of herbicide use	Harm beneficial insects
Higher yields	Lower genetic diversity

Figure 10-19 Trade-offs: projected advantages and disadvantages of *genetically modified crops and foods. Critical thinking: pick the single advantage and disadvantage that you think are the most important.*

Scientists also point out that continuing to increase fertilizer, water, and pesticide inputs eventually produces no additional increase in crop yields. For example, grain yields rose about 2.1% per year between 1950 and 1990, but then growth dropped to 1.1% per year between 1990 and 2000 and to 0.5% between 1997 and 2004. No one knows whether this downward trend will continue.

There is also concern that crop yields in some areas may decline as soil erodes and loses fertility, irrigated soil becomes salty and waterlogged, underground and surface water supplies become depleted and polluted with pesticides and nitrates from fertilizers, and populations of rapidly breeding pests develop genetic immunity to widely used pesticides. We do not know how close we are to such environmental limits.

Also, according to Indian economist Vandana Shiva, overall gains in crop yields from new green and gene revolution varieties may be much lower than claimed. The yields are based on comparisons between the output per hectare of old and new *monoculture* varieties rather than between the even higher yields per hectare for *polyculture* cropping systems and the new monoculture varieties that often replace polyculture crops.

There is also concern that the projected increased loss of biodiversity might limit the genetic raw material needed for future green and gene revolutions. The FAO estimates that two-thirds of all seeds planted in developing countries belong to uniform strains. Such genetic uniformity increases the vulnerability of food crops to pests, diseases, and harsh weather.

Science: Increasing Irrigation— A Limited Solution

The amount of irrigated land per person has been declining since 1978 and is projected to fall much more during the next few decades.

Approximately 40% of the world's food production comes from the 20% of the world's cropland that is irrigated. Between 1950 and 2004, the world's irrigated area tripled, with most of the growth occurring from 1950 to 1978.

However, the amount of irrigated land per person has been declining since 1978 and is projected to fall much more between 2005 and 2050. One reason is that since 1978, the world's population has grown faster than irrigated agriculture. Other factors are depletion of underground water supplies (aquifers), inefficient use of irrigation water, and salt buildup in soil on irrigated cropland. In addition, the majority of the world's farmers do not have enough money to irrigate their crops.

Science: Cultivating More Land—Another Limited Solution

Significant expansion of cropland is unlikely over the next few decades because of poor soils, limited water, high costs, and harmful environmental effects.

Theoretically, clearing tropical forests and irrigating arid land could more than double the world's cropland. In reality, much of this area is *marginal land* with poor soil fertility, steep slopes, or both. Cultivation of such land is unlikely to be sustainable.

Much of the world's potentially cultivable land lies in dry areas, especially in Australia and Africa. Large-scale irrigation in these areas would require expensive dam projects, use large inputs of fossil fuel to pump water over long distances, and deplete groundwater supplies by removing water faster than it is replenished. It would also require expensive efforts to prevent erosion, groundwater contamination, salinization, and waterlogging, all of which reduce crop productivity.

Furthermore, these potential increases in cropland would not offset the projected loss of almost one-third of today's cultivated cropland caused by erosion, overgrazing, waterlogging, salinization, and urbanization. Such cropland expansion would also reduce wildlife habitats and thus the world's biodiversity. Bottom line: *Significant expansion of cropland is unlikely over the next few decades.*

Environmental Science ⊛ Now™
Find out how much of the world's land is suitable for farming and how that land is now used at Environmental ScienceNow.

Science: Producing More Meat

Meat and meat products are important sources of protein, but meat production has many harmful environmental effects.

Meat and meat products are good sources of high-quality protein. Between 1950 and 2004, world meat production increased more than fivefold, and per capita meat production more than doubled. It is likely to more than double again by 2050 as affluence rises in middle-income developing countries (such as China and India) and people begin consuming more meat.

Some analysts expect most future increases in meat production to come from densely populated *feedlots*, where animals are fattened for slaughter by feeding on grain grown on cropland or meal produced from fish. Feedlots account for about 43% of the world's beef production, half of pork production, and almost three-fourths of poultry production.

In the United States, most production of cattle, pigs, and poultry is concentrated in increasingly large, factory-like production facilities in only a few areas. As many as 100,000 cattle may be confined to a singe feedlot complex, and 10,000 hogs may be crowded almost shoulder to shoulder in a giant barn.

Expanding feedlot production of meat will increase pressure on the world's grain supply because feedlot livestock consume grain produced on cropland instead of feeding on natural grasses. It will also increase pressure on the world's fish supply because about one-third of the world's fish catch goes to feed livestock. This industrialized approach increases meat productivity, but it has a number of harmful environmental effects, as discussed next.

Science Case Study: Some Environmental Consequences of Meat Production

Industrialized meat production has an enormous environmental impact.

The meat-based diet of affluent people in developed and developing countries has a number of harmful environmental effects. More than half of the world's cropland (19% in the United States) is used to grow livestock feed grain (mostly field corn, sorghum, and soybeans). Livestock and fish raised for food also consume 37% of the world's grain production and 70% of U.S. grain production.

Meat production uses more than half of the water withdrawn from the world's rivers and aquifers each year. Most of this water is used to irrigate crops fed to livestock and to wash away animal wastes.

About 14% of U.S. topsoil loss is directly associated with livestock grazing. Cattle belch out 16% of the methane (a greenhouse gas that is 25 times more potent than carbon dioxide) released into the atmosphere. Also, some of the nitrogen in commercial inorganic fertilizer used to grow livestock feed is converted to nitrous oxide, a greenhouse gas released from the soil into the atmosphere.

Livestock in the United States produce 20 times more waste (manure) than is produced by the country's human population. A single cow produces as much waste as 16 humans. Manure washing off the land or leaking from lagoons used to store animal wastes can kill fish by depleting dissolved oxygen.

Animal wastes from feedlot facilities are typically stored in enormous open lagoons, which can rupture or leak and contaminate groundwater and nearby streams and rivers. In 1999, torrential rains from Hurricane Floyd caused a number of hog and poultry waste lagoons in southeastern North Carolina to overflow and spill their wastes into local rivers. Living near a feedlot or animal waste lagoon is also a nasal assault. According to the Environmental Protection Agency, livestock wastes have contaminated

Figure 10-20 Natural capital degradation: overgrazed rangeland (left) and lightly grazed rangeland (right).

groundwater and polluted more than 43,000 kilometers (27,000 miles) of streams in about half of U.S. states.

Overgrazing occurs when too many animals graze for too long and exceed the carrying capacity of a grassland area (Figure 10-20, left). It lowers the NPP of grassland vegetation, reduces grass cover, and when combined with prolonged drought can cause desertification. It also exposes soil to erosion by water and wind and compacts soil (which diminishes its capacity to hold water). Limited data from surveys in various countries by the FAO indicate that overgrazing by livestock has caused as much as one-fifth of the world's rangeland to lose productivity, mostly as a result of desertification (Figure 10-11). Overgrazing by cattle can destroy the vegetation on and near stream banks (Figure 8-28, left, p. 178). Protecting such land from further grazing can eventually lead to its ecological restoration (Figure 8-28, right, p. 178).

Producing meat can also endanger wildlife species. According to a 2002 report by the National Public Lands Grazing Campaign, livestock grazing in the United States has contributed to population declines in almost one-fourth of the country's threatened and endangered species.

Science: Producing Meat More Sustainably

We can reduce the environmental impacts of meat production by relying more on fish and chicken and less on beef and pork.

Livestock and fish vary widely in the efficiency with which they convert grain into animal protein (Figure 10-21, p. 224). A more sustainable form of meat

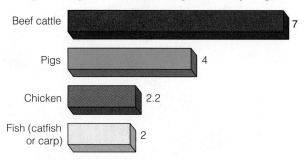

Kilograms of grain needed per kilogram of body weight

Beef cattle 7

Pigs 4

Chicken 2.2

Fish (catfish or carp) 2

Figure 10-21 Efficiency of converting grain into animal protein. Data in kilograms of grain per kilogram of body weight added. (Data from U.S. Department of Agriculture)

production and consumption would involve shifting from less grain-efficient forms of animal protein, such as beef and pork, to more grain-efficient ones, such as poultry and herbivorous farmed fish.

Some environmentalists have called for reducing livestock production (especially cattle) to decrease its environmental effects and to feed more people. This move would decrease the environmental impact of livestock production, but it would not free up much land or grain to feed more of the world's hungry people.

Cattle and sheep that graze on rangeland use a resource (grass) that humans cannot eat, and most of this land is not suitable for growing crops. Moreover, because of poverty, insufficient economic aid, and the nature of global economic and food distribution systems, very little (if any) additional grain grown on land once used to raise livestock or livestock feed would reach the world's hungry people.

Science: Harvesting Fish and Shellfish

After spectacular increases, the world's fish catch has leveled off.

The world's third major food-producing system consists of **fisheries:** concentrations of particular aquatic species suitable for commercial harvesting in a given ocean area or inland body of water. The world's commercial marine fishing industry is dominated by industrial fishing fleets that use global satellite positioning equipment, sonar, huge nets and long fishing lines, spotter planes, and large factory ships that can process and freeze their catches.

Approximately 55% of the annual commercial catch of fish and shellfish comes from the ocean using harvesting methods shown in Figure 10-22, mostly from plankton-rich coastal waters. Many commercially valuable species are being overfished by these increasingly efficient methods that "vacuum" the seas of fish and shellfish. About one-third of the world's marine fish harvest is used as animal feed, fishmeal, and oils. The rest of the catch comes from using *aquaculture* to raise fish much like livestock animals in feedlots in

ponds and underwater cages, and from inland freshwater fishing from lakes, rivers, reservoirs, and ponds.

Figure 10-23 shows the effects of the global efforts to boost the seafood harvest. After increasing fourfold between 1960 and 1982, the annual commercial fish catch (marine plus freshwater harvest, but excluding aquaculture) has declined and leveled off (Figure 10-23, left). After doubling between 1950 and 1956, the per capita catch leveled off until 1983. Since then, it has been declining (Figure 10-23, right) and may continue to fall because of overfishing, pollution, habitat loss, and population growth.

Science: Effects of Overfishing and Habitat Degradation on Fish Harvests

About three-fourths of the world's commercially valuable marine fish species are overfished or fished at their biological limit.

Fish are renewable resources as long as the annual harvest leaves enough breeding stock to renew the species for the next year. **Overfishing** is the taking of so many fish that too little breeding stock is left to maintain the species' numbers.

Prolonged overfishing leads to *commercial extinction,* when the population of a species declines to the point at which it is no longer profitable to hunt for them. Fishing fleets then move to a new species or a new region, hoping that the overfished species will eventually recover.

Overfishing is not new. Historical studies indicate that some species were overfished beginning centuries ago. However, this trend has greatly accelerated with the expansion of today's large and efficient global fishing fleets.

According to the FAO, three-fourths of the world's 200 commercially valuable marine fish species are either overfished or fished to their estimated maximum sustainable yield. The Ocean Conservancy states simply, "We are spending the principal of our marine fish resources rather than living off the interest they provide." Some fisheries are so depleted that even if all fishing stopped immediately, it would take as long as 20 years for stocks to recover.

Studies by the U.S. National Fish and Wildlife Foundation show that 14 major commercial fish species in U.S. waters have been severely depleted. Also, degradation, destruction, and pollution of wetlands, estuaries, coral reefs, salt marshes, and mangroves threaten populations of fish and shellfish.

Good news. In 1995, fisheries biologists studied population data for 128 depleted fish stocks and concluded that 125 of them could recover with careful management. This involves establishing fishing quotas, restricting use of certain types of fishing gear and methods, limiting the number of fishing boats, closing fisheries during spawning periods, and setting aside

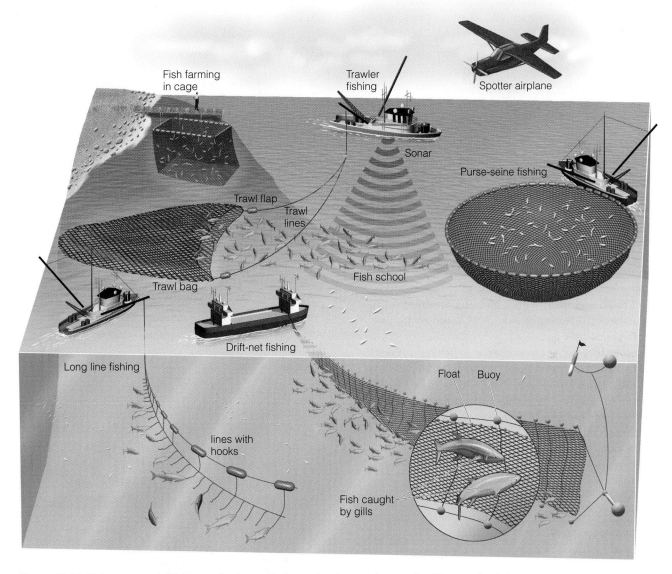

Figure 10-22 Major commercial fishing methods used to harvest various marine species. These methods have become so effective that many fish have become commercially extinct.

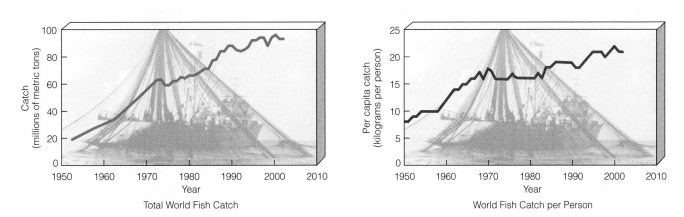

Total World Fish Catch

World Fish Catch per Person

Figure 10-23 Natural capital degradation: world fish catch (left) and world fish catch per person (right), 1950–2002. The total catch and per capita catches since 1990 may be about 10% lower than shown here because in 2000 it was discovered that China had been inflating its fish catches since 1990. (Data from UN Food and Agriculture Organization and Worldwatch Institute)

networks of no-take reserves. However, implementing such strategies costs money and is often politically unpopular.

Economics and Politics: Should Governments Continue Subsidizing Fishing Fleets?

Government subsidies given to the fishing industry are a major cause of overfishing.

Overfishing is a big and growing problem because too many commercial fishing boats and fleets are trying to hunt and gather a dwindling supply of the most desirable fish.

It costs the global fishing industry about $120 billion per year to catch $70 billion worth of fish. Government subsidies such as fuel tax exemptions, price controls, low-interest loans, and grants for fishing gear make up most of the $50 billion annual deficit of the industry. Without such subsidies, some of the world's fishing boats and fleets would go out of business and the number of fish caught would approach their sustainable yield.

Continuing to subsidize excess fishing allows some fishers to keep their jobs and boats a little longer while making less and less money until the fishery collapses. Then all jobs are gone, and fishing communities suffer even more—another example of the tragedy of the commons in action. Critics call for shifting some of the money from subsidies to programs to buy out some fishing boats and retrain their crews for other occupations.

X *HOW WOULD YOU VOTE?* Should governments eliminate most fishing subsidies? Cast your vote online at http://biology.brookscole.com/miller11.

Science: Aquaculture—Aquatic Feedlots

Aquaculture, the world's fastest-growing type of food production, has advantages and disadvantages.

Aquaculture involves raising fish and shellfish for food like crops instead of going out in fishing boats and hunting and gathering them. The world's fastest-growing type of food production, it accounts for about one-third of the fish and shellfish we eat. China, the world leader in this field, produces more than two-thirds of the world's aquaculture output.

There are two basic types of aquaculture: fish farming and fish ranching. **Fish farming** involves cultivating fish in a controlled environment (often a coastal or inland pond, lake, reservoir, or rice paddy) and harvesting them when they reach the desired size.

Fish ranching involves holding anadromous species, such as salmon that live part of their lives in fresh water and part in salt water, in captivity for the first few years of their lives, usually in fenced-in areas or floating cages in coastal lagoons and estuaries. The

fish are then released, and adults are harvested when they return to spawn.

Figure 10-24 lists the major advantages and disadvantages of aquaculture. Some analysts project that freshwater and saltwater aquaculture production could provide at least half of the world's seafood by 2020. Other analysts warn that the harmful environmental effects of aquaculture could limit future production.

Figure 10-25 lists some ways to make aquaculture more sustainable and to reduce its harmful environmental effects. However, even under the most optimistic projections, increasing both the wild catch and aquaculture will not increase world food supplies significantly. The reason: Fish and shellfish supply only about 1% of the calories and 6% of the protein in the human diet.

Economics and Politics: Government Agricultural and Food Production

Governments can use price controls to keep food prices artificially low, give farmers subsidies to encourage food production, or eliminate food price controls and subsidies and let farmers and fishers respond to market demand.

Figure 10-24 Trade-offs: advantages and disadvantages of *aquaculture. Critical thinking: pick the single advantage and disadvantage that you think are the most important.*

- Reduce use of fishmeal as a feed to reduce depletion of other fish

- Improve pollution management of aquaculture wastes

- Reduce escape of aquaculture species into the wild

- Restrict location of fish farms to reduce loss of mangrove forests and estuaries

- Farm some aquaculture species (such as salmon and cobia) in deeply submerged cages to protect them from wave action and predators and allow dilution of wastes into the ocean

- Set up a system for certifying sustainable forms of aquaculture

Figure 10-25 Solutions: ways to make aquaculture more sustainable and reduce its harmful environmental effects. *Critical thinking: which two of these solutions do you believe are the most important?*

Agriculture is a financially risky business. Whether farmers have a good year or a bad year depends on factors over which they have little control: weather, crop prices, crop pests and diseases, interest rates, and the global market. Because reliable food supplies are needed despite fluctuations in these factors, most governments provide assistance to farmers and consumers.

Governments use three main approaches to provide this assistance. One strategy is to use price controls to *keep food prices artificially low.* Consumers are happy, but farmers may not be able to make a living.

Another strategy is to *give farmers subsidies and tax breaks to keep them in business and encourage them to increase food production.* Globally, government price supports and other subsidies for agriculture total more than $350 billion per year (about $100 billion per year in the United States)—more than $666,000 per hour! If government subsidies are too generous and the weather is good, farmers may produce more food than can be sold. The resulting surplus depresses food prices, which reduces the financial incentive for farmers in developing countries to increase domestic food production—those connections again.

A third approach is to *eliminate most or all price controls and subsidies and let farmers and fishers respond to market demand without government interference.* Some analysts urge that any phaseout of farm and fishery subsidies should be coupled with increased aid for the poor and the lower middle class, who would suffer the most from any increase in food prices. Some environmental scientists say that instead of eliminating all subsidies, we should use them to reward farmers and ranchers

who protect the soil, conserve water, reforest degraded land, protect and restore wetlands, conserve wildlife, and practice more sustainable agriculture and fishing.

10-6 PROTECTING FOOD RESOURCES: PEST MANAGEMENT

Science: Natural Pest Control

Predators, parasites, and disease organisms found in nature control populations of most pest species as part of the earth's free ecological services.

A **pest** is any species that competes with us for food, invades lawns and gardens, destroys wood in houses, spreads disease, invades ecosystems, or is simply a nuisance. Worldwide, only about 100 species of plants ("weeds"), animals (mostly insects), fungi, and microbes (which can infect crop plants and livestock animals) cause about 90% of the damage to the crops we grow.

In natural ecosystems and many polyculture agroecosystems, *natural enemies* (predators, parasites, and disease organisms) control the populations of about 98% of the potential pest species as part of the earth's free ecological services. They help keep any one species from taking over for very long.

For example, the world's 30,000 known species of spiders, such as the wolf spider (Figure 10-26), kill far more insects every year than insecticides do. As we seek new ways to coexist with the insect rulers of the planet, we would do well to keep spiders on our side.

When we clear forests and grasslands, plant monoculture crops, and douse fields with pesticides, we upset many of these natural population checks and balances. Then we must devise ways to protect our monoculture crops, tree plantations, and lawns from insects and other pests that nature once controlled at no charge.

Figure 10-26 Natural capital: Spiders are insects' worst enemies. Most spiders, such as this ferocious-looking wolf spider, do not harm humans.

Science: Pesticides

We use chemicals to repel or kill pest organisms as plants have done for millions of years to defend themselves against hungry herbivores.

To help control pest organisms, we have developed a variety of **pesticides**—chemicals to kill or control populations of organisms we consider undesirable. Common types of pesticides include *insecticides* (insect killers), *herbicides* (weed killers), *fungicides* (fungus killers), and *rodenticides* (rat and mouse killers). *Biocide* is a more accurate name for such a chemical because most pesticides kill other organisms as well as their pest targets.

We did not invent the use of chemicals to repel or kill other species. Indeed, plants have been producing chemicals to ward off, deceive, or poison herbivores that feed on them for nearly 225 million years. This battle produces a never-ending, ever-changing coevolutionary process: Herbivores overcome various plant defenses through natural selection; then new plant defenses are favored by natural selection in this ongoing cycle of evolutionary punch and counterpunch.

Since 1950, pesticide use has increased more than 50-fold, and most of today's pesticides are more than ten times as toxic as those used in the 1950s. Three-fourths of these chemicals are used in developed countries, but their use in developing countries is soaring.

One-fourth of pesticide use in the United States is devoted to ridding houses, gardens, lawns, parks, playing fields, swimming pools, and golf courses of pests. According to the U.S. Environmental Protection Agency (EPA), the average lawn in the United States is doused with ten times more synthetic pesticides per hectare than U.S. cropland. Each year, more than 250,000 people in the United States become ill because of household pesticide use, and such pesticides are a major source of accidental poisonings and deaths for young children.

Broad-spectrum agents are toxic to many species. *Selective,* or *narrow-spectrum, agents* are effective against a narrowly defined group of organisms. Pesticides vary in their *persistence,* the length of time they remain deadly in the environment. In 1962, biologist Rachel Carson warned against relying on synthetic organic chemicals to kill insects and other species we deem pests (see Individuals Matter, right).

Environmental Science⊕Now™
Study descriptions of three major pesticides along with their chemical formulas at Environmental ScienceNow.

Science and Economics: Advantages of Modern Synthetic Pesticides

Modern pesticides save lives, increase food supplies, increase profits for farmers, work fast, and are safe if used properly.

Proponents of conventional chemical pesticides contend that their benefits outweigh their harmful effects. Conventional pesticides have a number of important benefits.

They save human lives. Since 1945, DDT and other chlorinated hydrocarbon and organophosphate insecticides probably have prevented the premature deaths of at least 7 million people (some say as many as 500 million) from insect-transmitted diseases such as malaria (carried by the *Anopheles* mosquito), bubonic plague (carried by rat fleas), and typhus (carried by body lice and fleas).

They increase food supplies. According to the FAO, 55% of the world's potential human food supply is lost to pests—about two-thirds of that before harvest and the rest after (Figure 10-27). Without pesticides, these losses would be worse, and food prices would rise.

They increase profits for farmers. Pesticide companies estimate that every $1 spent on pesticides leads to an increase in U.S. crop yields worth approximately $4. (Studies have shown this benefit drops to about $2 if the harmful effects of pesticides are included.)

They work faster and better than alternatives. Pesticides control most pests quickly and at a reasonable cost, have a long shelf life, are easily shipped and applied, and are safe when handled correctly by farm workers. When genetic resistance occurs, farmers can use stronger doses or switch to other pesticides.

When used properly, their health risks are very low compared with their benefits. According to Elizabeth Whelan, director of the American Council on Science and Health (ACSH), which presents the position of the pesticide industry, "The reality is that pesticides, when used in the approved regulatory manner, pose no risk to either farm workers or consumers."

Figure 10-27 Global outlook: rats, such as these caught by a farmer in India, destroy much of the world's wheat and rice before and after harvest.

Rachel Carson

Rachel Carson (1907–1964) was a pioneer in increasing public awareness of the importance of nature and the threat of pollution from pesticides. She began her professional career as a biologist for the Bureau of U.S. Fisheries (later the U.S. Fish and Wildlife Service). In that capacity, she carried out research on oceanography and marine biology and wrote articles about the oceans and topics related to the environment.

In 1951, Carson wrote *The Sea Around Us*, which described in easily understandable terms the natural history of oceans and how human activities were harming them. Her book sold more than 2 million copies, was translated into 32 languages, and won a National Book Award.

During the late 1940s and throughout the 1950s, DDT and related compounds were increasingly used to kill insects that ate food crops, attacked trees, bothered people, and transmitted diseases such as malaria.

In 1958, DDT was sprayed to control mosquitoes near the home and private bird sanctuary of one of Carson's friends. After the spraying, her friend witnessed the agonizing deaths of several birds. She begged Carson to find someone to investigate the effects of pesticides on birds and other wildlife.

Carson decided to look into the issue herself. She found that independent research on the environmental effects of pesticides was almost nonexistent. A well-trained scientist, she surveyed the scientific literature, became convinced that pesticides could harm wildlife and humans, and methodically developed information about the harmful effects of widespread use of pesticides.

In 1962, she published her findings in popular form in *Silent Spring*, whose title alluded to the silencing of "robins, catbirds, doves, jays, wrens, and scores of other bird voices" because of their exposure to pesticides. Many scientists, politicians, and policy makers read *Silent Spring*, and the public embraced it.

Chemical manufacturers viewed the book as a serious threat to their booming pesticide sales and mounted a campaign to discredit Carson. A parade of critical reviewers and industry scientists claimed her book was full of inaccuracies, made selective use of research findings, and failed to give a balanced account of the benefits of pesticides.

Some critics even claimed that, as a woman, Carson was incapable of understanding such a highly scientific and technical subject. Others charged that she was a hysterical woman and a radical nature lover trying to scare the public in an effort to sell books.

During these intense attacks, Carson was suffering from terminal cancer. Yet she strongly defended her research and countered her critics. She died in 1964—about 18 months after the publication of *Silent Spring*—without knowing that many historians consider her work an important contribution to the modern environmental movement then emerging in the United States.

U.S Fish and Wildlife Service

Figure 10-A Biologist Rachel Carson (1907–64) greatly increased our understanding of the importance of nature and the potential harmful effects of widespread use of pesticides. She died without knowing that her efforts were a key in beginning the modern environmental movement in the United States.

Newer pesticides are safer and more effective than many older ones. Greater use is being made of botanicals and microbotanicals. Derived originally from plants, they are safer for users and less damaging to the environment than many older pesticides. Genetic engineering is also being used to develop pest-resistant crop strains and genetically altered crops that produce pesticides.

Many new pesticides are used at very low rates per unit area compared to older products. Application amounts per hectare for many new herbicides are 1/100 the rates for older ones, and genetically engineered crops could reduce the use of toxic insecticides.

Science: The Ideal Pesticide and Pest

Scientists work to develop more effective and safer pesticides, but through coevolution pests find ways to combat the pesticides we throw at them.

Scientists continue to search for the *ideal pest-killing chemical*, which would have these qualities:

- Kill only the target pest

- Not cause genetic resistance in the target organism

- Disappear or break down into harmless chemicals after doing its job

- Be more cost-effective than doing nothing

The search continues, but so far no known natural or synthetic pesticide chemical meets all—or even most—of these criteria.

The *ideal insect pest* would attack a variety of plants, be highly prolific, have a short generation time and few natural predators, and be genetically resistant to a number of pesticides. The silverleaf whitefly has these characteristics, and farmers who have encountered it call it a *superbug.* This tiny white insect escaped from poinsettia greenhouses in Florida in 1986 and has become established in Florida, Arizona, California, and Texas.

It is known to eat at least 500 species of plants but does not like onions and asparagus. The silverleaf whitefly has no natural enemies. Dense swarms of these tiny insects attack plants, suck them dry, and leave them withered and dying.

U.S. crop losses from this insect are greater than $200 million per year—and growing. Scientists are scouring the world looking for natural enemies of this superbug. Stay tuned.

Science: Disadvantages of Modern Synthetic Pesticides

Pesticides can promote genetic resistance to their effects, wipe out natural enemies of pest species, create new pest species, end up in the environment, and sometimes harm wildlife and people.

Opponents of widespread pesticide use believe that the harmful effects of these chemicals outweigh their benefits. They cite several serious problems with the use of conventional pesticides.

They accelerate the development of genetic resistance to pesticides by pest organisms. Insects breed rapidly, and within five to ten years (much sooner in tropical areas) they can develop immunity to widely used pesticides through natural selection and come back stronger than before. Weeds and plant disease organisms also de-velop genetic resistance, but at a slower rate than insects. Since 1945, about 1,500 species of insects, mites, weeds, plant diseases, and rodents (mostly rats) have developed genetic resistance to one or more pesticides.

Because of genetic resistance, many insecticides (such as DDT) no longer do a good job of protecting people from insect-transmitted diseases in some parts of the world. Genetic resistance can also put farmers on a *pesticide treadmill,* whereby they pay more and more for a pest control program that often becomes less and less effective.

Some insecticides kill natural predators and parasites that help control the populations of pest species. Wiping out natural predators, such as spiders, can unleash new pests, whose populations their predators had previously held in check, and cause other unexpected effects (Connections, p. 125). Of the 300 most destructive insect pests in the United States, 100 were once minor pests that became major pests after widespread use of insecticides. Mostly because of genetic resistance and reduction of natural predators, pesticide use has not reduced U.S. crop losses to pests (Science Spotlight, below).

Pesticides do not stay put. According to the USDA, only 0.1–2% of the insecticide applied to crops by aerial spraying or ground spraying reaches the target pests. Also, less than 5% of herbicides applied to crops reach the target weeds. In other words, 98–99.9% of the pesticides and more than 95% of the herbicides we apply end up in the air, surface water, groundwater, bottom sediments, food, and nontarget organisms, including humans and wildlife (Figure 9-16, p. 197).

How Successful Have Synthetic Pesticides Been in Reducing Crop Losses in the United States?

SCIENCE SPOTLIGHT

Pesticides have not been as effective in reducing crop losses in the United States as agricultural experts had hoped, mostly because of genetic resistance and reductions in natural predators.

When David Pimentel, an expert in insect ecology, evaluated data from more than 300 agricultural scientists and economists, he reached three major conclusions.

First, although the use of synthetic pesticides has increased 33-fold since 1942, 37% of the U.S. food supply is lost to pests today compared to 31% in the 1940s. Since 1942, losses attributed to insects almost doubled from 7% to 13%, despite a 10-fold increase in the use of synthetic insecticides.

Second, the estimated environmental, health, and social costs of pesticide use in the United States total $4–10 billion per year. The International Food Policy Research Institute puts this figure much higher, at $100–200 billion per year, or $5–10 in damages for every dollar spent on pesticides.

Third, alternative pest management practices could halve the use of chemical pesticides on 40 major U.S. crops without reducing crop yields.

Numerous studies and experience show that pesticide use can be reduced sharply without reducing yields. In fact, yields may actually increase. Sweden has cut pesticide use in half with almost no decrease in crop yields. Campbell Soup uses no pesticides on tomatoes it grows in Mexico, and yields have not dropped. After a two-thirds cut in pesticide use on rice in Indonesia, yields increased by 15%.

Critical Thinking

Pesticide proponents argue that although crop losses to pests are higher today than in the past, without the widespread use of pesticides losses would be even higher. Explain why you agree or disagree with this position.

Crops that have been genetically altered to release small amounts of pesticides directly to pests can help overcome this problem. Unfortunately, they can also promote genetic resistance to the pesticides.

Some pesticides harm wildlife. According to the USDA and the U.S. Fish and Wildlife Service, each year pesticides applied to cropland in the United States wipe out about 20% of U.S. honeybee colonies and damage another 15%. Farmers lose at least $200 million per year from the reduced pollination of vital crops. Pesticides also kill more than 67 million birds and 6–14 million fish, and menace one of every five endangered and threatened species in the United States.

Some pesticides threaten human health. According to the WHO and the UN Environment Programme, each year pesticides seriously poison at least 3 million agricultural workers in developing countries and at least 300,000 people in the United States. They cause 20,000–40,000 deaths (about 25 in the United States) per year. Health officials believe the actual number of pesticide-related illnesses and deaths among the world's farm workers probably is greatly underestimated because of poor record-keeping, lack of doctors, inadequate reporting of illnesses, and faulty diagnoses.

According to studies by the National Academy of Sciences, exposure to legally allowed pesticide residues in food causes 4,000–20,000 cases of cancer per year in the United States. Roughly half of these individuals will die prematurely. Some scientists are becoming increasingly concerned about possible genetic mutations, birth defects, nervous system disorders (especially behavioral disorders), and effects on the immune and endocrine systems from long-term exposure to low levels of various pesticides. The pesticide industry disputes these claims.

Politics: Pesticide Protection Laws in the United States

Government regulation has banned a number of harmful pesticides but some scientists call for strenghtening pesticide laws.

How well the public in the United States is protected from the harmful effects of pesticides remains a controversial topic. The EPA banned or severely restricted the use of 57 active pesticide ingredients between 1972 and 2004. The 1996 Food Quality Protection Act (FQPA) also increased public protection from pesticides.

According to studies by the National Academy of Sciences, federal laws regulating pesticide use in the United States are inadequate and poorly enforced by the EPA, the Food and Drug Administration (FDA), and the USDA. One study by the National Academy of Sciences found that as much as 98% of the potential risk of developing cancer from pesticide residues on food grown in the United States would be eliminated

SCIENCE SPOTLIGHT

What Goes Around Can Come Around

U.S. pesticide companies make and export to other countries pesticides that have been banned or severely restricted—or never even approved—in the United States. Other industrial countries also export banned and unapproved pesticides.

But what goes around can come around. In what environmental scientists call a *circle of poison*, residues of some of these banned or unapproved chemicals exported to other countries can return to the exporting countries on imported food. The wind can also carry persistent pesticides such as DDT from one country to another.

Environmentalists have urged Congress—without success—to ban such exports. Supporters of the exports argue that such sales increase economic growth and provide jobs, and that banned pesticides are exported only with the consent of the importing countries. They also contend that if the United States did not export pesticides, other countries would.

In 1998, more than 50 countries developed an international treaty that requires exporting countries to have informed consent from importing counties for exports of 22 pesticides and 5 industrial chemicals. In 2000, more than 100 countries developed an international agreement to ban or phase out the use of 12 especially hazardous persistent organic pollutants (POPs)—9 of them persistent hydrocarbon pesticides such as DDT and other chemically similar pesticides.

Critical Thinking

Should U.S. companies be allowed to export pesticides that have been banned, severely restricted, or not approved for use in the United States? Explain.

if EPA standards were as strict for pre-1972 pesticides as they are for later ones. Another problem is that banned or unregistered pesticides may be manufactured in the United States and exported to other countries (Science Spotlight, above).

The pesticide industry disputes these findings, stating that eating food grown by using pesticides for the past 50 years has never harmed anyone in the United States. The industry also claims that the benefits of pesticides far outweigh their disadvantages.

X *HOW WOULD YOU VOTE?* Do the advantages of using synthetic chemical pesticides outweigh their disadvantages? Cast your vote online at http://biology.brookscole.com/miller11.

Science: Other Ways to Control Pests

A mix of cultivation practices and biological and ecological alternatives to conventional chemical pesticides can help control pests.

Many scientists believe we should greatly increase the use of biological, ecological, and other alternative methods for controlling pests and diseases that affect crops and human health. A number of methods are available.

A variety of *cultivation practices* can be employed to fake out pest species. Examples include rotating the types of crops planted in a field each year, adjusting planting times so major insect pests either starve or get eaten by their natural predators, and growing crops in areas where their major pests do not exist. Also, farmers can increase the use of polyculture, which uses plant diversity to reduce losses to pests. Homeowners can reduce weed invasions by cutting grass no lower than 8 centimeters (3 inches) high. This height provides a dense enough cover to keep out crabgrass and many other undesirable weeds.

Genetic engineering can be used to speed up the development of *pest- and disease-resistant crop strains* (Figure 10-28). Controversy persists over whether the projected advantages of increased use of genetically modified plants and foods outweigh their projected disadvantages (Figure 10-19).

We can increase the use of *biological control* by importing natural predators (Figures 10-26 and 10-29), parasites, and disease-causing bacteria and viruses to help regulate pest populations. This approach is nontoxic to other species, minimizes genetic resistance, and can save large amounts of money—about $25 for

Figure 10-29 Natural capital: biological pest control. The wasp pupae (white) will kill this tobacco hornworm.

every $1 invested in controlling 70 pests in the United States. However, biological control agents cannot always be mass-produced, are often slower acting and more difficult to apply than conventional pesticides, can sometimes multiply and become pests themselves, and must be protected from pesticides sprayed in nearby fields.

Sex attractants (called *pheromones*) can lure pests into traps or attract their natural predators into crop fields (usually the more effective approach). These chemicals attract only one species, work in trace amounts, have little chance of causing genetic resistance, and are not harmful to nontarget species. However, it is costly and time-consuming to identify, isolate, and produce the specific sex attractant for each pest or predator.

We can also use *hormones that disrupt an insect's normal life cycle* (Figure 10-30), thereby preventing it from reaching maturity and reproducing. Insect hormones have the same advantages as sex attractants. But they take weeks to kill an insect, often are ineffective with large infestations of insects, and sometimes break down before they can act. In addition, they must be applied at exactly the right time in the target insect's life cycle, can sometimes affect the target's predators and other nonpest species, and are difficult and costly to produce.

Some farmers have controlled certain insect pests by *spraying them with hot water.* This approach has worked well on cotton, alfalfa, and potato fields and in citrus groves in Florida, and its cost is roughly equal to that of using chemical pesticides.

Figure 10-28 Science: the results of one example of using *genetic engineering* to reduce pest damage. Both tomato plants were exposed to destructive caterpillars. The normal plant's leaves are almost gone (left), whereas the genetically altered plant shows little damage (right).

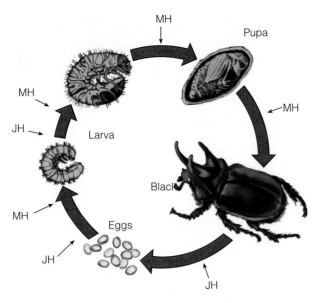

MH

Pupa

MH

MH

MH

Larva

JH

Black

MH

Eggs

JH

JH

Figure 10-30 Science: for normal insect growth, development, and reproduction to occur, certain juvenile hormones (JH) and molting hormones (MH) must be present at genetically determined stages in the insect's life cycle. If applied at the proper time, synthetic hormones disrupt the life cycles of insect pests and help control their populations.

Science: Integrated Pest Management

An ecological approach to pest control uses an integrated mix of cultivation and biological methods, and small amounts of selected chemical pesticides as a last resort.

Many pest control experts and farmers believe the best way to control crop pests is a carefully designed **integrated pest management (IPM)** program. In this approach, each crop and its pests are evaluated as parts of an ecological system. Then farmers develop a control program that includes cultivation, biological, and chemical methods applied in the proper sequence and with the proper timing.

The overall aim of IPM is not to eradicate pest populations but rather to reduce crop damage to an economically tolerable level. Fields are monitored carefully. When an economically damaging level of pests is reached, farmers first use biological methods (natural predators, parasites, and disease organisms) and cultivation controls, including vacuuming up harmful bugs. Small amounts of insecticides—mostly based on natural insecticides produced by plants—are applied only as a last resort. Also, different chemicals are used to slow the development of genetic resistance and to avoid killing predators of pest species.

In 1986, the Indonesian government banned 57 of the 66 pesticides used on rice, and phased out pesticide subsidies over a 2-year period. It also launched a nationwide education program to help farmers switch to IPM. The results were dramatic: Between 1987 and 1992, pesticide use dropped by 65%, rice production rose by 15%, and more than 250,000 farmers were trained in IPM techniques. Sweden and Denmark have used IPM to cut their pesticide use in half.

The experiences of these and other countries show that a well-designed IPM program can reduce pesticide use and pest control costs by at least half, cut preharvest pest-induced crop losses by half, and improve crop yields. It can also reduce inputs of fertilizer and irrigation water, and slow the development of genetic resistance because pests are assaulted less often and with lower doses of pesticides. IPM is an important form of *pollution prevention* that reduces risks to wildlife and human health.

Despite its promise, IPM—like any other form of pest control—has some disadvantages. It requires expert knowledge about each pest situation and acts more slowly than conventional pesticides. Methods developed for a crop in one area might not apply to areas with even slightly different growing conditions. Initial costs may be higher, although long-term costs typically are lower than those of using conventional pesticides.

Widespread use of IPM is hindered by government subsidies for conventional chemical pesticides and opposition by pesticide manufacturers, whose sales would drop sharply. There is also a lack of experts to help farmers shift to IPM.

A 1996 study by the National Academy of Sciences recommended that the United States shift from chemically based approaches to ecologically based pest management approaches. Within 5–10 years, such a shift could cut U.S. pesticide use in half, as it has in several other countries.

A growing number of scientists urge the USDA to use three strategies to promote IPM in the United States:

- Add a 2% sales tax on pesticides and use the revenue to fund IPM research and education.

- Set up a federally supported IPM demonstration project on at least one farm in every county.

- Train USDA field personnel and county farm agents in IPM so they can help farmers use this alternative.

The pesticide industry has successfully opposed such measures.

X *How Would You Vote?* Should governments heavily subsidize a switch to integrated pest management? Cast your vote online at http://biology.brookscole.com/miller11.

Several UN agencies and the World Bank have joined together to establish an IPM facility. Its goal is to promote the use of IPM by disseminating information and establishing networks among researchers, farmers, and agricultural extension agents involved in IPM.

10-7 SOLUTIONS: SUSTAINABLE AGRICULTURE

Science: Sustainable Organic Agriculture

We can produce food more sustainably by reducing resource throughput and working with nature.

There are three main ways to reduce hunger and malnutrition and the harmful environmental effects of agriculture. First, we can *slow population growth*. Second, we can *reduce poverty* so that people can grow or buy enough food for their survival and good health.

Third, we can develop and phase in systems of **sustainable** or **low-input agriculture**—also called **organic agriculture**—over the next few decades. Figure 10-31

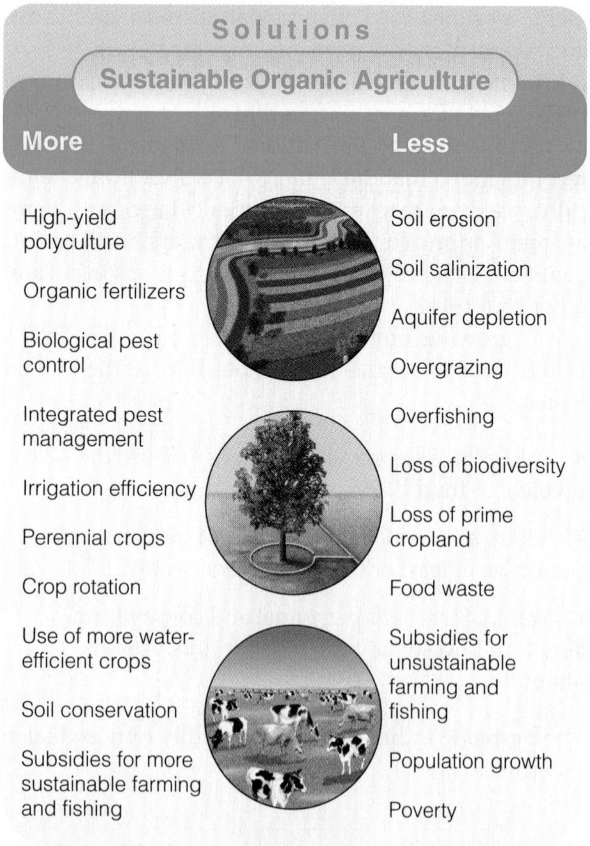

Figure 10-31 Solutions: components of more sustainable, low-throughput, or organic, agriculture. *Critical thinking: which two of the solutions in each column do you believe are the most important?*

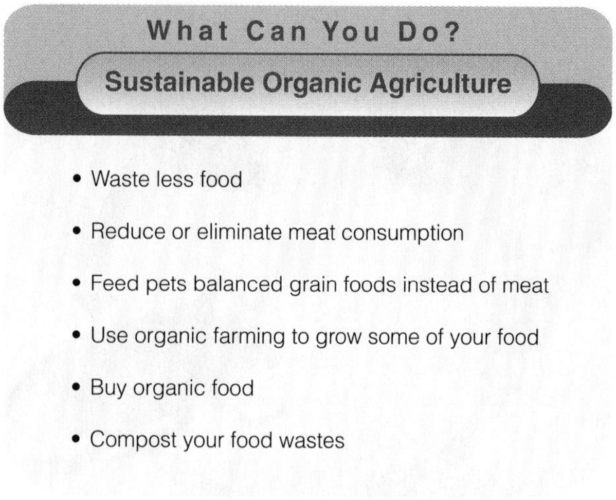

Figure 10-32 Individuals matter: ways to promote more sustainable agriculture. *Critical thinking: which three of these actions do you believe are the most important? Which of these actions in this list do you plan to do?*

lists the major components of more sustainable agriculture. Low-input organic agriculture produces roughly equivalent yields with lower carbon dioxide emissions, uses 30-50% less energy per unit of yield, improves soil fertility, reduces soil erosion, and generally is more profitable for the farmer than high-input farming.

Most proponents of more sustainable agriculture are not opposed to high-yield agriculture. Instead, they see it as vital for protecting the earth's biodiversity by reducing the need to cultivate new and often marginal land. They call for using environmentally sustainable forms of both high-yield polyculture and high-yield monoculture for growing crops with increasing emphasis on using more sustainable methods for producing food (Figure 10-31, left).

Solutions: Making the Transition to More Sustainable Agriculture

More research, demonstration projects, government subsidies, and training can promote a shift to more sustainable agriculture.

Analysts suggest four major strategies to help farmers make the transition to more sustainable organic agriculture. *First*, greatly increase research on sustainable agriculture and improving human nutrition. *Second*, set up demonstration projects so farmers can see how more sustainable organic agricultural systems work. *Third*, provide subsidies and increased foreign aid to encourage its use. *Fourth*, establish training programs in sustainable organic agriculture for farmers and government agricultural officials, and encourage the creation of college curricula in sustainable organic agriculture and human nutrition.

Phasing in more sustainable organic agriculture involves applying the four principles of sustainability

(Figure 6-18, p. 126) to producing food. The goal is to feed the world's people while sustaining and restoring the earth's natural capital and living off the natural income it provides. This will not be easy, but it can be done. Figure 10-32 lists some ways that you can promote more sustainable organic agriculture.

The sector of the economy that seems likely to unravel first is food. Eroding soils, deteriorating rangelands, collapsing fisheries, falling water tables, and rising temperatures are converging to make it difficult to expand food production fast enough to keep up with the demand.

LESTER R. BROWN

CRITICAL THINKING

1. Summarize the major economic and ecological advantages and limitations of each of the following proposals for increasing world food supplies and reducing hunger over the next 30 years:
   (a) Cultivating more land by clearing tropical forests and irrigating arid lands
   (b) Catching more fish in the open sea
   (c) Producing more fish and shellfish with aquaculture
   (d) Increasing the yield per area of cropland

2. List five ways in which your lifestyle directly or indirectly contributes to soil erosion.

3. What could happen to energy-intensive agriculture in the United States and other industrialized countries if world oil prices rose sharply?

4. What are the three most important actions you would take to reduce hunger **(a)** in the country where you live and **(b)** in the world?

5. Should governments phase in agricultural tax breaks and subsidies to encourage farmers to switch to more sustainable organic agriculture? Explain your answer.

6. According to physicist Albert Einstein, "Nothing will benefit human health and increase chances of survival of life on Earth as much as the evolution to a vegetarian diet." Are you willing to eat less meat or not eat any meat? Explain.

7. Explain how widespread use of a pesticide can **(a)** increase the damage done by a particular pest and **(b)** create new pest organisms.

8. Congratulations! You are in charge of the world. List the three most important features of your **(a)** your agricultural policy, **(b)** your policy to reduce soil erosion, and **(c)** your policy for pest management.

LEARNING ONLINE

The website for this book includes review questions for the entire chapter, flash cards for key terms and concepts, a multiple-choice practice quiz, interesting Internet sites, references, and a guide for accessing thousands of InfoTrac® College Edition articles.
   Visit

http://biology.brookscole.com/miller11

Then choose Chapter 10, and select a learning resource. For access to animations, additional quizzes, chapter outlines and summaries, register and log in to

Environmental Science ⊛ Now™

at **esnow.brookscole.com/miller11** using the access code card in the front of your book.

Active Graphing

Visit http://esnow.brookscole.com/miller11 to explore the graphing exercise for this chapter.

CASE STUDY

Water Conflicts in the Middle East

In the near future, water-short countries in the Middle East are likely to engage in conflicts over access to water resources. Most water in this dry region comes from three shared river basins: the Nile, Jordan, and Tigris–Euphrates (Figure 11-1).

Three countries—Ethiopia, Sudan, and Egypt—use most of the water that flows in Africa's Nile River, with Egypt being last in line along the river. To meet the water needs of its rapidly growing population, Ethiopia plans to divert more water from the Nile. So does Sudan. Such upstream diversions would reduce the amount of water available to Egypt, which cannot exist without irrigation water from the Nile.

Egypt could go to war with Sudan and Ethiopia for more water, cut population growth, or improve irrigation efficiency. Other options are to import more grain to reduce the need for irrigation water, work out water-sharing agreements with other countries, or suffer the harsh human and economic consequences of hydrological poverty.

The Jordan basin is by far the most water-short region, with fierce competition for its water among Jordan, Syria, Palestine (Gaza and the West Bank), and Israel. Syria plans to build dams and withdraw more water from the Jordan River, decreasing the downstream water supply for Jordan and Israel. Israel warns that it may destroy the largest dam that Syria plans to build.

Turkey, located at the headwaters of the Tigris and Euphrates rivers, controls how much water flows downstream to Syria and Iraq before emptying into the Persian Gulf. Turkey is building 24 dams along the upper Tigris and Euphrates to generate electricity and irrigate a large area of land.

If completed, these dams will reduce the flow of water downstream to Syria and Iraq by as much as 35% in normal years and by much more in dry years. Syria also plans to build a large dam along

Figure 11-1 Threatened natural capital: many countries in the Middle East, with some of the world's highest population growth, face water shortages and conflicts over access to water because they share water from three major river basins.

the Euphrates to divert water arriving from Turkey. This will leave little water for Iraq and could lead to a water war between that country and Syria.

Resolving these water distribution problems will require a combination of regional cooperation in allocating water supplies, slowed population growth, improved efficiency in water use, higher water prices to help improve irrigation efficiency, and increased grain imports to reduce water needs. Finding a solution in this and other water-short areas will not be easy. Currently there are no cooperative agreements for use of 158 of the world's 263 water basins that are shared by two or more countries.

To many analysts, emerging water shortages in many parts of the world—along with the related problems of biodiversity loss and climate change—are the three most serious environmental problems the world faces during this century.

Our liquid planet glows like a soft blue sapphire in the hard-edged darkness of space. There is nothing else like it in the solar system. It is because of water.

JOHN TODD

This chapter discusses the water supply and pollution problems we face and ways to use water—an irreplaceable resource—more sustainably. It addresses the following questions:

- Why is water so important, how much fresh water is available to us, and how much of it are we using?

- What causes freshwater shortages, and what can we do about these problems?

- What causes flooding, and what can we do about it?

- What pollutes water, where do these pollutants come from, and what effects do they have?

- What are the major water pollution problems affecting streams, lakes, and groundwater?

- What are the major water pollution problems affecting oceans?

- How can we prevent and reduce water pollution?

- How can we use the earth's water more sustainably?

KEY IDEAS

- We currently use more than half of the world's reliable runoff of surface water and could be using 70–90% by 2025.

- One of every six people does not have regular access to an adequate, safe, and affordable supply of clean water, and this number could increase to at least one of every four people by 2050.

- We can use water more sustainably by cutting waste, raising water prices, preserving forests in water basins, and slowing population growth.

- Stream pollution in most developing countries is a serious and growing problem.

- Groundwater pollution is a serious and growing problem in parts of the world.

- We can reduce pollution of coastal waters near heavily populated areas by preventing or reducing the flow of pollution from the land and from streams emptying into the ocean.

11-1 WATER'S IMPORTANCE, USE, AND RENEWAL

Science: Importance and Availability of Fresh Water

Water keeps us alive, moderates climate, sculpts the land, removes and dilutes wastes and pollutants, and is recycled by the hydrologic cycle.

We live on the water planet, with a precious film of water—most of it salt water—covering about 71% of the earth's surface. All organisms are made up of mostly water. Look in the mirror. What you see is about 60% water, most of it inside your cells.

You could survive for several weeks without food but only a few days without water. It takes huge amounts of water to supply you with food, provide shelter, and meet your other needs and wants. Water also plays a key role in sculpting the earth's surface, moderating climate, and removing and diluting water-soluble wastes and pollutants.

Despite its importance, water is one of our most poorly managed resources. We waste it and pollute it. We also charge too little for making it available. This encourages still greater waste and pollution of this renewable resource, for which we have no substitute. As Benjamin Franklin said many decades ago, "It is not until the well runs dry that we know the worth of water."

Only a tiny fraction of the planet's abundant water supply is readily available to us as fresh water (Figure 11-2, p. 238). If the world's water supply amounted to only 100 liters (26 gallons), our usable supply of fresh water would be only about 0.014 liter, or 2.5 teaspoons!

Fortunately, the world's freshwater supply is continuously collected, purified, recycled, and distributed in the solar-powered *hydrologic cycle* (Figure 3-24, p. 54). This magnificent water recycling and purification system works well as long as we do not overload water systems with slowly degradable and nondegradable wastes or withdraw water from underground supplies faster than it is replenished. In parts of the world, we are doing both of these things.

Differences in average annual precipitation divide the world's countries and people into water *haves* and *have-nots*. Some places get lots of rain (the dark and light green areas in Figure 5-2, p. 80), whereas others get very little (the yellow-green areas in Figure 5-2). For example, Canada, with only 0.5% of the world's population, has one-fifth of the world's fresh water. China, with one-fifth of the world's people, has only 7% of the supply.

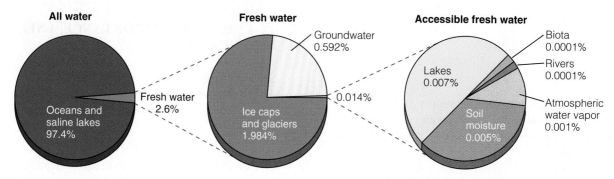

Figure 11-2 Natural capital: the planet's water budget. Only a tiny fraction by volume of the world's water supply is fresh water available for human use.

Science: Surface Water

Water that does not evaporate into the air or sink into the ground runs off the land into bodies of water.

One of our most precious resources is fresh water that flows across the earth's land surface and into rivers, streams, lakes, wetlands, and estuaries. The precipitation that does not return to the atmosphere by evaporation or infiltrate the ground is called **surface runoff.** The region from which surface water drains into a river, lake, wetland, or other body of water is called its **watershed** or **drainage basin.**

Two-thirds of the world's annual runoff is lost by seasonal floods and is not available for human use. The remaining one-third is **reliable runoff:** the amount of runoff that we can generally count on as a stable source of fresh water from year to year.

Science: Groundwater

Some precipitation infiltrates the ground and is stored in spaces in soil and rock.

Some precipitation infiltrates the ground and percolates downward through voids (pores, fractures, crevices, and other spaces) in soil and rock (Figure 11-3). The water in these spaces is called **groundwater**—and it is one of our most important sources of fresh water.

Close to the surface, the spaces in soil and rock hold little moisture. Below a certain depth, in the **zone of saturation,** these spaces are completely filled with water. The **water table** is located at the top of the zone of saturation. It falls in dry weather or when we remove groundwater faster than it is replenished, and it rises in wet weather.

Deeper down are geological layers called **aquifers:** porous, water-saturated layers of sand, gravel, or bedrock through which groundwater flows. They are like large elongated sponges through which groundwater seeps. Fairly watertight layers of rock or clay below an aquifer keep the water from escaping. About one of every three people on the earth depends on water pumped out of aquifers for drinking and other uses.

Most aquifers are replenished naturally by precipitation that percolates downward through soil and rock, a process called **natural recharge.** Others are recharged from the side by *lateral recharge* from nearby streams. Most aquifers recharge extremely slowly.

Groundwater normally moves from points of high elevation and pressure to points of lower elevation and pressure. This movement is quite slow, typically only a meter or so (about 3 feet) per year and rarely more than 0.3 meter (1 foot) per day.

Some aquifers get very little, if any, recharge and on a human time scale are nonrenewable resources. They are found fairly deep underground and were formed tens of thousands of years ago. Withdrawals from them amount to *water mining.* If kept up, they will deplete these ancient deposits of natural capital.

Tapping the World's Reliable Surface Water Supply

We currently use more than half of the world's reliable runoff of surface water and could be using 70–90% by 2025.

During the last century, the human population tripled, global water withdrawal increased sevenfold, and per capita withdrawal quadrupled. As a result, we now withdraw about 34% of the world's reliable runoff. We leave another 20% of this runoff in streams to transport goods by boats, dilute pollution, and sustain fisheries and wildlife. In total, *we directly or indirectly use about 54% of the world's reliable runoff of surface water.*

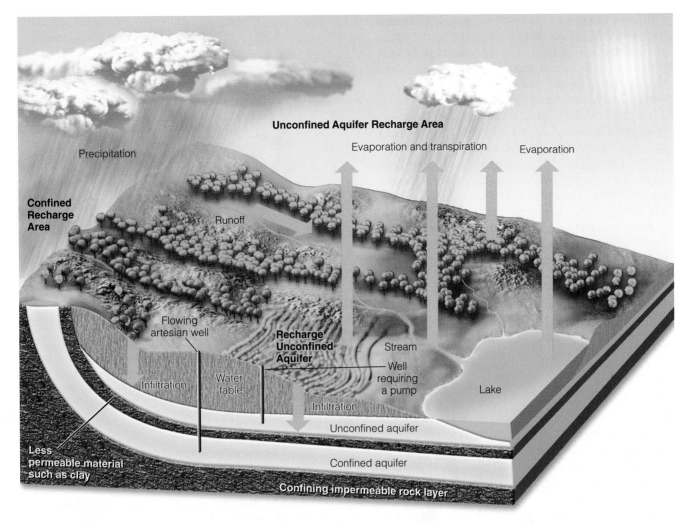

Figure 11-3 **Natural capital:** groundwater system. An *unconfined aquifer* is an aquifer with a water table. A *confined aquifer* is bounded above and below by less permeable beds of rock. Groundwater in this type of aquifer is confined under pressure. Some aquifers are replenished by precipitation; others are not.

To meet the demands of our growing population, global withdrawal rates of surface water could reach more than 70% of the reliable runoff by 2025 and 90% if per capita withdrawal of water continues increasing at the current rate. This is a global average. Withdrawal rates already exceed the reliable runoff in some areas.

Global Outlook: Uses of the World's Fresh Water

Irrigation is the biggest user of water (70%), followed by industries (20%) and cities and residences (10%).

Worldwide, we use 70% of the water we withdraw each year from rivers, lakes, and aquifers to irrigate one-fifth of the world's cropland. This land produces about 40% of the world's food, including two-thirds of the world's rice and wheat. Industry uses 20% of the

water withdrawn each year, and cities and residences use the remaining 10%.

Manufacturing and agriculture use large amounts of water. For example, it takes 400,000 liters (106,000 gallons) of water to produce an automobile, 9,000 liters (2,800 gallons) to produce 1 kilogram (2.2 pounds) of aluminum, and 7,000 liters (1,900 gallons) to produce 1 kilogram (2.2 pounds) of grain-fed beef. You could save more water by not eating a half a kilogram (1 pound) of grain-fed beef than by not showering for six months.

Science Case Study: Freshwater Resources in the United States

The United States has plenty of fresh water, but supplies vary in different areas depending on climate.

The United States has more than enough renewable fresh water. Unfortunately, much of it is in the wrong place at the wrong time or is contaminated by agricultural and industrial practices. The eastern states usually have ample precipitation, whereas many western states have too little (Figure 11-4, top).

In the East, the major uses for water are for energy production, cooling, and manufacturing. In the West, the largest use by far (85%) is for irrigation.

In many parts of the eastern United States, the most serious water problems are flooding, occasional urban shortages, and pollution. The major water problem in the arid and semiarid areas of the western half of the country (Figure 11-4, bottom) is a shortage of runoff, caused by low precipitation (Figure 11-4, top), high evaporation, and recurring prolonged drought.

Highly likely conflict potential
Substantial conflict potential
Moderate conflict potential
Unmet rural water needs

Figure 11-5 Water *hot spot* areas in 17 western states that by 2025 could face intense conflicts and "water wars" over scarce water needed for urban growth, irrigation, recreation, and wildlife. Some analysts suggest that this is a map of places not to live over the next 25 years. (Data from U.S. Department of the Interior)

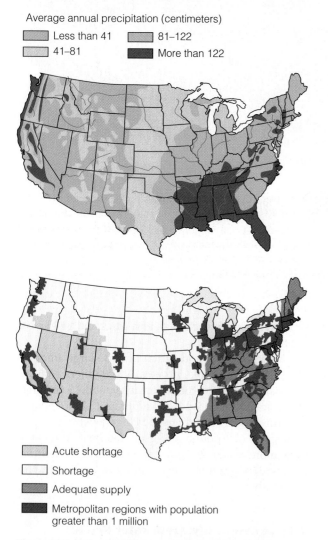

Average annual precipitation (centimeters)

Less than 41   81–122
41–81   More than 122

Acute shortage
Shortage
Adequate supply
Metropolitan regions with population greater than 1 million

Figure 11-4 Natural capital: average annual precipitation and major rivers (top) and water-deficit regions in the continental United States and their proximity to metropolitan areas having populations greater than 1 million (bottom). (Data from U.S. Water Resources Council and U.S. Geological Survey)

Water tables in many areas are dropping rapidly as farmers and cities deplete aquifers faster than they are recharged. Many U.S. urban centers (especially in the West and Midwest, purple in Figure 11-4, bottom) are located in areas that do not have enough water. In 2003, the U.S. Department of the Interior mapped out *water hot spots* in 17 western states (Figure 11-5).

11-2 SUPPLYING MORE WATER

Global Outlook: Freshwater Shortages

One of every six people does not have regular access to an adequate and affordable supply of clean water. This number could increase to at least one of every four people by 2050.

The two main factors causing water scarcity are a dry climate and too many people using the reliable supply of water. Figure 11-6 shows the degree of stress faced by the world's major river systems, based on a comparison of the amount of fresh water available with the amount used by humans.

A 2003 study by the United Nations found that one of every six people does not have regular access to an adequate and affordable supply of clean water. By 2050, this number could increase to at least one of every four people.

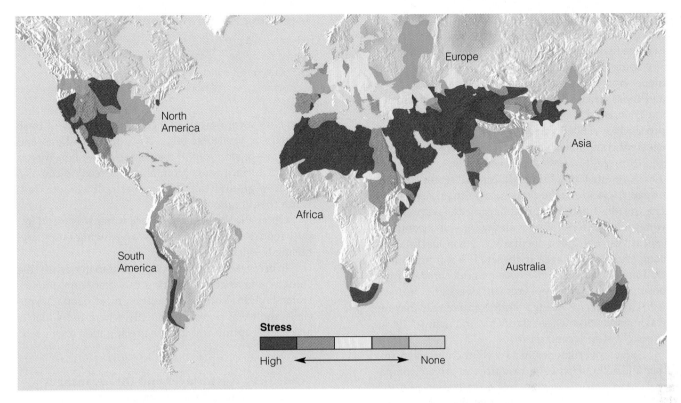

Figure 11-6 Natural capital degradation: stress on the world's major river basins, based on a comparison of the amount of water available with the amount used by humans. (Data from World Commission on Water Use in the 21st Century)

Poverty also governs access to water. Even when a plentiful supply of water exists, most of the world's 1.1 billion poor people living on less than $1 (U.S.) per day cannot afford a safe supply of drinking water and so live in *hydrological poverty.* Most are cut off from municipal water supplies and must collect water from unsafe sources or buy water—often taken from polluted rivers—from private vendors at high prices. In water-short rural areas in developing countries, many women and children must walk long distances each day, carrying heavy jars or cans, to get a meager and sometimes contaminated supply of water (Figure 11-7).

Politics and Ethics Case Study: Who Should Own and Manage Freshwater Resources?

Controversy exists regarding whether water supplies should be owned and managed by governments or by private corporations.

Most people believe that everyone should have access to clean water. But who will pay for making this water available to everyone?

Most water resources are owned by governments and managed as publicly owned resources for their citizens. An increasing number of governments are retaining ownership of these public resources but are hiring private companies to manage them. In addition, three large European companies—Vivendi, Suez, and RWE—have a long-range strategy to buy up as much of the world's water supplies as possible, especially in Europe and North America.

Currently, 85% of Americans get their water from publicly owned utilities. This may soon change. Within 10 years the three European-based water companies aim to control 70% of the water supply in the

Figure 11-7 Global outlook: villager in Mozambique, Africa, carrying water and an infant.

United States by buying up American water companies and entering into agreements with most cities to manage their water supplies.

Some argue that private companies have the money and expertise to manage these resources better and more efficiently than government bureaucracies. Experience with this public–private partnership approach has yielded mixed results. Some water management companies have improved efficiency, done a good job, and in a few cases lowered rates.

In the late 1980s, Prime Minister Margaret Thatcher placed England's water management in private hands. The result: financial mismanagement, skyrocketing water rates, deteriorating water quality, and company executives who gave themselves generous financial compensation packages. In the late 1990s, Prime Minister Tony Blair brought the system under control by imposing much stricter government oversight. *The message:* Governments hiring private companies to manage water resources must set standards and maintain strict oversight of such contracts.

Some government officials want to go even further and sell public water resources to private companies. Many people oppose full privatization of water resources because they believe that water is a public resource too important to be left solely in private hands. Also, once a city's water systems have been taken over by a foreign-based corporation, efforts to return the systems to public control can lead to severe economic penalties under the rules of the World Trade Organization (WTO).

In the Bolivian town of Cochabamba, 60% of the water was being lost through leaky pipes. With no money to fix the pipes, the Bolivian government sold the town's water system to a subsidiary of Bechtel Corporation. Within 6 months, the company doubled water rates and began seizing and selling the houses of people who did not pay their water bills. A general strike ensued, and violent street clashes between protesters and government troops led to 10,000 injured people and 7 deaths. The Bolivian government ended up tearing up the contract. Today's the town's government–private cooperative water management system is in shambles and most of the leaks persist.

Some analysts point to two other potential problems in a fully privatized water system. *First,* because private companies make money by delivering water, they have an incentive to sell as much water as they can rather than to conserve it. *Second,* because of lack of money to pay water bills, the poor will continue to be left out. There are no easy answers for managing the water that everyone needs.

✘ *HOW WOULD YOU VOTE?* Should private companies own and manage most of the world's water resources? Cast your vote online at http://biology.brookscole.com/miller11.

Increasing Freshwater Supplies

We can increase water supplies by building dams, bringing in water from somewhere else, withdrawing groundwater, converting salt water to fresh water, wasting less water, and importing food.

There are several ways to increase the supply of fresh water in a particular area. We can build dams and reservoirs to store runoff for release as needed. We can bring in surface water from another area. We can also withdraw groundwater and convert salt water to fresh water (desalination).

Other strategies are to reduce water waste and import food to reduce water use in growing crops and raising livestock.

In *developed countries,* people tend to live where the climate is favorable and bring in water from another watershed. In *developing countries,* most people (especially the rural poor) must settle where the water is and try to capture and use the precipitation they need.

Trade-Offs: Advantages and Disadvantages of Large Dams and Reservoirs

Large dams and reservoirs can produce cheap electricity, reduce downstream flooding, and provide year-round water for irrigating cropland, but they also displace people and disrupt aquatic systems.

Large dams and reservoirs have both benefits and drawbacks (Figure 11-8). Their main purpose is to capture and store runoff and release it as needed to control floods, generate electricity, and supply water for irrigation and for towns and cities. Reservoirs also provide recreational activities such as swimming, fishing, and boating.

The more than 45,000 large dams built on the world's 227 largest rivers have increased the annual reliable runoff available for human use by nearly one-third. At the same time, a series of dams on a river, especially in arid areas, can reduce downstream flow to a trickle and prevent it from reaching the sea as a part of the hydrologic cycle. According to the World Commission on Water in the 21st Century, half of the world's major rivers go dry part of the year because of flow reduction by dams, especially during drought years.

This engineering approach to river management has displaced 40–80 million people from their homes and flooded an area of mostly productive land roughly equal to the area of California. In addition, this approach often impairs some of the important ecological services rivers provide (Figure 11-9). In 2003, the World Resources Institute estimated that dams and reservoirs have strongly or moderately fragmented and disturbed 60% of the world's major river basins.

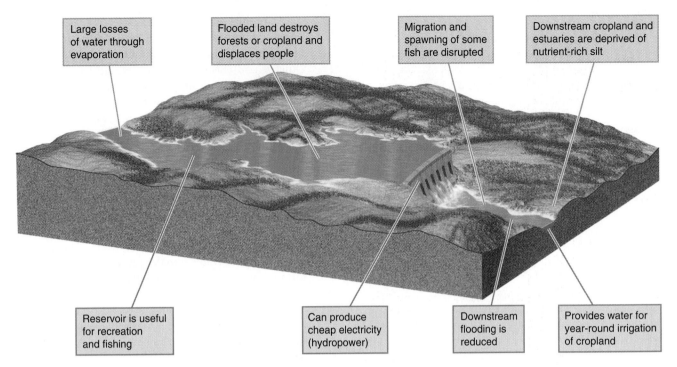

Large losses of water through evaporation	Flooded land destroys forests or cropland and displaces people
Migration and spawning of some fish are disrupted	Downstream cropland and estuaries are deprived of nutrient-rich silt

Reservoir is useful for recreation and fishing	Can produce cheap electricity (hydropower)
Downstream flooding is reduced	Provides water for year-round irrigation of cropland

Figure 11-8 Trade-offs: advantages (green) and disadvantages (orange) of large dams and reservoirs. The world's 45,000 large dams (higher than 15 meters or 50 feet) capture and store 14% of the world's runoff, provide water for 45% of irrigated cropland, and supply more than half the electricity used by 65 countries. The United States has more than 70,000 large and small dams, capable of capturing and storing half of the country's entire river flow. *Critical thinking: pick the single advantage and disadvantage that you think are the most important.*

Since 1960, the Colorado River in the United States has rarely made it to the Gulf of California because of a combination of multiple dams, large-scale water withdrawal, and prolonged drought. Its dwindling water supply threatens the survival of species that spawn in the river, destroys estuaries that serve as breeding grounds for numerous aquatic species, and increases saltwater contamination of aquifers near the coast.

X *How Would You Vote?* Do the advantages of large dams outweigh their disadvantages? Cast your vote online at http://biology.brookscole.com/miller11.

Trade-Offs: Advantages and Disadvantages of Water Transfers—the California Experience

The massive transfer of water from water-rich northern California to water-poor southern California has brought many benefits, but remains controversial.

Tunnels, aqueducts, and underground pipes can transfer stream runoff collected by dams and reservoirs from water-rich areas to water-poor areas. They also create environmental problems. Indeed, most of the world's dam projects and large-scale water transfers illustrate an important ecological principle: *You cannot do just one thing.* A number of unintended environmental consequences almost always occur.

Natural Capital

Ecological Services of Rivers

- Deliver nutrients to sea to help sustain coastal fisheries

- Deposit silt that maintains deltas

- Purify water

- Renew and renourish wetlands

- Provide habitats for wildlife

Figure 11-9 Natural capital: important ecological services provided by rivers. Currently, the services are assigned little or no monetary value when the costs and benefits of dam and reservoir projects are assessed. According to environmental economists, attaching even crudely estimated monetary values to these ecosystem services would help sustain them.

One of the world's largest water transfer projects is the *California Water Project* (Figure 11-10, p. 244). It uses a maze of giant dams, pumps, and aqueducts (cement-lined artificial rivers) to transport water from water-rich northern California to southern California's

Figure 11-10 Solutions: California Water Project and the Central Arizona Project. These projects involve large-scale water transfers from one watershed to another. Arrows show the general direction of water flow.

heavily populated, arid and semiarid agricultural regions and cities. In effect, this project supplies massive amounts of water to areas that without such water would be mostly desert.

For decades, northern and southern Californians have feuded over how the state's water should be allocated under this project. Southern Californians want more water from the north to grow more crops and to support Los Angeles, San Diego, and other growing urban areas. Agriculture consumes three-fourths of the water withdrawn in California, much of it used inefficiently for water-thirsty crops growing in desert-like conditions.

Northern Californians counter that sending more water south would degrade the Sacramento River, threaten fisheries, and reduce the flushing action that helps clean San Francisco Bay of pollutants. They also argue that much of the water sent south is wasted. They point to studies showing that making irrigation just 10% more efficient would provide enough water for domestic and industrial uses in southern California.

According to a 2002 joint study by a group of scientists and engineers, projected global warming will sharply reduce water availability in California (especially southern California) and other water-short states in the western United States even under the best-case scenario. Some analysts project that sometime during this century, many of the people living in arid southern California cities (such as Los Angeles and San Diego), as well as farmers in this area, will have to move somewhere else because of a lack of water.

Pumping out more groundwater is not the answer—groundwater is already being withdrawn faster than it is replenished throughout much of California. Quicker and cheaper solutions would be to improve irrigation efficiency, stop growing water-thirsty crops in a desert climate, and allow farmers to sell cities their legal rights to withdraw certain amounts of water from rivers.

Science Case Study: The Aral Sea Disaster

Diverting water from the Aral Sea and its two feeder rivers mostly for irrigation has created a major ecological, economic, and health disaster.

The shrinking of the Aral Sea (Figure 11-11) is a result of a large-scale water transfer project in an area of the former Soviet Union with the driest climate in central Asia. Since 1960, enormous amounts of irrigation water have been diverted from the inland Aral Sea and its two feeder rivers to create one of the world's largest irrigated areas, mostly for raising cotton and rice. The irrigation canal, the world's longest, stretches more than 1,300 kilometers (800 miles).

This large-scale water diversion project, coupled with droughts and high evaporation rates due to the area's hot and dry climate, has caused a regional ecological, economic, and health disaster. Since 1960, the sea's salinity has tripled, its surface area has decreased by 58%, and it has lost 83% of its volume of water. Water withdrawal for agriculture has reduced the sea's two supply rivers to mere trickles.

About 85% of the area's wetlands have been eliminated and roughly half the local bird and mammal species have disappeared. In addition, a huge area of former lake bottom has been converted to a human-made desert covered with glistening white salt. The increased salt concentration caused the presumed extinction of 20 of the area's 24 native fish species. This has devastated the area's fishing industry, which once provided work for more than 60,000 people. Fishing villages and boats once located on the sea's coastline now sit abandoned in the middle of a salt desert (Figure 11-12).

Winds pick up the salty dust that encrusts the lake's now-exposed bed and blow it onto fields as far as 300 kilometers (190 miles) away. As the salt spreads, it pollutes water and kills wildlife, crops, and other vegetation. Aral Sea dust settling on glaciers in the Himalayas is causing them to melt at a faster than normal rate—another example of connections and unintended consequences.

To raise yields, farmers have increased their inputs of herbicides, insecticides, fertilizers, and irrigation water on some crops. Many of these chemicals have percolated downward and accumulated to dan-

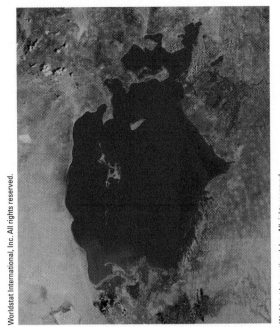

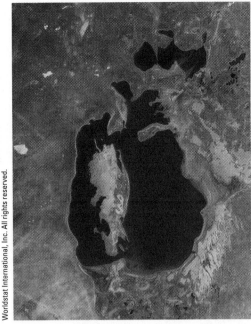

Figure 11-11 Natural capital degradation: the *Aral Sea* was once the world's fourth largest freshwater lake. Since 1960, it has been shrinking and getting saltier because most of the water from the rivers that replenish it has been diverted to grow cotton and food crops. These satellite photos show the sea in 1976 and in 1997. As the lake shrinks, it leaves behind a salty desert, economic ruin, increasing health problems, and severe ecological disruption.

gerous levels in the groundwater—the source of most of the region's drinking water.

Shrinkage of the Aral Sea has altered the area's climate. The once-huge sea acted as a thermal buffer that

Figure 11-12 Natural and economic capital degradation: as the Aral Sea shrank it left ships stranded in a newly formed desert.

moderated the heat of summer and the extreme cold of winter. Now there is less rain, summers are hotter and drier, winters are colder, and the growing season is shorter. The combination of such climate change and severe salinization has reduced crop yields by 20–50% on almost one-third of the area's cropland.

Finally, many of the 58 million people living in the Aral Sea's watershed have experienced increasing health problems from a combination of toxic dust, salt, and contaminated water.

Can the Aral Sea be saved, and can the area's serious ecological and human health problems be reversed? Since 1999, the United Nations and the World Bank have spent about $600 million to purify drinking water and upgrade irrigation and drainage systems, which improves irrigation efficiency and flushes salts from croplands. In addition, some artificial wetlands and lakes have been constructed to help restore aquatic vegetation, wildlife, and fisheries.

The five countries surrounding the lake and its two feeder rivers have worked to improve irrigation efficiency and to partially replace crops such as rice and cotton, which have high water requirements, with other crops requiring less irrigation water. As a result, the total annual volume of water in the Aral Sea basin has stabilized. Nevertheless, experts expect the largest portion of the Aral Sea to the south to continue shrinking.

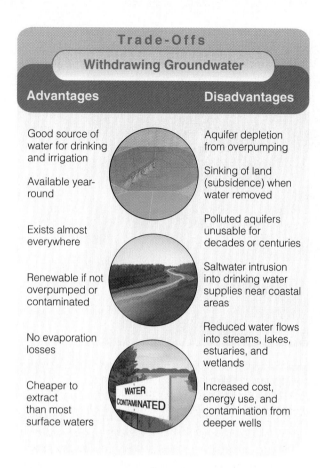

Trade-Offs

Withdrawing Groundwater

Advantages	Disadvantages
Good source of water for drinking and irrigation	Aquifer depletion from overpumping
Available year-round	Sinking of land (subsidence) when water removed
Exists almost everywhere	Polluted aquifers unusable for decades or centuries
Renewable if not overpumped or contaminated	Saltwater intrusion into drinking water supplies near coastal areas
No evaporation losses	Reduced water flows into streams, lakes, estuaries, and wetlands
Cheaper to extract than most surface waters	Increased cost, energy use, and contamination from deeper wells

Figure 11-13 Trade-offs: advantages and disadvantages of withdrawing groundwater. *Critical thinking: pick the single advantage and disadvantage that you think are the most important.*

Trade-Offs: Advantages and Disadvantages of Withdrawing Groundwater

Most aquifers are renewable sources unless water is removed faster than it is replenished or becomes contaminated.

Aquifers provide drinking water for about one-fourth of the world's people. In the United States, water pumped from aquifers supplies almost all of the drinking water in rural areas, one-fifth of that in urban areas, and 43% of irrigation water.

Relying more on groundwater has advantages and disadvantages (Figure 11-13). Aquifers are widely available and are renewable sources of water as long as the water is not withdrawn faster than it is replaced and as long as the aquifers do not become contaminated.

But water tables are falling in many areas of the world because the rate of pumping out water (mostly to irrigate crops) exceeds the rate of natural recharge from precipitation. The world's three largest grain-producing countries—China, India, and the United States—are overpumping many of their aquifers.

Saudi Arabia is as water-poor as it is oil-rich. It gets about 70% of its drinking water at a high cost from the world's largest desalination complex on its eastern coast. The rest of the country's water is pumped from deep aquifers, most as nonrenewable as the country's oil. This water-short nation wastes much of its scarcest resource to grow irrigated crops on desert land and to fill large numbers of fountains and swimming pools that let precious water evaporate into the hot, dry desert air. Hydrologists estimate that because of the rapid depletion of its fossil aquifers, most irrigated agriculture in Saudi Arabia may disappear within 10–20 years.

In the United States, groundwater is being withdrawn at four times its replacement rate. The most serious overdrafts are occurring in parts of the huge Ogallala Aquifer, extending from southern South Dakota to central Texas, and in parts of the arid Southwest (Figure 11-14). Serious groundwater depletion is also taking place in California's water-short Central Valley, which supplies half the country's vegetables and fruits.

Groundwater overdrafts near coastal areas can contaminate groundwater supplies by allowing salt water to intrude into freshwater aquifers used to supply water for irrigation and domestic purposes (Figure 11-15). This problem is especially serious in coastal areas in Florida, California, South Carolina, and Texas. Figure 11-16 lists ways to prevent or slow the problem of groundwater depletion.

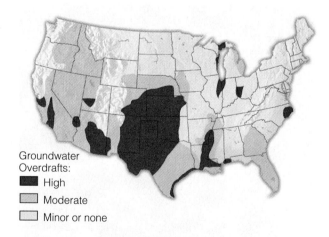

Groundwater Overdrafts:
- ■ High
- ▨ Moderate
- □ Minor or none

Environmental Science⊕Now™

Active Figure 11-14 Natural capital degradation: areas of greatest aquifer depletion from groundwater overdraft in the continental United States. Aquifer depletion is also high in Hawaii and Puerto Rico (not shown on map). This practice causes the land below the aquifer to subside or sink in most of these areas. *See an animation based on this figure and take a short quiz on the concept.* (Data from U.S. Water Resources Council and U.S. Geological Survey)

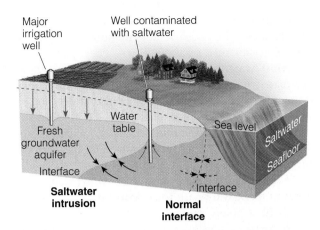

Figure 11-15 **Natural capital degradation:** *saltwater intrusion* along a coastal region. When the water table is lowered, the normal interface (dashed line) between fresh and saline groundwater moves inland, making groundwater drinking supplies unusable.

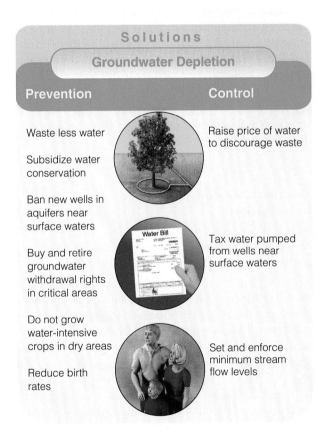

Figure 11-16 **Solutions:** ways to prevent or slow groundwater depletion. *Critical thinking: which two of these solutions do you believe are the most important?*

With global water shortages looming, scientists are evaluating deep aquifers—found at depths of 0.8 kilometer (0.5 mile) or more—as future water sources. Preliminary results suggest that some of these aquifers hold enough water to support billions of people for centuries. The quality of water in these aquifers may also be much higher than the quality of the water in most rivers and lakes.

Two major concerns arise regarding tapping these mostly one-time deposits of water. *First,* little is known about the geological and ecological impacts of pumping water from deep aquifers. *Second,* no international water treaties exist governing rights to, and ownership of, water found under several different countries. Without such treaties, legal and physical conflicts could ensue over who has the right to tap into and use these valuable resources.

Environmental Science ⊛ Now™

Find out where in the United States aquifers are being depleted, polluted, and contaminated by saltwater at Environmental ScienceNow.

Science and Economics: Desalination

Removing salt from seawater will probably not be done widely because of high costs and questions about what to do with the resulting salt.

Desalination involves removing dissolved salts from ocean water or from brackish (slightly salty) water in aquifers or lakes. It represents another way to increase supplies of fresh water.

One method for desalinating water is *distillation*—heating salt water until it evaporates, leaves behind salts in solid form, and condenses as fresh water. Another method is *reverse osmosis*—pumping salt water at high pressure through a thin membrane with pores that allow water molecules, but not most dissolved salts, to pass through. In effect, high pressure is used to push fresh water out of salt water.

Today about 13,500 desalination plants operate in 120 countries, especially in the arid, desert nations of the Middle East, North Africa, the Caribbean, and the Mediterranean. These plants meet less than 0.3% of the world's water needs.

There are two major problems with the widespread use of desalination. One is the high cost, because it takes a lot of energy to desalinate water. Currently, desalinating water costs two to three times as much as the conventional purification of fresh water, although recent advances in reverse osmosis have reduced the energy costs somewhat.

The second problem is that desalination produces large quantities of briny wastewater that contain lots of salt and other minerals. Dumping concentrated brine into a nearby ocean increases the salinity of the ocean water, threatening food resources and aquatic life in the vicinity. Dumping it on land could contaminate groundwater and surface water.

Bottom line: Currently, significant desalination is practical only for wealthy and water-short countries and cities that can afford its high cost.

Scientists are working to develop new membranes for reverse osmosis that can separate water from salt more efficiently and under less pressure. If successful, this strategy could bring down the cost of using desalination to produce drinking water. Even so, desalinated water probably will not be cheap enough to irrigate conventional crops or meet much of the world's demand for fresh water unless scientists can figure out how to use solar energy or other means to desalinate seawater cheaply and how to safely dispose of the salt left behind.

11-3 REDUCING WATER WASTE

Economics: Benefits of Reducing Water Waste

We waste about two-thirds of the water we use, but we could cut this waste to 15%.

Mohamed El-Ashry of the World Resources Institute estimates that *65–70% of the water people use throughout the world is lost through evaporation, leaks, and other losses.* The United States, the world's largest user of water, does slightly better but still loses about half of the water it withdraws. El-Ashry believes it is economically and technically feasible to reduce such water losses to 15%, thereby meeting most of the world's water needs for the foreseeable future.

This win-win solution would decrease the burden on wastewater plants and reduce the need for expensive dams and water transfer projects that destroy wildlife habitats and displace people. It would also slow depletion of groundwater aquifers and save both energy and money.

According to water resource experts, the main cause of water waste is that *we charge too little for water.* Such *underpricing* is mostly the result of government subsidies that provide irrigation water, electricity, and diesel fuel for farmers to pump water from rivers and aquifers at below-market prices.

Because these subsidies keep the price of water low, users have little or no financial incentive to invest in water-saving technologies. According to water resource expert Sandra Postel, "By heavily subsidizing water, governments give out the false message that it is abundant and can afford to be wasted—even as rivers are drying up, aquifers are being depleted, fisheries are collapsing, and species are going extinct."

Farmers, industries, and others benefiting from government water subsidies offer a counter-argument: They promote settlement and agricultural production in arid and semiarid areas, stimulate local economies, and help keep the prices of food, manufactured goods, and electricity low.

Most water resource experts believe that when water scarcity afflicts many areas in this century, governments will have to make the unpopular decision to raise water prices. China did so in 2002 because it faced water shortages in most of its major cities, rivers running dry, and falling water tables in key agricultural areas.

Higher water prices encourage water conservation but make it difficult for low-income farmers and city dwellers to buy enough water to meet their needs. South Africa has found a novel solution to this problem. When the country raised water prices, it established *lifeline* rates that give each household a set amount of water at a low price to meet basic needs. When users exceed this amount, the price rises.

The second major cause of water waste is *lack of government subsidies for improving the efficiency of water use.* A basic rule of economics is that you get more of what you reward. Withdrawing subsidies that encourage water waste and providing subsidies for efficient water use would sharply reduce water waste.

X *HOW WOULD YOU VOTE?* Should water prices be raised sharply to help reduce water waste? Cast your vote online at http://biology.brookscole.com/miller11.

Solutions: Wasting Less Water in Irrigation

Although 60% of the world's irrigation water is currently wasted, improved irrigation techniques could reduce this proportion to 5–20%.

About 60% of the irrigation water applied throughout the world does not reach the targeted crops and does not contribute to food production. Most irrigation systems obtain water from a groundwater well or a surface water source. The water then flows by gravity through unlined ditches in crop fields so the crops can absorb it (Figure 11-17, left). This *flood irrigation* method delivers far more water than is needed for crop growth and typically loses 40% of the water through evaporation, seepage, and runoff.

More efficient and environmentally sound irrigation technologies can greatly reduce water demands and waste on farms by delivering water more precisely to crops. For example, the *center-pivot low-pressure sprinkler* (Figure 11-17, right) uses pumps to spray water on a crop. Typically, it allows 80% of the water to reach crops. *Low-energy precision application (LEPA) sprinklers,* another form of center-pivot irrigation, put 90–95% of the water where crops need it by

Figure 11-17 Major *irrigation systems.* Because of high initial costs, center-pivot irrigation and drip irrigation are not widely used. The development of new low-cost drip-irrigation systems may change this situation.

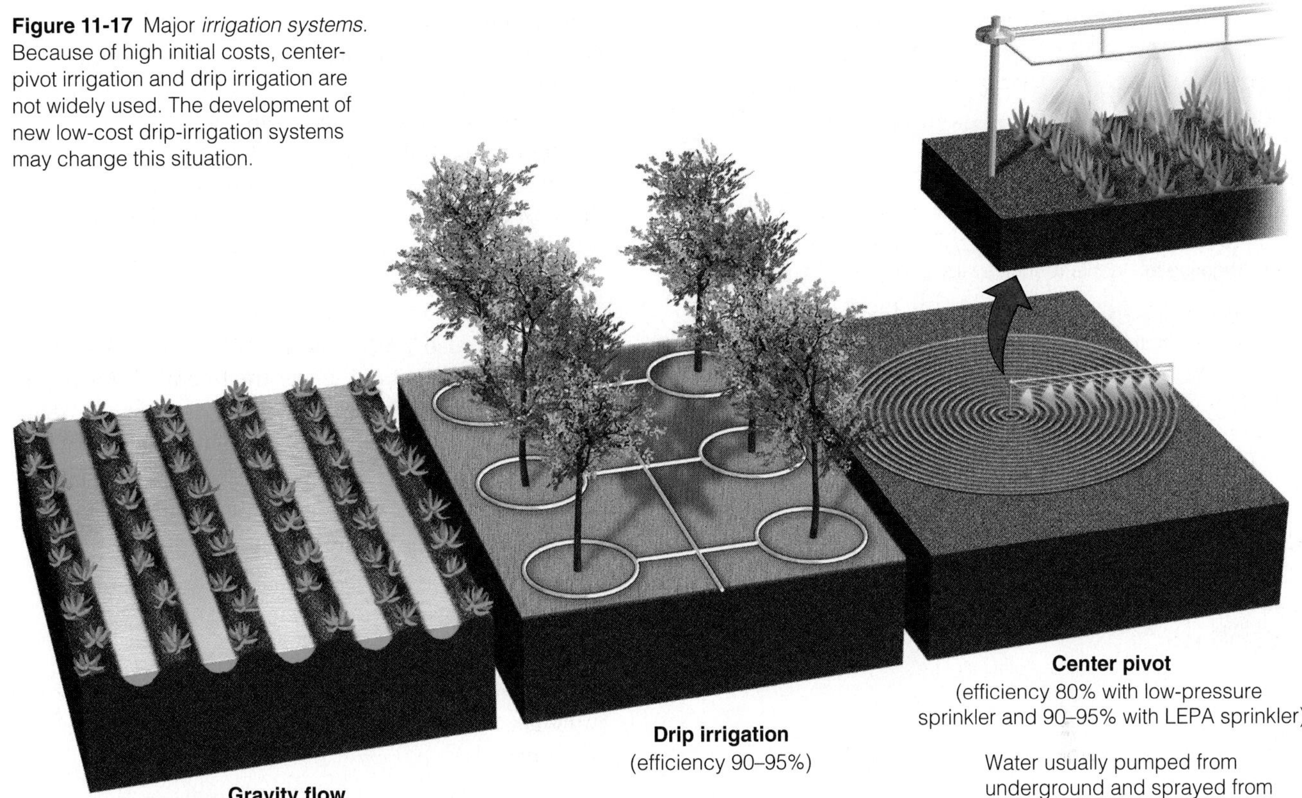

Center pivot
(efficiency 80% with low-pressure sprinkler and 90–95% with LEPA sprinkler)

Water usually pumped from underground and sprayed from mobile boom with sprinklers.

Drip irrigation
(efficiency 90–95%)

Above- or below-ground pipes or tubes deliver water to individual plant roots.

Gravity flow
(efficiency 60% and 80% with surge valves)

Water usually comes from an aqueduct system or a nearby river.

spraying the water closer to the ground and in larger droplets than the center-pivot, low-pressure system. Another method is to use *soil moisture detectors* to water crops only when they need it.

Drip irrigation or *microirrigation* (Figure 11-17, center) is the most efficient way to deliver small amounts of water precisely to crops. It consists of a network of perforated plastic tubing installed at or below the ground level. Small holes or emitters in the tubing deliver drops of water at a slow and steady rate, close to the plant roots.

Drip irrigation is very efficient, with 90–95% of the water input reaching the crops. The flexible and lightweight tubing system can easily be fitted to match the patterns of crops in a field and left in place or moved around.

Currently, drip irrigation is used on slightly more than 1% of the world's irrigated crop fields and 4% of those in the United States. This percentage rises to 90% in Cyprus, 66% in Israel, and 13% in California. Unfortunately, the capital cost of conventional drip irrigation systems remains too high for most poor farmers and for use on low-value row crops.

As noted earlier, irrigation water is underpriced. Raise water prices enough and drip irrigation would

quickly be adopted to irrigate most of the world's crops. *Good news.* The capital cost of a new type of drip irrigation system is one-tenth as much per hectare as that for conventional drip systems.

Figure 11-18 (p. 250) lists other ways to reduce water waste in irrigating crops. Since 1950, water-short Israel has used many of these techniques to slash irrigation water waste by 84% while irrigating 44% more land. Israel now treats and reuses 30% of its municipal sewage water for crop production and plans to increase this percentage to 80% by 2025. The government also gradually eliminated most water subsidies to raise the price of irrigation water to one of the highest in the world. Israelis also import most of their wheat and meat and concentrate on growing fruits, vegetables, and flowers that need less water.

Many of the world's poor farmers cannot afford most of the modern technological methods for increasing irrigation and irrigation efficiency. Instead, they resort to small-scale and low-cost traditional technologies.

Some (in Bangladesh, for example) use pedal-powered treadle pumps to move water through irrigation ditches. Others use buckets or small tanks with holes for drip irrigation.

- Lining canals bring water to irrigation ditches

- Leveling fields with lasers

- Irrigating at night to reduce evaporation

- Using soil and satellite sensors and computer systems to monitor soil moisture and add water only when necessary

- Polyculture

- Organic farming

- Growing water-efficient crops using drought-resistant and salt-tolerant crop varieties

- Irrigating with treated urban waste water

- Importing water-intensive crops and meat

Figure 11-18 Solutions: methods for reducing water waste in irrigation. *Critical thinking: which two of these solutions do you believe are the most important?*

Solutions: Wasting Less Water in Industry, Homes, and Businesses

We can save water by changing to yard plants that need little water, using drip irrigation, raising water prices, fixing leaks, and using water-saving toilets and other appliances.

Figure 11-19 lists ways to use water more efficiently in industries, homes, and businesses. Many homeowners and businesses in water-short areas are replacing green lawns with vegetation adapted to a dry climate. This win-win approach, called *xeriscaping* (pronounced "ZER-i-scaping"), reduces water use by 30–85% and sharply reduces inputs of labor, fertilizer, and fuel. It also reduces polluted runoff, air pollution, and yard wastes.

About one-fifth of all U.S. public water systems do not have water meters and charge a single low rate for almost unlimited use of high-quality water. In Boulder, Colorado, introducing water meters reduced water use by more than one-third. Many apartment dwellers have little incentive to conserve water because water use is included in their rent.

We can also save water by replacing the current system in which we use large amounts of water good enough to drink to dilute and wash or flush away industrial, animal, and household wastes with one that mimics the way nature deals with wastes. According to the FAO, if current trends continue, within 40 years we will need the world's entire reliable flow of river water just to dilute and transport the wastes we produce.

One potential solution is to ban the discharge of industrial toxic wastes into municipal sewer systems. Another is to rely more on waterless composting toilets that convert human fecal matter into a small amount of dry and odorless soil-like humus material that can be removed from a composting chamber every year or so and returned to the soil as fertilizer. They work. I used one for 15 years without any problems.

We can also return the nutrient-rich sludge produced by conventional waste treatment plants to the soil as a fertilizer. Banning the input of toxic industrial chemicals into sewage treatment plants will make this feasible.

Solutions: Using Water More Sustainably

We can use water more sustainably by cutting waste, raising water prices, preserving forests in water basins, and slowing population growth.

Sustainable water use is based on the commonsense principle stated in an old Inca proverb: "The frog does

- Redesign manufacturing processes

- Landscape yards with plants that require little water

- Use drip irrigation

- Fix water leaks

- Use water meters and charge for all municipal water use

- Raise water prices

- Use waterless composting toilets

- Require water conservation in water-short cities

- Use water-saving toilets, showerheads, and front-loading clothes washers

- Collect and reuse household water to irrigate lawns and nonedible plants

- Purify and reuse water for houses, apartments, and office buildings

Figure 11-19 Solutions: methods of reducing water waste in industries, homes, and businesses. *Critical thinking: which two of these solutions do you believe are the most important?*

not drink up the pond in which it lives." Figure 11-20 lists ways to implement this principle.

The challenge in encouraging such a *blue revolution* is to implement a mix of strategies. One strategy involves using technology to irrigate crops more efficiently and to save water in industries and homes. A second approach uses economic and political policies to remove subsidies that cause water to be underpriced and thus wasted, while guaranteeing low prices for low-income consumers and adding subsidies that reward reduced water waste.

A third component is to switch to new waste-treatment systems that accept only nontoxic wastes, use less or no water to treat wastes, return nutrients in plant and animal wastes to the soil, and mimic the ways that nature decomposes and recycles organic wastes. A fourth strategy is to leave enough water in rivers to protect wildlife, ecological processes, and the natural ecological services provided by rivers.

We can all help bring about this blue revolution by using and wasting less water. We can also support government policies that result in more sustainable use of the world's water and better ways to treat our

Solutions
Sustainable Water Use

- Not depleting aquifers
- Preserving ecological health of aquatic systems
- Preserving water quality
- Integrated watershed management
- Agreements among regions and countries sharing surface water resources
- Outside party mediation of water disputes between nations
- Marketing of water rights
- Raising water prices
- Wasting less water
- Decreasing government subsidies for supplying water
- Increasing government subsidies for reducing water waste
- Slowing population growth

Figure 11-20 Solutions: methods for achieving more sustainable use of the earth's water resources. *Critical thinking: which two of these solutions do you believe are the most important?*

What Can You Do?
Water Use and Waste

- Use water-saving toilets, showerheads, and faucet aerators.
- Shower instead of taking baths, and take short showers.
- Repair water leaks.
- Turn off sink faucets while brushing teeth, shaving, or washing.
- Wash only full loads of clothes or use the lowest possible water-level setting for smaller loads.
- Wash a car from a bucket of soapy water, and use the hose for rinsing only.
- If you use a commercial car wash, try to find one that recycles its water.
- Replace your lawn with native plants that need little if any watering.
- Water lawns and gardens in the early morning or evening.
- Use drip irrigation and mulch for gardens and flowerbeds.
- Use recycled (gray) water for watering lawns and houseplants and for washing cars.

Figure 11-21 Individuals matter: ways you can reduce your use and waste of water. *Critical thinking: which four of these actions do you believe are the most important? Which actions on this list do you do or plan to do?*

industrial and household wastes. Figure 11-21 lists ways you can reduce your water use and waste.

11-4 TOO MUCH WATER

Science: Flooding

Heavy rainfall, rapid snowmelt, removal of vegetation, and destruction of wetlands cause flooding.

Whereas some areas have too little water, others sometimes have too much because of natural flooding by streams, caused mostly by heavy rain or rapid melting of snow. A flood happens when water in a stream overflows its normal channel and spills into the adjacent area, called a **floodplain**. Floodplains, which include highly productive wetlands, help provide natural flood and erosion control, maintain high water quality, and recharge groundwater.

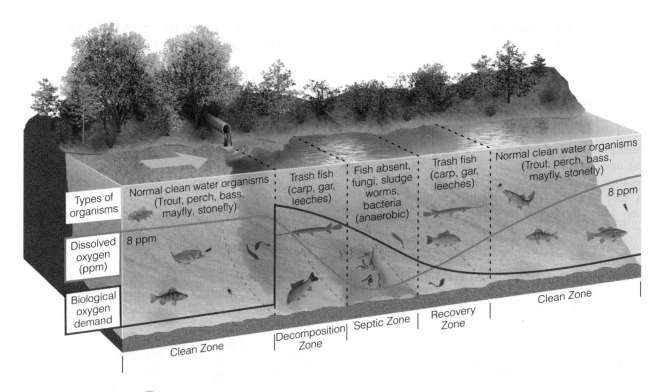

Environmental Science ⊕ Now™

Active Figure 11-24 **Natural capital:** dilution and decay of degradable, oxygen-demanding wastes and heat in a stream, showing the oxygen sag curve and the curve of oxygen demand. Depending on flow rates and the amount of pollutants, streams can recover from such pollution if they are given enough time and are not overloaded. *See an animation based on this figure and take a short quiz on the concept.*

it flowed through Cleveland. The highly publicized image of this burning river prompted elected officials to enact laws limiting the discharge of industrial wastes into the river and sewage systems, and to provide funds to upgrade sewage treatment facilities. Today the river is cleaner and is widely used by boaters and anglers. This accomplishment illustrates the power of bottom-up pressure by citizens, who prodded elected officials to change a severely polluted river into an economically and ecologically valuable public resource. Individuals matter!

On the other hand, large fish kills and drinking water contamination still occur in parts of developed countries. Two causes of these problems are accidental or deliberate releases of toxic inorganic and organic chemicals by industries or mines and malfunctioning sewage treatment plants. A third cause is nonpoint runoff of pesticides and excess plant nutrients from cropland and animal feedlots.

Global Outlook: Stream Pollution in Developing Countries

Stream pollution in most developing countries is a serious and growing problem.

Stream pollution from discharges of untreated sewage and industrial wastes is a serious and growing problem in most developing countries. According to a 2003 report by the World Commission on Water in the 21st Century, half of the world's 500 rivers are heavily polluted; most of them running through developing countries. Most of these countries cannot afford to build waste treatment plants and do not have—or do not enforce—laws for controlling water pollution.

Industrial wastes and sewage pollute more than two-thirds of India's water resources and 54 of the 78 streams monitored in China. Only 10% of the sewage produced in Chinese cities is treated. In Latin America and Africa, most streams passing through urban or industrial areas suffer from severe pollution.

Science: Pollution Problems of Lakes

Dilution of pollutants in lakes is less effective than in most streams because most lake water is not mixed well and has little flow.

In lakes and reservoirs, dilution of pollutants often is less effective than in streams for two reasons. *First,* lakes and reservoirs often contain stratified layers

(Figure 5-35, p. 105) that undergo little vertical mixing. *Second,* they have little flow. The flushing and changing of water in lakes and large artificial reservoirs can take from 1 to 100 years, compared with several days to several weeks for streams.

As a result, lakes and reservoirs are more vulnerable than streams to contamination by runoff or discharge of plant nutrients, oil, pesticides, and toxic substances such as lead, mercury, and selenium. These contaminants can kill bottom life and fish and birds that feed on contaminated aquatic organisms. Many toxic chemicals and acids also enter lakes and reservoirs from the atmosphere.

Science: Cultural Eutrophication

Various human activities can overload lakes with plant nutrients, which decrease dissolved oxygen and kill some aquatic species.

Eutrophication is the name given to the natural nutrient enrichment of lakes, mostly from runoff of plant nutrients such as nitrates and phosphates from surrounding land. An *oligotrophic lake* is low in nutrients and its water is clear (Figure 11-25, left). Over time, some oligotrophic lakes become more eutrophic as nutrients are added from the surrounding watershed and the atmosphere. Others do not because of differences in the surrounding drainage basin.

Near urban or agricultural areas, human activities can greatly accelerate the input of plant nutrients to a lake, a process called **cultural eutrophication.** Nitrate- and phosphate-containing effluents mostly cause this change. They come from sources such as runoff from farmland, animal feedlots, urban areas, and mining sites, and from the discharge of treated and untreated municipal sewage. Some nitrogen also reaches lakes by deposition from the atmosphere.

During hot weather or drought, this nutrient overload produces dense growths or "blooms" of organisms such as algae and cyanobacteria (Figure 11-25, right) and thick growths of water hyacinth, duckweed, and other aquatic plants. These dense colonies of plant life can reduce lake productivity and fish growth by decreasing the input of solar energy needed for photosynthesis by the phytoplankton that support fish.

In addition, when the algae die, their decomposition by swelling populations of aerobic bacteria depletes dissolved oxygen in the surface layer of water near the shore and in the bottom layer. This oxygen depletion can kill fish and other aerobic aquatic animals. If excess nutrients continue to flow into a lake, anaerobic bacteria will take over and produce gaseous decomposition products such as smelly, highly toxic hydrogen sulfide and flammable methane.

According to the U.S. Environmental Protection Agency (EPA), one-third of the 100,000 medium to large lakes and 85% of the large lakes near major population centers in the United States have some degree of cultural eutrophication. One-fourth of the lakes in China also suffer from cultural eutrophication.

Figure 11-25 Natural capital degradation: the effect of nutrient enrichment on a lake. Crater Lake in Oregon (left) is an *oligotrophic lake;* it is low in nutrients. Because of the low density of plankton, its water is quite clear. The lake on the right, found in western New York, is a *eutrophic lake.* Because of an excess of plant nutrients, its surface is covered with mats of algae and cyanobacteria.

There are several ways to *prevent* or *reduce* cultural eutrophication. For example, we can use advanced (but expensive) waste treatment to remove nitrates and phosphates before wastewater enters lakes, ban or limit the use of phosphates in household detergents and other cleaning agents, and employ soil conservation and land-use control to reduce nutrient runoff.

There are also several ways to *clean up* lakes suffering from cultural eutrophication. We can mechanically remove excess weeds, control undesirable plant growth with herbicides and algicides, and pump air through lakes and reservoirs to prevent oxygen depletion (an expensive and energy-intensive method).

As usual, pollution prevention is more effective and usually cheaper in the long run than cleanup. If excessive inputs of plant nutrients stop, a lake usually can return to its previous state.

Science: Groundwater Pollution

Groundwater can become contaminated with a variety of chemicals because it cannot effectively cleanse itself and dilute and disperse pollutants.

A serious threat to human health is the out-of-sight pollution of groundwater, a prime source of water for drinking and irrigation. Groundwater pollution comes from numerous sources (Figure 11-26). People who dump or spill gasoline, oil, and paint thinners and other organic solvents onto the ground also contaminate groundwater.

When groundwater becomes contaminated, it cannot cleanse itself of *degradable wastes* as flowing surface water does (Figure 11-24). Groundwater flows so slowly—usually less than 0.3 meter (1 foot) per day—that contaminants are not diluted and dispersed effectively. In addition, groundwater usually has much

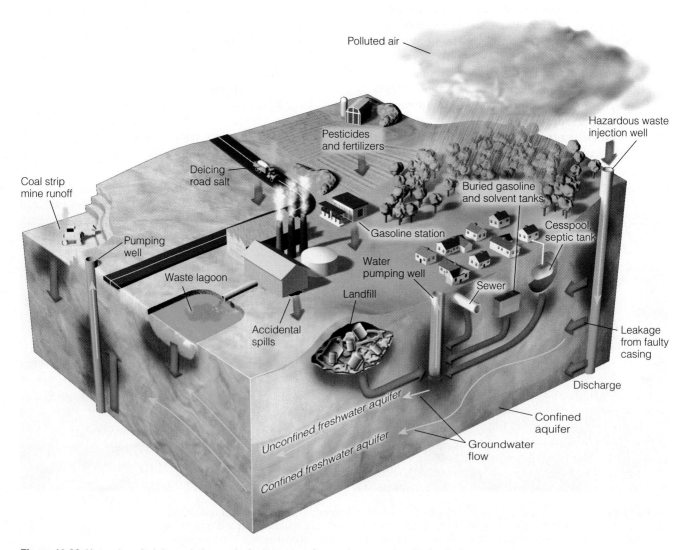

Figure 11-26 Natural capital degradation: principal sources of groundwater contamination in the United States.

lower concentrations of dissolved oxygen (which helps decompose many contaminants) and smaller populations of decomposing bacteria. The usually cold temperatures of groundwater also slow down chemical reactions that decompose wastes.

It can take hundreds to thousands of years for contaminated groundwater to cleanse itself of *degradable* wastes. On a human time scale, *nondegradable wastes* (such as toxic lead, arsenic, and fluoride) are there permanently.

Global Outlook: Extent of Groundwater Pollution

Leaks from chemical storage ponds, underground storage tanks, piping used to inject hazardous waste underground, and seepage of agricultural fertilizers can contaminate groundwater.

On a global scale, we do not know much about groundwater pollution because few countries go to the great expense of locating, tracking, and testing aquifers. Nevertheless, the results of scientific studies in scattered parts of the world are alarming.

According to the EPA and the U.S. Geological Survey, one or more organic chemicals contaminate about 45% of *municipal* groundwater supplies in the United States. An EPA survey of 26,000 industrial waste ponds and lagoons in the United States found that one-third of them had no liners to prevent toxic liquid wastes from seeping into aquifers. One-third of these sites are within 1.6 kilometers (1 mile) of a drinking water well. In 2002, the U.S. General Accounting Office estimated that at least 76,000 underground tanks storing gasoline, diesel fuel, home heating oil, or toxic solvents were leaking their contents into groundwater in the United States.

During this century, scientists expect many of the millions of such tanks installed around the world to corrode, leak, contaminate groundwater, and become a major global health problem. Determining the extent of a leak from a single underground tank can cost $25,000–250,000, and cleanup costs range from $10,000 to more than $250,000. If the chemical reaches an aquifer, effective cleanup is often not possible or is too costly. *Bottom line*: Wastes we think we have thrown away or stored safely can escape and come back to haunt us.

Toxic *arsenic* contaminates drinking water when a well is drilled into aquifers where soils and rock are naturally rich in arsenic. According to the WHO, more than 112 million people are drinking water with arsenic levels 5–100 times the WHO standard, mostly in Bangladesh, China, and West Bengal, India.

There is also concern over arsenic levels in drinking water in parts of the United States. The EPA plans to lower the acceptable level of arsenic in drinking wa-

Solutions	
Groundwater Pollution	
Prevention	**Cleanup**
Find substitutes for toxic chemicals	Pump to surface, clean, and return to aquifer (very expensive)
Keep toxic chemicals out of the environment	
Install monitoring wells near landfills and underground tanks	Inject microorganisms to clean up contamination (less expensive but still costly)
Require leak detectors on underground tanks	
Ban hazardous waste disposal in landfills and injection wells	Pump nanoparticles of inorganic compounds to remove pollutants (may be the cheapest, easiest, and most effective method but is still being developed)
Store harmful liquids in aboveground tanks with leak detection and collection systems	

Figure 11-27 Solutions: methods for preventing and cleaning up contamination of ground water. *Critical Thinking: which two of these solutions do you believe are the most important?*

ter from 50 to 10 parts per billion (ppb). According to the WHO and other scientists, even the 10 ppb standard is not safe. Many scientists call for lowering the standard to 3–5 ppb, which would be very expensive.

Solutions: Protecting Groundwater

Prevention is the most effective and affordable way to protect groundwater from pollutants.

Figure 11-27 lists ways to prevent and clean up groundwater contamination. Pumping polluted groundwater to the surface, cleaning it up, and returning it to the aquifer is very expensive. *Preventing contamination is the most effective and cheapest way to protect groundwater resources.*

11-7 OCEAN POLLUTION

Science: How Much Pollution Can the Oceans Tolerate?

Oceans, if they are not overloaded, can disperse and break down large quantities of degradable pollutants.

The oceans can dilute, disperse, and degrade large amounts of raw sewage, sewage sludge, oil, and some types of degradable industrial waste, especially in deep-water areas. Also, some forms of marine life have been affected less by several pollutants than expected.

Some scientists have suggested that it is safer to dump sewage sludge and most other harmful wastes into the deep ocean than to bury them on land or burn them in incinerators. Other scientists disagree, pointing out we know less about the deep ocean than we do about the moon. They add that dumping harmful wastes in the ocean would delay urgently needed pollution prevention and promote further degradation of this vital part of the earth's life-support system.

Science: Pollution of Coastal Waters

Pollution of coastal waters near heavily populated areas is a serious problem.

Coastal areas—especially wetlands and estuaries, coral reefs, and mangrove swamps—bear the brunt of our enormous inputs of wastes into the ocean (Figure 11-28). This is not surprising because 45% of the world's population lives on or near the coast and 14 of the world's 15 largest metropolitan areas (each with 10 million people or more) are near coastal waters (see Figure 7-13, p. 141, and Figure 7-15, p. 142).

In most coastal developing countries and in some coastal developed countries, municipal sewage and industrial wastes are dumped into the sea without treat-

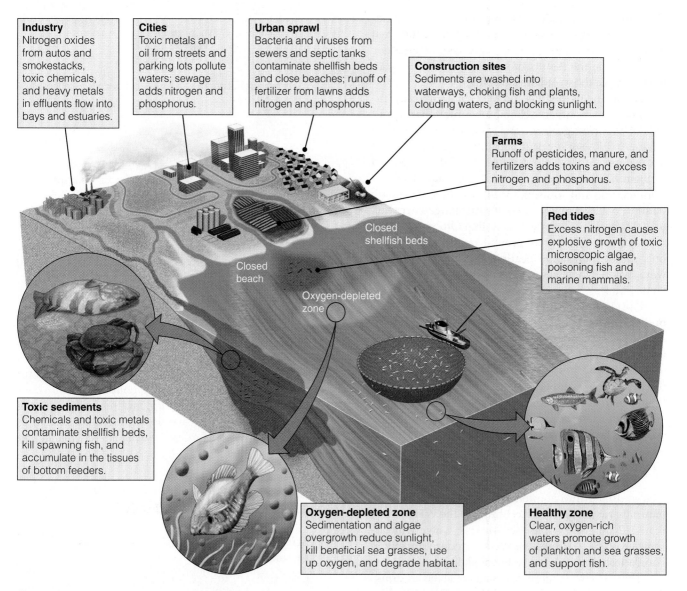

Industry
Nitrogen oxides from autos and smokestacks, toxic chemicals, and heavy metals in effluents flow into bays and estuaries.

Cities
Toxic metals and oil from streets and parking lots pollute waters; sewage adds nitrogen and phosphorus.

Urban sprawl
Bacteria and viruses from sewers and septic tanks contaminate shellfish beds and close beaches; runoff of fertilizer from lawns adds nitrogen and phosphorus.

Construction sites
Sediments are washed into waterways, choking fish and plants, clouding waters, and blocking sunlight.

Farms
Runoff of pesticides, manure, and fertilizers adds toxins and excess nitrogen and phosphorus.

Red tides
Excess nitrogen causes explosive growth of toxic microscopic algae, poisoning fish and marine mammals.

Closed shellfish beds

Closed beach

Oxygen-depleted zone

Toxic sediments
Chemicals and toxic metals contaminate shellfish beds, kill spawning fish, and accumulate in the tissues of bottom feeders.

Oxygen-depleted zone
Sedimentation and algae overgrowth reduce sunlight, kill beneficial sea grasses, use up oxygen, and degrade habitat.

Healthy zone
Clear, oxygen-rich waters promote growth of plankton and sea grasses, and support fish.

Figure 11-28 Natural capital degradation: residential areas, factories, and farms all contribute to the pollution of coastal waters and bays.

ment. For example, 85% of the sewage from large cities along the Mediterranean Sea (with a coastal population of 200 million people during tourist season) is discharged into the sea untreated. It causes widespread beach pollution and shellfish contamination.

Recent studies of some U.S. coastal waters have found vast colonies of viruses thriving in raw sewage, effluents from sewage treatment plants (which do not remove viruses), and leaking septic tanks. According to one study, one-fourth of the people using coastal beaches in the United States develop ear infections, sore throats, eye irritations, respiratory disease, or gastrointestinal disease.

Runoffs of sewage and agricultural wastes into coastal waters introduce large quantities of nitrate and phosphate plant nutrients, which can cause explosive growth of harmful algae. These *harmful algal blooms* (HABs) are called red, brown, or green toxic tides, depending on their color. They can release waterborne and airborne toxins that damage fisheries, kill some fish-eating birds, reduce tourism, and poison seafood.

According to the U.N. Environment Programme, each year some 150 large *oxygen-depleted zones* (sometimes inaccurately called dead zones) form in the world's coastal waters and in landlocked seas such as the Baltic and Black Seas. These zones result from excessive nonpoint inputs of fertilizers and animal wastes from land runoff and deposition of nitrogen compounds from the atmosphere.

This cultural eutrophication depletes dissolved oxygen. Without oxygen, most of the aquatic life (except bacteria) dies or moves elsewhere. The biggest such zone in U.S. waters and the third largest in the world forms every summer in a narrow stretch of the Gulf of Mexico off the mouth of the Mississippi River (Figure 11-29).

Preventive measures for reducing the number and size of these oxygen-depleted zones include reducing nitrogen inputs from various sources, planting forests and grasslands to soak up excess nitrogen and keep it out of waterways, improving sewage treatment to reduce discharges of nitrates into waterways, further reducing emissions of nitrogen oxides from motor vehicles, and phasing in forms of renewable energy to replace the burning of fossil fuels.

Science Case Study: The Chesapeake Bay

Pollutants from six states contaminate the shallow Chesapeake Bay estuary, but cooperative efforts have reduced some of the pollution inputs.

Since 1960, the Chesapeake Bay—the United States' largest estuary—has been in serious trouble from water pollution, mostly because of human activities. One problem is that between 1940 and 2004, the number of people living in the Chesapeake Bay area grew

Figure 11-29 Natural capital degradation: a large zone of oxygen-depleted water (less than 2 ppm dissolved oxygen) forms for half of the year in the Gulf of Mexico as a result of oxygen-depleting algal blooms. It is created mostly by huge inputs of nitrate (NO^{3-}) and phosphate (PO_4^{3-}) plant nutrients from the massive Mississippi River basin. In 2002, this area (shown in green) was roughly equivalent to the area of the U.S. state of New Jersey or the Central American country of Belize.

from 3.7 million to 17 million, and it may soon reach 18 million.

The estuary receives wastes from point and nonpoint sources scattered throughout a huge drainage basin that includes 9 large rivers and 141 smaller streams and creeks in parts of six states (Figure 11-30).

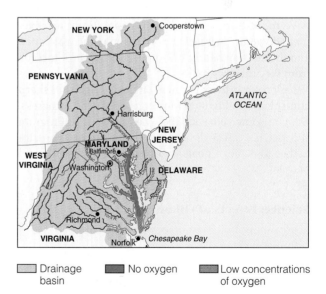

| Drainage basin | No oxygen | Low concentrations of oxygen |

Figure 11-30 Natural capital degradation: *Chesapeake Bay,* the largest estuary in the United States, is severely degraded as a result of water pollution from point and nonpoint sources in six states and from deposition of air pollutants.

The shallow bay has become a huge pollution sink, and only 1% of the waste entering it is flushed into the Atlantic Ocean.

Phosphate and nitrate levels have risen sharply in many parts of the bay, causing algal blooms and oxygen depletion. Commercial harvests of its once-abundant oysters, crabs, and several important fish have fallen sharply since 1960 because of a combination of pollution, overfishing, and disease.

Point sources—primarily sewage treatment plants and industrial plants (often in violation of their discharge permits)—account for 60% by weight of the phosphates. Nonpoint sources—mostly runoff of fertilizer and animal wastes from urban, suburban, and agricultural land and deposition from the atmosphere—account for 60% by weight of the nitrates. According to 2004 study by the Chesapeake Bay Foundation, animal manure is the largest source of nitrates and phosphates from agricultural pollution.

In 1983, the United States implemented the Chesapeake Bay Program. In the country's most ambitious attempt at *integrated coastal management,* citizens' groups, communities, state legislatures, and the federal government are working together to reduce pollution inputs into the bay. Strategies include establishing land-use regulations in the bay's six watershed states to reduce agricultural and urban runoff, banning phosphate detergents, upgrading sewage treatment plants, and better monitoring of industrial discharges. In addition, wetlands are being restored and large areas of the bay are being replanted with sea grasses to help filter out nutrients and other pollutants.

This hard work has paid off. Between 1985 and 2000, phosphorus levels declined 27%, nitrogen levels dropped 16%, and grasses growing on the bay's floor have made a comeback. This is a significant achievement given the increasing population in the watershed and the fact that nearly 40% of the nitrogen inputs come from the atmosphere.

Of course, there is still a long way to go, and sharp cuts in state and federal funding have slowed progress. Despite some setbacks, the Chesapeake Bay Program shows what can be done when diverse groups work together to achieve goals that benefit both wildlife and people.

Science: Effects of Oil on Ocean Life

Most aquatic oil pollution comes from human activities on land.

Crude petroleum (oil as it comes out of the ground) and *refined petroleum* (fuel oil, gasoline, and other processed petroleum products) reach the ocean from a number of sources.

Tanker accidents and blowouts at offshore drilling rigs (when oil escapes under high pressure from a borehole in the ocean floor) get most of the publicity because of their high visibility. In fact, *most ocean oil pollution comes from human activities on land.* According to a 2004 study by the Pew Oceans Commission, every 8 months an amount of oil equal to that spilled by the *Exxon Valdez* tanker into Alaska's Prince William Sound in 1989 drains from the land into the oceans. Almost half (some experts estimate 90%) of the oil reaching the oceans is waste oil dumped, spilled, or leaked onto the land or into sewers by cities, industries, and people changing their own motor oil.

Volatile organic hydrocarbons in oil immediately kill a number of aquatic organisms, especially in their vulnerable larval forms. Other chemicals in oil form tar-like globs that float on the surface and coat the feathers of birds (especially diving birds) and the fur of marine mammals. This oil coating destroys their natural insulation and buoyancy, causing many of them to drown or die of exposure from loss of body heat.

Heavy oil components that sink to the ocean floor or wash into estuaries can smother bottom-dwelling organisms such as crabs, oysters, mussels, and clams or make them unfit for human consumption. Some oil spills have killed coral reefs.

Research shows that most (but not all) forms of marine life recover from exposure to large amounts of *crude oil* within 3 years. Recovery from exposure to *refined oil,* especially in estuaries and salt marshes, can take 10–15 years. Oil slicks that wash onto beaches can have a serious economic impact on coastal residents, who lose income from fishing and tourist activities.

If they are not too large, oil spills can be partially cleaned up by mechanical (floating booms, skimmer boats, and absorbent devices such as large pillows filled with feathers or hair), chemical, fire, and natural methods (such as using cocktails of natural bacteria to speed up oil decomposition).

But scientists estimate that current methods can recover no more than 15% of the oil from a major spill. Thus preventing oil pollution is the most effective and, in the long run, the least costly approach.

Solutions: Protecting Coastal Waters

Preventing or reducing the flow of pollution from land and from streams emptying into the ocean is the key to protecting the oceans.

Figure 11-31 list ways for preventing and reducing excessive pollution of coastal waters. The key to protecting the oceans is to reduce the flow of pollution from land and from streams emptying into these waters. Ocean pollution control must be linked with land-use and air pollution policies because about one-third of all pollutants entering the ocean worldwide come from emissions of air pollutants.

Solutions

Coastal Water Pollution

Prevention	Cleanup
Reduce input of toxic pollutants	Improve oil-spill cleanup capabilities
Separate sewage and storm lines	
Ban dumping of wastes and sewage by maritime and cruise ships in coastal waters	Sprinkle nanoparticles over an oil or sewage spill to dissolve the oil or sewage without creating harmful by-products (still under development)
Ban ocean dumping of sludge and hazardous dredged material	
Protect sensitive areas from development, oil drilling, and oil shipping	Require at least secondary treatment of coastal sewage
Regulate coastal development	
Recycle used oil	Use wetlands, solar-aquatic, or other methods to treat sewage
Require double hulls for oil tankers	

Figure 11-31 Solutions: methods for preventing and cleaning up excessive pollution of coastal waters. *Critical thinking: which two of these solutions do you believe are the most important?*

11-8   PREVENTING AND REDUCING SURFACE WATER POLLUTION

Solutions: Reducing Surface Water Pollution from Nonpoint Sources

The key to reducing nonpoint pollution—most of it from agriculture—is to prevent it from reaching bodies of surface water.

There are a number of ways to reduce nonpoint water pollution, most of which comes from agriculture. Farmers can reduce soil erosion by keeping cropland covered with vegetation and by reforesting critical watersheds. They can also reduce the amount of fertilizer that runs off into surface waters and leaches into aquifers by using slow-release fertilizer, using none on steeply sloped land, and planting buffer zones of vegetation between cultivated fields and nearby surface water.

Applying pesticides only when needed and relying more on integrated pest management (p. 233) can reduce pesticide runoff. Farmers can control runoff and infiltration of manure from animal feedlots by planting buffers and locating feedlots and animal waste sites away from steeply sloped land, surface water, and flood zones.

☒ HOW WOULD YOU VOTE? Should we greatly increase efforts to reduce water pollution from nonpoint sources even though this could be quite costly? Cast your vote online at http://biology.brookscole.com/miller11.

Politics: Laws for Reducing Water Pollution from Point Sources

Most developed countries use laws to set water pollution standards. In most developing countries, such laws do not exist or are poorly enforced.

The Federal Water Pollution Control Act of 1972 (renamed the Clean Water Act when it was amended in 1977) and the 1987 Water Quality Act form the basis of U.S. efforts to control pollution of the country's surface waters. The Clean Water Act sets standards for allowed levels of key water pollutants and requires polluters to get permits limiting how much of various pollutants they can discharge into aquatic systems.

The EPA is also experimenting with a *discharge trading policy* that uses market forces to reduce water pollution in the United States. Under this program, a water pollution source is allowed to pollute at higher levels than allowed in its permits if it buys credits from permit holders with pollution levels below their allowed levels.

Some environmentalists support discharge trading. But they warn that such a system is no better than the caps set for total pollution levels in various areas, and call for careful scrutiny of the cap levels. They also warn that discharge trading could allow pollutants to build up to dangerous levels in areas where credits are bought. In addition, they call for gradually lowering the caps to encourage prevention of water pollution and development of better technology for controlling water pollution, neither of which is a part of the current EPA discharge trading system.

According to Sandra Postel, director of the Global Water Policy Project, most cities in developing countries discharge 80–90% of their untreated sewage directly into rivers, streams, and lakes. These waters are then used for drinking water, bathing, and washing clothes.

Science: Reducing Water Pollution from Point Sources

Septic tanks and various levels of sewage treatment can reduce point-source water pollution.

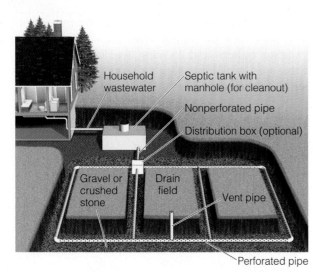

Figure 11-32 Solutions: *septic tank system* used for disposal of domestic sewage and wastewater in rural and suburban areas. This system separates solids from liquids, digests organic matter and large solids, and discharges the liquid wastes in a network of buried pipes with holes located over a large drainage or absorption field. As these wastes drain from the pipes and percolate downward, the soil filters out some potential pollutants, and soil bacteria decompose biodegradable materials. To be effective, septic tank systems must be properly installed in soils with adequate drainage, not placed too close together or too near well sites, and pumped out when the settling tank becomes full.

In rural and suburban areas with suitable soils, sewage from each house usually is discharged into a **septic tank** (Figure 11-32). One-fourth of all homes in the United States are served by septic tanks.

In U.S. urban areas, most waterborne wastes from homes, businesses, factories, and storm runoff flow through a network of sewer pipes to *wastewater* or *sewage treatment plants*. Raw sewage reaching a treatment plant typically undergoes one or two levels of wastewater treatment. The first level, **primary sewage treatment**, is a *physical* process that uses screens and a grit tank to remove large floating objects and solids such as sand and rock, and a settling tank that allows suspended solids to settle out as sludge (Figure 11-33). A second level, **secondary sewage treatment,** is a *biological* process in which aerobic bacteria remove as much as 90% of dissolved and biodegradable, oxygen-demanding organic wastes.

Because of the Clean Water Act, most U.S. cities have combined primary and secondary sewage treatment plants. According to the EPA, however, at least two-thirds of these plants have sometimes violated water pollution regulations. Also, 500 cities have failed to meet federal standards for sewage treatment plants, and 34 East Coast cities simply screen out large floating objects from their sewage before discharging it into coastal waters.

Before discharge, water from primary, secondary, or more advanced treatment undergoes *bleaching* to remove water coloration and *disinfection* to kill disease-carrying bacteria and some (but not all) viruses. The usual method for accomplishing this is *chlorination.* However, chlorine can react with organic materials in water to form small amounts of chlorinated hydrocarbons. Some of these chemicals cause cancers in test animals and may damage the human nervous, immune, and endocrine systems.

Use of other disinfectants, such as ozone and ultraviolet light, is increasing. These options cost more and their effects do not last as long as chlorination.

Science: Improving Sewage Treatment

Preventing toxic chemicals from reaching sewage treatment plants would eliminate such chemicals from the sludge and water discharged from such plants.

Environmental scientist Peter Montague calls for redesigning the sewage treatment system. The idea is to prevent toxic and hazardous chemicals from reaching sewage treatment plants and thus from getting into sludge and the water discharged from such plants.

Montague suggests several ways to do this. We could require industries and businesses to remove toxic and hazardous wastes before sending water to municipal sewage treatment plants. We could also encourage industries to reduce or eliminate their use and waste of toxic chemicals.

Another suggestion is to have more households, apartment buildings, and offices eliminate sewage outputs by switching to waterless *composting toilet systems.* Such systems would be cheaper to install and maintain than current sewage systems because they do not require vast systems of underground pipes connected to centralized sewage treatment plants. They also save large amounts of water.

Solutions: Treating Sewage by Working with Nature

Natural and artificial wetlands and other ecological systems can be used to treat sewage.

John Todd has developed an ecological approach to treating sewage, which he calls *living machines* (Figure 11-34). His purification process begins when sewage flows into a passive solar greenhouse or outdoor sites containing rows of large open tanks populated by an increasingly complex series of organisms. In the first set of tanks, algae and microorganisms decompose organic wastes, with sunlight speeding up the process. Water hyacinths, cattails, bulrushes, and other aquatic plants growing in the tanks take up the resulting nutrients.

After flowing though several of these natural purification tanks, the water passes through an artificial

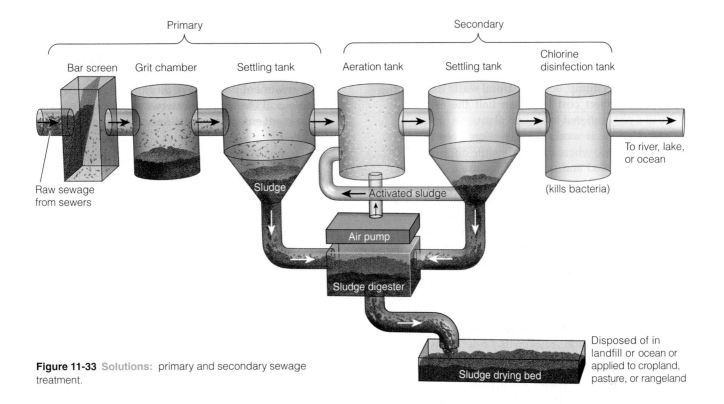

Primary

Bar screen   Grit chamber   Settling tank

Secondary

Aeration tank   Settling tank   Chlorine disinfection tank

Raw sewage from sewers

Sludge

← Activated sludge

Air pump

Sludge digester

To river, lake, or ocean

(kills bacteria)

Sludge drying bed

Disposed of in landfill or ocean or applied to cropland, pasture, or rangeland

Figure 11-33 Solutions: primary and secondary sewage treatment.

marsh of sand, gravel, and bulrushes, which filters out algae and remaining organic waste. Some of the plants also absorb (sequester) toxic metals such as lead and mercury and secrete natural antibiotic compounds that kill pathogens.

Next, the water flows into aquarium tanks. Snails and zooplankton consume microorganisms and are in turn consumed by crayfish, tilapia, and other fish that can be eaten or sold as bait. After 10 days, the clear water flows into a second artificial marsh for final filtering and cleansing.

The water can be made pure enough to drink by using ultraviolet light or by passing the water through an ozone generator, usually immersed out of sight in an attractive pond or wetland habitat. Operating costs are about the same as for a conventional sewage treatment plant.

Some communities work with nature by using nearby natural wetlands to treat sewage, and others create artificial wetlands for such purposes (see the following Case Study).

Science Case Study: Using Wetlands to Treat Sewage

Some communities use natural or artificially created wetlands to treat their sewage.

More than 150 cities and towns in the United States use natural and artificial wetlands to treat sewage as a low-tech, low-cost alternative to expensive waste

Ocean Arks International

Figure 11-34 Solutions: ecological wastewater purification by a *living machine.* At the Providence, Rhode Island, Solar Sewage Treatment Plant, biologist John Todd demonstrates how ecological waste engineering in a greenhouse can purify wastewater in an ecological process he invented. Todd and others are conducting research to perfect solar-aquatic sewage treatment systems based on working with nature.

of nonrenewable mineral resources. Instead, they claim that the major problem is the environmental damage caused by their extraction, processing, and conversion to products (Figure 12-12).

The environmental impacts from mining an ore are affected by its percentage of metal content, or *grade.* The more accessible and higher-grade ores are usually exploited first. As they are depleted, it takes more money, energy, water, and other materials to exploit lower-grade ores. This, in turn, increases land disruption, mining waste, and pollution.

For example, gold miners typically remove an amount of ore equal to the weight of 50 automobiles to extract a piece of gold that would fit inside your clenched fist. Most newlyweds would be surprised to know that about 5.5 metric tons (6 tons) of mining waste was created to make their two gold wedding rings. In Australia and North America, a mining technology called *cyanide heap leaching* is cheap enough to allow mining companies to level entire mountains containing very low-grade gold ore.

Currently, most of the harmful environmental costs of mining and processing minerals are not included in the prices for processed metals and the resulting consumer products. Instead, these costs are passed on to society and future generations, which gives mining companies and manufacturers little incentive to reduce resource waste and pollution. Environmentalists and some economists call for phasing in *full-cost pricing*—including the cost of environmental harm done in the price of goods made from minerals.

✘ *HOW WOULD YOU VOTE?* Should market prices of goods made from minerals include their harmful environmental costs? Cast your vote online at http://biology.brookscole.com /miller11.

12-6 SUPPLIES OF MINERAL RESOURCES

Science and Economics: Supplies of Nonrenewable Mineral Resources

The future supply of a resource depends on how available and affordable it is and how rapidly that supply is used.

The future supply of nonrenewable minerals depends on two factors: the actual or potential supply of the mineral and the rate at which we use it.

We never completely run out of any mineral. However, a mineral becomes *economically depleted* when it costs more to find, extract, transport, and process the remaining deposit than it is worth. At that point, there are five choices: *recycle or reuse existing supplies, waste less, use less, find a substitute,* or *do without.*

Depletion time is how long it takes to use up a certain proportion—usually 80%—of the reserves of a mineral at a given rate of use. When experts disagree about depletion times, it is often because they are using different assumptions about supply and rate of use (Figure 12-14).

The shortest depletion time assumes no recycling or reuse and no increase in reserves (curve A, Figure 12-14). A longer depletion time assumes that recycling will stretch existing reserves and that better mining technology, higher prices, and new discoveries will increase reserves (curve B, Figure 12-14). An even longer depletion time assumes that new discoveries will further expand reserves and that recycling, reuse, and reduced consumption will extend supplies (curve C, Figure 12-14). Finding a substitute for a resource leads to a new set of depletion curves for the new resource.

We can use geological methods to make fairly good estimates of the reserves of most resources (Figure 12-7, blue) and less accurate measurements of potential other supplies of mineral resources (Figure 12-7, red). Rising prices and improved mining technology can convert some of the other resources to reserves, but it is difficult to project how much this conversion will add to the usable supply.

The demand for mineral resources is increasing at a rapid rate as more people consume more stuff. For example, since 1940 Americans alone have used up as

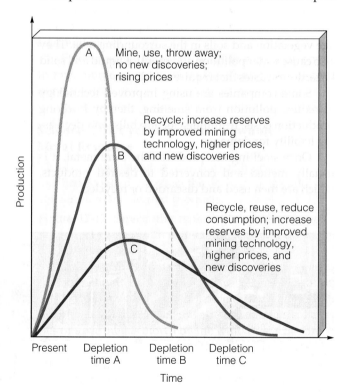

Figure 12-14 Natural capital depletion: *depletion curves* for a nonrenewable resource (such as aluminum or copper) using three sets of assumptions. Dashed vertical lines represent times when 80% depletion occurs.

large a share of the earth's mineral resources as all previous generations put together, and the resource-use treadmill is accelerating.

Will we run out of affordable supplies of a particular mineral resource? No one knows. If we do, can we find an acceptable substitute? Some think we can. Do environmental limits apply to the use of mineral resources? Some environmental scientists think so unless we can use microorganisms, or other less environmentally harmful ways, to extract and process minerals, or nanotechnology to construct materials we need from atoms and molecules.

Economics and Politics: Prices and Supplies of Nonrenewable Minerals

A rising price for a scarce mineral resource can increase supplies and encourage more efficient use.

Geologic processes determine the quantity and location of a mineral resource in the earth's crust. Economics determines what part of the known supply is actually extracted and used.

According to standard economic theory, in a competitive free market a plentiful mineral resource is cheap when its supply exceeds demand. When a resource becomes scarce, its price rises. This can encourage exploration for new deposits, stimulate development of better mining technology, and make it profitable to mine lower-grade ores. It can also encourage a search for substitutes and promote resource conservation.

According to some economists, this price effect may no longer apply very well in most developed countries. Industry and government in such countries often control the supply, demand, and prices of minerals to such an extent that a truly competitive free market does not exist.

Most mineral prices are kept artificially low because governments subsidize development of their domestic mineral resources to help promote economic growth and national security. In the United States, for instance, mining companies get depletion allowances amounting to 5–22% of their gross income (depending on the mineral). They can also reduce their taxes by deducting much of their costs for finding and developing mineral deposits. In addition, hardrock mining companies operating in the United States can buy public land at 1872 prices and pay no royalties to the government on the minerals they extract (see the Case Study at the beginning of the chapter).

Between 1982 and 2004, U.S. mining companies received more than $6 billion in government subsidies. Critics argue that taxing—rather than subsidizing—the extraction of nonfuel mineral resources would provide governments with revenue, create incentives for more efficient resource use, promote waste reduction and pollution prevention, and encourage recycling and reuse of mineral resources.

Mining company representatives insist that they need subsidies and low taxes to keep the prices of minerals low for consumers. They also claim that the subsidies encourage the companies not to move their mining operations to other countries with no such taxes and less stringent mining regulations.

Economic problems can also hinder the development of new supplies of mineral resources because finding them takes increasingly scarce investment capital and is financially risky. Typically, if geologists identify 10,000 possible deposits of a given resource, only 1,000 sites are worth exploring; only 100 justify drilling, trenching, or tunneling; and only 1 becomes a producing mine or well. If you had lots of financial capital, would you invest it in developing a nonrenewable mineral resource?

Science: Mining Lower-Grade Ores

New technologies can increase the mining of low-grade ores at affordable prices, but harmful environmental effects can limit this endeavor.

Some analysts contend that all we need to do to increase supplies of a mineral is to extract lower grades of ore. They point to the development of new earth-moving equipment, improved techniques for removing impurities from ores, and other technological advances in mineral extraction and processing.

In 1900, the average copper ore mined in the United States was about 5% copper by weight. Today that ratio is 0.5%, and copper costs less (adjusted for inflation). New methods of mineral extraction may allow even lower-grade ores of some metals to be used.

Several factors can limit the mining of lower-grade ores. One is the increased cost of mining and processing larger volumes of ore. Another is the availability of fresh water needed to mine and process some minerals—especially in arid and semiarid areas. A third limiting factor is the environmental impacts of the increased land disruption, waste material, and pollution produced during mining and processing (Figure 12-12).

One way to improve mining technology is to use microorganisms for in-place (*in situ*, pronounced "in SY-too") mining. This biological approach removes desired metals from ores while leaving the surrounding environment undisturbed. It also reduces the air pollution associated with the smelting of metal ores and the water pollution associated with using hazardous chemicals such as cyanides and mercury to extract gold.

Once a commercially viable ore deposit has been identified, wells are drilled into it and the ore is fractured. The ore is then inoculated with bacteria to extract the desired metal. Next, the well is flooded with water, which is pumped to the surface, where the desired metal is removed. The water is recycled.

Currently, more than 30% of all copper produced worldwide, worth more than $1 billion per year, comes from such *biomining*. If naturally occurring bacteria cannot be found to extract a particular metal, genetic engineering techniques could be used to produce such bacteria.

On the down side, microbiological ore processing is slow. It can take decades to remove the same amount of material that conventional methods can remove within months or years. So far, biological mining methods are economically feasible only with low-grade ores for which conventional techniques are too expensive.

Science Case Study: Using Nanotechnology to Produce New Materials

Building new materials from the bottom up by assembling atoms and molecules has enormous potential but could have potentially harmful unintended effects.

Nanotechnology uses science and engineering at the atomic and molecular levels to build materials with specified properties. It involves finding ways to manipulate atoms and molecules as small as 1–100 nanometers—billionths of a meter—wide. For comparison, your unaided eye cannot see things smaller than 10,000 nanometers across and the width of a typical human hair is 50,000 nanometers.

This atomic and molecular approach to manufacturing uses abundant atoms such as carbon, oxygen, and hydrogen as raw materials and arranges them to create everything from medicines and solar cells to automobile bodies. Ideally, this bottom-up process occurs with little environmental harm and without depleting nonrenewable resources. One example is the creation of soccer ball–shaped forms of carbon called *buckyballs*.

Nanotechnology scientists entice us with visions of a *molecular economy*. They ask us to imagine a supercomputer the size of a sugar cube that could store all the information in the U.S. Library of Congress, biocomposite materials smaller than a human cell that would make your bones and tendons super strong, designer molecules that could seek out and kill only cancer cells, and windows, kitchens, and bathrooms that never need cleaning. The list could go on.

This research is in its early stages and tangible results remain a decade away. Nevertheless, nanotechnology has already been used to develop stain-resistant, wrinkle-free materials for pants and sunscreens that block ultraviolet light.

Nobel laureate Horst Stormer says, "Nanotechnology has given us the tools . . . to play with the ultimate toy box of nature—atoms and molecules. . . . The possibilities to create new things appear limitless." You might want to consider this rapidly emerging field as a career choice.

So what is the catch? What are some possible unintended harmful consequences of nanotechnology? As particles get smaller, they become more reactive and potentially more toxic because they have large surface areas relative to their mass. Also, nanosize particles could breach some of the natural defenses of our bodies. They could easily reach the lungs and from there migrate to other organs, including possibly the central nervous system and the bloodstream.

In 2004, Eva Olberdorster, an environmental toxicologist at Southern Methodist University, found that fish swimming in water loaded with buckyballs experienced brain damage within 48 hours. Little is known about how buckyballs and other nanoparticles behave in the human body. Even so, factories are churning out buckyballs and these and other nanoparticles are starting to show up in products ranging from cosmetics to sunscreens and in the environment.

Many analysts say we need to take two steps before unleashing nanotechnology more broadly. *First*, we must carefully investigate its potential ecological, health, and societal risks. *Second*, we must develop guidelines and regulations for controlling and guiding its spread until we have better answers to many of the "What happens next?" questions about this technology.

Science, Economics, and Politics: Getting More Minerals from the Ocean

Most minerals in seawater cost too much to extract, and mineral resources found on the deep ocean floor are not being removed because of high costs and squabbles over who owns them.

Ocean mineral resources are found in seawater, sediments and deposits on the shallow continental shelf, hydrothermal ore deposits (Figure 12-15), and manganese-rich nodules on the deep-ocean floor.

Most of the chemical elements found in seawater occur in such low concentrations that recovering them takes more energy and money than they are worth. Only magnesium, bromine, and sodium chloride are abundant enough to be extracted profitably at current prices with existing technology.

Deposits of minerals (mostly sediments) along the continental shelf and near shorelines are significant sources of sand, gravel, phosphates, sulfur, tin, copper, iron, tungsten, silver, titanium, platinum, and diamonds.

Rich hydrothermal deposits of gold, silver, zinc, and copper are found as sulfide deposits in the deep-

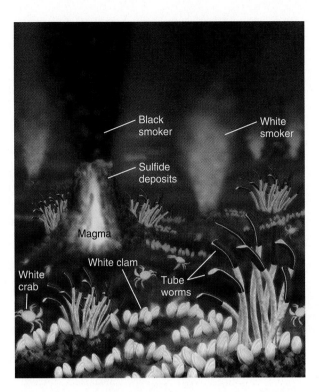

Figure 12-15 Natural capital: *hydrothermal ore deposits* form when mineral-rich superheated water shoots out of vents in solidified magma on the ocean floor. After mixing with cold seawater, black particles of metal ore precipitate out and build up as chimneylike ore deposits around the vents. A variety of organisms, supported by bacteria that produce food by chemosynthesis, exist in the dark ocean around these black smokers.

ocean floor and around black smokers (Figure 12-15). Currently, it costs too much to extract these minerals, even though some deposits contain large concentrations of important metals.

Manganese-rich nodules found on the deep-ocean floor may be a future source of manganese and other key metals. They might be sucked up from the ocean floor by giant vacuum pipes or scooped up by buckets on a continuous cable operated by a mining ship.

So far these nodules and resource-rich mineral beds in international waters have not been developed because of high costs and squabbles over who owns them and how any profits from extracting them should be distributed among the world's nations.

Some environmental scientists believe seabed mining probably would cause less environmental harm than mining on land. They remain concerned that removing seabed mineral deposits and dumping back unwanted material will stir up ocean sediments, destroy seafloor organisms, and have potentially harmful effects on poorly understood ocean food webs and marine biodiversity. They call for more research to help evaluate such possible effects.

Science: Finding Substitutes for Scarce Mineral Resources

Scientists and engineers are developing new types of materials that can serve as substitutes for many metals.

Some analysts believe that even if supplies of key minerals become too expensive or scarce, human ingenuity will find substitutes. They point to the current *materials revolution* in which silicon and new materials, particularly ceramics and plastics, are being developed and used as replacements for metals.

Ceramics offer many advantages over conventional metals. They are harder, stronger, lighter, and longer lasting than many metals, and they can withstand intense heat and do not corrode. Within a few decades, scientists may develop high-temperature ceramic superconductors in which electricity flows without resistance. The result: faster computers, more efficient power transmission, and affordable electromagnets for propelling high-speed magnetic levitation trains.

High-strength plastics and composite materials strengthened by lightweight carbon and glass fibers are beginning to transform the automobile and aerospace industries. They cost less to produce than metals because they take less energy, do not need painting, and can be molded into any shape. New plastics and gels are also being developed to provide superinsulation without taking up much space. Nanotechnology may also lead to the development of materials that can serve as substitutes for various minerals.

Substitutes exist for many scarce mineral resources. Unfortunately, finding substitutes for some key materials may prove difficult or impossible. Examples include helium, phosphorus for phosphate fertilizers, manganese for making steel, and copper for wiring motors and generators.

In addition, some substitutes are inferior to the minerals they replace. For example, aluminum could replace copper in electrical wiring. But producing aluminum takes much more energy than producing copper, and aluminum wiring is a greater fire hazard than copper wiring.

Clearly, there are a number of exciting possibilities and a number of environmental hazards involved in extracting and using the nonrenewable mineral resources that support our economies and provide us with many useful materials and products.

Mineral resources are the building blocks on which modern society depends. Knowledge of their physical nature and origins, the web they weave between all aspects of human society and the physical earth, can lay the foundations for a sustainable society.

ANN DORR

CRITICAL THINKING

1. What would probably happen if **(a)** plate tectonics stopped and **(b)** erosion and weathering stopped? If you could, would you eliminate either group of processes? Explain.

2. You are an igneous rock. Act as a microscopic reporter and send in a written report on what you experience as you move through the rock cycle (Figure 12-6, p. 275). Repeat this experience, assuming you are a sedimentary rock and then a metamorphic rock.

3. Use the second law of thermodynamics (p. 30) to analyze the scientific and economic feasibility of each of the following processes:
  (a) Extracting most minerals dissolved in seawater
  (b) Mining increasingly lower-grade deposits of minerals
  (c) Using inexhaustible solar energy to mine minerals
  (d) Continuing to mine, use, and recycle minerals at increasing rates

4. What mineral resources are extracted in your local area? What mining methods are used, and what have been their environmental impacts? How has mining these resources benefited the local economy?

5. Congratulations! You are in charge of the world. What are the three most important features of your policy for developing and sustaining the world's nonrenewable mineral resources?

LEARNING ONLINE

The website for this book includes review questions for the entire chapter, flash cards for key terms and concepts, a multiple-choice practice quiz, interesting Internet sites, references, and a guide for accessing thousands of InfoTrac® College Edition articles.
    Visit

http://biology.brookscole.com/miller11

Then choose Chapter 12, and select a learning resource. For access to animations, additional quizzes, chapter outlines and summaries, register and log in to

Environmental Science ⊛ Now ™

at **esnow.brookscole.com/miller11** using the access code card in the front of your book.

13 Energy

Nonrenewable Energy

Renewable Energy

The Coming Energy-Efficiency and Renewable-Energy Revolution

Energy analyst Amory Lovins built a large, solar-heated, superinsulated, partially earth-sheltered home and office in Snowmass, Colorado (Figure 13-1), which experiences severely cold winter temperatures.

The same structure also houses the research center for the Rocky Mountain Institute. This office–home gets 99% of its space and water heating and 95% of its daytime lighting from the sun, and uses one-tenth the usual amount of electricity for a structure of its size.

With today's superinsulating windows, a house can have many windows without experiencing much heat loss in cold weather or heat gain in hot weather. Thinner insulation now being developed will allow roofs and walls to be protected far better than in today's best superinsulated houses.

Some of today's green and smart buildings have lights run by sensors that dim when its sunny and brighten when its cloudy and air conditioners that shut off when a window is open. A new building at the Natural Energy Lab of Hawaii uses no energy from the electricity grid. It is cooled with piped in sea-water with the condensation from the cooling pipes used for irrigation.

A small (but growing) number of people in developed and developing countries get their electricity from *solar cells* that convert sunlight directly into electricity. These devices can be attached like shingles to a roof, used as roofing, or applied to window glass as a coating. Solar-cell prices are high but falling.

According to many scientists and executives of oil and automobile companies, we are in the beginning stages of a *hydrogen revolution* to be phased in during this century as the Age of Oil begins winding down. Because little hydrogen gas (H_2) is readily available, we must use another energy resource to produce it from water or various organic compounds such as methane. We could do so by passing electricity produced by renewable energy from wind turbines, hydroelectric power plants, solar cells, biomass, and geothermal energy from the earth's interior through water to make H_2 gas. Energy-efficient *fuel cells* could produce electricity by combining hydrogen and oxygen gas to run cars and appliances, heat water, and heat and cool buildings.

Burning hydrogen in a fuel cell by combining it with oxygen produces water vapor and no carbon dioxide. Shifting to hydrogen as our primary energy source would therefore eliminate most air pollution and greatly slow global warming—as long as the hydrogen is produced from water and not carbon-containing fossil fuels and the nuclear fuel cycle that emit the greenhouse gas CO_2 into the atmosphere.

Figure 13-1 The Rocky Mountain Institute in Colorado. This facility is a home and a center for the study of energy efficiency and sustainable use of energy and other resources. It is also an example of energy-efficient passive solar design.

Robert Millman/Rocky Mountain Institute

Typical citizens of advanced industrialized nations each consume as much energy in six months as typical citizens in developing countries consume during their entire life.

MAURICE STRONG

This chapter evaluates the use of nonrenewable fossil fuel and nuclear power energy resources, ways to improve energy efficiency, and use of renewable energy resources. It addresses the following questions:

- How should we evaluate energy resources?

- What are the advantages and disadvantages of conventional oil, heavy oils, natural gas, coal, and conversion of coal to gaseous and liquid fuels?

- What are the advantages and disadvantages of conventional nuclear fission, breeder nuclear fission, and nuclear fusion?

- How can we improve energy efficiency and what are the advantages of doing so?

- What are the advantages and disadvantages of using renewable energy in forms such as solar energy, flowing water, wind, biomass, geothermal energy, and hydrogen?

- How can we make a transition to a more sustainable energy future?

KEY IDEAS

- About 78% of the commercial energy we use comes from nonrenewable fossil fuels. These fuels produce much of the world's air and water pollution and add carbon dioxide to the troposphere, which can lead to global warming.

- The United States—the world's largest oil user—has only 2.9% of the world's proven oil reserves and only a small percentage of its unproven reserves. It imports 62% of its oil.

- The nuclear power fuel cycle has a fairly low environmental impact, an ample supply of fuel, and a very low risk of an accident, but costs are high, radioactive wastes must be stored safely for thousands of years, and facilities are vulnerable to terrorist attack.

- About 43% of the energy used in the United States is wasted unnecessarily.

- During this century, we will need to make a transition from dependence on nonrenewable fossil fuels to greatly improved energy efficiency and much greater dependence on renewable energy resources.

- Making the transition to a more sustainable energy future depends mainly on citizens insisting that elected officials and businesses implement policies for such a shift.

13-1 EVALUATING ENERGY RESOURCES

Science: Solar and Commercial Energy

About 99% of the energy that heats the earth and our buildings comes from the sun; the remaining 1% comes mostly from burning fossil fuels.

Almost all of the energy that heats the earth and our buildings comes from the sun at no cost to us. Without this essentially inexhaustible solar energy (*solar capital*), the earth's average temperature would be −240°C −400°F), and life as we know it would not exist.

This direct input of solar energy produces several other *indirect forms of renewable solar energy*. Examples include wind, falling and flowing water (hydropower), and biomass (solar energy converted to chemical energy and stored in the chemical bonds of organic compounds in trees and other plants).

Commercial energy sold in the marketplace makes up the remaining 1% of the energy we use to supplement the earth's direct input of solar energy. Most commercial energy comes from extracting and burning *nonrenewable mineral resources* obtained from the earth's crust, primarily carbon-containing fossil fuels—oil, natural gas, and coal (Figure 13-2).

Global Outlook: Types of Commercial Energy the World Depends On

About 76% of the commercial energy we use comes from nonrenewable fossil fuels.

About 82% of the commercial energy consumed in the world comes from *nonrenewable* energy resources— 76% from fossil fuels and 6% from nuclear power (Figure 13-3, left). The remaining 18% comes from *renewable* energy resources—biomass (11%), hydropower (4.5%), and a combination of geothermal, wind, and solar energy (1.5%).

Roughly half the world's people in developing countries burn wood and charcoal to heat their dwellings and cook their food. This *biomass energy* is renewable as long as wood supplies are not harvested faster than they are replenished.

Unfortunately, many of these individuals face a *fuelwood shortage* that is expected to worsen because of unsustainable harvesting of fuelwood. Also, at least 2 million people die prematurely each year from breathing particles emitted by burning wood indoors on open fires and in poorly designed primitive stoves.

Environmental Science⊛Now™
Examine and compare energy sources used in developing and developed countries at Environmental ScienceNow.

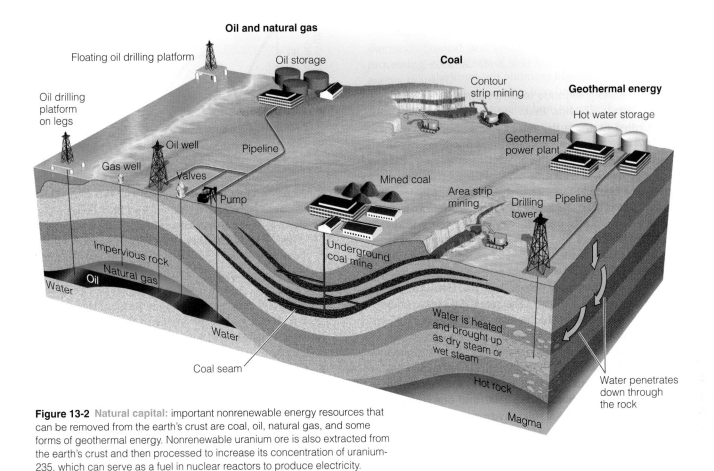

Oil and natural gas

Floating oil drilling platform

Oil storage

Coal

Contour strip mining

Geothermal energy

Hot water storage

Oil drilling platform on legs

Oil well

Pipeline

Geothermal power plant

Gas well

Valves

Pump

Mined coal

Area strip mining

Drilling tower

Pipeline

Impervious rock

Natural gas

Underground coal mine

Oil

Water

Water

Coal seam

Water is heated and brought up as dry steam or wet steam

Hot rock

Magma

Water penetrates down through the rock

Figure 13-2 Natural capital: important nonrenewable energy resources that can be removed from the earth's crust are coal, oil, natural gas, and some forms of geothermal energy. Nonrenewable uranium ore is also extracted from the earth's crust and then processed to increase its concentration of uranium-235, which can serve as a fuel in nuclear reactors to produce electricity.

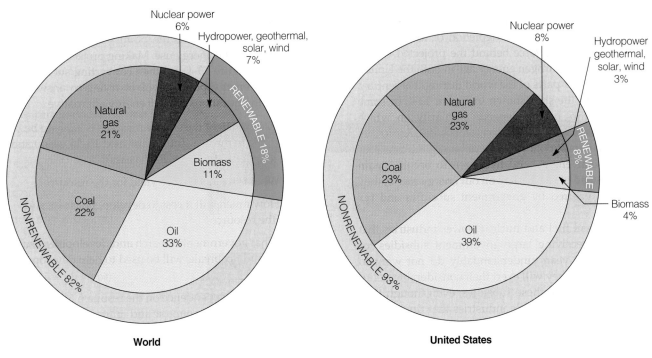

Nuclear power 6%

Hydropower, geothermal, solar, wind 7%

Natural gas 21%

RENEWABLE 18%

Biomass 11%

Coal 22%

Oil 33%

NONRENEWABLE 82%

World

Nuclear power 8%

Hydropower geothermal, solar, wind 3%

Natural gas 23%

RENEWABLE 8%

Coal 23%

Biomass 4%

Oil 39%

NONRENEWABLE 93%

United States

Figure 13-3 Natural capital: commercial energy use by source for the world (left) and the United States (right) in 2003. Commercial energy amounts to only 1% of the energy used in the world; the other 99% is direct solar energy received from the sun and is not sold in the marketplace. (Data from U.S. Department of Energy, British Petroleum, Worldwatch Institute, and International Energy Agency)

Conventional nuclear energy has a low net energy ratio because large amounts of energy are needed to extract and process uranium ore, convert it into nuclear fuel, build and operate nuclear power plants, dismantle the highly radioactive plants after their 15–60 years of useful life, and store the resulting highly radioactive wastes safely for 10,000–240,000 years. Each of these steps in the *nuclear fuel cycle* uses energy and costs money. Some analysts estimate that ultimately the conventional nuclear fuel cycle will lead to a net energy loss because we will have to put more energy into it than we will ever get out of it. Nuclear power proponents disagree.

13-2 NONRENEWABLE FOSSIL FUELS

Science: Crude Oil

Crude oil is a thick liquid containing hydrocarbons that we extract from underground deposits and separate into products such as gasoline, heating oil, and asphalt.

Petroleum, or **crude oil** (oil as it comes out of the ground), is a thick and gooey liquid consisting of hundreds of combustible hydrocarbons along with small amounts of sulfur, oxygen, and nitrogen impurities. It is also known as *conventional oil* or *light oil*.

Deposits of crude oil and natural gas often are trapped together under a dome deep within the earth's crust on land or under the seafloor (Figure 13-2). The crude oil is dispersed in pores and cracks in underground rock formations, somewhat like water saturating a sponge. To extract the oil, a well is drilled into the deposit. Then oil, drawn by gravity out of the rock

pores and into the bottom of the well, is pumped to the surface.

Drilling for oil causes only moderate damage to land because the wells occupy fairly little land area. However, drilling for oil and transporting it around the world inevitably results in some oil spills on land and in aquatic systems. In addition, harmful environmental effects are associated with the extraction, processing, and use of any nonrenewable resource from the earth's crust (Figure 12-12, p. 279).

After it is extracted, crude oil is transported to a *refinery* by pipeline, truck, or ship (oil tanker). There it is heated and distilled to separate it into components with different boiling points (Figure 13-6)—a technological marvel based on complex chemistry and engineering. Some of the products of oil distillation, called

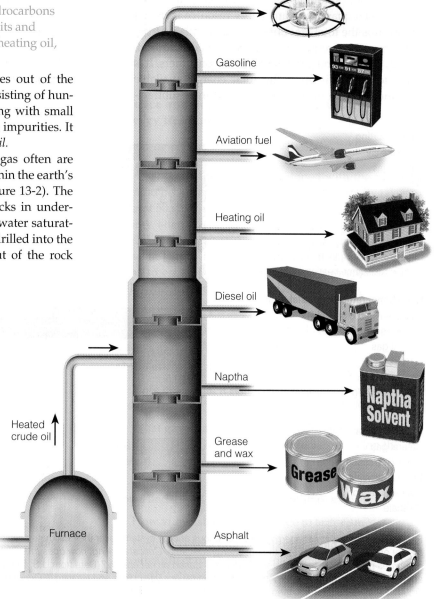

Figure 13-6 Science: refining crude oil. Based on their boiling points, components are removed at various levels in a giant distillation column. The most volatile components with the lowest boiling points are removed at the top of the column.

petrochemicals, are used as raw materials in industrial organic chemicals, pesticides, plastics, synthetic fibers, paints, medicines, and many other products.

Geology and Economics: Global Oil Supplies—OPEC Rules

Eleven OPEC countries—most of them in the Middle East—have 78% of the world's proven oil reserves and most of the world's unproven reserves.

The oil industry is the world's largest business. Control of the world's current and future oil reserves is the single greatest source of global economic and political power.

Oil *reserves* are identified deposits from which oil can be extracted profitably at current prices with current technology. The 11 countries that make up the Organization of Petroleum Exporting Countries (OPEC) have 78% of the world's crude oil reserves. This explains why OPEC is expected to have long-term control over the supplies and prices of the world's conventional oil. Today OPEC's members are Algeria, Indonesia, Iran, Iraq, Kuwait, Libya, Nigeria, Qatar, Saudi Arabia, the United Arab Emirates, and Venezuela.

Saudi Arabia has by far the largest proportion of the world's crude oil reserves (25%). It is followed by Canada (15%), whose huge supply of oil sand was recently classified as a conventional source of oil. Other countries with large proven reserves are Iraq (11%), the United Arab Emirates (9.3%), Kuwait (9.2%), and Iran (8.6%).

Science and Economics Case Study: U.S. Oil Supplies

The United States—the world's largest oil user—has only 2.9% of the world's proven oil reserves and only a small percentage of its unproven reserves.

Figure 13-7 (p. 292) shows the locations of the major known deposits of fossil fuels in the United States and Canada and ocean areas where more crude oil and natural gas might be found. About one-fourth of U.S. domestic oil production comes from offshore drilling (mostly off the coasts of Texas and Louisiana, Figure 13-7, bottom) and 17% from Alaska's North Slope.

The United States has only 2.9% of the world's oil reserves. But it uses about 26% of the crude oil extracted worldwide each year (more than two-thirds of that for transportation), mostly because oil is an abundant, convenient, and cheap fuel. Indeed, when adjusted for inflation, oil costs about as much today as it did in 1975 (Figure 13-8, p. 292). Although low oil prices have stimulated economic growth, they have discouraged improvements in energy efficiency and renewable energy resources that might make the United States less dependent on oil imports from the Middle East and other countries.

Despite an upsurge in exploration and test drilling, U.S. oil extraction has declined since 1985, and most geologists do not expect a significant increase in domestic supplies. The United States produces most of its dwindling supply of oil at a high cost, about $7.50–$10 per barrel compared to about $2.50 per barrel in Saudi Arabia. Thus, *opening all U.S. coastal waters, forests, and wild places to drilling would hardly put a dent in world oil prices or meet much of the U.S. demand for oil.*

In 2004, the United States imported about 62% of the oil it used (up from 36% in 1973, when OPEC imposed an oil embargo against the U.S. and other nations). It got most of its oil from four nations (in order of importance): the non-OPEC nations of Canada and Mexico and the OPEC nations of Venezuela and Saudi Arabia.

According to the DOE, in the not-too-distant future the United States will have to depend more on the Middle East for oil, because it contains by far most of the world's discovered and undiscovered oil. Reasons for this high dependence on imported oil are declining domestic oil reserves, higher production costs for domestic oil than for most oil imports, and increased oil use. The United States could be importing as much as 70% of the oil it uses by 2020. But it will be facing stiff competition for world oil supplies from rapidly industrializing China.

Some analysts favor depending on oil imports. They argue that using up limited and declining domestic oil supplies is a drain-America-first policy that will increase future dependence on foreign oil supplies. Indeed, if the United States stopped importing oil and depended totally on domestic supplies, its proven reserves would last only about a decade at current consumption rates.

Bottom line: If you think of U.S. oil reserves as a six-pack of oil, four of the cans are empty. Geologists estimate that if the country opens up virtually all of its public lands and coastal regions to oil exploration, it may find at best about half a can of new oil at a high cost (compared to much cheaper OPEC oil) and with serious harmful environmental effects.

Global Outlook: How Long Will Conventional Oil Supplies Last?

Known and projected global oil reserves should last for 42–93 years and U.S. reserves for 10–48 years, depending on how rapidly we use oil.

The world is not yet running out of oil. But like all nonrenewable resources, oil supplies are eventually expected

Figure 13-7 Natural capital: locations of the major known deposits of oil, natural gas, and coal in North America and offshore areas where more crude oil and natural gas might be found. Geologists do not expect to find very much new oil and natural gas in North America. Offshore drilling for oil accounts for about one-fourth of U.S. oil production. Nine of every 10 barrels of this oil comes from the Gulf of Mexico, where there are 4,000 oil drilling platforms and 53,000 kilometers (33,000 miles) of underwater pipeline (see insert). (Data from Council on Environmental Quality and U.S. Geological Survey)

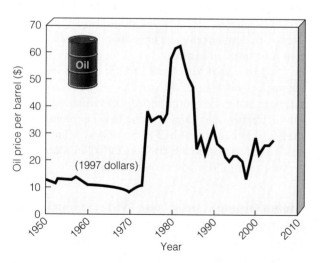

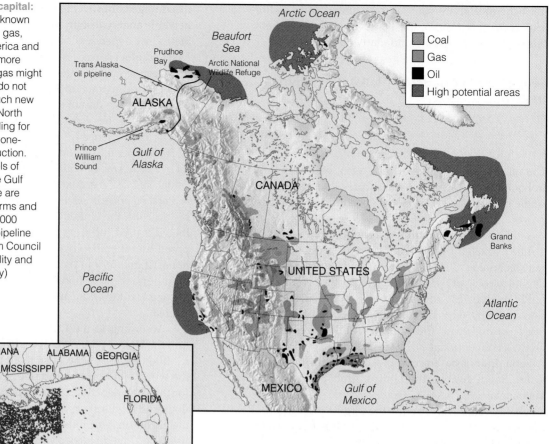

Figure 13-8 Economics: inflation-adjusted price of oil in the United States, 1950–2005. When adjusted for inflation, oil costs about the same as it did in 1975. (Data from U.S. Department of Energy and Department of Commerce)

to decline. At some point, prices will begin a steady rise as consumers begin competing for the world's dwindling oil reserves.

According to geologists, known and projected global reserves of oil are expected to be 80% depleted within 42–93 years and U.S. reserves in *10–48 years*, depending on how rapidly we use oil. If these estimates are correct, oil should be reaching its sunset years sometime during this century. (See Science Supplement 8 at the end of this book for a brief history of the Age of Oil.)

Oil geologist Colin J. Campbell warns that estimates of remaining oil may be too high because oil-producing countries often inflate their estimated oil reserves so that they can borrow money from the World Bank using their supposed oil supplies as collateral. Although Saudi Arabia vigorously denies it, some geologists believe that the country's estimated oil reserves are not as high as its leaders say.

Basically, we have three options: look for more oil, use or waste less oil, or use something else. Some analysts—mostly economists—contend that rising oil prices (when oil consumption exceeds oil production) will stimulate exploration and lead to enough new reserves to meet future demand through the next century or longer and that new technology will allow us to recover more oil from existing oil wells. Others argue that even if much more oil is somehow found, we are ignoring the consequences of the high (2–5% per year) exponential growth in global oil consumption.

It is hard to get a grip on the incredible amount of oil we consume. Maybe this will help: *Stretched end to end, the number of barrels of oil the world used in 2004 would circle the equator 636 times!*

Suppose we continue to use oil reserves at the current rate of about 2.8% per year with no increase in oil consumption—a highly unlikely assumption. Here are some of the results of using oil under this conservative no growth estimate:

- Saudi Arabia, with the world's largest known crude oil reserves, could supply the world's entire oil needs for about 10 years.

- The estimated reserves under Alaska's North Slope—the largest ever found in North America—would meet current world demand for only 6 months or U.S. demand for 3 years.

- The estimated reserves in Alaska's Arctic National Wildlife Refuge (ANWR) would met current oil demand for only 1–5 months and U.S. oil demand for 7–24 months.

Just for the world to keep using conventional oil at the current rate, we must discover global oil reserves that are the equivalent to a new Saudi Arabian supply every 10 years. According to most geologists, this is highly unlikely.

Many developing countries such as China and India are rapidly expanding their use of oil, putting them on a collision course with the United States. China is now the world's second largest oil importer after the United States. By 2025, that country could be using as much oil as the United States, and the two countries could be competing to import dwindling supplies of increasingly expensive oil. Indeed, if everyone in the world consumed as much oil as the average American, the world's proven oil reserves would be gone in a decade.

Exponential growth is an incredibly powerful force. If oil use grows by 5% per year, doubling the size of the world's oil reserves will add only 12 years to the life expectancy of these reserves. Even if we had a 1,000-year supply of oil, using it at a rate of 5% per year would deplete it in only 79 years.

Here is the problem in a nutshell: Oil is the most widely used energy resource in the world and its rate of use is growing exponentially. Most people in developed countries are *oilaholics*, and the world's largest suppliers for this addiction are Saudi Arabia, Canada, and Iraq.

Trade-Offs: Advantages and Disadvantages of Conventional Oil

Conventional oil is a versatile fuel that can last for at least 50 years, but burning it produces air pollution and releases the greenhouse gas carbon dioxide.

Figure 13-9 lists the advantages and disadvantages of using conventional crude oil as an energy resource. A serious problem associated with the use of conventional crude oil is that burning oil or any carbon-containing fossil fuel releases CO_2 into the atmosphere and thus can help promote climate change from global warming. Currently, burning oil mostly as gasoline and diesel fuel for transportation accounts for 43% of global CO_2 emissions.

Figure 13-10 (p. 294) compares the relative amounts of CO_2 emitted per unit of energy by the major fossil fuels and nuclear power.

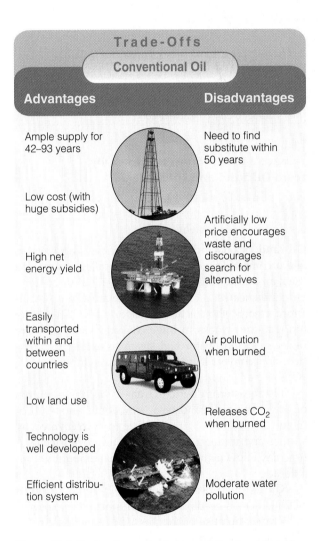

Trade-Offs

Conventional Oil

Advantages	Disadvantages
Ample supply for 42–93 years	Need to find substitute within 50 years
Low cost (with huge subsidies)	Artificially low price encourages waste and discourages search for alternatives
High net energy yield	
Easily transported within and between countries	Air pollution when burned
Low land use	
Technology is well developed	Releases CO_2 when burned
Efficient distribution system	Moderate water pollution

Figure 13-9 Trade-offs: advantages and disadvantages of using conventional crude oil as an energy resource. *Critical thinking: pick the single advantage and disadvantage that you think are the most important.*

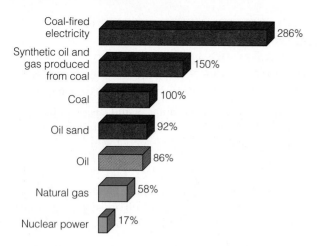

Coal-fired electricity — 286%
Synthetic oil and gas produced from coal — 150%
Coal — 100%
Oil sand — 92%
Oil — 86%
Natural gas — 58%
Nuclear power — 17%

Figure 13-10 Natural capital degradation: CO_2 emissions per unit of energy produced by various fuels, expressed as percentages of emissions produced by burning coal directly. These emissions can enhance the earth's natural greenhouse effect (Figure 5-5, p. 82) and lead to warming of the troposphere. (Data from U.S. Department of Energy)

✗ HOW WOULD YOU VOTE? Do the advantages of relying on conventional oil as the world's major energy resource outweigh its disadvantages? Cast your vote online at http://biology.brookscole.com/miller11.

Science and Economics: Heavy Oils from Oil Sand and Oil Shale

Heavy oils from oil sand and oil shale could supplement conventional oil, but there are environmental problems.

Oil sand, or **tar sand,** is a mixture of clay, sand, water, and a combustible organic material called *bitumen*—a thick and sticky heavy oil with a high sulfur content. Oil sands nearest the earth's surface are dug up by gigantic electric shovels and transported by 270-metric-ton (300-ton) trucks to huge cookers. There they are mixed with hot water and steam to extract the bitumen and convert it into a low-sulfur synthetic crude oil suitable for refining.

Northeastern Alberta in Canada has three-fourths of the world's oil sand resources, about one-tenth of them close enough to the surface to be recovered by surface and underground mining. Improved technology may allow extraction of twice that amount within a decade.

Currently, these deposits supply about one-fifth of Canada's oil needs, and this proportion is expected to increase. Because of the dramatic reductions in development and production costs, in 2003 the oil industry began counting Canada's oil sands as reserves of conventional oil. As a consequence, Canada has 15% of the world's oil reserves, second only to Saudi

Arabia. If a pipeline is built to transfer some of this synthetic crude oil from western Canada to the northwestern United States, Canada could greatly reduce future U.S. dependence on oil imports from the Middle East and add to Canadian income. However, even if everything goes right, this will not provide a significant amount of oil for the United States for 10–20 years.

Extracting and processing oil sands has a severe impact on the land and produces much more water pollution, much more air pollution (especially sulfur dioxide), and more CO_2 per unit of energy than exploiting conventional crude oil.

Oily rocks are another potential supply of heavy oil. Such rocks, called *oil shales* (Figure 13-11, left), contain a solid combustible mixture of hydrocarbons called *kerogen.* It can be extracted from crushed oil shales by heating them in a large container, a process that yields a distillate called **shale oil** (Figure 13-11, right). Before the thick shale oil can be sent by pipeline to a refinery, it must be heated to increase its flow rate and processed to remove sulfur, nitrogen, and other impurities.

Estimated potential global supplies of shale oil are about 240 times larger than estimated global supplies of conventional oil. But most deposits are of such a low grade that with current oil prices and technology, it takes more energy and money to mine and convert the kerogen to crude oil than the resulting fuel is worth. Producing and using shale oil also has a much higher environmental impact than exploiting conventional oil.

Figure 13-11 Natural capital: oil shale rock (left) and the shale oil (right) extracted from it. Big U.S. oil shale projects have been canceled because of excessive costs.

U.S. Department of Energy

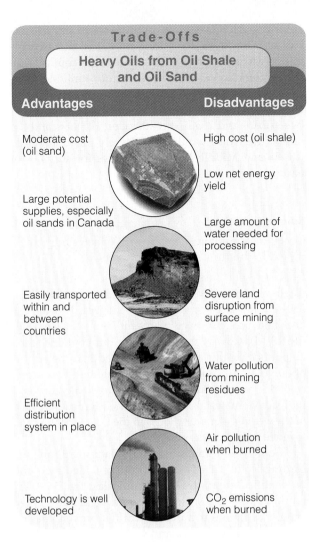

Trade-Offs

Heavy Oils from Oil Shale and Oil Sand

Advantages	Disadvantages
Moderate cost (oil sand)	High cost (oil shale)
	Low net energy yield
Large potential supplies, especially oil sands in Canada	Large amount of water needed for processing
Easily transported within and between countries	Severe land disruption from surface mining
	Water pollution from mining residues
Efficient distribution system in place	
	Air pollution when burned
Technology is well developed	CO_2 emissions when burned

Figure 13-12 Trade-offs: advantages and disadvantages of using heavy oils from oil sand and oil shale as energy resources. *Critical thinking: pick the single advantage and disadvantage that you think are the most important.*

Figure 13-12 lists the advantages and disadvantages of using heavy oil from oil sand and oil shales as energy resources.

Science: Natural Gas

Natural gas, consisting primarily of methane, is often found above reservoirs of crude oil.

In its underground gaseous state, **natural gas** is a mixture of 50–90% by volume of methane (CH_4), the simplest hydrocarbon. It also contains smaller amounts of heavier gaseous hydrocarbons such as ethane (C_2H_6), propane (C_3H_8), and butane (C_4H_{10}), and small amounts of highly toxic hydrogen sulfide (H_2S).

Conventional natural gas lies above most reservoirs of crude oil (Figure 13-2). However, unless a natural gas pipeline has been built, these deposits cannot be used. Indeed, the natural gas found above oil reservoirs in deep-sea and remote land areas is often viewed as an unwanted by-product and is burned off. This wastes a valuable energy resource and releases carbon dioxide into the atmosphere.

Unconventional natural gas is found by itself in other underground sources. So far, it costs too much to get natural gas from such unconventional sources, but the extraction technology is evolving rapidly.

When a natural gas field is tapped, propane and butane gases are liquefied and removed as **liquefied petroleum gas (LPG)**. LPG is stored in pressurized tanks for use mostly in rural areas not served by natural gas pipelines. The rest of the gas (mostly methane) is dried to remove water vapor, cleansed of poisonous hydrogen sulfide and other impurities, and pumped into pressurized pipelines for distribution. At a very low temperature, natural gas can be converted to **liquefied natural gas (LNG)**. This highly flammable liquid can then be shipped to other countries in refrigerated tanker ships. Some predict greatly increased use of imported LNG in the United States if enough shipping terminals can be built.

Russia has about 31% of the world's proven natural gas reserves, followed by Iran (15%) and Qatar (9%). The United States has only 3% of the world's proven natural gas reserves.

The long-term global outlook for natural gas supplies is better than that for conventional oil. At the current consumption rate, known reserves and undiscovered, potential reserves of conventional natural gas should last the world for 62–125 years and the United States for 55–80 years, depending on how rapidly they are used.

In total, *conventional* and *unconventional* supplies of natural gas (the latter available at higher prices) should last at least 200 years at the current consumption rate and 80 years if usage rates rise 2% per year.

Trade-Offs: Advantages and Disadvantages of Natural Gas

Natural gas is a versatile and clean-burning fuel, but it releases the greenhouse gases carbon dioxide (when burned) and methane (from leaks) into the atmosphere.

Figure 13-13 (p. 296) lists the advantages and disadvantages of using conventional natural gas as an energy resource. Because of its advantages over oil, coal, and nuclear energy, some analysts see natural gas as the best fuel to help make the transition to improved energy efficiency and greater use of solar energy and hydrogen over the next 50 years.

Natural gas heats about two-thirds of American homes, and it generates about 15% of the country's

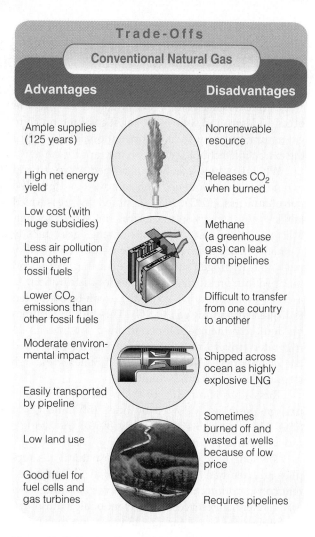

Trade-Offs

Conventional Natural Gas

Advantages	Disadvantages
Ample supplies (125 years)	Nonrenewable resource
High net energy yield	Releases CO_2 when burned
Low cost (with huge subsidies)	Methane (a greenhouse gas) can leak from pipelines
Less air pollution than other fossil fuels	
Lower CO_2 emissions than other fossil fuels	Difficult to transfer from one country to another
Moderate environmental impact	Shipped across ocean as highly explosive LNG
Easily transported by pipeline	
Low land use	Sometimes burned off and wasted at wells because of low price
Good fuel for fuel cells and gas turbines	Requires pipelines

Figure 13-13 Trade-offs: advantages and disadvantages of using conventional natural gas as an energy resource. *Critical thinking: pick the single advantage and disadvantage that you think are the most important.*

electricity. However, U.S. production of natural gas has been declining for a long time, and this trend is unlikely to be reversed. More natural gas could be imported from Canada, but it will require building a major pipeline between the two countries. Also, natural gas production in Canada is expected to peak between 2020 and 2030. Then the United States and the rest of the world would have to rely increasingly on Russia and Middle Eastern countries for supplies of natural gas.

Science: Coal

Coal is an abundant energy resource that is burned mostly to produce electricity and steel.

Coal is a solid fossil fuel that was formed in several stages as the buried remains of land plants that lived 300–400 million years ago were subjected to intense heat and pressure over many millions of years (Figure 13-14). Coal is mostly carbon but contains small amounts of sulfur, which are released into the atmosphere as sulfur dioxide when the coal burns. Burning coal also releases trace amounts of toxic mercury and radioactive materials. Coal is burned to generate 62% of the world's electricity (52% in the United States) and to make three-fourths of its steel.

Coal is the world's most abundant fossil fuel. According to the U.S. Geological Survey, identified and unidentified supplies of coal could last for 214–1,125 years, depending on our rate of usage. The United States has one-fourth of the world's proven coal reserves, Russia has 16%, and China has 12%. In 2004, slightly more than half of global coal consumption was split almost evenly between China and the United States.

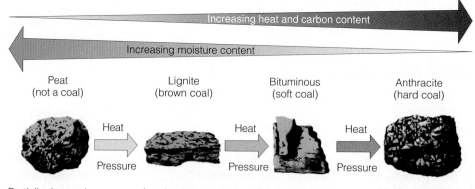

Figure 13-14 Natural capital: stages in coal formation over millions of years. Peat is a soil material made of moist, partially decomposed organic matter. Lignite and bituminous coal are sedimentary rocks, whereas anthracite is a metamorphic rock (Figure 12-6, p. 275).

Increasing heat and carbon content

Increasing moisture content

Peat (not a coal)	Lignite (brown coal)	Bituminous (soft coal)	Anthracite (hard coal)
Partially decayed plant matter in swamps and bogs; low heat content	Low heat content; low sulfur content; limited supplies in most areas	Extensively used as a fuel because of its high heat content and large supplies; normally has a high sulfur content	Highly desirable fuel because of its high heat content and low sulfur content; supplies are limited in most areas

Heat / Pressure

China has enough proven coal reserves to last 300 years at its current rate of consumption. According to the U.S. Geological Survey, identified U.S. coal reserves should also last about 300 years at the current consumption rate, and unidentified U.S. coal resources could extend those supplies for perhaps another 100 years, although at a higher cost. If U.S. coal use should increase by 4% per year—as the coal industry projects—the country's proven coal reserves would last only 64 years.

Trade-Offs: Advantages and Disadvantages of Coal

Coal is the most abundant fossil fuel, but compared to oil and natural gas it is not as versatile, has a much higher environmental impact, and releases more carbon dioxide into the atmosphere.

Figure 13-15 lists the advantages and disadvantages of using coal as an energy resource. *Bottom line:* Coal is the world's most abundant fossil fuel, but mining and burning coal has a severe environmental impact on the earth's air, water, and land, and accounts for more than one-third of the world's annual CO_2 emissions.

Each year in the United States alone, air pollutants from coal burning kill thousands of people prematurely (estimates range from 65,000 to 200,000), cause at least 50,000 cases of respiratory disease, and result in several billion dollars of property damage. Many people are unaware that burning coal is responsible for one-fourth of atmospheric mercury pollution in the United States, and it releases far more radioactive particles into the air than normally operating nuclear power plants.

Many analysts project a decline in coal use over the next 40–50 years because of its high CO_2 emissions (Figure 13-10) and harmful health effects, and the availability of safer and cheaper ways to produce electricity such as wind energy and burning natural gas in gas turbines.

X *HOW WOULD YOU VOTE?* Should coal use be phased out over the next 20 years? Cast your vote online at http//:biology.brookscole.com/miller11.

Trade-Offs: Advantages and Disadvantages of Converting Solid Coal into Gaseous and Liquid Fuels

Coal can be converted to gaseous and liquid fuels that burn cleaner than coal, but the costs are high, and producing and burning them add more carbon dioxide to the atmosphere than burning coal.

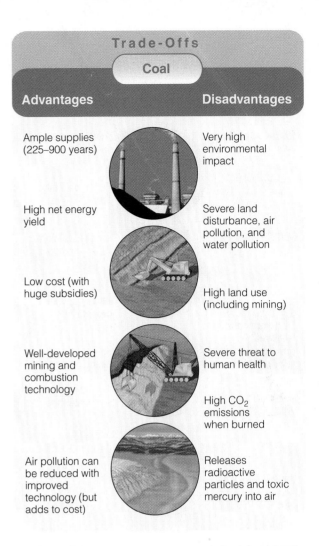

Trade-Offs

Coal

Advantages	Disadvantages
Ample supplies (225–900 years)	Very high environmental impact
High net energy yield	Severe land disturbance, air pollution, and water pollution
Low cost (with huge subsidies)	High land use (including mining)
Well-developed mining and combustion technology	Severe threat to human health
	High CO_2 emissions when burned
Air pollution can be reduced with improved technology (but adds to cost)	Releases radioactive particles and toxic mercury into air

Figure 13-15 Trade-offs: advantages and disadvantages of using coal as an energy resource. *Critical thinking: pick the single advantage and disadvantage that you think are the most important.*

Solid coal can be converted into **synthetic natural gas (SNG)** by **coal gasification** or into a liquid fuel such as methanol or synthetic gasoline by **coal liquefaction.** Figure 13-16 (p. 298) lists the advantages and disadvantages of using these *synfuels.*

Without huge government subsidies, most analysts expect synfuels to play a minor role as energy resources in the next 20–50 years. Compared to burning conventional coals, they require mining 50% more coal and their production and burning add 50% more carbon dioxide to the atmosphere. Also, they cost more to produce than coal.

Currently, the DOE and a consortium of major oil companies are working on ways to reduce CO_2 emissions during the coal gasification process. If these efforts prove successful, burning gasified coal could be a cheaper and cleaner way to produce electricity than burning coal, oil, or natural gas. Stay tuned.

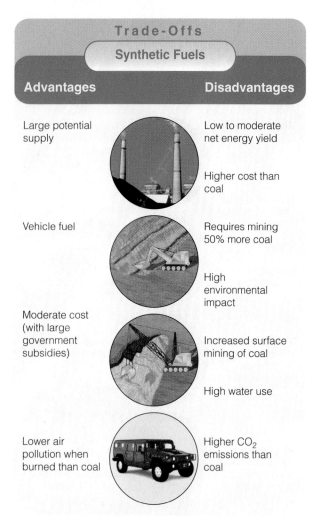

Advantages	Disadvantages
Large potential supply	Low to moderate net energy yield
	Higher cost than coal
Vehicle fuel	Requires mining 50% more coal
	High environmental impact
Moderate cost (with large government subsidies)	Increased surface mining of coal
	High water use
Lower air pollution when burned than coal	Higher CO_2 emissions than coal

Figure 13-16 Trade-offs: advantages and disadvantages of using synthetic natural gas (SNG) and liquid synfuels produced from coal. *Critical thinking: pick the single advantage and disadvantage that you think are the most important.*

13-3 NONRENEWABLE NUCLEAR ENERGY

Science: How Does a Nuclear Fission Reactor Work?

In a conventional nuclear reactor, isotopes of uranium and plutonium undergo controlled nuclear fission. The resulting heat produces steam that in turn spins turbines to generate electricity.

To evaluate the advantages and disadvantages of nuclear power, we must know how a conventional nuclear power plant and its accompanying nuclear fuel cycle work. In the reactor of a nuclear power plant, the rate of fission in a nuclear chain reaction (Figure 2-6, p. 28) is controlled and the heat generated is used to produce high-pressure steam, which spins turbines that generate electricity.

Light-water reactors (LWRs) like the one diagrammed in Figure 13-17 produce 85% of the world's nuclear-generated electricity (100% in the United States). *Control rods* are moved in and out of the reactor core to absorb neutrons, thereby regulating the rate of fission and amount of power produced. A *coolant*, usually water, circulates through the reactor's core to remove heat to keep fuel rods and other materials from melting and to produce steam for generating electricity. The greatest danger in water-cooled reactors is a loss of coolant, which would allow the nuclear fuel to overheat, melt down, and possibly release radioactive materials to the environment. An LWR includes an emergency core cooling system as a backup to help prevent such meltdowns.

A *containment vessel* with strong, thick walls surrounds the reactor core. It is designed to keep radioactive materials from escaping into the environment in case of an internal explosion or core meltdown within the reactor, and to protect the core from external threats such as a plane crash.

Water-filled pools or *dry casks* with thick steel walls are used for on-site storage of highly radioactive spent fuel rods, which are removed when reactors are refueled. Spent-fuel pools or casks are located in a separate building that is not nearly as well protected as the reactor core and are much more vulnerable to a head-on crash from a plane or terrorist attack. The long-term goal is to transport spent fuel rods and other long-lived radioactive wastes to an underground facility for long-term storage.

The overlapping and multiple safety features of a modern nuclear reactor greatly reduce the chance of a serious nuclear accident. At the same time, these safety features make nuclear power plants very expensive to build and maintain.

Nuclear power plants, each with one or more reactors, are merely one part of the nuclear fuel cycle (Figure 13-18, p. 300). Unlike other energy resources, nuclear energy produces highly radioactive materials that must be stored safely for 10,000–240,000 years until their radioactivity falls to safe levels. In addition, once a nuclear reactor comes to the end of its useful life (after 40–60 years), it cannot be shut down and abandoned like a coal-burning plant. It contains large quantities of intensely radioactive materials that must be kept out of the environment for many thousands of years.

In evaluating the safety and economic feasibility of nuclear power, energy experts and economists caution us to look at the entire nuclear fuel cycle, not just the nuclear plant itself.

What Happened to Nuclear Power?

After more than 50 years of development and enormous government subsidies, nuclear power has not lived up to its promise.

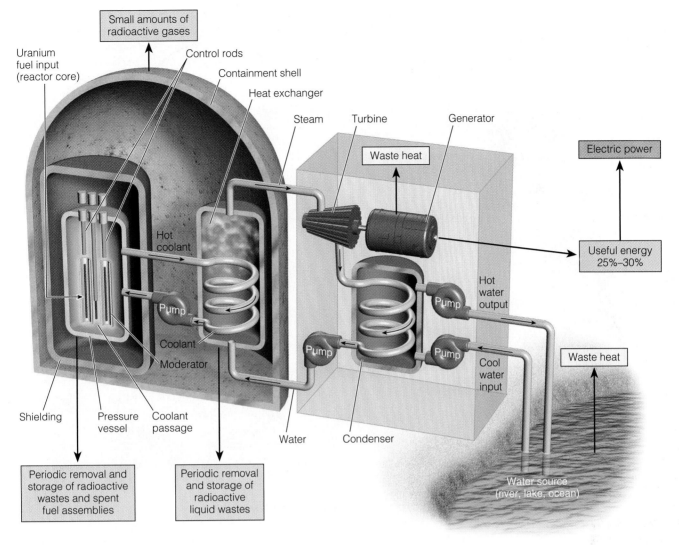

Figure 13-17 Science: light-water–moderated and –cooled nuclear power plant with a pressurized water reactor. Some plants use huge cooling towers to transfer some of the waste heat to the atmosphere.

In the 1950s, researchers predicted that by 2000 at least 1,800 nuclear power plants would supply 21% of the world's commercial energy (25% in the United States) and most of the world's electricity.

After almost 50 years of development, enormous government subsidies, and an investment of $2 trillion, these goals have not been met. Instead, in 2004, 439 commercial nuclear reactors in 30 countries were producing only 6% of the world's commercial energy and 16% of its electricity.

Since 1989, electricity production from nuclear power has increased only slightly and is now the world's slowest-growing energy source. According to the DOE, the percentage of the world's electricity produced by nuclear power is projected to fall to 12% by 2025, compared to 16% in 2003, because the retirement of aging plants is expected to exceed the construction of new ones.

No new nuclear power plants have been ordered in the United States since 1978, and all 120 plants ordered since 1973 have been canceled. In 2004, there were 103 licensed commercial nuclear power reactors in 31 states—most of them located in the eastern half of the country. These reactors generate about 21% of the country's electricity and 8% of its total energy. This percentage is expected to decline over the next two decades as existing plants wear out and are retired.

According to energy analysts and economists, several reasons explain the failure of nuclear power to grow as projected. They include multibillion-dollar construction cost overruns, higher operating costs and more malfunctions than expected, and poor management. Two other obstacles have been public concerns about safety and stricter government safety regulations, especially after the accidents in 1979 at the Three Mile Island nuclear plant in Pennsylvania and in 1986

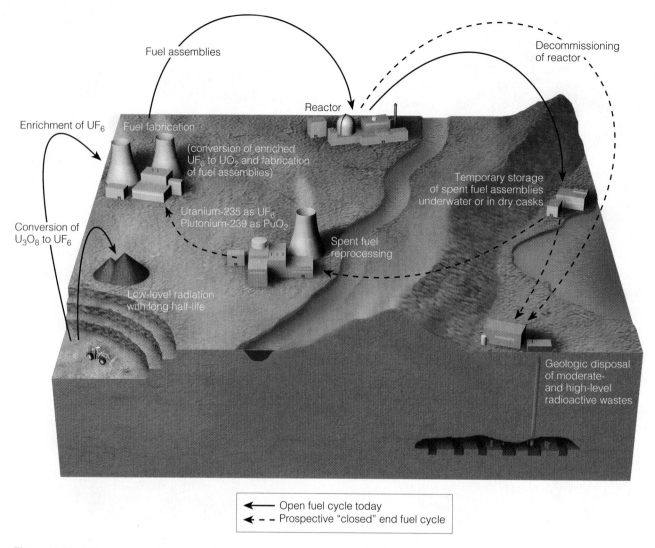

Fuel assemblies

Decommissioning
of reactor

Enrichment of UF$_6$

Reactor

Fuel fabrication

(conversion of enriched
UF$_6$ to UO$_2$ and fabrication
of fuel assemblies)

Temporary storage
of spent fuel assemblies
underwater or in dry casks

Conversion of
U$_3$O$_8$ to UF$_6$

Uranium-235 as UF$_6$
Plutonium-239 as PuO$_2$

Spent fuel
reprocessing

Low-level radiation
with long half-life

Geologic disposal
of moderate-
and high-level
radioactive wastes

← Open fuel cycle today
←- - Prospective "closed" end fuel cycle

Figure 13-18 Science: the nuclear fuel cycle.

at the Chernobyl nuclear plant in Ukraine (see Case Study, below)

Another problem is investor concerns about the economic feasibility of nuclear power, taking into account the entire nuclear fuel cycle. At Three Mile Island, investors lost more than $1 billion in one hour from damaged equipment and repair, even though no human lives were lost. Also, concern has been voiced about the vulnerability of nuclear power plants to terrorist attack after the destruction of New York's World Trade Center buildings and Washington, D.C.'s Pentagon on September 11, 2001. Experts are especially concerned about the vulnerability of poorly protected and intensely radioactive spent fuel rods stored in water pools or casks outside of reactor buildings.

Science Case Study: The Chernobyl Nuclear Power Plant Accident

The world's worst nuclear power plant accident occurred in 1986 in Ukraine.

Chernobyl is known around the globe as the site of the world's most serious nuclear power plant accident. On April 26, 1986, a series of explosions in one of the reactors in a nuclear power plant in Ukraine—then part of the Soviet Union—blew the massive roof off a reactor building and flung radioactive debris and dust high into the atmosphere. A huge radioactive cloud spread over much of Belarus, Russia, Ukraine, and other parts of Europe and eventually encircled the planet.

Clouds of radioactive material escaped into the atmosphere for 10 days. The surrounding environment and people were exposed to radiation levels 100 times higher than those caused by the atomic bomb dropped on Hiroshima, Japan, near the end of World War II.

According to various UN studies, the disaster, which was caused by poor reactor design and human error, had dire consequences. At least 31 people near the accident site were killed directly. More than half a million people were exposed to dangerous levels of radioactivity, which ultimately may cause 8,000–15,000 premature deaths.

More than 100,000 people had to leave their homes. Most were not evacuated until at least 10 days after the accident. In 2003, Ukraine officials downgraded the 27-kilometer (17-mile) area surrounding the reactor to a "zone with high risk" to allow those willing to accept the health risk to return to their homes.

Chernobyl taught us a hard lesson: *A major nuclear accident anywhere has effects that reverberate throughout much of the world.*

Environmental Science ⊕ Now™
Watch how winds carried radioactive fallout around the world after the Chernobyl meltdown at Environmental ScienceNow.

Trade-Offs: Advantages and Disadvantages of the Nuclear Power Fuel Cycle

The nuclear power fuel cycle has a fairly low environmental impact, an ample supply of fuel, and a very low risk of an accident, but costs are high, radioactive wastes must be stored safely for thousands of years, and facilities are vulnerable to terrorist attack.

Figure 13-19 lists the major advantages and disadvantages of the nuclear fuel cycle. Using nuclear power to produce electricity has some important advantages over coal-burning power plants (Figure 13-20, p. 302).

Because of the built-in safety features, the risk of exposure to radioactivity from nuclear power plants in the United States and most other developed countries is extremely low. However, a partial or complete meltdown or explosion is possible, as the accidents at the Chernobyl nuclear power plant in Ukraine and the Three Mile Island plant in Pennsylvania taught us.

The U.S. Nuclear Regulatory Commission (NRC) estimates there is a 15–45% chance of a complete core meltdown at a U.S. reactor during the next 20 years. The NRC also found that 39 U.S. reactors have an 80% chance of containment shell failure from a meltdown or an explosion of gases inside the containment structures.

In the United States, there is widespread public distrust of government agencies' ability to enforce nuclear safety in commercial (NRC) and military (DOE) nuclear facilities. In 1996, George Galatis, a respected senior nuclear engineer, said, "I believe in nuclear power but after seeing the NRC in action I'm convinced a serious accident is not just likely, but inevitable. . . . They're asleep at the wheel."

The 2001 terrorist attacks on the World Trade Center and Pentagon raised fears that a similar attack by a large plane loaded with fuel could break open a reactor's containment shell and set off a reactor meltdown that could create a major radioactive disaster.

Nuclear officials contend that these concerns are overblown and that U.S. nuclear plants could survive

Trade-Offs	
Conventional Nuclear Fuel Cycle	
Advantages	**Disadvantages**
Large fuel supply	High cost even with large subsidies
Low environmental impact (without accidents)	Low net energy yield
Emits 1/6 as much CO_2 as coal	High environmental impact (with major accidents)
Moderate land disruption and water pollution (without accidents)	Catastrophic accidents can happen (Chernobyl)
Moderate land use	No widely acceptable solution for long-term storage of radioactive wastes and decommissioning worn-out plants
Low risk of accidents because of multiple safety systems (except in 35 poorly designed and run reactors in former Soviet Union and eastern Europe)	Subject to terrorist attacks
	Spreads knowledge and technology for building nuclear weapons

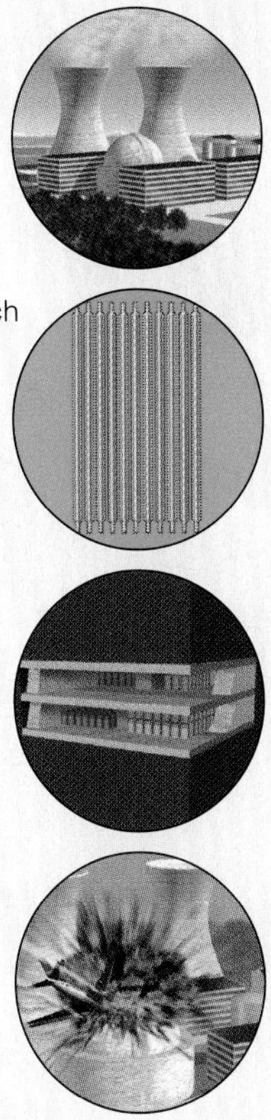

Figure 13-19 Trade-offs: advantages and disadvantages of using the conventional nuclear fuel cycle (Figure 13-18) to produce electricity. *Critical thinking: pick the single advantage and disadvantage that you think are the most important.*

such an attack because of the thickness and strength of the containment walls. But a 2002 study by the Nuclear Control Institute found that the plants were not designed to withstand the crash of a large jet traveling at the impact speed of the two hijacked airliners that hit the World Trade Center.

An even greater concern is insufficient security at U.S. nuclear power plants against ground-level attacks by terrorists. During a series of security exercises performed by the NRC between 1991 and 2001, mock attackers were able to simulate the destruction of enough equipment to cause a meltdown at nearly half of U.S. nuclear plants. A 2002 study also found that many security guards at nuclear power plants have low morale and are overworked, underpaid, undertrained, and not equipped with sufficient firepower to repel a serious ground attack by terrorists.

The NRC contends that the security weaknesses revealed by earlier mock tests have been corrected.

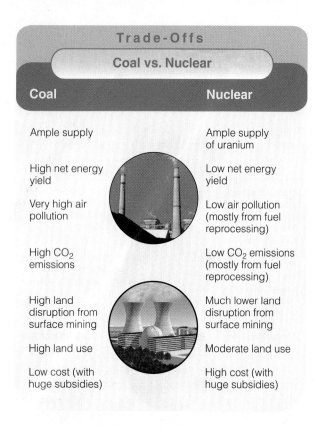

Figure 13-20 Trade-offs: comparison of the risks of using nuclear power and coal-burning plants to produce electricity. *Critical thinking: would you rather live next door to a coal-fired power plant or a nuclear power plant?*

Many nuclear power analysts remain unconvinced and note that since September 2001 the NRC has stopped staging such tests.

According to critics, the NRC is reluctant to require utilities to make significant improvements in plant security because it would increase the already high costs of the nuclear power fuel cycle and make it less competitive in the marketplace.

Throughout the world, nuclear scientists and government officials have urged the shutdown of 35 poorly designed and poorly operated nuclear reactors in some republics of the former Soviet Union and in eastern Europe. This is unlikely without economic aid from the world's developed countries.

Science: High-Level Radioactive Waste

Scientists disagree about the best methods for long-term storage of high-level radioactive waste.

Each part of the nuclear power fuel cycle produces low-level and high-level solid, liquid, and gaseous radioactive wastes. *High-level radioactive wastes* must be stored safely for at least 10,000 years, or 240,000 years if plutonium-239 is not removed by reprocessing as part of the nuclear fuel cycle. They consist mainly of spent fuel

rods from commercial nuclear power plants and assorted wastes from the production of nuclear weapons.

There is concern that some of the pools and casks used to stored spent fuel rods at various nuclear power plants in the United States are vulnerable to sabotage by terrorist attack. A spent-fuel pool typically holds 5–10 times more long-lived radioactivity than the radioactive core inside a plant's reactor. Unlike the reactor core with its thick concrete protective dome, spent-fuel pools and dry casks have little protective cover—although casks offer more protection than pools.

An earthquake, a deliberate crash by a small airplane, or an attack by a group of suicidal terrorists could drain a pool or rupture a dry cask containing spent fuel rods. According to the NRC, this would release significant amounts of radioactive materials into the atmosphere, contaminate large areas for decades, and create economic and psychological havoc.

U.S. nuclear power officials consider such events to be highly unlikely, worst-case scenarios and question some of the estimates. They also contend that nuclear power facilities are safe from attack. Critics are not convinced and call for constructing much more secure structures to protect spent-fuel storage sites. A 2005 study by the National Academy of Sciences recommends storing fuel rods now in pools in casks as soon as possible. Critics accuse the NRC of failing to improve the security of spent fuel rods because it would impose additional costs on utility companies and raise the cost of relying on the nuclear power fuel cycle.

After more than 50 years of research, scientists still do not agree on whether there is a safe way of storing high-level radioactive waste. Some believe the long-term safe storage or disposal of high-level radioactive wastes is technically possible. Others disagree, pointing out that it is impossible to demonstrate that any method will work for 10,000–240,000 years.

Here are some of the proposed methods and their possible drawbacks:

- *Bury it deep underground.* This favored strategy is under study by all countries producing nuclear waste and is the option being pursued in the United States (Science and Politics Case Study, p. 303) and at a slower pace in Finland and Sweden.

- *Shoot it into space or into the sun.* Costs would be very high, and a launch accident—like the explosion of the space shuttle *Challenger*—could disperse high-level radioactive wastes over large areas of the earth's surface. This strategy has been abandoned for now.

- *Bury it under the Antarctic ice sheet or the Greenland ice cap.* The long-term stability of the ice sheets is not known. They could be destabilized by heat from the wastes, and retrieving the wastes would be difficult or impossible if the method failed. This strategy is prohibited by international law.

- *Dump it into descending zones of the earth's crust in the deep ocean.* Wastes eventually might be spewed out somewhere else by volcanic activity, and containers might leak and contaminate the ocean before being carried downward. Also, retrieval would be impossible if the method did not work. This strategy is prohibited by international law.

- *Bury it in thick deposits of mud on the deep-ocean floor in areas that tests show have remained geologically stable for 65 million years.* The waste containers eventually would corrode and release their radioactive contents. This approach is prohibited by international law.

- *Change it into harmless, or less harmful, isotopes.* Currently, no way exists to do this. Even if a method was developed, costs would probably be very high, and the resulting toxic materials and low-level (but very long-lived) radioactive wastes would need to be disposed of safely.

Science and Politics Case Study: High-Level Radioactive Wastes in the United States

Scientists disagree over the decision to store high-level nuclear wastes at an underground storage site in Nevada.

In 1985, the DOE announced plans to build a repository for underground storage of high-level radioactive wastes from commercial nuclear reactors on federal land in the Yucca Mountain desert region, 160 kilometers (100 miles) northwest of Las Vegas, Nevada.

The proposed facility is expected to cost at least $58 billion to build (financed partly by a tax on nuclear power). It is scheduled to open by 2010, but may not begin operation until 2015.

Some scientists argue that it should never be allowed to open, mostly because rock fractures and tiny cracks may allow water to leak into the site and eventually corrode radioactive waste storage casks. Geologists also point out a nearby active volcano and 32 active earthquake fault lines running through the site—an unusually high number. In 1998, Jerry Szymanski, formerly the DOE's top geologist at Yucca Mountain and now an outspoken opponent of the site, said that if water flooded the site it could cause an explosion so large that "Chernobyl would be small potatoes."

In 2002, the U.S. National Academy of Sciences, in collaboration with Harvard and University of Tokyo scientists, urged the U.S. government to slow down and rethink its nuclear waste storage process. These scientists contend that storing spent fuel rods in dry-storage casks in well-protected buildings at nuclear plant sites is an adequate solution for at least 100 years in terms of safety and national security. This approach would buy time to carry out more research on this complex problem and to evaluate other sites and storage methods.

Opponents also contend that the Yucca Mountain waste site should not be opened because *it might decrease national security.* The plan calls for wastes to be put into specially designed casks and shipped by truck or rail cars to the Nevada site. This would require about 19,600 shipments of wastes over much of the country for the estimated 38 years before the site is filled. At the end of this period, the amount of newly collected radioactive waste stored at nuclear power plant sites would be about the same as before the Yucca Mountain repository opened. Critics contend that it will be much more difficult to protect such a large number of shipments from terrorist attacks than to provide more secure ways to store such wastes at nuclear power plant sites.

The DOE and proponents of nuclear power counter that the risks of an accident or sabotage of nuclear waste shipments are negligible. Opponents disagree. Utility companies oppose improved storage at plant sites because it would add to their costs and make nuclear power a more unattractive energy option.

Despite these and other objections from scientists and citizens, in 2002 Congress approved Yucca Mountain as the official site for storing the country's commercial nuclear wastes. Opponents want the law repealed. In 2005, Congress temporarily halted efforts to get final approval for the project after learning that some of the data used to justify the project were made up.

X *HOW WOULD YOU VOTE?* Should highly radioactive spent fuel be stored in casks in well-protected buildings at nuclear power plant sites instead of shipping them to a single site for underground burial? Cast your vote online at http://biology.brookscole.com/miller11.

Science: Dealing with Worn-Out Nuclear Plants

When a nuclear reactor reaches the end of its useful life, we must keep its highly radioactive materials from reaching the environment for thousands of years.

When a nuclear power plant comes to the end of its useful life, it must be *decommissioned,* or retired—the last step in the nuclear power fuel cycle. Scientists have proposed three ways to do this.

One strategy is to dismantle the plant after it closes and store its large volume of highly radioactive materials in a high-level, nuclear waste storage facility. Many scientists question the safety of such facilities. A second approach is to install a physical barrier around the plant and set up full-time security for 30–100 years before the plant is dismantled. A third option is to enclose the entire plant in a tomb that must last and be monitored for several thousand years. Regardless of the method chosen, decommissioning adds to the total costs of nuclear power as an energy option.

At least 228 large commercial reactors worldwide (20 in the United States) are scheduled for retirement by 2012. However, the NRC has approved extending the 40-year expected life expectancy of at least 20 U.S. reactors to 60 years. Opponents contend this could increase the risk of nuclear accidents in aging reactors.

Science and Terrorism: "Dirty" Bombs

Terrorists could wrap conventional explosives around small amounts of various radioactive materials that are fairly easy to get, detonate such bombs, and contaminate fairly large areas with radioactivity for decades.

Since the terrorist attacks in the United States on September 11, 2001, concern has been growing about the threats posed by *"dirty" bombs*. Such a bomb consists of an explosive such as dynamite mixed with or wrapped around an amount of radioactive material small enough fit in a coffee cup.

Such radioactive materials can be stolen from any of thousands of poorly guarded and difficult-to-protect sources or bought on the black market. For example, hospitals use radioisotopes (such as cobalt-60) to treat cancer and diagnose various diseases. Other potential sources include university research laboratories, and industries that use radioisotopes to detect leaks in underground pipes, irradiate food, examine mail and other materials, and detect flaws in pipe welds and boilers.

Since 1986, the NRC has recorded 1,700 incidents in the United States in which radioactive materials used by industrial, medical, or research facilities have been stolen or lost. Since 1991, the International Atomic Energy Agency (IAEA) has detected 671 incidents of illicit trafficking in dirty-bomb materials.

Detonating a dirty bomb at street level or on a rooftop does not cause a nuclear blast. Nevertheless, it could kill as many as 1,000 people in densely populated cities and spread radioactive material over several blocks. This would pose cancer risks for decades and cause widespread terror and panic—the primary objective of terrorists.

Science: Can Nuclear Power Reduce Dependence on Imported Oil and Help Reduce Global Warming?

Because so little oil is burned to produce electricity , building more nuclear power plants is neither a way to reduce dependence on imported oil nor a major way to reduce carbon dioxide emissions.

Some proponents of nuclear power in the United States claim it will help reduce the country's dependence on imported oil. Other analysts point out that use of nuclear power has little effect on U.S. oil use because burning oil typically produces only 2–3% of the electricity in the United States (and in most other countries). Also, the major use for oil is as gasoline and diesel fuel in transportation, which would not be affected by increasing the use of nuclear power plants to produce electricity.

Nuclear power advocates also contend that increased use of nuclear power would reduce the threat of global warming by eliminating emissions of CO_2 compared to burning coal. Scientists point out that this argument is only partially correct. Nuclear plants themselves are not emitters of CO_2, but the nuclear fuel cycle does produce some CO_2. However, such emissions are much less than that produced by burning coal or natural gas to produce the same amount of electricity (Figure 13-10). Environmentalists and many energy experts argue that reducing energy waste and increasing the use of wind turbines, solar cells, and hydrogen to produce electricity are better ways to reduce CO_2 emissions.

Economics of Nuclear Power

Even with massive government subsidies, the nuclear power fuel cycle is an expensive way to produce electricity compared to several other energy alternatives.

Experience has shown that the nuclear power fuel cycle is an expensive way to produce electricity, even when huge government subsidies partially shield it from free-market competition with other energy sources. In 1995, the World Bank said nuclear power is too costly and risky. *Forbes* magazine has called the failure of the U.S. nuclear power program "the largest managerial disaster in U.S. business history, involving $1 trillion in wasted investment and $10 billion in direct losses to stockholders."

In recent years, the operating costs of many U.S. nuclear power plants have dropped, mostly because of less downtime. Environmentalists and economists counter that the cost of nuclear power must reflect the entire nuclear power fuel cycle, not merely the operating costs of individual plants. According to these analysts, when these costs (including nuclear waste disposal and decommissioning of worn-out plants) are included, the overall cost of nuclear power is very high (even with huge government subsidies) compared to many other energy alternatives.

Partly to address these concerns, the U.S. nuclear industry hopes to persuade the Congress and utility companies to build hundreds of smaller, second-generation plants using standardized designs. The industry claims these plants are safer and can be built more quickly (in 3–6 years).

These *advanced light-water reactors (ALWRs)* have built-in *passive safety features* designed to make explosions or the release of radioactive emissions almost impossible. However, according to *Nucleonics Week,* an important nuclear industry publication, "Experts are flatly unconvinced that safety has been achieved—or even substantially increased—by the new designs." In addition, these new designs do not eliminate the threats and the expense and hazards of long-term radioactive waste storage and power plant decommissioning.

Science: Breeder Nuclear Fission

Because of very high costs and bad safety experiences with several nuclear breeder reactors, this technology has essentially been abandoned.

Some nuclear power proponents urge the development and widespread use of **breeder nuclear fission reactors,** which generate more nuclear fuel than they consume by converting nonfissionable uranium-238 into fissionable plutonium-239. Because breeders would use more than 99% of the uranium in ore deposits, the world's known uranium reserves would last at least 1,000 years, and perhaps several thousand years.

However, if the safety system of a breeder reactor fails, the reactor could lose some of its liquid sodium coolant, which ignites when exposed to air and reacts explosively if it comes into contact with water. The potential result: a runaway fission chain reaction and perhaps a nuclear explosion powerful enough to blast open the containment building and release a cloud of highly radioactive gases and particles into the atmosphere. Leaks of flammable liquid sodium can also cause fires, as has happened with all experimental breeder reactors built so far.

Existing experimental breeder reactors also produce plutonium so slowly that it would take 100–200 years for them to produce enough plutonium to fuel a significant number of other breeder reactors. In 1994, the United States ended government-supported research of breeder technology after providing about $9 billion in research and development funding.

In December 1986, France opened a commercial-size breeder reactor. It was so expensive to build and operate that after spending $13 billion the government spent another $2.75 billion to shut it down permanently in 1998. Because of this experience, other countries have abandoned their plans to build full-size commercial breeder reactors.

Science: Nuclear Fusion

Nuclear fusion has a number of advantages, but after more than five decades of research and billions of dollars in government research and development subsidies, this technology remains at the laboratory stage.

Nuclear fusion is a nuclear change in which two isotopes of light elements, such as hydrogen, are forced together at extremely high temperatures until they fuse to form a heavier nucleus, releasing energy in the process. Scientists hope that controlled nuclear fusion will provide an almost limitless source of high-temperature heat and electricity. Research has focused on the D–T nuclear fusion reaction, in which two isotopes of hydrogen—deuterium (D) and tritium (T)—fuse at about 100 million degrees (Figure 2-7, p. 28).

With nuclear fusion, there would be no risk of meltdown or release of large amounts of radioactive materials from a terrorist attack and little risk from additional proliferation of nuclear weapons because bomb-grade materials (such as enriched uranium-235 and plutonium-239) are not required for fusion energy. Fusion power might also be used to destroy toxic wastes, supply electricity for ordinary use, and decompose water to produce the hydrogen gas needed to run a hydrogen economy by the end of this century.

This sounds great. So what is holding up fusion energy? After more than 50 years of research and huge expenditures of mostly government funds, controlled nuclear fusion remains in the laboratory stage. None of the approaches tested so far has produced more energy than it uses.

If researchers can eventually get more energy out of nuclear fusion than they put in, the next step would be to build a small fusion reactor and then scale it up to commercial size. This is an extremely difficult engineering problem. Also, the estimated cost of a building and operating a commercial fusion reactor (even with huge government subsidies) is several times the costs for a comparable conventional fission reactor.

Proponents contend that with greatly increased federal funding, a commercial nuclear fusion power plant might be built by 2030 or perhaps by 2020. However, many energy experts do not expect nuclear fusion to be a significant energy source until 2100, if then.

Politics: Nuclear Power's Future in the United States

There is disagreement over whether the United States should phase out nuclear power or keep this option open in case other alternatives do not pan out.

Since 1948, nuclear energy (fission and fusion) has received about 58% of all federal energy research and development funds in the United States—compared to 22% for fossil fuels, 11% for renewable energy, and 8% for energy efficiency and conservation. Because the results of this huge investment of taxpayer dollars have been largely disappointing, some analysts call for phasing out all or most government subsidies and tax breaks for nuclear power and using the money to accelerate the development of other, more promising energy technologies.

To these critics, conventional nuclear power is a complex, expensive, inflexible, and centralized way to produce electricity that is too vulnerable to terrorist attack. They believe it is a technology whose time has passed in a world where electricity will increasingly be provided by small, decentralized, easily expandable power plants such as natural gas turbines, farms of wind turbines, arrays of solar cells, and hydrogen-powered fuel cells. According to investors and World Bank economic analysts, conventional nuclear power simply cannot compete in today's increasingly open, decentralized, and unregulated energy market unless it is artificially shielded from free-market competition by huge government subsidies.

Proponents of nuclear power argue that governments should continue funding research and development and pilot-plant testing of potentially safer and cheaper reactor designs along with breeder fission and nuclear fusion. They insist we need to keep these nuclear options available for use in the future if renewable energy options fail to keep up with electricity demands and reduce CO_2 emissions to acceptable levels. Germany does not buy these arguments—it has plans to phase out nuclear power over the next two decades. However, China plans to build a number of nuclear power plants.

☒ *HOW WOULD YOU VOTE?* Should nuclear power be phased out in the country where you live over the next 20–30 years? Cast your vote online at http://biology.brookscole.com/miller11.

13-4 IMPROVING ENERGY EFFICIENCY

Science: Wasting Energy

The United States unnecessarily wastes about 43% of the energy it uses.

You may be surprised to learn that 84% of all commercial energy used in the United States goes to waste (Figure 13-21). About 41% of this energy is lost automatically because of the degradation of energy quality imposed by the second law of thermodynamics (p. 30). Another 43% is wasted unnecessarily, largely as the result of using fuel-wasting motor vehicles, furnaces, and other devices, and living and working in leaky, poorly insulated, poorly designed buildings. (See the Guest Essay on this topic by Amory Lovins on the website for this chapter.)

Energy efficiency in the United States has improved since the oil price shock in 1979. Even so, ac-

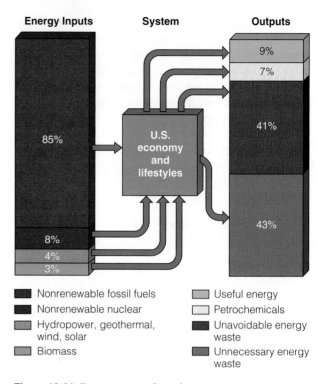

| Energy Inputs | System | Outputs |

85%

U.S. economy and lifestyles

9%

7%

41%

43%

8%
4%
3%

■ Nonrenewable fossil fuels
■ Nonrenewable nuclear
■ Hydropower, geothermal, wind, solar
■ Biomass

□ Useful energy
□ Petrochemicals
■ Unavoidable energy waste
■ Unnecessary energy waste

Figure 13-21 Energy waste: flow of commercial energy through the U.S. economy. Only 16% of all commercial energy used in the United States ends up performing useful tasks or being converted to petrochemicals; the rest is unavoidably wasted because of the second law of thermodynamics (41%) or is wasted unnecessarily (43%). (Data from U.S. Department of Energy)

cording to the DOE, the United States still wastes as much energy as two-thirds of the world's population consumes.

Reducing energy waste requires improving *energy efficiency* by using less energy to do more useful work. In other words, we must learn how to do more with less. Reducing such energy waste has numerous economic and environmental advantages (Figure 13-22).

Improvements in energy efficiency since 1973 have cut U.S. energy bills by $200 billion per year. Nevertheless, unnecessary energy waste costs the United States about $300 billion per year—an average of $570,000 per minute.

Four widely used devices waste large amounts of energy:

- An *incandescent light bulb* wastes 95% of its energy input of electricity. In other words, it is a *heat bulb.*

- A *nuclear power plant* producing electricity for space heating or water heating wastes about 86% of the energy in its nuclear fuel and probably 92% when we include the energy needed to deal with its radioactive wastes and to retire the plant.

- A motor vehicle with an *internal combustion engine* wastes 75–80% of the energy in its fuel.

Solutions

Reducing Energy Waste

Prolongs fossil fuel supplies

Reduces oil imports

Very high net energy

Low cost

Reduces pollution and environmental degradation

Buys time to phase in renewable energy

Less need for military protection of Middle East oil resources

Improves local economy by reducing flow of money out to pay for energy

Creates local jobs

Figure 13-22 Solutions: advantages of reducing energy waste. Global improvements in energy efficiency could save the world about $1 trillion per year—an average of $114 million per hour! *Critical thinking: which two of these advantages do you believe are the most important?*

■ In a *coal-burning power plant,* two-thirds of the energy released by burning coal ends up as waste heat in the environment.

Energy experts call for us to replace these devices or greatly improve their energy efficiency over the next few decades.

According to a 2004 study by David Pimentel and other scientists, the U.S. government could within a decade implement energy conservation and efficiency measures that would reduce current energy consumption by one-third and save consumers $438 billion per year.

Science: Saving Energy and Money in Industry

Industries can save energy and money by producing both heat and electricity from an energy source and by using more energy-efficient electric motors and lighting.

Some industries save energy and money by using **cogeneration,** or *combined heat and power (CHP)* systems. In such a system, two useful forms of energy (such as steam and electricity) are produced from the same fuel source. These systems have an efficiency as high as 80% (compared to 30–40% for coal-fired boilers and nuclear power plants) and emit two-thirds less CO_2 per unit of energy produced than conventional coal-fired boilers.

Cogeneration has been widely used in Western Europe for years. Its use in the United States (where it now produces 9% of the country's electricity) and China is growing.

Another way to save energy and money in industry is to *replace energy-wasting electric motors,* which consume one-fourth of the electricity produced in the United States. Most of these motors are inefficient because they run only at full speed with their output throttled to match the task—somewhat like driving a car very fast with your foot on the brake pedal. Each year, a heavily used electric motor consumes 10 times its purchase cost in electricity—equivalent to using $200,000 worth of gasoline each year to fuel a $20,000 car! The costs of replacing such motors with adjustable-speed drive motors would be paid back in about 1 year and save an amount of energy equal to that generated by 150 large (1,000-megawatt) power plants.

A third way to save energy is to *switch from low-efficiency incandescent lighting to higher-efficiency fluorescent lighting.*

Science: Saving Energy in Transportation

The best way to save energy in transportation is to increase the fuel efficiency of motor vehicles.

Good news. Between 1973 and 1985, average fuel efficiency for new vehicles sold in the United States rose 37% because of government-mandated *corporate average fuel economy (CAFE)* standards. *Bad news.* Between 1985 and 2004, the average fuel efficiency of new cars sold in the United States leveled off or declined slightly and in 2004 reached a 23-year low. According to energy expert Amory Lovins, the average fuel efficiency of Ford cars and trucks is now worse than when the company started 100 years ago with the Model A.

Fuel-efficient cars are available but account for less than 1% of all car sales. One reason is that the inflation-adjusted price of gasoline today in the United States is low (Figure 13-23 and Connections, both on p. 308). A second reason is that two-thirds of U.S. consumers prefer SUVs, pickup trucks, minivans, and other large, inefficient vehicles. A third reason is the failure of elected officials to raise CAFE standards since 1985 because of opposition from automakers and oil companies.

Suppose that Congress required the average car in the United States get 17 kilometers per liter (40 miles per gallon) within 10 years. According to energy analysts, this change would cut gasoline consumption in half, save more than three times the amount of oil in

Economics and Politics: The Real Cost of Gasoline in the United States

CONNECTIONS

Many Americans complain about high and rising gasoline prices. But economists and environmentalists point out that gasoline costs U.S. consumers much more than it appears. This is because *most of the real cost of gasoline is not paid directly at the pump.*

According to a 1998 study by the International Center for Technology Assessment, the hidden costs of gasoline to U.S. consumers approximate $1.30–3.70 per liter ($5–14 per gallon), depending on how estimates are constructed. These hidden costs include government subsidies and tax breaks for oil companies and road builders, pollution control and cleanup, military protection of oil supplies in the Middle East, and environmental, health,

and social costs. The latter costs include increased medical bills and insurance premiums, time wasted in traffic jams, noise pollution, increased mortality from air and water pollution, urban sprawl, and harmful effects on wildlife species and habitats.

If these harmful costs were included as taxes in the market price of gasoline, we would have much more energy-efficient and less polluting cars. This would also increase the country's military and economic security by sharply reducing U.S. dependence on imported oil. But gasoline and car companies benefit financially by being able to pass these hidden costs on to consumers, future generations, and the environment.

This political stalemate could be broken if two things happened. *First,* enough informed and commit-

ted voters could band together to educate people and elected officials about the need to impose much higher gasoline taxes as an important part of the country's national and economic security. *Second,* these voters could insist that the government use most of the revenue from higher gasoline taxes to reduce payroll and income taxes and to provide an energy safety net for poor and lower middle-class citizens.

Critical Thinking

Would you support drastically higher gasoline taxes if taxes on your wages, income, and wealth were reduced to the point where the gasoline taxes would not increase your living expenses? Explain. Would you become involved in a political movement to bring about such a tax-shift policy? Explain.

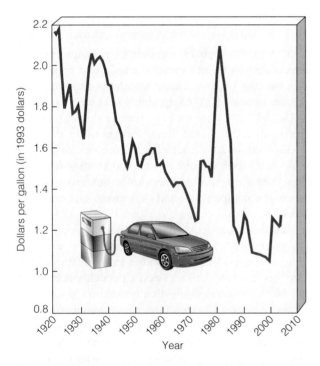

Figure 13-23 Economics: inflation-adjusted price of gasoline (in 1993 dollars) in the United States, 1920–2005. The 225 million motor vehicles in the United States use 40% of the world's gasoline. Gasoline is one of the cheapest items American consumers buy—it costs less per liter than bottled water. (Data from U.S. Department of Energy)

the nation's current proven oil reserves, and save enough oil to eliminate all current oil imports to the United States from the Middle East. In 2003, China announced plans to impose much stricter fuel-efficiency standards than the United States in an effort to reduce the country's dependence on oil imports and reduce carbon dioxide emissions.

X *How Would You Vote?* Should the government greatly increase fuel efficiency standards for all vehicles in the United States or the country where you live? Cast your vote online at http://biology.brookscole.com/miller11.

Science: Hybrid and Fuel-Cell Cars

Fuel-efficient vehicles powered by a hybrid gas-electric engine and electric vehicles powered by fuel cells running on hydrogen are being developed.

There is growing interest in developing *superefficient cars* that could eventually get 34–128 kilometers per liter (80–300 miles per gallon). Amory Lovins developed and promoted this concept in the 1980s. See his Guest Essay on the website for this chapter.

One type of energy-efficient car uses a *hybrid-electric internal combustion engine.* It runs on gasoline, diesel fuel, or natural gas and uses a small battery (recharged by the internal combustion engine) to provide the energy needed for acceleration and hill climbing (Figure 13-24).

Toyota introduced its first hybrid vehicle in 1997. Carmakers plan to introduce as many as 20 hybrid models, including cars, trucks, SUVs, and vans, in the next 4–5 years. Sales of hybrid motor vehicles are projected to grow rapidly and will probably dominate motor vehicle sales between 2010 and 2030.

Another type of efficient car uses a *fuel cell*—a device that combines hydrogen gas (H_2) and oxygen gas (O_2) in the air to produce electrical energy to power the car and water vapor (H_2O), which is emitted into the atmosphere (Figure 13-25, p. 310).

Fuel cells are at least twice as efficient as internal combustion engines, have no moving parts, require little maintenance, and produce little or no pollution,

depending on how their hydrogen fuel is produced. Most major automobile companies have developed prototype fuel-cell cars and plan to market a variety of such vehicles by 2020 (with a few models available by 2010) and greatly increase their use by 2050. Until then, hybrids will probably retain an advantage because they are available now and get their fuel from gasoline filling stations instead of having to depend on building a new network of hydrogen filling stations.

Science: Designing Buildings to Save Energy

We can save energy in buildings by getting heat from the sun, superinsulating them, and using plant-covered eco-roofs.

Atlanta's 13-story Georgia Power Company building uses 60% less energy than conventional office buildings of the same size. The largest surface of the building faces south to capture solar energy. Each floor extends out over the one below it. This blocks out the higher summer sun to reduce air conditioning costs but allows warming by the lower winter sun. Energy-efficient lights focus on desks rather than illuminating entire rooms. In contrast, the conventional Sears Tower building in Chicago consumes more energy in a day than does a city of 150,000 people.

Another energy-efficient design is a *superinsulated house* (Figure 13-26, p. 310). Such houses typically cost 5% more to build than conventional houses of the same size. The extra cost is paid back by energy savings within about 5 years and can save a homeowner $50,000–100,000 over a 40-year period. Superinsulated houses in Sweden use 90% less energy for heating and cooling than the typical American home.

Since the mid-1980s, interest has been growing in *strawbale houses* (Figure 13-27, p. 311). The walls of these superinsulated houses are made by stacking compacted bales of low-cost straw and then covering the bales on the outside and inside with plaster or adobe. The main problem is getting banks and other moneylenders to recognize the potential of this and other unconventional types of housing and to provide homeowners with construction loans. See the Guest Essay about strawbale and solar energy houses by Nancy Wicks on the website for this chapter.

Eco-roofs or *green roofs* covered with plants have been used in Germany, in other parts of Europe, and in Iceland for decades. With proper design, these plant-covered roof gardens provide good insulation, absorb storm water and release it slowly, outlast conventional roofs, and make a building or home more energy efficient. Designing and installing such systems could be an interesting career.

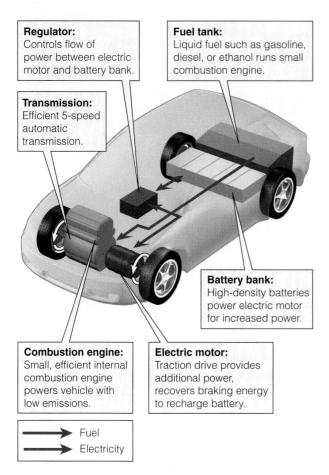

Regulator:
Controls flow of power between electric motor and battery bank.

Fuel tank:
Liquid fuel such as gasoline, diesel, or ethanol runs small combustion engine.

Transmission:
Efficient 5-speed automatic transmission.

Battery bank:
High-density batteries power electric motor for increased power.

Combustion engine:
Small, efficient internal combustion engine powers vehicle with low emissions.

Electric motor:
Traction drive provides additional power, recovers braking energy to recharge battery.

→ Fuel
→ Electricity

Figure 13-24 Solutions: general features of a car powered by a *hybrid gas–electric engine*. The bodies of future models will probably be made of lightweight composite plastics that offer more protection in crashes, do not need to be painted, do not rust, can be recycled, and have fewer parts than conventional car bodies. (Concept information from DaimlerChrysler, Ford, Honda, and Toyota)

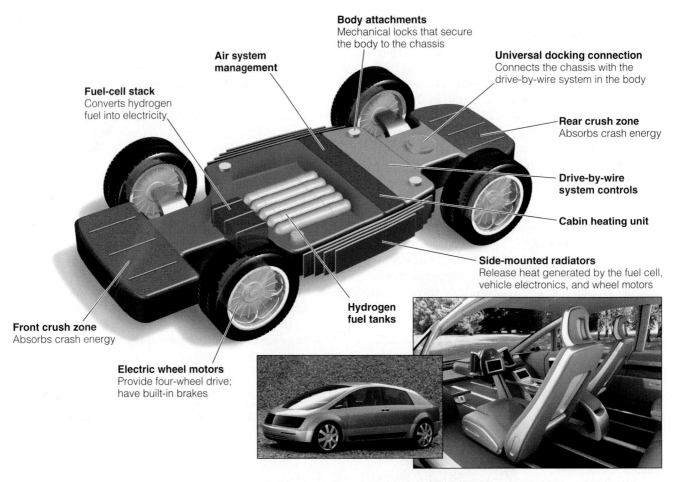

Body attachments
Mechanical locks that secure the body to the chassis

Air system management

Fuel-cell stack
Converts hydrogen fuel into electricity

Universal docking connection
Connects the chassis with the drive-by-wire system in the body

Rear crush zone
Absorbs crash energy

Drive-by-wire system controls

Cabin heating unit

Side-mounted radiators
Release heat generated by the fuel cell, vehicle electronics, and wheel motors

Front crush zone
Absorbs crash energy

Hydrogen fuel tanks

Electric wheel motors
Provide four-wheel drive; have built-in brakes

Figure 13-25 Solutions: prototype Hy-Wire car developed by General Motors. It combines a hydrogen fuel cell with drive-by-wire technology. The Hy-Wire consists of a skateboard-like chassis and a variety of snap-on bodies. General Motors claims the car could be on the road within a decade, but some analysts believe that it will be 2020 before similar cars from various manufacturers will be mass produced. (Basic information from General Motors)

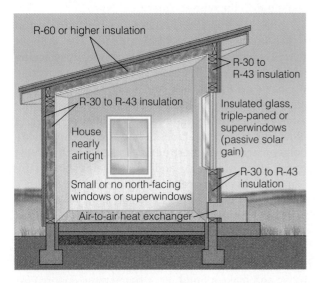

R-60 or higher insulation

R-30 to R-43 insulation

R-30 to R-43 insulation

Insulated glass, triple-paned or superwindows (passive solar gain)

House nearly airtight

R-30 to R-43 insulation

Small or no north-facing windows or superwindows

Air-to-air heat exchanger

Figure 13-26 Solutions: major features of a *superinsulated house*. The house is so heavily insulated and airtight that heat from direct sunlight, appliances, and human bodies can warm it with little or no need for a backup heating system. An air-to-air heat exchanger prevents buildup of indoor air pollution.

Science: Saving Energy in Existing Buildings

We can save energy in existing buildings by insulating them, plugging leaks, and using energy-efficient heating and cooling systems, appliances, and lighting.

Here are some ways to save energy in existing buildings:

- *Insulate and plug leaks.* One-third of the heated air in typical U.S. homes and buildings escapes through closed windows and holes and cracks (Figure 13-28) —roughly equal to the energy in all the oil flowing through the Alaska pipeline every year. During hot weather, these windows and cracks let heat in, increasing the use of air conditioning. Although not very sexy, adding insulation and plugging leaks in a house are two of the quickest, cheapest, and best ways to save energy and money.

- *Use energy-efficient windows.* Replacing all windows in the United States with energy-efficient windows would cut expensive heat losses from houses by two-

Figure 13-27 Solutions: energy-efficient, environmentally healthy, and affordable Victorian-style *strawbale house* designed and built by Alison Gannett in Crested Butte, Colorado. The left photo was taken during construction; the right photo shows the completed house. Depending on the thickness of the bales, plastered strawbale walls have an insulating value of R-35 to R-60, compared to R-12 to R-19 in a conventional house. (The R-value is a measure of resistance to heat flow.) Such houses are also great sound insulators.

thirds, lessen cooling costs in the summer, and reduce CO_2 emissions. Widely available superinsulating windows insulate as well as 8–12 sheets of glass. Although they cost 10–15% more than double-glazed windows, this cost is paid back rapidly by the energy they save. Even better windows will reach the market soon.

■ *Stop other heating and cooling losses.* Leaky heating and cooling ducts in attics and unheated basements allow 20–30% of a home's heating and cooling energy to escape and draw unwanted moisture and heat into the home. Careful sealing can reduce this loss. Some designs for new homes keep the ducts inside the home's thermal envelope so that escaping hot or cool air is fed back into the living space. Also, using white instead of dark roofs can reduce city temperatures and cut electricity for air conditioning.

■ *Heat houses more efficiently.* In order, the most energy-efficient ways to heat space are superinsulation, a geothermal heat pump, passive solar heating, a conventional heat pump (in warm climates only), small

Figure 13-28 Unnecessary energy waste: an infrared photo (thermogram) showing heat loss (red, white, and orange) around the windows, doors, roofs, and foundations of houses and stores in Plymouth, Michigan. Many homes and buildings in the United States and other countries are so full of leaks that their heat loss in cold weather and heat gain in hot weather are equivalent to having a large window-sized hole in the wall of the house. How leaky is the dwelling where you live?

cogenerating microturbines, and a high-efficiency (85–98%) natural gas furnace. The most wasteful and expensive way is to use electric resistance heating with the electricity produced by a coal-fired or nuclear power plant.

■ *Heat water more efficiently.* One approach is to use a *tankless instant water heater* (about the size of a small suitcase) fired by natural gas or LPG but not by electricity. These devices, which are widely used in many parts of Europe, heat water instantly as it flows through a small burner chamber, provide hot water only when it is needed, and use less energy than traditional water heaters.* A well-insulated, conventional natural gas or LPG water heater is fairly efficient. Nevertheless, all conventional natural gas and electric resistance heaters waste energy by keeping a large tank of water hot all day and night and can run out after a long shower or two.

■ *Use energy-efficient appliances and lighting.* If all households in the United States used the most efficient frost-free refrigerator now available, 18 large (1,000-megawatt) power plants could close. Microwave ovens can cut electricity use for cooking by 25–50% (but not if used for defrosting food). Clothes dryers with moisture sensors cut energy use by 15%, and front-loading washers use 50% less energy than top-loading models but cost about the same. Replacing 21 energy-wasting incandescent light bulbs in a house or building with energy-efficient fluorescent bulbs typically saves $1,125. What a great investment payoff.

■ *Set strict energy-efficiency standards for new buildings.* Building codes could require that new houses use 60–80% less energy than conventional houses of the same size, as has been done in Davis, California. Because of tough energy-efficiency standards, the average Swedish home consumes about one-third as much energy as the average American home of the same size.

Economics: Why Are We Still Wasting So Much Energy?

Low-priced fossil fuels and lack of government tax breaks for saving energy promote energy waste.

With an impressive array of benefits (Figure 13-22), why is so little emphasis placed on improving energy efficiency? One reason is a glut of low-cost gasoline (Figure 13-23) and other fossil fuels. As long as energy remains artificially cheap because its market price does not include its harmful costs, people are more

likely to waste it and not make investments in improving energy efficiency.

Another reason is a lack of tax breaks and other economic incentives for consumers and businesses to invest in improving energy efficiency compared to government subsidies and tax breaks for energy alternatives such as fossil fuels and nuclear power..

Would you like to earn about 20% per year on your money, tax free and risk free? Invest it in improving the energy efficiency of your home and in energy-efficient lights and appliances. You will get your investment back in a few years and then make about 20% per year by having lower heating, cooling, and electricity bills. This is a win-win deal for you and the earth.

✗ *HOW WOULD YOU VOTE?* Should the United States or the country where you live greatly increase its emphasis on improving energy efficiency? Cast your vote online at http://biology.brookscole.com/miller11.

13-5 USING RENEWABLE ENERGY TO PROVIDE HEAT AND ELECTRICITY

Science and Economics: Types of Renewable Energy

Six types of renewable energy are solar, flowing water, wind, biomass, hydrogen, and geothermal.

One of the four keys to sustainability (Figure 6-18, p. 126), based on learning from nature, is to *rely mostly on renewable solar energy.* We can get renewable solar energy directly from the sun or indirectly from moving water, wind, and biomass. Two other forms of renewable energy are geothermal energy from the earth's interior and use of renewable energy to produce hydrogen fuel from water. Like fossil fuels and nuclear power, each of these alternatives has advantages and disadvantages.

If renewable energy is so great, why is it that it provides only 18% of the world's energy and 6% of the United States' energy? One reason is that renewable energy resources have received and continue to receive much lower government tax breaks, subsidies, and research and development funding than fossil fuels and nuclear power have received for decades. The other reason is that the prices we pay for fossil fuels and nuclear power do not include the costs of their harm to the environment and to human health.

In other words, the economic dice have been loaded against solar, wind, and other forms of renewable energy. If the economic playing field was made more even, energy analysts say that many of these forms of renewable energy would take over—another example of the *you-get-what-you-reward* economic principle in action.

*They work great. I used them in a passive solar office and living space for 15 years. For information on available models visit **www.foreverhotwater.com**.

Science: Passive and Active Solar Energy Heating

We can heat buildings by orienting them toward the sun (passive solar heating) or by pumping a liquid such as water through rooftop collectors (active solar heating).

Buildings and water can be heated by solar energy using two methods: passive and active (Figure 13-29). A **passive solar heating system** absorbs and stores heat from the sun directly within a structure (Figure 13-29, left, and Figure 13-30, p. 314). Energy-efficient windows and attached greenhouses face the sun to collect solar energy by direct gain. Walls and floors of concrete, adobe, brick, stone, salt-treated timber, and water in metal or plastic containers store much of the collected solar energy as heat and release it slowly throughout the day and night. A small backup heating system such as a vented natural gas or propane heater may be used but is not necessary in many climates. (See the Guest Essay by Nancy Wicks on this topic on the website for this chapter.)

On a life-cycle cost basis, good passive solar and superinsulated design is the cheapest way to heat a home or small building in regions with access to ample sunlight. Such a system usually adds 5–10% to the construction cost, but the life-cycle cost of operating such a house is 30–40% lower. The typical payback time for passive solar features is 3–7 years.

An **active solar heating system** absorbs energy from the sun by pumping a heat-absorbing fluid (such as water or an antifreeze solution) through special collectors usually mounted on a roof or on special racks to face the sun (Figure 13-29, right). Some of the collected heat can be used directly. The rest can be stored in a large insulated container filled with gravel, water, clay, or a heat-absorbing chemical for release as needed. Active solar collectors can also supply hot water and are widely used in areas of the world with sunny climates.

Figure 13-31 (p. 314) lists the major advantages and disadvantages of using passive or active solar heating systems for heating buildings. Passive solar energy is great for new homes in areas with adequate sunlight. But it cannot be used to heat existing homes and buildings not oriented to receive sunlight or where other buildings or trees block access to sunlight. Active solar collectors are good for heating water in sunny areas. Most analysts do not expect widespread use of active solar collectors for heating houses because of their high costs, maintenance requirements, and unappealing appearance.

Science: Cooling Houses Naturally

We can cool houses by superinsulating them, taking advantage of breezes, shading them, having light-colored or garden roofs, and using geothermal cooling.

Here are some ways to have a cooler house. Use superinsulation and superinsulating windows, open windows to take advantage of breezes, and use fans to keep air moving. Block the high summer sun with deciduous trees and window overhangs (Figure 13-30, top), or with awnings.

Use a light-colored roof to reflect as much as 80% of the sun's heat, compared to only 8% for a dark-gray roof. Alternatively, use insulating garden-top roofs (Figure 13-30, bottom). Suspend reflective insulating foil in an attic to block heat from radiating down into the house.

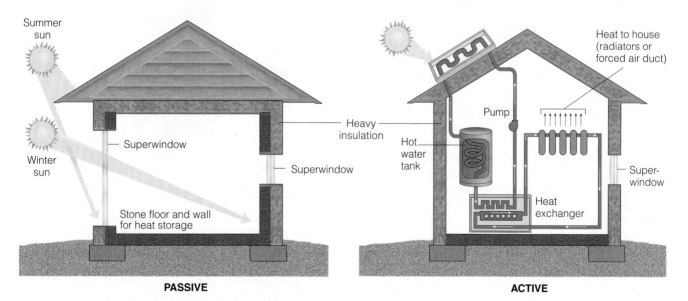

Figure 13-29 Solutions: passive and active solar heating for a home.

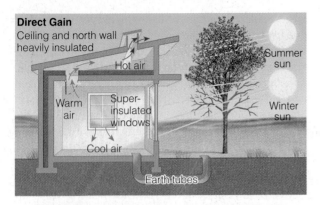

Direct Gain
Ceiling and north wall heavily insulated

Summer sun

Winter sun

Hot air

Warm air

Super-insulated windows

Cool air

Earth tubes

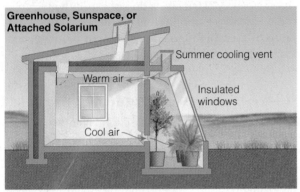

Greenhouse, Sunspace, or Attached Solarium

Summer cooling vent

Warm air

Insulated windows

Cool air

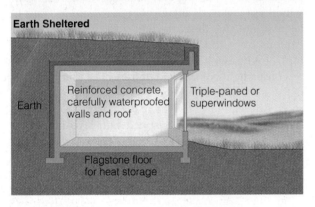

Earth Sheltered

Earth

Reinforced concrete, carefully waterproofed walls and roof

Triple-paned or superwindows

Flagstone floor for heat storage

Figure 13-30 Solutions: three examples of *passive solar design* for houses.

Another option is to place plastic *earth tubes* underground where the earth is cool year-round. In this geothermal cooling system, a tiny fan can pipe cool and partially dehumidified air into an energy-efficient house (Figure 13-30, top).* In warm climates, you can also use high-efficiency heat pumps for air conditioning. Toronto, Canada's largest city, cools downtown buildings by pumping cold water from the depths of Lake Ontario and passing it through building air conditioning systems.

---

*They work. I used them in a passively heated and cooled office and home for 15 years. People allergic to pollen and molds should add an air purification system, but this is also necessary with a conventional cooling system.

Science: Using Solar Energy to Generate High-Temperature Heat and Electricity

Large arrays of solar collectors in sunny deserts can produce high-temperature heat to spin turbines and produce electricity, but costs are high.

Several *solar thermal systems* can collect and transform radiant energy from the sun into high-temperature thermal energy (heat), which can then be used directly or converted to electricity. These approaches are used mostly in desert areas with ample sunlight. Figure 13-32 lists the advantages and disadvantages of concentrating solar energy to produce high-temperature heat or electricity.

One method uses a *central receiver system*, called a *power tower*. Huge arrays of computer-controlled mirrors called *heliostats* track the sun and focus sunlight on a central heat collection tower (top drawing in Figure 13-32).

Another approach is a *solar thermal plant*, in which sunlight is collected and focused on oil-filled pipes running through the middle of a large area of curved solar collectors (bottom drawing in Figure 13-32). This concentrated sunlight can generate temperatures high enough to produce steam for running turbines and

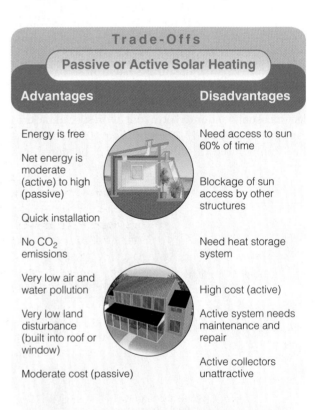

Trade-Offs

Passive or Active Solar Heating

Advantages	Disadvantages
Energy is free	Need access to sun 60% of time
Net energy is moderate (active) to high (passive)	Blockage of sun access by other structures
Quick installation	
No CO$_2$ emissions	Need heat storage system
Very low air and water pollution	High cost (active)
Very low land disturbance (built into roof or window)	Active system needs maintenance and repair
Moderate cost (passive)	Active collectors unattractive

Figure 13-31 Trade-offs: advantages and disadvantages of heating a house with passive or active solar energy. *Critical thinking: pick the single advantage and the single disadvantage that you think are the most important.*

Trade-Offs

Solar Energy for High-Temperature Heat and Electricity

Advantages	Disadvantages
Moderate net energy	Low efficiency
	High costs
Moderate environmental impact	Needs backup or storage system
No CO_2 emissions	Need access to sun most of the time
Fast construction (1–2 years)	High land use
Costs reduced with natural gas turbine backup	May disturb desert areas

Figure 13-32 Trade-offs: advantages and disadvantages of using solar energy to generate high-temperature heat and electricity. *Critical thinking: pick the single advantage and the single disadvantage that you think are the most important.*

generating electricity. At night or on cloudy days, high-efficiency, combined-cycle, natural gas turbines can supply backup electricity as needed.

Most analysts do not expect widespread use of such technologies over the next few decades because of their high costs, the limited number of suitable sites, and availability of much cheaper ways to produce electricity such as combined-cycle natural gas turbines and wind turbines.

On an individual scale, inexpensive *solar cookers* can focus and concentrate sunlight to cook food, especially in rural villages in sunny developing countries. They can be made by fitting an insulated box big enough to hold three or four pots with a transparent, removable top. Solar cookers reduce deforestation for fuelwood as well as the time and labor needed to collect firewood. They also reduce indoor air pollution from smoky fires.

Science: Producing Electricity with Solar Cells

Solar cells that convert sunlight to electricity can be incorporated into roofing materials or windows, and the currently high costs of doing so are expected to fall.

Solar energy can be converted directly into electrical energy by **photovoltaic (PV) cells,** commonly called **solar cells** (Figure 13-33). A typical solar cell is a transparent wafer that contains a semiconductor with a thickness ranging from less than that of a human hair to a sheet of paper. Sunlight energizes and causes electrons in the semiconductor to flow, creating an electrical current. These devices have no moving parts, require little maintenance, produce no pollution during operation, and last as long as a conventional fossil fuel or nuclear power plant. The semiconductor material

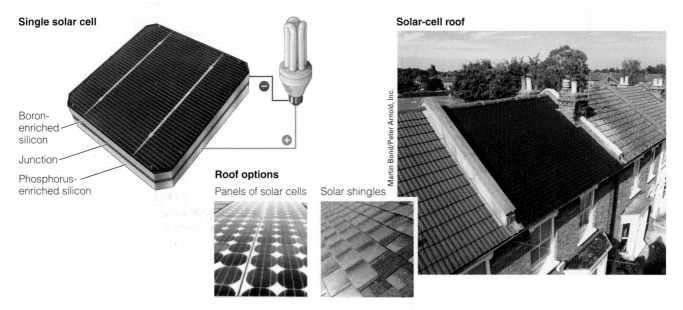

Single solar cell

Boron-enriched silicon

Junction

Phosphorus-enriched silicon

Roof options

Panels of solar cells

Solar shingles

Solar-cell roof

Martin Bond/Peter Arnold, Inc.

Figure 13-33 Solutions: photovoltaic (PV) cells can provide electricity for a house or building using solar-cell roof shingles, as shown in the house on the right in Richmond, Surrey, England. PV panel roof systems that look like a metal roof are also available. In addition, new thin-film solar cells can be applied to windows and outside glass walls.

Figure 13-34 Global solutions: solar cells used to provide electricity for a remote village in Niger, Africa.

Peter Arnold, Inc.

used in solar cells can be made into paper-thin rigid or flexible sheets that can incorporated into traditional-looking roofing materials (Figure 13-33, right). Glass walls and windows of buildings can also have built-in solar cells.

Easily expandable banks of solar cells could be used in developing countries to provide electricity for the 1.7 billion people in rural villages who now lack it (Figure 13-34). With financing from the World Bank, India (the world's largest market for solar cells) is installing solar-cell systems in 38,000 villages. Zimbabwe is bringing solar electricity to 2,500 villages, mostly because they are located long distances from power grids.

Figure 13-35 lists the advantages and disadvantages of solar cells. Currently, costs of producing electricity from solar cells are high. In the future, they will likely drop thanks to mass production and new designs.

Currently, solar cells supply only 0.05% of the world's electricity. With increased government and private research and development, plus greater government tax breaks and other subsidies, they could provide one-fourth of the world's electricity by 2040. If these projections prove correct, the production, sale, and installation of solar cells could become one of the world's largest and fastest-growing businesses.

How Would You Vote? Should we greatly increase our dependence on solar cells for producing electricity? Cast your vote online at http://biology.brookscole.com/miller11.

Trade-Offs

Solar Cells

Advantages	Disadvantages
Fairly high net energy	Need access to sun
Work on cloudy days	Low efficiency
Quick installation	
Easily expanded or moved	Need electricity storage system or backup
No CO$_2$ emissions	
Low environmental impact	High land use (solar-cell power plants) could disrupt desert areas
Last 20–40 years	
Low land use (if on roof or built into walls or windows)	High costs (but should be competitive in 5–15 years)
Reduces dependence on fossil fuels	DC current must be converted to AC

Figure 13-35 Trade-offs: advantages and disadvantages of using solar cells to produce electricity. *Critical thinking: pick the single advantage and the single disadvantage that you think are the most important.*

Science: Producing Electricity from Flowing Water—Dams, Tides, and Waves

Water flowing in rivers and streams can be trapped in reservoirs behind dams and released as needed to spin turbines and produce electricity.

Solar energy evaporates water and deposits it as water and snow in other areas through the water cycle. Water flowing from higher to lower elevations in rivers and streams can be controlled by dams and reservoirs and used to produce electricity (Figure 11-8, p. 243). This indirect form of renewable solar energy is called *hydropower.*

The most popular approach is to build a high dam across a large river to create a reservoir. Some of the water stored in the reservoir is allowed to flow through huge pipes at controlled rates, spinning turbines and producing electricity. Smaller and lower dams can also be used.

In 2003, hydropower supplied about one-fifth of the world's electricity, including 99% of that used in Norway, 75% in New Zealand, 25% in China, and 7% in the United States (but about 50% on the West Coast).

Figure 13-36 lists the advantages and disadvantages of using large-scale hydropower plants to produce electricity.

Because of increasing concern about the harmful environmental and social consequences of large dams, the World Bank and other development agencies have been pressured to stop funding new large-scale hydropower projects. Also, according to a 2000 study by the World Commission on Dams, hydropower in tropical countries is a major emitter of greenhouse gases. The reservoirs that power the dams can trap rotting vegetation, which can emit greenhouse gases such as carbon dioxide and methane.

X *HOW WOULD YOU VOTE?* Should we greatly increase our dependence on renewable large-scale dams for producing electricity? Cast your vote online at **http://biology.brookscole .com/miller11**.

Small-scale hydropower projects eliminate most of the harmful environmental effects of large-scale projects. But their electrical output can vary with seasonal changes in stream flow.

We can also produce electricity from water flows by tapping into the energy from *tides* and *waves*. Most analysts expect these sources to make little contribution to world electricity production because of high costs and lack of enough areas with the right conditions.

Science: Producing Electricity from Wind

Since 1995, the use of wind turbines to produce electricity has increased almost sevenfold.

Trade-Offs

Large-Scale Hydropower

Advantages	Disadvantages
Moderate to high net energy	High construction costs
High efficiency (80%)	High environmental impact from flooding land to form a reservoir
Large untapped potential	
Low-cost electricity	High CO_2 emissions from biomass decay in shallow tropical reservoirs
Long life span	
No CO_2 emissions during operation in temperate areas	Floods natural areas behind dam
	Converts land habitat to lake habitat
May provide flood control below dam	Danger of collapse
Provides water for year-round irrigation of cropland	Uproots people
	Decreases fish harvest below dam
Reservoir is useful for fishing and recreation	Decreases flow of natural fertilizer (silt) to land below dam

Figure 13-36 Trade-offs: advantages and disadvantages of using large dams and reservoirs to produce electricity. *Critical thinking: pick the single advantage and the single disadvantage that you think are the most important.*

The greater heating of the earth at the equator than at the poles and the earth's rotation set up flows of air called *wind*. This indirect form of solar energy can be captured by wind turbines and converted into electricity (Figure 13-37, p. 318).

Since 1990, wind power has been the world's fastest-growing source of energy, with its use increasing almost sevenfold between 1995 and 2004. Europe is leading the way into the *age of wind energy*. About three-fourths of the world's wind-generated power is produced in Europe in inland and offshore wind farms or parks. European companies manufacture 80% of the wind turbines sold in the global marketplace. Wind power also is being developed rapidly in India and to a lesser extent in China.

Much of the world's potential for wind power remains untapped. According to a 2003 Wind Force 12 report, wind parks on only one-tenth of the earth's land could produce twice the world's projected demand for electricity by 2020. This estimate does not include the establishment of wind parks at sea.

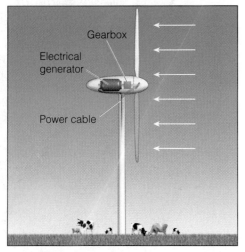

Wind turbine

Wind farm

Figure 13-37 Solutions: Wind turbines can be used to produce electricity individually or in clusters, called *wind farms* or *wind parks* that can be located on land or offshore. Since 1990, wind power has been the world's fastest-growing source of energy.

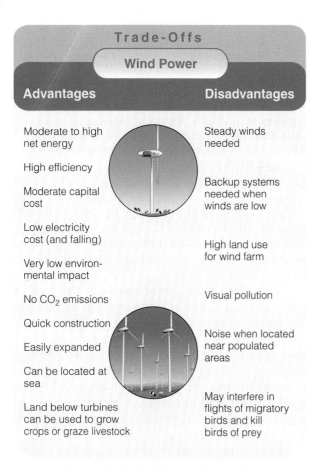

Trade-Offs	
Wind Power	
Advantages	**Disadvantages**
Moderate to high net energy	Steady winds needed
High efficiency	
Moderate capital cost	Backup systems needed when winds are low
Low electricity cost (and falling)	
Very low environmental impact	High land use for wind farm
No CO$_2$ emissions	Visual pollution
Quick construction	
Easily expanded	Noise when located near populated areas
Can be located at sea	
Land below turbines can be used to grow crops or graze livestock	May interfere in flights of migratory birds and kill birds of prey

Figure 13-38 Trade-offs: advantages and disadvantages of using wind to produce electricity. By 2020, wind power could supply more than 10% of the world's electricity and 10–25% of the electricity used in the United States. *Critical thinking: pick the single advantage and the single disadvantage that you think are the most important.*

The DOE calls the Great Plains states of North Dakota, South Dakota, Nebraska, Kansas, Oklahoma, and Texas the "Saudi Arabia of wind." In theory, these states have enough wind resources to more than meet all of the nation's electricity needs. According to the American Wind Energy Association, with increased and consistent government subsidies and tax breaks, wind power in the United States could produce almost one-fourth of the country's electricity by 2025. This electricity could be passed through water to produce hydrogen gas, which could in turn run fuel cells.

Some U.S. farmers and ranchers already make more money by using their land for wind power production than by growing crops or raising cattle. In the 1980s, the United States was the world leader in wind technology. It later lost that lead to Europe and Japan, mostly because of insufficient and irregular government subsidies and tax breaks for wind power compared to those regularly received for decades by fossil fuels and nuclear power.

Figure 13-38 lists the advantages and disadvantages of using wind to produce electricity. According to energy analysts, wind power has more benefits and fewer serious drawbacks than any other energy resource. Many governments and corporations now recognize that wind is a vast, climate-benign, renewable energy resource that can supply both electricity and hydrogen fuel for running fuel cells at an affordable cost. They understand that *there is money in wind* and that our energy future may be *blowing in the wind*.

🗙 *HOW WOULD YOU VOTE?* Should we greatly increase our dependence on wind power? Cast your vote online at http://biology.brookscole.com/miller11.

Science: Burning Solid Biomass

Plant materials and animal wastes can be burned to provide heat or electricity or converted into gaseous or liquid biofuels.

Biomass consists of plant materials (such as wood and agricultural waste) and animal wastes that can be burned directly as a solid fuel or converted into gaseous or liquid **biofuels** (Figure 13-39). Biomass is an indirect form of solar energy because it consists of combustible organic compounds produced by photosynthesis.

Most biomass is burned *directly* for heating, cooking, and industrial processes or *indirectly* to drive turbines and produce electricity. Burning wood and animal manure for heating and cooking supplies 11% of the world's energy and 30% of the energy used in developing countries.

Almost 70% of the people living in developing countries heat their homes and cook their food by burn-

Figure 13-40 Natural biomass capital: making fuel briquettes from cow dung in India. The scarcity of fuelwood causes people to collect and burn such dung. However, this practice deprives the soil of an important source of plant nutrients from dung decomposition.

ing wood or charcoal (derived from wood). However, 2.7 billion people in these countries cannot find—or are too poor to buy—enough fuelwood to meet their needs.

One way to produce biomass fuel is to plant, harvest, and burn large numbers of fast-growing trees (such as cottonwoods, poplars, and sycamores), shrubs, perennial grasses (such as switchgrass), and water hyacinths in *biomass plantations*.

In agricultural areas, *crop residues* (such as sugarcane residues, rice husks, cotton stalks, and coconut shells) and *animal manure* can be collected and burned or converted into biofuels (Figure 13-40). Some ecologists argue that it makes more sense to use animal manure as a fertilizer and crop residues to feed livestock, retard soil erosion, and fertilize the soil.

Figure 13-41 (p. 320) lists the general advantages and disadvantages of burning solid biomass as a fuel. Burning biomass produces CO_2. However, if the rate of use of biomass does not exceed the rate at which it is replenished by new plant growth (which takes up CO_2), no net increase in CO_2 emissions occurs. Nevertheless, repeated cycles of growing and harvesting biomass plantations can deplete the soil of key nutrients.

According to a 2004 study by the World Wide Fund for Nature and the European Biomass Industry Association, burning biomass currently provides about 1% of the power needs of developed countries but could provide 15% by 2020.

X *HOW WOULD YOU VOTE?* Should we greatly increase our dependence on burning solid biomass to provide heat and produce electricity? Cast your vote online at http://biology.brookscole.com/miller11.

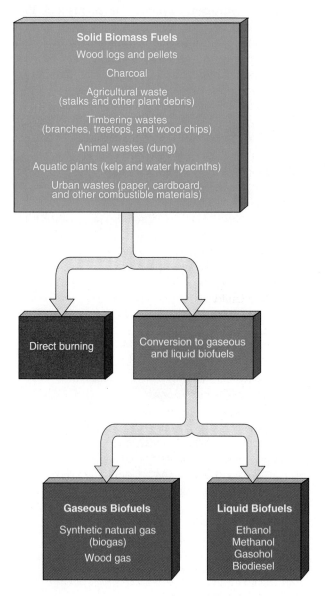

Figure 13-39 Natural capital: principal types of biomass fuel.

Solid Biomass Fuels

Wood logs and pellets

Charcoal

Agricultural waste
(stalks and other plant debris)

Timbering wastes
(branches, treetops, and wood chips)

Animal wastes (dung)

Aquatic plants (kelp and water hyacinths)

Urban wastes (paper, cardboard, and other combustible materials)

Direct burning

Conversion to gaseous and liquid biofuels

Gaseous Biofuels

Synthetic natural gas (biogas)

Wood gas

Liquid Biofuels

Ethanol
Methanol
Gasohol
Biodiesel

Trade-Offs

Solid Biomass

Advantages	Disadvantages
Large potential supply in some areas	Nonrenewable if harvested unsustainably
Moderate costs	Moderate to high environmental impact
No net CO_2 increase if harvested and burned sustainably	CO_2 emissions if harvested and burned unsustainably
Plantation can be located on semiarid land not needed for crops	Low photosynthetic efficiency
	Soil erosion, water pollution, and loss of wildlife habitat
Plantation can help restore degraded lands	Plantations could compete with cropland
Can make use of agricultural, timber, and urban wastes	Often burned in inefficient and polluting open fires and stoves

Figure 13-41 Trade-offs: general advantages and disadvantages of burning solid biomass as a fuel. *Critical thinking: pick the single advantage and single disadvantage that you think are the most important.*

Producing Gaseous and Liquid Fuels from Solid Biomass

Some forms of solid biomass can be converted into gaseous and liquid biofuels.

Bacteria and various chemical processes can convert some forms of biomass into gaseous and liquid biofuels. Examples include *biogas* (a mixture of 60% methane and 40% CO_2), *liquid ethanol,* and *liquid methanol.*

In rural China, anaerobic bacteria in more than 500,000 *biogas digesters* convert plant and animal wastes into methane gas that is used for heating and cooking. After the biogas has been removed, the almost odorless solid residue serves as fertilizer on food crops or on trees. When they work, biogas digesters are very efficient. Unfortunately, they are slow and unpredictable, a problem that could be corrected by developing more reliable models. They also add CO_2 to the atmosphere.

Some analysts believe that liquid ethanol and methanol produced from biomass could replace gasoline and diesel fuel when oil becomes too scarce and expensive. *Ethanol* can be made from sugar and grain crops (sugarcane, sugar beets, sorghum, sunflowers, and corn) by fermentation and distillation. Gasoline mixed with 10–23% pure ethanol makes *gasohol,* which can be burned in conventional gasoline engines. Figure 13-42 lists the advantages and disadvantages of using ethanol as a vehicle fuel compared to gasoline.

Methanol is typically made from natural gas but can be produced at a higher cost from carbon dioxide, coal, and biomass such as wood, wood wastes, agricultural wastes, sewage sludge, and garbage. Figure 13-43 lists the advantages and disadvantages of using methanol as a vehicle fuel compared to gasoline.

Chemist George A. Olah believes that establishing a *methanol economy* is preferable to pursuing the highly publicized hydrogen economy. He points out that methanol can be produced chemically from carbon dioxide in the atmosphere, which could slow projected global warming. In addition, methanol can be converted to other hydrocarbon compounds that can be

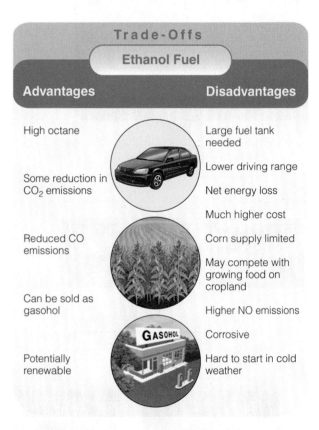

Trade-Offs

Ethanol Fuel

Advantages	Disadvantages
High octane	Large fuel tank needed
	Lower driving range
Some reduction in CO_2 emissions	Net energy loss
	Much higher cost
Reduced CO emissions	Corn supply limited
	May compete with growing food on cropland
Can be sold as gasohol	Higher NO emissions
	Corrosive
Potentially renewable	Hard to start in cold weather

Figure 13-42 Trade-offs: general advantages and disadvantages of using ethanol as a vehicle fuel compared to gasoline. *Critical thinking: pick the single advantage and single disadvantage that you think are the most important.*

Trade-Offs

Methanol Fuel

Advantages	Disadvantages
High octane	Large fuel tank needed
Some reduction in CO_2 emissions	Half the driving range
Lower total air pollution (30–40%)	Corrodes metal, rubber, plastic
Can be made from natural gas, agricultural wastes, sewage sludge, garbage, and CO_2	High CO_2 emissions if made from coal
	Expensive to produce
Can be used to produce H_2 for fuel cells	Hard to start in cold weather

Figure 13-43 Trade-offs: general advantages and disadvantages of using methanol as a vehicle fuel compared to gasoline. *Critical thinking: pick the single advantage and the single disadvantage that you think are the most important.*

used to produce useful chemicals like the petrochemicals made from petroleum and natural gas.

✗ *How Would You Vote?* Should we greatly increase our dependence on ethanol and methanol as vehicle fuels? Cast your vote online at http://biology.brookscole.com/miller11.

13-6 GEOTHERMAL ENERGY

Science: Tapping the Earth's Internal Heat

We can use geothermal energy stored in the earth's mantle to heat and cool buildings and to produce electricity.

Geothermal energy consists of heat stored in soil, underground rocks, and fluids in the earth's mantle. We can tap into this stored energy to heat and cool buildings and to produce electricity.

Geothermal heat pumps can exploit the difference between underground and surface temperatures in most places and use a system of pipes and ducts to heat or cool a building. These devices extract heat from the earth in winter. In summer, they can store heat removed from a house in the earth. They are a very efficient and cost-effective way to heat or cool a space.

A related way to heat or cool a building is *geothermal exchange* or *geoexchange*. Buried pipes filled with a fluid move heat in or out of the ground or from nearby bodies of water, depending on the season and the heating or cooling requirements. In winter, for example, heat is removed from the fluid in the buried pipes and blown through house ducts. In summer, this process is reversed. According to the EPA, geothermal exchange is the most energy-efficient, cost-effective, and environmentally clean way to heat or cool a building.

We have also learned to tap into deeper, more concentrated underground reservoirs of geothermal energy. One type of reservoir contains *dry steam* with water vapor but no water droplets. Another consists of *wet steam*, a mixture of steam and water droplets. A third is *hot water* trapped in fractured or porous rock at various places in the earth's crust.

If such geothermal sites are close to the surface, wells can be drilled to extract the dry steam, wet steam, or hot water (Figure 13-2). It can then be used to heat homes and buildings or to spin turbines and produce electricity.

Three other nearly nondepletable sources of geothermal energy exist. One is *molten rock* (magma). Another is *hot dry-rock zones*, where molten rock that has penetrated the earth's crust heats subsurface rock to high temperatures. A third source is low- to moderate-temperature *warm-rock reservoir deposits*. Heat from such deposits could be used to preheat water and run heat pumps for space heating and air conditioning. Hot dry-rock zones can be found almost anywhere 8–10 kilometers (5–6 miles) below the earth's surface. Research is being carried out in several countries to see whether these zones can provide affordable geothermal energy.

Currently, about 22 countries (most of them in the developing world) extract enough energy from geothermal sites to produce about 1% of the world's electricity. Geothermal energy is used to heat 85% of Iceland's buildings, produce electricity, and grow most of that country's fruits and vegetables in greenhouses heated by geothermal energy.

Geothermal electricity meets the electricity needs of 6 million Americans and supplies 6% of California's electricity. The world's largest operating geothermal system, called *The Geysers*, extracts energy from a dry-steam reservoir north of San Francisco, California. Currently, heat is being withdrawn from this geothermal site about 80 times faster than it is being replenished, converting this renewable resource to a nonrenewable source of energy. In 1999, Santa Monica, California, became the first city to get all its electricity from geothermal energy. Figure 13-44 (p. 322) lists the advantages and disadvantages of using geothermal energy.

Trade-Offs

Geothermal Energy

Advantages	Disadvantages

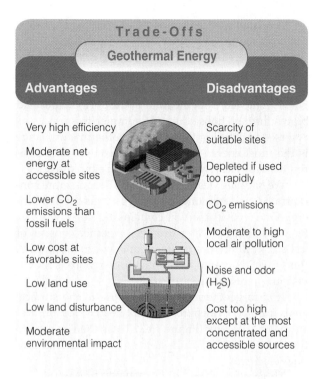

Advantages	Disadvantages
Very high efficiency	Scarcity of suitable sites
Moderate net energy at accessible sites	Depleted if used too rapidly
Lower CO_2 emissions than fossil fuels	CO_2 emissions
Low cost at favorable sites	Moderate to high local air pollution
Low land use	Noise and odor (H_2S)
Low land disturbance	Cost too high except at the most concentrated and accessible sources
Moderate environmental impact	

Figure 13-44 Trade-offs: advantages and disadvantages of using geothermal energy for space heating and to produce electricity or high-temperature heat for industrial processes. *Critical thinking: pick the single advantage and the single disadvantage that you think are the most important.*

✗ *HOW WOULD YOU VOTE?* Should we greatly increase our dependence on geothermal energy to provide heat and to produce electricity? Cast your vote online at http://biology .brookscole.com/miller11.

13-7 HYDROGEN

Science: Can Hydrogen Replace Oil?

Some energy analysts view hydrogen gas as the best fuel to replace oil during the last half of this century.

When oil is gone or when the remaining oil costs too much to use, how will we fuel our vehicles, industry, and buildings? Many scientists and executives of major oil companies and automobile companies say the fuel of the future is hydrogen gas (H_2). Figure 13-45 lists the advantages and disadvantages of using hydrogen as an energy resource.

When hydrogen gas burns in air or in fuel cells, it combines with oxygen gas in the air to produce non-polluting water vapor. Widespread use of hydrogen as a fuel would eliminate most of our current air pollution problems and greatly reduce the threats from

global warming because it does not emit CO_2—as long as the hydrogen is not produced from fossil fuels or other carbon-containing compounds.

So what is the catch? Three problems arise in turning the vision of widespread use of hydrogen as a fuel into reality. *First,* hydrogen is chemically locked up in water and in organic compounds such as methane and gasoline. *Second,* it takes energy and money to produce hydrogen from water and organic compounds. In other words, *hydrogen is not a source of energy—it is a fuel produced by using energy. Third,* fuel cells are the best way to use hydrogen to produce electricity, but current versions are expensive.

Trade-Offs

Hydrogen

Advantages	Disadvantages

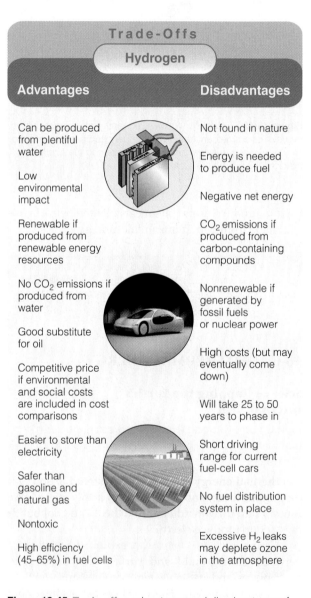

Advantages	Disadvantages
Can be produced from plentiful water	Not found in nature
Low environmental impact	Energy is needed to produce fuel
	Negative net energy
Renewable if produced from renewable energy resources	CO_2 emissions if produced from carbon-containing compounds
No CO_2 emissions if produced from water	Nonrenewable if generated by fossil fuels or nuclear power
Good substitute for oil	
Competitive price if environmental and social costs are included in cost comparisons	High costs (but may eventually come down)
	Will take 25 to 50 years to phase in
Easier to store than electricity	Short driving range for current fuel-cell cars
Safer than gasoline and natural gas	No fuel distribution system in place
Nontoxic	
High efficiency (45–65%) in fuel cells	Excessive H_2 leaks may deplete ozone in the atmosphere

Figure 13-45 Trade-offs: advantages and disadvantages of using hydrogen as a fuel for vehicles and for providing heat and electricity. *Critical thinking: pick the single advantage and the single disadvantage that you think are the most important.*

We could use electricity from coal-burning and conventional nuclear power plants to decompose water into hydrogen and oxygen gas. This approach is expensive, however, and does not avoid the harmful environmental effects associated with using these fuels (Figures 13-15 and 13-19). We can also make hydrogen from coal and strip it from organic compounds found in fuels such as natural gas (methane), methanol, and gasoline. However, according to a 2002 study by physicist Marin Hoffer and a team of other scientists, producing hydrogen from organic compounds will add more CO_2 to the atmosphere per unit of heat generated than does burning these carbon-containing fuels directly. Using this approach could accelerate projected global warming unless we can develop affordable ways to store (sequester) the CO_2 underground or in the deep ocean.

Most proponents of hydrogen believe that if we are to receive its very low pollution and low CO_2 emission benefits, the energy used to produce H_2 by decomposing water must come from low-polluting, renewable sources that also emit little or no CO_2. The most likely sources are electricity generated by wind farms, geothermal energy, solar cells, or biological processes in bacteria and algae (Science Spotlight, right). Other analysts insist that the best and most efficient ways to reduce the greenhouse gas emissions from burning coal to produce electricity are to rely more on renewable wind, hydropower, geothermal energy, and solar cells to produce electricity instead of using this energy to produce hydrogen that is then burned to produce electricity.

In 1999, DaimlerChrysler, Royal Dutch Shell, Norsk Hydro, and Icelandic New Energy announced government-approved plans to turn the tiny country of Iceland, with nearly 300,000 residents, into the world's first "hydrogen economy" and make the country self-sufficient in energy by 2050. Hydrogen would be produced mostly by using Iceland's ample supplies of hydropower and geothermal energy to produce hydrogen from water. The world's first hydrogen fueling station, built in the capital city of Reykjavik, is supplying hydrogen to power several fuel cell buses. The next step is to test a fleet of hydrogen-powered cars for use by corporate or government employees.

Some analysts urge the United States to institute an *Apollo*-type program to spur the rapid development of a renewable-energy–hydrogen revolution that would be phased in during the last half of this century.

Once produced, hydrogen can be stored in a pressurized tank, as liquid hydrogen, and in solid metal hydride compounds, which when heated release hydrogen gas. Scientists are also evaluating ways to store H_2 by absorbing it onto the surfaces of activated charcoal or graphite nanofibers, which when heated re-

Producing Hydrogen from Green Algae Found in Pond Scum

SCIENCE SPOTLIGHT

In a few decades we may be able to use large-scale cultures of green algae to produce hydrogen gas. This simple plant grows all over the world and is commonly found in pond scum.

When living in air and sunlight, green algae carry out photosynthesis like other plants and produce carbohydrates and oxygen gas. In 2000, Tasios Melis, a researcher at the University of California, Berkeley, found a way to make these algae produce bubbles of hydrogen rather than oxygen.

First, Melis grew cultures of hundreds of billions of the algae in the normal way with plenty of sunlight, nutrients, and water. Then he cut off their supply of two key nutrients: sulfur and oxygen. Within 20 hours, the plant cells underwent a metabolic change and switched from an oxygen-producing to a hydrogen-producing metabolism, allowing the researcher to collect hydrogen gas bubbling from the culture.

Melis believes he can increase the efficiency of this hydrogen-producing process tenfold. If so, eventually a *biological hydrogen factory* might cycle a mixture of algae and water through a system of clear tubes exposed to sunlight to produce hydrogen. The gene responsible for producing the hydrogen might even be transferred to other plants to allow them to produce hydrogen.

Critical Thinking

What might be some ecological problems related to the widespread use of this method for producing hydrogen?

lease hydrogen gas. Another possibility is to store it inside tiny glass microspheres. Much more research is needed to convert these dreams into reality.

Some good news. Metal hydrides, charcoal powders, graphite nanofibers, and glass microspheres containing hydrogen will not explode or burn if a vehicle's fuel tank or system is ruptured in an accident. This makes hydrogen a much safer fuel than the highly volatile gasoline, diesel fuel, methanol, and natural gas.

X **HOW WOULD YOU VOTE?** Do the advantages of producing and burning hydrogen as an energy resource outweigh the disadvantages? Cast your vote online at http://biology .brookscole.com/miller11.

13-8 A SUSTAINABLE ENERGY STRATEGY

Global Outlook: What Are the Best Energy Alternatives?

A more sustainable energy policy would improve energy efficiency, rely more on renewable energy, and reduce the harmful effects of using fossil fuels and nuclear energy

In 1999, the International Energy Agency noted that "the world is in the early stages of an inevitable transition to a sustainable energy system that will be largely dependent on renewable resources." Scientists and energy experts who have evaluated energy alternatives have come to three general conclusions.

First, *there will be a gradual shift from large, centralized macropower systems to smaller, decentralized micropower systems* such as small natural gas turbines for commercial buildings, wind turbines, fuel cells, and household solar panels and solar roofs (Figure 13-46). This shift from centralized *macropower* to dispersed *micropower* is analogous to the computer industry's shift from large, centralized mainframes to increasingly smaller, widely dispersed PCs, laptops, and handheld computers.

Second, *the best alternatives combine improved energy efficiency and the use of natural gas as a fuel to make the transition to small-scale, decentralized, locally available, renewable energy resources and possibly nuclear fusion (if it proves feasible)*. Figure 13-47 lists strategies for making the transition to a more sustainable energy future over the next 50 years.

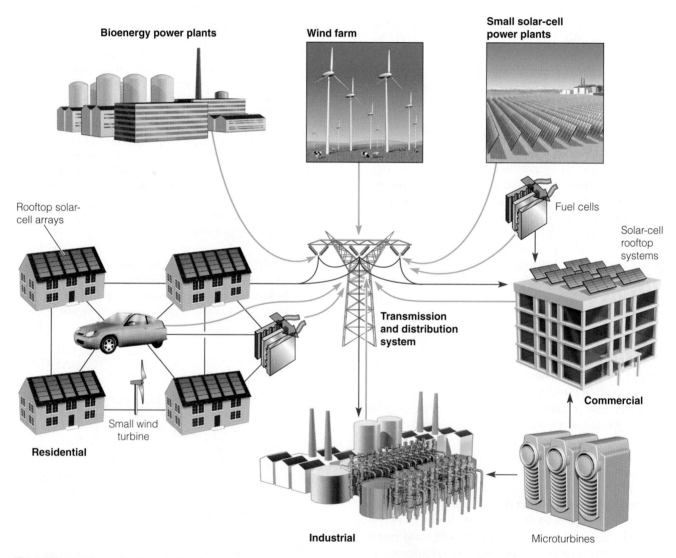

Figure 13-46 Solutions: *decentralized power system* in which electricity is produced by a large number of dispersed, small-scale *micropower systems*. Some would produce power on site; others would feed the power they produce into a conventional electrical distribution system. Over the next few decades, many energy and financial analysts expect a shift to this type of power system. Proponents say that such a dispersed power system is less vulnerable to terrorism and blackouts from hurricanes, floods, and other disasters.

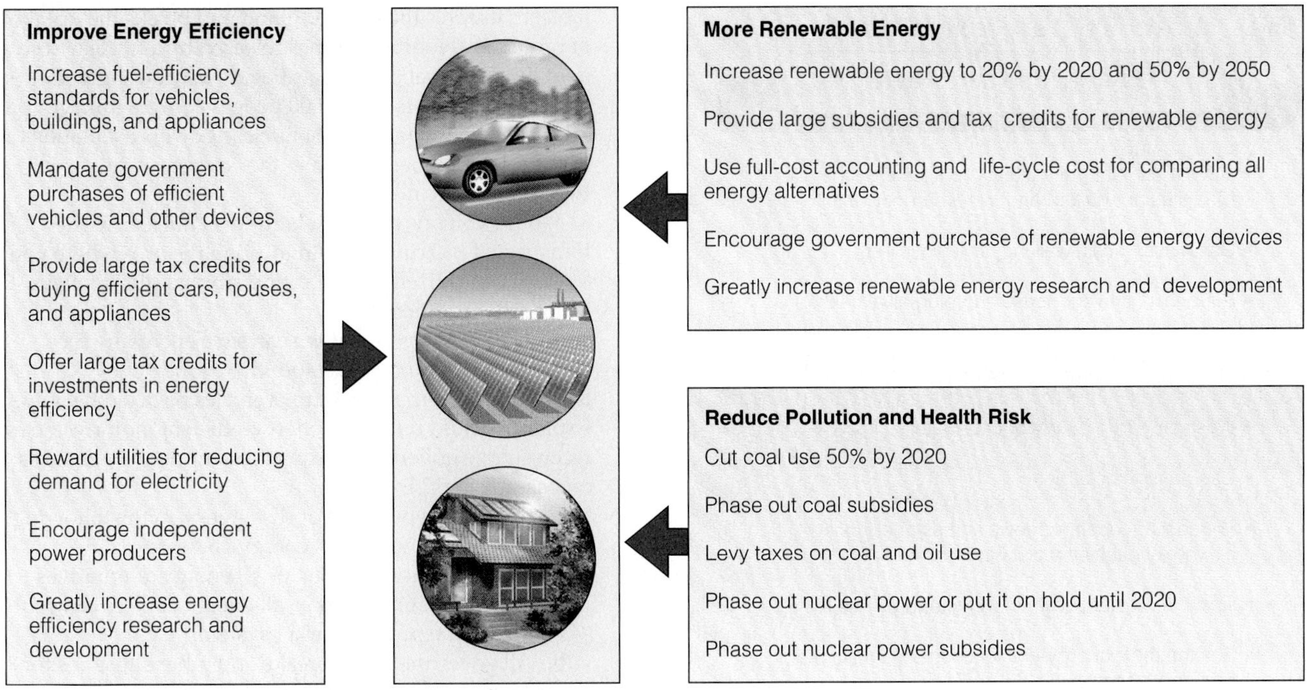

Improve Energy Efficiency

Increase fuel-efficiency standards for vehicles, buildings, and appliances

Mandate government purchases of efficient vehicles and other devices

Provide large tax credits for buying efficient cars, houses, and appliances

Offer large tax credits for investments in energy efficiency

Reward utilities for reducing demand for electricity

Encourage independent power producers

Greatly increase energy efficiency research and development

More Renewable Energy

Increase renewable energy to 20% by 2020 and 50% by 2050

Provide large subsidies and tax credits for renewable energy

Use full-cost accounting and life-cycle cost for comparing all energy alternatives

Encourage government purchase of renewable energy devices

Greatly increase renewable energy research and development

Reduce Pollution and Health Risk

Cut coal use 50% by 2020

Phase out coal subsidies

Levy taxes on coal and oil use

Phase out nuclear power or put it on hold until 2020

Phase out nuclear power subsidies

Figure 13-47 Solutions: suggestions of various energy analysts to help make the transition to a more sustainable energy future. *Critical thinking: which five of these solutions do you believe are the most important?*

Third, *over the next 50 years, the choice is not between using nonrenewable fossil fuels and renewable energy sources.* Because of their supplies and low prices, fossil fuels will continue to be used in large quantities. The challenge is to find ways to reduce the harmful environmental impacts of widespread fossil fuel use, with special emphasis on reducing air pollution and emissions of greenhouse gases as less harmful alternatives are phased in.

Economics, Politics, and Energy Resources

Governments can use a combination of subsidies, tax breaks, and taxes to promote or discourage use of various energy alternatives.

To most analysts, the key to making a shift to more sustainable energy resources and societies lies in economics and politics. Governments can use two basic economic and political strategies to help stimulate or dampen the short-term and long-term use of a particular energy resource.

First, they can *keep energy prices artificially low to encourage use of selected energy resources.* They can provide research and development subsidies and tax breaks, and enact regulations that help stimulate the development and use of energy resources receiving such support. For decades, this approach has been employed to stimulate the development and use of fossil fuels and nuclear power in the United States and in most other developed countries. It has created an uneven economic playing field that encourages energy waste and rapid depletion of nonrenewable energy resources. It also discourages improvements in energy efficiency and the development of renewable energy.

Second, governments can *keep energy prices artificially high to discourage use of a resource.* They can raise the price of an energy resource by eliminating existing tax breaks and other subsidies, enacting restrictive regulations, or adding taxes on its use. This approach will increase government revenues, encourage improvements in energy efficiency, reduce dependence on imported energy, and decrease use of an energy resource that has a limited future supply.

Making this transition to a more sustainable energy future depends primarily on *politics*, which depends largely on the pressure individuals put on elected officials by voting with their ballots and on companies by voting with their pocket books. Figure 13-48 (p. 326) lists some ways you can contribute to making this transition by reducing the amount of energy you use and waste.

A transition to renewable energy is inevitable, not because fossil fuel supplies will run out—large reserves of oil, coal, and gas remain in the world—but because the costs and risks of using these supplies will continue to increase relative to renewable energy.

MOHAMED EL-ASHRY

- Drive a car that gets at least 15 kilometers per liter (35 miles per gallon) and join a carpool.

- Use mass transit, walking, and bicycling.

- Superinsulate your house and plug all air leaks.

- Turn off lights, TV sets, computers, and other electronic equipment when they are not in use.

- Wash laundry in warm or cold water.

- Use passive solar heating.

- For cooling, open windows and use ceiling fans or whole-house attic or window fans.

- Turn thermostats down in winter and up in summer.

- Buy the most energy-efficient homes, lights, cars, and appliances available.

- Turn down the thermostat on water heaters to 43–49°C (110–120°F) and insulate hot water heaters and pipes.

Figure 13-48 Individuals matter: ways to reduce your use and waste of energy. *Critical thinking: which three of these actions do you believe are the most important? Which things in the list do you do or plan to do?*

CRITICAL THINKING

1. To continue using oil at the current rate (not the projected higher exponential increase in its annual use), we must discover and add to global oil reserves the equivalent of a new Saudi Arabian supply (the world's largest) *every 10 years.* Do you believe this is possible? If not, what effects might this failure have on your life and on the lives of your children or grandchildren?

2. List five actions you can take to reduce your dependence on oil and the gasoline derived from it. Which of these things do you actually plan to do?

3. Are you for or against continuing to increase oil imports in the United States or in the country where you live? If you favor reducing dependence on oil imports, list the three best ways to do so.

4. Explain why you agree or disagree with the following proposals made by various energy analysts as means to solve energy problems:

   a. Find and develop more domestic supplies of oil
   b. Place a heavy federal tax on gasoline and imported oil to help reduce the waste of oil resources
   c. Increase dependence on nuclear power
   d. Phase out all nuclear power plants by 2025

5. After a nuclear accident or terrorist attack, exposed children and adults can take iodine tablets to help prevent thyroid cancer. The stable isotope of iodine in the tablets saturates the thyroid gland and blocks the uptake of radioactive iodine isotopes released by a nuclear accident. In 1997, French officials began distributing potassium iodide tablets to 600,000 people living within 10 kilometers (6 miles) of its 24 nuclear power installations. Has such action been taken in the country where you live? If not, why not?

6. Would you favor having high-level nuclear waste transported by truck or train through the area where you live to a centralized underground storage site? Explain. What are the options?

7. Someone tells you that we can save energy by recycling it. How would you respond?

8. Should gas-guzzling motor vehicles be taxed heavily? Explain. Should buyers of energy-efficient motor vehicles receive large government subsidies, funded by the taxes on gas-guzzlers? Explain.

9. Explain why you agree or disagree with the following proposals made by various energy analysts:

   a. Government subsidies for all energy alternatives should be eliminated so all energy choices can compete in a true free-market system.
   b. All government tax breaks and other subsidies for conventional fuels (oil, natural gas, and coal), synthetic natural gas and oil, and nuclear power (fission and fusion) should be phased out. They should be replaced with subsidies and tax breaks for improving energy efficiency and developing solar, wind, geothermal, hydrogen, and biomass energy alternatives.
   c. Development of solar, wind, and hydrogen energy should be left to private enterprise and receive little or no help from the federal government, but nuclear energy and fossil fuels should continue to receive large federal subsidies.

10. Congratulations! You are in charge of the world. List the five most important features of your energy policy.

LEARNING ONLINE

The website for this book includes review questions for the entire chapter, flash cards for key terms and concepts, a multiple-choice practice quiz, interesting Internet sites, references, and a guide for accessing thousands of InfoTrac® College Edition articles.
   Visit

http://biology.brookscole.com/miller11

Then choose Chapter 13, and select a learning resource. For access to animations, additional quizzes, chapter outlines and summaries, register and log in to

Environmental Science ⊕ Now™

at **esnow.brookscole.com/miller11** using the access code card in the front of your book.

> **Active Graphing**
>
> Visit http://esnow.brookscole.com/miller11 to explore the graphing exercise for this chapter.

14 Risk, Human Health, and Toxicology

CASE STUDY
The Big Killer

What is roughly the diameter of a 30-caliber bullet, can be bought almost anywhere, is highly addictive, and kills about 13,700 people every day, or one every 6 seconds? A cigarette. *Cigarette smoking is the world's most preventable major cause of suffering and premature death among adults.*

According to the World Health Organization (WHO), tobacco helped kill 85 million people between 1950 and 2005—almost three times the 30 million people killed in battle in all wars since 1900!

The WHO estimates that each year tobacco contributes to the premature deaths of at least 5 million people (about half from developed countries and half from developing countries) from 34 illnesses including *heart disease, lung cancer, other cancers, bronchitis, emphysema,* and *stroke.* By 2030, the annual death toll from smoking-related diseases is projected to reach 10 million—an average of 27,400 preventable deaths per day or 1 death every 3 seconds. About 70% of these deaths are expected to occur in developing countries.

According to a 2002 study by the Centers for Disease Control and Prevention (CDC), smoking kills about 440,000 Americans per year prematurely—an average of 1,205 deaths per day (Figure 14-1). This death toll is roughly equivalent to three fully loaded 400-passenger jumbo jets crashing *every day* with no survivors! Yet, this ongoing major human tragedy rarely makes the news.

The overwhelming consensus in the scientific community is that the nicotine inhaled in tobacco smoke is highly addictive. Only 1 in 10 people who try to quit smoking succeeds, about the same relapse rate as for recovering alcoholics and those addicted to heroin or crack cocaine. A British government study showed that adolescents who smoke more than one cigarette have an 85% chance of becoming smokers. People can also be exposed to secondhand smoke from others, called *passive smoking.*

A 50-year study published in 2004 by Richard Doll and Richard Peto found that cigarette smokers die on average 10 years earlier than nonsmokers but that kicking the habit—even at 50 years old—can cut a person's risk in half. If people quit smoking by the age of 30, they can avoid nearly all the risk of dying prematurely.

Many health experts urge that a $3–5 federal tax be added to the price of a pack of cigarettes in the United States. The users of cigarettes (and other tobacco products)—not the rest of society—would then pay a much greater share of the $158 billion per year in health, economic, and social costs associated with their smoking.

Other suggestions for reducing the death toll and the health effects of smoking in the United States (and in other countries) include banning all cigarette advertising, prohibiting the sale of cigarettes and other tobacco products to anyone younger than 21 (with strict penalties for violators), and banning cigarette vending machines. Analysts also call for classifying and regulating the use of nicotine as an addictive and dangerous drug, eliminating all federal subsidies and tax breaks to tobacco farmers and tobacco companies, and using cigarette tax income to finance an aggressive antitobacco advertising and education program. So far, the U.S. Congress has not enacted such reforms.

X *HOW WOULD YOU VOTE?* Do you favor classifying and regulating nicotine as an addictive and dangerous drug? Cast your vote online at http://biology.brookscole.com/miller11.

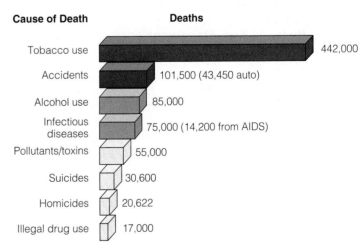

Figure 14-1 Annual deaths in the United States from tobacco use and other causes in 2003. Smoking is by far the nation's leading cause of preventable death, causing more premature deaths each year than all the other categories in this figure combined. (Data from U.S. National Center for Health Statistics and Centers for Disease Control and Prevention and U.S. Surgeon General)

The dose makes the poison.
PARACELSUS, 1540

In this chapter you will learn the risks of harm from disease and chemicals, how such risks are determined, and how well we perceive risks. It addresses the following questions:

- What types of hazards do people face?

- What types of disease (biological hazards) threaten people in developing countries and developed countries?

- What chemical hazards do people face, and how can they be measured?

- How can risks be estimated and perceived?

KEY IDEAS

- Rapidly producing infectious bacteria can undergo natural selection and become genetically resistant to widely used antibiotics.

- In terms of deaths caused, the three most serious hazards are poverty/malnutrition, smoking, and pneumonia/flu and the three most serious infectious diseases are AIDS, malaria, and diarrhea.

- It is difficult and costly to determine the risk of harm from chemicals.

- Most of us have a distorted view of the risks we face.

14-1   RISKS AND HAZARDS

Science: Risk and Risk Assessment

Risk is a measure of the likelihood that you will suffer harm from a hazard.

Risk is the *possibility* of suffering harm from a hazard that can cause injury, disease, death, economic loss, or environmental damage. It is usually expressed in terms of *probability*: a mathematical statement about how likely it is to suffer harm from a hazard. Scientists often state probability in terms such as "The lifetime probability of developing lung cancer from smoking a pack of cigarettes per day is 1 in 250." This means that 1 of every 250 people who smoke a pack of cigarettes per day will develop lung cancer over a typical lifetime (usually considered 70 years).

It is important to distinguish between *possibility* and *probability*. When we say that it is *possible* that a smoker can get lung cancer, we are saying that this event could happen. *Probability* gives us an estimate of the likelihood of such an event.

Figure 14-2 Science: *Risk assessment* and *risk management*.

Risk assessment is the scientific process of estimating how much harm a particular hazard can cause to human health. **Risk management** involves deciding whether or how to reduce a particular risk to a certain level and at what cost. Figure 14-2 summarizes how risks are assessed and managed.

Science: Types of Hazards

We can suffer harm from cultural hazards, biological hazards, chemical hazards, and physical hazards, but determining the risks involved is difficult.

We can suffer harm from four major types of hazards:

- *Cultural hazards* such as smoking, unsafe working conditions, poor diet, drugs, drinking, driving, criminal assault, unsafe sex, and poverty

- *Biological hazards* from pathogens (bacteria, viruses, and parasites) that cause infectious disease

- *Chemical hazards* from harmful chemicals in the air, water, soil, and food

- *Physical hazards* such as a fire, earthquake, volcanic eruption, flood, tornado, and hurricane

---

14-2   BIOLOGICAL HAZARDS: DISEASE IN DEVELOPED AND DEVELOPING COUNTRIES

Science: Nontransmissible and Transmissible Diseases

Diseases not caused by living organisms do not spread from one person to another, while those caused by living organisms such as bacteria and viruses can spread from person to person.

A **nontransmissible disease** is not caused by living organisms and does not spread from one person to another. Such diseases tend to develop slowly and have multiple causes. Examples include cardiovascular (heart and blood vessel) disorders, asthma, emphysema, and malnutrition.

In an infection, a pathogen in the form of a bacterium, virus, or parasite invades your body and multiplies in its cells and tissues. This can lead to an **infectious** or **transmissible disease** if your body cannot mobilize its defenses fast enough to keep the pathogen from interfering with your bodily functions.

Pathogens or infectious agents can be spread from one person to another by air, water, food, and body fluids such as pathogen-loaded droplets present in the sneezes, coughs, feces, urine, or blood of infected people. Pathogen-containing organisms such as mosquitoes can spread some infectious diseases. Figure 14-3 shows the annual death toll from the world's seven deadliest infectious diseases.

Great news. Since 1900, and especially since 1950, the incidences of infectious diseases and the death rates from such diseases have been greatly reduced. This has been achieved mostly by a combination of better health care, the use of antibiotics to treat infectious disease caused by bacteria, and the development of vaccines to prevent the spread of some infectious viral diseases.

Bad news. Many disease-carrying bacteria have developed genetic immunity to widely used antibiotics.

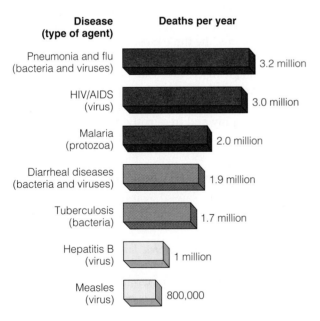

Disease (type of agent)	Deaths per year
Pneumonia and flu (bacteria and viruses)	3.2 million
HIV/AIDS (virus)	3.0 million
Malaria (protozoa)	2.0 million
Diarrheal diseases (bacteria and viruses)	1.9 million
Tuberculosis (bacteria)	1.7 million
Hepatitis B (virus)	1 million
Measles (virus)	800,000

Figure 14-3 Global outlook: the World Health Organization estimates that each year the world's seven deadliest infectious diseases kill 1.9 million people—most of them poor people in developing countries. This amounts to about 38,000 mostly preventable deaths every day. *Critical thinking: what three things would you do to reduce the death toll?* (Data from The World Health Organization)

Also, many disease-transmitting species of insects such as mosquitoes have become immune to widely used pesticides that once helped control their populations.

Science Case Study: Growing Germ Resistance to Antibiotics

Rapidly producing infectious bacteria can undergo natural selection and become genetically resistant to widely used antibiotics.

We may be falling behind in our efforts to prevent infectious bacterial diseases because of the astounding reproductive rate of bacteria, some of which can produce 16,777,216 offspring in 24 hours. Their high reproductive rate allows these organisms to become genetically resistant to an increasing number of antibiotics through natural selection. They can also transfer such resistance to nonresistant bacteria.

Other factors play a key role in fostering resistance. One is the spread of bacteria (some beneficial and some harmful) around the globe by human travel and international trade. Another is the overuse of pesticides, which increases populations of pesticide-resistant insects and other carriers of bacterial diseases.

Yet another factor is overuse of antibiotics by doctors. According to a 2000 study by Richard Wenzel and Michael Edward, at least half of all antibiotics used to treat humans are prescribed unnecessarily. In many countries, antibiotics are available without a prescription, which also promotes unnecessary use. Resistance to some antibiotics has also increased due to their widespread use in livestock and diary animals to control disease and to promote growth.

As a result of these factors acting together, every major disease-causing bacterium now has strains that resist at least one of the roughly 160 antibiotics we use to treat bacterial infections, such as those from tuberculosis.

Science Case Study: The Growing Global Threat from Tuberculosis

Tuberculosis kills 1.7 million people per year and could kill 26 million people between 2005 and 2020.

Since 1990, one of the world's most underreported stories has been the rapid spread of tuberculosis (TB). According to the WHO, this highly infectious bacterial disease strikes 9 million people per year and kills 1.7 million of them—about 84% of them in developing countries. The WHO projects that between 2005 and 2020, 26 million people will die of this disease unless current efforts and funding to control TB are greatly strengthened and expanded.

Many TB-infected people do not appear to be sick. Indeed, half of them do not know they are infected.

Several factors account for the recent increase in TB incidence. One is the lack of TB screening and control programs, especially in developing countries, where

95% of the new cases occur. A second problem is that most strains of the TB bacterium have developed genetic resistance to almost all effective antibiotics.

Population growth and urbanization have also led to increased contacts between people and spread TB, especially in areas where large numbers of the poor crowd together. In addition, the spread of AIDS is a factor because the disease greatly weakens the immune system and allows TB bacteria to multiply in people who have AIDS.

Slowing the spread of the disease requires early identification and treatment of people with active TB, especially those with a chronic cough. Treatment with a combination of four inexpensive drugs can cure 90% of individuals with active TB. To be effective, the drugs must be taken every day for 6–8 months. Because the symptoms disappear after a few weeks, many patients think they are cured and stop taking the drugs. This allows the disease to recur in a drug-resistant form. It then spreads to other people, and drug-resistant strains of TB bacteria develop. *Critical thinking: What three things would you do to reduce the spread of TB?*

Science: Viral Diseases

HIV, flu, and hepatitis B viruses infect and kill many more people each year than the highly publicized Ebola, West Nile, and SARS viruses.

What are the world's three most widespread and dangerous viruses? The biggest killer is the *human immunodeficiency virus (HIV)*, which is transmitted by unsafe sex, sharing of needles by drug users, infected mothers who pass the virus to their offspring before or during birth, and exposure to infected blood. On a global scale, HIV infects at least 5 million new people each year (about 41,000 in the United States), half of them women. The resulting complications from AIDS kill about 3 million people annually (Case Study, right).

The second biggest killer is the *influenza* or *flu* virus, which is transmitted by the body fluids or airborne emissions of an infected person, and kills about 1 million people per year. In 1918, an especially dangerous flu virus spread rapidly around the globe and killed an estimated 30 million people.

The third largest killer is the *hepatitis B virus (HBV)*, which damages the liver and kills about 1 million people each year. Like HIV, it is transmitted by unsafe sex, sharing of needles by drug users, infected mothers who pass the virus to their offspring before or during birth, and exposure to infected blood.

In recent years, three other viruses that cause previously unknown diseases have received widespread coverage in the media. The *Ebola virus* is transmitted by the blood or other body fluids of an infected person.

The *West Nile virus* is transmitted by the bite of a common mosquito that became infected by feeding on birds that carry the virus. It can infect 230 species of animals, including at least 130 bird species. In the United States, it was first reported in 1999, when it was apparently introduced by a mosquito or bird from the Middle East. Since then, the virus has spread throughout most of the lower 48 U.S. states, infecting thousands, and killing several hundred people. The chance of being infected and killed by this disease is low (about 1 in 2,500). In 2004, the flu killed more Americans in two days than the West Nile virus killed during the entire year.

The *severe acute respiratory syndrome (SARS) virus* was first transmitted to humans in China in 2002 from the flesh of wild animals valued as delicacies. From these wild hosts, this highly infectious disease it was transmitted to pigs, chickens, and ducks eaten by humans. It spread to chefs and animal handlers and then to other humans in close contact with SARS-infected people.

Its flu-like symptoms include chills, fever, a dry cough, headaches, and muscle pains; in the elderly and sick, SARS can quickly turn into life-threatening pneumonia. In 2003, the disease began spreading to other countries, mostly by global travelers. Within six months it had reached 31 other countries, infecting at least 8,500 people and causing 812 deaths. Swift local action by the WHO and other health agencies helped contain the spread of this disease by July 2003. Without careful vigilance, it might break out again.

Health officials are concerned about the emergence and spread of these three and other *emerging viral diseases* and are working hard to control their spread. Nevertheless, in terms of annual infection rates and deaths, the three most dangerous viruses by far remain HIV, flu, and HBV.

You can greatly reduce your chances of getting infectious diseases that spread from person to person by practicing good old-fashioned hygiene. Wash your hands thoroughly and frequently, and avoid touching your mouth, nose, and eyes.

Science Case Study: HIV and AIDS

The spread of AIDS, caused by infection with HIV, is one of the world's most serious and rapidly growing health threats.

The global spread of *acquired immune deficiency syndrome (AIDS)*, caused by infection with HIV, is considered the world's most serious and rapidly growing health threat. The virus itself is not deadly, but it kills immune cells and leaves the body defenseless against infectious bacteria and other viruses.

According to the WHO, by the beginning of 2004 some 39.4 million people worldwide (96% of them in developing countries, especially African countries south of the Sahara Desert) were infected with HIV.

Every day about 13,000 more people—most of them between the ages of 15 and 24—become infected with HIV. According to former U.S. Secretary of State Colin Powell, "AIDS is . . . now more destructive than any army, any conflict, and any weapon of mass destruction."

Within 7–10 years, at least half of all HIV-infected people will develop AIDS. This long incubation period means that infected people often spread the virus for several years without knowing they are infected. *There is no vaccine to prevent HIV and no cure for AIDS. Once you get AIDS, you will eventually die from it,* although drugs may help some infected people live longer. Unfortunately, only a tiny fraction of those suffering from AIDS can afford to use these costly drugs. Between 1980 and 2005, about 23 million people (502,000 in the United States) died of AIDS-related diseases.

AIDS has caused the life expectancy of the 700 million people living in sub-Saharan Africa to drop from 62 to 47 years. The premature deaths of teachers, health-care workers, and other young productive adults in such countries leads to diminished education and health care, decreased food production and economic development, and disintegrating families. Such deaths drastically alter a country's age structure diagram (Figure 14-4). Between 2004 and 2020, the WHO estimates 60 million more deaths from AIDS. By 2020, its death toll could reach 5 million per year.

According to the WHO, a global strategy to slow the spread of AIDS should have five major priorities. *First,* shrink the number of people capable of infecting others by quickly reducing the number of new infections below the number of deaths. *Second,* concentrate on the groups in a society that are most likely to spread the disease, such as truck drivers, sex workers, and soldiers. *Third,* provide free HIV testing and pressure people to get tested.

Fourth, implement a mass-advertising and education program geared toward adults and schoolchildren to help prevent the disease, emphasizing abstinence and condom use. *Fifth,* provide free or low-cost drugs to slow the progress of the disease.

Implementing such a program will require action by government officials and a massive aid program by developed nations. These countries will need to send both monetary aid and workers to help rebuild devastated societies.

Environmental Science ⊕ Now™
Examine the HIV virus and how it replicates itself by using a host cell at Environmental ScienceNow.

___

✗ *How Would You Vote?* Should developed and developing countries mount a global crash program to reduce the spread of AIDS and to help countries affected by this disease? Cast your vote online at http://biology.brookscole.com/miller11.

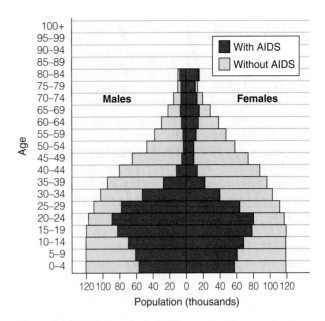

Figure 14-4 Global outlook: AIDS can affect the age structure of a population. This figure shows the projected age structure of Botswana's population in 2020 with and without AIDS. (Data from U.S. Census Bureau)

Science Case Study: Malaria

Malaria kills about 3 million people per year and has probably killed more people than all of the wars ever fought.

About one of every five people in the world—most of them living in poor African countries—is at risk from malaria (Figure 14-5). Worldwide, an estimated 300–500 million people are infected with the protozoan parasites that cause this disease, and 270–500 million new cases are reported each year. Malaria is not just a concern for the people living in the areas where it occurs, but also for anyone traveling to these areas—including

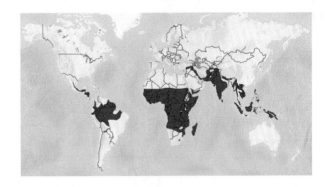

Figure 14-5 Global outlook: distribution of malaria. About 40% of the world's population live in areas in which malaria is present. Malaria kills about 3 million people each year. (Data from the World Health Organization and U.S. Centers for Disease Control and Prevention)

many unsuspecting tourists—because there is no vaccine for this disease.

Malaria is caused by a parasite that is spread by the bites of certain mosquito species. It infects and destroys red blood cells, causing fever, chills, drenching sweats, anemia, severe abdominal pain, headaches, vomiting, extreme weakness, and greater susceptibility to other diseases. It kills about 3 million people each year—on a par with AIDS (Figure 14-3)—and causes an average of 8,200 deaths per day. Approximately 90% of those dying are children younger than age 5. Many children who survive bouts of severe malaria have brain damage or impaired learning ability.

Four species of protozoan parasites in the genus *Plasmodium* cause malaria. Most cases of the disease occur when an uninfected female of any of about 60 *Anopheles* mosquito species bites a person infected with *Plasmodium,* ingests blood that contains the parasite, and later bites an uninfected person (Figure 14-6). *Plasmodium* parasites then move out of the mosquito and into the human's bloodstream, multiply in the liver, and enter blood cells to continue multiplying. Malaria can also be transmitted by blood transfusions or by sharing needles.

The malaria cycle repeats itself until immunity develops, treatment is given, or the victim dies. *Over the course of human history, malarial protozoa probably have killed more people than all the wars ever fought.*

During the 1950s and 1960s, the spread of malaria was sharply curtailed by draining swamplands and marshes, spraying breeding areas with insecticides, and using drugs to kill the parasites in the bloodstream. Since 1970, however, malaria has come roaring back. Most species of the *Anopheles* mosquito have become genetically resistant to most insecticides. Even worse, the *Plasmodium* parasites have become genetically resistant to common antimalarial drugs.

Researchers are working to develop new antimalarial drugs, vaccines, and biological controls for *Anopheles* mosquitoes. They are also using artemisinins derived from qinghaosu, an ancient Chinese herbal remedy that cures 90% of malaria patients in three days. Unfortunately, such approaches receive too little funding and have proved more difficult to implement than originally thought.

For now, prevention is the best approach to slowing the spread of malaria. Methods include increasing water flow in irrigation systems to prevent mosquito larvae from developing (an expensive and wasteful use of water), fixing leaking water pipes, and providing poor people in malarial regions with window screens for their dwellings and insecticide-treated bed nets.

Other approaches include cultivating fish that feed on mosquito larvae (biological control), clearing vegetation around houses, planting trees that soak up water in low-lying marsh areas where mosquitoes thrive (a method that can degrade or destroy ecologically important wetlands), and using zinc and vitamin A supplements to boost resistance to malaria in children.

Spraying the insides of homes with low concentrations of the pesticide DDT twice a year greatly reduces the number of malaria cases. Under an international treaty enacted in 2002, DDT and five of its chlorinated-hydrocarbon cousins are being phased out in developing countries. However, the treaty allows 25 countries to continue using DDT for malaria control until other alternatives become available. Health officials in developing countries call for much greater funding of research geared toward finding ways to prevent and treat malaria and of schemes to provide poor people in malaria-prone countries with window screens and insecticide-treated bed nets.

Environmental Science ⊕ Now™
Watch through a microscope what happens when a mosquito infects a human with malaria at Environmental ScienceNow.

Merozoites enter bloodstream and develop into gametocytes causing malaria and making infected person a new reservoir

Female mosquito bites infected human, ingesting blood that contains *Plasmodium* gametocytes

Plasmodium develop in mosquito

Sporozoites penetrate liver and develop into merozoites

Female mosquito injects *Plasmodium* sporozoites into human host.

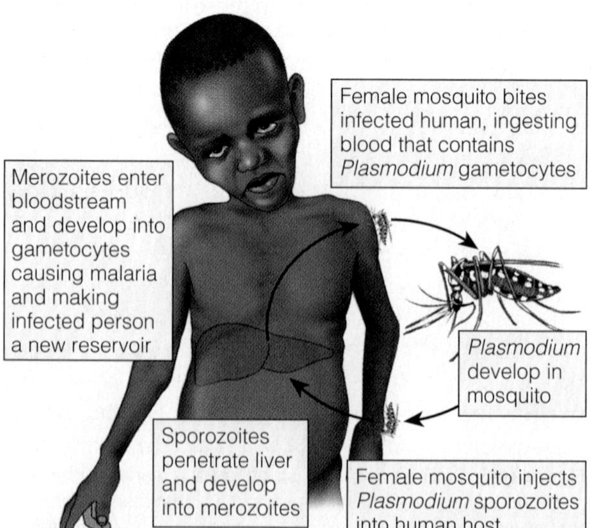

Figure 14-6 Science: the life cycle of malaria. *Plasmodium* circulates from mosquito to human and back to mosquito.

Solutions: Reducing the Incidence of Infectious Diseases

There are a number of ways to reduce the incidence of infectious diseases if the world is willing to provide the necessary funds and assistance.

Good news. According to the WHO, the global death rate from infectious diseases decreased by about two-thirds between 1970 and 2000 and is projected to continue dropping. Also, between 1971 and 2000, the percentage of children in developing countries immunized with vaccines to prevent tetanus, measles, diphtheria, typhoid fever, and polio increased from 10% to 84%—saving about 10 million lives each year.

Solutions

Infectious Diseases

Increase research on tropical diseases and vaccines

Reduce poverty

Decrease malnutrition

Improve drinking water quality

Reduce unnecessary use of antibiotics

Educate people to take all of an antibiotic prescription

Reduce antibiotic use to promote livestock growth

Careful hand washing by all medical personnel

Immunize children against major viral diseases

Oral rehydration for diarrhea victims

Global campaign to reduce HIV/AIDS

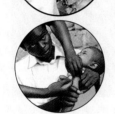

Figure 14-7 Solutions: ways to prevent or reduce the incidence of infectious diseases, especially in developing countries. *Critical thinking: which three of these approaches do you believe are the most important?*

Figure 14-7 lists measures health scientists and public health officials suggest to help prevent or reduce the incidence of infectious diseases that affect humanity—especially in developing countries. An important breakthrough has been the development of simple *oral rehydration therapy* to help prevent death from dehydration for victims of diarrheal diseases, which cause about one-fourth of all deaths of children younger than age 5. It involves administering a simple solution of boiled water, salt, and sugar or rice, at a cost of only a few cents per person. It has been the major factor in reducing the annual number of deaths from diarrhea from 4.6 million in 1980 to 1.9 million in 2002. Few investments have saved so many lives at such a low cost.

Bad news. The WHO estimates that only 10% of global medical research and development money goes toward preventing infectious diseases in developing countries, even though more people worldwide suffer and die from these diseases than from all other infectious diseases combined.

Science, Politics, and Ethics: Bioterrorism

Bioterrorism that involves releasing infectious organisms into the air, water supply, or food supply is a serious and growing threat.

Bioterrorism involves the deliberate release of disease-causing bacteria or viruses into the air, water supply, or food supply of concentrated urban populations. According to antiterrorism experts, it is a much easier, cheaper, and more effective way to cause illness, death, and mass terror and chaos than crashing planes into buildings or setting off dirty nuclear weapons. The materials and tools to make biological weapons are inexpensive and easy to get. A state-of-the-art laboratory for making biological warfare agents requires only $10,000 of off-the shelf equipment that can be set up in a space about the size of a small bathroom. Now that the sequencing of the flu virus's genome is nearly complete, for example, bioterrorists might develop more lethal flu viruses and easily transmit them through the air in tiny droplets.

Since the end of World War II, the United States and the former Soviet Union have spent billions of dollars developing, producing, and stockpiling large quantities of biological weapons of mass destruction. Figure 14-8 (p. 334) provides information about some of the common bacterial and viral agents these countries have investigated.

Both countries have used recombinant DNA techniques to produce more dangerous versions of these organisms that act more rapidly, are more virulent, and are resistant to antibiotics used to treat them. They have also created new and even more dangerous infectious organisms with properties that are classified as top secret.

In 1995, the U.S. Central Intelligence Agency (CIA) identified 16 nations suspected of having programs to develop and stockpile biological warfare agents. In addition, thousands of molecular biologists and graduate-school students around the world have enough knowledge about recombinant DNA and cloning technology to design and mass-produce biological warfare agents.

Once made, the bacteria or viruses can be carried in a small vial or aerosol container not detectable by conventional security equipment. They could be released in a crowded subway car, into a public water supply, or into the unprotected, ground-level air intakes found in most office buildings. A terrorist organization with volunteers willing to die for their cause could infect volunteers with a normally fatal disease organism that is easily transmitted from one human to another. After waiting until they are contagious, the volunteers could be sent on airplane trips throughout the world. Millions could die, and the social and economic fabric of affected societies would unravel.

Agent	Contagious	Symptoms	Mortality (if untreated)	Existence of vaccine	Treatment
Smallpox (virus)	Yes	Fever, aches, headache, red spots on face and torso	30%	Yes	Vaccination within 4 days after exposure, IV hydration
Hemorrhagic fever (viruses)	Yes	Vary but include fever, bleeding, shock, and coma	Varies	No	Ebola has no cure; antiviral riboflavin and some antibiotics may help
Inhalation anthrax (bacterium)	No	Fever, chest pain, difficulty breathing, respiratory failure	90–100%	Yes	Early treatment with Cipro and other antibiotics
Botulism (bacterium)	No	Blurred vision, progressive paralysis, death within 24 hours if not treated	60–100%	Yes	Equine antitoxin given early. Intensive care, respirator
Pneumonic plague (bacterium)	Yes	High fever, chills, headache, coughing blood, difficulty breathing, respiratory failure	90–100%	No	Antibiotics
Tularemia (bacterium)	No	Fever, sore throat, weakness, respiratory stress, pneumonia	30–60%	Yes (in testing)	Antibiotics

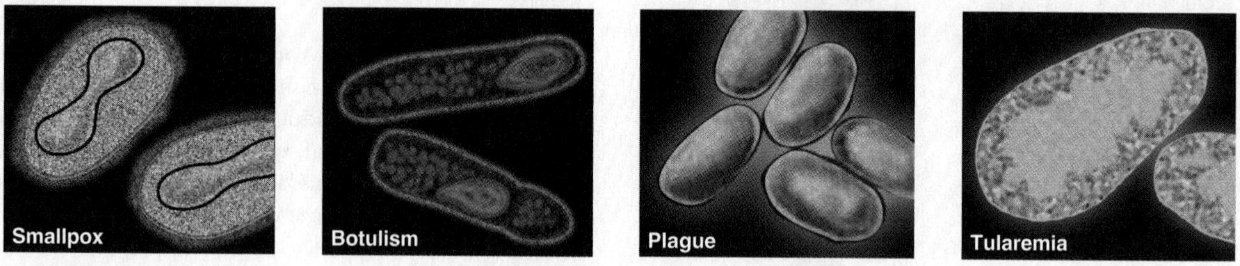

Figure 14-8 Bioterrorist weapons: characteristics of common agents that might be used by terrorists as biological weapons.

Perhaps you are thinking, "Whoa, enough already. This stuff is depressing and scary." But it is a reality in today's world. Let us look at some more hopeful news about bioterrorism.

Early detection of biological agents is the key in treating exposed victims and preventing the spread of diseases to others. Some scientists are trapping common insects such as bees, beetles, moths, and crickets to see whether they can be used as environmental monitors of chemical and biological agents. Others are trying to develop inexpensive and easy-to-use DNA detectors to quickly and accurately diagnose any infectious disease such as smallpox. For example, MIT biologist Todd Rider has developed a biological sensor that can detect dangerous biological agents such as anthrax within minutes. He made the sensor out of mouse immune cells by inserting a gene for antibodies to a particular biological agent (such as anthrax) along with a gene that causes a jellyfish to glow. When a biological agent activates the antibody, the immune cells of the mouse light up.

Treatments are available for the most common biological agents (Figure 14-8)—unless they have been genetically modified to make such treatments fail. Outbreaks can be kept under control if hospitals stock large supplies of antibiotics and vaccines for treatment of common diseases, provide emergency and hospital workers with detection systems and protective gear, and alert doctors to the symptoms of the most common biological warfare agents. *Critical thinking: what three things would you do to reduce the threat of bioterrorism?*

14-3 CHEMICAL HAZARDS

Science: Toxic and Hazardous Chemicals

Toxic chemicals can harm or kill, and hazardous chemicals can cause various types of harm.

A **toxic chemical** can cause temporary or permanent harm or death to humans or animals. A **hazardous chemical** can harm humans or other animals because it is flammable or explosive or because it can irritate or damage the skin or lungs, interfere with oxygen uptake, or induce allergic reactions.

There are three major types of potentially toxic agents: mutagens, teratogens, and carcinogens.

Mutagens include chemicals or ionizing radiation that cause or increase the frequency of mutations, or changes, in DNA molecules. An example is nitrous acid (HNO_2), which is formed by the digestion of nitrite preservatives in foods. Harmful mutations occurring in reproductive cells can be passed on to offspring and to future generations. There is no safe threshold for exposure to mutagens.

Teratogens are chemicals that cause harm or birth defects to a fetus or embryo. Ethyl alcohol is an example of a teratogen. Drinking during pregnancy can lead to offspring with low birth weight and a number of physical, developmental, and mental problems. Another potent teratogen is thalidomide, which was developed in the 1950s as a sleeping pill and a means to help prevent nausea during pregnancy. Unfortunately, experience showed that even a single dose of this chemical during pregnancy could cause limb deformities and organ defects in offspring.

Carcinogens are chemicals or ionizing radiation that can cause or promote cancer—the growth of a malignant (cancerous) tumor, in which certain cells multiply uncontrollably. One example is benzene, a widely used chemical solvent. Another is formaldehyde, a major indoor air pollutant that is used to manufacture common household materials such as plywood, particleboard, flooring, furniture, and adhesives in carpeting and wallpaper.

Many cancerous tumors spread by **metastasis,** in which malignant cells break off from tumors and travel in body fluids to other parts of the body. There they start new tumors, making treatment much more difficult.

Typically, 10–40 years may elapse between the initial exposure to a carcinogen and the appearance of detectable symptoms. Partly because of this time lag, many healthy teenagers and young adults have trouble believing that their smoking, drinking, eating, and other lifestyle habits today could lead to some form of cancer before they reach age 50.

Science: Effects of Chemicals on the Immune, Nervous, and Endocrine Systems

Long-term exposure to some chemicals at low doses may disrupt the body's immune, nervous, and endocrine systems.

Since the 1970s, a growing body of research on wildlife and laboratory animals, along with some epidemiological studies of humans, suggests that long-term exposure to some chemicals in the environment can disrupt the body's immune, nervous, and endocrine systems.

The *immune system* consists of specialized cells and tissues that protect the body against disease and harmful substances by forming antibodies that render invading agents harmless. Ionizing radiation and some chemicals such as arsenic and dioxins can weaken the human immune system and leave the body vulnerable to attacks by allergens, infectious bacteria, viruses, and protozoans.

Some natural and synthetic chemicals in the environment, called **neurotoxins,** can harm the human *nervous system* (brain, spinal cord, and peripheral nerves). For example, many poisons and the venom of some snakes are neurotoxins. They inhibit, damage, or destroy nerve cells (neurons) that transmit electrochemical messages throughout the body. Effects can include behavioral changes, paralysis, and death. Examples of chemical neurotoxins include PCBs, methyl mercury, lead, and certain pesticides.

The *endocrine system* is a complex network of glands that releases tiny amounts of *hormones* into the bloodstream of humans and other vertebrate animals. Low levels of these chemical messengers turn on and off bodily systems that control sexual reproduction, growth, development, learning ability, and behavior.

Examples of hormone disrupters or *hormonally active agents (HAAs)* include DDT, PCBs, and certain herbicides. Exposure to low levels of HAAs may disrupt the effects of natural hormones in some animals, but more research is needed to verify their effects on humans. Some scientists say we need to wait for the results of more research before banning or severely restricting HAAs—but this will take decades. Other scientists believe that as a precaution, we should sharply reduce our use of potential hormone disrupters.

Critical thinking: should we ban or severely restrict the use of potential hormone disrupters? Why hasn't this been done?

14-4 TOXICOLOGY: ASSESSING CHEMICAL HAZARDS

Science: What Determines Whether a Chemical Is Harmful?

The harm caused by exposure to a chemical depends on the amount of exposure (dose), the frequency of exposure, the person who is exposed, the effectiveness of the body's detoxification systems, and one's genetic makeup.

Toxicity—a measure of how harmful a substance is in causing injury, illness, or death to a living organism—depends on several factors. One is the **dose,** the amount of a substance a person has ingested, inhaled, or absorbed through the skin. Other factors are how often the exposure occurred, who is exposed (adult or child, for example), and how well the body's detoxification systems (such as the liver, lungs, and kidneys) work.

Toxicity also depends on *genetic makeup,* which determines an individual's sensitivity to a particular toxin. Some individuals are sensitive to a number of toxins—a condition known as *multiple chemical sensitivity (MCS)*. This genetic variation in individual responses to exposure to various toxins raises a difficult ethical, political, and economic question: When regulating levels of a toxic substance in the environment, should the allowed level be set to protect the most sensitive individuals (at great cost) or the average person?

Five major factors can affect the harm caused by a substance. One is its *solubility*. *Water-soluble toxins* (which are often inorganic compounds) can move throughout the environment and get into water supplies and the aqueous solutions that surround the cells in our bodies. *Oil- or fat-soluble toxins* (which are usually organic compounds) can penetrate the membranes surrounding cells because the membranes allow similar oil-soluble chemicals to pass through them. Thus, oil- or fat-soluble toxins can accumulate in body tissues and cells.

A second factor is a substance's *persistence*. Many chemicals, such as DDT, are used precisely because of their persistence or resistance to breakdown. Of course, this persistence also means they can have long-lasting harmful effects on the health of wildlife and people.

A third factor for some substances is *bioaccumulation*, in which some molecules are absorbed and stored in specific organs or tissues at higher than normal levels. As a consequence, a chemical found at a fairly low concentration in the environment can build up to a harmful level in certain organs and tissues.

A related factor is *biomagnification*, in which levels of some potential toxins in the environment become magnified as they pass through food chains and webs. Organisms at low trophic levels might ingest only small amounts of a toxin, but each animal on the next trophic level up that eats many of those organisms will take in increasingly larger amounts of that toxin (Figure 9-16, p. 197). Examples of chemicals that can be biomagnified include long-lived, fat-soluble organic compounds such as DDT, PCBs (oily chemicals used in electrical transformers), and some radioactive isotopes (such as strontium-90).

A fifth factor is *chemical interactions* that can decrease or multiply the harmful effects of a toxin. An *antagonistic interaction* can reduce harmful effects. For example, there is preliminary evidence that vitamins E and A can interact to reduce the body's response to some cancer-causing chemicals.

A *synergistic interaction* multiplies harmful effects. For instance, workers exposed to tiny fibers of asbestos increase their chances of getting lung cancer 20-fold. But asbestos workers who also smoke have a 400-fold increase in lung cancer rates. In such cases, one plus one can be a lot greater than two.

The type and amount of health damage that result from exposure to a chemical or other agent are called the response. An *acute effect* is an immediate or rapid harmful reaction to an exposure—ranging from dizziness to death. A *chronic effect* is a permanent or long-lasting consequence (kidney or liver damage, for example) from exposure to a single dose or to repeated lower doses of a harmful substance.

A basic concept of toxicology is that *any synthetic or natural chemical can be harmful if ingested in a large enough quantity*. For example, drinking 100 cups of strong coffee one after another would expose most people to a lethal dosage of caffeine. Similarly, downing 100 tablets of aspirin or 1 liter (1.1 quarts) of pure alcohol (ethanol) would kill most people.

The critical question is this: *How much exposure to a particular toxic chemical causes a harmful response?* This is the meaning of the chapter-opening quote by the German scientist Paracelsus about the dose making the poison.

People vary in terms of the dose of a toxin they can tolerate without incurring significant harm, because of differences in their genetic makeup. For this reason, a better way to state Paracelsus's principle of toxicology is this: The dose makes the poison, *but differently for different individuals*.

Your body has three major mechanisms for reducing the harmful effects of some chemicals. *First*, it can break down (usually by enzymes found in the liver), dilute, or excrete (for example, in your breath, sweat, and urine) small amounts of most toxins to keep them from reaching harmful levels. However, accumulations of high levels of toxins can overload the ability of your liver and kidneys to degrade and excrete such substances.

Second, your cells have enzymes that can sometimes repair damage to DNA and protein molecules. *Third*, cells in some parts of your body (such as your skin and the linings of your gastrointestinal tract, lungs, and blood vessels) can reproduce fast enough to replace damaged cells.

Science: Effects of Trace Levels of Toxic Chemicals

Trace amounts of chemicals in the environment or your body may or may not be harmful.

Should we be concerned about trace amounts of various chemicals in air, water, food, and our bodies? The honest answer is that, in most cases, we do not know because of a lack of data and the difficulty of determining the effects of exposures to low levels of chemicals.

Some scientists think that trace levels of most chemicals are not harmful. They point to the dramatic increase in average life expectancy in the United States since 1950. Other scientists are not so sure and suggest that much more research is needed to evaluate the possible long-term harm caused by exposure to low levels of the thousands of new synthetic chemicals that we have put into the environment during the past few decades.

Chemists are able to detect increasingly small amounts of potentially toxic chemicals in air, water, and food. This is good news, but it can give the false impression that dangers from toxic chemicals are increasing. In reality, we may simply be uncovering levels of chemicals that have been around for a long time.

Some people also have the mistaken idea that natural chemicals are safe and synthetic chemicals are harmful. In fact, many synthetic chemicals are quite safe if used as intended, and many natural chemicals are deadly.

Science: Estimating the Toxicity of a Chemical

Chemicals vary widely in their toxicity to humans and other animals.

A **poison** or **toxin** is a chemical that adversely affects the health of a living human or animal by causing injury, illness, or death. One method for determining the relative toxicities of various chemicals is to measure their effects on test animals. For example, we can determine a chemical's *lethal dose (LD)*. A chemical's median lethal dose (LD50) means that the amount received in that dose kills 50% of the animals (usually rats and mice) in a test population within a 14-day period (Figure 14-9).

Chemicals vary widely in their toxicity (Table 14-1). Some poisons can cause serious harm or death after a single acute exposure at very low dosages. Others cause such harm only at dosages so huge that it is nearly impossible to get enough into the body to cause injury or death. Most chemicals fall between these two extremes.

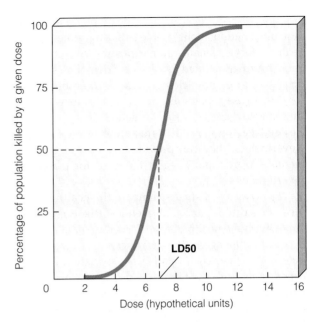

Figure 14-9 Science: hypothetical *dose-response curve* showing determination of the LD50, the dosage of a specific chemical that kills 50% of the animals in a test group. Toxicologists use this method to compare the toxicities of different chemicals.

Science: Using Case Reports and Epidemiological Studies to Estimate Toxicity

We can estimate toxicity by using case reports about the harmful effects of chemicals on human health and by comparing the health of a group of people exposed to a chemical with the health of a similar group not exposed to the chemical.

Scientists use several methods to get information about the harmful effects of chemicals on human health. For

Table 14-1 Toxicity Ratings and Average Lethal Doses for Humans			
Toxicity Rating	**LD50 (milligrams per kilogram of body weight)***	**Average Lethal Dose†**	**Examples**
Supertoxic	Less than 0.01	Less than 1 drop	Nerve gases, botulism toxin, mushroom toxins, dioxin (TCDD)
Extremely toxic	Less than 5	Less than 7 drops	Potassium cyanide, heroin, atropine, parathion, nicotine
Very toxic	5–50	7 drops to 1 teaspoon	Mercury salts, morphine, codeine
Toxic	50–500	1 teaspoon to 1 ounce	Lead salts, DDT, sodium hydroxide, sodium fluoride, sulfuric acid, caffeine, carbon tetrachloride
Moderately toxic	500–5,000	1 ounce to 1 pint	Methyl (wood) alcohol, ether, phenobarbital, amphetamines (speed), kerosene, aspirin
Slightly toxic	5,000–15,000	1 pint to 1 quart	Ethyl alcohol, Lysol, soaps
Essentially nontoxic	15,000 or greater	More than 1 quart	Water, glycerin, table sugar

*Dosage that kills 50% of individuals exposed
†Amounts of substances in liquid form at room temperature that are lethal when given to a 70.4-kilogram (155-pound) human

example, *case reports,* usually made by physicians, provide information about people suffering some adverse health effect or death after exposure to a chemical. Such information often involves accidental poisonings, drug overdoses, homicides, or suicide attempts.

Most case reports are not reliable sources for estimating toxicity because the actual dosage and the exposed person's health status are often not known. Nevertheless, they can provide clues about environmental hazards and suggest the need for laboratory investigations.

Epidemiological studies compare the health of people exposed to a particular chemical (the *experimental group*) with the health of a similar group of people not exposed to the agent (the *control group*). The goal is to determine whether the statistical association between exposure to a toxic chemical and a health problem is strong, moderate, weak, or undetectable.

Three factors can limit the usefulness of epidemiological studies. *First,* in many cases, too few people have been exposed to high enough levels of a toxic agent to detect statistically significant differences. *Second,* conclusively linking an observed effect with exposure to a particular chemical is difficult because people are exposed to many different toxic agents throughout their lives and can vary in their sensitivity to such chemicals. *Third,* we cannot use epidemiological studies to evaluate hazards from new technologies or chemicals to which people have not yet been exposed.

Science, Ethics, and Economics: Using Laboratory Experiments to Estimate Toxicity

Exposing a population of live laboratory animals (especially mice and rats) to known amounts of a chemical is the most widely used method for determining its toxicity.

The most widely used method for determining toxicity is to expose a population of live laboratory animals (especially mice and rats) to measured doses of a specific substance under controlled conditions. Animal tests take 2–5 years and cost $200,000–2,000,000 per substance tested. Such tests can be painful to the test animals and can kill or harm them. The goal is to develop data on the responses of the test animals to various doses of a chemical, but estimating the effects of low doses is difficult.

Animal welfare groups want to limit or ban the use of test animals or ensure that they are treated in the most humane manner possible. More humane methods for carrying out toxicity tests are available. They include computer simulations and using tissue cultures of cells and bacteria, chicken egg membranes, and measurements of changes in the electrical properties of individual animal cells.

These alternatives can greatly decrease the use of animals for testing toxicity. Some scientists point out that some animal testing is needed because the alternative methods cannot adequately mimic the complex biochemical interactions taking place in a live animal. *Critical thinking: should animal testing be banned?*

Acute toxicity tests are run to develop a dose-response curve, which shows the effects of various dosages of a toxic agent on a group of test organisms. In *controlled experiments,* the effects of the chemical on a *test group* are compared with the responses of a *control group* of organisms not exposed to the chemical. Care is taken that organisms in each group are as identical as possible in terms of age, health status, and genetic makeup, and that all are exposed to the same environmental conditions.

Fairly high dosages are used to reduce the number of test animals needed, obtain results quickly, and lower costs. Otherwise, tests would have to be run on millions of laboratory animals for many years, and manufacturers could not afford to test most chemicals.

For the same reasons, scientists often use mathematical models to extrapolate the results of high-dose exposures to low-dose levels. Then they extrapolate the low-dose results from the test organisms to humans to estimate LD50 values for acute toxicity (Table 14-1).

Two general types of dose-response curves exist (Figure 14-10). With the *nonthreshold dose-response model* (Figure 14-10, left), any dosage of a toxic chemical or ionizing radiation causes harm that increases with the dosage. With the *threshold dose-response model* (Figure 14-10, right), a threshold dosage must be reached before any detectable harmful effects occur, presumably because the body can repair the damage caused by low dosages of some substances.

Establishing which of these models applies at low dosages is extremely difficult and controversial. To be on the safe side, the nonthreshold dose-response model often is assumed.

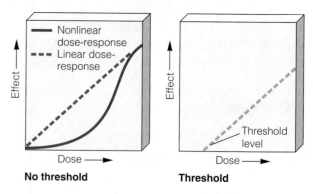

Figure 14-10 Science: two types of *dose-response curves.* The linear and nonlinear curves in the left graph apply if even the smallest dosage of a chemical or ionizing radiation has a harmful effect that increases with the dosage. The curve on the right applies if a harmful effect occurs only when the dosage exceeds a certain *threshold level.* Which model is better for a specific harmful agent is uncertain because of the difficulty in estimating the response to very low dosages.

Some scientists challenge the validity of extrapolating data from test animals to humans because human physiology and metabolism often differ from those of the test animals. Other scientists say that such tests and models work fairly well (especially for revealing cancer risks) when the correct experimental animal is chosen or when a chemical is toxic or harmful to several different test-animal species.

The problem of estimating toxicities gets worse. In real life, each of us is exposed to a variety of chemicals, some of which can interact in ways that decrease or enhance their individual effects over the short and long term.

Toxicologists have great difficulty in estimating the toxicity of a single substance. But adding the problem of evaluating mixtures of potentially toxic substances, separating out which ones are the culprits, and determining how they can interact with one another is overwhelming from a scientific and economic standpoint. For example, just studying the interactions of three of the 500 most widely used industrial chemicals would take 20.7 million experiments—a physical and financial impossibility.

Solutions: Protecting Children from Toxic Chemicals

Children tend to be much more vulnerable than adults to toxic chemicals.

Children are usually much more susceptible to the effects of toxic substances than are adults for three major reasons. *First,* children breathe more air, drink more water, and eat more food per unit of body weight than do adults. *Second,* they are exposed to toxins in dust or soil when they put their fingers, toys, or other objects in their mouths (as they frequently do). *Third,* immune systems and bodily processes for degrading or excreting toxins and repairing damage are usually less well developed in children than in adults.

In 2003, the U.S. Environmental Protection Agency (EPA) proposed that in determining risk, regulators should assume children have 10 times the exposure risk of adults to cancer-causing chemicals. Some health scientists contend that these guidelines are too weak. They suggest that, to be on the safe side, we should assume that the risk of harm from toxins for children is 100 times that of adults.

Science, Politics, and Economics: Why Do We Know So Little about the Harmful Effects of Chemicals?

Under existing laws, most chemicals are considered innocent until proven guilty, and estimating their toxicity to establish guilt is difficult, uncertain, and expensive.

As discussed previously, all methods for estimating toxicity levels and risks have serious limitations. But they are all we have. To take this uncertainty into account and minimize harm, scientists and regulators typically set allowed exposure levels to toxic substances and ionizing radiation at 1/100 or even 1/1,000 of the estimated harmful levels.

According to risk assessment expert Joseph V. Rodricks, "Toxicologists know a great deal about a few chemicals, a little about many, and next to nothing about most." The U.S. National Academy of Sciences estimates that only 10% of at least 80,000 chemicals in commercial use have been thoroughly screened for toxicity, and only 2% have been adequately tested to determine whether they are carcinogens, teratogens, or mutagens. Hardly any of the chemicals in commercial use have been screened for possible damage to humans' nervous, endocrine, and immune systems.

Currently, federal and state governments do not regulate about 99.5% of the commercially used chemicals in the United States. Several reasons explain this lack of regulation. For example, under existing U.S. laws, most chemicals are *considered innocent until proven guilty.* Some analysts think this is the opposite of the way it should be. Why, they ask, should chemicals have the same legal rights as people? *Critical thinking: should chemicals be considered guilty until proven innocent?*

In addition, there are not enough funds, personnel, facilities, and test animals available to provide such information for more than a small fraction of the many individual chemicals we encounter in our daily lives. It is also too difficult and expensive to analyze the combined effects of multiple exposures to various chemicals and their possible interactions.

Science and Economics: Pollution Prevention and the Precautionary Principle

Preliminary but not conclusive evidence that a chemical causes significant harm should spur preventive action, some say.

So where does this leave us? We do not know a lot about the potentially toxic chemicals around us and inside of us, and estimating their effects is very difficult, time-consuming, and expensive. Is there a way out of this dilemma?

Some scientists and health officials, especially those in European Union countries, are pushing for much greater emphasis on *pollution prevention.* They say we should not release into the environment chemicals that we know or suspect can cause significant harm. This means looking for harmless or less harmful substitutes for toxic and hazardous chemicals or recycling them within production processes so they do not reach the environment.

This prevention approach is based on the **precautionary principle:** When there is plausible but incomplete scientific evidence (frontier science evidence) of significant harm to humans or the environment from a proposed or existing chemical or technology, we should take action to prevent or reduce the risk instead of waiting for more conclusive (sound or consensus science) evidence.

Under this approach, those proposing to introduce a new chemical or technology would bear the burden of establishing its safety—a *guilty-until-proven-innocent approach*. This means two major changes in the way we evaluate risks. *First,* new chemicals and technologies would be assumed to be harmful until scientific studies can show otherwise. *Second,* existing chemicals and technologies that appear to have a strong chance of causing significant harm would be removed from the market until their safety can be established.

Some movement is being made in this direction, especially in the European Union. In 2000, negotiators agreed to a global treaty that would ban or phase out use of 12 of the most notorious *persistent organic pollutants (POPs)*, also called the *dirty dozen*. The list included DDT and eight other persistent pesticides, PCBs, and dioxins and furans. New chemicals would be added to the list when the harm they could potentially cause is seen as outweighing their usefulness. This treaty went into effect in 2004.

Manufacturers and businesses contend that widespread application of the precautionary principle would make it too expensive and almost impossible to introduce any new chemical or technology. They argue that we can never have a risk-free society.

Conversely, proponents of increased reliance on the precautionary principle say that it will encourage innovation in developing less harmful alternative chemicals and technologies, and in finding ways to prevent as much pollution as possible instead of relying mostly on pollution control. We can never have a risk-free society. But proponents believe we should make greater use of the precautionary principle to reduce many of the risks we face.

X *HOW WOULD YOU VOTE?* Should we rely more on the precautionary principle as a way to reduce the risks from chemicals and technologies? Cast your vote online at http://biology .brookscole.com/miller11.

14-5 RISK ANALYSIS

Science, Poverty, and Lifestyles: Estimating Risks

Scientists have developed ways to evaluate and compare risks, decide how much risk is acceptable, and find affordable ways to reduce it.

Risk analysis involves identifying hazards and evaluating their associated risks (*risk assessment;* Figure 14-2, left), ranking risks (*comparative risk analysis*), determin-

ing options and making decisions about reducing or eliminating risks (*risk management;* Figure 14-2, right), and informing decision makers and the public about risks (*risk communication*).

Statistical probabilities based on past experience, animal testing and other tests, and epidemiological studies are used to estimate risks from older technologies and chemicals. To evaluate new technologies and products, risk evaluators use more uncertain statistical probabilities, based on models rather than actual experience and testing.

Figure 14-11 lists the results of a *comparative risk analysis,* summarizing the greatest ecological and health risks identified by a panel of scientists acting as advisers to the EPA.

Comparative Risk Analysis

Most Serious Ecological and Health Problems

High-Risk Health Problems
- Indoor air pollution
- Outdoor air pollution
- Worker exposure to industrial or farm chemicals
- Pollutants in drinking water
- Pesticide residues on food
- Toxic chemicals in consumer products

High-Risk Ecological Problems
- Global climate change
- Stratospheric ozone depletion
- Wildlife habitat alteration and destruction
- Species extinction and loss of biodiversity

Medium-Risk Ecological Problems
- Acid deposition
- Pesticides
- Airborne toxic chemicals
- Toxic chemicals, nutrients, and sediment in surface waters

Low-Risk Ecological Problems
- Oil spills
- Groundwater pollution
- Radioactive isotopes
- Acid runoff to surface waters
- Thermal pollution

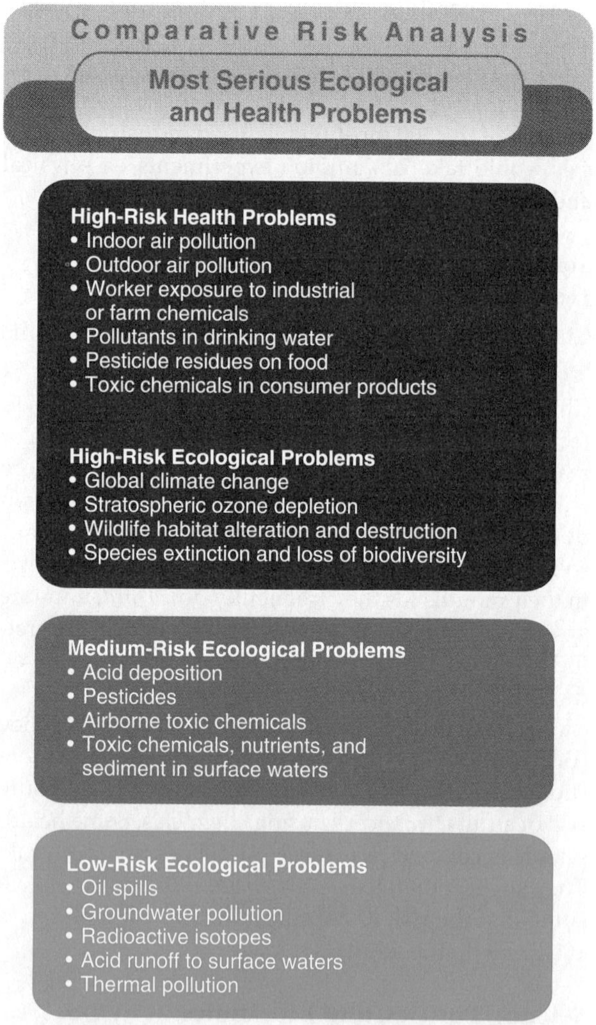

Figure 14-11 Science: *comparative risk analysis* of the most serious ecological and health problems according to scientists acting as advisers to the EPA. Risks under each category are not listed in rank order. *Critical thinking: which two risks in each of the two high-risk problems do you believe are the most serious?* (Data from Science Advisory Board, *Reducing Risks,* Washington, D.C.: Environmental Protection Agency, 1990)

The greatest risks many people face today are rarely dramatic enough to make the daily news. In terms of the number of premature deaths per year (Figure 14-12) and reduced life span (Figure 14-13, p. 342), *the greatest risk by far is poverty.* Its high death toll is a result of malnutrition, increased susceptibility to normally nonfatal infectious diseases, and often-fatal infectious diseases from lack of access to a safe water supply.

A sharp reduction in or elimination of poverty would do far more to improve longevity and human health than any other measure. It would also greatly improve human rights, provide more people with income to stimulate economic development, and reduce environmental degradation and the threat of terrorism. Sharply reducing poverty is a win-win situation for people, economies, and the environment.

After the health risks associated with poverty and gender, the greatest risks of premature death mostly result from voluntary choices people make about their lifestyles (Figures 14-12 and 14-13). The best ways to reduce one's risk of premature death and serious health risks are to avoid smoking and exposure to smoke, lose excess weight, reduce consumption of foods containing cholesterol and saturated fats, eat a variety of fruits and vegetables, exercise regularly, not drink alcohol or drink no more than two drinks in a single day, avoid excess sunlight (which ages skin and causes skin cancer), and practice safe sex (including abstinence).

Science: Estimating Risks from Technologies

Estimating risks from using certain technologies is difficult because of the unpredictability of human behavior, chance, and sabotage.

The more complex a technological system and the more people needed to design and run it, the more difficult it is to estimate the risks. The overall *reliability* or the probability (expressed as a percentage) that a person or device will perform without failure or error is the product of two factors:

$$\text{System reliability (\%)} = \frac{\text{Technology}}{\text{reliability}} \times \frac{\text{Human}}{\text{reliability}}$$

With careful design, quality control, maintenance, and monitoring, a highly complex system such as a nuclear power plant or space shuttle can achieve a high degree of technological reliability. But human reliability usually is much lower than technological reliability and almost impossible to predict: To err is human.

Suppose the technological reliability of a nuclear power plant is 95% (0.95) and human reliability is 75% (0.75). Then the overall system reliability is 71% (0.95 × 0.75 = 71%). Even if we could make the technology 100% reliable (1.0), the overall system reliability would still be only 75% (1.0 × 0.75 × 100 = 75%). The crucial dependence of even the most carefully designed systems on unpredictable human reliability helps explain

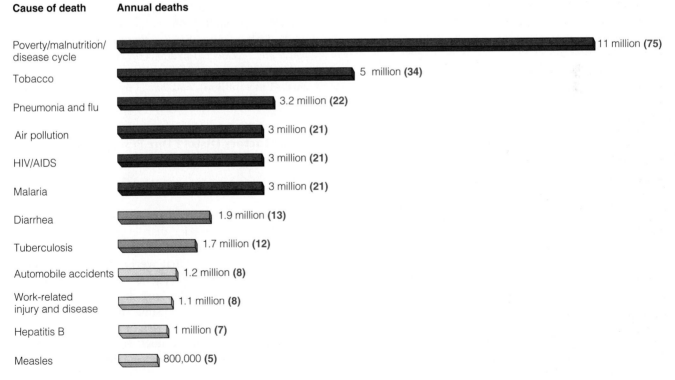

Cause of death   **Annual deaths**

- Poverty/malnutrition/disease cycle — 11 million **(75)**
- Tobacco — 5 million **(34)**
- Pneumonia and flu — 3.2 million **(22)**
- Air pollution — 3 million **(21)**
- HIV/AIDS — 3 million **(21)**
- Malaria — 3 million **(21)**
- Diarrhea — 1.9 million **(13)**
- Tuberculosis — 1.7 million **(12)**
- Automobile accidents — 1.2 million **(8)**
- Work-related injury and disease — 1.1 million **(8)**
- Hepatitis B — 1 million **(7)**
- Measles — 800,000 **(5)**

Figure 14-12 Global outlook: number of deaths per year in the world from various causes. Numbers in parentheses give these deaths in terms of the number of fully loaded 400-passenger jumbo jets crashing *every day of the year* with no survivors. Because of sensational media coverage, most people have a distorted view of the largest annual causes of death. *Critical thinking: which three of these items are most likely to shorten your life span?*

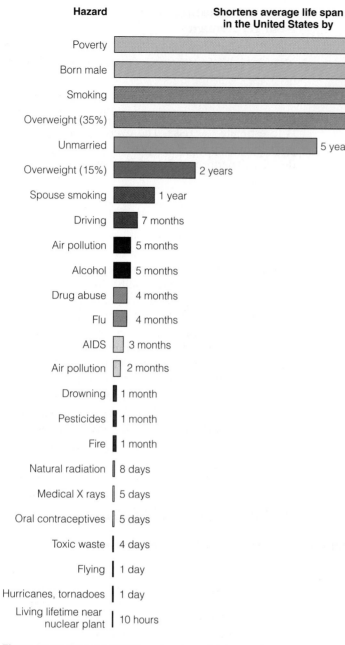

Hazard	Shortens average life span in the United States by
Poverty	7–10 years
Born male	7.5 years
Smoking	6–10 years
Overweight (35%)	6 years
Unmarried	5 years
Overweight (15%)	2 years
Spouse smoking	1 year
Driving	7 months
Air pollution	5 months
Alcohol	5 months
Drug abuse	4 months
Flu	4 months
AIDS	3 months
Air pollution	2 months
Drowning	1 month
Pesticides	1 month
Fire	1 month
Natural radiation	8 days
Medical X rays	5 days
Oral contraceptives	5 days
Toxic waste	4 days
Flying	1 day
Hurricanes, tornadoes	1 day
Living lifetime near nuclear plant	10 hours

Figure 14-13 Global outlook: comparison of risks people face, expressed in terms of shorter average life span. After poverty and gender, the greatest risks people face come mostly from the lifestyle choices they make. These are merely generalized relative estimates. Individual responses to these risks can differ because of factors such as genetic variation, family medical history, emotional makeup, stress, and social ties and support. *Critical thinking: which three of these items are most likely to shorten your life span?* (Data from Bernard L. Cohen)

allegedly almost "impossible" tragedies such as the Chernobyl nuclear power plant accident and the *Challenger* and *Columbia* space shuttle accidents.

One way to make a system more foolproof or fail-safe is to move more of the potentially fallible elements from the human side to the technical side. However, chance events such as a lightning bolt can knock out an automatic control system, and no machine or computer program can completely replace human judgment. Also, the parts in any automated control system are manufactured, assembled, tested, certified, and maintained by fallible human beings. In addition, computer software programs used to monitor and control complex systems can be flawed because of human error or can be deliberately modified by computer viruses to malfunction.

Perceiving Risks

Most individuals are poor at evaluating the relative risks they face, mostly because of misleading information, denial, and irrational fears.

Most of us are not good at assessing the relative risks from the hazards that surround us. Also, many people deny or shrug off the high-risk chances of death (or injury) from voluntary activities they enjoy, such as *motorcycling* (1 death in 50 participants), *smoking* (1 in 250 by age 70 for a pack-a-day smoker), *hang gliding* (1 in 1,250), and *driving* (1 in 3,300 without a seatbelt and 1 in 6,070 with a seatbelt). Indeed, the most dangerous thing most people in many countries do each day is drive or ride in a car.

Yet some of these same people may be terrified about the possibility of being killed by a *gun* (1 in 28,000 in the United States), *flu* (1 in 130,000), *nuclear power plant accident* (1 in 200,000), *West Nile virus* (1 in 1 million), *lightning* (1 in 3 million), *commercial airplane crash* (1 in 9 million), *snakebite* (1 in 36 million), or *shark attack* (1 in 281 million).

What Factors Distort Our Perceptions of Risk?

Several factors can give people a distorted sense of risk.

Four factors can cause people to see a technology or a product as being riskier than experts judge it to be. First is the *degree of control* we have. Most of us have a greater fear of things that we do not have personal control over. For example, some individuals feel safer driving their own car for long distances through bad traffic than traveling the same distance on a plane. But look at the math. The risk of dying in a car accident while using your seatbelt is 1 in 6,070 whereas the risk of dying in a commercial airliner crash is 1 in 9 million.

Second is *fear of the unknown*. Most people have greater fear of a new, unknown product or technology than they do of an older, more familiar one. Examples include a greater fear of genetically modified food than of food produced by traditional plant-breeding techniques, and a greater fear of nuclear power plants than of more familiar coal-fired power plants.

Third is *whether we voluntarily take the risk.* For example, we might perceive that the risk from driving, which is largely voluntary, is less than that from a nuclear power plant, which is mostly imposed on us whether we like it or not.

Fourth is *whether a risk is catastrophic,* not chronic. We usually have a much greater fear of a well-publicized death toll from a single catastrophic accident than from the same or an even larger death toll spread out over a longer time. Examples include a severe nuclear power plant accident, an industrial explosion, or an accidental plane crash, as opposed to coal-burning power plants, automobiles, or smoking. *Critical thinking: what three things do you fear that are not very risky?*

There is also concern over the *unfair distribution of risks* from the use of a technology or chemical. Citizens are outraged when government officials decide to put a hazardous waste landfill or incinerator in or near their neighborhood. Even when the decision is based on careful risk analysis, it is usually seen as politics, not science. Residents will not be satisfied by estimates that the lifetime risks of cancer death from the facility are not greater than, say, 1 in 100,000. Instead, they point out that living near the facility means that they will have a much higher risk of dying from cancer than would people living farther away.

Becoming Better at Risk Analysis

To become better at risk analysis, you can carefully evaluate or tune out the barrage of bad news covered in the media, compare risks, and concentrate on reducing risks over which you have some control.

You can do three things to become better at estimating risks.

First, carefully evaluate news reports. Recognize that the media often give an exaggerated view of risks to capture our interest and sell newspapers or gain TV viewers.

Second, compare risks. Do you risk getting cancer by eating a charcoal-broiled steak once or twice a week? Yes, because in theory anything can harm you. The question is whether this danger is great enough for you to worry about. In evaluating a risk, the question is not "Is it safe?" but rather "How risky is it compared to other risks?"

Third, concentrate on the most serious risks to your life and health that you have some control over and stop worrying about smaller risks and those over which you have no control. When you worry about

something, the most important question to ask is, "Do I have any control over this?"

You have control over major ways to reduce risks from heart attack, stroke, and many forms of cancer because you can decide whether you smoke, what you eat, how much exercise you get, how much alcohol you consume, how often you expose yourself to the sun's ultraviolet rays, and whether you practice safe sex. Concentrate on evaluating these important choices, and you will have a much greater chance of living a longer, healthier, happier, and less fearful life.

The burden of proof imposed on individuals, companies, and institutions should be to show that pollution prevention options have been thoroughly examined, evaluated, and used before lesser options are chosen.

JOEL HIRSCHORN

CRITICAL THINKING

1. Explain why you agree or disagree with the proposals for reducing the death toll and other harmful effects of smoking listed on p. 327. Do you believe that there should be a ban on smoking indoors in all public places? Explain.

2. Should we have zero pollution levels for all toxic and hazardous chemicals? Explain. What are the alternatives?

3. Do you believe that health and safety standards in the workplace should be strengthened and enforced more vigorously, even if this causes a loss of jobs when companies transfer operations to countries with weaker standards? Explain.

4. Evaluate the following statements:
   a. We should not get worked up about exposure to toxic chemicals because almost any chemical can cause some harm at a large enough dosage.
   b. We should not worry so much about exposure to toxic chemicals because through genetic adaptation we can develop immunity to such chemicals.
   c. We should not worry so much about exposure to toxic chemicals because we can use genetic engineering to reduce or eliminate such problems.

5. How can changes in the age structure of a human population increase the spread of infectious diseases? How can the spread of infectious diseases affect the age structure of human populations?

6. Should laboratory-bred animals be used in laboratory experiments in toxicology? Explain. What are the alternatives?

7. What are the five major risks you face from **(a)** your lifestyle, **(b)** where you live, and **(c)** what you do for a living? Which of these risks are voluntary and which are involuntary? List the five most important things you can do to reduce these risks. Which of these things do you actually plan to do?

8. Congratulations! You are in charge of the world. List the three most important features of your program to reduce the risk from exposure to **(a)** infectious disease organisms and **(b)** toxic and hazardous chemicals.

LEARNING ONLINE

The website for this book includes review questions for the entire chapter, flash cards for key terms and concepts, a multiple-choice practice quiz, interesting Internet sites, references, and a guide for accessing thousands of InfoTrac® College Edition articles.

Visit

http://biology.brookscole.com/miller11

Then choose Chapter 14, and select a learning resource. For access to animations, additional quizzes, chapter outlines and summaries, register and log in to

Environmental Science ⊕ Now™

at **esnow.brookscole.com/miller11** using the access code card in the front of your book.

Active Graphing

Visit http://esnow.brookscole.com/miller11 to explore the graphing exercise for this chapter.

CASE STUDY
When Is a Lichen Like a Canary?

Nineteenth-century coal miners took canaries with them into the mines—not to enjoy their songs but to listen for the moment when they stopped singing. Then the miners knew it was time to get out of the mine because the air contained methane, which could ignite and explode.

Today we use sophisticated equipment to monitor air quality, but living things such as lichens (Figure 15-1) can also warn us of bad air. Lichens consist of a fungus and an alga living together, usually in a mutually beneficial (mutualistic) partnership. You have probably seen lichens growing as crusts or leafy growths on rocks (Figure 15-1, right), walls, tombstones, and tree trunks or hanging down from twigs and branches (Figure 15-1, left).

These hardy pioneer species are good biological indicators of air pollution because they continually absorb air as a source of nourishment. A highly polluted area around an industrial plant may have no lichens or only gray-green crusty lichen. An area with moderate air pollution may have orange crusty lichens on walls. Walls and trees in areas with fairly clean air may support leafy lichens.

Some lichen species are sensitive to specific air-polluting chemicals. Old man's beard (*Usnea trichodea*, Figure 15-1, left) and yellow *Evernia* lichens, for example, sicken or die in the presence of excess sulfur dioxide.

Because lichens are widespread, long lived, and anchored in place, they can also help track pollution to its source. Isle Royale in Lake Superior is a place where no car or smokestack has ever intruded. The scientist who discovered sulfur dioxide pollution there used *Evernia* lichens to point the finger northward to coal-burning facilities at Thunder Bay, Canada.

In 1986, the Chernobyl nuclear power plant in Ukraine exploded and spewed radioactive particles into the atmosphere. Some of these particles fell to the ground over northern Scandinavia and were absorbed by the lichens that carpet much of Lapland. The area's Saami people depend on reindeer meat for food, and the reindeer feed on lichens. After Chernobyl, more than 70,000 reindeer had to be killed and the meat discarded because it was too radioactive to eat. Scientists helped the Saami identify which of the remaining reindeer to move by analyzing lichens to pinpoint the most contaminated areas.

We all must breathe air from a global atmospheric commons in which air currents and winds can transport some pollutants over long distances. Lichens can alert us to the danger, but as with all forms of pollution, the best solution is prevention.

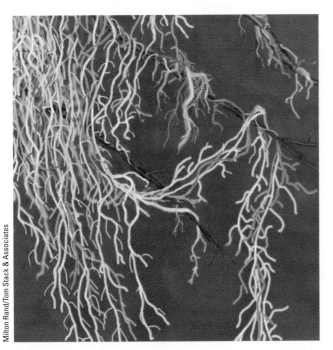

Figure 15-1 Natural capital: red and yellow crustose lichens growing on tundra rock in the foothills of the Sierra Nevada near Merced, California (right), and *Usnea trichodea* lichen growing on a branch of a larch tree in Gifford Pinchot National Park, Washington (left). The vulnerability of various lichen species to specific air pollutants can help researchers detect levels of these pollutants and track down their sources.

I thought I saw a blue jay this morning. But the smog was so bad that it turned out to be a cardinal holding its breath.

MICHAEL J. COHEN

This chapter discusses the types, sources, and effects of chemicals that pollute the outdoor and indoor air that we breathe and presents solutions for reducing these threats to our health and to ecosystems. It addresses the following questions:

- What layers are found in the atmosphere?

- What are the major outdoor air pollutants, and where do they come from?

- What are two types of smog?

- What is acid deposition, and how can it be reduced?

- What are the harmful effects of air pollutants?

- How can we prevent and control air pollution?

KEY IDEAS

- Burning fossil fuels in motor vehicles and power and industrial plants is the major source of air pollution from human activities.

- Photochemical smog is a mixture of air pollutants formed by the reaction of nitrogen oxides and volatile organic hydrocarbons under the influence of sunlight.

- Industrial smog is a mixture of sulfur dioxide, droplets of sulfuric acid, and suspended solid particles emitted or formed in the atmosphere when coal and oil are burned.

- Sulfur dioxide, nitrogen oxides, and particulates produced mostly by burning coal can react in the atmosphere to produce acidic chemicals that can travel long distances before returning to the earth's surface as harmful acid deposition.

- Indoor air pollution usually poses a much greater threat to human health than outdoor air pollution, especially for the poor in developing countries.

- At least 3 million people (most of them in Asia) die prematurely each year from the effects of air pollution—mostly from burning wood or coal inside dwellings in developing countries.

- The Clean Air Acts in the United States have greatly reduced outdoor air pollution from six major pollutants but these laws can be made more effective.

- We need to focus on preventing air pollution, with emphasis on sharply reducing indoor air pollution in developing countries.

15-1 STRUCTURE AND SCIENCE OF THE ATMOSPHERE

Science: The Troposphere

The atmosphere's innermost layer consists mostly of nitrogen and oxygen, plus smaller amounts of water vapor and carbon dioxide.

We live at the bottom of a thin layer of gases surrounding the earth, called the *atmosphere*. It is divided into several spherical layers (Figure 15-2), each of which is characterized by abrupt changes in temperature as a result of differences in the absorption of incoming solar energy.

About 75–80% of the earth's air mass is found in the **troposphere,** the atmospheric layer closest to the earth's surface. This layer extends only 17 kilometers (11 miles) above sea level at the equator and 8 kilometers (5 miles) over the poles. If the earth were the size of an apple, this lower layer containing the air we breathe would be no thicker than the apple's skin.

Take a deep breath. About 99% of the volume of the air you inhaled from the troposphere consists of two gases: nitrogen (78%) and oxygen (21%). The remainder consists of water vapor (varying from 0.01% at the frigid poles to 4% in the humid tropics), slightly less than 1% argon (Ar), 0.038% carbon dioxide (CO_2), and trace amounts of several other gases.

The troposphere is also involved in the chemical cycling of the earth's vital nutrients. In addition, this thin and turbulent layer of rising and falling air currents and winds is largely responsible for the planet's short-term *weather* and long-term *climate*.

Science: The Stratosphere

Ozone in the atmosphere's second layer filters out most of the sun's UV radiation that is harmful to us and most other species.

The atmosphere's second layer is the **stratosphere,** which extends 17–48 kilometers (11–30 miles) above the earth's surface (Figure 15-2). Although the stratosphere contains less matter than the troposphere, its composition is similar, with two notable exceptions: its volume of water vapor is about 1/1,000 as much and its concentration of ozone (O_3) is much higher.

Stratospheric ozone is produced when oxygen molecules there interact with ultraviolet (UV) radiation emitted by the sun ($3\ O_2 + UV \longrightarrow 2\ O_3$). This "global sunscreen" of ozone in the stratosphere prevents 95% of the sun's harmful UV radiation from reaching the earth's surface.

The UV filter of "good" ozone in the lower stratosphere allows us and other forms of life to exist on land and helps protect us from sunburn, skin and eye cancer, cataracts, and damage to our immune systems. It

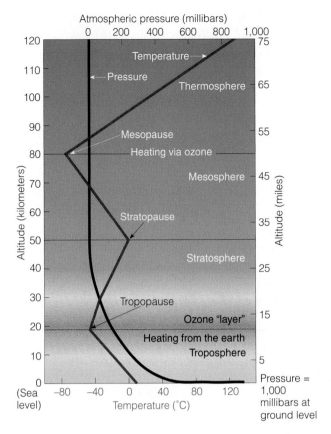

Atmospheric pressure (millibars)

Temperature

Pressure

Thermosphere

Mesopause

Heating via ozone

Mesosphere

Stratopause

Stratosphere

Tropopause

Ozone "layer"

Heating from the earth

Troposphere

Altitude (kilometers)

Altitude (miles)

Temperature (°C)

Pressure = 1,000 millibars at ground level

Figure 15-2 Natural capital: the earth's atmosphere consists of several layers. The average temperature of the atmosphere varies with altitude (red line). Most UV radiation from the sun is absorbed by ozone (O_3), found primarily in the stratosphere in the *ozone layer* 17–26 kilometers (10–16 miles) above sea level.

also prevents much of the oxygen in the troposphere from being converted to photochemical ozone, a harmful air pollutant.

Some human activities are *decreasing* the amount of beneficial or "good" ozone in the stratosphere and *increasing* the amount of harmful or "bad" ozone in the troposphere—especially in some urban areas. Ozone in this portion of the atmosphere near the earth's surface damages plants, lung tissues, and some materials such as rubber.

15-2 OUTDOOR AIR POLLUTION

Science: Types and Sources of Air Pollution

Outdoor air pollutants come mostly from natural sources and burning fossil fuels in motor vehicles and power and industrial plants.

Air pollution is the presence of chemicals in the atmosphere in concentrations high enough to harm or-

ganisms and materials (such as metals and stone used in buildings and statues) and to alter climate. The effects of air pollution range from annoying to lethal.

Air pollutants come from both natural and human sources. Natural sources include dust blowing off the earth's surface (Figure 5-1, p. 78), forest fires, volcanic eruptions volatile organic chemicals released by some plants, the decay of plants, and sea spray. Most natural sources of air pollution are spread out and, except for those from volcanic eruptions and some forest fires, rarely reach harmful levels.

Air pollution from human activities is not new (Science Spotlight, p. 350). Since our discovery of fire, we have added various types of pollutants to the troposphere. Our inputs increased when we began extracting and burning coal, first for heat and later for generating electricity and producing materials such as steel.

Most outdoor pollutants from human activities in today's urban areas enter the atmosphere from the burning of fossil fuels in power plants and factories (*stationary sources*, Figure 1-8, p. 12) and in motor vehicles (*mobile sources*). Pollutants from such activities can reach harmful levels in the troposphere, especially in urban areas where people, cars, and industrial activities are concentrated.

Scientists classify outdoor air pollutants into two categories. **Primary pollutants** are emitted directly into the troposphere in a potentially harmful form. Examples include soot and carbon monoxide. While in the troposphere, some primary pollutants may react with one another or with the basic components of air to form new pollutants, called **secondary pollutants** (Figure 15-3, p. 348).

Because of their high concentrations of cars and factories, cities normally have higher air pollution levels than rural areas. However, prevailing winds can spread long-lived primary and secondary air pollutants from urban and industrial areas to the countryside as well as to other urban areas.

Indoor air pollutants come from infiltration of polluted outdoor air and chemicals used or produced inside buildings. Experts in risk analysis rate indoor and outdoor air pollution as high-risk human health problems.

According to the World Health Organization (WHO), one of every six people on the earth (more than 1.1 billion people) lives in an urban area where the outdoor air is unhealthy to breathe. Most of these individuals live in densely populated cities in developing countries where air pollution control laws do not exist or are poorly enforced. The biggest health threat comes from indoor air pollution when the poor must burn wood, charcoal, coal, or dung in open fires or poorly designed stoves to heat their dwellings and cook their food. In other words, poverty can mean poor air for poor people.

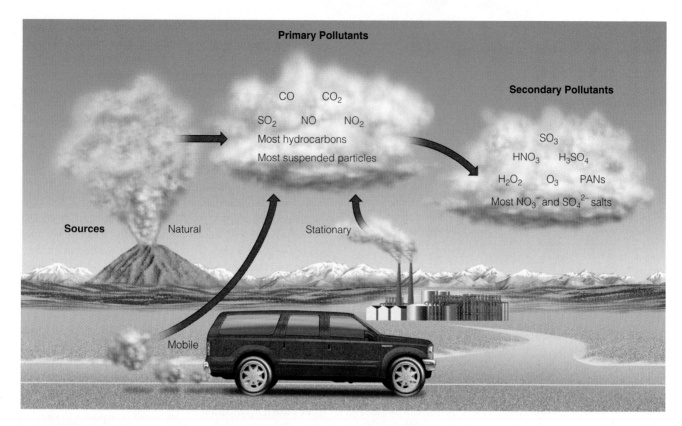

Primary Pollutants

CO   CO$_2$

SO$_2$ NO   NO$_2$

Most hydrocarbons

Most suspended particles

Secondary Pollutants

SO$_3$

HNO$_3$ H$_3$SO$_4$

H$_2$O$_2$ O$_3$ PANs

Most NO$_3^-$ and SO$_4^{2-}$ salts

Sources   Natural   Stationary

Mobile

Figure 15-3 Natural capital degradation: sources and types of air pollutants. Human inputs of air pollutants may come from *mobile sources* (such as cars) and *stationary sources* (such as industrial and power plants). Some *primary air pollutants* may react with one another or with other chemicals in the air to form *secondary air pollutants.*

In the United States and most other developed countries, government-mandated standards set maximum allowable atmospheric concentrations, or criteria, for six *criteria or conventional air pollutants* commonly found in outdoor air. Table 15-1 lists the properties, sources, and effects of these six types of pollutants. Study this table carefully.

Most scientists would add two other chemicals to the six criteria pollutants shown in Table 15-1. *Volatile organic compounds (VOCs),* mostly hydrocarbons, play a role in the formation of the photochemical smog that plagues many cities. *Carbon dioxide (CO₂)* has the ability to increase the temperature of the troposphere and thus change climate, as discussed in Chapter 16.

Many oil and coal companies do not want carbon dioxide classified as a pollutant and have lobbied U.S. elected officials successfully to prevent such classification. Otherwise, we might use less of these fuels and the companies would have to spend more money controlling CO₂ emissions. Twelve U.S. states have sued the EPA for failure to regulate CO₂ emissions. Some states are also passing their own laws that regulate CO₂ emissions. *Critical thinking: do you believe that carbon dioxide should be classified as a pollutant?*

15-3 PHOTOCHEMICAL AND INDUSTRIAL SMOG

Science: Photochemical Smog

Photochemical smog is a mixture of air pollutants formed by the reaction of nitrogen oxides and volatile organic hydrocarbons under the influence of sunlight.

A *photochemical reaction* is any chemical reaction activated by light. Air pollution known as **photochemical smog** forms when a mix of nitrogen oxides (NO and NO$_2$, collectively called NO$_x$) and volatile organic hydrocarbon compounds from natural and human sources react under the influence of UV radiation from the sun to produce a mixture of more than 100 primary and secondary pollutants. In greatly simplified terms,

VOCs + NO$_x$ + heat + sunlight ⟶ ground level ozone (O$_3$)
+ other photochemical oxidants
+ aldehydes
+ other secondary air pollutants

CARBON MONOXIDE (CO)

Description: Colorless, odorless gas that is poisonous to air-breathing animals; forms during the incomplete combustion of carbon-containing fuels ($2\,C + O_2 \longrightarrow 2\,CO$).

Major human sources: Cigarette smoking, incomplete burning of fossil fuels. About 77% (95% in cities) comes from motor vehicle exhaust.

Health effects: Reacts with hemoglobin in red blood cells and reduces the ability of blood to bring oxygen to body cells and tissues. This impairs perception and thinking; slows reflexes; causes headaches, drowsiness, dizziness, and nausea; can trigger heart attacks and angina; damages the development of fetuses and young children; and aggravates chronic bronchitis, emphysema, and anemia. At high levels it causes collapse, coma, irreversible brain cell damage, and death.

NITROGEN DIOXIDE (NO₂)

Description: Reddish-brown irritating gas that gives photochemical smog its brownish color; in the atmosphere can be converted to nitric acid (HNO_3), a major component of acid deposition.

Major human sources: Fossil fuel burning in motor vehicles (49%) and power and industrial plants (49%).

Health effects: Lung irritation and damage; aggravates asthma and chronic bronchitis; increases susceptibility to respiratory infections such as the flu and common colds (especially in young children and older adults).

Environmental effects: Reduces visibility; acid deposition of HNO_3 can damage trees, soils, and aquatic life in lakes.

Property damage: HNO_3 can corrode metals and eat away stone on buildings, statues, and monuments; NO_2 can damage fabrics.

SULFUR DIOXIDE (SO₂)

Description: Colorless, irritating; forms mostly from the combustion of sulfur-containing fossil fuels such as coal and oil ($S + O_2 \longrightarrow SO_2$); in the atmosphere can be converted to sulfuric acid (H_2SO_4), a major component of acid deposition.

Major human sources: Coal burning in power plants (88%) and industrial processes (10%).

Health effects: Breathing problems for healthy people; restriction of airways in people with asthma; chronic exposure can cause a permanent condition similar to bronchitis. According to the WHO, at least 625 million people are exposed to unsafe levels of sulfur dioxide from fossil fuel burning.

Environmental effects: Reduces visibility; acid deposition of H_2SO_4 can damage trees, soils, and aquatic life in lakes.

Property damage: SO_2 and H_2SO_4 can corrode metals and eat away stone on buildings, statues, and monuments; SO_2 can damage paint, paper, and leather.

SUSPENDED PARTICULATE MATTER (SPM)

Description: Variety of particles and droplets (aerosols) small and light enough to remain suspended in atmosphere for short periods (large particles) to long periods (small particles; cause smoke, dust, and haze.

Major human sources: Burning coal in power and industrial plants (40%), burning diesel and other fuels in vehicles (17%), agriculture (plowing, burning off fields), unpaved roads, construction.

Health effects: Nose and throat irritation, lung damage; aggravates bronchitis and asthma; shortens life; toxic particulates (such as lead, cadmium, PCBs, and dioxins) can cause mutations, reproductive problems, cancer.

Environmental effects: Reduces visibility; acid deposition of H_2SO_4 droplets can damage trees, soils, and aquatic life in lakes.

Property damage: Corrodes metal; soils and discolors buildings, clothes, fabrics, and paints.

OZONE (O₃)

Description: Highly reactive, irritating gas with an unpleasant odor that forms in the troposphere as a major component of photochemical smog.

Major human sources: Chemical reaction with volatile organic compounds (VOCs, emitted mostly by cars and industries) and nitrogen oxides to form photochemical smog.

Health effects: Breathing problems; coughing; eye, nose, and throat irritation; aggravates chronic diseases such as asthma, bronchitis, emphysema, and heart disease; reduces resistance to colds and pneumonia; may speed up lung tissue aging.

Environmental effects: Ozone can damage plants and trees; smog can reduce visibility.

Property damage: Damages rubber, fabrics, and paints.

LEAD

Description: Solid toxic metal and its compounds, emitted into the atmosphere as particulate matter.

Major human sources: Paint (old houses), smelters (metal refineries), lead manufacture, storage batteries, leaded gasoline (being phased out in developed countries).

Health effects: Accumulates in the body; brain and other nervous system damage and mental retardation (especially in children); digestive and other health problems; some lead-containing chemicals cause cancer in test animals.

Environmental effects: Can harm wildlife.

*Data from U.S. Environmental Protection Agency.

The formation of photochemical smog involves a complex series of chemical reactions. It begins inside automobile engines and in the boilers of coal-burning power and industrial plants. At the high temperatures found there, nitrogen and oxygen in air react to produce colorless nitric oxide ($N_2 + O_2 \longrightarrow 2\,NO$). In the atmosphere, some of the NO is converted to nitrogen dioxide (NO_2), a yellowish-brown gas with a choking odor. NO_2 is the cause of the brownish haze that hangs over many cities during the afternoons of sunny days, explaining why photochemical smog sometimes is called *brown-air smog*.

When exposed to UV radiation from the sun, some of the NO_2 engages in a complex series of reactions with hydrocarbons (mostly released by vegetation, motor vehicles, and other human activities) that

Air Pollution in the Past: The Bad Old Days

Modern civilization did not invent air pollution. It probably began when humans discovered fire, began to burn wood in poorly ventilated caves for warmth and cooking, and inhaled unhealthy smoke and soot.

During the Middle Ages, a haze of wood smoke hung over densely packed urban areas. The Industrial Revolution brought even worse air pollution as coal was burned to power factories and heat homes.

By the 1850s, London had become famous for its "pea soup" fog—a mixture of coal smoke and fog that blanketed the city. In 1880, a prolonged coal fog killed an estimated 2,200 people. Another in 1911 killed more than 1,100 Londoners. The authors of a report on this disaster coined the word *smog* to describe the deadly mixture of smoke and fog that enveloped the city.

In 1952, an even worse yellow fog lasted for 5 days and killed 4,000–12,000 Londoners, prompting Parliament to pass the Clean Air Act of 1956. Air pollution disasters in 1956, 1957, and 1962 killed 2,500 more people. Because of strong air pollution laws, London's air today

is much cleaner, and "pea soup" fogs are a thing of the past. Instead, the major threat comes from air pollutants emitted by motor vehicles.

The Industrial Revolution, powered by coal-burning factories and homes, brought air pollution to the United States. Large industrial cities such as Pittsburgh, Pennsylvania, and St. Louis, Missouri, were notorious for their smoky air. By the 1940s, the air over some cities was so polluted that people had to use their automobile headlights during the day.

The first documented air pollution disaster in the United States occurred on October 29, 1948, at the small industrial town of Donora in Pennsylvania's Monongahela River Valley. Pollutants from the area's coal-burning industries became trapped in a dense fog that stagnated over the valley for 5 days. About 6,000 of the town's 14,000 inhabitants became sick, and 22 of them died. This killer fog resulted from a combination of mountainous terrain surrounding the valley and weather conditions that trapped and concentrated deadly pollutants emitted by the community's steel mill, zinc smelter, and sulfuric acid plant.

In 1963, high concentrations of air pollutants accumulated in the

air over New York City, killing about 300 people and injuring thousands. Other episodes in New York, Los Angeles, and other large cities in the 1960s resulted in much stronger air pollution control programs in the 1970s.

In 1952, Oregon became the first state to pass a law controlling air pollution. The U.S. Congress passed the original version of the Clean Air Act in 1963. It did not have much effect until a stronger version was enacted in 1970 and the Environmental Protection Agency (EPA) was created and empowered to set and enforce national air pollution standards. Even stricter emission standards were imposed by amendments to the Clean Air Act in 1977 and 1990. Mostly as a result of these laws and actions by states and local areas, air quality has dramatically improved throughout the United States.

Critical Thinking

Explain why you agree or disagree with the following statement: "Air pollution in the United States is no longer a major concern because of the significant progress made in reducing outdoor air pollution since 1970."

---

produce photochemical smog—a mixture of ozone, nitric acid, aldehydes, peroxyacyl nitrates (PANs), and other secondary pollutants.

Hotter days lead to higher levels of ozone and other components of smog. As traffic increases on a sunny day, photochemical smog (dominated by O_3) usually builds up to peak levels by early afternoon, irritating people's eyes and respiratory tracts.

All modern cities suffer from some photochemical smog, but it is much more common in cities with sunny, warm, dry climates and lots of motor vehicles—for example, Los Angeles, Denver, and Salt Lake City in the United States; Sydney, Australia; São Paulo, Brazil; Buenos Aires, Argentina; and Mexico City, Mexico (Figure 15-4). According to a 1999 study, if 400 million people in China drive gasoline-powered cars by 2050 as projected, the resulting photochemical

smog could cover the entire western Pacific with ozone, extending to the United States.

Environmental Science ⊕ Now™

See how photochemical smog forms and how it affects us at Environmental ScienceNow.

Science: Industrial Smog

Industrial smog is a mixture of sulfur dioxide, droplets of sulfuric acid, and suspended solid particles emitted by burning coal and oil.

Fifty years ago, cities such as London, England, and Chicago and Pittsburgh in the United States burned large amounts of coal and heavy oil (which contain sulfur impurities) in power plants and factories and for heating homes and cooking food. During winter,

Mark Edwards/Peter Arnold, Inc.

Figure 15-4 Global outlook: photochemical smog in Mexico City, Mexico. *Critical thinking: why is photochemical smog worse in cities with sunny, warm climages?*

people in such cities were exposed to **industrial smog** consisting mostly of sulfur dioxide, aerosols containing suspended droplets of sulfuric acid formed from sulfur dioxide, and a variety of suspended solid particles that gave the resulting smog a gray color (hence the alternative name *gray-air smog*).

Today, urban industrial smog is rarely a problem in most developed countries. Coal and heavy oil in those nations are now burned only in large boilers with reasonably good pollution control or with tall smokestacks that transfer the pollutants to downwind rural areas.

Unfortunately, industrial smog remains a problem in industrialized urban areas of China, India (Figure 15-5), Ukraine, and some eastern European countries (especially the "black triangle" region of Slovakia, Poland, Hungary, and the Czech Republic), where large quantities of coal are burned in factories and in houses with inadequate pollution controls.

A 2002 study by the UN Environment Programme found that a huge blanket of mostly industrial smog—called the *Asian brown cloud*—stretches nearly continuously across much of India, Bangladesh, and the industrial heart of China and parts of the open sea in this area. This cloud—a whopping 3 kilometers (2 miles) thick—results from enormous emissions of ash, smoke, dust, and acidic compounds produced by people burning coal in industries and homes and clearing and burning forests for planting crops, along with dust blowing off deserts in western Asia. As the cloud travels, it picks up many toxic pollutants.

The rapid industrialization of parts of southeastern Asia, especially China and India, is repeating on a

much larger scale the smoky, unhealthy, coal-burning past of the Industrial Revolution in Europe and the United States during the 19th and early 20th centuries (Science Spotlight, p. 350).

Science: Factors Influencing the Formation of Photochemical and Industrial Smog

Outdoor air pollution can be reduced by precipitation and winds and increased by urban buildings, mountains, and high temperatures.

Three natural factors help *reduce* outdoor air pollution. First, *rain and snow* help cleanse the air of pollutants. This helps explain why cities with dry climates are more prone to photochemical smog than cities with wet climates. Second, *salty sea spray from the oceans* can wash out much of the particulates and water-soluble pollutants from air that flows from land over the oceans. Third, *winds* can help sweep pollutants away, dilute them by mixing them with cleaner air, and bring in fresh air. Ultimately, these pollutants are blown somewhere else or are deposited from the sky onto surface waters, soil, and buildings. There is no away.

Four other factors can *increase* outdoor air pollution. First, *urban buildings* can slow wind speed and reduce dilution and removal of pollutants. Second, *hills and mountains* can reduce the flow of air in valleys below them and allow pollutant levels to build up at ground level. Third, *high temperatures* promote the chemical reactions leading to photochemical smog formation.

Deb Kushal/Peter Arnold, Inc.

Figure 15-5 Global outlook: industrial smog from an industrial plant in India. *Critical thinking: why does industrial smog tend to be more serious in cold climates?*

A fourth factor—the so-called *grasshopper effect*—is based on atmospheric distillation that transfers volatile air pollutants from tropical and temperate areas to the earth's poles. It occurs when volatile compounds evaporate from warm terrestrial areas at low latitudes and rise high into the atmosphere. Then they are deposited in the oceans or carried to higher latitudes at or near the earth's poles by atmospheric currents and oceanic currents of water.

Science: Effects of Temperature Inversions on Outdoor Air Pollution

A layer of warm air sitting on top of a layer of cool air near the ground can prevent outdoor pollutants from rising and dispersing.

During daylight, the sun warms the air near the earth's surface. Normally, this warm air and most of the pollutants it contains rise to mix with the cooler air above it. The mixing of warm and cold air creates turbulence, which disperses the pollutants.

Under certain atmospheric conditions, however, a layer of warm air can lie atop a layer of cooler air nearer the ground, creating a **temperature inversion.** Because the cooler air is denser than the warmer air above it, the air near the surface does not rise and mix with the air above it. Pollutants can concentrate in this stagnant layer of cool air near the ground.

Areas with certain types of topography and weather conditions are especially susceptible to prolonged temperature inversions. One such area is a town or city located in a valley surrounded by mountains that experiences cloudy and cold weather during part of the year. In such cases, the surrounding mountains and the clouds block much of the winter sun that causes air to heat and rise, and the mountains block air from being blown away. As long as these stagnant conditions persist, concentrations of pollutants in the valley below will build up to harmful and even lethal concentrations. This is what happened during the 1948 air pollution disaster in the valley town of Donora, Pennsylvania (Science Spotlight, p. 350).

Another type of area vulnerable to temperature inversion is a city with several million people and motor vehicles in an area with a sunny climate, light winds, mountains on three sides, and the ocean on the other side. Here, the conditions are ideal for photochemical smog worsened by frequent thermal inversions.

This description applies to California's heavily populated Los Angeles basin, which experiences prolonged subsidence temperature inversions at least half of the year, mostly during the warm summer and fall. When a thermal inversion persists throughout the day, the surrounding mountains prevent the polluted surface air from being blown away by sea breezes.

Environmental Science ⊕ Now™
Learn more about thermal inversions and what they can mean for people in some cities at Environmental ScienceNow.

15-4 REGIONAL OUTDOOR AIR POLLUTION FROM ACID DEPOSITION

Science: Acid Deposition

Sulfur dioxide, nitrogen oxides, and particulates can react in the atmosphere to produce acidic chemicals that can travel long distances before returning to the earth's surface.

Most coal-burning power plants, ore smelters, and other industrial plants in developed countries use tall smokestacks to emit sulfur dioxide, suspended particles, and nitrogen oxides high into the troposphere where wind can mix, dilute, and disperse them.

These tall smokestacks reduce *local* air pollution, but they can increase *regional* air pollution downwind. The primary pollutants (sulfur dioxide and nitrogen oxides) emitted into the atmosphere above the inversion layer may be transported as far as 1,000 kilometers (600 miles) by prevailing winds. During their trip, they form secondary pollutants such as nitric acid vapor, droplets of sulfuric acid, and particles of acid-forming sulfate and nitrate salts.

These acidic substances remain in the atmosphere for 2–14 days, depending mostly on prevailing winds, precipitation, and other weather patterns. During this period they descend to the earth's surface in two forms. *Wet deposition* consists of acidic rain, snow, fog, and cloud vapor with a pH less than 5.6 (Figure 3-23, p. 53). *Dry deposition* consists of acidic particles. The resulting mixture is called **acid deposition** (Figure 15-6), sometimes termed *acid rain*. Most dry deposition occurs within 2–3 days fairly near the emission sources, whereas most wet deposition takes place within 4–14 days in more distant downwind areas.

Acid deposition is a *regional* air pollution problem in most parts of the world that lie downwind from coal-burning facilities and from urban areas with large numbers of cars. Such areas include the eastern United States (Figure 15-7, p. 354) and other parts of the world (Figure 15-8, p. 355).

In the United States, older coal-burning power and industrial plants without adequate pollution controls in the Ohio Valley emit the largest quantities of sulfur dioxide and other pollutants that can cause acid deposition. Mostly as a result of these emissions along with those by other industries and motor vehicles in urban areas, typical precipitation in the eastern United States

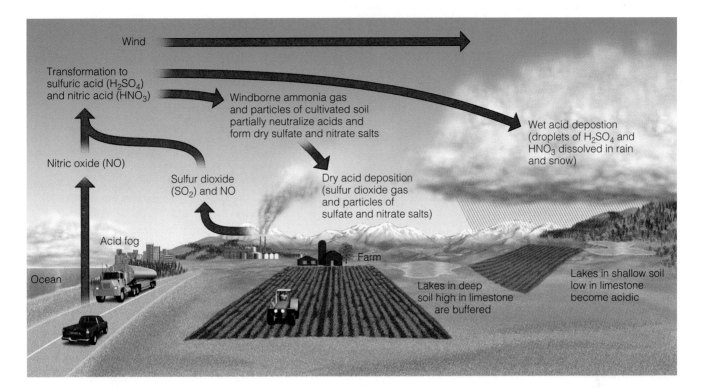

Environmental Science⊕Now™

Active Figure 15-6 Natural capital degradation: *acid deposition*, which consists of rain, snow, dust, or gas with a pH lower than 5.6, is commonly called acid rain. Soils and lakes vary in their ability to buffer or remove excess acidity. *See an animation based on this figure and take a short quiz on the concept.*

has a pH of 4.4–4.7 (Figure 15-7). This is 10 or more times the acidity of natural precipitation, which has a pH of 5.6. Some mountaintop forests in the eastern United States and east of Los Angeles are bathed in fog and dews as acidic as lemon juice, with a pH of 2.3—about 1,000 times the acidity of normal precipitation.

In some areas, soils contain basic compounds such as calcium carbonate ($CaCO_3$) or limestone that can react with and neutralize, or *buffer*, some inputs of acids. The areas most sensitive to acid deposition are those containing thin, acidic soils without such natural buffering (Figure 15-8, green and most red areas) and those in which the buffering capacity of soils has been depleted by decades of acid deposition.

Many acid-producing chemicals generated in one country are exported to other countries by prevailing winds. For example, acidic emissions from industrialized areas of western Europe (especially the United Kingdom and Germany) and eastern Europe blow into Norway, Switzerland, Austria, Sweden, the Netherlands, and Finland. Some SO_2 and other emissions from coal-burning power and industrial plants in the Ohio Valley of the United States (Figure 15-7, red dots, p. 354) end up in southeastern Canada. Some acidic emissions in China end up in Japan and North and South Korea.

The worst acid deposition occurs in Asia, especially China, which gets about 59% of its energy from burning coal. Scientists estimate that by 2025, China will emit more sulfur dioxide than the United States, Canada, and Japan combined.

Environmental Science⊕Now™
Learn more about the sources of acid deposition, how it forms, and what it can do to lakes and soils at Environmental ScienceNow.

Science: Harmful Effects of Acid Deposition

Acid deposition can cause or worsen respiratory disease, attack metallic and stone objects, decrease atmospheric visibility, and kill fish.

Acid deposition has a number of harmful effects. It contributes to human respiratory diseases such as bronchitis and asthma, and can leach toxic metals (such as lead and copper) from water pipes into drinking water. It also damages statues, national monuments, buildings, metals, and car finishes.

Acid deposition decreases atmospheric visibility, mostly because of the sulfate particles it contains. On some days, the air in Grand Canyon National Park and

http://biology.brookscole.com/miller11   353

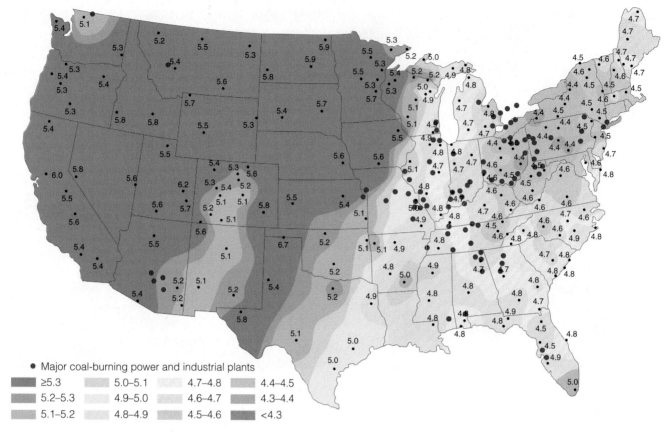

●	Major coal-burning power and industrial plants

≥5.3	5.0–5.1	4.7–4.8	4.4–4.5
5.2–5.3	4.9–5.0	4.6–4.7	4.3–4.4
5.1–5.2	4.8–4.9	4.5–4.6	<4.3

Figure 15-7 Natural capital degradation: pH values from field measurements at 250 sites (red dots) in 48 states in 2002. Yellow, tan, and orange indicate the areas with lowest pH (highest acidity). Red dots show major sources of sulfur dioxide (SO_2) emissions, mostly large coal-burning power plants. The damaging acidic rain and snow that fall in the Northeast result from coal-fired power and industrial plants and cars in the region and from pollution that blows in from such plants in the Midwest. In the East, the primary component of acid deposition is H_2SO_4 (formed from SO_2 emitted by coal-burning plants). In the West, HNO_3 predominates (formed mostly from NO_x emissions from motor vehicles). According to the EPA, two-thirds of the SO_2 and one-fourth of the NO_x that are the primary causes of acid deposition come from coal-burning power plants. (Data from National Atmospheric Deposition Program/National Trends Network, 2003)

certain other national parks is so smoggy from sulfate particles and other pollutants that visitors cannot see the magnificent vistas.

Acid deposition has harmful ecological effects on aquatic systems as well. Most fish cannot survive in water with a pH less than 4.5. Acid deposition can also release aluminum ions (Al^{3+}) attached to minerals in nearby soil into lakes. These ions asphyxiate many kinds of fish by stimulating excessive mucus formation, which clogs their gills.

Because of excess acidity, several thousand lakes in Norway and Sweden contain no fish, and many more lakes there have lost most of their acid-neutralizing capacity. In Canada, at least 1,200 acidified lakes contain few if any fish, and fish populations in thousands more lakes are declining because of increased acidity. In the United States, several hundred lakes (most in the Northeast) are threatened with excess acidity. Acid deposition is not always the main culprit, however. Some

lakes are acidic because they are surrounded by naturally acidic soils.

Acid deposition (often along with other air pollutants such as ozone) can harm forests and crops—especially when the soil pH falls below 5.1—by leaching essential plant nutrients such as calcium and magnesium salts from soils. This reduces plant productivity and the soils' ability to buffer or neutralize acidic inputs.

Acid deposition rarely kills trees directly, but can weaken them and leave them vulnerable to stresses such as severe cold, diseases, insect attacks, drought, and harmful mosses. Effects of acid deposition on trees and other plants are caused partly by chemical interactions in forest and cropland soils (Figure 15-9, p. 356).

Mountaintop forests are the terrestrial areas hardest hit by acid deposition (Figure 15-10, p. 356). They tend to have thin soils without much buffering capacity, and trees on mountaintops (especially conifers

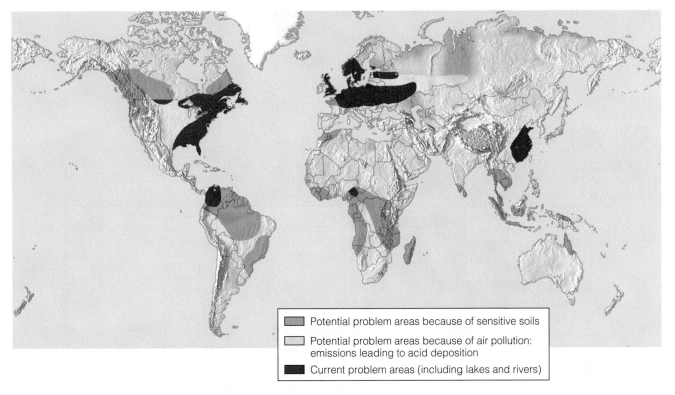

Figure 15-8 Natural capital degradation: regions where acid deposition is now a problem and regions with the potential to develop this problem. Such regions have large inputs of air pollution (mostly from power plants, industrial plants, and ore smelters) or are sensitive areas with soils and bedrock that cannot neutralize (buffer) inputs of acidic compounds. (Data from World Resources Institute and U.S. Environmental Protection Agency)

- Potential problem areas because of sensitive soils
- Potential problem areas because of air pollution: emissions leading to acid deposition
- Current problem areas (including lakes and rivers)

such as red spruce and balsam fir that keep their leaves year-round) are bathed almost continuously in very acidic fog and clouds.

Most of the world's forests and lakes are not being destroyed or seriously harmed by acid deposition. Rather, this regional problem is harming forests and lakes that lie downwind from coal-burning facilities and from large car-dominated cities without adequate pollution controls.

Environmental Science ⊛ Now™
Examine how acid deposition can harm a pine forest and what it means to surrounding land and waters at Environmental ScienceNow.

Solutions: Reducing Acid Deposition

Acid deposition can be prevented and cleaned up, but this is politically difficult.

Figure 15-11 (p. 357) summarizes ways to reduce acid deposition. According to most scientists studying the problem, the best solutions are *prevention approaches* that reduce or eliminate emissions of sulfur dioxide, nitrogen oxides, and particulates.

Currently, controlling acid deposition is a political hot potato. One problem is that the people and eco-

systems it affects often are quite distant from those that cause the problem. Also, countries with large supplies of coal (such as China, India, Russia, and the United States) have a strong incentive to use it as a major energy resource. Owners of coal-burning power plants say the costs of adding pollution control equipment, using low-sulfur coal, or removing sulfur from coal are too high and would increase the cost of electricity for consumers.

Environmental scientists respond that affordable and much cleaner ways are available to produce electricity—including wind turbines and burning natural gas in turbines. They also point out that the largely hidden health and environmental costs of burning coal amount to roughly twice its market cost. Including these costs would spur prevention of acid deposition.

Large amounts of limestone or lime can be used to neutralize acidified lakes or surrounding soil—the only cleanup approach now being used. Liming creates several problems, however. But this expensive and temporary remedy usually must be repeated annually. Also, it can kill some types of plankton and aquatic plants and can harm wetland plants that need acidic water. Finally, it is difficult to know how much lime to put where (in the water or at selected places on the ground).

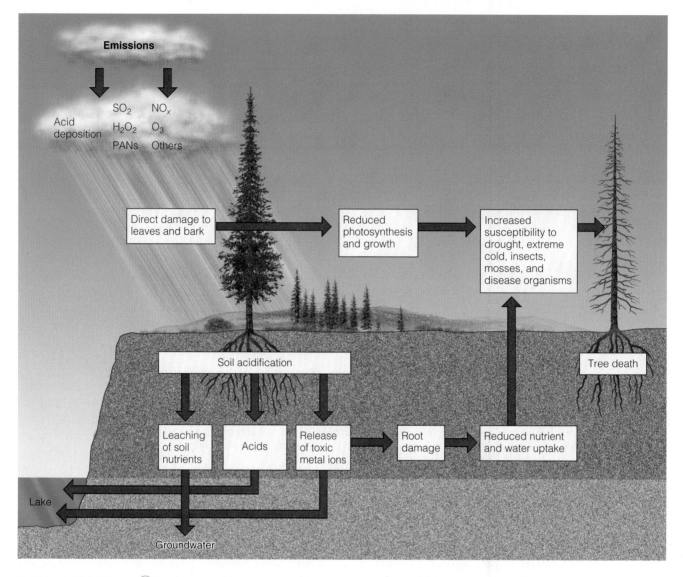

Active Figure 15-9 Natural capital degradation: air pollutants are one of several interacting stresses that can damage, weaken, or kill trees and pollute surface and groundwater. *See an animation based on this figure and take a short quiz on the concept.*

Figure 15-10 Natural capital degradation: damage (mostly from acid deposition) to high elevation Fraser fir and red spruce trees in Mount Mitchell State Park, North Carolina. Trees on the left (windward) side of the mountain are the most damaged from prolonged exposure to polluted air blowing in mostly from older coal and industrial plants in the midwestern United States without adequate pollution controls (red dots in Figure 15-7).

Carolina Biological/Visuals Unlimited

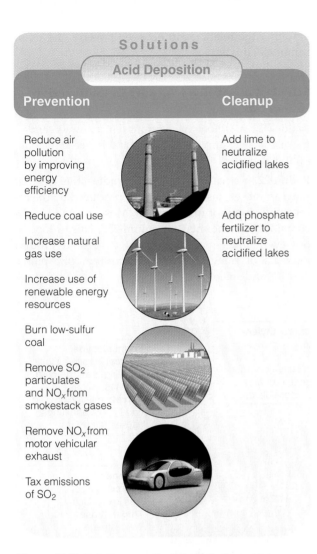

Solutions

Acid Deposition

Prevention	Cleanup
Reduce air pollution by improving energy efficiency	Add lime to neutralize acidified lakes
Reduce coal use	Add phosphate fertilizer to neutralize acidified lakes
Increase natural gas use	
Increase use of renewable energy resources	
Burn low-sulfur coal	
Remove SO_2 particulates and NO_x from smokestack gases	
Remove NO_x from motor vehicular exhaust	
Tax emissions of SO_2	

Figure 15-11 Solutions: methods for reducing acid deposition and its damage. *Critical thinking: which two of these solutions do you believe are the most important?*

15-5 INDOOR AIR POLLUTION

Science: Indoor Air Pollution

Indoor air pollution usually poses a much greater threat to human health than outdoor air pollution.

If you are reading this book indoors, you may be inhaling more air pollutants with each breath than if you were outside. Figure 15-12 (p. 358) shows some typical sources of indoor air pollution in a modern home.

EPA studies have revealed some alarming facts about indoor air pollution in the United States. *First,* levels of 11 common pollutants generally are two to five times higher inside homes and commercial buildings than outdoors and as much as 100 times higher in some cases. *Second,* pollution levels inside cars in traffic-clogged urban areas can be as much as 18 times higher than outside. *Third,* the health risks from exposure to such chemicals are magnified because most people in developed countries typically spend 70–98% of their time indoors or inside vehicles.

As a result of these studies, in 1990 the EPA placed indoor air pollution at the top of the list of 18 sources of cancer risk—causing as many as 6,000 premature cancer deaths per year in the United States. At greatest risk are smokers, infants and children younger than age 5, the old, the sick, pregnant women, people with respiratory or heart problems, and factory workers.

Danish and U.S. EPA studies have linked air pollutants found in buildings to dizziness, headaches, coughing, sneezing, shortness of breath, nausea, burning eyes, chronic fatigue, irritability, skin dryness and irritation, and flu-like symptoms, known as the *sick-building syndrome.* New buildings are more commonly "sick" than old ones because of reduced air exchange (to save energy) and chemicals released from new carpeting and furniture. EPA studies indicate that almost one in five of the 4 million commercial buildings in the United States is considered "sick." Mold spores that grow in damp places are probably the greatest cause of allergic reactions to indoor air pollution. Such molds can flourish in air ducts and when moisture becomes trapped in the walls of airtight houses.

The solution is not to give up on improving energy efficiency in buildings. Instead, air heat exchangers should be used to provide a healthy flow of air inside homes and other buildings without affecting overall heat gains and losses in such structures. Having outdoor vents in kitchens and bathrooms also helps reduce the indoor buildup of moisture and air pollutants from cooking. Gas and oil furnaces should be checked for carbon monoxide production, and homeowners should install carbon monoxide detectors and warning devices near bedrooms. Homeowners can also put a plastic cover over dirt in crawlspaces to reduce inputs of moisture from the ground and use vents to ensure that crawlspaces are well ventilated.

According to the EPA and public health officials, the four most dangerous indoor air pollutants in developed countries are *cigarette smoke* (Case Study, p. 327), *formaldehyde, radon-222 gas* (Case Study, p. 358), and *very small fine and ultrafine particles.*

The chemical that causes most people in developed countries difficulty is *formaldehyde,* a colorless, extremely irritating gas widely used to manufacture common household materials. According to the EPA and the American Lung Association, 20–40 million Americans suffer from chronic breathing problems, dizziness, rash, headaches, sore throat, sinus and eye irritation, wheezing, and nausea caused by daily exposure to low levels of formaldehyde emitted from common household materials.

The many sources of *formaldehyde* include building materials (such as plywood, particleboard, paneling, and high-gloss wood used in floors and cabinets), furniture, drapes, upholstery, adhesives in carpeting and wallpaper, urethane-formaldehyde insulation, fingernail hardener, and wrinkle-free coating on permanent-press clothing (Figure 15-12). The EPA estimates that as many as 1 of every 5,000 people who live in manufactured homes for more than 10 years will develop cancer from formaldehyde exposure.

In developing countries, the indoor burning of wood, charcoal, dung, crop residues, and coal in open fires or in unvented or poorly vented stoves for cooking and heating exposes inhabitants to high levels of particulate air pollution. According to the World Bank, as many as 2.8 million people (most of them women and children) in developing countries die prematurely each year from breathing elevated levels of such indoor smoke. *Indoor air pollution for the poor is by far the world's most serious air pollution problem*—a glaring ex-

ample of the relationship between poverty and environmental quality.

Science Case Study: Exposure to Radioactive Radon Gas

Radon-222, a radioactive gas found in some soil and rocks, can seep into some houses and increase the risk of lung cancer.

Radon-222—a naturally occurring radioactive gas that you cannot see, taste, or smell—is produced by the radioactive decay of uranium-238. Most soils and rocks contain small amounts of uranium-238. This isotope is, however, much more concentrated in underground deposits of minerals such as uranium, phosphate, granite, and shale.

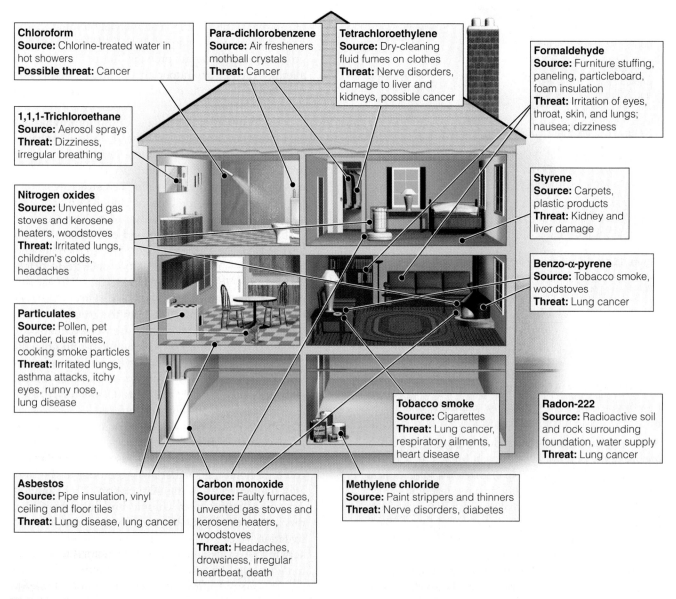

Figure 15-12 Science: Some important indoor air pollutants. *Critical thinking: which of these are you probably exposed to?* (Data from U.S. Environmental Protection Agency)

When colorless and odorless radon gas from such deposits seeps upward through the soil and is released outdoors, it disperses quickly in the atmosphere and decays to harmless levels. However, in buildings above such deposits, radon gas can enter through cracks in foundations and walls, openings around sump pumps and drains, and hollow concrete blocks (Figure 15-13). Once inside, it can build up to high levels, especially in unventilated lower levels of homes and buildings.

Radon-222 gas quickly decays into solid particles of other radioactive elements, such as polonium-210, that if inhaled expose lung tissue to a large amount of ionizing radiation from alpha particles. This exposure can damage lung tissue and lead to lung cancer over the course of a 70-year lifetime. Such exposure makes radon the second-leading cause of lung cancer (after smoking) in the United States. Your chances of getting lung cancer from radon depend mostly on how much radon is in your home, how much time you spend in your home, and whether you are a smoker or have ever smoked.

According to the EPA, nearly 1 in every 15 homes in the United States has a potentially dangerous level of indoor radon. Ideally, radon levels should be monitored continuously in the main living areas (not basements or crawlspaces) for 2 months to a year. Only 6% of U.S. households had followed the EPA's recommendation to conduct radon tests (most lasting only 2–7 days and costing $20–100 per home).

For information about radon testing, visit the EPA website at **http://www.epa.gov/iaq/radon**. According to the EPA, radon control could add $350–500 to the cost of a new home, and correcting a radon problem in an existing house could run $800–2,500. Remedies include sealing cracks in the foundation and walls, increasing ventilation by cracking a window or installing vents, and using a fan to create cross ventilation.

15-6   HARMFUL EFFECTS OF AIR POLLUTION

Science: How Does Your Respiratory System Help Protect You from Air Pollution?

Your respiratory system has several ways to help protect you from air pollution, but some air pollutants can overcome these defenses.

Your respiratory system (Figure 15-14, p. 360) has a number of mechanisms that help protect you from air

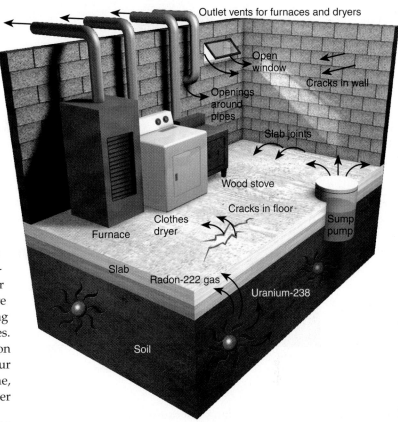

Figure 15-13 Science: sources and paths of entry for radon-222 gas. *Have you monitored the indoor air where you live for radon-222?* (Data from U.S. Environmental Protection Agency)

pollution. Hairs in your nose filter out large particles. Sticky mucus in the lining of your upper respiratory tract captures smaller (but not the smallest) particles and dissolves some gaseous pollutants. Sneezing and coughing expel contaminated air and mucus when pollutants irritate your respiratory system.

In addition, hundreds of thousands of tiny mucus-coated hairlike structures called *cilia* line your upper respiratory tract. They continually wave back and forth and transport mucus and the pollutants they trap to your throat (where they are swallowed or expelled).

Prolonged or acute exposure to air pollutants, including tobacco smoke, can overload or break down these natural defenses. This can cause or contribute to various respiratory diseases. An example is *asthma*—typically an allergic reaction causing sudden episodes of muscle spasms in the bronchial walls—results in acute shortness of breath. More than 17 million Americans have asthma, about 5 million of them children younger than age 5.

Years of smoking and breathing air pollutants can lead to *lung cancer* and *chronic bronchitis*, which involves persistent inflammation and damage to the cells lining the bronchi and bronchioles. The results are mucus buildup, painful coughing, and shortness of

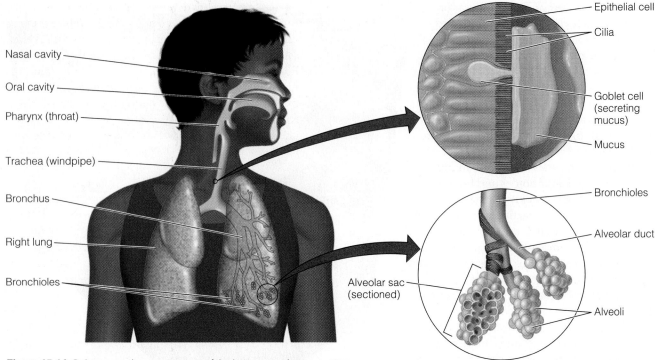

Nasal cavity

Oral cavity

Pharynx (throat)

Trachea (windpipe)

Bronchus

Right lung

Bronchioles

Epithelial cell

Cilia

Goblet cell (secreting mucus)

Mucus

Bronchioles

Alveolar duct

Alveolar sac (sectioned)

Alveoli

Figure 15-14 Science: major components of the human respiratory system.

breath. Damage deeper in the lung can cause *emphysema*, in which irreversible damage to air sacs or alveoli leads to abnormal dilation of air spaces, loss of lung elasticity, and acute shortness of breath (Figure 15-15).

People with respiratory diseases are especially vulnerable to air pollution, as are older adults, infants, pregnant women, and people with heart disease.

Science: Harmful Health Effects of Key Air Pollutants

Air pollutants damage materials and human lungs and worldwide prematurely kill at least 3 million people each year.

Table 15-1 lists some of the harmful health effects of prolonged or chronic exposure to six major air pollu-

tants. According to the WHO, worldwide at least 3 million people (most of them in Asia) die prematurely each year from the effects of air pollution—an average of 8,200 deaths per day. About 2.8 million of these deaths (93%) result from indoor air pollution, typically from heart attacks, respiratory diseases, and lung cancer related to prolonged breathing of polluted air. *The WHO and the World Bank consider indoor air pollution to be one of the world's most serious environmental problems.*

In the United States, the EPA estimates that annual deaths related to indoor and outdoor air pollution range from 150,000 to 350,000 people—equivalent to one to two fully loaded 400-passenger jumbo jets crashing *each day* with no survivors. Millions more become ill and lose work time. Most of these deaths are related to inhalation of fine and ultrafine particulates

Figure 15-15 Science: normal human lungs (left) and the lungs of a person who died of emphysema (right). Prolonged smoking and exposure to air pollutants can cause emphysema in anyone, but about 2% of emphysema cases result from a defective gene that reduces the elasticity of the air sacs in the lungs. Anyone with this hereditary condition, for which testing is available, should not smoke and should not live or work in a highly polluted area.

O. Auerbach/Visuals Unlimited

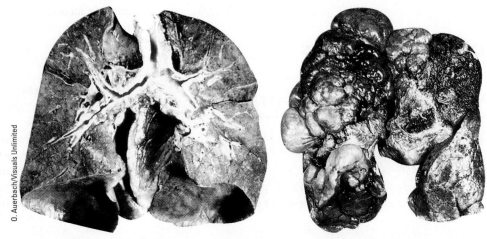

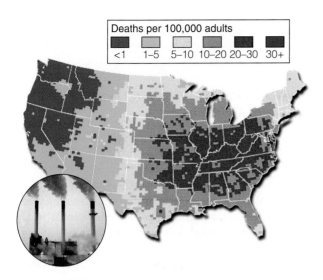

Figure 15-16 Premature deaths from air pollution in the United States, mostly from coal-burning power plants. *Critical thinking: what is the risk where you live?* (Data from U.S. Environmental Protection Agency)

Deaths per 100,000 adults
<1 1–5 5–10 10–20 20–30 30+

in indoor air and in outdoor air from coal-burning power plants, mostly in the eastern half of the United States (Figure 15-16).

According to recent studies by the EPA, each year more than 125,000 Americans (120,000 of them in urban areas) get cancer from breathing soot-laden diesel fumes from buses and trucks. Other sources of these fumes include tractors, bulldozers and other construction equipment, portable generators, and other off-road diesel vehicles. In one year, a large diesel-powered bulldozer can produce as much air pollution as 26 cars.

15-7 PREVENTING AND REDUCING AIR POLLUTION

Politics: Air Pollution Laws in the United States

The Clean Air Acts in the United States have greatly reduced outdoor air pollution from six major pollutants.

The U.S. Congress passed the Clean Air Acts in 1970, 1977, and 1990. With these laws, the federal government established key air pollution regulations that are enforced by each state and by major cities.

Congress directed the EPA to establish *national ambient air quality standards (NAAQS)* for six outdoor criteria pollutants (Table 15-1). The EPA regulates these chemicals by using *criteria* developed from risk assessment methods to set maximum permissible levels in outdoor air.

One limit, called a *primary standard*, is set to protect human health. Another limit, called a *secondary standard*, is intended to prevent environmental and property damage. Each standard specifies the maximum allowable level, averaged over a specific period, for a certain pollutant in outdoor (ambient) air. The EPA has also established national emission standards for more than 188 *hazardous air pollutants (HAPs)* that may cause serious *health* and *ecological effects*. These chemicals include neurotoxins, carcinogens, mutagens, teratogens, endocrine system disrupters, and other toxic compounds. Most of these chemicals are chlorinated hydrocarbons, VOCs, or compounds of toxic metals.

One of the best sources of information about HAPs in a local area is the annual *Toxic Release Inventory (TRI)*, which is collected and released to the public as part of community "right to know" laws enacted by Congress in 1986. The TRI law requires 23,000 refineries, power plants, hardrock mines, chemical manufacturers, and factories to report their releases above certain minimum amounts and their waste management methods for 667 toxic chemicals.

Great news. According to a 2003 EPA report, aggregate emissions of the six criteria air pollutants decreased by 48% between 1970 and 2002 even with significant increases in gross domestic product, vehicle miles traveled, energy consumption, and population.

Bad news. After dropping in the 1980s, smog levels did not decline between 1993 and 2002 mostly because reducing smog requires much bigger cuts in emissions of nitrogen oxides from power and industrial plants and motor vehicles. Also, according to the EPA, in 2003 more than 170 million people lived in 470 of the nation's 2,700 counties in 31 states where air is unhealthy to breathe during part of the year because of high levels of air pollutants—primarily ozone and fine particles. In 2004, the EPA estimated that airborne fine particles cause 15,000 premature deaths and 95,000 cases of chronic or acute bronchitis per year. And according to a 2004 joint study by the Environmental Integrity Project and the Galveston Houston Association for Smog Prevention, refinery and chemical plant emissions of major toxic pollutants may actually be four to five times higher than is currently being reported by the EPA and nearly all state governments.

Politics: Improving U.S. Air Pollution Laws

Environmental scientists applaud the success of U.S. air pollution control laws but have suggested several ways to make them more effective.

The reduction of outdoor air pollution in the United States since 1970 has been a remarkable success story. It occurred because of two factors. *First*, U.S. citizens insisted that laws be passed and enforced to improve air quality. *Second*, the country was affluent enough to afford such controls and improvements.

But more can be done. Environmental scientists point to several deficiencies in the Clean Air Act. First, *we continue to rely mostly on pollution cleanup rather than prevention.* The power of prevention is clear: In the United States, the air pollutant with the largest drop (98% between 1970 and 2002) in its atmospheric level was lead, which was largely banned in gasoline. This change in the lead level is viewed as one the greatest environmental success stories in the country's history.

Second, *Congress has failed to increase fuel-efficiency standards for cars, sport utility vehicles (SUVs), and light trucks.* According to environmental scientists, increased fuel efficiency would reduce air pollution from motor vehicles more quickly and effectively than any other method, reduce CO_2 emissions and help slow global warming, save energy, and save consumers enormous amounts of money. *Critical thinking: why hasn't this been done?*

Third, *regulation of emissions from inefficient two-cycle gasoline engines remains inadequate.* These engines are used in lawn mowers, leaf blowers, chain saws, jet skis, outboard motors, and snowmobiles. According to the California Air Resources Board, a 1-hour ride on a typical jet ski creates more air pollution than the average U.S. car does in a year, and operating a 100-horsepower boat engine for 7 hours emits more air pollutants than driving a new car for 160,000 kilometers (100,000 miles). In 2001, the EPA announced plans to reduce emissions from most of these sources by 2007. Manufacturers have pushed for extending these deadlines.

Fourth, *the acts have done little to reduce emissions of carbon dioxide and other greenhouse gases*—mostly because CO_2 has not been classified as an air pollutant.

Fifth, *the acts have failed to deal seriously with indoor air pollution,* even though it is by far the most serious air pollution problem in terms of poorer health, premature death, and economic losses from lost work time and increased health costs. *Critical thinking: why hasn't this been done?*

Finally, there is a need for *better enforcement of the Clean Air Acts.* According to a 2002 government study, more rigorous enforcement would save about 6,000 lives and prevent 140,000 asthma attacks each year in the United States.

Executives of companies affected by implementing such policies claim that correcting these deficiencies in the Clean Air Acts would cost too much, harm economic growth, and cost jobs. Proponents contend that history has shown that most industry estimates of the cost of implementing various air pollution control standards in the United States were many times the actual cost. In addition, implementing such standards has boosted economic growth and created jobs by stimulating companies to develop new technologies for reducing air pollution emissions, as has happened in Germany. Many of these technologies are sold in the international marketplace.

X *HOW WOULD YOU VOTE?* Should the 1990 U.S. Clean Air Act be strengthened? Cast your vote online at http://biology.brookscole.com/miller11.

Economics Case Study: Using the Marketplace to Reduce Air Pollution

Allowing producers of air pollutants to buy and sell government air pollution allotments in the marketplace can help reduce emissions.

To help reduce SO_2 emissions, the Clean Air Act of 1990 authorizes an *emissions trading program,* which enables the 110 most polluting power plants in 21 states (primarily in the Midwest and East, red dots in Figure 15-7) to buy and sell SO_2 pollution rights.

Each year, a coal-burning power plant is given a certain number of pollution credits, or rights that allow it to emit a certain amount of SO_2. A utility that emits less SO_2 than its limit has a surplus of pollution credits. It can use these credits to avoid reductions in SO_2 emissions at another of its plants, keep them for future plant expansions, or sell them to other utilities, private citizens, or environmental groups.

Proponents argue that this system allows the marketplace to determine the cheapest, most efficient way to get the job done instead of having the government dictate how to control air pollution. Some environmentalists see this *cap-and-trade* market approach as an improvement over the regulatory *command-and-control* approach, as long as it achieves a net reduction in SO_2 pollution. That goal would be accomplished by limiting the total number of credits and gradually lowering the *emissions cap* or annual number of credits, as has been done since 2000.

One of the neat things about the SO_2 emissions market is that anyone can participate. Environmental groups can buy up such rights to pollute and not use them. You could personally reduce air pollution by buying a certificate allowing you to add 0.9 metric ton (1 ton) of SO_2 to the atmosphere and hanging it on the wall. You can purchase these certificates and give them away as birthday or holiday gifts. See **www.epa.gov/airmarkets/** for a list of brokers and other sellers of SO_2 permits.

Some environmentalists criticize the cap-and-trade program. They contend that it allows utilities with older, dirtier power plants to buy their way out of their environmental responsibilities and continue emitting unacceptable levels of SO_2. This practice could lead (and, indeed, already has led) to continuing high levels of air pollution in certain areas or "hot spots."

The cap-and-trade program also creates incentives to cheat because air quality regulation is based largely on self-reporting of emissions. Environmentalists call for unannounced spot monitoring by the government and large fines for cheaters.

Ultimately, the success of any emissions trading approach depends on how low the initial cap is set and then on how much it is reduced annually to promote continuing innovation in air pollution prevention and control. Without these elements, emissions trading programs mostly move air pollutants from one area to another without achieving any overall reduction in air quality.

Good news. Between 1990 and 2002, the emissions trading system helped reduce SO_2 emissions from electric power plants in the United States by 35%. The cost of doing so amounted to less than one-tenth of the cost projected by industry, because the market-based system motivated companies to reduce emissions in more efficient ways.

The EPA has also created an emissions trading program for smog-forming nitrogen oxides in a number of states in the East and Midwest. Emissions trading may also be implemented for particulate emissions and VOCs and for the combined emissions of SO_2, NO_x, and mercury from coal-burning power plants. Environmental and health scientists are particularly opposed to using a cap-and-trade program to control emissions of mercury by coal-burning power plants and industries because this pollutant is highly toxic, falls out of the atmosphere adjacent to such facilities, and does not break down in the environment. Coal-burning plants choosing to buy permits instead of sharply reducing their mercury emissions would create toxic hot spots with unacceptably high levels of mercury.

Bad news. In 2002, the EPA reported results from the country's oldest and largest emissions trading program, in effect since 1993 in southern California. According to the EPA, this cap-and-trade model "produced far less emissions reductions than were either projected for the program or could have been expected from" the command-and-control system it replaced. The same study also found accounting abuses, including emissions caps set 60% higher than current emissions levels. This highlights the need for more careful monitoring of all cap-and-trade programs.

X *HOW WOULD YOU VOTE?* Should emissions trading be used as the primary way to control emissions of all major air pollutants? Cast your vote online at http://biology.brookscole .com/miller11.

Solutions: Reducing Outdoor Air Pollution

There are a number of ways to prevent and control air pollution from coal-burning facilities and motor vehicles.

Figure 15-17 summarizes ways to reduce emissions of sulfur oxides, nitrogen oxides, and particulate matter from stationary sources such as electric power plants and industrial plants that burn coal.

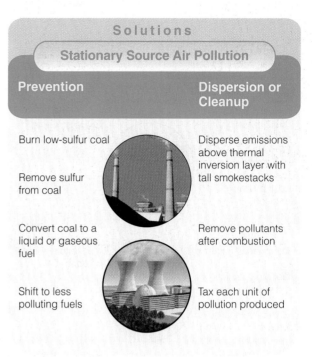

Figure 15-17 Solutions: methods for reducing emissions of sulfur oxides, nitrogen oxides, and particulate matter from stationary sources such as coal-burning electric power plants and industrial plants. *Critical thinking: which two of these solutions do you believe are the most important?*

Approximately 20,000 older coal-burning plants, industrial plants, and oil refineries in the United States have not been required to meet the air pollution standards required for new facilities under the Clean Air Acts. Environmentalists and officials of states subject to pollution from such plants have been trying to get Congress reverse this situation since 1970. To date, they have been not been successful because of strong lobbying efforts by U.S. coal and oil industries.

In 2004, Abt Associates, a consulting firm hired by the EPA, estimated that air pollution from the nation's 1,100 largest coal-fired power plants causes 24,000 premature deaths, 38,200 nonfatal heart attacks, 554,000 asthma attacks, and 3 million lost workdays each year. Approximately 90% of these deaths could be prevented by rigid enforcement of the existing Clean Air Acts and by requiring older coal-burning plants to install currently available air-pollution control technology systems.

Figure 15-18 (p. 364) lists ways to reduce emissions from motor vehicles, the primary culprits in producing photochemical smog.

Good news. Over the next 10–20 years, air pollution from motor vehicles should decrease from increased use of *partial zero-emission vehicles (PZEVs)* that emit almost no air pollutants thanks to their improved engine and emission systems, *hybrid-electric vehicles* (Figure 13-24, p. 309), and vehicles powered by fuel cells running on hydrogen (Figure 13-25, p. 310).

Solutions

Motor Vehicle Air Pollution

Prevention	Cleanup
Mass transit	Emission control devices
Bicycles and walking	
Less polluting engines	
Less polluting fuels	
Improve fuel efficiency	Car exhaust inspections twice a year
Get older, polluting cars off the road	
Give buyers large tax write-offs for buying low-polluting, energy-efficient vehicles	
Restrict driving in polluted areas	Stricter emission standards

Figure 15-18 Solutions: methods for reducing emissions from motor vehicles. *Critical thinking: which two of these solutions do you believe are the most important?*

✗ How Would You Vote? Should older coal-burning power and industrial plants have to meet the same air pollution standards as new facilities? Cast your vote online at http://biology.brookscole.com/miller11.

Bad news. The growing number of motor vehicles in urban areas of many developing countries is worsening the already poor air quality there. Many of these vehicles are 10 or more years old, have no pollution control devices, and burn leaded gasoline.

Solutions: Reducing Indoor Air Pollution

Little effort has been devoted to reducing indoor air pollution even though it poses a much greater threat to human health than outdoor air pollution.

Reducing indoor air pollution does not require setting indoor air quality standards and monitoring the more than 100 million homes and buildings in the United States (or the buildings in any country). Instead, air pollution experts suggest several ways to prevent or reduce indoor air pollution, as shown in Figure 15-19. Another possibility to ensure cleaner indoor air in high-rise buildings is to build rooftop greenhouses through which building air can be circulated.

In developing countries, indoor air pollution from open fires and leaky and inefficient stoves that burn wood, charcoal, or coal could be reduced if governments gave people inexpensive clay or metal stoves that burn biofuels more efficiently while venting their exhaust to the outside, or stoves that use solar energy to cook food (solar cookers). This policy would also reduce deforestation by using less fuelwood and charcoal.

What Is the Next Step?

We need to focus on preventing air pollution, with emphasis on sharply reducing indoor air pollution in developing countries.

Encouraging news. Since 1970, most of the world's developed countries have enacted laws and regulations

Solutions

Indoor Air Pollution

Prevention	Cleanup or Dilution
Cover ceiling tiles and lining of AC ducts to prevent release of mineral fibers	Use adjustable fresh air vents for work spaces
Ban smoking or limit it to well-ventilated areas	Increase intake of outside air
Set stricter formaldehyde emissions standards for carpet, furniture, and building materials	Change air more frequently
	Circulate a building's air through rooftop greenhouses
Prevent radon infiltration	
Use office machines in well-ventilated areas	Use exhaust hoods for stoves and appliances burning natural gas
Use less polluting substitutes for harmful cleaning agents, paints, and other products	Install efficient chimneys for wood-burning stoves

Figure 15-19 Solutions: ways to prevent and reduce indoor air pollution. *Critical thinking: which two of these solutions do you believe are the most important?*

that have significantly reduced outdoor air pollution. Without individuals and organized groups putting strong political pressure on elected officials in the 1970s and 1980s, these laws and regulations would not have been enacted, funded, and implemented. In turn, these legal requirements spurred companies, scientists, and engineers to come up with better ways to control outdoor pollution.

The current laws represent a useful *output approach* to controlling pollution. To environmental scientists, the next step is to shift to *preventing air pollution.* With this approach, the question is not *What can we do about the air pollutants we produce?* but rather *How can we avoid producing these pollutants in the first place?*

Figure 15-20 shows ways to prevent outdoor and indoor air pollution over the next 30–40 years. Like the shift to *controlling outdoor air pollution* between 1970 and 2000, this new shift to *preventing outdoor and indoor air pollution* will not take place without political pressure on elected officials by individual citizens and groups. Figure 15-21 lists some ways that you can reduce your exposure to indoor air pollution.

Solutions

Air Pollution

Outdoor	Indoor
Improve energy efficiency to reduce fossil fuel use	Reduce poverty
Rely more on lower-polluting natural gas	Distribute cheap and efficient cookstoves to poor families in developing countries
Rely more on renewable energy (especially solar cells, wind, and solar-produced hydrogen)	Reduce or ban indoor smoking
Transfer technologies for latest energy efficiency, renewable energy, and pollution prevention to developing countries	Develop simple and cheap tests for indoor pollutants such as particulates, radon, and formaldehyde

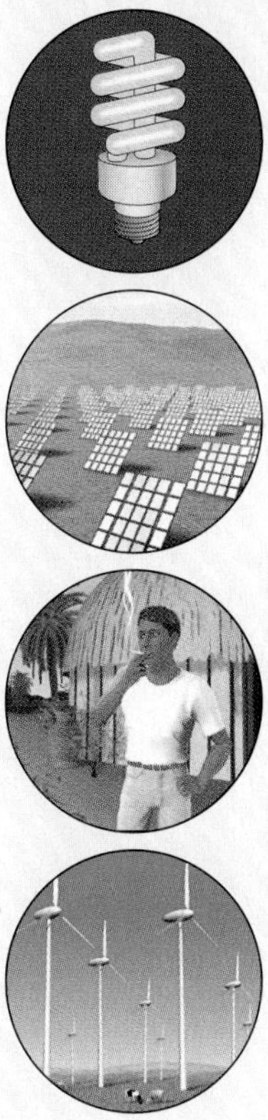

Figure 15-20 Solutions: ways to prevent outdoor and indoor air pollution over the next 30–40 years. *Critical thinking: which two of these solutions do you believe are the most important?*

What Can You Do?

Indoor Air Pollution

- Test for radon and formaldehyde inside your home and take corrective measures as needed.
- Do not buy furniture and other products containing formaldehyde.
- Remove your shoes before entering your house to reduce inputs of dust, lead, and pesticides.
- Test your house or workplace for asbestos fiber levels and for any crumbling asbestos materials if it was built before 1980.
- Don't live in a pre-1980 house without having its indoor air tested for asbestos and lead.
- Do not store gasoline, solvents, or other volatile hazardous chemicals inside a home or attached garage.
- If you smoke, do it outside or in a closed room vented to the outside.
- Make sure that wood-burning stoves, fireplaces, and kerosene- and gas-burning heaters are properly installed, vented, and maintained.
- Install carbon monoxide detectors in all sleeping areas.

Figure 15-21 Individuals matter: ways to reduce your exposure to air pollution. *Critical thinking: which three of these actions do you believe are the most important? Which things in this list do you do or plan to do?*

Turning the corner on air pollution requires moving beyond patchwork, end-of-pipe approaches to confront pollution at its sources. This will mean reorienting energy, transportation, and industrial structures toward prevention.

HILARY F. FRENCH

CRITICAL THINKING

1. Explain why you agree or disagree with the following statement: "Because we have not proved absolutely that anyone has died or suffered serious disease from nitrogen oxides, current federal emission standards for this pollutant should be relaxed."

2. Identify climate and topographic factors in your local community that **(a)** intensify air pollution and **(b)** help reduce air pollution.

3. Should all tall smokestacks be banned in an effort to promote greater emphasis on preventing air pollution? Explain.

4. Have buildings at your school been tested for radon? If so, what were the results? What has been done about areas with unacceptable levels? If this testing has not been done, talk with school officials about having it done. Has your house been tested for radon?

5. Do you agree or disagree with the possible weaknesses of the U.S. Clean Air Acts listed on p. 362? Defend each of your choices.

6. Explain why you agree or disagree with each of the proposals listed in Figure 15-20 (p. 365) for shifting the emphasis to preventing air pollution over the next several decades.

7. Congratulations! You are in charge of reducing air pollution in the country where you live. List the three most important features of your policy for **(a)** outdoor air pollution and **(b)** indoor air pollution.

LEARNING ONLINE

The website for this book includes review questions for the entire chapter, flash cards for key terms and concepts, a multiple-choice practice quiz, interesting Internet sites, references, and a guide for accessing thousands of InfoTrac® College Edition articles.

Visit

http://biology.brookscole.com/miller11

Then choose Chapter 15, and select a learning resource. For access to animations, additional quizzes, chapter outlines and summaries, register and log in to

Environmental Science Now™

at **esnow.brookscole.com/miller11** using the access code card in the front of your book.

Active Graphing

Visit http://esnow.brookscole.com/miller11 to explore the graphing exercise for this chapter.

CASE STUDY

Studying a Volcano to Understand Climate Change

In 1991, NASA scientist James Hansen announced that the recent explosion of a volcano in the Philippines would probably cool the average temperature of the earth by 0.5°C (1°F) over a 15-month period. The earth would then begin to warm, returning to the temperatures observed before the explosion by 1995. His predictions turned out to be correct.

The volcano in question was Mount Pinatubo (Figure 16-1). After 600 years of slumber, in June 1991 it exploded in the second-largest volcanic eruption of the 20th century (the largest took place in Alaska in 1912). A huge amount of volcanic material blasted out of the mountain, sending a cloud of gas and ash to a height of 35 kilometers (22 miles). Simultaneously, avalanches of hot gas and ash roared down the sides of the mountain, killing hundreds of people and filling valleys with volcanic deposits.

The eruption of Mount Pinatubo was a terrible catastrophe, killing many people, destroying homes and farmland, and causing hundreds of millions of dollars in damage. At the same time, the tragedy enabled scientists to test whether they understood the global climate well enough to estimate how the eruption would affect temperatures on earth.

By the late 1980s most of the world's climate scientists had become concerned that human actions (such as fossil fuel use) were contributing to *global warming*—a rise in the temperature of the entire earth. By the late 1980s, some were so worried that they stated publicly that global warming was likely to occur and could have disastrous ecological and economic effects. Their concerns were based in part on results from computer models of the global climate. But were these models reliable?

Although the complex global climate models mimicked past and present climates well, scientists wanted to perform a more rigorous test. Mount Pinatubo provided just such a test. To make his forecasts, Hansen added the estimated amount of sulfur dioxide released by the volcano's eruption to a global climate model and then used the model to forecast

how the earth's temperature would change. His model passed the test with flying colors. Its success helped convince most scientists and policy makers that climate model projections—including the impact of human actions—should be taken seriously.

Hansen's model and more than a dozen other climate models indicate that global temperatures are likely to rise several degrees over the next hundred years—in part because of human actions—and affect the earth's global and regional climates, economies, and our way of life. To many scientists, climate change represents the biggest challenge humanity faces during this century. The primary question now is this: What should we do about it?

U.S. Geological Survey

Figure 16-1 Science: an enormous cloud of gas and ash rises above Mount Pinatubo in the Philippines on June 12, 1991. Three days later, the volcano exploded in a cataclysmic eruption, killing hundreds. Sulfur dioxide and other gases emitted into the atmosphere by the eruption circled the globe, reduced sunlight reaching the earth's surface, and cooled the atmosphere for 15 months. Scientists used this event to test global climate models.

We are in the middle of a large, uncontrolled experiment on the only planet we have.

DONALD KENNEDY

Mount Pinatubo's eruption temporarily altered the temperature of the entire earth. This chapter discusses the considerable body of evidence indicating that the earth's troposphere is warming (partly because of human activities), how our activities are depleting ozone in the stratosphere, and what we can do about these threats. It addresses the following questions:

- How have the earth's temperature and climate changed in the past?

- How might the earth's temperature change in the future?

- What factors influence the earth's average temperature?

- What are some possible beneficial and harmful effects of a warmer earth?

- How can we slow or adapt to projected increases in the earth's temperature?

- How have human activities depleted ozone in the stratosphere, and why should we care?

- How can we slow and eventually reverse ozone depletion in the stratosphere caused by human activities?

KEY IDEAS

- During its long history, the earth has experienced prolonged periods of global warming and global cooling. Considerable evidence indicates that the earth's troposphere is warming, partly because of human activities, and that this trend will lead to significant climate change during this century.

- Global warming is one of the most serious environmental threats faced by humanity.

- Although we cannot stop climate change, we can use existing technological and policy options to help slow the rate of climate change and adapt to its effects.

- Widespread use of certain long-lived chemicals has reduced ozone levels in the stratosphere, which allows more harmful ultraviolet radiation to reach the earth's surface.

- If countries adhere to its terms, an international treaty phasing out ozone-depleting chemicals should return ozone concentrations in the stratosphere to 1980 levels by 2050 and to 1950 levels by 2100.

16-1 PAST CLIMATE CHANGE AND THE NATURAL GREENHOUSE EFFECT

Science: Past Historic Changes in the Earth's Temperature

The earth has experienced prolonged periods of global warming and global cooling.

Changes in the earth's climate are neither new nor unusual. Over the past 4.7 billion years, the planet's climate has been altered by volcanic emissions, changes in solar input, continents moving as a result of shifting tectonic plates (Figure 4-8, p. 72), strikes by large meteors, and other factors.

Over the past 900,000 years, the average temperature of the troposphere has experienced prolonged periods of *global cooling* and *global warming* (Figure 16-2, top left). These alternating cycles of freezing and thawing are known as *glacial and interglacial* (between ice ages) *periods.*

During each cold period, thick glacial ice covered much of the earth's surface for about 100,000 years. For roughly 12,000 years, we have had the good fortune to live in an interglacial period characterized by a fairly stable climate and average global surface temperature (Figure 16-2, top, right). In other words, since agriculture began, the global climate has been favorable to life as we know it. However, even during this generally stable period, regional climates have changed significantly.

Past temperature changes such as those depicted in Figure 16-2 are estimated by analysis of radioisotopes in rocks and fossils, plankton and radioisotopes in ocean sediments, ice cores from ancient glaciers (Figure 16-3), temperature measurements taken at different depths from boreholes drilled deep into the earth's surface, pollen from lake bottoms and bogs, tree rings, historical records, and temperature measurements (since 1860).

Science: The Natural Greenhouse Effect

Certain gases in the atmosphere absorb heat and warm the lower atmosphere.

In addition to incoming sunlight, a natural process called the *greenhouse effect* (Figure 5-5, p. 82) warms the earth's lower troposphere and surface. Swedish chemist Svante Arrhenius first recognized this natural tropospheric heating effect in 1896. Since then, numerous laboratory experiments and measurements of atmospheric temperatures at different altitudes have confirmed this relationship. It is now one of the most widely accepted theories in the atmospheric sciences.

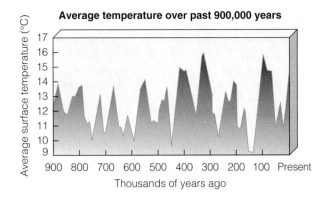
Average temperature over past 900,000 years

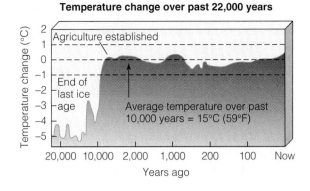
Temperature change over past 22,000 years

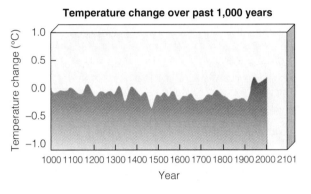
Temperature change over past 1,000 years

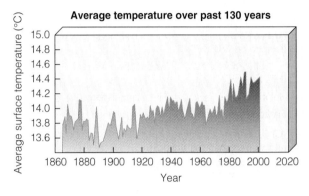
Average temperature over past 130 years

Figure 16-2 Science: estimated changes in the average global temperature of the atmosphere near the earth's surface over different periods of time. Although a particular place might have much lower or much higher readings than the troposphere's average temperature, such averages provide a valuable way to measure long-term trends. (Data from Goddard Institute for Space Studies, Intergovernmental Panel on Climate Change, National Academy of Sciences, National Aeronautics and Space Agency, National Center for Atmospheric Research, and National Oceanic and Atmospheric Administration)

The earth's natural greenhouse effect is one of the reasons you are alive to read these words.

Environmental Science ⊕ Now™
See how greenhouse gases trap heat in the atmosphere and raise the earth's temperature at Environmental ScienceNow.

Science: Major Greenhouse Gases

The two major greenhouse gases are water vapor and carbon dioxide.

The two greenhouse gases with the largest concentrations are *water vapor,* controlled by the hydrologic cycle, and *carbon dioxide* (CO_2), controlled by the carbon cycle. Carbon dioxide is the major greenhouse gas that humans have added to the troposphere (Table 16-1, p. 370).

Scientists have analyzed the concentrations of greenhouse gases such as CO_2 and CH_4 (methane) in bubbles trapped at various depths in ancient glacial ice (Figure 16-3). According to these measurements, the

Figure 16-3 Science: ice cores such as this one extracted by drilling deep holes in ancient glaciers at various sites in Antarctica and Greenland can be analyzed to obtain information about past climates.

changes in tropospheric CO_2 levels correlate fairly closely with variations in the average global temperature near the earth's surface during the past 160,000 years (Figure 16-4, p. 370).

Table 16-1 Science: Major Greenhouse Gases from Human Activities

Greenhouse Gas	Human Sources	Average Time in the Troposphere	Relative Warming Potential (compared to CO_2)
Carbon dioxide (CO_2)	Fossil fuel burning, especially coal (70–75%), deforestation, and plant burning	100–120 years	1
Methane (CH_4)	Rice paddies, guts of cattle and termites, landfills, coal production, coal seams, and natural gas leaks from oil and gas production and pipelines	12–18 years	23
Nitrous oxide (N_2O)	Fossil fuel burning, fertilizers, livestock wastes, and nylon production	114–120 years	296
Chlorofluorocarbons (CFCs)*	Air conditioners, refrigerators, plastic foams	11–20 years (65–110 years in the stratosphere)	900–8,300
Hydrochloro-fluorocarbons (HCFCs)	Air conditioners, refrigerators, plastic foams	9–390	470–2,000
Hydrofluorocarbons (HFCs)	Air conditioners, refrigerators, plastic foams	15–390	130–12,700
Halons	Fire extinguishers	65	5,500
Carbon tetrachloride	Cleaning solvent	42	1,400

*CFC use is being phased out, but these compounds remain in the troposphere for 1–2 decades and then enter the stratosphere.

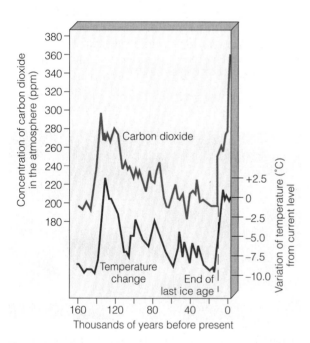

Figure 16-4 Science: atmospheric carbon dioxide levels and global temperature. Estimated long-term variations in average global temperature of the atmosphere near the earth's surface are graphed along with average tropospheric CO_2 levels over the past 160,000 years. The rough correlation between CO_2 levels in the troposphere and temperature shown in these estimates based on ice core data suggests a connection between the two variables, although no definitive causal link has been established. In 1999, the world's deepest ice core sample revealed a similar correlation between air temperatures and the greenhouse gases CO_2 and CH_4 going back for 460,000 years. (Data from Intergovernmental Panel on Climate Change and National Center for Atmospheric Research)

16-2 CLIMATE CHANGE AND HUMAN ACTIVITIES

Science: Signs That the Troposphere Is Warming

A considerable body of scientific evidence indicates that the earth's troposphere is warming, partly because of human activities.

In 1988, the United Nations and the World Meteorological Organization established the Intergovernmental Panel on Climate Change (IPCC) to document past climate changes and project future climate changes. The IPCC network includes more than 2,000 climate experts from 70 nations.

Since 1861, the concentrations of the greenhouse gases CO_2, CH_4, and N_2O in the troposphere have risen sharply, especially since 1950 (Figure 16-5). According to studies by IPCC and the U.S. National Academy of Sciences, humans have increased concentrations of these greenhouse gases in the troposphere by burning fossil fuels (which adds CO_2 and CH_4), clearing and burning forests and grasslands (which add CO_2 and N_2O), and planting rice and using inorganic fertilizers (which release N_2O into the troposphere). Since 1980, the concentration of CO_2 in the troposphere has increased from 280 parts per million (ppm) to 380 ppm, the highest estimated level in 420,000 years. Within a few decades, CO_2 levels are projected to exceed 500 ppm and lead to significant warming of the planet.

Although the United States has only 4.6% of the world's population it is by far the world's largest emit-

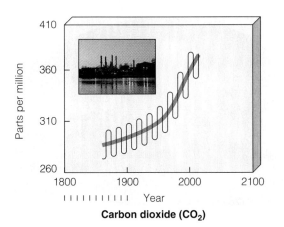

Carbon dioxide (CO₂)

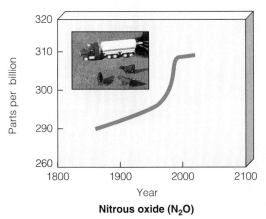

Methane (CH₄)

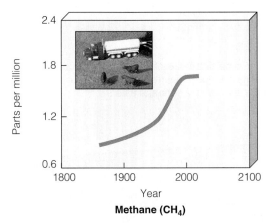

Nitrous oxide (N₂O)

Active Figure 16-5 Science: Increases in average concentrations of three greenhouse gases—carbon dioxide, methane, and nitrous oxide—in the troposphere between 1860 and 2004. The fluctuations in the CO_2 curve reflect seasonal changes in photosynthetic activity that cause small differences between summer and winter concentrations of CO_2. *See an animation based on this figure and take a short quiz on the concept.* (Data from Intergovernmental Panel on Climate Change, National Center for Atmospheric Research, and World Resources Institute)

ter of CO_2—producing almost one-fourth of the annual global emissions of this gas. It is followed by the European Union (12%), China (11%), Russia (7%), Japan (5%), and India (5%). China's emissions are expected to surpass U.S. emissions by 2025. The United States

also has the world's highest per capita CO_2 emissions, followed by Australia, Canada, and the Netherlands.

In its 2001 report, the IPCC listed a number of findings indicating that is it *very likely* (90–99% probability) that the troposphere is getting warmer (Figure 16-2, bottom right).

First, the 20th century was the hottest century in the past 1,000 years (Figure 16-2, bottom left).

Second, since 1861 the average global temperature of the troposphere near the earth's surface has risen 0.6°C (1.1°F) over the entire globe (Figure 16-2, bottom right) and about 0.8°C (1.4°F) over the continents. Most of this increase has taken place since 1980.

Third, the 16 warmest years on record have occurred since 1980 and the 10 warmest years have occurred since 1990.

Fourth, glaciers and floating sea ice in some parts of the world are melting and shrinking (Figure 16-6, p. 372). This process exposes darker and less reflective surfaces of water and land, resulting in a warmer troposphere. As more ice melts, the troposphere can become warmer, which melts more ice and increases the tropospheric temperature even more. According to a 2004 study by nearly 300 scientists, the Arctic "is now experiencing some of the most rapid and severe climate change on Earth." The study concluded that the accelerated melting of Greenland's ice sheets will cause sea levels to rise around the world.

Fifth, warmer temperatures in Alaska and in other parts of the Arctic are melting permafrost. This is releasing large amounts of CO_2 and CH_4 into the troposphere, which can accelerate tropospheric warming.

Sixth, sea levels are rising. During the last century, the world's average sea level rose by 0.1–0.2 meters (4–8 inches), mostly because of runoff from melting land-based ice and because of the expansion of ocean water as its temperature increases.

Examine how four major greenhouse gases have increased in recent decades and find out why at Environmental ScienceNow.

Science: The Scientific Consensus about Future Climate Change

Most climate scientists agree that human activities have influenced recent climate changes and will lead to further significant climate change during this century.

To project the effects of increases in greenhouse gases on average global temperature, scientists develop complex *mathematical models* of interactions among the earth's sunlight, clouds, landmasses, oceans and ocean currents, ice, and the atmospheric concentration of gases like carbon dioxide and sulfur dioxide. Then they run the models on supercomputers and use the results to forecast future changes in the earth's average

Figure 16-6 Science: satellite data showing Arctic sea ice in 1979 (left) and in 2003 (right). According to NASA, the ice cover shrunk by 9% during this period. *Critical thinking: why should you care if ice melts in the artic sea?* (Data collected by Defense Satellite Program [DMSP] Special Sensor Microwave Imager[SSMI], Image: NASA)

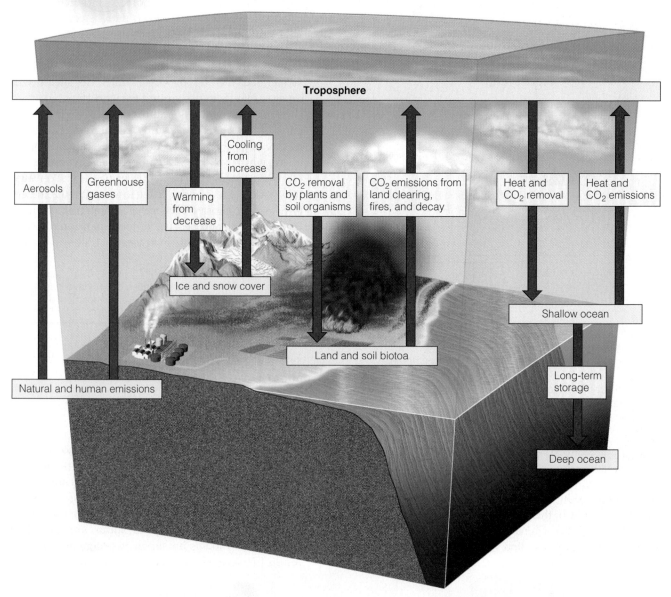

Figure 16-7 Science: simplified model of some of major processes that interact to determine the average temperature and greenhouse gas content of the troposphere and thus the earth's climate.

temperature. Figure 16-7 (p. 372) gives a greatly simplified summary of some of the interactions in the global climate system.

Such models provide scenarios of what is *very likely* (90–99% probability) or *likely* (66–89% probability) to happen to the average temperature of the troposphere based on various assumptions and data fed into the model. How well the results correspond to the real world depends on the assumptions of the model (based on current knowledge about the systems making up the earth, oceans, and atmosphere) and the accuracy of the data used.

In 1990, 1995, and 2001, the IPCC published reports that evaluated how global temperatures have changed in the past (Figure 16-2) and made forecasts of how they are likely to change during this century. The 2001 report included two important findings. *First,* "there is new and stronger evidence that most of the warming observed over the last 50 years is attributable to human activities." *Second,* it is *very likely* (90–99% probability) that the earth's mean surface temperature will increase by 1.4–5.8°C (2.5–10.4°F) between 2000 and 2100 (Figure 16-8). Other major scientific advisory bodies, such as the U.S. National Academy of Sciences, have reached similar conclusions.

More recent runs of various climate models suggest that the most likely temperature increase during this century will be in the range of 2.5°C (4.5°F) to 3.5°C (6.3°F)—a major increase in such a short period.

Science, Economics, and Ethics: Why Should We Be Concerned about a Warmer Earth?

A rapid increase in the temperature of the troposphere during this century would give us little time to deal with its harmful effects.

Climate scientists warn that the concern is not just about a temperature change but how rapidly it occurs. Most past changes in the temperature of the troposphere took place over thousands of years to a hundred thousand years. The problem we face now is a fairly sharp projected increase in temperature of the troposphere during this century or less than the typical span of a human lifetime (Figure 16-8). According to the IPPC, there is a 90–99% chance that this will be the fastest temperature change of the past 1,000 years.

Such rapid change can affect the availability of water resources by altering rates of evaporation and precipitation. It can shift areas where crops can be grown; alter some ocean currents; increase average sea levels and flood some coastal wetlands, cities, and low-lying islands; alter the structure and location of some of the world's biomes; and affect the economic and social fabric of societies throughout the world. The projected increase in the earth's temperature within a few decades or a century gives us little time to deal with its effects.

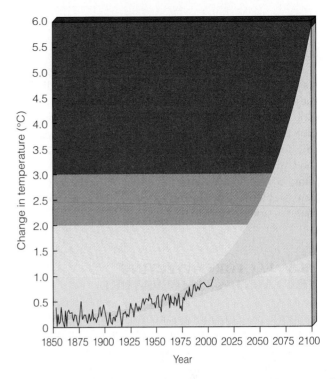

Figure 16-8 Science: comparison of measured changes in the average temperature of the atmosphere at the earth's surface between 1860 and 2004 and the projected range of temperature increase during the rest of this century. Climate models project a 90–99% probability that such changes will occur. (Data from U.S. National Academy of Sciences, National Center for Atmospheric Research, and Intergovernmental Panel on Climate Change)

In addition, climate scientists warn that climate rarely changes in a linear fashion. Large, abrupt changes in climate have repeatedly occurred in past. For example, temperature changes as great as 10°C (18°F) in a decade have occurred in some locales, often spurred by sudden shifts in oceanic circulation.

In 2002, a National Academy of Sciences study suggested that the temperature of the troposphere could increase drastically in only a decade or two. This conclusion was based on analysis of greenhouse gases, such as carbon dioxide and methane, in ice cores and analysis of pollen and zooplankton buried in ocean sediment that revealed sudden climate flips throughout the last ice age—from about 70,000 to 11,500 years ago—when vast sheets of ice blanketed much of the northern hemisphere.

The National Academy of Sciences report lays out a nightmarish worst-case scenario in which human activities or a combination of human activities and natural climate changes succeed in flipping one of the earth's climate switches and triggering a new abrupt change. It describes ecosystems suddenly collapsing, low-lying cities being flooded, forests being consumed in vast fires, grasslands dying out and turning into dust bowls, much wildlife disappearing, more frequent and intensified coastal storms and hurricanes, and tropical waterborne and insect-transmitted infectious diseases spreading rapidly beyond their current ranges.

These possibilities were supported by a 2003 analysis carried out by Peter Schwartz and Doug Randall for the U.S. Department of Defense. They concluded that global warming "must be viewed as a serious threat to global stability and should be elevated beyond a scientific debate to a U.S. national security concern." In 2004, the United Kingdom's chief science adviser, David A. King, wrote, "In my view, climate change is the most severe problem we are facing today—more serious even than the threat of terrorism."

16-3 FACTORS AFFECTING THE EARTH'S TEMPERATURE

Scientists have identified a number of natural and human-influenced factors that might *amplify* or *dampen* projected changes in the average temperature of the troposphere. The fairly wide range of projected future temperature changes shown in Figure 16-8 results from including what is known about these factors in climate models. Let us examine some possible wild cards that could make matters worse or better during this century.

Science: Can the Oceans Store More CO_2 and Heat?

There is uncertainty about how much CO_2 and heat the oceans can remove from the troposphere and how long they might remain in the oceans.

According to a 2004 study by Christopher L. Sabine and other researchers, the oceans help moderate the earth's average surface temperature by removing about 48% of the excess CO_2 we pumped into the atmosphere as part of the global carbon cycle between 1800 and 1994. They also absorb heat from the atmosphere and slowly transfer some of it to the deep ocean, where it is removed from the climate system for unknown periods.

Ocean currents on the surface and deep down are connected and act like a gigantic conveyor belt to store CO_2 and heat in the deep sea and to transfer hot and cold water from the tropics to the poles (Figure 16-9). This loop of water flow is propelled by winds and differences in water density, which changes with the temperature and salinity of seawater. In a warmer world, an influx of fresh water from increased rain in the North Atlantic and thawing ice in the Arctic region might slow or disrupt this conveyor belt, causing drastic climate changes in as little as a decade.

Large changes in the speed of this conveyor belt and its stopping and starting may have contributed to wild swings in northern hemisphere temperatures during past ice ages. Scientists are trying to learn more about how this belt operates to evaluate the likelihood

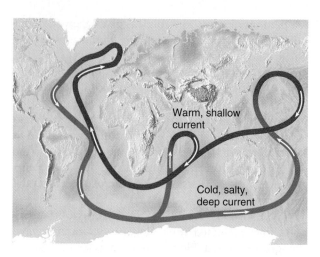

Figure 16-9 Natural capital: A connected loop of shallow and deep ocean currents stores CO_2 in the deep sea and transmits warm and cool water to various parts of the earth. It occurs when ocean water in the North Atlantic near Iceland is dense enough (because of its salt content and cold temperature) to sink to the ocean bottom, flow southward, and then move eastward to well up in the warmer Pacific. A shallower return current aided by winds then brings warmer and less salty—and thus less dense—water to the Atlantic. This water can cool and sink to begin the cycle again. A warmer planet would be a rainier one, which, coupled with melting glaciers, would increase the amount of fresh water flowing into the North Atlantic. This could slow or even jam the loop by diluting the saltwater and making it more buoyant (less dense) and less prone to sinking.

of it slowing down or stalling during this century, and the effects this might have on regional and global atmospheric temperatures.

The large loop of shallow and deep ocean currents shown in Figure 16-9 helps keep much of the northern hemisphere fairly warm by pulling warm tropical water north, pushing cold water south, and releasing much of the heat stored in the water into the troposphere.

If this loop of currents should slow sharply or shut down, northern Europe and the northeast coast of North America would experience severe regional cooling. In other words, *global warming can lead to significant global cooling in some parts of the world*. Disruption or significant slowing of the loop would also disrupt other parts of the world with floods, droughts, severe storms, and searing heat.

Science: Effects of Cloud Cover

Warmer temperatures create more clouds that could warm or cool the troposphere, but we do not know which effect might dominate.

A major unknown in global climate models is how changes in the global distribution of clouds might alter the temperature of the troposphere. Warmer temperatures increase evaporation of surface water and create more clouds. These additional clouds can have a *warm-*

ing effect, by absorbing and releasing heat into the troposphere, or a *cooling effect*, by reflecting more sunlight back into space.

An increase in thick and continuous clouds at low altitudes can *decrease* surface warming by reflecting and blocking more sunlight. In contrast, an increase in thin and discontinuous cirrus clouds at high altitudes can warm the lower troposphere and *increase* surface warming.

In addition, infrared satellite images indicate that the wispy condensation trails (contrails) left behind by jet planes might have a greater impact on the temperature of the troposphere than scientists once thought. NASA scientists found that jet contrails expand and turn into large cirrus clouds that tend to release heat into the upper troposphere. If these preliminary results are confirmed, emissions from jet planes could be responsible for as much as half of the tropospheric warming in the northern hemisphere.

Science: Effects of Outdoor Air Pollution

Aerosol pollutants and soot produced by human activities can warm or cool the atmosphere, but such effects will decrease with any decline in outdoor air pollution.

Aerosols (microscopic droplets and solid particles) of various air pollutants are released or formed in the troposphere by volcanic eruptions (Figure 16-1) and human activities. They can either warm or cool the air depending on factors such as their size and the reflectivity of the underlying surface.

Most tropospheric aerosols, such as sulfate particles produced by fossil fuel combustion, tend to cool the atmosphere and thus temporarily slow global warming. However, a recent study by Mark Jacobson of Stanford University indicated that tiny particles of *soot* or *black carbon aerosols*—produced mainly from incomplete combustion in coal burning, diesel engines, and open fires—may be the second biggest contributor to global warming after the greenhouse gas CO_2.

Climate scientists do not expect aerosol pollutants to counteract or enhance projected global warming very much in the next 50 years for two reasons. *First,* aerosols and soot fall back to the earth or are washed out of the lower atmosphere within weeks or months, whereas CO_2 and other greenhouse gases remain in the atmosphere for decades to several hundred years. *Second,* aerosol inputs into the atmosphere are being reduced—especially in developed countries.

Science: Effects of Higher CO_2 Levels on Photosynthesis

Increased CO_2 in the troposphere could increase plant photosynthesis, but several factors can limit or offset this effect.

Some studies suggest that larger amounts of CO_2 in the atmosphere could increase the rate of photosynthesis in some areas with adequate water and soil nutrients. This would remove more CO_2 from the troposphere and help slow global warming.

However, recent studies indicate that this effect would be temporary for two reasons. *First,* the increase in photosynthesis would slow as the plants reach maturity and take up less CO_2 from the troposphere. *Second,* carbon stored by the plants would be returned to the atmosphere as CO_2 when the plants die and decompose or burn.

Science: Effects of a Warmer Troposphere on Methane Emissions

Warmer air can release methane gas stored in bogs, wetlands, and tundra soils and make the air even warmer.

Global warming could be accelerated by an increased release of methane (a potent greenhouse gas; see Table 16-1) from two major sources: bogs and other wetlands and ice-like compounds called *methane hydrates* trapped beneath arctic permafrost. Significant amounts of methane would be released into the troposphere if the permafrost in tundra and boreal forest soils melts, as is currently occurring in parts of Canada, Alaska, China, and Mongolia. The resulting tropospheric warming could lead to more methane release and still more warming.

16-4 POSSIBLE EFFECTS OF A WARMER WORLD

Science: Projected Effects of a Warmer Troposphere

A warmer climate would have beneficial and harmful effects, but poor nations in the tropics would suffer the most.

A warmer global climate could have a number of harmful and beneficial effects (Figure 16-10, p. 376) for humans, other species, and ecosystems, depending mostly on location and the rate at which the climate changes.

Some areas will benefit because of less severe winters, more precipitation in some dry areas, less precipitation in wet areas, and increased food production. Also, some plant and animal species adapted to higher temperatures may be able to expand their populations and range.

Other areas will suffer harm from excessive heat, lack of water, and decreased food production. Wildfires are likely to increase in forest and grassland areas where the climate becomes drier. Also, tree deaths would increase from larger disease and pest populations that would thrive in areas with a warmer climate. Many

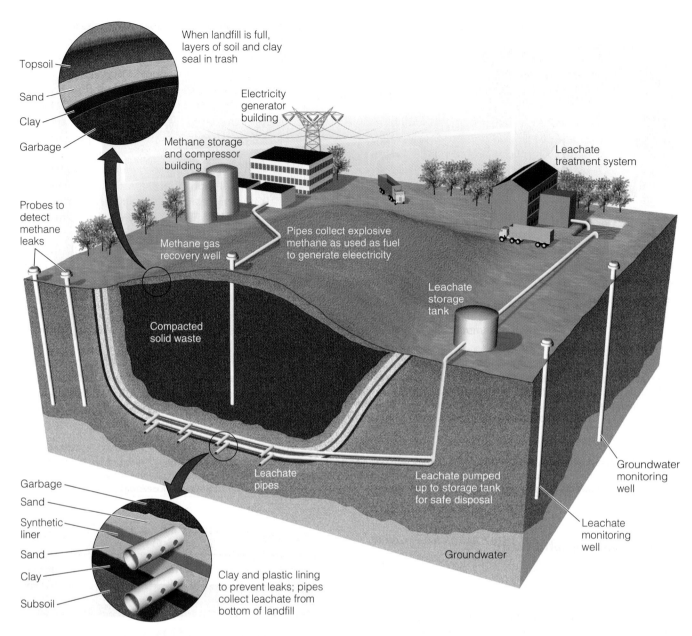

Topsoil

Sand

Clay

Garbage

When landfill is full, layers of soil and clay seal in trash

Electricity generator building

Methane storage and compressor building

Leachate treatment system

Probes to detect methane leaks

Methane gas recovery well

Pipes collect explosive methane as used as fuel to generate eleectricity

Leachate storage tank

Compacted solid waste

Leachate pumped up to storage tank for safe disposal

Groundwater monitoring well

Leachate pipes

Leachate monitoring well

Groundwater

Garbage

Sand

Synthetic liner

Sand

Clay

Subsoil

Clay and plastic lining to prevent leaks; pipes collect leachate from bottom of landfill

Figure 17-11 Solutions: state-of-the-art *sanitary landfill*, which is designed to eliminate or minimize environmental problems that plague older landfills. Even these landfills are expected to leak eventually, passing both the effects of contamination and the cleanup costs on to future generations. Since 1997, only sanitary landfills have been able to operate in the United States. As a result, many older and small landfills have been closed and replaced with larger local and regional modern landfills.

before being filled with garbage, as shown in Figure 17-11. The landfill bottom is covered with a second impermeable liner, usually made of several layers of clay, thick plastic, and sand. This liner collects *leachate* (rainwater contaminated as it percolates through the solid waste) and is intended to prevent its leakage into groundwater. Wells are drilled around the landfill to monitor any leakage.

The leachate is pumped from the bottom of the landfill, stored in tanks, and sent to a regular sewage treatment plant or an on-site treatment plant. When it becomes full, the landfill is covered with clay, sand, gravel, and topsoil to prevent water from seeping in.

Sanitary landfills have a network of vent pipes to collect landfill gases (consisting mostly of two greenhouse gases, methane and carbon dioxide) released by the underground decomposition of wastes. The methane is filtered out and burned in small gas turbines to produce steam or electricity for nearby facilities or sold to utilities for use as a fuel.

400 CHAPTER 17 Solid and Hazardous Waste

Trade-Offs

Sanitary Landfills

Advantages	Disadvantages
No open burning	Noise and traffic
Little odor	Dust
Low groundwater pollution if sited properly	Air pollution from toxic gases and volatile organic compounds
Can be built quickly	Releases greenhouse gases (methane and CO_2) unless they are collected
Low operating costs	Groundwater contamination
Can handle large amounts of waste	Slow decomposition of wastes
Filled land can be used for other purposes	Discourages recycling and waste reduction
No shortage of landfill space in many areas	Eventually leaks and can contaminate groundwater

Figure 17-12 Trade-offs: advantages and disadvantages of using sanitary landfills to dispose of solid waste. *Critical thinking: pick the single advantage and disadvantage that you think are the most important.*

Figure 17-12 lists the advantages and disadvantages of using sanitary landfills to dispose of solid waste. According to the EPA, all landfills eventually leak.

✗ *How Would You Vote?* Do the advantages of burying solid waste in sanitary landfills outweigh the disadvantages? Cast your vote online at http://biology.brookscole.com/miller11.

17-7 HAZARDOUS WASTE

Science and Politics: Hazardous Waste

Developed countries produce 80–90% of the world's solid and liquid wastes that can harm people, and most such wastes are not regulated.

Hazardous waste is any discarded solid or liquid material that is *toxic, ignitable, corrosive,* or *reactive* enough to explode or release toxic fumes. According to the UN

Environment Programme, developed countries produce 80–90% of these wastes. Figure 17-13 (p. 402) lists some the harmful chemicals found in many homes.

In the United States, about 5% of all hazardous waste is regulated under the Resource Conservation and Recovery Act (RCRA, pronounced "RICK-ra") and is often referred to as *RCRA*. The EPA sets standards for management of several types of hazardous waste and issues firms permits to produce and dispose of a certain amount of wastes in acceptable ways. Permit holders must use a *cradle-to-grave* system to keep track of waste they transfer from a point of generation (cradle) to an approved off-site disposal facility (grave) and submit proof of this disposal to the EPA.

RCRA is a good start, but it and other laws do not regulate about 95% of these wastes. In most other countries, especially developing countries, even less hazardous waste is regulated.

In 1980, the U.S. Congress passed the *Comprehensive Environmental Response, Compensation, and Liability Act*, commonly known as the *CERCLA* or *Superfund* program. Its goals are to identify hazardous waste sites and clean them up on a priority basis. The worst sites that represent an immediate and severe threat to human health are put on a *National Priorities List (NPL)* and scheduled for total cleanup using the most cost-effective method. Currently, about 1,250 sites are on this list and cleanup has essentially been completed on about 72% of the sites at an average cost of $20 million per site. The Waste Management Research Institute estimates that the Superfund list could eventually include at least 10,000 priority sites, with cleanup costs of as much as $1 trillion, not counting legal fees.

The Superfund law was designed to have polluters pay for cleaning up abandoned hazardous waste sites. However, facing pressure from polluters, Congress did away with this *polluter-pays principle* in 1995 and converted it to a *taxpayers-pay* program. As a result, the amount of funds appropriated for cleanup has dropped sharply and the pace of cleanup has slowed.

✗ *How Would You Vote?* Should the U.S. Congress reinstate the polluter-pays principle by using taxes from chemical, oil, mining, and smelting companies to reestablish a fund for cleaning up existing and new Superfund sites? Register your vote online at http://biology.brookscole.com/miller11.

Both the U.S. Congress and several state legislatures have also passed laws that encourage the cleanup of *brownfields*—abandoned industrial and commercial sites that are often contaminated with hazardous wastes. Examples include factories, junkyards, older landfills, and gas stations.

What Harmful Chemicals Are in Your Home?

Cleaning

- Disinfectants
- Drain, toilet, and window cleaners
- Spot removers
- Septic tank cleaners

Paint

- Latex and oil-based paints
- Paint thinners, solvents, and strippers
- Stains, varnishes, and lacquers
- Wood preservatives
- Artist paints and inks

General

- Dry-cell batteries (mercury and cadmium)
- Glues and cements

Gardening

- Pesticides
- Weed killers
- Ant and rodent killers
- Flea powders

Automotive

- Gasoline
- Used motor oil
- Antifreeze
- Battery acid
- Solvents
- Brake and transmission fluid
- Rust inhibitor and rust remover

Figure 17-13 Science: harmful chemicals found in many homes. Congress has exempted disposal of these materials from government regulation. *Critical thinking: which of these chemicals are in your home? Which ones could you do without?*

Figure 17-14 lists the priorities that prominent scientists believe we should follow in dealing with hazardous waste. Denmark is following these priorities but most countries are not.

Science and Politics: How Safe Are U.S. Chemical Plants from Terrorist Attacks?

Large amounts of hazardous wastes could be released into the environment by terrorist attacks on major chemical plants in the United States.

Managers of industrial plants that manufacture and use chemicals work hard to prevent accidental release of chemicals that can harm workers or nearby residents. But accidents can happen, as thousands of people living near a pesticide manufacturing plant in Bhopal, India, learned in 1984 (see Case Study, p. 403).

Roughly 15,000 chemical plants, refineries, and other sites in the United States contain large quantities of hazardous chemicals. According to the EPA, at about 790 sites the toll of death or injury from a catastrophic disaster at a chemical plant could reach from 100,00 to more than 1 million people. The 2001 attacks on New York City's World Trade Center Towers and the Pentagon have heightened concerns about terrorist acts against such plants.

Analysts view such plants as easy targets for acts of sabotage. There are no federal laws establishing minimum security at these chemical facilities. In addition, slow-moving railcars carrying hazardous chemicals to and from such plants regularly pass through urban areas. And some of the railcars are parked near residential areas for extended periods. Barges carrying hazardous chemicals that move up and down the country's waterways are also largely unprotected.

In 2004, a study by the Working Group on the Community Right-to-Know pointed out that terrorist acts or accidents could release large quantities of gaseous ammonia or chlorine used at some 275 of the nation's power plants. Such releases pose a potential danger to 3.5 million Americans. The study noted that less harmful substitutes for these chemicals are available. *Critical thinking: what three things would you do to reduce such risks?*

Science, Economics, and Ethics Case Study: A Black Day in Bhopal, India

The world's worst industrial accident occurred in 1984 at a pesticide plant in Bhopal, India.

December 2, 1984, will long stand as a black day in Indian history. On that date, the world's worst industrial accident occurred at a Union Carbide pesticide plant in Bhopal, India.

An explosion in an underground storage tank released a large quantity of highly toxic methyl isocyanate (MIC) gas, used to produce carbamate pesticides. Water leaking into the tank through faulty valves and corroded pipes caused the explosive chemical reaction.

Once in the atmosphere, some of the toxic MIC was converted to more deadly hydrogen cyanide gas. The toxic cloud of gas settled over an area of 78 square kilometers (30 square miles), exposing as many as 600,000 people. Many were illegal squatters living near the plant because they had no other place to go. The deadly cloud spread through Bhopal without warning because the plant's warning sirens had been turned off to save money.

Indian officials say that at least 10,000 people died within a few days after the accident. Union Carbide put the death toll at 3,800. The International Campaign for Justice in Bhopal estimates that by 2001 the accident had killed 20,000. It also puts the number of people suffering from chronic illnesses from the accident at 120,000–150,000. An international team of medical specialists estimated in 1996 that 50,000–60,000 people sustained permanent injuries such as blindness, lung damage, and neurological problems. Clearly, this was the world's largest industrial tragedy.

Indian officials claim that Union Carbide might have prevented the tragedy by spending no more than $1 million to upgrade plant equipment and improve safety. According to an investigation by India's Central Bureau of Investigation (CBI), the company's U.S. corporate managers decided to save money by cutting back on maintenance and safety because the plant had proven to be a financial disappointment for the firm. The CBI found that on the night of the disaster six safety measures designed to prevent a leak of toxic materials were inadequate, shut down, or malfunctioning.

In 1989, Union Carbide agreed to pay $470 million to compensate the victims without admitting any guilt or negligence concerning the accident. The Indian government has distributed only part of these funds to the victims or their families, with the average payment amounting to $500.

In 1992, the Court of the Chief Judicial Magistrate for Bhopal charged Warren Anderson, CEO of Union Carbide at the time of the accident, with "culpable homicide" (the equivalent of manslaughter) and issued a warrant for his arrest. He has refused to appear in court, and the U.S. government has not responded to India's request to extradite him to India to stand trial.

Science: Detoxifying Hazardous Waste

Chemical and biological methods can be used to remove hazardous wastes or to reduce their toxicity.

In Denmark, all hazardous and toxic waste from industries and households is delivered to 21 transfer stations throughout the country. The waste is then transferred to a large treatment facility. There, three-fourths of the waste is detoxified by physical, chemical, and biological methods; the rest is buried in a carefully designed and monitored landfill.

Some scientists and engineers consider biological treatment of hazardous waste as the wave of the future for cleaning

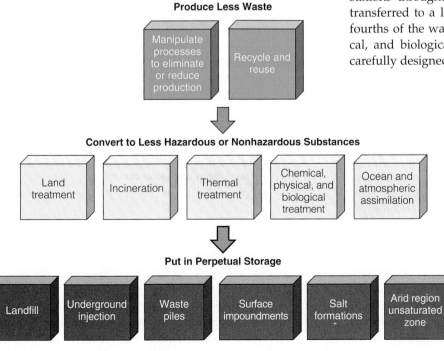

Figure 17-14 Solutions: priorities for dealing with hazardous waste. To date, these priorities have not been followed in the United States or most other countries. *Critical thinking: why do most countries not follow these priorities?* (Data from U.S. National Academy of Sciences)

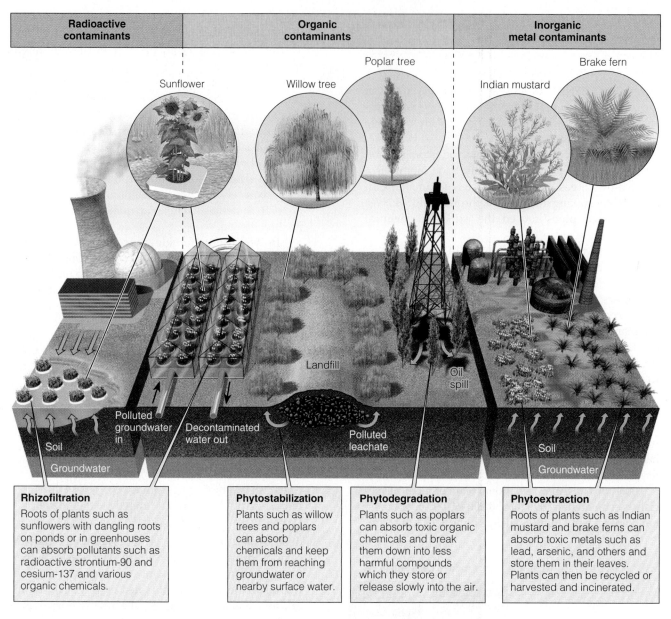

Radioactive contaminants	Organic contaminants	Inorganic metal contaminants

Sunflower

Poplar tree

Willow tree

Brake fern

Indian mustard

Landfill

Oil spill

Polluted groundwater in

Decontaminated water out

Polluted leachate

Soil

Groundwater

Soil

Groundwater

Rhizofiltration

Roots of plants such as sunflowers with dangling roots on ponds or in greenhouses can absorb pollutants such as radioactive strontium-90 and cesium-137 and various organic chemicals.

Phytostabilization

Plants such as willow trees and poplars can absorb chemicals and keep them from reaching groundwater or nearby surface water.

Phytodegradation

Plants such as poplars can absorb toxic organic chemicals and break them down into less harmful compounds which they store or release slowly into the air.

Phytoextraction

Roots of plants such as Indian mustard and brake ferns can absorb toxic metals such as lead, arsenic, and others and store them in their leaves. Plants can then be recycled or harvested and incinerated.

Figure 17-15 Solutions: *phytoremediation.* Various types of plants can be used as pollution sponges to clean up soil and water and radioactive substances (left), organic compounds (center), and toxic metals (right). (Data from American Society of Plant Physiologists, U.S. Environmental Protection Agency, and Edenspace)

up some types of toxic and hazardous waste. In *bioremediation,* for example, bacteria and enzymes help destroy toxic or hazardous substances or convert them to harmless compounds. See the Guest Essay by John Pichtel on this topic on the website for this chapter.

Phytoremediation involves using natural or genetically engineered plants to absorb, filter, and remove contaminants from polluted soil and water, as shown in Figure 17-15. Various plants have been identified as "pollution sponges" to help clean up soil and water contaminated with chemicals such as pesticides, organic solvents, radioactive metals, and toxic metals such as lead, mercury, and arsenic. Figure 17-16 lists advantages and disadvantages of phytoremediation.

✗ *HOW WOULD YOU VOTE?* Do the advantages of phytoremediation outweigh the disadvantages? Cast your vote online at http://biology.brookscole.com/miller11.

Science: Burning and Burying Solid Waste

Hazardous waste can be incinerated or disposed of on or underneath the earth's surface, but this practice can pollute the air and water.

Hazardous waste can be incinerated. Burning such wastes has the same mixture of advantages and disadvantages as burning solid wastes (Figure 17-10). Unfortunately, incinerating hazardous waste releases

Trade-Offs

Phytoremediation

Advantages	Disadvantages

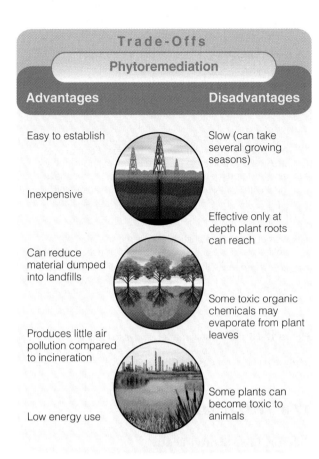

Advantages

Easy to establish

Inexpensive

Can reduce material dumped into landfills

Produces little air pollution compared to incineration

Low energy use

Disadvantages

Slow (can take several growing seasons)

Effective only at depth plant roots can reach

Some toxic organic chemicals may evaporate from plant leaves

Some plants can become toxic to animals

Figure 17-16 Trade-offs: advantages and disadvantages of using *phytoremediation* to remove or detoxify hazardous waste. *Critical thinking: pick the single advantage and disadvantage that you think are the most important.*

air pollutants such as toxic dioxins and produces a highly toxic ash that must be safely and permanently stored.

Most hazardous waste in the United States is disposed of on land in deep underground wells; surface impoundments such as ponds, pits, or lagoons; and state-of-the-art landfills. In *deep-well disposal,* liquid hazardous wastes are pumped under pressure through a pipe into dry, porous geologic formations or zones of rock far beneath the aquifers tapped for drinking and irrigation water. Theoretically, these liquids soak into the porous rock material and are isolated from overlying groundwater by essentially impermeable layers of rock.

Figure 17-17 lists the advantages and disadvantages of deep-well disposal of liquid hazardous wastes. Many scientists believe that current regulations for deep-well disposal are inadequate and should be improved.

✘ *How Would You Vote?* Do the advantages of deep-well disposal of hazardous waste from soil and water outweigh the disadvantages? Cast your vote online at http://biology .brookscole.com/miller11.

Trade-Offs

Deep Underground Wells

Advantages	Disadvantages

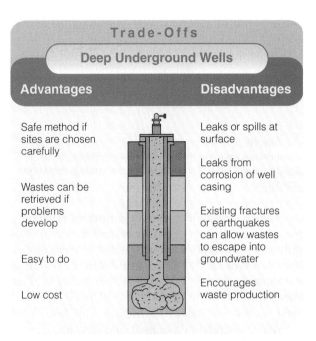

Advantages

Safe method if sites are chosen carefully

Wastes can be retrieved if problems develop

Easy to do

Low cost

Disadvantages

Leaks or spills at surface

Leaks from corrosion of well casing

Existing fractures or earthquakes can allow wastes to escape into groundwater

Encourages waste production

Figure 17-17 Trade-offs: advantages and disadvantages of injecting liquid hazardous wastes into deep underground wells. *Critical thinking: pick the single advantage and disadvantage that you think are the most important.*

Trade-Offs

Surface Impoundments

Advantages	Disadvantages

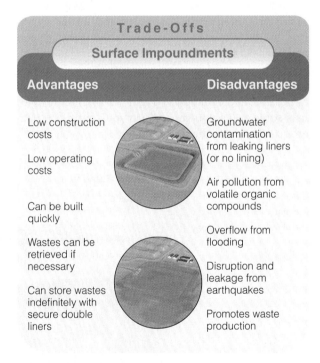

Advantages

Low construction costs

Low operating costs

Can be built quickly

Wastes can be retrieved if necessary

Can store wastes indefinitely with secure double liners

Disadvantages

Groundwater contamination from leaking liners (or no lining)

Air pollution from volatile organic compounds

Overflow from flooding

Disruption and leakage from earthquakes

Promotes waste production

Figure 17-18 Trade-offs: advantages and disadvantages of storing liquid hazardous wastes in surface impoundments. *Critical thinking: pick the single advantage and disadvantage that you think are the most important.*

Surface impoundments are excavated depressions such as ponds, pits, or lagoons into which liquid hazardous wastes are drained and stored (Figure 11-26, p. 258). As water evaporates, the waste settles and becomes more concentrated. Figure 17-18 lists the advantages and disadvantages of this method of disposal. EPA studies found that 70% of these storage basins in the United States have no liners, and as many as 90% may threaten groundwater. According to the EPA, all liners are likely to leak eventually and can contaminate groundwater.

How Would You Vote? Do the advantages of storing hazardous waste in surface impoundments outweigh the disadvantages? Cast your vote online at http://biology.brookscole.com/miller11.

Sometimes liquid and solid hazardous wastes are put into drums or other containers and buried in carefully designed and monitored *secure hazardous waste landfills* (Figure 17-19). Sweden buries its concentrated hazardous wastes in underground vaults made of reinforced concrete. By contrast, in the United Kingdom most hazardous wastes are mixed with household garbage and stored in hundreds of conventional landfills throughout the country.

Hazardous wastes can also be stored in carefully designed *aboveground buildings*—a good option in areas where the water table is close to the surface or areas that are above aquifers used for drinking water. These structures are built to withstand storms and to prevent the release of toxic gases. Leaks are monitored and any leakage is collected and treated.

To most environmental scientists, the real solution to the hazardous waste problem is to produce as little as possible in the first place (Figure 17-14). Figure 17-20 lists some ways you can reduce your output of hazardous waste into the environment.

What Can You Do?

Hazardous Waste

- Use pesticides in the smallest amount possible.
- Use less harmful substances instead of commercial chemicals for most household cleaners. For example, use liquid ammonia to clean appliances and windows; vinegar to polish metals, clean surfaces, and remove stains and mildew; baking soda to clean household utensils, deodorize, and remove stains; borax to remove stains and mildew.
- Do not dispose of pesticides, paints, solvents, oil, antifreeze, or other products containing hazardous chemicals by flushing them down the toilet, pouring them down the drain, burying them, throwing them into the garbage, or dumping them down storm drains.

Figure 17-20 Individuals matter: ways to reduce your input of hazardous waste into the environment. *Critical thinking: which of the things in this list do you do or plan to do?*

17-8 TOXIC METALS

Science Case Study: Lead

Lead is especially harmful to children and is still used in leaded gasoline and household paints in about 100 countries.

Because it is a chemical element, lead (Pb) does not break down in the environment. This potent neurotoxin can harm the nervous system, especially in young

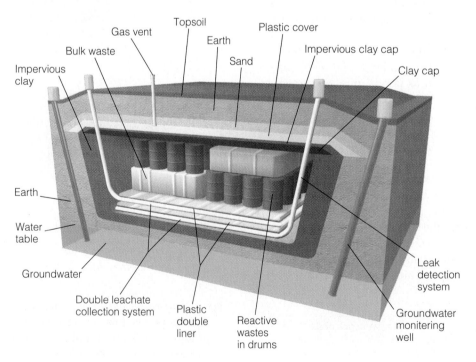

Figure 17-19 Solutions: secure hazardous waste landfill.

children. Each year in the United States, 12,000–16,000 children younger than age 9 are treated for acute lead poisoning, and about 200 die. About 30% of the survivors suffer from palsy, partial paralysis, blindness, and mental retardation.

Children younger than age 6 and unborn fetuses even with low blood levels of lead are especially vulnerable to nervous system impairment, lowered IQ (by an average of 7.4 points), shortened attention span, hyperactivity, hearing damage, and various behavior disorders.

Good news. Between 1976 and 2000, the percentage of U.S. children ages 1–5 with blood lead levels above the safety standard dropped from 85% to 2.2%, so that at least 9 million childhood lead poisonings were prevented. The primary reason for this trend was that government regulations banned leaded gasoline in 1976 (with complete phaseout by 1986) and lead-based paints in 1970 (but illegal use continued until about 1978). This is an excellent example of the power of pollution prevention.

Bad news. The U.S. Centers for Disease Control and Prevention (CDC) estimates that at least 400,000

U.S. children still have unsafe blood levels of lead caused by exposure from a number of sources. A major source is inhalation or ingestion of lead particles from peeling lead-based paint found in about 38 million houses built before 1960. Lead can also leach from water lines and pipes and faucets containing lead. In addition, a 1993 study by the U.S. National Academy of Sciences and numerous other studies indicate *there is no safe level of lead in children's blood.*

Health scientists have proposed a number of ways to help protect children from lead poisoning, as listed in Figure 17-21. Although the threat from lead has been reduced in the United States, it remains a danger in many developing countries. About 80% of the gasoline sold in the world today is unleaded, but about 100 countries still use leaded gasoline. The World Health Organization (WHO) estimates that 130–200 million children around the world are at risk from lead poisoning, and 15–18 million children in developing countries have permanent brain damage because of lead poisoning—mostly from use of leaded gasoline. *Good news.* China recently phased out leaded gasoline in less than three years.

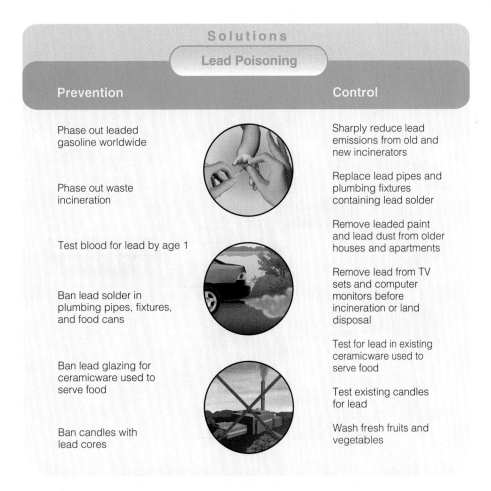

Solutions

Lead Poisoning

Prevention

- Phase out leaded gasoline worldwide
- Phase out waste incineration
- Test blood for lead by age 1
- Ban lead solder in plumbing pipes, fixtures, and food cans
- Ban lead glazing for ceramicware used to serve food
- Ban candles with lead cores

Control

- Sharply reduce lead emissions from old and new incinerators
- Replace lead pipes and plumbing fixtures containing lead solder
- Remove leaded paint and lead dust from older houses and apartments
- Remove lead from TV sets and computer monitors before incineration or land disposal
- Test for lead in existing ceramicware used to serve food
- Test existing candles for lead
- Wash fresh fruits and vegetables

Figure 17-21 Solutions: ways to help protect children from lead poisoning. *Critical thinking: which three of the solutions do you believe are the most important?*

Science Case Study: Mercury

Mercury is released into the environment mostly by burning coal and incinerating wastes and can build to high levels in some types of fish consumed by humans.

Mercury—a toxic metal—is released into the atmosphere from rocks, soil, and volcanoes and by vaporization from the ocean. Such natural sources account for about one-third of the mercury released into the atmosphere each year. According to the EPA, the remaining two-thirds comes from human activities—mostly coal burning and to a lesser extent waste incineration, gold and silver mining, and smelting of metal ores. Once in the atmosphere, elemental mercury may be converted to more toxic inorganic and organic mercury compounds, as shown in Figure 17-22.

Humans are exposed to mercury in two ways. *First*, they may inhale vaporized elemental mercury (Hg) or particulates of inorganic mercury (Hg^{2+}) salts (such as HgS and $HgCl_2$). *Second*, they may eat fish contaminated with methylmercury (CH_3Hg^+). The greatest risk is brain damage from exposure to low levels of methylmercury in fetuses and young children whose nervous systems are still developing.

Once moderately harmful inorganic mercury ions (Hg^{2+}) enter an aquatic system, bacteria may convert them to highly toxic methylmercury, which can be biologically magnified in food chains and webs (Figure 17-22). As a consequence, high levels of methylmercury are often found in the tissues of sharks, swordfish, king mackerel, tilefish, and albacore (white) tuna feeding at high trophic levels in food chains and webs. In 2004, the U.S. Food and Drug Administration (FDA) and the EPA advised women who may become pregnant, pregnant women, and nursing mothers not to eat shark, swordfish, king

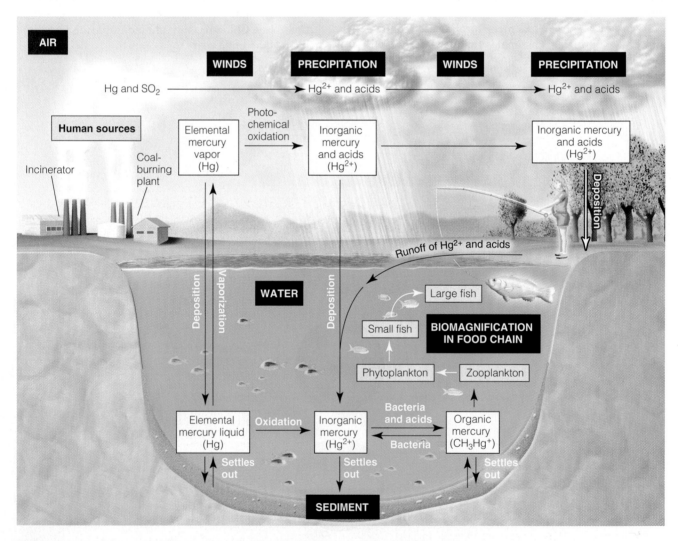

Figure 17-22 Science: cycling of mercury in aquatic environments, in which mercury is converted from one form to another. The most toxic form to humans is methylmercury (CH_3Hg^+), which can be biologically magnified in aquatic food chains. Some mercury is also released back into the atmosphere as mercury vapor.

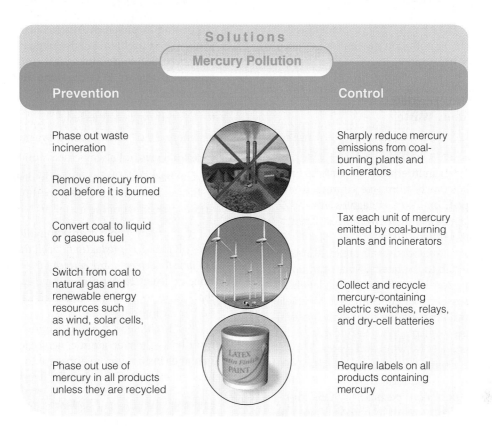

Solutions

Mercury Pollution

Prevention	Control
Phase out waste incineration	Sharply reduce mercury emissions from coal-burning plants and incinerators
Remove mercury from coal before it is burned	
Convert coal to liquid or gaseous fuel	Tax each unit of mercury emitted by coal-burning plants and incinerators
Switch from coal to natural gas and renewable energy resources such as wind, solar cells, and hydrogen	Collect and recycle mercury-containing electric switches, relays, and dry-cell batteries
Phase out use of mercury in all products unless they are recycled	Require labels on all products containing mercury

Figure 17-23 Solutions: ways to prevent or control inputs of mercury into the environment from human activities—mostly through control of coal-burning plants and incinerators. *Critical thinking: which four of these solutions do you believe are the most important?*

mackerel, or tilefish and to limit their consumption of albacore tuna to no more than 170 grams (6 ounces) per week. They also advised these populations to check local advisories about the safety of fish caught in local lakes, rivers, and coastal areas.

In 2004, the EPA warned that one-fourth of the nation's rivers, one-third of its lakes (including all of the Great Lakes), and three-fourths of its coastal waters are contaminated with mercury and other pollutants that could cause health problems for children and pregnant women who eat too much fish caught in them.

Figure 17-23 lists ways to prevent or control human exposure to mercury. In its 2003 report on global mercury pollution, the UN Environment Programme recommended phasing out coal burning and waste incineration as rapidly as possible.

17-9 ACHIEVING A LOW-WASTE SOCIETY

Politics: Grassroots Action for Better Solid and Hazardous Waste Management

In the United States, citizens have kept large numbers of incinerators, landfills, and hazardous waste treatment plants from being built in their local areas.

In the United States, individuals have organized to prevent hundreds of incinerators, landfills, and treatment plants for hazardous and radioactive wastes from being built in or near their communities. Opposition has grown as numerous studies have shown that such facilities have traditionally been located in communities populated mostly by African Americans, Asian Americans, Latinos, and poor whites. This practice has been cited as an example of *environmental injustice*. See the Guest Essay on this subject by Robert Bullard on the website for this chapter.

Health risks from incinerators and landfills, when averaged over the entire country, are quite low. However, the risks for people living near these facilities are much higher. They are the ones whose health, lives, and property values are being threatened.

Manufacturers and waste industry officials point out that something must be done with the toxic and hazardous wastes produced while providing people with certain goods and services. They contend that if local citizens adopt a "not in my back yard" (NIMBY) approach, the waste still ends up in someone's back yard.

Many citizens do not accept this argument. To them, the best way to deal with most toxic or hazardous wastes is to produce much less of them, as suggested by the U.S. National Academy of Sciences (Figure 17-14). For such materials, their goal is "not in anyone's back yard" (NIABY) or "not on planet Earth" (NOPE) by emphasizing pollution prevention and use of the precautionary principle.

Global Outlook: International Action to Reduce Hazardous Waste

An international treaty calls for phasing out the use of harmful persistent organic pollutants.

Between 1989 and 1994, an international treaty to limit transfer of hazardous waste from one country to another was developed. In 2000, delegates from 122 countries completed a global treaty to control 12 *persistent organic pollutants (POPs).*

These widely used toxic chemicals are insoluble in water and soluble in fat. In the fatty tissues of humans and other organisms feeding at high trophic levels in food webs, they can become concentrated to levels hundreds of thousand times higher than in the general environment (Figure 9-16, p. 197). POPs can also be transported long distances by wind and water.

The list of 12 chemicals covered by the 2000 treaty—the *dirty dozen*—includes DDT and 8 other chlorine-containing persistent pesticides, PCBs, dioxins, and furans. The treaty seeks to ban or phase out use of these chemicals and detoxify or isolate stockpiles of them. About 25 countries can continue using DDT to combat malaria until safer alternatives become available.

Environmentalists consider the POPs treaty to be an important milestone in international environmental law because it uses the *precautionary principle* to manage and reduce the risks from toxic chemicals. This list is expected to grow as scientific studies uncover evidence of toxic and environmental damage from other chemicals.

In 2000, the Swedish Parliament enacted a law that by 2020 will ban all chemicals that are persistent and can bioaccumulate in living tissue. This law also requires an industry to perform risk assessments on all old and new chemicals and show that these chemicals are safe to use, as opposed to requiring the government to show that they are dangerous. In other words, chemicals are assumed to be guilty until proven innocent—the reverse of the current policy in the United States and most countries. There is strong opposition to this approach in the United States, especially by the industries producing potentially dangerous chemicals.

Solutions: Making the Transition to a Low-Waste Society

A number of the principles and programs discussed in this chapter can be used to make the transition to a low-waste society during this century.

According to physicist Albert Einstein, "A clever person solves a problem, a wise person avoids it." To prevent pollution and reduce waste, many environmental scientists urge us to understand and live by four key principles:

- Everything is connected.
- There is no "away" for the wastes we produce.
- Dilution is not always the solution to pollution.
- The best and cheapest way to deal with waste and pollution is to produce fewer pollutants and to reuse and recycle most of the materials we use.

Good news. There is growing interest in and greater use of *resource productivity, pollution prevention, eco-industrial systems,* and *service-flow businesses.* In addition, at least 24 countries have *eco-labeling programs* that certify a product or service as having met specified environmental standards.

Such changes start off slowly but can accelerate rapidly as their economic, ecological, and health advantages become more apparent to investors, business leaders, elected officials, and citizens.

Environmental Science⊕Now™
Learn more about how shifting to a low-waste (low-throughput) economy would be the best long-term solution to environmental and resource problems at Environmental ScienceNow.

The key to addressing the challenge of toxics use and wastes rests on a fairly straightforward principle: harness the innovation and technical ingenuity that has characterized the chemicals industry from its beginning and channel these qualities in a new direction that seeks to detoxify our economy.
ANNE PLATT MCGINN

CRITICAL THINKING

1. Collect all of the trash (excluding food waste) that you generate in a typical week. Measure its total weight and volume. Sort it into major categories such as paper, plastic, metal, and glass. Then weigh each category and calculate the percentage by weight of the trash in each category. What percentage by weight of this waste consists of materials that could be recycled or reused? What percentage by weight of the items could you have done without? Tally and compare the results for your entire class.

2. Are you for or against bringing about an *ecoindustrial revolution* in the country and community where you live? Explain. Do you believe it will be possible to phase in such a revolution over the next two to three decades? Explain. What are the three most important strategies for doing this?

3. Are you for or against shifting to a *service-flow economy* based on eco-leasing in the country and community where you live? Explain. Do you believe it will be possible to shift to such an economy over the next two to three decades? Explain. What are the three most important strategies for doing this?

4. Would you oppose having a hazardous waste landfill, waste treatment plant, deep-injection well, or incinerator in your community? Explain. If you oppose these disposal facilities, how do you believe the hazardous waste generated in your community and your state should be managed?

5. Give your reasons for agreeing or disagreeing with each of the following proposals for dealing with hazardous waste:

  a. Reduce the production of hazardous waste and encourage recycling and reuse of hazardous materials by charging producers a tax or fee for each unit of waste generated.

  b. Ban all land disposal and incineration of hazardous waste to encourage recycling, reuse, and treatment, and to protect air, water, and soil from contamination.

  c. Provide low-interest loans, tax breaks, and other financial incentives to encourage industries producing hazardous waste to reduce, recycle, reuse, treat, and decompose such waste.

6. Congratulations! You are in charge of the world. List the three most important components of your strategy for dealing with **(a)** solid waste and **(b)** hazardous waste.

LEARNING ONLINE

The website for this book includes review questions for the entire chapter, flash cards for key terms and concepts, a multiple-choice practice quiz, interesting Internet sites, references, and a guide for accessing thousands of InfoTrac® College Edition articles.

Visit

http://biology.brookscole.com/miller11

Then choose Chapter 17, and select a learning resource. For access to animations, additional quizzes, chapter outlines and summaries, register and log in to

Environmental Science ⊛ Now™

at **esnow.brookscole.com/miller11** using the access code card in the front of your book.

Active Graphing

Visit http://esnow.brookscole.com/miller11 to explore the graphing exercise for this chapter.

Environmental Economics, Politics, and Worldviews

CASE STUDY
Biosphere 2: A Lesson in Humility

In 1991, eight scientists (four men and four women) were sealed into Biosphere 2, a $200 million facility designed to be a self-sustaining life-support system (Figure 18-1).

The 1.3-hectare (3.2-acre) sealed system of interconnected domes was constructed in the desert near Tucson, Arizona. It had a variety of natural living systems, each built from scratch. They included a tropical rain forest, savanna, desert, lakes, streams, freshwater and saltwater wetlands, and a mini-ocean with a coral reef.

Biosphere 2 was designed to mimic the earth's natural chemical recycling systems. Water evaporated from its ocean and other aquatic systems, then condensed to provide rainfall over the tropical rain forest. This water trickled through soil filters into the marshes and the ocean to provide fresh water for the crew and other living organisms before evaporating

again. Human and animal excrement and other wastes were treated and recycled to help support plant growth.

The facility was stocked with more than 4,000 species of organisms selected to maintain life-support functions. Sunlight and external natural gas–powered generators provided energy. The Biospherians were supposed to be isolated for 2 years and to raise their own food using intensive organic agriculture, breathe air recirculated by plants, and drink water cleansed by natural recycling processes.

From the beginning, many unexpected problems cropped up. The life-support system began unraveling. Large amounts of oxygen disappeared when soil organisms converted it to carbon dioxide. Additional oxygen had to be pumped in from the outside to keep the Biospherians from suffocating.

The nitrogen and carbon recycling systems also failed to function properly. Levels of nitrous oxide rose high enough to threaten the occupants with brain damage and had to be controlled by outside intervention. Carbon dioxide levels skyrocketed to heights that threatened to poison the humans and spurred the growth of weedy vines that choked out food crops.

Tropical birds died after the first freeze. An Arizona ant species got into the enclosure, proliferated, and killed off most of the system's introduced insect species. As a result, the facility was overrun with cockroaches and katydids. In total, 19 of Biosphere 2's 25 small animal species became extinct. Before the 2-year period was up, all plant-pollinating insects had died, thereby dooming to extinction most of the plant species.

Despite many problems, the facility's waste and wastewater were recycled. The Biospherians were also able to produce 80% of their food supply.

Scientists Joel Cohen and David Tilman, who evaluated the project, concluded, "No one yet knows how to engineer systems that provide humans with life-supporting services that natural ecosystems provide for free." In other words, an expenditure of $200 million failed to maintain a life-support system for eight people.

Columbia University took over Biosphere 2 as a research facility for a few years, but abandoned it in 2003.

James Marshall/CORBIS

Figure 18-1 Biosphere 2, constructed near Tucson, Arizona, was designed to be a self-sustaining life-support system for eight people sealed into the facility in 1991. The experiment failed because of a breakdown in its nutrient cycling systems.

The main ingredients of an environmental ethic are caring about the planet and all of its inhabitants, allowing unselfishness to control the immediate self-interest that harms others, and living each day so as to leave the lightest possible footprints on the planet.

ROBERT CAHN

This chapter examines the basic principles underlying environmental economics and environmental politics and looks at three environmental worldviews that guide human behavior toward the environment. It addresses the following questions:

- How do neoclassical and ecological economists differ in their view of the earth's economic systems?

- How can we monitor environmental progress?

- What economic tools can we use to shift to full-cost pricing?

- How does poverty reduce environmental quality, and how can we reduce poverty?

KEY IDEAS

- Economists differ in their views about the importance of natural capital and the long-term sustainability of economic growth.

- Environmentalists and many economists call for prices of goods and services to include indirect environmental, health, and other harmful costs associated with their production and use. They suggest removing environmentally harmful government subsidies and tax breaks and shifting taxes from wages and profits to pollution and waste.

- Sharply cutting poverty can improve environmental quality and human health.

- To shift to more environmentally sustainable economies, we can design them to use the four principles of sustainability found in nature.

- Most improvements in environmental quality result from millions of citizens putting pressure on elected officials and individuals developing innovative solutions to environmental problems.

- Planetary management, stewardship, and earth wisdom are three major environmental worldviews

- The message of environmentalism is not gloom and doom, fear, and catastrophe, but rather hope, a positive vision of the future, and a call for responsibility in dealing with the environmental challenges we face.

- How can we shift to more environmentally sustainable economies over the next few decades?

- How is environmental policy formulated in the United States?

- What are some guidelines for making environmental policy? How can people affect such decisions?

- What are three major environmental worldviews?

- How can we live more sustainably?

18-1 ECONOMIC SYSTEMS AND SUSTAINABILITY

Economic Resources

An economic system produces goods and services by using natural, human, and manufactured resources.

An **economic system** is a social institution through which goods and services are produced, distributed, and consumed to satisfy people's needs and unlimited wants in the most efficient possible way. In a *market-based economic system,* buyers (demanders) and sellers (suppliers) interact in *markets* to make economic decisions about which goods and services are produced, distributed, and consumed.

Three types of resources are used to produce goods and services (Figure 18-2, p. 414). **Natural resources,** or **natural capital,** includes the goods and services produced by the earth's natural processes, which support all economies and all life. (See Figure 1-3, p. 7, and the Guest Essay on natural capital by Paul Hawken on the website for this chapter.) **Human resources,** or **human capital,** includes people's physical and mental talents that provide labor, innovation, culture, and organization. **Manufactured resources,** or **manufactured capital,** includes items such as machinery, equipment, and factories made from natural resources with the help of human resources.

Neoclassical, Environmental, and Ecological Economists

Economists differ in their opinion of the importance of natural capital and the long-term sustainability of economic growth.

Neoclassical economists such as Milton Friedman and Robert Samuelson view natural resources as important but not vital because of our ability to find substitutes for scarce resources and ecosystem services. They also contend that continuing economic growth is necessary, desirable, and essentially unlimited.

Ecological economists such as Herman Daly and Robert Costanza view economic systems as subsystems

Natural Resources   Manufactured Resources   Human Resources   Goods and Services

Figure 18-2 Three types of resources are used to produce goods and services.

of the environment that depend heavily on the earth's irreplaceable natural resources (Figure 18-3). They point out that there are no substitutes for many natural resources, such as air, water, fertile soil, and biodiversity. They also believe that conventional economic growth eventually will become unsustainable because it can deplete or degrade many of the natural resources on which economic systems depend.

Taking the middle ground in this debate are *environmental economists*. They generally agree with ecological economists that some forms of economic growth are not sustainable. At the same time, they believe we can modify the principles of neoclassical economics and reform current economic systems, rather than having to redesign them to provide more environmentally sustainable economic development.

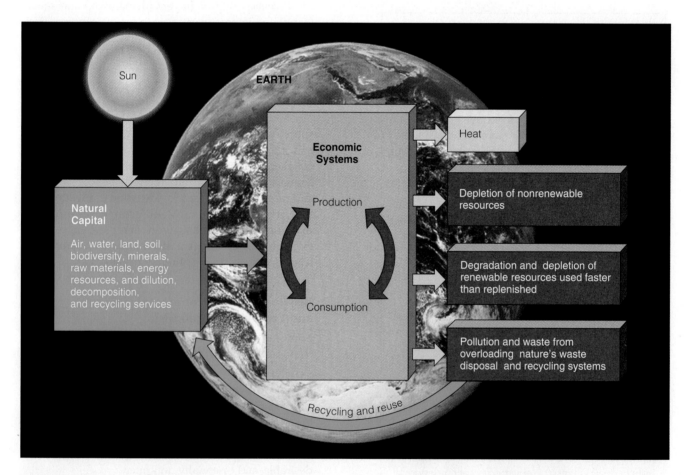

Environmental Science⊕Now™

Active Figure 18-3 Solutions: *ecological economists* see all economies as human subsystems that depend on resources and services provided by the sun and the earth's natural resources.

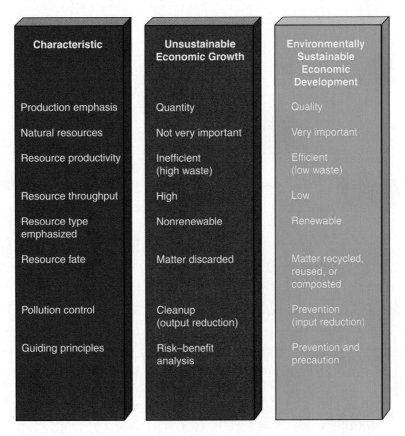

Characteristic	Unsustainable Economic Growth	Environmentally Sustainable Economic Development
Production emphasis	Quantity	Quality
Natural resources	Not very important	Very important
Resource productivity	Inefficient (high waste)	Efficient (low waste)
Resource throughput	High	Low
Resource type emphasized	Nonrenewable	Renewable
Resource fate	Matter discarded	Matter recycled, reused, or composted
Pollution control	Cleanup (output reduction)	Prevention (input reduction)
Guiding principles	Risk–benefit analysis	Prevention and precaution

Figure 18-4 Comparison of unsustainable economic growth and environmentally sustainable economic development according to ecological economists and many environmental economists.

Ecological and environmental economists distinguish between unsustainable economic growth and environmentally sustainable economic development (Figure 18-4).

Ecological and environmental economists call for making a shift from our current economy based on unlimited economic growth to a more *environmentally sustainable economy,* or *eco-economy,* over the next several decades. See the Guest Essay on this topic by Herman Daly on the website for this chapter. Figure 18-5 (p. 416) shows some of the goods and services these economists would encourage in such an eco-economy.

Environmental Science ⊕ Now™
Learn more about how ecological economists view market-based systems, and contrast those views with conventional economics at Environmental ScienceNow.

Ecological and environmental economists have suggested several strategies to help make the transition to a more sustainable eco-economy over the next several decades:

■ Use indicators that monitor economic and environmental health.

■ Include in the market prices of goods and services their estimated harmful effects on the environment and human health (*full-cost pricing*).

■ Phase out environmentally harmful government subsidies and tax breaks while increasing such subsidies and breaks for environmentally friendly activities, goods, and services.

■ Lower taxes on income and wealth but increase *green taxes and fees*— those levied on pollution, resource waste, and environmentally harmful goods and services.

■ Pass laws and regulations to prevent pollution and resource depletion in certain areas.

■ Use tradable permits or rights to pollute or use resources to limit overall pollution and resource use in given areas.

■ Reduce poverty, one of the basic causes of resource depletion and pollution.

Let's look at these proposed solutions in more detail.

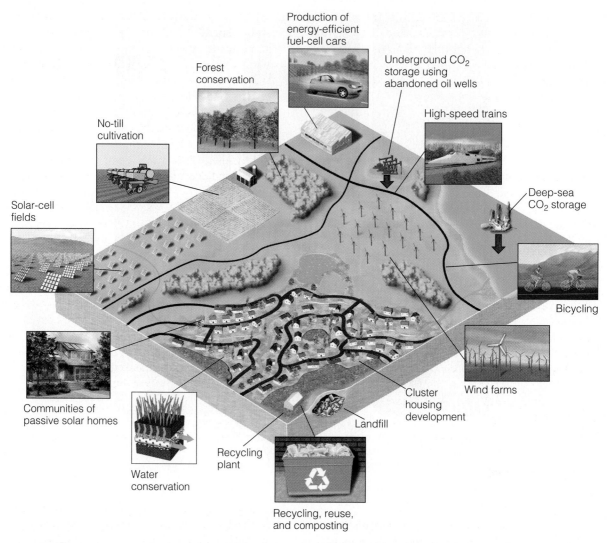

Production of
energy-efficient
fuel-cell cars

Underground CO_2
storage using
abandoned oil wells

Forest
conservation

High-speed trains

No-till
cultivation

Deep-sea
CO_2 storage

Solar-cell
fields

Bicycling

Communities of
passive solar homes

Wind farms

Cluster
housing
development

Landfill

Recycling
plant

Water
conservation

Recycling, reuse,
and composting

Figure 18-5 Solutions: Some components of the more environmentally sustainable economic development favored by ecological economists, environmental economists, and environmental scientists. The goal is to have economic systems that put more emphasis on conserving and sustaining the air, water, soil, biodiversity, and other natural resources that sustain all life and all economies.

18-2 USING ECONOMICS TO IMPROVE ENVIRONMENTAL QUALITY

Environmental and Economic Indicators

We need new indicators that reflect changing levels of environmental quality and human health.

Gross domestic product (GDP) and *per capita GDP* indicators provide a standardized and useful method for measuring and comparing the economic outputs of nations. The GDP is deliberately designed to measure the annual economic value of all goods and services produced within a country without attempting to distinguish between those goods and services that are environmentally or socially beneficial and those that are harmful. Economists who developed the GDP many decades ago never intended it to be used for measuring environmental quality or human well-being.

Environmental and ecological economists and environmental scientists call for the development and adoption of new indicators to help monitor environmental quality and human well-being. One approach is to develop indicators that *add* to the GDP the costs of things not counted in the marketplace that enhance environmental quality and human well-being. They would also *subtract* from the GDP the costs of things that lead to a lower quality of life and deplete natural resources.

One such indicator is the *genuine progress indicator (GPI)*, introduced in 1995 by Redefining Progress, a nonprofit organization that develops economics and policy tools to help evaluate and promote sustainability. (The same group developed the concept of ecological footprints, Figure 1-7, p. 11) Within the GPI, the estimated value of beneficial transactions that meet basic needs, but in which no money changes hands, are added to the GDP. Examples include unpaid volunteer work, health care for family members, child care, and housework. Then the estimated harmful environmental costs (such as pollution and resource depletion and degradation) and social costs (such as crime) are subtracted from the GDP.

Genuine
progress = GDP + benefits not included in − harmful environmental
indicator   market transactions   and social costs

Figure 18-6 compares the per capita GDP and GPI for the United States between 1950 and 2002. While the per capita GDP rose sharply over this period, the per capita GPI stayed nearly flat and even declined slightly between 1975 and 2002.

The GPI and other environmental indicators under development are far from perfect and include many crude estimates. Nevertheless, without such indicators, we would not know much about what is happening to people, the environment, and the planet's natural capital. With them, we have a way to determine which policies work. In essence, according to ecological and environmental economists, if we ignore these indicators, we attempt to guide national and global economies through treacherous economic and environmental waters at ever-increasing speeds using faulty radar.

Internal and External Costs

The direct price you pay for something does not include indirect environmental, health, and other harmful costs associated with its production and use.

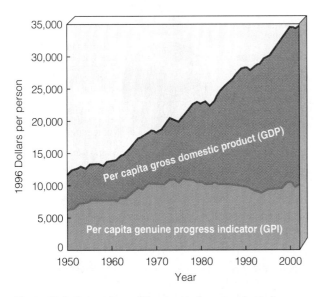

Figure 18-6 Comparison of the per capita gross domestic product (GDP) and per capita genuine progress indicator (GPI) in the United States between 1950 and 2002. (Data from Redefining Progress, 2004)

All economic goods and services have *internal* or *direct costs* associated with producing them. For example, if you buy a car, the direct price you pay includes the costs of raw materials, labor, and shipping, as well as a markup to allow the car company and its dealers to earn some profits. Once you buy the car, you must pay additional direct costs for gasoline, maintenance, and repair.

Making, distributing, and using economic goods or services also lead to *indirect* or *external costs* that are not included in their market prices and that affect people other than the buyer and seller. For example, extracting and processing raw materials to make a car use nonrenewable energy and mineral resources, produce solid and hazardous wastes, disturb land, and pollute the air and water. These external costs can have short- and long-term harmful effects on other people and on the earth's life-support systems.

Because these costs are not included in the market price, most people do not connect them with car ownership. Even so, the car buyer and other people in a society pay these hidden costs sooner or later, in the form of poorer health, higher costs of health care and insurance, higher taxes for pollution control, traffic congestion, and land used for highways and parking.

Full-Cost Pricing

Including external costs in market prices informs consumers about the impacts that their purchases have on the earth's life-support systems and human health.

For most economists, an *environmentally honest market system* provides a way to deal with the harmful costs of goods and services. In this system, the harmful indirect or external costs of goods and services are included in the market price of any good or service, so that its price comes as close as possible to its **full cost**—its internal costs plus its external costs. Such a system would allow consumers to make more informed choices because they would be aware of most or all of the costs involved when they buy something.

Full-cost pricing encourages producers to invent more resource-efficient and less-polluting methods of production, thereby cutting their production costs. Jobs would be lost in environmentally harmful businesses as consumers more often choose green products, whereas jobs would be created in environmentally beneficial businesses. If a shift to full-cost pricing took place over several decades, most environmentally harmful businesses would have time to transform themselves into environmentally beneficial businesses. Likewise, consumers would have time to adjust their buying habits and learn to purchase more environmentally friendly products and services.

Full-cost pricing seems to make a lot of sense. So why is it not used more widely? Two major reasons explain why.

First, many producers of harmful and wasteful goods would have to charge more, and some would go out of business. Naturally, they oppose such pricing.

Second, it is difficult to put a price tag on many environmental and health costs. But to ecological and environmental economists, making the best possible estimates is far better than not including such costs in what we pay for most goods and services.

☒ *HOW WOULD YOU VOTE?* Should full-cost pricing be used in setting the market prices of goods and services? Cast your vote online at http://biology.brookscole.com/miller11.

Phasing in such a system will require government action. Few companies will volunteer to reduce their short-term profits in an effort to become more environmentally responsible. Governments can use several strategies to encourage or force producers to work toward full-cost pricing, including phasing out environmentally harmful subsidies, levying taxes on environmentally harmful goods and services, passing laws to regulate pollution and resource depletion, and using tradable permits for reducing pollution or resource use.

Government Subsidies and Tax Breaks

We can improve environmental quality and help phase in full-cost pricing by removing environmentally harmful government subsidies and tax breaks.

One way to encourage a shift to full-cost pricing is to *phase out* environmentally harmful subsidies and tax breaks, which cost the world's governments about $1.9 trillion per year, according to studies by Norman Myers and other analysts. Examples include government depletion subsidies and tax breaks for extracting minerals and oil from the ground, cutting timber on public lands, and using low-cost water for farming, and measures that encourage overfishing by providing low-cost loans to buy fishing boats.

On paper, phasing out such subsidies may seem like a great idea. In reality, the powerful interests receiving such benefits strongly oppose these political decisions. They want to keep—and if possible increase—these measures and often lobby against subsidies and tax breaks for more environmentally beneficial competitors.

Some countries have begun reducing environmentally harmful government subsidies. Japan, France, and Belgium have phased out all coal subsidies. Germany has cut coal subsidies in half and plans to phase them out completely by 2010. China has cut coal subsidies by about 73% and has imposed a tax on high-sulfur coals.

Green Taxes and Fees and Tax Shifting

Taxes and fees on pollution and resource use can take us closer to full-cost pricing, and shifting taxes from wages and profits to pollution and waste helps makes this feasible.

Another way to discourage pollution and resource waste is to *use green taxes* to include many of the harmful environmental costs of production and consumption in market prices. Taxes can be levied on a per-unit basis on the amount of pollution produced and the amount of hazardous or nuclear waste produced, and on the use of fossil fuels, timber, and minerals. Higher fees can also be charged for extracting lumber and minerals from public lands, using water provided by government-financed projects, and using public lands for livestock grazing. Figure 18-7 lists advantages and disadvantages of using green taxes and fees.

☒ *HOW WOULD YOU VOTE?* Do the advantages of relying more on environmental or green taxes and fees to reduce pollution and resource waste outweigh their disadvantages? Cast your vote online at http://biology.brookscole.com/miller11.

Trade-Offs

Environmental Taxes and Fees

Advantages	Disadvantages
Helps bring about full-cost pricing	Penalizes low-income groups unless safety nets are provided
Provides incentive for businesses to do better to save money	Hard to determine optimal level for taxes and fees
Can change behavior of polluters and consumers if taxes and fees are set at a high enough level	Need to frequently readjust levels, which is technically and politically difficult
Easily administered by existing tax agencies	Governments may see this as a way of increasing general revenue instead of using funds to improve environmental quality and reduce taxes on income, payroll, and profits
Fairly easy to detect cheaters	

Figure 18-7 Trade-offs: advantages and disadvantages of using environmental or green taxes and fees to reduce pollution and resource waste. *Critical thinking: pick the single advantage and disadvantage that you think are the most important.*

- Decreases depletion and degradation of natural resources

- Improves environmental quality by full-cost pricing

- Encourages pollution prevention and waste reduction

- Stimulates creativity in solving environmental problems to avoid paying pollution taxes and thereby increases profits

- Rewards recycling and reuse

- Relies more on marketplace rather than regulation for environmental protection

- Provides jobs

- Can stimulate sustainable economic development

- Allows cuts in income, payroll, and sales taxes

Figure 18-8 Solutions: advantages of taxing wages and profits less and pollution and waste more. *Critical thinking: pick the two advantages that you think are the most important.*

Economists point out two requirements for successful implementation of green taxes. *First,* they should reduce or replace income, payroll, or other taxes. *Second,* the poor and middle class need a safety net to reduce the regressive nature of consumption taxes on essentials such as food, fuel, and housing.

To many analysts, the tax system in most countries is backward. It *discourages* what we want more of—jobs, income, and profit-driven innovation—and *encourages* what we want less of—pollution, resource waste, and environmental degradation. A more environmentally sustainable economic system would *lower* taxes on labor, income, and wealth and *raise* taxes on environmentally harmful activities.

Shifting more of the tax burden from wages and profits to pollution and waste has a number of advantages (Figure 18-8). Some 2,500 economists, including eight Nobel Prize winners, have endorsed the concept of such tax shifting.

Nine Western European countries have begun trial versions of such tax shifting, known as *environmental tax reform.* So far, only a small amount of revenue has been shifted by taxes on emissions of CO_2 and toxic metals, garbage production, and vehicles entering congested cities. But such experience shows that this idea works.

X *HOW WOULD YOU VOTE?* Do you favor shifting taxes on wages and profits to pollution and waste? Cast your vote online at http://biology.brookscole.com/miller11.

Environmental Laws and Regulations

Environmental laws and regulations work best if they motivate companies to find innovative ways to control and prevent pollution and reduce resource waste.

Regulation is a form of government intervention in the marketplace that is widely used to help control or prevent pollution and reduce resource waste. It involves enacting and enforcing laws and establishing regulations that set pollution standards, regulate harmful activities, ban the release of toxic chemicals into the environment, and require that certain irreplaceable or slowly replenished resources be protected from unsustainable use.

A number of environmental and business leaders agree that *innovation-friendly regulations* can motivate companies to develop eco-friendly products and industrial processes that increase their profits and competitiveness in national and international markets. But they also agree that some pollution control regulations are too costly and discourage innovation. Examples include regulations that concentrate on cleanup instead of prevention, are too prescriptive (for example, by mandating specific technologies), set compliance deadlines that are too short to allow companies to find innovative solutions, and discourage risk taking and experimentation.

An innovation-friendly regulatory process emphasizes pollution prevention and waste reduction and requires industry and environmental interests to work together in developing realistic standards and timetables. It sets goals, frees industries to meet them in any way that works, and allows enough time for innovation to emerge.

For many years, many companies mostly resisted environmental regulation and developed an adversarial relationship with government regulators. In recent years, a growing number of companies have realized the economic and competitive advantages of making environmental improvements and recognized that their shareholder value depends in part on having a good environmental record. As a result, some firms have begun looking for innovative and profitable ways to reduce resource use, pollution, and waste (Individuals Matter, p. 394). At the same time, many consumers have begun buying green products.

Tradable Pollution and Resource-Use Permits

The government can set limits on pollution emissions or use of a resource, issue pollution or resource-use permits, and allow holders to trade their permits in the marketplace.

One market approach would have the government *grant tradable pollution and resource-use permits.* For

Trade-Offs

Tradable Environmental Permits

Advantages	Disadvantages
Flexible	Big polluters and resource wasters can buy their way out
Easy to administer	May not reduce pollution at dirtiest plants
Encourages pollution prevention and waste reduction	Can exclude small companies from buying permits
	Caps can be too low
Can promote achievement of caps	Caps must be gradually reduced to encourage innovation
	Determining caps is difficult
Permit prices determined by market transactions	Must decide who gets permits and why
	Administrative costs high with many participants
Confronts ethical problem of how much pollution or resource waste is acceptable	Emissions and resource wastes must be monitored
	Self-monitoring can promote cheating
Confronts problem of how permits should be fairly distributed	Sets bad example by selling legal rights to pollute or waste resources

Figure 18-9 Trade-offs: advantages and disadvantages of using tradable resource-use permits to reduce pollution and resource waste. *Critical thinking: pick the single advantage and disadvantage that you think are the most important.*

example, the government could set a limit or cap on total emissions of a pollutant or use of a resource such as a fishery. It would then issue or auction permits that allocate the total among manufacturers or users.

A permit holder not using its entire allocation could use the permit as a credit against future expansion, use it in another part of its operation, or sell it to other companies. In the United States, this approach has been employed to reduce the emissions of sulfur dioxide and several other air pollutants. Tradable rights can also be established among countries to help preserve biodiversity and reduce emissions of greenhouse gases and other pollutants with harmful regional or global effects.

Figure 18-9 lists the advantages and disadvantages of tradable pollution and resource-use permits. The effectiveness of such programs depends on how high or low the initial cap is set and the rate at which the cap is reduced. However, this approach can allow major polluters and resource wasters to buy their way out instead of making technological improvements. The result: hot spots of pollution, land degradation, and resource depletion.

X *HOW WOULD YOU VOTE?* Do the advantages of using tradable pollution and resource-use permits to reduce pollution and resource waste outweigh the disadvantages? Cast your vote online at http://biology.brookscole.com/miller11.

Eco-Labeling

Labeling environmentally beneficial goods and resources extracted by more sustainable methods can help consumers decide which goods and services to buy.

Product eco-labeling can encourage companies to develop green products and services and help consumers select more environmentally beneficial products and services. Eco-labeling programs have been developed in Europe, Japan, Canada, and the United States (for example, the *Green Seal* labeling program that has certified more than 300 products; see Figure 18-10).

Germany: Blue Angel (1978)

Canada: Environmental Choice (1988)

United States: Green Seal (1989)

Nordic Council: White Swan (1989)

European Union: Eco-label (1992)

China: Environmental label (1993)

Figure 18-10 Solutions: symbols used in some *eco-labeling* programs that evaluate green or environmentally favorable products.

Eco-labels are also being used to identify fish caught by sustainable methods (certified by the Marine Stewardship Council) and to certify timber produced and harvested by sustainable methods (evaluated by organizations such as the Forestry Stewardship Council; see Solutions, p. 165).

18-3 REDUCING POVERTY TO IMPROVE ENVIRONMENTAL QUALITY AND HUMAN WELL-BEING

Distribution of the World's Wealth

Since 1960, most of the financial benefits of global economic growth have flowed up to the rich rather than down to the poor and middle class.

Poverty is defined as the inability to meet one's basic economic needs. According to a 2000 World Bank study, half of humanity is trying to live on less than $2 (U.S.) per day, and one of every five people on the planet is struggling to survive on an income of roughly $1 (U.S.) per day. Millions of people in developing countries are homeless and often must sleep on the streets (Figure 18-11).

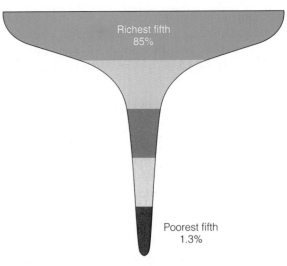

Figure 18-12 **Global outlook:** the *global distribution of income* shows that most of the world's income flows up; the richest 20% of the world's population receive more of the world's income than all of the remaining 80%. Each horizontal band in this diagram represents one-fifth of the world's population. This upward flow of global income has accelerated since 1960 and especially since 1980. This trend can increase environmental degradation by increasing average per capita consumption by the richest 20% of the population and causing the poorest 20% of the world's people to survive by using renewable resources faster than they are replenished. (Data from UN Development Programme and Ismail Serageldin, "World Poverty and Hunger—A Challenge for Science," *Science* 296 (2002): 54–58)

Figure 18-11 **Global outlook:** according to the United Nations, one-fifth of the world's people have inadequate housing, and at least 100 million have no housing at all. These homeless people in Calcutta, India, must sleep on the street. Many more poor people live in crowded and dangerous shantytowns and slums in major cities throughout developing countries.

Poverty has numerous harmful health and environmental effects (Figure 1-11, p. 14, and Figure 14-12, p. 341) and has been identified as one of the five major causes of the environmental problems we face.

According to most neoclassical economists, a growing economy can help the poor by creating more jobs, enable more of the wealth created by economic growth to reach workers, and provide greater tax revenues that can be used to help the poor help themselves. Economists call this the *trickle-down* effect. However, since 1960, most of the benefits of global economic growth as measured by income have flowed up to the rich, rather than down to the poor and middle class (Figure 18-12). Since 1980, this *wealth gap* has grown. According to Ismail Serageldin, the planet's three richest people have more wealth than the combined GDP of the world's 47 poorest countries. South African President Thabo Mbeki told delegates at the 2003 Johannesburg World Summit on Sustainable Development, "A global human society based on poverty for many and prosperity for a few, characterized by islands of wealth, surrounded by a sea of poverty, is unsustainable."

This upward flow of wealth does not mean that economic growth causes poverty. Instead, it indicates that rich nations and individuals have not been willing to devote very much of their wealth to helping reduce poverty and its harmful effects on the environment and human well-being. Poverty is also sustained by corruption, the absence of property rights, insufficient legal protection, and the inability of many people to borrow money to grow crops or start a small business.

Solutions: Reducing Poverty

We can sharply cut poverty by forgiving the international debts of the poorest countries and greatly increasing international aid and small individual loans to help the poor help themselves.

Analysts point out that reducing poverty will require the governments of most developing countries to make policy changes. For example, they need to shift more of the national budget to help the rural and urban poor work their way out of poverty. They also need to give villages, villagers, and the urban poor title to common lands and to crops and trees they plant on them.

One way to help reduce global poverty might be to forgive at least 60% of the $2.4 trillion debt that developing countries owe to developed countries and international lending agencies, and all of the $422 billion debt of the poorest and most heavily indebted countries. This would be done on the condition that the money saved on interest debt be devoted to meeting basic human needs. Currently, developing countries pay almost $300 billion per year in interest to developed countries to service their debt.

Developed countries can also take the following steps:

- Increase nonmilitary government and private aid, with mechanisms to assure that most of it goes directly to the poor to help them become more self-reliant and to help provide social safety nets such as welfare, unemployment payments, and pension benefits

- Mount a massive global effort to combat malnutrition and the infectious diseases that kill millions of people prematurely and help perpetuate poverty

- Have lending agencies make small loans to poor people who want to increase their income (see Solutions, p. 423)

- Make investments in small-scale infrastructure such as solar-cell power facilities in villages (Figure 13-34, p. 316), small-scale irrigation projects, and farm-to-market roads

According to the United Nations Development Program (UNDP), it will cost about $50 billion per year to provide universal access to basic services such as education, health, nutrition, family planning, safe water, and sanitation. This amount is less than 0.1% of the world's annual income and a mere fraction of what the world devotes each year to military spending (Figure 18-13).

Solutions: Making the Transition to an Eco-Economy

An eco-economy copies nature's four principles of sustainability and uses various economic strategies to help implement full-cost pricing.

An eco-economy mimics the processes that sustain the earth's natural systems (Figure 6-19, p. 126). Figure 18-14 lists principles that Paul Hawken and other business leaders and economists have suggested for using the sustainability strategies and economic tools discussed in this chapter to make the transition to more environmentally sustainable economies over the next several decades. Hawken has a simple golden rule for such an economy: *"Leave the world better than you found it, take no more than you need, try not to harm life or the environment, and make amends if you do."*

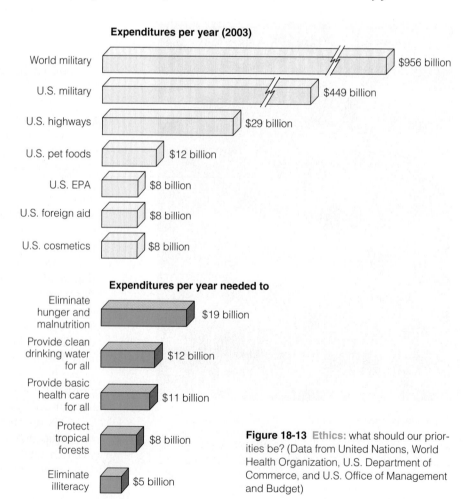

Expenditures per year (2003)

World military	$956 billion
U.S. military	$449 billion
U.S. highways	$29 billion
U.S. pet foods	$12 billion
U.S. EPA	$8 billion
U.S. foreign aid	$8 billion
U.S. cosmetics	$8 billion

Expenditures per year needed to

Eliminate hunger and malnutrition	$19 billion
Provide clean drinking water for all	$12 billion
Provide basic health care for all	$11 billion
Protect tropical forests	$8 billion
Eliminate illiteracy	$5 billion

Figure 18-13 Ethics: what should our priorities be? (Data from United Nations, World Health Organization, U.S. Department of Commerce, and U.S. Office of Management and Budget)

Global Outlook: Microloans for the Poor

SOLUTIONS

Most of the world's poor desperately want to earn more, become more self-reliant, and have a better life. But they have no credit record. Also, they have few assets to use for collateral to secure a loan to buy seeds and fertilizer for farming or tools and materials for a small business.

During the last 28 years, an innovation called *microlending*, or *microfinance*, has helped deal with this problem. For example, since economist Muhammad Yunus started it in 1976, the Grameen (Village) Bank in Bangladesh has provided more than $4 billion in microloans (ranging from $50 to $500) to several million mostly poor, rural, and landless women in 40,000 villages. About 94% of the loans go to women who start their own small businesses as seamstresses, weavers, bookbinders, or vendors.

To stimulate repayment and provide support, the Grameen Bank organizes microborrowers into five-member "solidarity" groups. If one member of the group misses a weekly payment or defaults on the loan, the other members of the group must make the payments.

The Grameen Bank's experience has shown that microlending is both successful and profitable. For example, fewer than 3% of microloan repayments to the Grameen Bank are late, and the repayment rate on its loans is an astounding 90–95%.

About half of Grameen's borrowers move above the poverty line within five years, and domestic violence, divorce, and birth rates are lower among borrowers. Microloans by the Grameen Bank are being used to develop day-care centers, health clinics, reforestation projects, drinking-water supply projects, literacy programs, and group insurance programs. Microloans are also being used to bring small-scale solar and wind power systems to rural villages.

Grameen's model has inspired the development of microcredit projects in more than 58 countries that have reached 36 million people (including dependents), and the number is growing rapidly.

Critical Thinking

Why do you think international development and lending agencies such as the World Bank and the International Monetary Fund have largely ignored microloans? How would you change this situation?

Economics

Reward (subsidize) earth-sustaining behavior

Penalize (tax and do not subsidize) earth-degrading behavior

Shift taxes from wages and profits to pollution and waste

Use full-cost pricing

Sell more services instead of more things

Do not deplete or degrade natural capital

Live off income from natural capital

Reduce poverty

Use environmental indicators to measure progress

Certify sustainable practices and products

Use eco-labels on products

Environmentally Sustainable Economy (Eco-Economy)

Resource Use and Pollution

Reduce resource use and waste by refusing, reducing, reusing, and recycling

Improve energy efficiency

Rely more on renewable solar and geothermal energy

Shift from a carbon-based (fossil fuel) economy to a renewable fuel–based economy

Ecology and Population

Mimic nature

Preserve biodiversity

Repair ecological damage

Stabilize population by reducing fertility

Figure 18-14 Solutions: principles for shifting to more environmentally sustainable economies or eco-economies during this century. *Critical thinking: which five of these solutions do you believe are the most important?*

18-4 POLITICS AND ENVIRONMENTAL POLICY

Factors Hindering the Ability of Democracies to Deal with Environmental Problems

Democracies are designed to deal mostly with short-term, isolated problems.

Politics is the process by which individuals and groups try to influence or control the policies and actions of governments at local, state, national, and international levels. Politics is concerned with who has power over the distribution of resources and who gets what, when, and how. Many people think of politics in national terms, but what directly affects most people is what happens in their local community.

Democracy is government by the people through elected officials and representatives. In a *constitutional democracy,* a constitution provides the basis of government authority, limits government power by mandating free elections, and guarantees free speech.

Political institutions in constitutional democracies are designed to allow gradual change to ensure economic and political stability. In the United States, for example, rapid and destabilizing change is curbed by a system of checks and balances that distributes power among the three branches of government—*legislative, executive,* and *judicial*—and among federal, state, and local governments.

In passing laws, developing budgets, and formulating regulations, elected and appointed government officials must deal with pressure from many competing *special-interest groups.* Each group advocates passing laws, providing subsidies or tax breaks, or establishing regulations favorable to its cause and weakening or repealing laws and regulations unfavorable to its position.

Some special-interest groups such as corporations are *profit-making organizations;* others are *nonprofit nongovernmental organizations (NGOs).* Examples of NGOs include labor unions and mainline and grassroots environmental organizations.

The deliberately stable design of democracies is highly desirable. At the same time, several features of democratic governments hinder their ability to deal with environmental problems. Many problems such as climate change, biodiversity loss, and long-lived hazardous waste have long-range effects, are related to one another, and require integrated, long-term solutions emphasizing prevention. But because elections are held every few years, most politicians seeking re-election tend to focus on short-term, isolated problems rather than on complex, interrelated, time-consuming, and long-term problems. Also, too many political leaders have too little understanding of how the earth's natural systems work and how they support all life, economies, and societies.

Principles for Making Environmental Policy Decisions

Several principles can guide us in making environmental decisions.

An **environmental policy** consists of laws, rules, and regulations related to an environmental problem that are developed, implemented, and enforced by a particular government agency. Analysts have suggested that legislators and individuals evaluating existing or proposed environmental policies should be guided by the following principles:

- *The humility principle:* Our understanding of nature and of the consequences of our actions is quite limited.

- *The reversibility principle:* Try not to do something that cannot be reversed later if the decision turns out to be wrong.

- *The precautionary principle:* When substantial evidence indicates an activity threatens human health or the environment, take precautionary measures to prevent or reduce such harm, even if some of the cause-and-effect relationships are not fully established scientifically. It is better to be safe than sorry.

- *The prevention principle:* Whenever possible, make decisions that help prevent a problem from occurring or becoming worse.

- *The polluter-pays principle:* Develop regulations and use economic tools such as full-cost pricing to ensure that polluters bear the costs of the pollutants and the wastes they produce.

- *The integrative principle:* Make decisions that involve integrated solutions to environmental and other problems.

- *The public participation principle:* Citizens should have ready access to environmental data and information and the right to participate in developing, criticizing, and modifying environmental policies.

- *The human rights principle:* All people have a right to an environment that does not harm their health and well-being.

- *The environmental justice principle:* Establish environmental policy so that no group of people bears an unfair share of the burden created by pollution, environmental degradation, or the execution of environmental laws. See the Guest Essay on this subject by Robert D. Bullard in the website for this chapter.

How Can Individuals Affect Environmental Policy?

Most improvements in environmental quality result from millions of citizens putting pressure on elected officials and individuals developing innovative solutions to environmental problems.

A major theme of this book is that *individuals matter.* History shows that significant change usually comes from the *bottom up* when individuals join with others to bring about change (Figure 18-15). Without grassroots political action by millions of individual citizens and organized groups, the air you breathe and the water you drink today would be much more polluted, and much more of the earth's biodiversity would have disappeared. Once environmental laws are passed, citizens must insist that they be implemented and work together to prevent them from being weakened and to improve them.

Figure 18-16 lists ways you can influence and change government policies in constitutional democracies. Active political participation is guided by Aldo Leopold's insight: "All ethics rest upon a single premise: that the individual is a member of a community of interdependent parts."

What Can You Do?

Influencing Environmental Policy

- Become informed on issues
- Run for office (especially at local level)
- Make your views known at public hearings
- Make your views known to elected representatives
- Contribute money and time to candidates for office
- Vote
- Form or join nongovernment organizations (NGOs) seeking change
- Support reform of election campaign financing

Figure 18-16 Individuals matter: ways you can influence environmental policy. *Critical thinking: which three of these actions do you believe are the most important? Which things in this list do you do or plan to do?*

Environmental Leadership

Each of us can play a leadership role in establishing and changing environmental policy.

Each of us can provide environmental leadership in four ways. First, we can *lead by example,* using our own lifestyle to show others that change is possible and beneficial.

Second, we can *work within existing economic and political systems to bring about environmental improvement.* We can influence political decisions by campaigning and voting for candidates and by communicating with elected officials. We can also send a message to companies making harmful environment products or policies by *voting with our wallets* and letting them know what choices we have made. In addition, we can work within the system by choosing environmental careers.

Third, we can *run for some sort of local office.* Look in the mirror. Maybe you are one who can make a difference as an office holder.

Fourth, we can *propose and work for better solutions to environmental problems.* Leadership is more than being against something. It also involves coming up with better ways to accomplish goals and persuading people to work together to achieve such goals. If we care enough, each of us can make a difference.

STOP
ILLEGAL
LOGGING
– ISLA VERDE SURVIVO

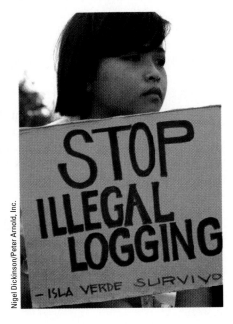

Nigel Dickinson/Peter Arnold, Inc.

Figure 18-15 Global outlook: democracy in action. Anti-logging demonstration in the Philippines. Deforestation caused by commercial and illegal logging led to flash floods that wreaked havoc on towns and communities near rivers. In one night, 7,000 people lost their lives from such floods.

Case Study: Environmental Policy in the United States

Formulating, legislating, and executing environmental policy in the United States is a complex, lengthy, and controversial process.

The major function of the federal government in the United States (and in other democratic countries) is to develop and implement *policy* for dealing with various issues. *Policy* is typically composed of various *laws* passed by the legislative branch, *regulations* instituted by the executive branch to put laws into effect, and *funding* to implement and enforce the laws and regulations and oversight of the process by the judicial branch. Figure 18-17 (p. 426) presents a greatly simplified overview of how individuals and lobbyists for and against a particular environmental law interact with the three branches of government in the United States.

Figure 18-18 (p. 427) lists some of the major environmental laws passed in the United States since 1969. There are continual efforts to weaken, abolish, or find ways to get around these laws.

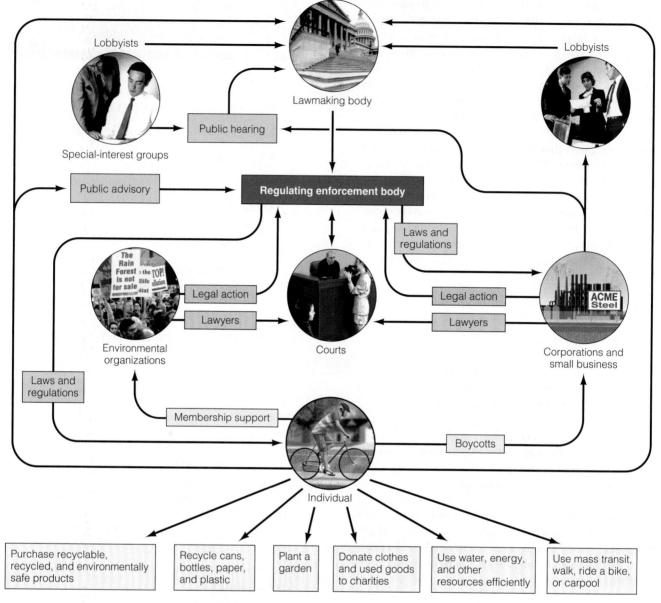

Figure 18-17 Individuals matter: greatly simplified overview of how individuals and lobbyists for and against a particular environmental law interact with the legislative, executive, and judicial branches of the U.S. government. The bottom of this diagram also shows some ways in which individuals can bring about environmental change through their own lifestyles. See the website for this book for details on contacting your elected representatives.

Passing an environmental law is not enough to make policy. The next step involves trying to get enough funds appropriated to implement and enforce each law. Indeed, developing and adopting a budget is the most important and controversial activity of the executive and legislative branches.

Once a law has been passed and funded, the appropriate government department or agency must draw up regulations for implementing it. An affected group may take the agency to court for failing to implement and enforce the regulations or for enforcing them too rigidly.

Politics plays an important role in the policies and staffing of environmental regulatory agencies—depending on which political party is in power and which environmental attitudes prevail at the time. Industries facing environmental regulations often put political pressure on regulatory agencies and lobby to have the president appoint people to high positions in such agencies who come from the industries being regulated. In other words, the regulated try to take over the agencies and become the regulators—described by some as "putting foxes in charge of the henhouse."

In addition, people in regulatory agencies work closely with and often develop friendships with officials in the industries they are regulating. Some industries offer regulatory agency employees high-paying jobs in an attempt to influence their regulatory decisions. This can lead to a *revolving door,* as employees move back and forth between industry and government.

According to social scientists, the development of public policy in democracies often goes through a *policy life cycle* consisting of four stages: *recognition, formulation, implementation,* and *control.* Figure 18-19 (p. 428) illustrates this cycle and shows the general positions of several major environmental problems in the policy life cycle in the United States and most other developed countries.

Mainline and Grassroots Environmental Groups

Environmental groups monitor environmental activities, work to pass and strengthen environmental laws, and work with corporations to find solutions to environmental problems.

The spearhead of the global conservation and environmental movement consists of more than 100,000 nonprofit NGOs working at the international, national, state, and local levels—up from about 2,000 such groups in 1970. The growing influence of these organizations is one of the most important changes related to environmental decisions and policies.

NGOs range from grassroots groups that have just a few members to global organizations like the 5-million-member World Wide Fund for Nature, which has offices in 48 countries. Other international groups with large memberships include Greenpeace, the World Wildlife Fund, the Nature Conservancy, Grameen Bank (see Solutions, p. 423), and Conservation International.

In the United States, more than 8 million citizens belong to more than 30,000 NGOs dealing with environmental issues. They range from small *grassroots* groups to large heavily *funded* groups, the latter staffed by expert lawyers, scientists, and economists.

The large groups have become powerful and important forces within the political system. They have persuaded Congress to pass and strengthen environmental laws (Figure 18-18) and work to fight off attempts to weaken or repeal such laws.

Some industries and environmental groups are working together to find solutions to environmental problems. For example, the Environmental Defense has worked with McDonald's to redesign its packaging system to eliminate its plastic hamburger containers, and with General Motors to remove high-pollution cars from the road.

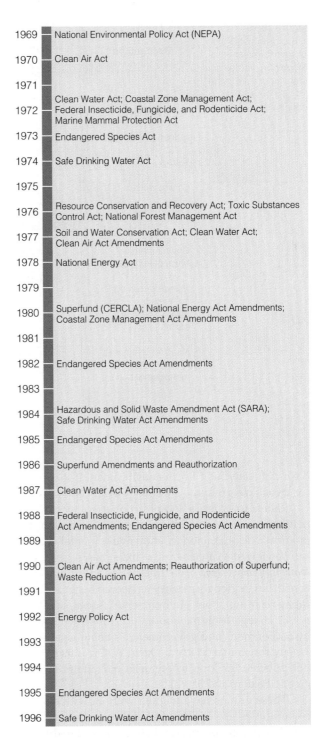

Figure 18-18 Solutions: some major environmental laws and their amended versions enacted in the United States since 1969. A more detailed list is found on the website for this chapter.

The base of the environmental movement in the United States and throughout the world consists of thousands of grassroots citizens' groups organized to improve environmental quality, often at the local level. According to political analyst Konrad von Moltke, "There isn't a government in the world that would

Figure 18-19
Solutions: positions
of several major envi-
ronmental problems
in the *policy life cycle*
in most developed
countries.

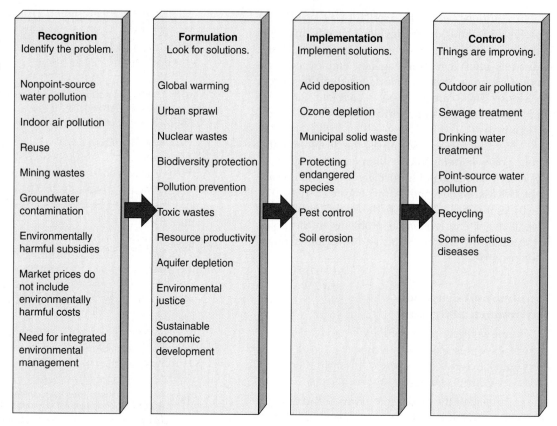

Recognition Identify the problem.	**Formulation** Look for solutions.	**Implementation** Implement solutions.	**Control** Things are improving.
Nonpoint-source water pollution	Global warming	Acid deposition	Outdoor air pollution
Indoor air pollution	Urban sprawl	Ozone depletion	Sewage treatment
Reuse	Nuclear wastes	Municipal solid waste	Drinking water treatment
Mining wastes	Biodiversity protection	Protecting endangered species	Point-source water pollution
Groundwater contamination	Pollution prevention	Pest control	Recycling
Environmentally harmful subsidies	Toxic wastes	Soil erosion	Some infectious diseases
Market prices do not include environmentally harmful costs	Resource productivity		
Need for integrated environmental management	Aquifer depletion		
	Environmental justice		
	Sustainable economic development		

have done anything for the environment if it weren't for the citizen groups."

Taken together, a loosely connected worldwide network of grassroots NGOs working for bottom-up political, social, economic, and environmental change can be viewed as an emerging citizen-based *global sustainability movement.* The Internet is informing and connecting a global citizenry as people begin collaborating to bring about environmental, social, and economic change from the bottom up. Top-down control by corporations and governments is being weakened. As environmental educator David W. Orr puts it, "It is not possible for long to organize our affairs around greed, illusion, and ill will."

These groups have worked with individuals and communities to oppose harmful projects such as landfills, waste incinerators, nuclear waste dumps, clear-cutting of forests (Figure 18-15), pollution from factories and refineries, and a variety of development projects. They have also taken action against environmental injustice. (See the Guest Essay on this topic by Robert D. Bullard on the website for this chapter.) Grassroots groups have formed land trusts and other local organizations to save wetlands, forests, farmland, and ranchland from development and have helped restore forests (see Individuals Matter, p. 172), degraded rivers, and wetlands.

Some grassroots environmental groups use the nonviolent and nondestructive tactics of protest marches, tree sitting (see Individuals Matter, p. 166), and other devices for generating publicity to help educate and sway members of the public to oppose various environmentally harmful activities. Much more controversial are militant environmental groups that use violent means to achieve their ends. Most environmentalists oppose such tactics.

✗ HOW WOULD YOU VOTE? Do you support the use of nonviolent and nondestructive civil disobedience tactics by some environmental groups? Cast your vote online at http://biology.brookscole.com/miller11.

Case Study: Environmental Action by Students in the United States

Many student environmental groups work to bring about environmental improvements in their schools and local communities.

Since 1988, there has been a boom in environmental awareness on college campuses and in public schools across the United States. Most student environmental groups work with members of the faculty and administration to bring about environmental improvements in their schools and local communities.

Many of these groups make *environmental audits* of their campuses or schools. They use the data gathered in this way to propose changes that will make their campus or school more environmentally sustainable, usually saving money in the process. As a result, students have helped convince almost 80% of universities and colleges in the United States to develop recycling programs.

Students at Oberlin College in Ohio helped design a more sustainable environmental studies building. At Northland College in Wisconsin, students helped design a "green" dorm that features a large wind generator, panels of solar cells, recycled furniture, and waterless (composting) toilets.

At Minnesota's St. Olaf College, students have carried out sustainable agriculture and ecological restoration projects. At Temple University in Philadelphia, students passed a resolution expressing a willingness to pay an extra fee if the school gets much of its electricity from wind power.

Students also recognize that college campuses are major polluters. A recent Yale University study reported that the school emits more greenhouse gases than 32 developing countries. Students at Columbia University are pressuring the New York legislature to cap carbon dioxide emissions and the university to make more socially and environmentally investments with its endowment funds.

Such student-spurred environmental activities and research studies are spreading to universities in at least 42 other countries. See Noel Perrin's Guest Essay on environmental activities at U.S. colleges on this chapter's website.

How Successful Have U.S. Environmental Groups and Their Opponents Been?

Environmental groups have helped educate the public and business and political leaders about environmental issues and pass environmental laws, but an organized movement has undermined many of these efforts.

Since 1970, a variety of environmental groups in the United States and other countries have helped increase public understanding of environmental issues and have gained public support for an array of environmental and resource-use laws in the United States and other countries. In addition, they have helped individuals deal with a number of local environmental problems.

Polls show that about 80% of the U.S. public strongly supports environmental laws and regulations and does not want them weakened. Polls also show that less than 10% of the U.S. public views the environment as one of the nation's most pressing problems. As a result, environmental concerns often do not get transferred to the ballot box. As one political scientist put it, "Environmental concerns are like the Florida Everglades, a mile wide but only a few inches deep." Since 1980, a well-organized and well-funded movement has also undermined much of the improve-

ments made in environmental understanding and support for environmental concerns in the United States. Three major groups are strongly opposed to many environmental proposals, laws, and regulations. *First,* leaders of some corporations and people in positions of economic and political power see environmental laws and regulations as threats to their wealth and power. *Second,* some citizens see environmental laws and regulations as threats to their private property rights and jobs. *Third,* some state and local government officials are tired of having to implement federal environmental laws and regulations without federal funding (unfunded mandates) or disagree with certain regulations.

One problem is that the focus of environmental issues has shifted from easy-to-see dirty smokestacks and burning rivers to more complex and controversial environmental problems such as climate change and biodiversity loss. Explaining such complex issues to the public and mobilizing support for often controversial, long-range solutions to such problems is difficult. See the Guest Essay on environmental reporting by Andrew C. Revkin on the website for this chapter.

Solutions: Developing Environmentally Sustainable Political and Economic Systems

We need to work together to find and implement innovative solutions to local, national, and global environmental, economic, and social problems.

Many environmentalists call for people from all political persuasions and walks of life to work together to develop a positive vision for a transition to more environmentally sustainable societies and economies throughout the world.

A major goal is to promote the development of creative experiments at local levels—such as the one in Curitiba, Brazil (p. 152)—that could be implemented in other areas over the next few decades. A second goal is to get citizens, business leaders, and elected officials to cooperate in trying to find and implement innovative solutions to local, national, and global environmental, economic, and social problems.

According to business leader Paul Hawken, making a cultural shift to more environmentally sustainable societies over the next 50 years

> *means thinking big and long into the future. It also means doing something now. It means electing people who really want to make things work, and who can imagine a better world. It means writing to companies and telling them what you think. It means never forgetting that the cash register is the daily voting booth in democratic capitalism.*

Several guidelines have been suggested for fostering cooperation instead of confrontation as we deal with important environmental problems.

First, recognize that business is not the enemy. Businesses want to make money for their investors and stockholders. Why not reward them while simultaneously encouraging new environmental innovations by shifting government subsidies from earth-degrading activities to earth-sustaining activities and by shifting taxes from income and wealth to pollution and resource waste? In this way, environmentalists and leaders of corporations could become partners in a joint quest for environmental and economic sustainability.

Second, shift the emphasis from dealing with environmental problems after the fact to preventing or minimizing them.

Third, use well-designed and carefully monitored marketplace solutions to prevent most environmental problems instead of relying primarily on laws, regulations, and litigation.

Fourth, cooperate and innovate to find *win-win* solutions to environmental problems instead of using confrontational tactics to come up with less effective *I-win-you-lose* solutions in which the earth always ends up losing.

Fifth, stop exaggerating. People on both sides of thorny environmental issues should take a vow not to exaggerate or distort their positions in attempts to play win-lose or winner-take-all environmental games. There are trade-offs in any environmental decision—as presented throughout this book—so they must work together to find balanced win-win solutions that are implemented in a flexible and adaptive manner.

In working to make the earth a better place to live, we should be guided by historian Arnold Toynbee's observation, "If you make the world ever so little better, you will have done splendidly, and your life will have been worthwhile," and by George Bernard Shaw's reminder that "indifference is the essence of inhumanity."

In the end, it comes back to *each of us taking responsibility.* We must decide whether we want to be part of the problem or part of the solution to the environmental challenges we face.

18-5 GLOBAL ENVIRONMENTAL POLICY

National and Global Security

Many analysts believe that environmental security is as important as military and economic security.

Countries are legitimately concerned with *military security* and *economic security.* However, ecologists and many economists point out that all economies are supported by the earth's natural capital (Figure 1-9, p. 13).

According to environmental expert Norman Myers,

If a nation's environmental foundations are degraded or depleted, its economy may well decline, its social

fabric deteriorate, and its political structure become destabilized as growing numbers of people seek to sustain themselves from declining resource stocks. Thus, national security is no longer about fighting forces and weaponry alone. It relates increasingly to watersheds, croplands, forests, genetic resources, climate, and other factors that, taken together, are as crucial to a nation's security as are military factors.

Proponents of Myers' view call for all countries to make environmental security a major focus of diplomacy and government policy at all levels. This perspective would be implemented by a council of advisers made up of highly qualified experts in environmental, economic, and military security who integrate all three security concerns in making major decisions. Proponents note that failure to protect chemical and nuclear plants and water supplies from terrorist acts can lead to serious environmental harm.

Likewise, acting to prevent pollution and reduce resource waste can improve economic and environmental security. For example, reducing dependence on imported oil by improving gas mileage saves consumers money, reduces air pollution, helps slow global warming, and improves military security by reducing the need for troops from oil-addicted countries to help protect Middle East oil supplies.

X *HOW WOULD YOU VOTE?* Is environmental security just as important as economic and military security? Cast your vote online at http://biology.brookscole.com/miller11.

Global Outlook: International Environmental Organizations

International environmental organizations gather and evaluate environmental data, help develop environmental treaties, and provide funds and loans for sustainable economic development.

A variety of international environmental organizations help shape and set environmental policy. Perhaps the most influential is the United Nations, the umbrella that shelters a large family of organizations such as the UN Environment Programme (UNEP), the World Health Organization (WHO), the UN Development Programme (UNDP), and the Food and Agriculture Organization (FAO).

Other organizations that make or influence environmental decisions include the World Bank, the Global Environment Facility (GEF), and the World Conservation Union (IUCN).

These and other organizations have played important roles in

- Expanding understanding of environmental issues
- Gathering and evaluating environmental data

- Developing and monitoring international environmental treaties
- Providing funds and loans for more sustainable economic development and reducing poverty
- Helping more than 100 nations develop environmental laws and institutions

Despite their often limited funding, these diverse organizations have made important contributions to global and national environmental progress since 1970.

International Environmental Cooperation and Policy

Earth summits and international environmental treaties have played important roles in dealing with global environmental problems, but most treaties are not effectively monitored and enforced.

Since the 1972 UN Conference on the Human Environment in Stockholm, Sweden, progress has been made in addressing environmental issues at the global level. Figure 18-20 lists some of the good and bad news about international efforts to deal with global environmental problems such as poverty, climate change, biodiversity loss, and ocean pollution.

In 2004, environmental leader Gus Speth argued that global environmental problems are getting worse and that international efforts to solve them are inadequate. He proposes the creation of a World Environmental Organization (WEO), on the order of the World Health Organization and the World Trade Organization, to deal with global challenges.

18-6 ENVIRONMENTAL WORLDVIEWS: CLASHING VALUES AND CULTURES

What Is an Environmental Worldview?

Your environmental worldview encompasses how you think the world works, what you believe your environmental role in the world should be, and what you believe is right and wrong environmental behavior.

People disagree about how serious our environmental problems are and what we should do about them. These conflicts arise mostly out of differing **environmental worldviews**: how people think the world works, what they believe their environmental role in the world should be, and what they believe is right and wrong environmental behavior **(environmental ethics)**. People with widely differing environmental worldviews can take the same data, be logically consistent, and yet arrive at quite different conclusions because they start with different assumptions and values.

Trade-Offs
Global Efforts on Environmental Problems

Good News	Bad News
Environmental protection agencies in 115 nations	 Most international environmental treaties lack criteria for monitoring and evaluating their effectiveness
Over 500 international environmental treaties and agreements	1992 Rio Earth Summit led to nonbinding agreements without enough funding to implement them
UN Environment Programme (UNEP) created in 1972 to negotiate and monitor international environmental treaties	 By 2003 there was little improvement in the major environmental problems discussed at the 1992 Rio summit
1992 Rio Earth Summit adopted key principles for dealing with global environmental problems	2002 Johannesburg Earth Summit failed to provide adequate goals, deadlines, and funding for dealing with global environmental problems such as climate change, biodiversity loss, and poverty
2002 Johannesburg Earth Summit attempted to implement policies and goals of 1992 Rio summit and find ways to reduce poverty	

Figure 18-20 Trade-offs: good and bad news about international efforts to deal with global environmental problems. *Critical thinking: pick the single piece of good news and bad news that you think are the most important.*

Many types of environmental worldviews exist. Some are *human centered* while others are *life centered*, with the primary focus on individual species or on sustaining the earth's natural life forms (biodiversity) and life-support systems (biosphere) for the benefit of humans and other forms of life.

Human-Centered Environmental Worldviews

According to the human-centered view, humans are the planet's most important species and should manage the earth mostly for our own benefit.

Many people in today's industrial consumer societies have a **planetary management worldview**. According

Environmental Worldviews

Planetary Management

- We are apart from the rest of nature and can manage nature to meet our increasing needs and wants.

- Because of our ingenuity and technology we will not run out of resources.

- The potential for economic growth is essentially unlimited.

- Our success depends on how well we manage the earth's life-support systems mostly for our benefit.

Stewardship

- We have an ethical responsibility to be caring managers, or *stewards*, of the earth.

- We will probably not run out of resources, but they should not be wasted.

- We should encourage environmentally beneficial forms of economic growth and discourage environmentally harmful forms.

- Our success depends on how well we manage the earth's life-support systems for our benefit and for the rest of nature.

Environmental Wisdom

- We are a part of and totally dependent on nature and nature exists for all species.

- Resources are limited, should not be wasted, and are not all for us.

- We should encourage earth-sustaining forms of economic growth and discourage earth-degrading forms.

- Our success depends on learning how nature sustains itself and integrating such lessons from nature into the ways we think

Figure 18-21 Comparison of three environmental worldviews.

to this worldview, humans are the planet's most important and dominant species, and we can and should manage the earth mostly for our own benefit. Other species and parts of nature are seen as having *instrumental value* based on how useful they are to us. According to environmental leader Gus Speth, "This view of the world—that nature belongs to us rather than we to nature—is powerful and pervasive, and it has led to much mischief." Figure 18-21 (left) summarizes the four major beliefs or assumptions of one version of this worldview.

Another increasingly popular human-centered environmental worldview is the **stewardship worldview.** It assumes that we have an ethical responsibility to be caring and responsible managers, or *stewards,* of the earth. Figure 18-21 (center) summarizes the major beliefs of this worldview.

According to the stewardship view, when we use the earth's natural capital we are borrowing from the earth and from future generations, and we have an ethical responsibility to pay the debt by leaving the earth in at least as good a condition as we now enjoy. Some analysts believe we should consider the wisdom of the 18th-century Iroquois Confederation of Native Americans: *In our every deliberation, we must consider the impact of our decisions on the next seven generations.*

Life-Centered Environmental Worldviews

Some believe that we have an ethical responsibility not to degrade the earth's ecosystems, biodiversity, and biosphere for all forms of life.

Some people believe any human-centered worldview will eventually fail because it wrongly assumes we now have or can gain enough knowledge to become effective managers or stewards of the earth. These critics point out that we do not know how many species live on the earth, much less what their roles are and how they interact with one another and their non-living environment. We have only an inkling of what goes on in a handful of soil, a meadow, or any other part of the earth.

These critics believe that human-centered environmental worldviews should be expanded to recognize the *inherent* or *intrinsic value* of all forms of life, regardless of their potential or actual use to humans. Most people with such a life-centered worldview believe we have an ethical responsibility to avoid causing the premature extinction of species through our activities for three reasons. *First,* each species is a unique storehouse of genetic information that should be respected and protected simply because it exists (*intrinsic value*). *Second,* each species is a potential economic good for human use (*instrumental value*). *Third,* populations of species are capable, through evolution and speciation, of adapting to changing environmental conditions.

Some people believe we must go beyond focusing mostly on species. According to these people, we have an ethical responsibility not to degrade the earth's ecosystems, biodiversity, and biosphere for this and future generations of humans and other species. This *earth-centered*, or *ecocentric*, environmental worldview is devoted to preserving the earth's biodiversity and the functioning of its life-support systems for all forms of life.

One earth-centered worldview is called the **environmental wisdom worldview**. Figure 18-21 (right) summarizes its major beliefs. In many respects, it is the opposite of the planetary management worldview (Figure 18-21, left). According to this worldview, we are part of—not apart from—the community of life and the ecological processes that sustain all life (Figure 18-22).

This worldview suggests that the earth does not need us managing it to go on, whereas we do need the earth to survive. We cannot save the earth because it does not need saving. What we need to save is the existence of our own species and other species that may become extinct because of our activities. See Lester Milbrath's Guest Essay on this topic on the website for this chapter.

X *HOW WOULD YOU VOTE?* Which of the following comes closest to your environmental worldview: planetary management, stewardship, or environmental wisdom? Cast your vote online at http://biology.brookscole.com/miller11.

Courtesy of Earth Flag Co.

Figure 18-22 The earth flag is a symbol of commitment to promoting environmental sustainability by working with the earth at the individual, local, national, and international levels.

18-7 LIVING MORE SUSTAINABLY

Environmental Literacy

Environmentally literate citizens and leaders are needed to build more environmentally sustainable and just societies.

Most environmentalists believe that learning how to live more sustainably requires a foundation of environmental education. They cite the key goals of environmental education or ecological literacy:

- Develop respect or reverence for all life.

- Understand as much as we can about how the earth works and sustains itself, and use such knowledge to guide our lives, communities, and societies.

- Look for connections within the biosphere and between our actions and the biosphere.

- Use critical thinking skills to become seekers of environmental wisdom instead of overfilled vessels of environmental information.

- Understand and evaluate our environmental worldview and see this as a lifelong process.

- Learn how to evaluate the beneficial and harmful consequences to the earth of our choices of lifestyle and profession, today and in the future.

- Foster a desire to make the world a better place and act on this desire.

Specifically, an ecologically literate person should have a basic comprehension of the following:

- Concepts such as environmental sustainability, natural capital, exponential growth, carrying capacity, and risks and risk analysis

- Environmental history (to help keep us from repeating past mistakes)

- The laws of thermodynamics and the law of conservation of matter

- Basic principles of ecology

- Human population dynamics

- Sustainable cities and design

- Ways to sustain biodiversity

- Sustainable agriculture and forestry

- Soil conservation

- Sustainable water use

- Nonrenewable mineral resources

- Nonrenewable and renewable energy resources

- Climate change and ozone depletion

- Pollution prevention and waste reduction

- Environmentally sustainable economic and political systems

- Environmental worldviews and ethics

If you have studied this book carefully and listened to your instructor, you are well on your way to becoming environmentally literate.

According to environmental educator Mitchell Thomashow, four basic questions lie at the heart of environmental literacy. *First,* where do the things I consume come from? *Second,* what do I know about the place where I live? *Third,* how am I connected to the earth and other living things? *Fourth,* what is my purpose and responsibility as a human being? How we answer these four questions determines our *ecological identity.* What are your answers?

Figure 18-23 summarizes guidelines and strategies that have been discussed throughout this book for achieving more sustainable societies.

Learning from the Earth

In addition to formal learning, we need to learn by experiencing nature directly.

Formal environmental education is important, but is it enough? Many analysts say no. They urge us to escape the cultural and technological body armor we use to insulate ourselves from nature and to experience nature directly.

They suggest we kindle a sense of awe, wonder, mystery, and humility by standing under the stars, sitting in a forest, or taking in the majesty and power of an ocean.

We might pick up a handful of soil and try to sense the teeming microscopic life in it that keeps us alive. We might look at a tree, mountain, rock, or bee and try to sense how they are a part of us and we are a part of them as interdependent participants in the earth's life-sustaining recycling processes.

Many psychologists believe that, consciously or unconsciously, we spend much of our lives searching for roots: something to anchor us in a bewildering and frightening sea of change. As philosopher Simone Weil observed, "To be rooted is perhaps the most important and least recognized need of the human soul."

Earth-focused philosophers say that to be rooted, each of us needs to find a *sense of place:* a stream, a mountain, a yard, a neighborhood lot, or any piece of the earth we feel at one with as a place we know, experience emotionally, and love. When we become part of a place, it becomes a part of us. Then we are driven to defend it from harm and to help heal its wounds. This quest might lead us to recognize that the healing of the earth and the healing of the human spirit are one and the same. We might discover and tap into what Aldo Leopold calls "the green fire that burns in our hearts" and use it as a force for respecting and working with the earth and with one another.

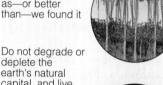

Solutions

Developing Environmentally Sustainable Societies

Guidleines

Leave the world in as good a shape as—or better than—we found it

Do not degrade or deplete the earth's natural capital, and live off the natural income it provides

Copy nature

Take no more than we need

Do not reduce biodiversity

Try not to harm life, air, water, soil

Do not change the world's climate

Do not overshoot the earth's carrying capacity

Help maintain the earth's capacity for self-repair

Repair past ecological damage

Strategies

Sustain biodiversity

Eliminate poverty

Develop eco-economies

Build sustainable communities

Do not use renewable resources faster than nature can replace them

Use sustainable agriculture

Depend more on locally available renewable energy from the sun, wind, flowing water, and sustainable biomass

Emphasize pollution prevention and waste reduction

Do not waste matter and energy resources

Recycle, reuse, and compost 60–80% of matter resources

Maintain a human population size such that needs are met without threatening life-support systems

Emphasize ecological restoration

Figure 18-23 Solutions: guidelines and strategies for achieving more sustainable societies. *Critical thinking: which five of these solutions do you believe are the most important?*

Living More Simply

Some affluent people are voluntarily adopting lifestyles in which they enjoy life more by consuming less.

Many analysts urge us to *learn how to live more simply.* Seeking happiness through the pursuit of material things is considered folly by almost every major religion and philosophy. Yet, this message is preached incessantly by modern advertising, which encourages us to buy more and more things. As humorist Will Rogers put it, "Too many people spend money they haven't earned to buy things they don't want, to impress people they don't like."

Some affluent people in developed countries are adopting a lifestyle of *voluntary simplicity,* doing and enjoying more with less by learning to live more simply. Voluntary simplicity is based on Mahatma Gandhi's *principle of enoughness:* "The earth provides enough to satisfy every person's need but not every person's greed. . . . When we take more than we need, we are simply taking from each other, borrowing from the future, or destroying the environment and other species." As environmental educator David W. Orr puts it, "We do not have to rob the world and steal from our children to live well."

Most of the world's major religions have similar teachings. "Why do you spend your money for that which is not bread, and your labor for that which does not satisfy?" (Christianity: Old Testament, Isaiah 55:2). "Eat and drink, but waste not by excess" (Islam: Koran 7.31). "One should abstain from acquisitiveness" (Hinduism: Acarangastura 2.119). "He who knows he has enough is rich" (Taoism: Tao Te Ching, Chapter 33).

Implementing this principle means asking yourself, "How much is enough?" This is not easy because people in affluent societies are conditioned to want more and more.

Becoming Better Environmental Citizens

We can help make the world a better place by not falling into mental traps that lead to denial and inaction and by keeping our empowering feelings of hope slightly ahead of our immobilizing feelings of despair.

We all make some direct or indirect contributions to the environmental problems we face. However, because we do not want to feel guilty about the environmental harm we may be creating, we try not to think about this issue too much—a path that can lead to denial and inaction.

Analysts suggest that we move beyond fear, denial, apathy, and guilt to more responsible environmental actions in our daily lives by recognizing and avoiding common mental traps that lead to denial, indifference, and inaction. These traps include *gloom-and-doom pessimism* (it is hopeless), *blind technological optimism* (science and technofixes will save us), *fatalism* (we have no control over our actions and the future), *extrapolation to infinity* (if I cannot change the entire world quickly, I will not try to change any of it), *paralysis by analysis* (searching for the perfect worldview,

philosophy, solutions, and scientific information before doing anything), and *faith in simple, easy answers.* They also urge us to keep our empowering feelings of hope slightly ahead of our immobilizing feelings of despair.

Recognizing that no single correct or best solution to the environmental problems we face exists is also important. Indeed, one of nature's most important lessons is that preserving diversity—in this case, being flexible and adaptable in trying a variety of solutions to our problems—is the best way to adapt to the earth's largely unpredictable, ever-changing conditions.

Finally, we should have fun and take time to enjoy life. Laugh every day and enjoy nature, beauty, friendship, and love. This empowers us to become good earth citizens who practice *good earthkeeping.*

Components of the Sustainability Revolution

The message of environmentalism is not gloom and doom, fear, and catastrophe, but rather hope, a positive vision of the future, and a call for commitment in dealing with the environmental challenges we face.

The *environmental* or *sustainability revolution* that many environmental scientists and environmentalists call for us to achieve during this century would have several interrelated components:

- A *biodiversity protection revolution* devoted to protecting and sustaining the genes, species, natural systems, and chemical and biological processes that make up the earth's biodiversity

- An *efficiency revolution* that minimizes the wasting of matter and energy resources

- An *energy revolution* based on decreasing our dependence on carbon-based, nonrenewable fossil fuels and increasing our dependence on forms of renewable energy such as the wind, sun, flowing water, and geothermal energy

- A *pollution prevention revolution* that reduces pollution and environmental degradation from harmful chemicals

- A *sufficiency revolution* dedicated to meeting the basic needs of all people on the planet while affluent societies learn to live more sustainably by living with less

- A *demographic revolution* based on bringing the size and growth rate of the human population into balance with the earth's ability to support humans and other species sustainably

- An *economic and political revolution* in which we use economic systems to reward environmentally beneficial behavior and discourage environmentally harmful behavior

We possess an incredible array of scientific, technological, and economic solutions to the environmental problems we face. Our challenge is to implement such solutions by converting environmental knowledge, wisdom, and beliefs into political action.

This requires understanding that *individuals matter.* Virtually all of the environmental progress we have made during the last few decades occurred because individuals banded together to insist that we can do better.

This journey begins in your own community, because in the final analysis *all sustainability is local.* We help make the world more sustainable by working to make our local communities more sustainable. This endeavor begins with your own lifestyle. It is the meaning of the motto: "Think globally, act locally."

It is an incredibly exciting time to be alive as we struggle to implement such ideals by entering into a new relationship with the earth that keeps us all alive and supports our economies.

Envision the earth's life-sustaining processes as a beautiful and diverse web of interrelationships—a kaleidoscope of patterns, rhythms, and connections whose very complexity and multitude of possibilities remind us that cooperation, sharing, honesty, humility, compassion, and love should be the guidelines for our behavior toward one another and toward the earth.

When there is no dream, the people perish.
PROVERBS 29:18

CRITICAL THINKING

1. Should we attempt to maximize economic growth by producing and consuming increasingly more economic goods? Explain. What are the alternatives?

2. Suppose that over the next 20 years the current harmful environmental and health costs of goods and services are internalized until their market prices reflect their total costs. What harmful and beneficial effects might such full-cost pricing have on your lifestyle?

3. List all the goods you use, and then identify those that meet your basic needs and those that satisfy your wants. Identify any economic wants **(a)** you would be willing to give up, **(b)** you believe you should give up but are unwilling to give up, and **(c)** you hope to give up in the future. Relate the results of this analysis to your personal impact on the environment. Compare your results with those of your classmates.

4. What are the greatest strengths and weaknesses of the system of government in your country with respect to **(a)** protecting the environment and **(b)** ensuring environmental justice for all? What three major changes, if any, would you make in this system?

5. This chapter summarized several different environmental worldviews. Go through these worldviews and find the beliefs you agree with to describe your own environmental worldview. Which of your beliefs were added or modified as a result of taking this course?

6. Would you (or do you already) use the principle of voluntary simplicity in your life? How?

7. Explain why you agree or disagree with the following ideas:
   a. Everyone has the right to have as many children as they want.
   b. Each member of the human species has a right to use as many resources as they want.
   c. Individuals should have the right to do anything they want with land they own.
   d. Other species exist to be used by humans.
   e. All forms of life have an intrinsic value and therefore have a right to exist.

Are your answers consistent with the beliefs making up your environmental worldview that you described in Question 5?

LEARNING ONLINE

The website for this book includes review questions for the entire chapter, flash cards for key terms and concepts, a multiple-choice practice quiz, interesting Internet sites, references, and a guide for accessing thousands of InfoTrac® College Edition articles.

Visit

http://biology.brookscole.com/miller11

Then choose Chapter 18, and select a learning resource. For access to animations, additional quizzes, chapter outlines and summaries, register and log in to

Environmental Science ⊛ Now™

at **esnow.brookscole.com/miller11** using the access code card in the front of your book.

> ### Active Graphing
> Visit http://esnow.brookscole.com/miller11 to explore the graphing exercise for this chapter.

LENGTH

Metric

1 kilometer (km) = 1,000 meters (m)
1 meter (m) = 100 centimeters (cm)
1 meter (m) = 1,000 millimeters (mm)
1 centimeter (cm) = 0.01 meter (m)
1 millimeter (mm) = 0.001 meter (m)

English

1 foot (ft) = 12 inches (in)
1 yard (yd) = 3 feet (ft)
1 mile (mi) = 5,280 feet (ft)
1 nautical mile = 1.15 miles

Metric–English

1 kilometer (km) = 0.621 mile (mi)
1 meter (m) = 39.4 inches (in)
1 inch (in) = 2.54 centimeters (cm)
1 foot (ft) = 0.305 meter (m)
1 yard (yd) = 0.914 meter (m)
1 nautical mile = 1.85 kilometers (km)

AREA

Metric

1 square kilometer (km^2) = 1,000,000 square meters (m^2)
1 square meter (m^2) = 1,000,000 square millimeters (mm^2)
1 hectare (ha) = 10,000 square meters (m^2)
1 hectare (ha) = 0.01 square kilometer (km^2)

English

1 square foot (ft^2) = 144 square inches (in^2)
1 square yard (yd^2) = 9 square feet (ft^2)
1 square mile (mi^2) = 27,880,000 square feet (ft^2)
1 acre (ac) = 43,560 square feet (ft^2)

Metric–English

1 hectare (ha) = 2.471 acres (ac)
1 square kilometer (km^2) = 0.386 square mile (mi^2)
1 square meter (m^2) = 1.196 square yards (yd^2)
1 square meter (m^2) = 10.76 square feet (ft^2)
1 square centimeter (cm^2) = 0.155 square inch (in^2)

VOLUME

Metric

1 cubic kilometer (km^3) = 1,000,000,000 cubic meters (m^3)
1 cubic meter (m^3) = 1,000,000 cubic centimeters (cm^3)
1 liter (L) = 1,000 milliliters (mL) = 1,000 cubic centimeters (cm^3)
1 milliliter (mL) = 0.001 liter (L)
1 milliliter (mL) = 1 cubic centimeter (cm^3)

English

1 gallon (gal) = 4 quarts (qt)
1 quart (qt) = 2 pints (pt)

Metric–English

1 liter (L) = 0.265 gallon (gal)
1 liter (L) = 1.06 quarts (qt)
1 liter (L) = 0.0353 cubic foot (ft^3)
1 cubic meter (m^3) = 35.3 cubic feet (ft^3)
1 cubic meter (m^3) = 1.30 cubic yards (yd^3)
1 cubic kilometer (km^3) = 0.24 cubic mile (mi^3)
1 barrel (bbl) = 159 liters (L)
1 barrel (bbl) = 42 U.S. gallons (gal)

MASS

Metric

1 kilogram (kg) = 1,000 grams (g)
1 gram (g) = 1,000 milligrams (mg)
1 gram (g) = 1,000,000 micrograms (μg)
1 milligram (mg) = 0.001 gram (g)
1 microgram (μg) = 0.000001 gram (g)
1 metric ton (mt) = 1,000 kilograms (kg)

English

1 ton (t) = 2,000 pounds (lb)
1 pound (lb) = 16 ounces (oz)

Metric–English

1 metric ton (mt) = 2,200 pounds (lb) = 1.1 tons (t)
1 kilogram (kg) = 2.20 pounds (lb)
1 pound (lb) = 454 grams (g)
1 gram (g) = 0.035 ounce (oz)

ENERGY AND POWER

Metric

1 kilojoule (kJ) = 1,000 joules (J)
1 kilocalorie (kcal) = 1,000 calories (cal)
1 calorie (cal) = 4.184 joules (J)

Metric–English

1 kilojoule (kJ) = 0.949 British thermal unit (Btu)
1 kilojoule (kJ) = 0.000278 kilowatt-hour (kW-h)
1 kilocalorie (kcal) = 3.97 British thermal units (Btu)
1 kilocalorie (kcal) = 0.00116 kilowatt-hour (kW-h)
1 kilowatt-hour (kW-h) = 860 kilocalories (kcal)
1 kilowatt-hour (kW-h) = 3,400 British thermal units (Btu)
1 quad (Q) = 1,050,000,000,000,000 kilojoules (kJ)
1 quad (Q) = 293,000,000,000 kilowatt-hours (kW-h)

TEMPERATURE CONVERSIONS

Fahrenheit ($^\circ$F) to Celsius ($^\circ$C):
$$^\circ C = (^\circ F - 32.0) \div 1.80$$
Celsius($^\circ$C) to Fahrenheit($^\circ$F):
$$^\circ F = (^\circ C \times 1.80) + 32.0$$

ECOLOGICAL FOOTPRINTS

NASA Goddard Space Flight Center Image by Reto Stöckli (land surface, shallow water, clouds). Enhancements by Robert Simmon (ocean color, compositing, 3D globes, animation)

Figure 1 Natural capital: composite satellite view of the earth showing its major terrestrial and aquatic features.

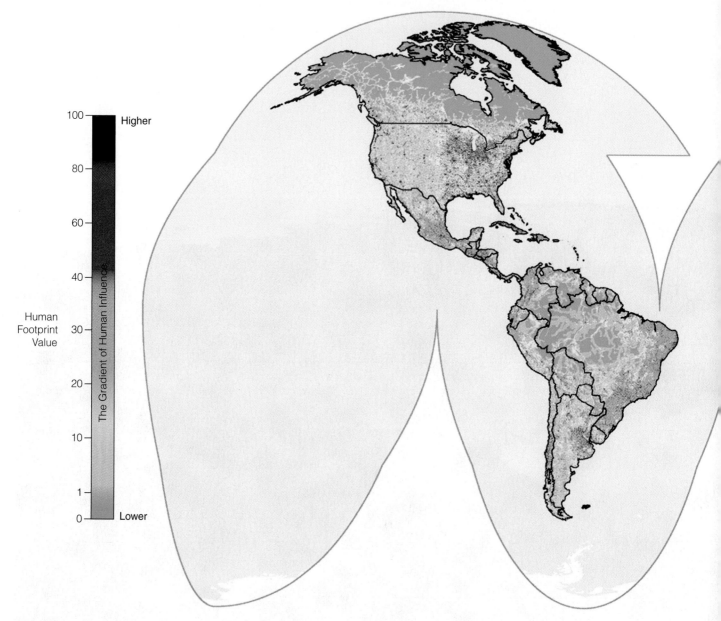

Figure 2 Natural capital degradation: the human footprint on the earth's land surface—in effect, the sum of all ecological footprints (Figure 1-7, p. 11) of the human population. Colors represent the percentage of each area influenced by human activities. Excluding Antarctica and Greenland, human activities have directly affected to some degree 83% of the earth's land surface and 98% of the area where it is possible to grow rice, wheat, or maize. *Critical thinking: how large is the human footprint where you live? What contribution does your lifestyle make to the environmental impacts?* (Data from Wildlife Conservation Society and the Center for International Earth Science Information Network at Columbia University)

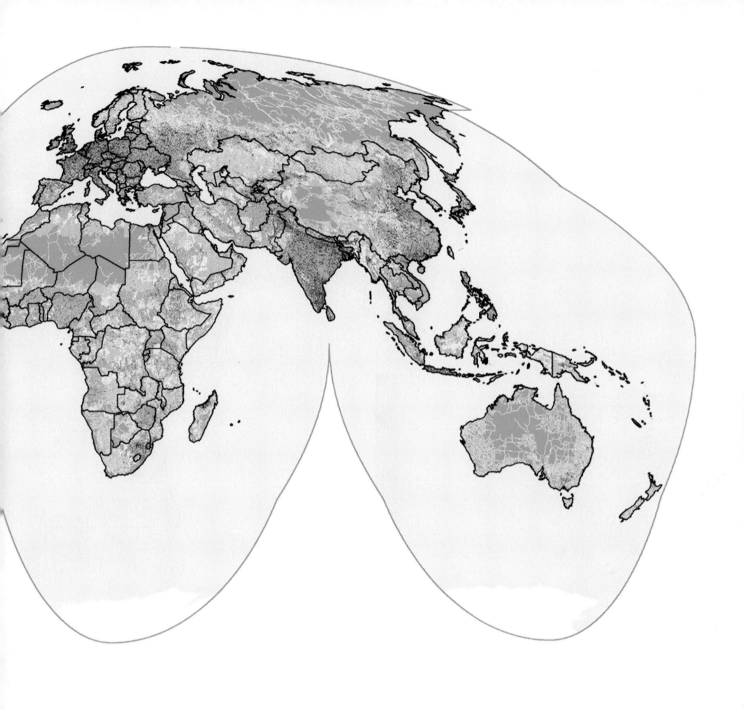

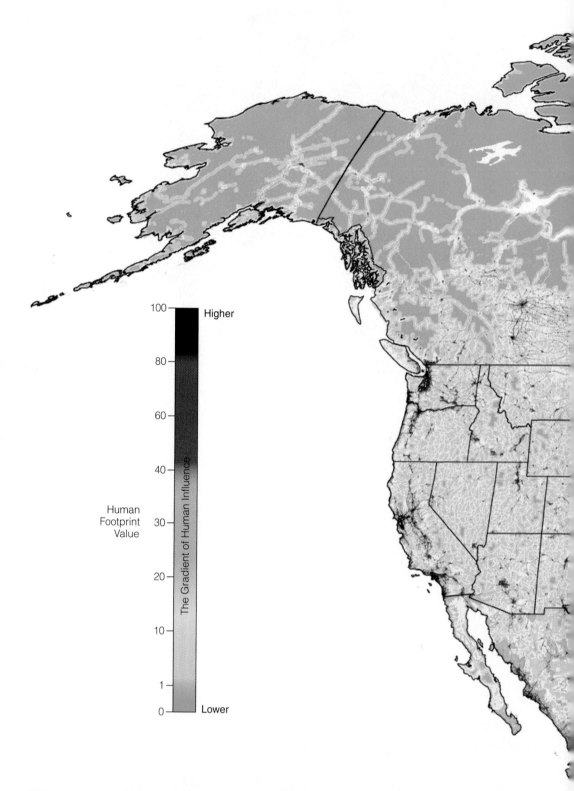

Figure 3 Natural capital degradation: the human ecological footprint in North America. Colors represent the percentage of each area influenced by human activities. This is an expanded portion of Figure 2 showing the human footprint on the earth's entire land surface. *Critical thinking: if you live in the United States, how large is the human ecological footprint where you live?* (Data from Wildlife Conservation Society and the Center for International Earth Science Information Network at Columbia University)

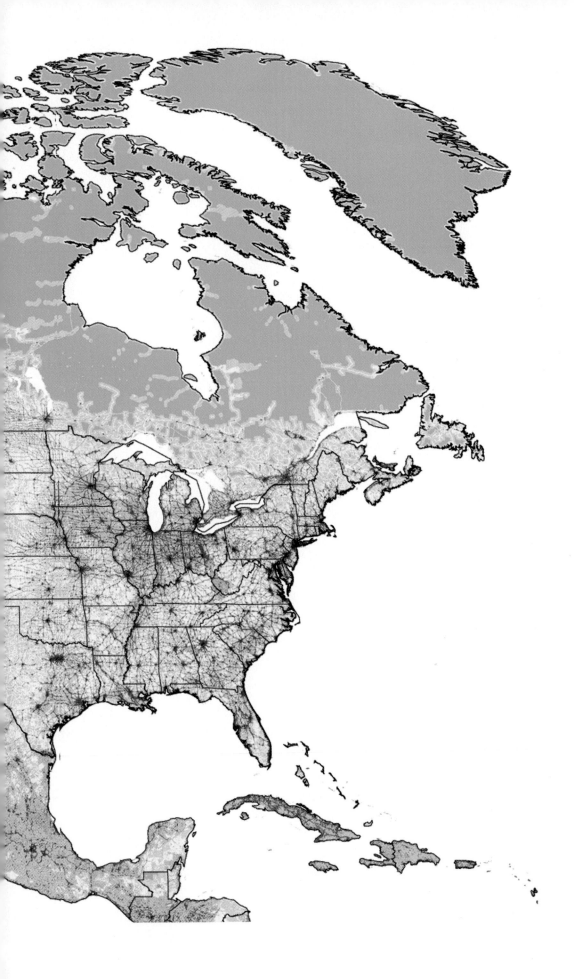

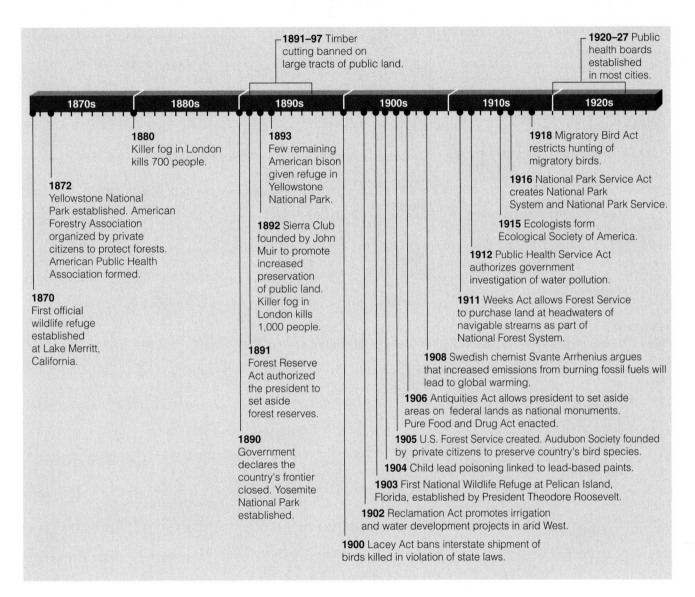

1891–97 Timber cutting banned on large tracts of public land.

1920–27 Public health boards established in most cities.

1870s   1880s   1890s   1900s   1910s   1920s

1880 Killer fog in London kills 700 people.

1893 Few remaining American bison given refuge in Yellowstone National Park.

1872 Yellowstone National Park established. American Forestry Association organized by private citizens to protect forests. American Public Health Association formed.

1892 Sierra Club founded by John Muir to promote increased preservation of public land. Killer fog in London kills 1,000 people.

1870 First official wildlife refuge established at Lake Merritt, California.

1891 Forest Reserve Act authorized the president to set aside forest reserves.

1890 Government declares the country's frontier closed. Yosemite National Park established.

1918 Migratory Bird Act restricts hunting of migratory birds.

1916 National Park Service Act creates National Park System and National Park Service.

1915 Ecologists form Ecological Society of America.

1912 Public Health Service Act authorizes government investigation of water pollution.

1911 Weeks Act allows Forest Service to purchase land at headwaters of navigable streams as part of National Forest System.

1908 Swedish chemist Svante Arrhenius argues that increased emissions from burning fossil fuels will lead to global warming.

1906 Antiquities Act allows president to set aside areas on federal lands as national monuments. Pure Food and Drug Act enacted.

1905 U.S. Forest Service created. Audubon Society founded by private citizens to preserve country's bird species.

1904 Child lead poisoning linked to lead-based paints.

1903 First National Wildlife Refuge at Pelican Island, Florida, established by President Theodore Roosevelt.

1902 Reclamation Act promotes irrigation and water development projects in arid West.

1900 Lacey Act bans interstate shipment of birds killed in violation of state laws.

1870–1930

Figure 1 Examples of the increased role of the federal government in resource conservation and public health and establishment of key private environmental groups, 1870–1930.

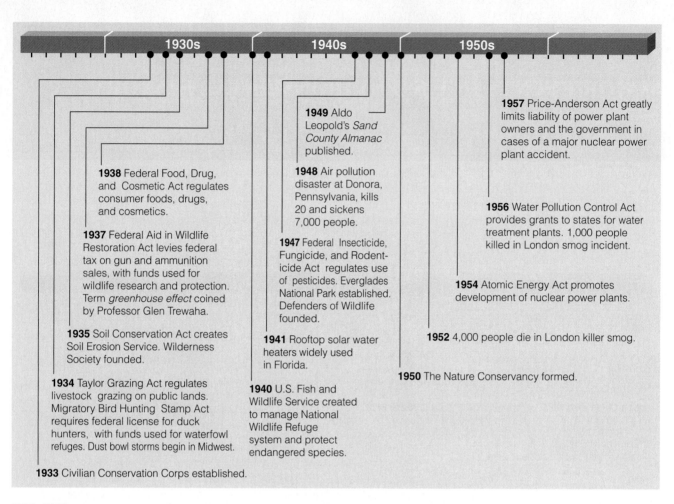

1949 Aldo Leopold's *Sand County Almanac* published.

1948 Air pollution disaster at Donora, Pennsylvania, kills 20 and sickens 7,000 people.

1947 Federal Insecticide, Fungicide, and Rodenticide Act regulates use of pesticides. Everglades National Park established. Defenders of Wildlife founded.

1941 Rooftop solar water heaters widely used in Florida.

1940 U.S. Fish and Wildlife Service created to manage National Wildlife Refuge system and protect endangered species.

1938 Federal Food, Drug, and Cosmetic Act regulates consumer foods, drugs, and cosmetics.

1937 Federal Aid in Wildlife Restoration Act levies federal tax on gun and ammunition sales, with funds used for wildlife research and protection. Term *greenhouse effect* coined by Professor Glen Trewaha.

1935 Soil Conservation Act creates Soil Erosion Service. Wilderness Society founded.

1934 Taylor Grazing Act regulates livestock grazing on public lands. Migratory Bird Hunting Stamp Act requires federal license for duck hunters, with funds used for waterfowl refuges. Dust bowl storms begin in Midwest.

1933 Civilian Conservation Corps established.

1957 Price-Anderson Act greatly limits liability of power plant owners and the government in cases of a major nuclear power plant accident.

1956 Water Pollution Control Act provides grants to states for water treatment plants. 1,000 people killed in London smog incident.

1954 Atomic Energy Act promotes development of nuclear power plants.

1952 4,000 people die in London killer smog.

1950 The Nature Conservancy formed.

1930–1960

Figure 2 Some important conservation and environmental events, 1930–1960.

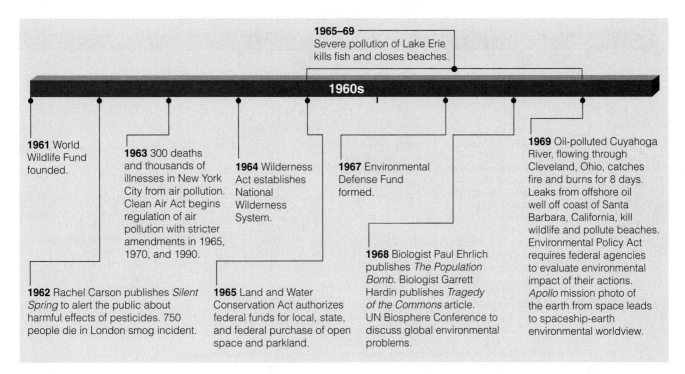

1965–69 Severe pollution of Lake Erie kills fish and closes beaches.

1960s

1961 World Wildlife Fund founded.

1962 Rachel Carson publishes *Silent Spring* to alert the public about harmful effects of pesticides. 750 people die in London smog incident.

1963 300 deaths and thousands of illnesses in New York City from air pollution. Clean Air Act begins regulation of air pollution with stricter amendments in 1965, 1970, and 1990.

1964 Wilderness Act establishes National Wilderness System.

1965 Land and Water Conservation Act authorizes federal funds for local, state, and federal purchase of open space and parkland.

1967 Environmental Defense Fund formed.

1968 Biologist Paul Ehrlich publishes *The Population Bomb*. Biologist Garrett Hardin publishes *Tragedy of the Commons* article. UN Biosphere Conference to discuss global environmental problems.

1969 Oil-polluted Cuyahoga River, flowing through Cleveland, Ohio, catches fire and burns for 8 days. Leaks from offshore oil well off coast of Santa Barbara, California, kill wildlife and pollute beaches. Environmental Policy Act requires federal agencies to evaluate environmental impact of their actions. *Apollo* mission photo of the earth from space leads to spaceship-earth environmental worldview.

1960s

Figure 3 Some important environmental events during the 1960s.

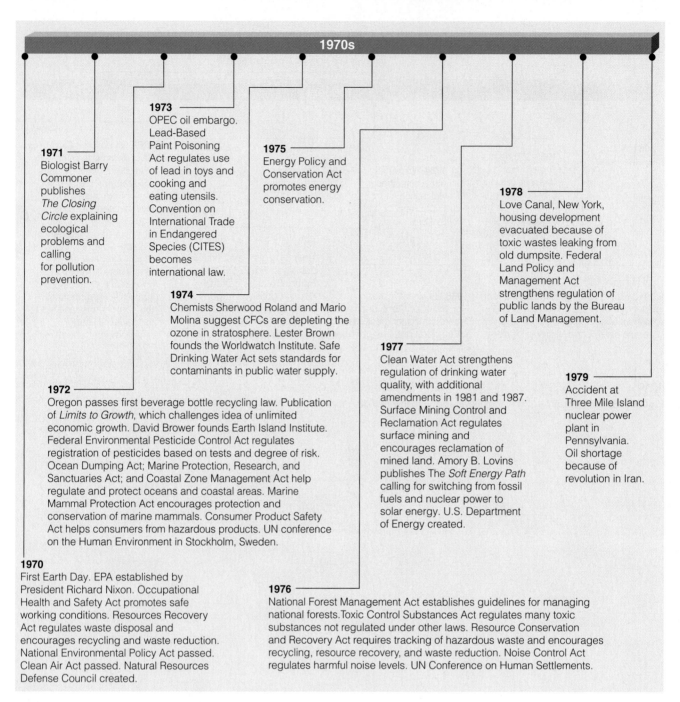

1970s

1971
Biologist Barry Commoner publishes *The Closing Circle* explaining ecological problems and calling for pollution prevention.

1973
OPEC oil embargo. Lead-Based Paint Poisoning Act regulates use of lead in toys and cooking and eating utensils. Convention on International Trade in Endangered Species (CITES) becomes international law.

1975
Energy Policy and Conservation Act promotes energy conservation.

1978
Love Canal, New York, housing development evacuated because of toxic wastes leaking from old dumpsite. Federal Land Policy and Management Act strengthens regulation of public lands by the Bureau of Land Management.

1974
Chemists Sherwood Roland and Mario Molina suggest CFCs are depleting the ozone in stratosphere. Lester Brown founds the Worldwatch Institute. Safe Drinking Water Act sets standards for contaminants in public water supply.

1977
Clean Water Act strengthens regulation of drinking water quality, with additional amendments in 1981 and 1987. Surface Mining Control and Reclamation Act regulates surface mining and encourages reclamation of mined land. Amory B. Lovins publishes The *Soft Energy Path* calling for switching from fossil fuels and nuclear power to solar energy. U.S. Department of Energy created.

1979
Accident at Three Mile Island nuclear power plant in Pennsylvania. Oil shortage because of revolution in Iran.

1972
Oregon passes first beverage bottle recycling law. Publication of *Limits to Growth*, which challenges idea of unlimited economic growth. David Brower founds Earth Island Institute. Federal Environmental Pesticide Control Act regulates registration of pesticides based on tests and degree of risk. Ocean Dumping Act; Marine Protection, Research, and Sanctuaries Act; and Coastal Zone Management Act help regulate and protect oceans and coastal areas. Marine Mammal Protection Act encourages protection and conservation of marine mammals. Consumer Product Safety Act helps consumers from hazardous products. UN conference on the Human Environment in Stockholm, Sweden.

1970
First Earth Day. EPA established by President Richard Nixon. Occupational Health and Safety Act promotes safe working conditions. Resources Recovery Act regulates waste disposal and encourages recycling and waste reduction. National Environmental Policy Act passed. Clean Air Act passed. Natural Resources Defense Council created.

1976
National Forest Management Act establishes guidelines for managing national forests. Toxic Control Substances Act regulates many toxic substances not regulated under other laws. Resource Conservation and Recovery Act requires tracking of hazardous waste and encourages recycling, resource recovery, and waste reduction. Noise Control Act regulates harmful noise levels. UN Conference on Human Settlements.

1970s

Figure 4 Some important environmental events during the 1970s, sometimes called the *environmental decade*.

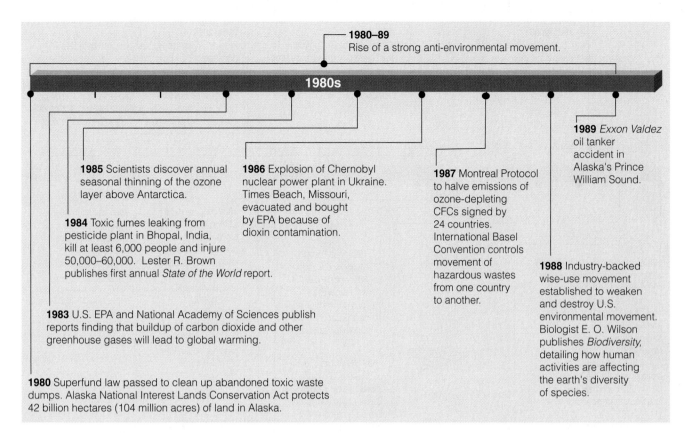

1980–89
Rise of a strong anti-environmental movement.

1980s

1985 Scientists discover annual seasonal thinning of the ozone layer above Antarctica.

1986 Explosion of Chernobyl nuclear power plant in Ukraine. Times Beach, Missouri, evacuated and bought by EPA because of dioxin contamination.

1987 Montreal Protocol to halve emissions of ozone-depleting CFCs signed by 24 countries. International Basel Convention controls movement of hazardous wastes from one country to another.

1989 *Exxon Valdez* oil tanker accident in Alaska's Prince William Sound.

1984 Toxic fumes leaking from pesticide plant in Bhopal, India, kill at least 6,000 people and injure 50,000–60,000. Lester R. Brown publishes first annual *State of the World* report.

1988 Industry-backed wise-use movement established to weaken and destroy U.S. environmental movement. Biologist E. O. Wilson publishes *Biodiversity*, detailing how human activities are affecting the earth's diversity of species.

1983 U.S. EPA and National Academy of Sciences publish reports finding that buildup of carbon dioxide and other greenhouse gases will lead to global warming.

1980 Superfund law passed to clean up abandoned toxic waste dumps. Alaska National Interest Lands Conservation Act protects 42 billion hectares (104 million acres) of land in Alaska.

1980s

Figure 5 Some important environmental events during the 1980s.

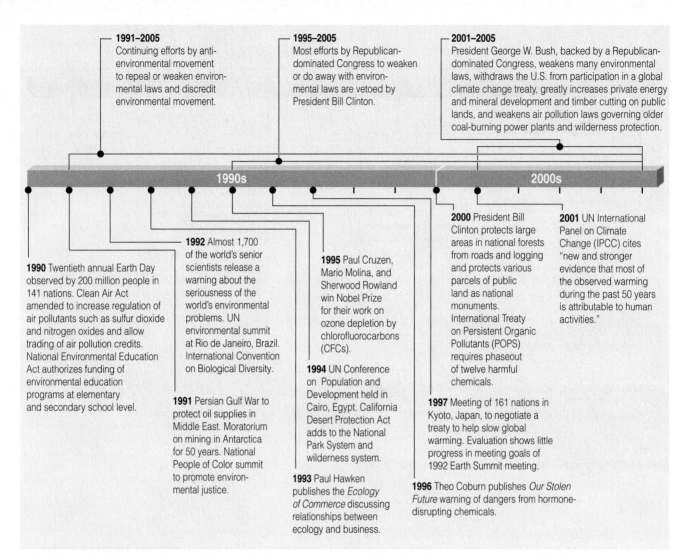

1991–2005 Continuing efforts by anti-environmental movement to repeal or weaken environmental laws and discredit environmental movement.

1995–2005 Most efforts by Republican-dominated Congress to weaken or do away with environmental laws are vetoed by President Bill Clinton.

2001–2005 President George W. Bush, backed by a Republican-dominated Congress, weakens many environmental laws, withdraws the U.S. from participation in a global climate change treaty, greatly increases private energy and mineral development and timber cutting on public lands, and weakens air pollution laws governing older coal-burning power plants and wilderness protection.

1990s   2000s

1990 Twentieth annual Earth Day observed by 200 million people in 141 nations. Clean Air Act amended to increase regulation of air pollutants such as sulfur dioxide and nitrogen oxides and allow trading of air pollution credits. National Environmental Education Act authorizes funding of environmental education programs at elementary and secondary school level.

1992 Almost 1,700 of the world's senior scientists release a warning about the seriousness of the world's environmental problems. UN environmental summit at Rio de Janeiro, Brazil. International Convention on Biological Diversity.

1991 Persian Gulf War to protect oil supplies in Middle East. Moratorium on mining in Antarctica for 50 years. National People of Color summit to promote environmental justice.

1993 Paul Hawken publishes the *Ecology of Commerce* discussing relationships between ecology and business.

1995 Paul Cruzen, Mario Molina, and Sherwood Rowland win Nobel Prize for their work on ozone depletion by chlorofluorocarbons (CFCs).

1994 UN Conference on Population and Development held in Cairo, Egypt. California Desert Protection Act adds to the National Park System and wilderness system.

2000 President Bill Clinton protects large areas in national forests from roads and logging and protects various parcels of public land as national monuments. International Treaty on Persistent Organic Pollutants (POPS) requires phaseout of twelve harmful chemicals.

1997 Meeting of 161 nations in Kyoto, Japan, to negotiate a treaty to help slow global warming. Evaluation shows little progress in meeting goals of 1992 Earth Summit meeting.

1996 Theo Coburn publishes *Our Stolen Future* warning of dangers from hormone-disrupting chemicals.

2001 UN International Panel on Climate Change (IPCC) cites "new and stronger evidence that most of the observed warming during the past 50 years is attributable to human activities."

1990–2005

Figure 6 Some important environmental events, 1990–2005.

SCIENCE SUPPLEMENT 4

BALANCING CHEMICAL EQUATIONS

Keeping Track of Atoms

In keeping with the law of conservation of matter, chemical equations are used as an accounting system to verify that no atoms are created or destroyed in a chemical reaction. As a consequence, each side of a chemical equation must have the same number of atoms of each element involved.

Ensuring that this condition is met leads to what chemists call a *balanced chemical equation.* The equation for the burning of carbon ($C + O_2 \longrightarrow CO_2$) is balanced because one atom of carbon and two atoms of oxygen are on both sides of the equation.

Consider the following chemical reaction: When electricity passes through water (H_2O), the latter can be broken down into hydrogen (H_2) and oxygen (O_2), as represented by the following equation:

$$H_2O \longrightarrow H_2 + O_2$$
2 H atoms   2 H atoms   2 O atoms
1 O atom

This equation is unbalanced because one atom of oxygen is on the left side of the equation but two atoms are on the right side.

We cannot change the subscripts of any of the formulas to balance this equation because that would change the arrangements of the atoms, leading to different substances. Instead, we must use different numbers of the molecules involved to balance the equation. For example, we could use two water molecules:

$$2 H_2O \longrightarrow H_2 + O_2$$
4 H atoms   2 H atoms   2 O atoms
2 O atoms

This equation is still unbalanced. Although the numbers of oxygen atoms on both sides of the equation are now equal, the numbers of hydrogen atoms are not.

We can correct this problem by having the reaction produce two hydrogen molecules:

$$2 H_2O \longrightarrow 2 H_2 + O_2$$
4 H atoms   4 H atoms   2 O atoms
2 O atoms

Now the equation is balanced, and the law of conservation of matter has been observed. For every two molecules of water through which we pass electricity, two hydrogen molecules and one oxygen molecule are produced.

If scientists and engineers can find economical ways to decompose water by using electricity or heat produced from solar energy, this reaction may be used as a way to produce hydrogen gas (H_2) for use as a fuel to help replace oil during this century. The hydrogen would be used in *fuel cells* where it would combine with oxygen gas to produce water and energy for heating houses and water and propelling motor vehicles and planes.

Bringing about such a *hydrogen revolution* would reduce the world's dependence on dwindling supplies of oil, eliminate most forms of air pollution because the major emission from a fuel cell is water vapor, and help slow global warming by not emitting the carbon dioxide gas that is released when any carbon-containing fuel is burned.

Practice Exercise

Balance the chemical equation for the reaction of nitrogen gas (N_2) with hydrogen gas (H_2) to form ammonia gas (NH_3).

Classifying Species

Biologists classify species into different *kingdoms*, on the basis of similarities and differences in characteristics such as their modes of nutrition, cell structure, appearance, and developmental features.

In this book, the earth's organisms are classified into six kingdoms: *eubacteria, archaebacteria, protists, fungi, plants,* and *animals.* Most bacteria, fungi, and protists are *microorganisms:* organisms so small that they cannot be seen with the naked eye.

Eubacteria consist of all single-celled prokaryotic bacteria except archaebacteria. Examples include various cyanobacteria and bacteria such as *Staphylococcus* and *Streptococcus.*

Archaebacteria are single-celled bacteria that are evolutionarily closer to eukaryotic cells than to eubacteria. Examples include methanogens that live in anaerobic sediments of lakes and swamps and in animal guts, halophiles that live in extremely salty water, and thermophiles that live in hot springs, hydrothermal vents, and acidic soil.

Protists (Protista) are mostly single-celled eukaryotic organisms such as diatoms, dinoflagellates, amoebas, golden brown and yellow-green algae, and protozoans. Some protists cause human diseases such as malaria and sleeping sickness.

Fungi are mostly many-celled, sometimes microscopic, eukaryotic organisms such as mushrooms, molds, mildews, and yeasts. Many fungi are decomposers. Other fungi kill various plants and cause huge losses of crops and valuable trees.

Plants (Plantae) are mostly many-celled eukaryotic organisms such as red, brown, and green algae and mosses, ferns, and flowering plants (whose flowers produce seeds that perpetuate the species). Some plants such as corn and marigolds are *annuals,* meaning that they complete their life cycles in one growing season. Others are *perennials,* which can live for more than 2 years, such as roses, grapes, elms, and magnolias.

Animals (Animalia) are also many-celled, eukaryotic organisms. Most have no backbones and hence are called *invertebrates.* Invertebrates include sponges, jellyfish, worms, arthropods (insects, shrimp, and spiders), mollusks (snails, clams, and octopuses), and echinoderms (sea urchins and sea stars). Insects play roles that are vital to our existence. *Vertebrates* (animals with backbones and a brain protected by skull bones) include fishes (sharks and tuna), amphibians (frogs and salamanders), reptiles (crocodiles and snakes), birds (eagles and robins), and mammals (bats, elephants, whales, and humans).

Naming Species

Within each kingdom, biologists have created subcategories based on anatomical, physiological, and behavioral characteristics. Kingdoms are divided into *phyla,* which are divided into subgroups called *classes.* Classes are subdivided into *orders,* which are further divided into *families.* Families consist of *genera* (singular, *genus*), and each genus contains one or more *species.* Note that the word *species* is both singular and plural. Figure 1 shows this detailed taxonomic classification for the current human species.

Most people call a species by its common name, such as robin or grizzly bear. Biologists use scientific names (derived from Latin) consisting of two parts (printed in italics, or underlined) to describe a species. The first word is the capitalized name (or abbreviation) for the genus to which the organism belongs. It is followed by a lowercase name that distinguishes the species from other members of the same genus. For example, the scientific name of the robin is *Turdus migratorius* (Latin for "migratory thrush") and the grizzly bear goes by the scientific name *Ursus horribilis* (Latin for "horrible bear").

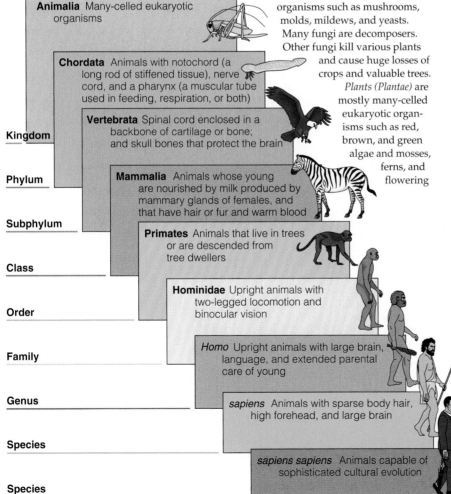

Figure 1 Taxonomic classification of the latest human species, *Homo sapiens sapiens.*

Animalia Many-celled eukaryotic organisms

Chordata Animals with notochord (a long rod of stiffened tissue), nerve cord, and a pharynx (a muscular tube used in feeding, respiration, or both)

Vertebrata Spinal cord enclosed in a backbone of cartilage or bone; and skull bones that protect the brain

Mammalia Animals whose young are nourished by milk produced by mammary glands of females, and that have hair or fur and warm blood

Primates Animals that live in trees or are descended from tree dwellers

Hominidae Upright animals with two-legged locomotion and binocular vision

Homo Upright animals with large brain, language, and extended parental care of young

sapiens Animals with sparse body hair, high forehead, and large brain

sapiens sapiens Animals capable of sophisticated cultural evolution

Kingdom

Phylum

Subphylum

Class

Order

Family

Genus

Species

Species

WEATHER BASICS

What Is Weather?

Weather is the result of the atmospheric conditions in a particular area.

Weather is an area's short-term atmospheric conditions—typically those occurring over hours or days. Examples of atmospheric conditions include temperature, pressure, moisture content, precipitation, sunshine, cloud cover, and wind direction and speed.

Meteorologists use equipment on weather balloons, aircraft, ships, and satellites, as well as radar and stationary sensors, to obtain data on weather variables. They then feed these data into computer models to draw weather maps. Other computer models project the weather for the next several days by calculating the probabilities that air masses, winds, and other factors will move and change in certain ways.

Much of the weather you experience results from interactions between the leading edges of moving masses of warm and cold air. Weather changes as one air mass replaces or meets another. The most dramatic changes in weather occur along a **front,** the boundary between two air masses with different temperatures and densities.

A **warm front** is the boundary between an advancing warm air mass and the cooler one it is replacing (Figure 1, top). Because warm air is less dense (weighs less per unit of volume) than cool air, an advancing warm front rises up over a mass of cool air.

As the warm front rises, its moisture begins condensing into droplets, forming layers of clouds at different altitudes. Gradually the clouds thicken, descend to a lower altitude, and often release their moisture as rainfall. A moist warm front can bring days of cloudy skies and drizzle.

A **cold front** (Figure 1, bottom) is the leading edge of an advancing mass of cold air. Because cold air is denser than warm air, an advancing cold front stays close to the ground and wedges underneath less dense warmer air. An approaching cold front produces rapidly moving, towering clouds called *thunderheads.*

As a cold front passes through, we may experience high surface winds and thunderstorms. After it leaves the area, we usually have cooler temperatures and a clear sky.

Near the top of the troposphere, hurricane-force winds circle the earth. These powerful winds, called *jet streams,* follow rising and falling paths that have a strong influence on weather patterns.

Highs and Lows

Weather is affected by up-and-down movements of air masses in conjunction with high and low atmospheric pressure.

Weather is also affected by changes in atmospheric pressure. *Air pressure* results from zillions of tiny molecules of gases (mostly nitrogen and oxygen) in the atmosphere zipping around at incredible speeds and hitting and bouncing off anything they encounter.

Atmospheric pressure is greater near the earth's surface because the molecules in the atmosphere are squeezed together under the weight of the air above them. An air mass with high pressure, called a **high,** contains cool, dense air that descends toward the earth's surface and becomes warmer. Fair weather follows as long as this high-pressure air mass remains over the area.

In contrast, a low-pressure air mass, called a **low,** produces cloudy and sometimes stormy weather. Because of its low pressure and low density, the center of a low rises, and its warm air expands and cools. When the temperature drops below a certain level where condensation takes place, called the *dew point,* moisture in the air condenses and forms clouds. If the

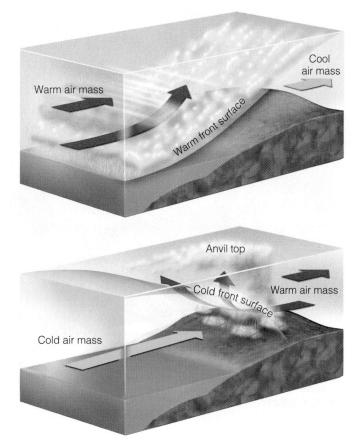

Figure 1 Natural capital: a *warm front* (top) arises when an advancing mass of warm air meets and rises up over a retreating mass of denser cool air. A *cold front* (bottom) forms when a mass of cold air wedges beneath a retreating mass of less dense warm air.

droplets in the clouds coalesce into large and heavy drops, then precipitation occurs. Recall that the condensation of water vapor into water drops usually requires that the air contain suspended tiny particles of material such as dust, smoke, sea salts, or volcanic ash. These so-called *condensation nuclei* provide surfaces on which the droplets of water can form and coalesce.

Tornadoes and Tropical Cyclones

Tornadoes and tropical storms are weather extremes that can cause tremendous damage but can sometimes have beneficial ecological effects.

Sometimes we experience *weather extremes.* Two examples are violent storms called *tornadoes* (which form over land) and *tropical cyclones* (which form over warm ocean waters and sometimes pass over coastal land).

Tornadoes or *twisters* are swirling funnel-shaped clouds that form over land. They can destroy houses and cause other serious damage in areas where they touch down on the earth's surface. The United States is the world's most tornado-prone country, followed by Australia.

Tornadoes in the plains of the Midwest usually occur when a large, dry, cold-air front moving southward from Canada runs into a large mass of humid air moving northward from the Gulf of Mexico. Most tornadoes occur in the spring when fronts of cold air from the north penetrate deeply into the midwestern plains.

As the large warm-air mass moves rapidly over the more dense cold-air mass, it rises rapidly and forms strong vertical convection currents that suck air upward, as shown in Figure 2. Scientists hypothesize that the rising vortex of air starts spinning because the air near the ground in the funnel is moving more slowly than the air above. This difference causes the air ahead of the advancing front to roll or spin in a vertically rising air mass or vortex.

Tropical cyclones are spawned by the formation of low-pressure cells of air over warm tropical seas. Figure 3 shows the formation and structure of a tropical cyclone. *Hurricanes* are tropical cyclones that form in the Atlantic Ocean; those forming in the Pacific Ocean usually are called *typhoons.*

Tropical cyclones take a long time to form and gain strength. As a result, meteo-

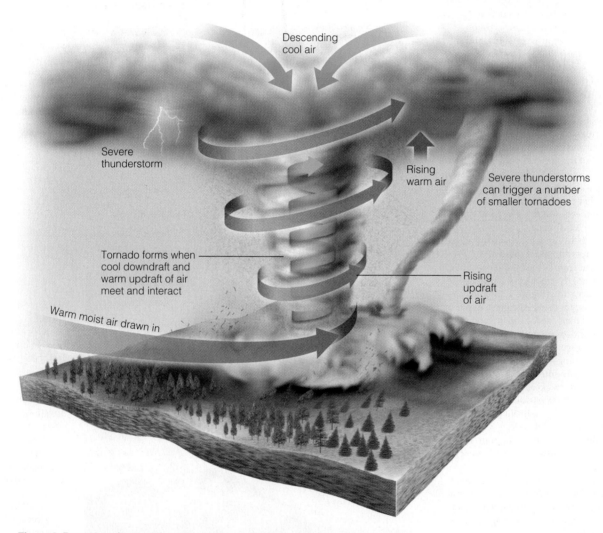

Figure 2 Formation of a *tornado* or *twister.* Although twisters can form at any time of the year, the most active tornado season in the United States is usually March through August. Meteorologists cannot tell us with great accuracy when and where most tornadoes will form.

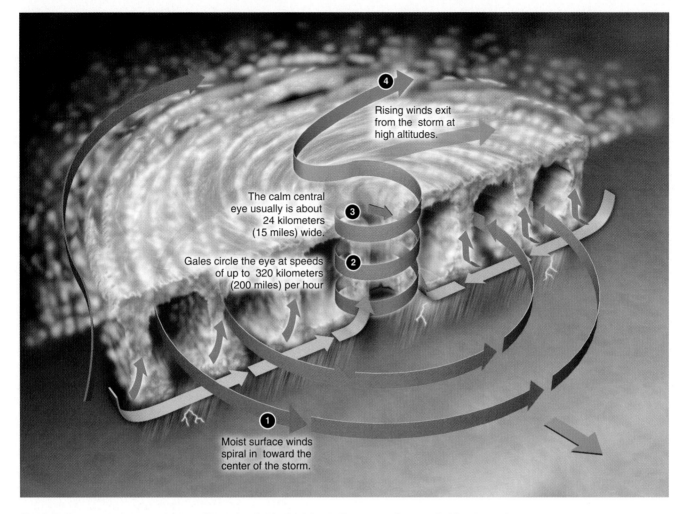

Rising winds exit from the storm at high altitudes.

The calm central eye usually is about 24 kilometers (15 miles) wide.

Gales circle the eye at speeds of up to 320 kilometers (200 miles) per hour

Moist surface winds spiral in toward the center of the storm.

Figure 3 Formation of a *tropical cyclone*. Those forming in the Atlantic Ocean usually are called *hurricanes;* those forming in the Pacific Ocean usually are called *typhoons*.

rologists can track their paths and wind speeds and warn people in areas likely to be hit by these violent storms.

Hurricanes and typhoons can kill and injure people and damage property and agricultural production. Sometimes, however, the long-term ecological and economic benefits of a tropical cyclone can exceed its short-term harmful effects.

For example, in parts of Texas along the Gulf of Mexico, coastal bays and marshes normally are closed off from freshwater and saltwater inflows. In August 1999, Hurricane Brett struck this coastal area. According to marine biologists, it flushed out excess nutrients from land runoff and swept dead sea grasses and rotting vegetation from the coastal bays and marshes. It also carved out 12 channels through the barrier islands along the coast, allowing huge quantities of fresh seawater to flood the bays and marshes.

This flushing out of the bays and marshes reduced brown tides consisting of explosive growths of algae feeding on excess nutrients. It also increased growth of sea grasses, which serve as nurseries for shrimp, crabs, and fish and provide food for millions of ducks wintering in Texas bays. Production of commercially important species of shellfish and fish also increased.

Earthquakes

Earthquakes occur when a part of the earth's crust suddenly fractures, shifts to relieve stress, and releases energy as shock waves.

Stress in the earth's crust can cause solid rock to deform until it suddenly fractures and shifts along the fracture, producing a fault (Figure 12-5, bottom, p.274). The faulting or a later abrupt movement on an existing fault causes an **earthquake** (Figure 1).

Relief of the earth's internal stress releases energy as shock waves, which move outward from the earthquake's focus like ripples in a pool of water. Scientists measure the severity of an earthquake by the *magnitude* of its shock waves. The magnitude is a measure of the amount of energy released in the earthquake, as indicated by the amplitude (size) of the vibrations when they reach a recording instrument (seismograph).

Scientists use the *Richter scale,* on which each unit represents an amplitude 10 times greater than the next smaller unit. Thus a magnitude 5.0 earthquake is 10 times greater than a magnitude 4.0 earthquake, and a magnitude 6.0 quake is 100 times greater than a magnitude 4.0 quake. Seismologists rate earthquakes as *insignificant* (less than 4.0 on the Richter scale), *minor* (4.0–4.9), *damaging* (5.0–5.9), *destructive* (6.0–6.9), *major* (7.0–7.9), and *great* (over 8.0).

Earthquakes often have *aftershocks* that gradually decrease in frequency over a period of as long as several months. Some also are preceded by *foreshocks* that occur from seconds to weeks before the main shock.

The *primary effects of earthquakes* include shaking and sometimes a permanent vertical or horizontal displacement of the ground. These effects may have serious consequences for people and for buildings, bridges, freeway overpasses, dams, and pipelines. An earthquake is a very large rock-and-roll event.

Secondary effects of earthquakes include rockslides, urban fires, and flooding caused by *subsidence* (sinking) of land. Coastal areas can be severely damaged by earthquakes at sea, which can generate huge water waves, called *tsunamis* (also called tidal waves, although they have nothing to do with tides), that travel as fast as 950 kilometers (590 miles) per hour. You cannot outrun a tsunami wave.

Large tsunamis start when part of the ocean bottom suddenly drops or rises, usually when a geological fault moves up or down and triggers huge waves. In December 2004, an earthquake in the Indian Ocean caused a large tsunami that killed 228,000 people (168,000 of them in Indonesia) and devastated large areas of Asia.

One way to reduce the loss of life and property damage from earthquakes is to examine historical records and make geologic measurements to locate active fault zones. We can also map high-risk areas, establish building codes that regulate the placement and design of buildings in such areas, and increase research geared toward predicting when and where earthquakes will occur. Then people can decide how high the risk might be and whether they want to accept that risk and live in areas subject to earthquakes.

Engineers know how to make homes, large buildings, bridges, and freeways more earthquake resistant. But this can be expensive, especially the reinforcement of existing structures.

Volcanoes

Some volcanoes erupt quietly with oozing flows of molten rock. Others erupt explosively and spew large chunks of lava rock, ash, and harmful gases into the atmosphere.

An active **volcano** occurs where magma (molten rock) reaches the earth's surface through a central vent or a long crack (*fissure;* Figure 2). Volcanic activity can release *ejecta* (debris ranging from large chunks of lava rock to glowing hot ash), liquid lava, and gases (such as water vapor, carbon dioxide, and sulfur dioxide) into the environment.

Volcanic activity is concentrated for the most part in the same areas as seismic activity. Some volcanoes erupt explosively and eject large quantities of gases and particulate matter (soot and mineral ash) high into the troposphere. Most of this soot and ash soon falls back to the earth's surface. Gases such as sulfur dioxide remain in the atmosphere, however, where they are converted to tiny droplets of sulfuric acid. This acid may remain above the clouds and not be washed out by rain for as long as 3 years. The tiny droplets reflect some of the sun's energy and can cool the atmosphere for 1–4 years.

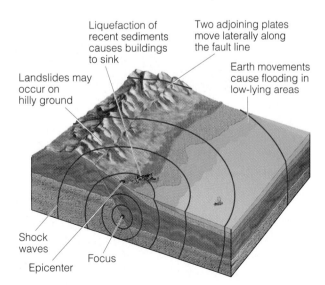

Liquefaction of recent sediments causes buildings to sink

Two adjoining plates move laterally along the fault line

Earth movements cause flooding in low-lying areas

Landslides may occur on hilly ground

Shock waves

Epicenter

Focus

Figure 1 Major features and effects of an *earthquake.*

Figure 2 A *volcano* erupts when molten magma in the partially molten asthenosphere rises in a plume through the lithosphere to erupt on the surface as lava that can spill over or be ejected into the atmosphere. Chains of islands can be created by eruptions of volcanoes that then become inactive.

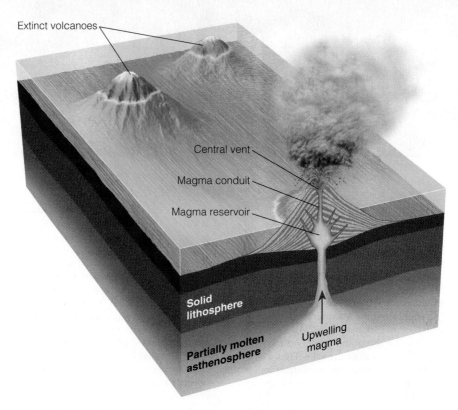

Extinct volcanoes

Central vent

Magma conduit

Magma reservoir

Solid lithosphere

Partially molten asthenosphere

Upwelling magma

Other volcanoes erupt more quietly. They involve primarily lava flows, which can cover roads and villages and ignite brush, trees, and homes.

We tend to think of volcanic activity as an undesirable event, but it does provide some benefits. For example, it creates outstanding scenery in the form of majestic mountains, some lakes (such as Crater Lake in Oregon; Figure 11-25, left, p. 257),

and other landforms. Perhaps the most important benefit of volcanism is the highly fertile soils produced by the weathering of lava.

We can reduce the loss of human life and sometimes property damage caused by volcanic eruptions in several ways. For example, we can use historical records and geologic measurements to identify high-risk areas so that people can try to avoid

living in them. We can also develop effective evacuation plans and measurements that warn us when volcanoes are likely to erupt.

Scientists continue to study the phenomena that precede an eruption. Examples include tilting or swelling of the cone, changes in magnetic and thermal properties of the volcano, changes in gas composition, and increased seismic activity.

SCIENCE SUPPLEMENT 8

BRIEF HISTORY OF THE AGE OF OIL

Some milestones in the Age of Oil:

- **1859:** First commercial oil well drilled near Titusville, Pennsylvania.

- **1905:** Oil supplies 10% of U.S. energy.

- **1925:** The United States produces 71% of the world's oil.

- **1930:** Because of an oil glut, oil sells for 10¢ per barrel.

- **1953:** U.S. oil companies account for about half of the world's oil production, and the United States is the world's leading oil exporter.

- **1955:** The United States has 20% of the world's estimated proven oil reserves.

- **1960:** OPEC is formed so that developing countries, with most of the world's known oil and projected oil reserves, can get a higher price for their oil.

- **1973:** The United States uses 30% of the world's oil, imports 36% of this oil, and has only 5% of the world's proven oil reserves.

- **1973–1974:** OPEC reduces oil imports to the West and bans oil exports to the United States because of its support for Israel in the 18-day Yom Kippur War with Egypt and Syria. World oil prices rise sharply and lead to double-digit inflation in the United States and many other countries and a global economic recession.

- **1975:** Production of estimated U.S. oil reserves peaks.

- **1979:** Iran's Islamic Revolution shuts down most of Iran's oil production and reduces world oil production.

- **1981:** The Iran–Iraq war pushes global oil prices to a historic high.

- **1983:** Facing an oil glut, OPEC cuts its oil prices.

- **1985:** U.S. domestic oil production begins to decline and is not expected to increase enough to affect the global price of oil or to reduce U.S. dependence on oil imports.

- **August 1990–June 1991:** The United States and its allies fight the Persian Gulf War to oust Iraqi invaders of Kuwait and to protect Western access to Saudi Arabian and Kuwaiti oil supplies.

- **2004:** The United States and a small number of allies fight a second Persian Gulf war to oust Sadam Hussein from power and to protect Western access to oil from Saudi Arabia, Kuwait, and Iraq.

- **2004:** OPEC has 67% of world oil reserves and produces 40% of the world's oil. The United States has only 2.9% of oil reserves, uses 26% of the world's oil production, and imports 62% of its oil.

- **2020:** The United States could be importing at least 70% of the oil it uses, as consumption continues to exceed production.

- **2010–2030:** Production of oil from the world's estimated oil reserves is expected to peak. Oil prices are expected to increase gradually as the demand for oil increasingly exceeds the supply—unless the world decreases its demand by wasting less energy and shifting to other sources of energy.

- **2010–2048:** Domestic U.S. oil reserves are projected to be 80% depleted.

- **2042–2083:** A gradual decline in dependence on oil is expected.

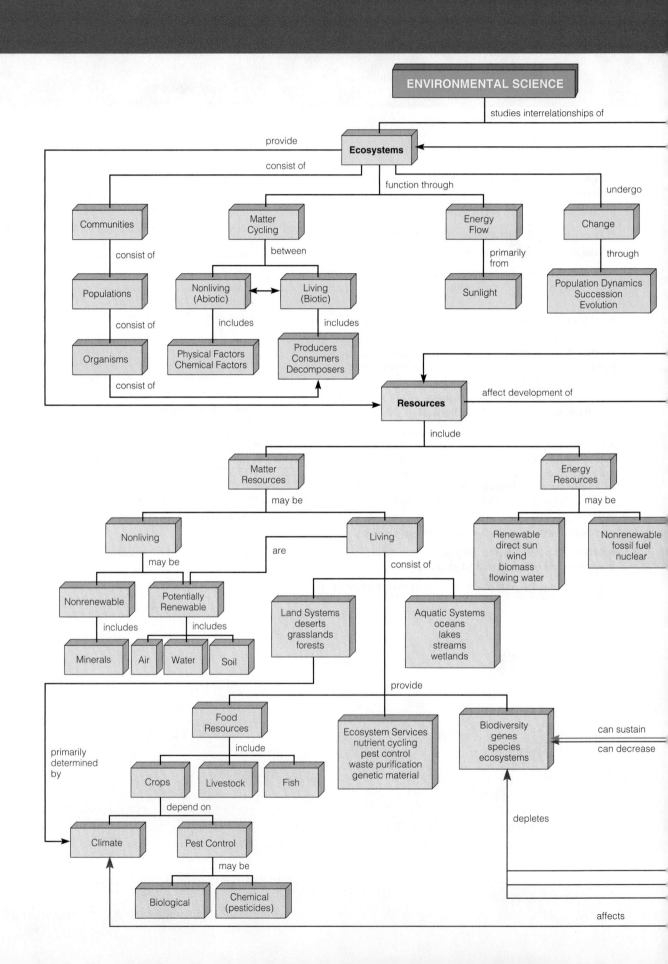

ENVIRONMENTAL SCIENCE: CONCEPTS AND CONNECTIONS

Developed by **Jane Heinze-Fry** with assistance from G. Tyler Miller, Jr.
(For assistance in creating your own concept maps, see the website for this book.)

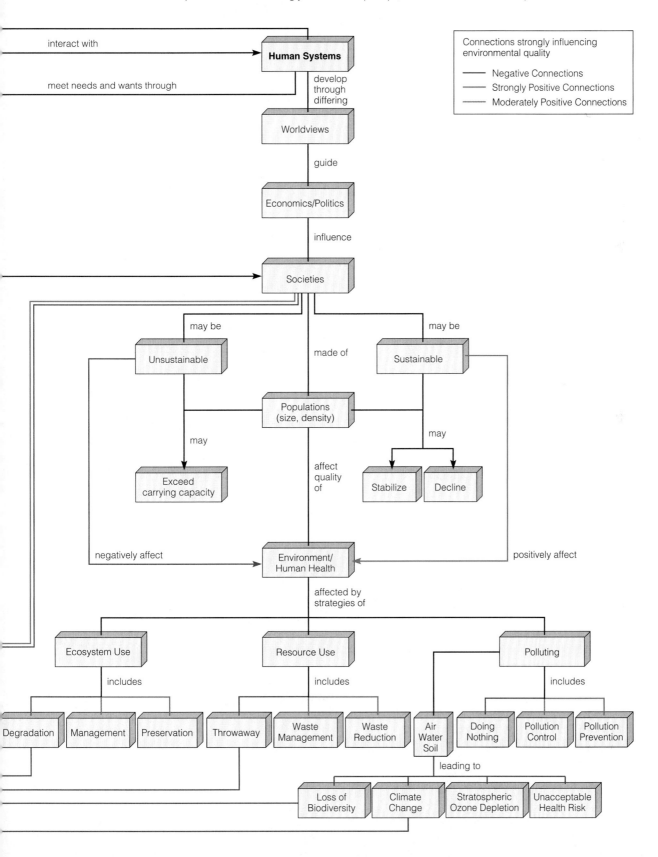

GLOSSARY

abiotic   Nonliving. Compare *biotic*.

acid   See *acid solution*.

acid deposition   The falling of acids and acid-forming compounds from the atmosphere to the earth's surface. Acid deposition is commonly known as *acid rain*, a term that refers to the wet deposition of droplets of acids and acid-forming compounds.

acid rain   See *acid deposition*.

acid solution   Any water solution that has more hydrogen ions (H$^+$) than hydroxide ions (OH$^-$); any water solution with a pH less than 7. Compare *basic solution, neutral solution*.

active solar heating system   System that uses solar collectors to capture energy from the sun and store it as heat for space heating and water heating. Liquid or air pumped through the collectors transfers the captured heat to a storage system, such as an insulated water tank or rock bed. Pumps or fans then distribute the stored heat or hot water throughout a dwelling as needed. Compare *passive solar heating system*.

adaptation   Any genetically controlled structural, physiological, or behavioral characteristic that helps an organism survive and reproduce under a given set of environmental conditions. It usually results from a beneficial mutation. See *biological evolution, differential reproduction, mutation, natural selection*.

adaptive management   Flexible management that views attempts to solve problems as experiments, analyzes failures to see what went wrong, and tries to modify and improve an approach before abandoning it. Because of the inherent unpredictability of complex systems, it often uses the precautionary principle as a management tool. See *precautionary principle*.

adaptive radiation   Process in which numerous new species evolve to fill vacant and new ecological niches in changed environments, usually after a mass extinction. Typically, this process takes millions of years.

adaptive trait   See *adaptation*.

aerobic respiration   Complex process that occurs in the cells of most living organisms,

in which nutrient organic molecules such as glucose (C$_6$H$_{12}$O$_6$) combine with oxygen (O$_2$) to produce carbon dioxide (CO$_2$), water (H$_2$O), and energy. Compare *photosynthesis*.

affluenza   Unsustainable addiction to overconsumption and materialism exhibited in the lifestyles of affluent consumers in the United States and other developed countries.

age structure   Percentage of the population (or number of people of each sex) at each age level in a population.

agricultural revolution   Gradual shift from small, mobile hunting and gathering bands to settled agricultural communities in which people survived by breeding and raising wild animals and cultivating wild plants near where they lived. It began 10,000–12,000 years ago. Compare *environmental revolution, hunter–gatherers, industrial–medical revolution, information and globalization revolution*.

agroforestry   Planting trees and crops together.

air pollution   One or more chemicals in high enough concentrations in the air to harm humans, other animals, vegetation, or materials. Excess heat and noise are also considered forms of air pollution. Such chemicals or physical conditions are called air pollutants. See *primary pollutant, secondary pollutant*.

albedo   Ability of a surface to reflect light.

alien species   See *nonnative species*.

allele   Slightly different molecular form found in a particular gene.

alley cropping   Planting of crops in strips with rows of trees or shrubs on each side.

alpha particle   Positively charged matter, consisting of two neutrons and two protons, that is emitted as radioactivity from the nuclei of some radioisotopes. See also *beta particle, gamma ray*.

altitude   Height above sea level. Compare *latitude*.

anaerobic respiration   Form of cellular respiration in which some decomposers

get the energy they need through the breakdown of glucose (or other nutrients) in the absence of oxygen. Compare *aerobic respiration*.

ancient forest   See *old-growth forest*.

animal manure   Dung and urine of animals used as a form of organic fertilizer. Compare *green manure*.

annual   Plant that grows, sets seed, and dies in one growing season. Compare *perennial*.

anthropocentric   Human-centered.

aquaculture   Growing and harvesting of fish and shellfish for human use in freshwater ponds, irrigation ditches, and lakes, or in cages or fenced-in areas of coastal lagoons and estuaries. See *fish farming, fish ranching*.

aquatic   Pertaining to water. Compare *terrestrial*.

aquatic life zone   Marine and freshwater portions of the biosphere. Examples include freshwater life zones (such as lakes and streams) and ocean or marine life zones (such as estuaries, coastlines, coral reefs, and the deep ocean).

aquifer   Porous, water-saturated layers of sand, gravel, or bedrock that can yield an economically significant amount of water.

arable land   Land that can be cultivated to grow crops.

area strip mining   Type of surface mining used where the terrain is flat. An earthmover strips away the overburden, and a power shovel digs a cut to remove the mineral deposit. The trench is then filled with overburden, and a new cut is made parallel to the previous one. The process is repeated over the entire site. Compare *mountaintop removal, open-pit mining, subsurface mining*.

arid   Dry. A desert or other area with an arid climate has little precipitation.

artificial selection   Process by which humans select one or more desirable genetic traits in the population of a plant or animal species and then use *selective breeding* to produce populations containing many individuals with the desired traits. Compare *genetic engineering, natural selection*.

asexual reproduction Reproduction in which a mother cell divides to produce two identical daughter cells that are clones of the mother cell. This type of reproduction is common in single-celled organisms. Compare *sexual reproduction*.

atmosphere The whole mass of air surrounding the earth. See *stratosphere, troposphere*.

atom Minute unit made of subatomic particles that is the basic building block of all chemical elements and thus all matter; the smallest unit of an element that can exist and still have the unique characteristics of that element. Compare *ion, molecule*.

atomic number Number of protons in the nucleus of an atom. Compare *mass number*.

autotroph See *producer*.

background extinction Normal extinction of various species as a result of changes in local environmental conditions. Compare *mass depletion, mass extinction*.

bacteria Prokaryotic, one-celled organisms. Some transmit diseases. Most act as decomposers and get the nutrients they need by breaking down complex organic compounds in the tissues of living or dead organisms into simpler inorganic nutrient compounds.

barrier islands Long, thin, low offshore islands of sediment that generally run parallel to the shore along some coasts.

basic solution Water solution with more hydroxide ions (OH^-) than hydrogen ions (H^+); water solution with a pH greater than 7. Compare *acid solution, neutral solution*.

benthos Bottom-dwelling organisms. Compare *decomposer, nekton, plankton*.

beta particle Swiftly moving electron emitted by the nucleus of a radioactive isotope. See also *alpha particle, gamma rays*.

bioaccumulation An increase in the concentration of a chemical in specific organs or tissues at a level higher than would normally be expected. Compare *biomagnification*.

biocentric Life-centered. Compare *anthropocentric*.

biodegradable Capable of being broken down by decomposers.

biodegradable pollutant Material that can be broken down into simpler substances (elements and compounds) by bacteria or other decomposers. Paper and most organic wastes such as animal manure are biodegradable but can take decades to biodegrade in modern landfills. Compare *degradable pollutant, nondegradable pollutant, slowly degradable pollutant*.

biodiversity Variety of different species (*species diversity*), genetic variability among individuals within each species (*genetic diversity*), variety of ecosystems (*ecological diversity*), and functions such as energy flow and matter cycling needed for the survival of species and biological communities (*functional diversity*).

biofuel Gas or liquid fuel (such as ethyl alcohol) made from plant material (biomass).

biogeochemical cycle Natural processes that recycle nutrients in various chemical forms from the nonliving environment to living organisms and then back to the nonliving environment. Examples include the carbon, oxygen, nitrogen, phosphorus, sulfur, and hydrologic cycles.

biological amplification See *biomagnification*.

biological community See *community*.

biological diversity See *biodiversity*.

biological evolution Change in the genetic makeup of a population of a species in successive generations. If continued long enough, it can lead to the formation of a new species. Note that populations—not individuals—evolve. See also *adaptation, differential reproduction, natural selection, theory of evolution*.

biological pest control Control of pest populations by natural predators, parasites, or disease-causing bacteria and viruses (pathogens).

biomagnification Increase in concentration of DDT, PCBs, and other slowly degradable, fat-soluble chemicals in organisms at successively higher trophic levels of a food chain or web. Compare *bioaccumulation*.

biomass Organic matter produced by plants and other photosynthetic producers; total dry weight of all living organisms that can be supported at each trophic level in a food chain or web; dry weight of all organic matter in plants and animals in an ecosystem; plant materials and animal wastes used as fuel.

biome Terrestrial regions inhabited by certain types of life, especially vegetation. Examples include various types of deserts, grasslands, and forests.

biopharming Use of genetically engineered animals to act as biofactories for producing drugs, vaccines, antibodies, hormones, industrial chemicals such as plastics and detergents, and human body organs.

biosphere Zone of the earth where life is found. It consists of parts of the atmosphere (the troposphere), hydrosphere (mostly surface water and groundwater), and lithosphere (mostly soil and surface rocks and sediments on the bottoms of oceans and other bodies of water) where life is found. Sometimes called the *ecosphere*.

biotic Living organisms. Compare *abiotic*.

biotic potential Maximum rate at which the population of a given species can increase when there are no limits on its rate of growth. See *environmental resistance*.

birth rate See *crude birth rate*.

bitumen Gooey, black, high-sulfur, heavy oil extracted from oil sand and then upgraded to synthetic fuel oil. See *oil sand*.

breeder nuclear fission reactor Nuclear fission reactor that produces more nuclear fuel than it consumes by converting nonfissionable uranium-238 into fissionable plutonium-239.

broadleaf deciduous plants Plants such as oak and maple trees that survive drought and cold by shedding their leaves and becoming dormant. Compare *broadleaf evergreen plants, coniferous evergreen plants*.

broadleaf evergreen plants Plants that keep most of their broad leaves year-round. Examples include the trees found in the canopies of tropical rain forests. Compare *broadleaf deciduous plants, coniferous evergreen plants*.

buffer Substance that can react with hydrogen ions in a solution and thus hold the acidity or pH of a solution fairly constant. See *pH*.

calorie Unit of energy; amount of energy needed to raise the temperature of 1 gram of water by 1°C (unit on Celsius temperature scale). See also *kilocalorie*.

cancer Group of more than 120 different diseases, one for each type of cell in the human body. Each type of cancer produces a tumor in which cells multiply uncontrollably and invade surrounding tissue.

carbon cycle Cyclic movement of carbon in different chemical forms from the environment to organisms and then back to the environment.

carcinogen Chemicals, ionizing radiation, and viruses that cause or promote the development of cancer. See *cancer*. Compare *mutagen, teratogen*.

carnivore Animal that feeds on other animals. Compare *herbivore, omnivore*.

carrying capacity (K) Maximum population of a particular species that a given habitat can support over a given period.

cell Smallest living unit of an organism. Each cell is encased in an outer membrane or wall and contains genetic material (DNA) and other parts to perform its life function. Organisms such as bacteria consist of only one cell, but most other organisms contain many cells.

CFCs See *chlorofluorocarbons*.

chain reaction Multiple nuclear fissions, taking place within a certain mass of a fissionable isotope, that release an enormous amount of energy in a short time.

chemical One of the millions of different elements and compounds found naturally and synthesized by humans. See *compound, element*.

chemical change Interaction between chemicals in which the chemical composition of the elements or compounds involved changes. Compare *nuclear change, physical change*.

chemical evolution Formation of the earth and its early crust and atmosphere, evolution of the biological molecules necessary for life, and evolution of systems of chemical reactions needed to produce the first living cells. These processes are believed to have occurred about 1 billion years before biological evolution. Compare *biological evolution*.

chemical formula Shorthand way to show the number of atoms (or ions) in the basic structural unit of a compound. Examples include H_2O, $NaCl$, and $C_6H_{12}O_6$.

chemical reaction See *chemical change*.

chemosynthesis Process in which certain organisms (mostly specialized bacteria) extract inorganic compounds from their environment and convert them into organic nutrient compounds without the presence of sunlight. Compare *photosynthesis*.

chlorinated hydrocarbon Organic compound made up of atoms of carbon, hydrogen, and chlorine. Examples include DDT and PCBs.

chlorofluorocarbons (CFCs) Organic compounds made up of atoms of carbon, chlorine, and fluorine. An example is Freon-12 (CCl_2F_2), which is used as a refrigerant in refrigerators and air conditioners and in making plastics such as Styrofoam. Gaseous CFCs can deplete the ozone layer when they slowly rise into the stratosphere and their chlorine atoms react with ozone molecules. Their use is being phased out.

chromosome A grouping of genes and associated proteins in plant and animal cells that carry certain types of genetic information. See *genes*.

clear-cutting Method of timber harvesting in which all trees in a forested area are removed in a single cutting. Compare *seed-tree cutting, selective cutting, shelterwood cutting, strip cutting*.

climate Physical properties of the troposphere of an area based on analysis of its weather records over a long period (at least 30 years). The two main factors determining an area's climate are the *temperature*, with its seasonal variations, and the amount and distribution of *precipitation*. Compare *weather*.

climax community See *mature community*.

coal Solid, combustible mixture of organic compounds with 30–98% carbon by weight, mixed with various amounts of water and small amounts of sulfur and nitrogen compounds. It forms in several stages as the remains of plants are subjected to heat and pressure over millions of years.

coal gasification Conversion of solid coal to synthetic natural gas (SNG).

coal liquefaction Conversion of solid coal to a liquid hydrocarbon fuel such as synthetic gasoline or methanol.

coastal wetland Land along a coastline, extending inland from an estuary that is covered with salt water all or part of the year. Examples include marshes, bays, lagoons, tidal flats, and mangrove swamps. Compare *inland wetland*.

coastal zone Warm, nutrient-rich, shallow part of the ocean that extends from the high-tide mark on land to the edge of a shelflike extension of continental landmasses known as the continental shelf. Compare *open sea*.

coevolution Evolution in which two or more species interact and exert selective pressures on each other that can lead each species to undergo adaptations. See *evolution, natural selection*.

cogeneration Production of two useful forms of energy, such as high-temperature heat or steam and electricity, from the same fuel source.

cold front Leading edge of an advancing mass of cold air. Compare *warm front*.

commensalism An interaction between organisms of different species in which one type of organism benefits and the other type is neither helped nor harmed to any great degree. Compare *mutualism*.

commercial extinction Depletion of the population of a wild species used as a resource to a level at which it is no longer profitable to harvest the species.

commercial inorganic fertilizer Commercially prepared mixture of plant nutrients such as nitrates, phosphates, and potassium applied to the soil to restore fertility and increase crop yields. Compare *organic fertilizer*.

common-property resource Resource that people normally are free to use; each user can deplete or degrade the available supply. Most such resources are renewable and owned by no one. Examples include clean air, fish in parts of the ocean not under the control of a coastal country, migratory birds, gases of the lower atmosphere, and the ozone content of the upper atmosphere (stratosphere). See *tragedy of the commons*.

community Populations of all species living and interacting in an area at a particular time.

competition Two or more individual organisms of a single species (*intraspecific competition*) or two or more individuals of different species (*interspecific competition*) attempting to use the same scarce resources in the same ecosystem.

compost Partially decomposed organic plant and animal matter used as a soil conditioner or fertilizer.

compound Combination of atoms, or oppositely charged ions, of two or more elements held together by attractive forces called chemical bonds. Compare *element*.

concentration Amount of a chemical in a particular volume or weight of air, water, soil, or other medium.

condensation nuclei Tiny particles on which droplets of water vapor can collect.

coniferous evergreen plants Cone-bearing plants (such as spruces, pines, and firs) that keep some of their narrow, pointed leaves (needles) all year. Compare *broadleaf deciduous plants, broadleaf evergreen plants*.

coniferous trees Cone-bearing trees, mostly evergreens, that have needle-shaped or scalelike leaves. They produce wood known commercially as softwood. Compare *deciduous plants*.

consensus science See *sound science*.

conservation Sensible and careful use of natural resources by humans. People with this view are called *conservationists*.

conservation biology Multidisciplinary science created to deal with the crisis of maintaining the genes, species, communities, and ecosystems that make up earth's biological diversity. Its goals are to investigate human impacts on biodiversity and to develop practical approaches to preserving biodiversity.

conservationist Person concerned with using natural areas and wildlife in ways that sustain them for current and future generations of humans and other forms of life.

conservation-tillage farming Crop cultivation in which the soil is disturbed little (minimum-tillage farming) or not at all (no-till farming) in an effort to reduce soil erosion, lower labor costs, and save energy. Compare *conventional-tillage farming*.

constancy Ability of a living system, such as a population, to maintain a certain size.

consumer Organism that cannot synthesize the organic nutrients it needs and gets its organic nutrients by feeding on the tissues of producers or of other consumers; generally divided into *primary consumers* (herbivores), *secondary consumers* (carnivores), *tertiary (higher-level) consumers, omnivores,* and *detritivores* (decomposers and detritus feeders). In economics, one who uses economic goods.

contour farming Plowing and planting across the changing slope of land, rather than in straight lines, to help retain water and reduce soil erosion.

contour strip mining Form of surface mining used on hilly or mountainous terrain. A power shovel cuts a series of terraces into the side of a hill. An earthmover removes the overburden, and a power shovel extracts the coal. The overburden from each new terrace is dumped onto the one below. Compare *area strip mining, dredging, mountaintop removal, open-pit mining, subsurface mining.*

controlled burning Deliberately set, carefully controlled surface fires that reduce flammable litter and decrease the chances of damaging crown fires. See *ground fire, surface fire.*

conventional-tillage farming Crop cultivation method in which a planting surface is made by plowing land, breaking up the exposed soil, and then smoothing the surface. Compare *conservation-tillage farming.*

convergent plate boundary Area where the earth's lithospheric plates are pushed together. See *subduction zone.* Compare *divergent plate boundary, transform fault.*

coral reef Formation produced by massive colonies containing billions of tiny coral animals, called polyps, that secrete a stony substance (calcium carbonate) around themselves for protection. When the corals die, their empty outer skeletons form layers and cause the reef to grow. Coral reefs are found in the coastal zones of warm tropical and subtropical oceans.

core Inner zone of the earth. It consists of a solid inner core and a liquid outer core. Compare *crust, mantle.*

critical mass Amount of fissionable nuclei needed to sustain a nuclear fission chain reaction.

crop rotation Planting a field, or an area of a field, with different crops from year to year to reduce soil nutrient depletion. A plant such as corn, tobacco, or cotton, which removes large amounts of nitrogen from the soil, is planted one year. The next year a legume such as soybeans, which adds nitrogen to the soil, is planted.

crown fire Extremely hot forest fire that burns ground vegetation and treetops. Compare *controlled burning, ground fire, surface fire.*

crude birth rate Annual number of live births per 1,000 people in the population of a geographic area at the midpoint of a given year. Compare *crude death rate.*

crude death rate Annual number of deaths per 1,000 people in the population of a geographic area at the midpoint of a given year. Compare *crude birth rate.*

crude oil Gooey liquid consisting mostly of hydrocarbon compounds and small amounts of compounds containing oxygen, sulfur, and nitrogen. Extracted from underground accumulations, it is sent to oil refineries, where it is converted to heating oil, diesel fuel, gasoline, tar, and other materials.

crust Solid outer zone of the earth. It consists of oceanic crust and continental crust. Compare *core, mantle.*

cultural eutrophication Overnourishment of aquatic ecosystems with plant nutrients (mostly nitrates and phosphates) because of human activities such as agriculture, urbanization, and discharges from industrial plants and sewage treatment plants. See *eutrophication.*

DDT Dichlorodiphenyltrichloroethane, a chlorinated hydrocarbon that has been widely used as an insecticide but is now banned in some countries.

death rate See *crude death rate.*

debt-for-nature swap Agreement in which a certain amount of foreign debt is canceled in exchange for local currency investments that will improve natural resource management or protect certain areas in the debtor country from harmful development.

deciduous plants Trees, such as oaks and maples, and other plants that survive during dry seasons or cold seasons by shedding their leaves. Compare *coniferous trees, succulent plants.*

decomposer Organism that digests parts of dead organisms and cast-off fragments and wastes of living organisms by breaking down the complex organic molecules in those materials into simpler inorganic compounds and then absorbing the soluble nutrients. Producers return most of these chemicals to the soil and water for reuse. Decomposers consist of various bacteria and fungi. Compare *consumer, detritivore, producer.*

deforestation Removal of trees from a forested area without adequate replanting.

degradable pollutant Potentially polluting chemical that is broken down com-

pletely or reduced to acceptable levels by natural physical, chemical, and biological processes. Compare *biodegradable pollutant, nondegradable pollutant, slowly degradable pollutant.*

degree of urbanization Percentage of the population in the world, or a country, living in areas with a population of more than 2,500 people (higher in some countries). Compare *urban growth.*

democracy Government by the people through their elected officials and appointed representatives. In a *constitutional democracy,* a constitution provides the basis of government authority and puts restraints on government power through free elections and freely expressed public opinion.

demographic transition Hypothesis that countries, as they become industrialized, have declines in death rates followed by declines in birth rates.

depletion time The time it takes to use a certain fraction (usually 80%) of the known or estimated supply of a nonrenewable resource at an assumed rate of use. Finding and extracting the remaining 20% usually costs more than it is worth.

desalination Purification of salt water or brackish (slightly salty) water by removal of dissolved salts.

desert Biome in which evaporation exceeds precipitation and the average amount of precipitation is less than 25 centimeters (10 inches) per year. Such areas have little vegetation or have widely spaced, mostly low vegetation. Compare *forest, grassland.*

desertification Conversion of rangeland, rain-fed cropland, or irrigated cropland to desertlike land, with a drop in agricultural productivity of 10% or more. It usually is caused by a combination of overgrazing, soil erosion, prolonged drought, and climate change.

detritivore Consumer organism that feeds on detritus, parts of dead organisms, and cast-off fragments and wastes of living organisms. The two principal types are *detritus feeders* and *decomposers.*

detritus Parts of dead organisms and cast-off fragments and wastes of living organisms.

detritus feeder Organism that extracts nutrients from fragments of dead organisms and their cast-off parts and organic wastes. Examples include earthworms, termites, and crabs. Compare *decomposer.*

deuterium (D; hydrogen-2) Isotope of the element hydrogen, with a nucleus containing one proton and one neutron and a mass number of 2.

developed country Country that is highly industrialized and has a high per capita GNP. Compare *developing country.*

developing country Country that has low to moderate industrialization and low to moderate per capita GNP. Most are located in Africa, Asia, and Latin America. Compare *developed country.*

dieback Sharp reduction in the population of a species when its numbers exceed the carrying capacity of its habitat. See *carrying capacity.*

differential reproduction Phenomenon in which individuals with adaptive genetic traits produce more living offspring than do individuals without such traits. See *natural selection.*

dioxins Family of 75 chlorinated hydrocarbon compounds formed as unwanted by-products in chemical reactions involving chlorine and hydrocarbons, usually at high temperatures.

dissolved oxygen (DO) content Amount of oxygen gas (O_2) dissolved in a given volume of water at a particular temperature and pressure, often expressed as a concentration in parts of oxygen per million parts of water.

disturbance A discrete event that disrupts an ecosystem or community. Examples of *natural disturbances* include fires, hurricanes, tornadoes, droughts, and floods. Examples of *human-caused disturbances* include deforestation, overgrazing, and plowing.

divergent plate boundary Area where the earth's lithospheric plates move apart in opposite directions. Compare *convergent plate boundary, transform fault.*

DNA (deoxyribonucleic acid) Large molecules in the cells of organisms that carry genetic information in living organisms.

domesticated species Wild species tamed or genetically altered by crossbreeding for use by humans for food (cattle, sheep, and food crops), pets (dogs and cats), or enjoyment (animals in zoos and plants in gardens). Compare *wild species.*

dose The amount of a potentially harmful substance an individual ingests, inhales, or absorbs through the skin. Compare *response.* See *dose-response curve, median lethal dose.*

dose-response curve Plot of data showing the effects of various doses of a toxic agent on a group of test organisms. See *dose, median lethal dose, response.*

doubling time The time it takes (usually in years) for the quantity of something growing exponentially to double. It can be calculated by dividing the annual percentage growth rate into 70.

drainage basin See *watershed.*

drift-net fishing Catching fish in huge nets that drift in the water.

drought Condition in which an area does not get enough water because of lower-than-normal precipitation or higher-than-normal temperatures that increase evaporation.

earthquake Shaking of the ground resulting from the fracturing and displacement of rock, which produces a fault, or from subsequent movement along the fault.

ecological diversity The variety of forests, deserts, grasslands, oceans, streams, lakes, and other biological communities interacting with one another and with their nonliving environment. See *biodiversity.* Compare *functional diversity, genetic diversity, species diversity.*

ecological efficiency Percentage of energy transferred from one trophic level to another in a food chain or web.

ecological footprint Amount of biologically productive land and water needed to supply a population with the renewable resources it uses and to absorb or dispose of the wastes from such resource use. It measures the average environmental impact of populations in different countries and areas. See *per capita ecological footprint*

ecological niche Total way of life or role of a species in an ecosystem. It includes all physical, chemical, and biological conditions that a species needs to live and reproduce in an ecosystem. See *fundamental niche, realized niche.*

ecological restoration Deliberate alteration of a degraded habitat or ecosystem to restore as much of its ecological structure and function as possible.

ecological succession Process in which communities of plant and animal species in a particular area are replaced over time by a series of different and often more complex communities. See *primary succession, secondary succession.*

ecologist Biological scientist who studies relationships between living organisms and their environment.

ecology Study of the interactions of living organisms with one another and with their nonliving environment of matter and energy; study of the structure and functions of nature.

economic decision Deciding which goods and services to produce, how to produce them, how much to produce, and how to distribute them to people.

economic depletion Exhaustion of 80% of the estimated supply of a nonrenewable resource. Finding, extracting, and processing the remaining 20% usually costs more

than it is worth. May also apply to the depletion of a renewable resource, such as a fish or tree species.

economic development Improvement of living standards by economic growth. Compare *economic growth, environmentally sustainable economic development.*

economic growth Increase in the capacity to provide people with goods and services produced by an economy; an increase in gross domestic product (GDP). Compare *economic development, environmentally sustainable economic development.* See *gross domestic product.*

economic resources Natural resources, capital goods, and labor used in an economy to produce material goods and services. See *natural resources.*

economic system Method that a group of people uses to choose which goods and services to produce, how to produce them, how much to produce, and how to distribute them to people.

economy System of production, distribution, and consumption of economic goods.

ecosphere See *biosphere.*

ecosystem Community of different species interacting with one another and with the chemical and physical factors making up its nonliving environment.

ecosystem services Natural services or natural capital that support life on the earth and are essential to the quality of human life and the functioning of the world's economies. See *natural resources.*

electromagnetic radiation Forms of kinetic energy traveling as electromagnetic waves. Examples include radio waves, TV waves, microwaves, infrared radiation, visible light, ultraviolet radiation, X rays, and gamma rays. Compare *ionizing radiation, nonionizing radiation.*

electron (e) Tiny particle moving around outside the nucleus of an atom. Each electron has one unit of negative charge and almost no mass. Compare *neutron, proton.*

element Chemical, such as hydrogen (H), iron (Fe), sodium (Na), carbon (C), nitrogen (N), or oxygen (O), whose distinctly different atoms serve as the basic building blocks of all matter. Two or more elements combine to form the compounds that make up most of the world's matter. Compare *compound.*

endangered species A wild species with so few individual survivors that the species could soon become extinct in all or most of its natural range. Compare *threatened species.*

endemic species Species that is found in only one area. Such species are especially vulnerable to extinction.

energy   Capacity to do work by performing mechanical, physical, chemical, or electrical tasks or to cause a heat transfer between two objects at different temperatures.

energy efficiency   Percentage of the total energy input that does useful work and is not converted into low-quality, generally useless heat in an energy conversion system or process. See *energy quality, net energy*. Compare *material efficiency*.

energy productivity   See *energy efficiency*.

energy quality   Ability of a form of energy to do useful work. High-temperature heat and the chemical energy in fossil fuels and nuclear fuels are concentrated high-quality energy. Low-quality energy, such as low-temperature heat, is dispersed or diluted and cannot do much useful work. See *high-quality energy, low-quality energy*.

environment   All external conditions and factors, living and nonliving (chemicals and energy), that affect an organism or other specified system during its lifetime.

environmental degradation   Depletion or destruction of a potentially renewable resource such as soil, grassland, forest, or wildlife that is used faster than it is naturally replenished. If such use continues, the resource becomes nonrenewable (on a human time scale) or nonexistent (extinct). See also *sustainable yield*.

environmental ethics   Human beliefs about what is right or wrong environmental behavior.

environmentalism   A social movement dedicated to protecting the earth's life support systems for us and other species.

environmentalist   Person who is concerned about the impact of people on environmental quality and believes that some human actions are degrading parts of the earth's life-support systems for humans and many other forms of life.

environmental justice   Fair treatment and meaningful involvement of all people regardless of race, color, sex, national origin, or income with respect to the development, implementation, and enforcement of environmental laws, regulations, and policies.

environmental policy   Laws, rules, and regulations related to an environmental problem that are developed, implemented, and enforced by a particular government agency.

environmentally sustainable economic development   Development that *encourages* forms of economic growth that meet the basic needs of the current generations of humans and other species without preventing future generations of humans and other species from meeting their basic needs and

discourages environmentally harmful and unsustainable forms of economic growth. It is the economic component of an *environmentally sustainable society*. Compare *economic development, economic growth*.

environmentally sustainable society   Society that satisfies the basic needs of its people without depleting or degrading its natural resources and thereby preventing current and future generations of humans and other species from meeting their basic needs.

environmental resistance   All of the limiting factors that act together to limit the growth of a population. See *biotic potential, limiting factor*.

environmental revolution   Cultural change involving halting population growth and altering lifestyles, political and economic systems, and the way we treat the environment so that we can help sustain the earth for ourselves and other species. It requires working with the rest of nature by learning more about how nature sustains itself. See *environmental wisdom worldview*.

environmental science   Interdisciplinary study that uses information from the physical sciences and social sciences to learn how the earth works, how we interact with the earth, and how to deal with environmental problems.

environmental scientist   Scientist who uses information from the physical sciences and social sciences to understand how the earth works, learn how humans interact with the earth, and develop solutions to environmental problems.

environmental wisdom worldview   Beliefs that (1) nature exists for all of the earth's species, not just for humans, and we are not in charge of the rest of nature; (2) there is not always more, and it is not all for us; (3) some forms of economic growth are beneficial and some are harmful, and our goals should be to design economic and political systems that encourage earth-sustaining forms of growth and discourage or prohibit earth-degrading forms; and (4) our success depends on learning to cooperate with one another and with the rest of nature instead of trying to dominate and manage earth's life-support systems primarily for our own use. Compare *planetary management worldview, stewardship worldview*.

environmental worldview   How people think the world works, what they think their role in the world should be, and what they believe is right and wrong environmental behavior (environmental ethics).

EPA   U.S. Environmental Protection Agency; responsible for managing federal efforts to control air and water pollution,

radiation and pesticide hazards, environmental research, hazardous waste, and solid waste disposal.

epidemiology   Study of the patterns of disease or other harmful effects from toxic exposure within defined groups of people to find out why some people get sick and some do not.

epiphyte   Plant that uses its roots to attach itself to branches high in trees, especially in tropical forests.

erosion   Process or group of processes by which loose or consolidated earth materials are dissolved, loosened, or worn away and removed from one place and deposited in another. See *weathering*.

estuary   Partially enclosed coastal area at the mouth of a river where its fresh water, carrying fertile silt and runoff from the land, mixes with salty seawater.

euphotic zone   Upper layer of a body of water through which sunlight can penetrate and support photosynthesis.

eutrophication   Physical, chemical, and biological changes that take place after a lake, estuary, or slow-flowing stream receives inputs of plant nutrients—mostly nitrates and phosphates—from natural erosion and runoff from the surrounding land basin. See *cultural eutrophication*.

eutrophic lake   Lake with a large or excessive supply of plant nutrients, mostly nitrates and phosphates. Compare *mesotrophic lake, oligotrophic lake*.

evaporation   Conversion of a liquid into a gas.

even-aged management   Method of forest management in which trees, sometimes of a single species in a given stand, are maintained at roughly the same age and size and are harvested all at once. Compare *uneven-aged management*.

evergreen plants   Plants that keep some of their leaves or needles throughout the year. Examples include ferns and cone-bearing trees (conifers) such as firs, spruces, pines, redwoods, and sequoias. Compare *deciduous plants, succulent plants*.

evolution   See *biological evolution*.

exhaustible resource   See *nonrenewable resource*.

exotic species   See *nonnative species*.

experiment   Procedure a scientist uses to study some phenomenon under known conditions. Scientists conduct some experiments in the laboratory and others in nature. The resulting scientific data or facts must be verified or confirmed by repeated observations and measurements, ideally by several different investigators.

exponential growth Growth in which some quantity, such as population size or economic output, increases at a constant rate per unit of time. An example is the growth sequence 2, 4, 8, 16, 32, 64, and so on. When the increase in quantity over time is plotted, this type of growth yields a curve shaped like the letter J. Compare *linear growth*.

external benefit Beneficial social effect of producing and using an economic good that is not included in the market price of the good. Compare *external cost, full cost*.

external cost Harmful social effect of producing and using an economic good that is not included in the market price of the good. Compare *external benefit, full cost, internal cost*.

externalities Social benefits ("goods") and social costs ("bads") not included in the market price of an economic good. See *external benefit, external cost*. Compare *full cost, internal cost*.

extinction Complete disappearance of a species from the earth. It happens when a species cannot adapt and successfully reproduce under new environmental conditions or when a species evolves into one or more new species. Compare *speciation*. See also *endangered species, mass depletion, mass extinction, threatened species*.

family planning Providing information, clinical services, and contraceptives to help people choose the number and spacing of children they want to have.

famine Widespread malnutrition and starvation in a particular area because of a shortage of food, usually caused by drought, war, flood, earthquake, or other catastrophic events that disrupt food production and distribution.

feedlot Confined outdoor or indoor space used to raise hundreds to thousands of domesticated livestock. Compare *rangeland*.

fermentation See *anaerobic respiration*.

fertilizer Substance that adds inorganic or organic plant nutrients to soil and improves its ability to grow crops, trees, or other vegetation. See *commercial inorganic fertilizer, organic fertilizer*.

first law of thermodynamics In any physical or chemical change, no detectable amount of energy is created or destroyed, but energy can be changed from one form to another; you cannot get more energy out of something than you put in; in terms of energy quantity, you cannot get something for nothing (there is no free lunch). This law does not apply to nuclear changes, in which energy can be produced from small amounts of matter. See *second law of thermodynamics*.

fishery Concentrations of particular aquatic species suitable for commercial harvesting in a given ocean area or inland body of water.

fish farming Form of aquaculture in which fish are cultivated in a controlled pond or other environment and harvested when they reach the desired size. See also *fish ranching*.

fish ranching Form of aquaculture in which members of a fish species such as salmon are held in captivity for the first few years of their lives, released, and then harvested as adults when they return from the ocean to their freshwater birthplace to spawn. See also *fish farming*.

fissionable isotope Isotope that can split apart when hit by a neutron at the right speed and thus undergo nuclear fission. Examples include uranium-235 and plutonium-239.

floodplain Flat valley floor next to a stream channel. For legal purposes, the term often applies to any low area that has the potential for flooding, including certain coastal areas.

flows See *throughput*.

flyway Generally fixed route along which waterfowl migrate from one area to another at certain seasons of the year.

food chain Series of organisms in which each eats or decomposes the preceding one. Compare *food web*.

food web Complex network of many interconnected food chains and feeding relationships. Compare *food chain*.

forest Biome with enough average annual precipitation (at least 76 centimeters, or 30 inches) to support the growth of tree species and smaller forms of vegetation. Compare *desert, grassland*.

fossil fuel Products of partial or complete decomposition of plants and animals that occur as crude oil, coal, natural gas, or heavy oils as a result of exposure to heat and pressure in the earth's crust over millions of years. See *coal, crude oil, natural gas*.

fossils Skeletons, bones, shells, body parts, leaves, seeds, or impressions of such items that provide recognizable evidence of organisms that lived long ago.

foundation species A species that plays a major role in shaping communities by creating and enhancing a habitat that benefits other species. Compare *indicator species, keystone species, native species, nonnative species*.

free-access resource See *common-property resource*.

Freons See *chlorofluorocarbons*.

freshwater life zones Aquatic systems where water with a dissolved salt concentration of less than 1% by volume accumulates on or flows through the surfaces of terrestrial biomes. Examples include *standing* (lentic) bodies of fresh water such as lakes, ponds, and inland wetlands and *flowing* (lotic) systems such as streams and rivers. Compare *biome*.

front The boundary between two air masses with different temperatures and densities. See *cold front, warm front*.

frontier science Preliminary scientific data, hypotheses, and models that have not been widely tested and accepted. Compare *junk science, sound science*.

frontier worldview View by European colonists settling North America in the 1600s that the continent had vast resources and was a wilderness to be conquered by settlers clearing and planting land.

full cost Cost of a good when its internal costs and its estimated short- and long-term external costs are included in its market price. Compare *external cost, internal cost*.

functional diversity Biological and chemical processes or functions such as energy flow and matter cycling needed for the survival of species and biological communities. See *biodiversity, ecological diversity, genetic diversity, species diversity*.

fundamental niche The full potential range of the physical, chemical, and biological factors a species can use if it does not face any competition from other species. See *ecological niche*. Compare *realized niche*.

fungicide Chemical that kills fungi.

game species Type of wild animal that people hunt or fish for, for sport and recreation and sometimes for food.

gamma ray A form of ionizing electromagnetic radiation with a high energy content emitted by some radioisotopes. It readily penetrates body tissues. See also *alpha particle, beta particle*.

GDP See *gross domestic product*.

gene mutation See *mutation*.

gene pool The sum total of all genes found in the individuals of the population of a particular species.

generalist species Species with a broad ecological niche. They can live in many different places, eat a variety of foods, and tolerate a wide range of environmental conditions. Examples include flies, cockroaches, mice, rats, and humans. Compare *specialist species*.

genes Coded units of information about specific traits that are passed from parents to offspring during reproduction. They consist of segments of DNA molecules found in chromosomes.

gene splicing See *genetic engineering.*

genetic adaptation Changes in the genetic makeup of organisms of a species that allow the species to reproduce and gain a competitive advantage under changed environmental conditions. See *differential reproduction, evolution, mutation, natural selection.*

genetically modified organism (GMO) Organism whose genetic makeup has been altered by genetic engineering.

genetic diversity Variability in the genetic makeup among individuals within a single species. See *biodiversity.* Compare *ecological diversity, functional diversity, species diversity.*

genetic engineering Insertion of an alien gene into an organism to give it a beneficial genetic trait. Compare *artificial selection, natural selection.*

genome Complete set of genetic information for an organism.

geographic isolation Separation of populations of a species for long times into different areas.

geology Study of the earth's dynamic history. Geologists study and analyze rocks and the features and processes of the earth's interior and surface.

geothermal energy Heat transferred from the earth's underground concentrations of dry steam (steam with no water droplets), wet steam (a mixture of steam and water droplets), or hot water trapped in fractured or porous rock.

globalization Broad process of global social, economic, and environmental change that leads to an increasingly integrated world.

global warming Warming of the earth's atmosphere because of increases in the concentrations of one or more greenhouse gases primarily as a result of human activities. See *greenhouse effect, greenhouse gases.*

grassland Biome found in regions where moderate annual average precipitation (25–76 centimeters, or 10–30 inches) is enough to support the growth of grass and small plants but not enough to support large stands of trees. Compare *desert, forest.*

greenhouse effect A natural effect that releases heat in the atmosphere (troposphere) near the earth's surface. Water vapor, carbon dioxide, ozone, and other gases in the lower atmosphere (troposphere) absorb some of the infrared radiation (heat) radiated by the earth's surface. Their molecules vibrate and transform the absorbed energy into longer-wavelength infrared radiation (heat) in the troposphere. If the

atmospheric concentrations of these greenhouse gases increase and other natural processes do not remove them, the average temperature of the lower atmosphere will increase gradually. Compare *global warming.* See also *natural greenhouse effect.*

greenhouse gases Gases in the earth's lower atmosphere (troposphere) that cause the greenhouse effect. Examples include carbon dioxide, chlorofluorocarbons, ozone, methane, water vapor, and nitrous oxide.

green manure Freshly cut or still-growing green vegetation that is plowed into the soil to increase the organic matter and humus available to support crop growth. Compare *animal manure.*

green revolution Popular term for the introduction of scientifically bred or selected varieties of grain (rice, wheat, maize) that, with adequate inputs of fertilizer and water, can greatly increase crop yields.

gross domestic product (GDP) Annual market value of all goods and services produced by all firms and organizations, foreign and domestic, operating within a country.

gross primary productivity (GPP) The rate at which an ecosystem's producers capture and store a given amount of chemical energy as biomass in a given length of time. Compare *net primary productivity.*

ground fire Fire that burns decayed leaves or peat deep below the ground surface. Compare *crown fire, surface fire.*

groundwater Water that sinks into the soil and is stored in slowly flowing and slowly renewed underground reservoirs called aquifers; underground water in the zone of saturation, below the water table. Compare *runoff, surface water.*

habitat Place or type of place where an organism or population of organisms lives. Compare *ecological niche.*

habitat fragmentation Breakup of a habitat into smaller pieces, usually as a result of human activities.

half-life Time needed for one-half of the nuclei in a radioisotope to emit their radiation. Each radioisotope has a characteristic half-life, which may range from a few millionths of a second to several billion years. See *radioisotope.*

hazard Something that can cause injury, disease, economic loss, or environmental damage. See also *risk.*

hazardous chemical Chemical that can cause harm because it is flammable or explosive, can irritate or damage the skin or lungs (such as strong acidic or alkaline substances), or can cause allergic reactions of the immune system (allergens). See also *toxic chemical.*

hazardous waste Any solid, liquid, or containerized gas that can catch fire easily, is corrosive to skin tissue or metals, is unstable and can explode or release toxic fumes, or has harmful concentrations of one or more toxic materials that can leach out. See also *toxic waste.*

heat Total kinetic energy of all randomly moving atoms, ions, or molecules within a given substance, excluding the overall motion of the whole object. Heat always flows spontaneously from a hot sample of matter to a colder sample of matter. This is one way to state the second law of thermodynamics. Compare *temperature.*

herbicide Chemical that kills a plant or inhibits its growth.

herbivore Plant-eating organism. Examples include deer, sheep, grasshoppers, and zooplankton. Compare *carnivore, omnivore.*

heterotroph See *consumer.*

high An air mass with a high pressure. Compare *low.*

high-input agriculture See *industrialized agriculture.*

high-quality energy Energy that is concentrated and has great ability to perform useful work. Examples include high-temperature heat and the energy in electricity, coal, oil, gasoline, sunlight, and nuclei of uranium-235. Compare *low-quality energy.*

high-quality matter Matter that is concentrated and contains a high concentration of a useful resource. Compare *low-quality matter.*

high-throughput economy The situation in most advanced industrialized countries, in which ever-increasing economic growth is sustained by maximizing the rate at which matter and energy resources are used, with little emphasis on pollution prevention, recycling, reuse, reduction of unnecessary waste, and other forms of resource conservation. Compare *low-throughput economy, matter-recycling-and-reuse economy.*

high-waste economy See *high-throughput economy.*

host Plant or animal on which a parasite feeds.

human resources People's physical and mental talents that support an economy by providing labor, innovation, culture, and organization. Compare *manufactured resources, natural resources.*

humus Slightly soluble residue of undigested or partially decomposed organic material in topsoil. This material helps retain water and water-soluble nutrients, which can be taken up by plant roots.

hunter–gatherers People who get their food by gathering edible wild plants and other materials and by hunting wild animals and fish.

hydrocarbon Organic compound of hydrogen and carbon atoms. The simplest hydrocarbon is methane (CH_4), the major component of natural gas.

hydroelectric power plant Structure in which the energy of falling or flowing water spins a turbine generator to produce electricity.

hydrologic cycle Biogeochemical cycle that collects, purifies, and distributes the earth's fixed supply of water from the environment to living organisms and then back to the environment.

hydropower Electrical energy produced by falling or flowing water. See *hydroelectric power plant*.

hydrosphere The earth's *liquid water* (oceans, lakes, other bodies of surface water, and underground water), *frozen water* (polar ice caps, floating ice caps, and ice in soil, known as permafrost), and *water vapor* in the atmosphere. See also *hydrologic cycle*.

identified resources Deposits of a particular mineral-bearing material of which the location, quantity, and quality are known or have been estimated from direct geological evidence and measurements. Compare *undiscovered resources*.

igneous rock Rock formed when molten rock material (magma) wells up from the earth's interior, cools, and solidifies into rock masses. Compare *metamorphic rock, sedimentary rock*. See *rock cycle*.

immature community Community at an early stage of ecological succession. It usually has a low number of species and ecological niches and cannot capture and use energy and cycle critical nutrients as efficiently as more complex, mature communities. Compare *mature community*.

immigrant species See *nonnative species*.

immigration Migration of people into a country or area to take up permanent residence.

indicator species Species that serve as early warnings that a community or ecosystem is being degraded. Compare *foundation species, keystone species, native species, nonnative species*.

industrialized agriculture Using large inputs of energy from fossil fuels (especially oil and natural gas), water, fertilizer, and pesticides to produce large quantities of crops and livestock for domestic and foreign sale. Compare *subsistence farming*.

industrial–medical revolution Use of new sources of energy from fossil fuels and later from nuclear fuels, and use of new technologies, to grow food and manufacture products. Compare *agricultural revolution, environmental revolution, hunter–gatherers, information and globalization revolution*.

industrial smog Type of air pollution consisting mostly of a mixture of sulfur dioxide, suspended droplets of sulfuric acid formed from some of the sulfur dioxide, and suspended solid particles. Compare *photochemical smog*.

infant mortality rate Number of babies out of every 1,000 born each year who die before their first birthday.

infectious disease See *transmissible disease*.

infiltration Downward movement of water through soil.

information and globalization revolution Use of new technologies such as the telephone, radio, television, computers, the Internet, automated databases, and remote sensing satellites to enable people to have increasingly rapid access to much more information on a global scale. Compare— *agricultural revolution, environmental revolution, hunter–gatherers, industrial–medical revolution*.

inherent value See *intrinsic value*.

inland wetland Land away from the coast, such as a swamp, marsh, or bog, that is covered all or part of the time with fresh water. Compare *coastal wetland*.

inorganic compounds All compounds not classified as organic compounds. See *organic compounds*.

inorganic fertilizer See *commercial inorganic fertilizer*.

input Matter, energy, or information entering a system. Compare *output, throughput*.

input pollution control See *pollution prevention*.

insecticide Chemical that kills insects.

instrumental value Value of an organism, species, ecosystem, or the earth's biodiversity based on its usefulness to humans. Compare *intrinsic value*.

integrated pest management (IPM) Combined use of biological, chemical, and cultivation methods in proper sequence and timing to keep the size of a pest population below the size that causes economically unacceptable loss of a crop or livestock animal.

intercropping Growing two or more different crops at the same time on a plot. For example, a carbohydrate-rich grain that depletes soil nitrogen and a protein-rich legume that adds nitrogen to the soil may be intercropped. Compare *monoculture, polyculture, polyvarietal cultivation*.

internal cost Direct cost paid by the producer and the buyer of an economic good. Compare *external benefit, external cost, full cost*.

interplanting Simultaneously growing a variety of crops on the same plot. See *agro-forestry, intercropping, polyculture, polyvarietal cultivation*.

interspecific competition Attempts by members of two or more species to use the same limited resources in an ecosystem. See *competition, intraspecific competition*.

intertidal zone The area of shoreline between low and high tides.

intraspecific competition Attempts by two or more organisms of a single species to use the same limited resources in an ecosystem. See *competition, interspecific competition*.

intrinsic rate of increase (r) Rate at which a population could grow if it had unlimited resources. Compare *environmental resistance*.

intrinsic value Value of an organism, species, ecosystem, or the earth's biodiversity based on its existence, regardless of whether it has any usefulness to humans. Compare *instrumental value*.

invasive species See *nonnative species*.

inversion See *temperature inversion*.

invertebrates Animals that have no backbones. Compare *vertebrates*.

ion Atom or group of atoms with one or more positive ($+$) or negative ($-$) electrical charges. Compare *atom, molecule*.

ionizing radiation Fast-moving alpha or beta particles or high-energy radiation (gamma rays) emitted by radioisotopes. They have enough energy to dislodge one or more electrons from atoms they hit, thereby forming charged ions in tissue that can react with and damage living tissue. Compare *nonionizing radiation*.

isotopes Two or more forms of a chemical element that have the same number of protons but different mass numbers because they have different numbers of neutrons in their nuclei.

J-shaped curve Curve with a shape similar to that of the letter J; can represent prolonged exponential growth. See *exponential growth*.

junk science Scientific results or hypotheses presented as sound science but not having undergone the rigors of the peer review process. Compare *frontier science, sound science*.

kerogen Solid, waxy mixture of hydrocarbons found in oil shale rock. Heating the rock to high temperatures causes the kerogen to vaporize. The vapor is condensed, purified, and then sent to a refinery to produce gasoline, heating oil, and other products. See also *oil shale, shale oil*.

keystone species Species that play roles affecting many other organisms in an ecosystem. Compare *foundation species, indicator species, native species, nonnative species*.

kilocalorie (kcal) Unit of energy equal to 1,000 calories. See *calorie*.

kilowatt (kW) Unit of electrical power equal to 1,000 watts. See *watt*.

kinetic energy Energy that matter has because of its mass and speed or velocity. Compare *potential energy*.

K-selected species Species that produce a few, often fairly large offspring but invest a great deal of time and energy to ensure that most of those offspring reach reproductive age. Compare *r-selected species*.

K-strategists See *K-selected species*.

lake Large natural body of standing fresh water formed when water from precipitation, land runoff, or groundwater flow fills a depression in the earth created by glaciation, earth movement, volcanic activity, or a giant meteorite. See *eutrophic lake*, *mesotrophic lake*, *oligotrophic lake*.

land degradation A decrease in the ability of land to support crops, livestock, or wild species in the future as a result of natural or human-induced processes.

landfill See *sanitary landfill*.

latitude Distance from the equator. Compare *altitude*.

law of conservation of energy See *first law of thermodynamics*.

law of conservation of matter In any physical or chemical change, matter is neither created nor destroyed but merely changed from one form to another; in physical and chemical changes, existing atoms are rearranged into different spatial patterns (physical changes) or different combinations (chemical changes).

law of tolerance The existence, abundance, and distribution of a species in an ecosystem are determined by whether the levels of one or more physical or chemical factors fall within the range tolerated by the species. See *threshold effect*.

LD50 See *median lethal dose*.

LDC See *developing country*.

leaching Process in which various chemicals in upper layers of soil are dissolved and carried to lower layers and, in some cases, to groundwater.

less developed country (LDC) See *developing country*.

life-cycle cost Initial cost plus lifetime operating costs of an economic good. Compare *full cost*.

life expectancy Average number of years a newborn infant can be expected to live.

limiting factor Single factor that limits the growth, abundance, or distribution of the population of a species in an ecosystem. See *limiting factor principle*.

limiting factor principle Too much or too little of any abiotic factor can limit or prevent growth of a population of a species in an ecosystem, even if all other factors are at or near the optimal range of tolerance for the species.

linear growth Growth in which a quantity increases by some fixed amount during each unit of time. An example is growth that increases in the sequence 2, 4, 6, 8, 10, and so on. Compare *exponential growth*.

liquefied natural gas (LNG) Natural gas converted to liquid form by cooling it to a very low temperature.

liquefied petroleum gas (LPG) Mixture of liquefied propane (C_3H_8) and butane (C_4H_{10}) gas removed from natural gas and used as a fuel.

lithosphere Outer shell of the earth, composed of the crust and the rigid, outermost part of the mantle outside the asthenosphere; material found in the earth's plates. See *crust*, *mantle*.

loams Soils containing a mixture of clay, sand, silt, and humus. Good for growing most crops.

logistic growth Pattern in which exponential population growth occurs when the population is small, and population growth decreases steadily with time as the population approaches the carrying capacity. See *S-shaped curve*.

low An air mass with a low pressure. Compare *high*.

low-input agriculture See *sustainable agriculture*.

low-quality energy Energy that is dispersed and has little ability to do useful work. An example is low-temperature heat. Compare *high-quality energy*.

low-quality matter Matter that is dilute or dispersed or contains a low concentration of a useful resource. Compare *high-quality matter*.

low-throughput economy Economy based on working with nature by recycling and reusing discarded matter; preventing pollution; conserving matter and energy resources by reducing unnecessary waste and use; not degrading renewable resources; building things that are easy to recycle, reuse, and repair; not allowing population size to exceed the carrying capacity of the environment; and preserving biodiversity and ecological integrity. Compare *high-throughput economy*, *matter-recycling-and-reuse economy*.

low-waste economy See *low-throughput economy*.

LPG See *liquefied petroleum gas*.

macroevolution Long-term, large-scale evolutionary changes among groups of species. Compare *microevolution*.

macronutrients Chemical elements that are needed in fairly large amounts. Examples are carbon, hydrogen, nitrogen, and oxygen found in protein, carbohydrates, and fats. Compare *micronutrients*.

magma Molten rock below the earth's surface.

malnutrition Faulty nutrition, caused by a diet that does not supply an individual with enough protein, essential fats, vitamins, minerals, and other nutrients needed for good health. Compare *overnutrition*, *undernutrition*.

mangrove swamps Swamps found on the coastlines in warm tropical climates. They are dominated by mangrove trees, any of about 55 species of trees and shrubs that can live partly submerged in the salty environment of coastal swamps.

mantle Zone of the earth's interior between its core and its crust. Compare *core*, *crust*. See *lithosphere*.

manufactured capital See *manufactured resources*.

manufactured resources Manufactured items made from natural resources and used to produce and distribute economic goods and services bought by consumers. They include tools, machinery, equipment, factory buildings, and transportation and distribution facilities. Compare *human resources*, *natural resources*.

manure See *animal manure*, *green manure*.

mass The amount of material in an object.

mass depletion Widespread, often global period during which extinction rates are higher than normal but not high enough to classify as a mass extinction. Compare *background extinction*, *mass extinction*.

mass extinction A catastrophic, widespread, often global event in which major groups of species are wiped out over a short time compared with normal (background) extinctions. Compare *background extinction*, *mass depletion*.

mass number Sum of the number of neutrons (n) and the number of protons (p) in the nucleus of an atom. It gives the approximate mass of that atom. Compare *atomic number*.

mass transit Buses, trains, trolleys, and other forms of transportation that carry large numbers of people.

material efficiency Total amount of material needed to produce each unit of goods or services. Also called *resource productivity*. Compare *energy efficiency*.

matter Anything that has mass (the amount of material in an object) and takes up space. On the earth, where gravity is present, we weigh an object to determine its mass.

matter quality Measure of how useful a matter resource is, based on its availability and concentration. See *high-quality matter, low-quality matter*.

matter-recycling-and-reuse economy Economy that emphasizes recycling the maximum amount of all resources that can be recycled. The goal is to allow economic growth to continue without depleting matter resources and without producing excessive pollution and environmental degradation. Compare *high-throughput economy, low-throughput economy*.

mature community Fairly stable, self-sustaining community in an advanced stage of ecological succession; usually has a diverse array of species and ecological niches; captures and uses energy and cycles critical chemicals more efficiently than simpler, immature communities. Compare *immature community*.

maximum sustainable yield See *sustainable yield*.

MDC See *developed country*.

median lethal dose (LD50) Amount of a toxic material per unit of body weight of test animals that kills half the test population in a certain time.

megacity City with 10 million or more people.

meltdown The melting of the core of a nuclear reactor.

mesotrophic lake Lake with a moderate supply of plant nutrients. Compare *eutrophic lake, oligotrophic lake*.

metabolism Ability of a living cell or organism to capture and transform matter and energy from its environment to supply its needs for survival, growth, and reproduction.

metamorphic rock Rock produced when a preexisting rock is subjected to high temperatures (which may cause it to melt partially), high pressures, chemically active fluids, or a combination of these agents. Compare *igneous rock, sedimentary rock*. See *rock cycle*.

metastasis Spread of malignant (cancerous) cells from a tumor to other parts of the body.

metropolitan area See *urban area*.

microevolution The small genetic changes a population undergoes. Compare *macroevolution*.

micronutrients Chemical elements that organisms need in small or even trace amounts to live, grow, or reproduce. Examples include sodium, zinc, copper, chlorine, and iodine. Compare *macronutrients*.

microorganisms Organisms such as bacteria that are so small that it takes a microscope to see them.

micropower systems Systems of small-scale decentralized units that generate 1–10,000 kilowatts of electricity. Examples include microturbines, fuel cells, and household solar panels and solar roofs.

mineral Any naturally occurring inorganic substance found in the earth's crust as a crystalline solid. See *mineral resource*.

mineral resource Concentration of naturally occurring solid, liquid, or gaseous material in or on the earth's crust in a form and amount such that extracting and converting it into useful materials or items is currently or potentially profitable. Mineral resources are classified as *metallic* (such as iron and tin ores) or *nonmetallic* (such as fossil fuels, sand, and salt).

minimum-tillage farming See *conservation-tillage farming*.

minimum viable population (MVP) Estimate of the smallest number of individuals necessary to ensure the survival of a population in a region for a specified time period, typically ranging from decades to 100 years.

mixture Combination of one or more elements and compounds.

model An approximate representation or simulation of a system being studied.

molecule Combination of two or more atoms of the same chemical element (such as O_2) or different chemical elements (such as H_2O) held together by chemical bonds. Compare *atom, ion*.

monoculture Cultivation of a single crop, usually on a large area of land. Compare *polyculture, polyvarietal cultivation*.

more developed country (MDC) See *developed country*.

mountaintop removal Type of surface mining that uses explosives, massive shovels, and even larger machinery called draglines to remove the top of a mountain to expose seams of coal underneath a mountain. Compare *area strip mining, contour strip mining*.

multiple use Use of an ecosystem such as a forest for a variety of purposes such as timber harvesting, wildlife habitat, watershed protection, and recreation. Compare *sustainable yield*.

municipal solid waste (MSW) Solid materials discarded by homes and businesses in or near urban areas. See *solid waste*.

mutagen Chemical or form of radiation that causes inheritable changes (mutations) in the DNA molecules in genes. See *carcinogen, mutation, teratogen*.

mutation A random change in DNA molecules making up genes that can alter anatomy, physiology, or behavior in offspring. See *mutagen*.

mutualism Type of species interaction in which both participating species generally benefit. Compare *commensalism*.

native species Species that normally live and thrive in a particular ecosystem. Compare *foundation species, indicator species, keystone species, nonnative species*.

natural capital See *natural resources*.

natural gas Underground deposits of gases consisting of 50–90% by weight methane gas (CH_4) and small amounts of heavier gaseous hydrocarbon compounds such as propane (C_3H_8) and butane (C_4H_{10}).

natural greenhouse effect Heat buildup in the troposphere because of the presence of certain gases, called greenhouse gases. Without this effect, the earth would be nearly as cold as Mars, and life as we know it could not exist. Compare *global warming*.

natural law See *scientific law*.

natural radioactive decay Nuclear change in which unstable nuclei of atoms spontaneously shoot out particles (usually alpha or beta particles) or energy (gamma rays) at a fixed rate.

natural rate of extinction See *background extinction*.

natural recharge Natural replenishment of an aquifer by precipitation, which percolates downward through soil and rock. See *recharge area*.

natural resources The earth's natural materials and processes that sustain life on the earth and our economies.

natural selection Process by which a particular beneficial gene (or set of genes) is reproduced in succeeding generations more than other genes. The result of natural selection is a population that contains a greater proportion of organisms better adapted to certain environmental conditions. See *adaptation, biological evolution, differential reproduction, mutation*.

nekton Strongly swimming organisms found in aquatic systems. Compare *benthos, plankton*.

net energy Total amount of useful energy available from an energy resource or energy system over its lifetime, minus the amount of energy *used* (the first energy law), *automatically wasted* (the second energy law), and *unnecessarily wasted* in finding, processing, concentrating, and transporting it to users.

net primary productivity (NPP) Rate at which all the plants in an ecosystem produce net useful chemical energy; equal to the difference between the rate at which the plants in an ecosystem produce useful chemical energy (gross primary productivity) and the rate at which they use some of that energy through cellular respiration. Compare *gross primary productivity*.

neurotoxins Chemicals that can harm the human *nervous system* (brain, spinal cord, peripheral nerves).

neutral solution Water solution containing an equal number of hydrogen ions (H^+) and hydroxide ions (OH^-); water solution with a pH of 7. Compare *acid solution, basic solution*.

neutron (n) Elementary particle in the nuclei of all atoms (except hydrogen-1). It has a relative mass of 1 and no electric charge. Compare *electron, proton*.

niche See *ecological niche*.

nitrogen cycle Cyclic movement of nitrogen in different chemical forms from the environment to organisms and then back to the environment.

nitrogen fixation Conversion of atmospheric nitrogen gas into forms useful to plants by lightning, bacteria, and cyanobacteria; it is part of the nitrogen cycle.

noise pollution Any unwanted, disturbing, or harmful sound that impairs or interferes with hearing, causes stress, hampers concentration and work efficiency, or causes accidents.

nondegradable pollutant Material that is not broken down by natural processes. Examples include the toxic elements lead and mercury. Compare *biodegradable pollutant, degradable pollutant, slowly degradable pollutant*.

nonionizing radiation Forms of radiant energy such as radio waves, microwaves, infrared light, and ordinary light that do not have enough energy to cause ionization of atoms in living tissue. Compare *ionizing radiation*.

nonnative species Species that migrate into an ecosystem or are deliberately or accidentally introduced into an ecosystem by humans. Compare *native species*.

nonpersistent pollutant See *degradable pollutant*.

nonpoint source Large or dispersed land areas such as crop fields, streets, and lawns that discharge pollutants into the environment over a large area. Compare *point source*.

nonrenewable resource Resource that exists in a fixed amount (stock) in various places in the earth's crust and has the potential for renewal by geological, physical, and chemical processes taking place over hundreds of millions to billions of years. Examples include copper, aluminum, coal, and oil. We classify these resources as exhaustible because we are extracting and using them at a much faster rate than they were formed. Compare *renewable resource*.

nontransmissible disease A disease that is not caused by living organisms and does not spread from one person to another. Examples include most cancers, diabetes, cardiovascular disease, and malnutrition. Compare *transmissible disease*.

no-till farming See *conservation-tillage farming*.

nuclear change Process in which nuclei of certain isotopes spontaneously change, or are forced to change, into one or more different isotopes. The three principal types of nuclear change are natural radioactivity, nuclear fission, and nuclear fusion. Compare *chemical change, physical change*.

nuclear energy Energy released when atomic nuclei undergo a nuclear reaction such as the spontaneous emission of radioactivity, nuclear fission, or nuclear fusion.

nuclear fission Nuclear change in which the nuclei of certain isotopes with large mass numbers (such as uranium-235 and plutonium-239) are split apart into lighter nuclei when struck by a neutron. This process releases more neutrons and a large amount of energy. Compare *nuclear fusion*.

nuclear fusion Nuclear change in which two nuclei of isotopes of elements with a low mass number (such as hydrogen-2 and hydrogen-3) are forced together at extremely high temperatures until they fuse to form a heavier nucleus (such as helium-4). This process releases a large amount of energy. Compare *nuclear fission*.

nucleus Extremely tiny center of an atom, making up most of the atom's mass. It contains one or more positively charged protons and one or more neutrons with no electrical charge (except for a hydrogen-1 atom, which has one proton and no neutrons in its nucleus).

nutrient Any food or element an organism must take in to live, grow, or reproduce.

nutrient cycle See *biogeochemical cycle*.

oil See *crude oil*.

oil sand Deposit of a mixture of clay, sand, water, and varying amounts of a tarlike heavy oil known as bitumen. Bitumen can be extracted from oil sand by heating. It is then purified and upgraded to synthetic crude oil. See *bitumen.*.

oil shale Fine-grained rock containing various amounts of kerogen, a solid, waxy mixture of hydrocarbon compounds. Heating the rock to high temperatures converts the kerogen into a vapor that can be condensed to form a slow-flowing heavy oil called shale oil. See *kerogen, shale oil*.

old-growth forest Virgin and old, second-growth forests containing trees that are often hundreds—sometimes thousands—of years old. Examples include forests of Douglas fir, western hemlock, giant sequoia, and coastal redwoods in the western United States. Compare *second-growth forest, tree plantation*.

oligotrophic lake Lake with a low supply of plant nutrients. Compare *eutrophic lake, mesotrophic lake*.

omnivore Animal that can use both plants and other animals as food sources. Examples include pigs, rats, cockroaches, and humans. Compare *carnivore, herbivore*.

open dump Fields or holes in the ground where garbage is deposited and sometimes covered with soil. They are rare in developed countries, but are widely used in many developing countries, especially to handle wastes from megacities. Compare *sanitary landfill*.

open-pit mining Removing minerals such as gravel, sand, and metal ores by digging them out of the earth's surface and leaving an open pit behind. Compare *area strip mining, contour strip mining, mountaintop removal, subsurface mining*.

open sea The part of an ocean that lies beyond the continental shelf. Compare *coastal zone*.

ore Part of a metal-yielding material that can be economically and legally extracted at a given time. An ore typically contains two parts: the ore mineral, which contains the desired metal, and waste mineral material (gangue).

organic agriculture Producing crops and livestock naturally by using organic fertilizer (manure, legumes, compost) and natural pest control (bugs that eat harmful bugs, plants that repel bugs, and environmental controls such as crop rotation) instead of using commercial inorganic fertilizers and synthetic pesticides and herbicides. See *sustainable agriculture*.

organic compounds Compounds containing carbon atoms combined with each other and with atoms of one or more other elements such as hydrogen, oxygen, nitrogen, sulfur, phosphorus, chlorine, and fluorine. All other compounds are called *inorganic compounds*.

organic fertilizer Organic material such as animal manure, green manure, and compost, applied to cropland as a source of plant nutrients. Compare *commercial inorganic fertilizer*.

organism Any form of life.

other resources Identified and undiscovered resources not classified as reserves. Compare *identified resources, reserves, undiscovered resources*.

output Matter, energy, or information leaving a system. Compare *input, throughput*.

output pollution control See *pollution cleanup*.

overburden Layer of soil and rock overlying a mineral deposit. Surface mining removes this layer.

overfishing Harvesting so many fish of a species, especially immature fish, that not enough breeding stock is left to replenish the species and it becomes unprofitable to harvest them.

overgrazing Destruction of vegetation when too many grazing animals feed too long and exceed the carrying capacity of a rangeland or pasture area.

overnutrition Diet so high in calories, saturated (animal) fats, salt, sugar, and processed foods and so low in vegetables and fruits that the consumer runs a high risk of developing diabetes, hypertension, heart disease, and other health hazards. Compare *malnutrition, undernutrition*.

oxygen-demanding wastes Organic materials that are usually biodegraded by aerobic (oxygen-consuming) bacteria if there is enough dissolved oxygen in the water.

ozone depletion Decrease in concentration of ozone (O_3) in the stratosphere. See *ozone layer*.

ozone layer Layer of gaseous ozone (O_3) in the stratosphere that protects life on earth by filtering out most harmful ultraviolet radiation from the sun.

PANs Peroxyacyl nitrates; group of chemicals found in photochemical smog.

parasite Consumer organism that lives on or in, and feeds on, a living plant or animal, known as the host, over an extended period. The parasite draws nourishment from and gradually weakens its host; it may or may not kill the host. See *parasitism*.

parasitism Interaction between species in which one organism, called the parasite, preys on another organism, called the host, by living on or in the host. See *host, parasite*.

parts per billion (ppb) Number of parts of a chemical found in 1 billion parts of a particular gas, liquid, or solid.

parts per million (ppm) Number of parts of a chemical found in 1 million parts of a particular gas, liquid, or solid.

parts per trillion (ppt) Number of parts of a chemical found in 1 trillion parts of a particular gas, liquid, or solid.

passive solar heating system System that captures sunlight directly within a structure and converts it into low-temperature heat for space heating or for heating water for domestic use without the use of mechanical devices. Compare *active solar heating system*.

pasture Managed grassland or enclosed meadow that usually is planted with domesticated grasses or other forage to be grazed by livestock. Compare *feedlot, rangeland*.

pathogen Organism that produces disease. Examples include bacteria, viruses, and parasites.

PCBs See *polychlorinated biphenyls*.

per capita ecological footprint Amount of biologically productive land and water needed to supply each person or population with the renewable resources they use and to absorb or dispose of the wastes from such resource use. It measures the average environmental impact of individuals or populations in different countries and areas.

per capita GDP Annual gross domestic product (GDP) of a country divided by its total population at midyear. It gives the average slice of the economic pie per person. Used to be called per capita gross national product (GNP). See *gross domestic product*.

percolation Passage of a liquid through the spaces of a porous material such as soil.

perennial Plant that can live for more than 2 years. Compare *annual*.

permafrost Perennially frozen layer of the soil that forms when the water there freezes. It is found in arctic tundra.

permeability The degree to which underground rock and soil pores are interconnected and thus a measure of the degree to which water can flow freely from one pore to another. Compare *porosity*.

perpetual resource An essentially inexhaustible resource on a human time scale. Solar energy is an example. Compare *nonrenewable resource, renewable resource*.

persistence How long a pollutant stays in the air, water, soil, or body.

persistent pollutant See *slowly degradable pollutant*.

pest Unwanted organism that directly or indirectly interferes with human activities.

pesticide Any chemical designed to kill or inhibit the growth of an organism that people consider undesirable. See *fungicide, herbicide, insecticide*.

petrochemicals Chemicals obtained by refining (distilling) crude oil. They are used as raw materials in manufacturing most industrial chemicals, fertilizers, pesticides, plastics, synthetic fibers, paints, medicines, and many other products.

petroleum See *crude oil*.

pH Numeric value that indicates the relative acidity or alkalinity of a substance on a scale of 0 to 14, with the neutral point at 7. Acid solutions have pH values lower than 7; basic or alkaline solutions have pH values greater than 7.

phosphorus cycle Cyclic movement of phosphorus in different chemical forms from the environment to organisms and then back to the environment.

photochemical smog Complex mixture of air pollutants produced in the lower atmosphere by the reaction of hydrocarbons and nitrogen oxides under the influence of sunlight. Especially harmful components include ozone, peroxyacyl nitrates (PANs), and various aldehydes. Compare *industrial smog*.

photosynthesis Complex process that takes place in cells of green plants. Radiant energy from the sun is used to combine carbon dioxide (CO_2) and water (H_2O) to produce oxygen (O_2), carbohydrates (such as glucose, $C_6H_{12}O_6$), and other nutrient molecules. Compare *aerobic respiration, chemosynthesis*.

photovoltaic (PV) cell Device that converts radiant (solar) energy directly into electrical energy. Also called a solar cell.

physical change Process that alters one or more physical properties of an element or a compound without changing its chemical composition. Examples include changing the size and shape of a sample of matter (crushing ice and cutting aluminum foil) and changing a sample of matter from one physical state to another (boiling and freezing water). Compare *chemical change, nuclear change*.

phytoplankton Small, drifting plants, mostly algae and bacteria, found in aquatic ecosystems. Compare *plankton, zooplankton*.

pioneer community First integrated set of plants, animals, and decomposers found in an area undergoing primary ecological succession. See *immature community, mature community*.

pioneer species First hardy species—often microbes, mosses, and lichens—that begin colonizing a site as the first stage of ecological succession. See *ecological succession, pioneer community*.

planetary management worldview Beliefs that (1) humans are the planet's most important species; (2) there is always more, and it is all for us; (3) all economic growth is good, more economic growth is better, and the potential for economic growth is limitless; and (4) our success depends on how well we can understand, control, and manage the earth's life-support systems for our own benefit. Compare *environmental wisdom worldview, stewardship worldview*.

plankton Small plant organisms (phytoplankton) and animal organisms (zooplankton) that float in aquatic ecosystems.

plantation agriculture Growing specialized crops such as bananas, coffee, and cacao in tropical developing countries, primarily for sale to developed countries.

plates Areas of the earth's lithosphere that move slowly around with the mantle's flowing asthenosphere. Most earthquakes and volcanoes occur around the boundaries of these plates. See *lithosphere, plate tectonics, tectonic plates*.

plate tectonics Theory of geophysical processes that explains the movements of lithospheric plates and the processes that occur at their boundaries. See *lithosphere, tectonic plates*.

point source Single identifiable source that discharges pollutants into the environment. Examples include the smokestack of a power plant or an industrial plant, drainpipe of a meatpacking plant, chimney of a house, or exhaust pipe of an automobile. Compare *nonpoint source*.

poison A chemical that adversely affects the health of a living human or animal by causing injury, illness, or death.

politics Process through which individuals and groups try to influence or control government policies and actions that affect the local, state, national, and international communities.

pollutant A particular chemical or form of energy that can adversely affect the health, survival, or activities of humans or other living organisms. See *pollution*.

pollution An undesirable change in the physical, chemical, or biological characteristics of air, water, soil, or food that can ad-versely affect the health, survival, or activities of humans or other living organisms.

pollution cleanup Device or process that removes or reduces the level of a pollutant after it has been produced or has entered the environment. Examples include automobile emission control devices and sewage treatment plants. Compare *pollution prevention*.

pollution prevention Device or process that prevents a potential pollutant from forming or entering the environment or sharply reduces the amount entering the environment. Compare *pollution cleanup*.

polychlorinated biphenyls (PCBs) Group of 209 toxic, oily, synthetic chlorinated hydrocarbon compounds that can be biologically amplified in food chains and webs.

polyculture Complex form of intercropping in which a large number of different plants maturing at different times are planted together. See also *intercropping*. Compare *monoculture, polyvarietal cultivation*.

polyvarietal cultivation Planting a plot of land with several varieties of the same crop. Compare *intercropping, monoculture, polyculture*.

population Group of individual organisms of the same species living in a particular area.

population change An increase or decrease in the size of a population. It is equal to (Births + Immigration) − (Deaths + Emigration).

population density Number of organisms in a particular population found in a specified area or volume.

population dispersion General pattern in which the members of a population are arranged throughout its habitat.

population distribution Variation of population density over a particular geographic area. For example, a country has a high population density in its urban areas and a much lower population density in its rural areas.

population dynamics Major abiotic and biotic factors that tend to increase or decrease the population size and affect the age and sex composition of a species.

population size Number of individuals making up a population's gene pool.

porosity Percentage of space in rock or soil occupied by voids, whether the voids are isolated or connected. Compare *permeability*.

potential energy Energy stored in an object because of its position or the position of its parts. Compare *kinetic energy*.

poverty Inability to meet basic needs for food, clothing, and shelter.

ppb See *parts per billion*.

ppm See *parts per million*.

ppt See *parts per trillion*.

precautionary principle When there is scientific uncertainty about potentially serious harm from chemicals or technologies, decision makers should act to prevent harm to humans and the environment. See *pollution prevention*.

precipitation Water in the form of rain, sleet, hail, and snow that falls from the atmosphere onto land and bodies of water.

predation Situation in which an organism of one species (the predator) captures and feeds on parts or all of an organism of another species (the prey).

predator Organism that captures and feeds on parts or all of an organism of another species (the prey).

predator–prey relationship Interaction between two organisms of different species in which one organism, called the *predator*, captures and feeds on parts or all of the other organism, called the *prey*.

prey Organism that is captured and serves as a source of food for an organism of another species (the predator).

primary consumer Organism that feeds on all or part of plants (herbivore) or on other producers. Compare *detritivore, omnivore, secondary consumer*.

primary pollutant Chemical that has been added directly to the air by natural events or human activities and occurs in a harmful concentration. Compare *secondary pollutant*.

primary productivity See *gross primary productivity, net primary productivity*.

primary sewage treatment Mechanical sewage treatment in which large solids are filtered out by screens and suspended solids settle out as sludge in a sedimentation tank. Compare *secondary sewage treatment*.

primary succession Ecological succession in a bare area that has never been occupied by a community of organisms. See *ecological succession*. Compare *secondary succession*.

probability A mathematical statement about how likely it is that something will happen.

producer Organism that uses solar energy (green plants) or chemical energy (some bacteria) to manufacture the organic compounds it needs as nutrients from simple inorganic compounds obtained from its environment. Compare *consumer, decomposer*.

proton (p) Positively charged particle in the nuclei of all atoms. Each proton has a relative mass of 1 and a single positive charge. Compare *electron, neutron*.

pyramid of energy flow Diagram representing the flow of energy through each trophic level in a food chain or food web. With each energy transfer, only a small part (typically 10%) of the usable energy entering one trophic level is transferred to the organisms at the next trophic level. Compare *pyramid of biomass, pyramid of numbers*.

radiation Fast-moving particles (particulate radiation) or waves of energy (electromagnetic radiation). See *alpha particle, beta particle, gamma ray*.

radioactive decay Change of a radioisotope to a different isotope by the emission of radioactivity.

radioactive isotope See *radioisotope*.

radioactive waste Waste products of nuclear power plants, research, medicine, weapon production, or other processes involving nuclear reactions. See *radioactivity*.

radioactivity Nuclear change in which unstable nuclei of atoms spontaneously shoot out "chunks" of mass, energy, or both at a fixed rate. The three principal types of radioactivity are gamma rays and fast-moving alpha particles and beta particles.

radioisotope Isotope of an atom that spontaneously emits one or more types of radioactivity (alpha particles, beta particles, gamma rays).

rain shadow effect Low precipitation on the far side (leeward side) of a mountain when prevailing winds flow up and over a high mountain or range of high mountains. This creates semiarid and arid conditions on the leeward side of a high mountain range.

rangeland Land that supplies forage or vegetation (grasses, grasslike plants, and shrubs) for grazing and browsing animals and is not intensively managed. Compare *feedlot, pasture*.

range of tolerance Range of chemical and physical conditions that must be maintained for populations of a particular species to stay alive and grow, develop, and function normally. See *law of tolerance*.

rare species A species that has naturally small numbers of individuals (often because of limited geographic ranges or low population densities) or that has been locally depleted by human activities.

realized niche Parts of the fundamental niche of a species that are actually used by that species. See *ecological niche, fundamental niche*.

recharge area Any area of land allowing water to pass through it and into an aquifer. See *aquifer, natural recharge*.

reconciliation ecology The science of inventing, establishing, and maintaining new habitats to conserve species diversity in places where people live, work, or play.

recycling Collecting and reprocessing a resource so that it can be made into new products. An example is collecting aluminum cans, melting them down, and using the aluminum to make new cans or other aluminum products. Compare *reuse*.

reforestation Renewal of trees and other types of vegetation on land where trees have been removed; can be done naturally by seeds from nearby trees or artificially by planting seeds or seedlings.

reliable runoff Surface runoff of water that generally can be counted on as a stable source of water from year to year. See *runoff*.

renewable resource Resource that can be replenished rapidly (hours to several decades) through natural processes. Examples include trees in forests, grasses in grasslands, wild animals, fresh surface water in lakes and streams, most groundwater, fresh air, and fertile soil. If such a resource is used faster than it is replenished, it can be depleted and converted into a nonrenewable resource. Compare *nonrenewable resource* and *perpetual resource*. See also *environmental degradation*.

replacement-level fertility Number of children a couple must have to replace them. The average for a country or the world usually is slightly higher than 2 children per couple (2.1 in the United States and 2.5 in some developing countries) because some children die before reaching their reproductive years. See also *total fertility rate*.

reproduction Production of offspring by one or more parents.

reproductive isolation Long-term geographic separation of members of a particular sexually reproducing species.

reproductive potential See *biotic potential*.

reserves Resources that have been identified and from which a usable mineral can be extracted profitably at present prices with current mining technology. See *identified resources, undiscovered resources*.

resource Anything obtained from the living and nonliving environment to meet human needs and wants. It can also be applied to other species.

resource partitioning Process of dividing up resources in an ecosystem so that species with similar needs (overlapping ecological niches) use the same scarce resources at different times, in different ways, or in different places. See *ecological niche, fundamental niche, realized niche*.

resource productivity See *material efficiency*.

respiration See *aerobic respiration*.

response The amount of health damage caused by exposure to a certain dose of a harmful substance or form of radiation. See *dose, dose-response curve, median lethal dose*.

restoration ecology Research and scientific study devoted to restoring, repairing, and reconstructing damaged ecosystems.

reuse Using a product over and over again in the same form. An example is collecting, washing, and refilling glass beverage bottles. Compare *recycling*.

riparian zones Thin strips and patches of vegetation that surround streams. They are very important habitats and resources for wildlife.

risk The probability that something undesirable will result from deliberate or accidental exposure to a hazard. See *risk analysis, risk assessment, risk management*.

risk analysis Identifying hazards, evaluating the nature and severity of risks (*risk assessment*), using this and other information to determine options and make decisions about reducing or eliminating risks (*risk management*), and communicating information about risks to decision makers and the public (*risk communication*).

risk assessment Process of gathering data and making assumptions to estimate short- and long-term harmful effects on human health or the environment from exposure to hazards associated with the use of a particular product or technology.

risk communication Communicating information about risks to decision makers and the public. See *risk, risk analysis*.

risk management Using risk assessment and other information to determine options and make decisions about reducing or eliminating risks. See *risk, risk analysis, risk communication*.

rock Any material that makes up a large, natural, continuous part of the earth's crust. See *mineral*.

rock cycle Largest and slowest of the earth's cycles, consisting of geologic, physical, and chemical processes that form and modify rocks and soil in the earth's crust over millions of years.

r-selected species Species that reproduce early in their life span and produce large numbers of usually small and short-lived offspring in a short period. Compare *K-selected species*.

r-strategists See *r-selected species.*

rule of 70 Doubling time (in years) = 70/(percentage growth rate). See *doubling time, exponential growth.*

runoff Fresh water from precipitation and melting ice that flows on the earth's surface into nearby streams, lakes, wetlands, and reservoirs. See *reliable runoff, surface runoff, surface water.* Compare *groundwater.*

salinity Amount of various salts dissolved in a given volume of water.

salinization Accumulation of salts in soil that can eventually make the soil unable to support plant growth.

saltwater intrusion Movement of salt water into freshwater aquifers in coastal and inland areas as groundwater is withdrawn faster than it is recharged by precipitation.

sanitary landfill Waste disposal site on land in which waste is spread in thin layers, compacted, and covered with a fresh layer of clay or plastic foam each day.

scavenger Organism that feeds on dead organisms that were killed by other organisms or died naturally. Examples include vultures, flies, and crows. Compare *detritivore.*

science Attempts to discover order in nature and use that knowledge to make predictions about what should happen in nature. See *frontier science, scientific data, scientific hypothesis, scientific law, scientific methods, scientific model, scientific theory, sound science.*

scientific data Facts obtained by making observations and measurements. Compare *scientific hypothesis, scientific law, scientific methods, scientific model, scientific theory.*

scientific hypothesis An educated guess that attempts to explain a scientific law or certain scientific observations. Compare *scientific data, scientific law, scientific methods, scientific model, scientific theory.*

scientific law Description of what scientists find happening in nature repeatedly in the same way, without known exception. See *first law of thermodynamics, law of conservation of matter, second law of thermodynamics.* Compare *scientific data, scientific hypothesis, scientific methods, scientific model, scientific theory.*

scientific methods The ways scientists gather data and formulate and test scientific hypotheses, models, theories, and laws. See *scientific data, scientific hypothesis, scientific law, scientific model, scientific theory.*

scientific model A simulation of complex processes and systems. Many are mathematical models that are run and tested using computers.

scientific theory A well-tested and widely accepted scientific hypothesis. Compare *scientific data, scientific hypothesis, scientific law, scientific methods, scientific model.*

secondary consumer Organism that feeds only on primary consumers. Compare *detritivore, omnivore, primary consumer.*

secondary pollutant Harmful chemical formed in the atmosphere when a primary air pollutant reacts with normal air components or other air pollutants. Compare *primary pollutant.*

secondary sewage treatment Second step in most waste treatment systems in which aerobic bacteria decompose as much as 90% of degradable, oxygen-demanding organic wastes in wastewater. It usually involves bringing sewage and bacteria together in trickling filters or in the activated sludge process. Compare *primary sewage treatment.*

secondary succession Ecological succession in an area in which natural vegetation has been removed or destroyed but the soil is not destroyed. See *ecological succession.* Compare *primary succession.*

second-growth forest Stands of trees resulting from secondary ecological succession. Compare *old-growth forest, tree farm.*

second law of energy See *second law of thermodynamics.*

second law of thermodynamics In any conversion of heat energy to useful work, some of the initial energy input is always degraded to lower-quality, more dispersed, less useful energy—usually low-temperature heat that flows into the environment; you cannot break even in terms of energy quality. See *first law of thermodynamics.*

sedimentary rock Rock that forms from the accumulated products of erosion and in some cases from the compacted shells, skeletons, and other remains of dead organisms. Compare *igneous rock, metamorphic rock.* See *rock cycle.*

seed-tree cutting Removal of nearly all trees on a site in one cutting, with a few seed-producing trees left uniformly distributed to regenerate the forest. Compare *clear-cutting, selective cutting, shelterwood cutting, strip cutting.*

selective cutting Cutting of intermediate-aged, mature, or diseased trees in an uneven-aged forest stand, either singly or in small groups. This practice encourages the growth of younger trees and maintains an uneven-aged stand. Compare *clear-cutting, seed-tree cutting, shelterwood cutting, strip cutting.*

septic tank Underground tank for treating wastewater from a home in rural and suburban areas. Bacteria in the tank decompose organic wastes, and the sludge settles to the bottom of the tank. The effluent flows out of the tank into the ground through a field of drainpipes.

sexual reproduction Reproduction in organisms that produce offspring by combining sex cells or *gametes* (such as ovum and sperm) from both parents. It produces offspring that have combinations of traits from their parents. Compare *asexual reproduction.*

shale oil Slow-flowing, dark brown, heavy oil obtained when kerogen in oil shale is vaporized at high temperatures and then condensed. Shale oil can be refined to yield gasoline, heating oil, and other petroleum products. See *kerogen, oil shale.*

shelterbelt See *windbreak.*

shelterwood cutting Removal of mature, marketable trees through a series of partial cuttings to allow regeneration of a new stand under the partial shade of older trees, which are later removed. Typically, this is done by making two or three cuts over a decade. Compare *clear-cutting, seed-tree cutting, selective cutting, strip cutting.*

shifting cultivation Clearing a plot of ground in a forest, especially in tropical areas, and planting crops on it for a few years (typically 2–5 years) until the soil is depleted of nutrients or the plot has been invaded by a dense growth of vegetation from the surrounding forest. Then a new plot is cleared and the process is repeated. The abandoned plot cannot successfully grow crops for 10–30 years. See also *slash-and-burn cultivation.*

slash-and-burn cultivation Cutting down trees and other vegetation in a patch of forest, leaving the cut vegetation on the ground to dry, and then burning it. The ashes that are left add nutrients to the nutrient-poor soils found in most tropical forest areas. Crops are planted between tree stumps. Plots must be abandoned after a few years (typically 2–5 years) because of loss of soil fertility or invasion of vegetation from the surrounding forest. See also *shifting cultivation.*

slowly degradable pollutant Material that is slowly broken down into simpler chemicals or reduced to acceptable levels by natural physical, chemical, and biological processes. Compare *biodegradable pollutant, degradable pollutant, nondegradable pollutant.*

sludge Gooey mixture of toxic chemicals, infectious agents, and settled solids removed from wastewater at a sewage treatment plant.

smart growth Form of urban planning which recognizes that urban growth will occur but uses zoning laws and other tools to prevent sprawl, direct growth to certain areas, protect ecologically sensitive and important lands and waterways, and develop urban areas that are more environmentally sustainable and more enjoyable places to live.

smelting Process in which a desired metal is separated from the other elements in an ore mineral.

smog Originally a combination of smoke and fog but now used to describe other mixtures of pollutants in the atmosphere. See *industrial smog, photochemical smog.*

soil Complex mixture of inorganic minerals (clay, silt, pebbles, and sand), decaying organic matter, water, air, and living organisms.

soil conservation Methods used to reduce soil erosion, prevent depletion of soil nutrients, and restore nutrients previously lost by erosion, leaching, and excessive crop harvesting.

soil erosion Movement of soil components, especially topsoil, from one place to another, usually by wind, flowing water, or both. This natural process can be greatly accelerated by human activities that remove vegetation from soil.

soil horizons Horizontal zones that make up a particular mature soil. Each horizon has a distinct texture and composition that vary with different types of soils. See *soil profile.*

soil permeability Rate at which water and air move from upper to lower soil layers. Compare *porosity.*

soil porosity See *porosity.*

soil profile Cross-sectional view of the horizons in a soil. See *soil horizons.*

soil structure How the particles that make up a soil are organized and clumped together. See also *soil permeability.*

solar capital Solar energy from the sun reaching the earth. Compare *natural resources.*

solar cell See *photovoltaic cell.*

solar collector Device for collecting radiant energy from the sun and converting it into heat. See *active solar heating system, passive solar heating system.*

solar energy Direct radiant energy from the sun and a number of indirect forms of energy produced by the direct input of such radiant energy. Principal indirect forms of solar energy include wind, falling and flowing water (hydropower), and biomass (solar energy converted into chemical energy stored in the chemical bonds of organic compounds in trees and other plants).

solid waste Any unwanted or discarded material that is not a liquid or a gas. See *municipal solid waste.*

sound science Scientific data, models, theories, and laws that are widely accepted by experts in a particular field of the natural or social sciences. These results of science are very reliable. Compare *frontier science, junk science.*

spaceship-earth worldview View of the earth as a spaceship: a machine that we can understand, control, and change at will by using advanced technology. See *planetary management worldview.* Compare *environmental wisdom worldview.*

specialist species Species with a narrow ecological niche. They may be able to live in only one type of habitat, tolerate only a narrow range of climatic and other environmental conditions, or use only one type or a few types of food. Compare *generalist species.*

speciation Formation of two species from one species because of divergent natural selection in response to changes in environmental conditions; usually takes thousands of years. Compare *extinction.*

species Group of organisms that resemble one another in appearance, behavior, chemical makeup and processes, and genetic structure. Organisms that reproduce sexually are classified as members of the same species only if they can actually or potentially interbreed with one another and produce fertile offspring.

species diversity Number of different species and their relative abundances in a given area. See *biodiversity.* Compare *ecological diversity, genetic diversity.*

spoils Unwanted rock and other waste materials produced when a material is removed from the earth's surface or subsurface by mining, dredging, quarrying, and excavation.

S-shaped curve Leveling off of an exponential, J-shaped curve when a rapidly growing population exceeds the carrying capacity of its environment and ceases to grow.

stewardship worldview View that because of our superior intellect and power or because of our religious beliefs, humans have an ethical responsibility to manage and care for domesticated plants and animals and the rest of nature. Compare *environmental wisdom worldview, planetary management worldview.*

stratosphere Second layer of the atmosphere, extending about 17–48 kilometers (11–30 miles) above the earth's surface. It contains small amounts of gaseous ozone (O_3), which filters out about 95% of the incoming harmful ultraviolet (UV) radiation emitted by the sun. Compare *troposphere.*

stream Flowing body of surface water. Examples are creeks and rivers.

strip cropping Planting regular crops and close-growing plants, such as hay or nitrogen-fixing legumes, in alternating rows or bands to help reduce depletion of soil nutrients.

strip cutting A variation of clear-cutting in which a strip of trees is clear-cut along the contour of the land, with the corridor being narrow enough to allow natural regeneration within a few years. After regeneration, another strip is cut above the first, and so on. Compare *clear-cutting, seed-tree cutting, selective cutting, shelterwood cutting.*

strip mining Form of surface mining in which bulldozers, power shovels, or stripping wheels remove large chunks of the earth's surface in strips. See *area strip mining, contour strip mining, surface mining.* Compare *subsurface mining.*

subatomic particles Extremely small particles—electrons, protons, and neutrons—that make up the internal structure of atoms.

subduction zone Area in which oceanic lithosphere is carried downward (subducted) under an island arc or continent at a convergent plate boundary. A trench ordinarily forms at the boundary between the two converging plates. See *convergent plate boundary.*

subsidence Slow or rapid sinking of part of the earth's crust that is not slope-related.

subsistence farming Supplementing solar energy with energy from human labor and draft animals to produce enough food to feed oneself and family members; in good years enough food may be left over to sell or put aside for hard times. Compare *industrialized agriculture.*

subsurface mining Extraction of a metal ore or fuel resource such as coal from a deep underground deposit. Compare *surface mining.*

succession See *ecological succession, primary succession, secondary succession.*

succulent plants Plants, such as desert cacti, that survive in dry climates by having no leaves, thus reducing the loss of scarce water. They store water and use sunlight to produce the food they need in the thick, fleshy tissue of their green stems and branches. Compare *deciduous plants, evergreen plants.*

sulfur cycle Cyclic movement of sulfur in various chemical forms from the environment to organisms and then back to the environment.

superinsulated house House that is heavily insulated and extremely airtight. Typically, active or passive solar collectors are used to heat water, and an air-to-air heat exchanger prevents buildup of excessive moisture and indoor air pollutants.

surface fire   Forest fire that burns only undergrowth and leaf litter on the forest floor. Compare *crown fire, ground fire*. See *controlled burning*.

surface mining   Removing soil, subsoil, and other strata and then extracting a mineral deposit found fairly close to the earth's surface. See *area strip mining, contour strip mining, mountaintop removal, open-pit mining*. Compare *subsurface mining*.

surface runoff   Water flowing off the land into bodies of surface water. See *reliable runoff*.

surface water   Precipitation that does not infiltrate the ground or return to the atmosphere by evaporation or transpiration. See *runoff*. Compare *groundwater*.

survivorship curve   Graph showing the number of survivors in different age groups for a particular species.

sustainability   Ability of a system to survive for some specified (finite) time.

sustainable agriculture   Method of growing crops and raising livestock based on organic fertilizers, soil conservation, water conservation, biological pest control, and minimal use of nonrenewable fossil fuel energy.

sustainable development   See *environmentally sustainable economic development*.

sustainable living   Taking no more potentially renewable resources from the natural world than can be replenished naturally and not overloading the capacity of the environment to cleanse and renew itself by natural processes.

sustainable society   A society that manages its economy and population size without doing irreparable environmental harm by overloading the planet's ability to absorb environmental insults, replenish its resources, and sustain human and other forms of life over a specified period, usually hundreds to thousands of years. During this period, the society satisfies the needs of its people without depleting natural resources and thereby jeopardizing the prospects of current and future generations of humans and other species.

sustainable yield (sustained yield)   Highest rate at which a potentially renewable resource can be used without reducing its available supply throughout the world or in a particular area. See also *environmental degradation*.

synergistic interaction   Interaction of two or more factors or processes so that the combined effect is greater than the sum of their separate effects.

synfuels   Synthetic gaseous and liquid fuels produced from solid coal or sources other than natural gas or crude oil.

synthetic natural gas (SNG)   Gaseous fuel containing mostly methane produced from solid coal.

system   A set of components that function and interact in some regular and theoretically predictable manner.

tailings   Rock and other waste materials removed as impurities when waste mineral material is separated from the metal in an ore.

tar sand   See *oil sand*.

tectonic plates   Various-sized areas of the earth's lithosphere that move slowly around with the mantle's flowing asthenosphere. Most earthquakes and volcanoes occur around the boundaries of these plates. See *lithosphere, plate tectonics*.

temperature   Measure of the average speed of motion of the atoms, ions, or molecules in a substance or combination of substances at a given moment. Compare *heat*.

temperature inversion   Layer of dense, cool air trapped under a layer of less dense, warm air. It prevents upward-flowing air currents from developing. In a prolonged inversion, air pollution in the trapped layer may build up to harmful levels.

teratogen   Chemical, ionizing agent, or virus that causes birth defects. Compare *carcinogen, mutagen*.

terracing   Planting crops on a long, steep slope that has been converted into a series of broad, nearly level terraces with short vertical drops from one to another that run along the contour of the land to retain water and reduce soil erosion.

terrestrial   Pertaining to land. Compare *aquatic*.

tertiary (higher-level) consumers   Animals that feed on animal-eating animals. They feed at high trophic levels in food chains and webs. Examples include hawks, lions, bass, and sharks. Compare *detritivore, primary consumer, secondary consumer*.

tertiary sewage treatment   See *advanced sewage treatment*.

theory of evolution   Widely accepted scientific idea that all life forms developed from earlier life forms. Although this theory conflicts with the creation stories of many religions, it is the way biologists explain how life has changed over the past 3.6–3.8 billion years and why it is so diverse today.

thermal inversion   See *temperature inversion*.

threatened species   A wild species that is still abundant in its natural range but is likely to become endangered because of a decline in numbers. Compare *endangered species*.

threshold effect   The harmful or fatal effect of a small change in environmental conditions that exceeds the limit of tolerance of an organism or population of a species. See *law of tolerance*.

throughput   Rate of flow of matter, energy, or information through a system. Compare *input, output*.

throwaway society   See *high-throughput economy*.

tolerance limits   Minimum and maximum limits for physical conditions (such as temperature) and concentrations of chemical substances beyond which no members of a particular species can survive. See *law of tolerance*.

total fertility rate (TFR)   Estimate of the average number of children who will be born alive to a woman during her lifetime if she passes through all her childbearing years (ages 15–44) conforming to age-specific fertility rates of a given year. More simply, it is an estimate of the average number of children a woman will have during her childbearing years.

toxic chemical   See *poison, carcinogen, hazardous chemical, mutagen, teratogen*.

toxicity   Measure of how harmful a substance is.

toxicology   Study of the adverse effects of chemicals on health.

toxic waste   Form of hazardous waste that causes death or serious injury (such as burns, respiratory diseases, cancers, or genetic mutations). See *hazardous waste*.

toxin See *poison*.

traditional intensive agriculture   Producing enough food for a farm family's survival and perhaps a surplus that can be sold. This type of agriculture uses higher inputs of labor, fertilizer, and water than traditional subsistence agriculture. See *traditional subsistence agriculture*. Compare *industrialized agriculture*.

traditional subsistence agriculture   Production of enough crops or livestock for a farm family's survival and, in good years, a surplus to sell or put aside for hard times. Compare *industrialized agriculture, traditional intensive agriculture*.

tragedy of the commons   Depletion or degradation of a potentially renewable resource to which people have free and unmanaged access. An example is the depletion of commercially desirable fish species in the open ocean beyond areas controlled by coastal countries. See *common-property resource*.

transform fault   Area where the earth's lithospheric plates move in opposite but parallel directions along a fracture (fault) in the lithosphere. Compare *convergent plate boundary, divergent plate boundary*.

transmissible disease   A disease that is caused by living organisms (such as bacteria, viruses, and parasitic worms) and can spread from one person to another by air, water, food, or body fluids (or in some cases by insects or other organisms). Compare *nontransmissible disease*.

transpiration   Process in which water is absorbed by the root systems of plants, moves up through the plants, passes through pores (stomata) in their leaves or other parts, and evaporates into the atmosphere as water vapor.

tree farm   See *tree plantation*.

tree plantation   Site planted with one or only a few tree species in an even-aged stand. When the stand matures, it is usually harvested by clear-cutting and then replanted. These farms normally raise rapidly growing tree species for fuelwood, timber, or pulpwood. See *even-aged management*. Compare *old-growth forest, second-growth forest, uneven-aged management*.

trophic level   All organisms that are the same number of energy transfers away from the original source of energy (for example, sunlight) that enters an ecosystem. For example, all producers belong to the first trophic level, and all herbivores belong to the second trophic level in a food chain or a food web.

troposphere   Innermost layer of the atmosphere. It contains about 75% of the mass of earth's air and extends about 17 kilometers (11 miles) above sea level. Compare *stratosphere*.

true cost   See *full cost*.

undernutrition   Consuming insufficient food to meet one's minimum daily energy needs for a long enough time to cause harmful effects. Compare *malnutrition, overnutrition*.

undiscovered resources   Potential supplies of a particular mineral resource, believed to exist because of geologic knowledge and theory, although their specific locations, quality, and amounts are unknown. Compare *identified resources, reserves*.

uneven-aged management   Method of forest management in which trees of different species in a given stand are maintained at many ages and sizes to permit continuous natural regeneration. Compare *even-aged management*.

upwelling   Movement of nutrient-rich bottom water to the ocean's surface. It can occur far from shore but usually takes place along certain steep coastal areas where the surface layer of ocean water is pushed away from shore and replaced by cold, nutrient-rich bottom water.

urban area   Geographic area with a population of 2,500 or more. The number of people used in this definition may vary, with some countries setting the minimum number of people at 10,000–50,000.

urban growth   Rate of growth of an urban population. Compare *degree of urbanization*.

urbanization   See *degree of urbanization*.

urban sprawl   Growth of low-density development on the edges of cities and towns. See *smart growth*.

utilitarian value   See *instrumental value*.

volcano   Vent or fissure in the earth's surface through which magma, liquid lava, and gases are released into the environment.

warm front   The boundary between an advancing warm air mass and the cooler one it is replacing. Because warm air is less dense than cool air, an advancing warm front rises over a mass of cool air. Compare *cold front*.

water cycle   See *hydrologic cycle*.

waterlogging   Saturation of soil with irrigation water or excessive precipitation so that the water table rises close to the surface.

water pollution   Any physical or chemical change in surface water or groundwater that can harm living organisms or make water unfit for certain uses.

watershed   Land area that delivers water, sediment, and dissolved substances via small streams to a major stream (river).

water table   Upper surface of the zone of saturation, in which all available pores in the soil and rock in the earth's crust are filled with water.

watt   Unit of power, or rate at which electrical work is done. See *kilowatt*.

weather   Short-term changes in the temperature, barometric pressure, humidity, precipitation, sunshine, cloud cover, wind direction and speed, and other conditions in the troposphere at a given place and time. Compare *climate*.

weathering   Physical and chemical processes in which solid rock exposed at earth's surface is changed to separate solid particles and dissolved material, which can then be moved to another place as sediment. See *erosion*.

wetland   Land that is covered all or part of the time with salt water or fresh water, excluding streams, lakes, and the open ocean. See *coastal wetland, inland wetland*.

wilderness   Area where the earth and its community of life have not been seriously disturbed by humans and where humans are only temporary visitors.

wildlife   All free, undomesticated species. Sometimes the term is used to describe animals only.

wildlife resources   Wildlife species that have actual or potential economic value to people.

wild species   Species found in the natural environment. Compare *domesticated species*.

windbreak   Row of trees or hedges planted to partially block wind flow and reduce soil erosion on cultivated land.

wind farm   Cluster of small to medium-sized wind turbines in a windy area to capture wind energy and convert it into electrical energy.

worldview   How people think the world works and what they think their role in the world should be. See *environmental wisdom worldview, planetary management worldview, stewardship worldview*.

zone of aeration   Zone in soil that is not saturated with water and that lies above the water table. See *water table, zone of saturation*.

zone of saturation   Area where all available pores in soil and rock in the earth's crust are filled by water. See *water table, zone of aeration*.

zoning   Regulating how various parcels of land can be used.

zooplankton   Animal plankton; small floating herbivores that feed on plant plankton (phytoplankton). Compare *phytoplankton*.

INDEX

Note: Page numbers in **boldface** type indicate definitions of key terms. Page numbers followed by italicized *f*, *t*, or *b* indicate figures, tables, and boxed material.

Abiotic components in ecosystems, 41, 42*f*, 43*f*
Aboveground buildings, hazardous waste storage in, 406
Abyssal zone, ocean, 98, 103*f*
Acid deposition (acid rain), **352**
  formation of, 353*f*
  global regions affected by, 355*f*
  harmful effects of, 353–55
  human interference in sulfur cycle resulting in, 59–60
  strategies for reducing, 355, 357*f*
  in United States, 352, 354*f*, 356*f*
Acid mine drainage, 278
Acquired immune deficiency syndrome (AIDS), 330–31
Active solar heating system, **313**, 314*f*
Acute effects of toxic substances, 336
Adaptation, **67**
  evolution and, 65–67
  limits on, 69–70
Adaptive ecosystem management, 175
Adaptive radiations, 73*f*
Adaptive trait, **67**
Advanced light-water reactors (ALWRs), 305
Aerobic respiration, **44**
  carbon cycle and, 55
Aesthetic value of biodiversity, 156, 157*f*
Affluence
  potential positive effects of, 15
  resource consumption and negative effects of, 14–15
Affluenza, **14**–15
Africa
  AIDS epidemic in, and population age structure of, 331
  consumption of bushmeat in, 196
Age structure, human population, **134**–37
  effect of, on population growth, 120, 135, 136*f*
  effect of AIDS on, 331*f*
  making population and economic projections based on, 135
  of select countries, 135*f*
Agribusiness, 209
Agricultural residue, paper produced from 168

Agricultural revolution, 16
  green revolution and, 209, 210*f*, 221–22
Agriculture. *See also* Food production
  effects of climate change on, 376*f*
  environmental effects of, 219–20
  government policies on, 226–27
  industrialized (high-input), 208, 209–11
  irrigation and, 214–16, 222
  monoculture, 159, 160, 222
  plantation, 208
  polyculture, 211
  soil conservation and methods of, 216, 217*f*
  sustainable, 234–35
  traditional intensive, 208
  traditional subsistence, 208–9
  in United States, 209–11
Agroforestry, **211**, 216
AIDS (acquired immune deficiency syndrome), 330–31
Air circulation, climate and global, 80, 81*f*
Air plants, 118
Air pollutants
  effects of, on human health, 360–61
  effects of, on trees and water, 356*f*
  primary, and secondary, 347
  radon gas, 358, 359*f*
  types of, and effects, 349*t*
Air pollution, 345–66
  acid deposition and regional outdoor, 352–56
  atmospheric structure and, 346–47
  coal burning as source of, 297
  defined, **347**
  disasters involving, 350*b*
  harmful effects of, 359–61
  indoor, 357–59
  lichen as indicator of, 345
  major types and sources of outdoor, 347–48, 349*t*
  photochemical and industrial smog as, 348–52
  premature deaths from, in U.S., 361*f*
  preventing and reducing, 361–65
Alaska, oil reserves in, 293
Algae, production of hydrogen from green, 323*b*
Algal blooms, 78, 257*f*
Alien (nonnative) species, **111**. *See also* Nonnative species
Alley cropping, **211**, 216, 217*f*
Alligators as keystone species, 108
Alpine tundra, 87

Altitude, climate, biomes, and effects of, **84**, 85*f*
Ammonification, nitrogen fixation and, 55
Amphibians as indicator species, 111–12
Anderson, Ray, development of sustainable green corporation by, 394*b*
Animal(s)
  endangered (*see* Endangered species)
  gray wolf, 154
  keystone species, 108, 112
  manure from, as biofuel, 319
  sharks, 113
  zoos and aquariums for protection of endangered, 202–3
Animal manure, **216**
  as fuel, 319
Antarctic, food web in, 48*f*
Antibiotics, pathogen resistance to, 329
Aquaculture, 224, **226**
  advantages and disadvantages of, 226*f*
  sustainable, 227*f*
Aquariums, protecting endangered species in, 202–3
Aquatic life zones, **40**, 95–106
  biodiversity in, 156, 179–80
  effects of acid deposition on, 354
  food web in, 48*f*
  freshwater, 41, 42*f*, 104–6
  human impacts on, 179–80
  limiting factors at different depths in, 95
  net primary productivity of, 50*f*
  organisms living in, 95
  protecting and sustaining, 180
  range of tolerance for temperature in, 43*f*
  saltwater, 41, 96–103
  types of, 95
Aquifers, 54, **238**. *See also* Groundwater
  confined and unconfined, 239*f*
  depletion of, 246–47
Aral Sea water diversion project, and ecological disaster, 244, 245*f*
Arboretum, 202
Arctic tundra, 87, 89*f*
Area strip mining, **277**
Argentine fire ant, 194, 195*f*
Arsenic as water pollutant, 259
Artificial selection, **74**
El-Ashry, Mohamed, 248
Asian brown cloud, 351
Asthma, 359

Atmosphere, **38**, 346–47
    changes in average temperature of,
        projected and measured average, 369*f*,
        373*f*
    climate and (*see* Climate; Climate
        change)
    layers and temperature of, 347*f*
    ozone in, 383, 386–87 (*see also* Ozone
        depletion)
    stratosphere of, 38, 346–47
    troposphere of, 38, 346
Atom(s), **24**
    structure of, 24–25
Atomic number, **24**
Australia, 96*f*
Autotrophs, **43**

Background extinction, **71**
Bacon, Francis, 127
Bacteria
    antibiotic resistant, 329
    ecological role of, 36–37
    as pathogen (*see* Pathogens)
Balance of nature, 119
Bangladesh
    flooding in, 252–53
    microlending to end poverty in, 423*b*
Barrier beaches, 100, 102*f*
Barrier islands, 100
Basal cell carcinoma, 384, 385*f*
Baseline data, ecological, 62
Bathyal zone, ocean, 98*f*, 103
Bats, 67
Bees, African "killer," 111
Behavioral adaptations, 67
Behavioral strategies against predation,
    116
Benthic zones, lake, 104, 105*f*
Benthos, **95**
Bequest value of biodiversity, 156
Berry, Wendell, 177
Bhopal, India, toxic chemical release at, 402,
    403
Bicycles as transportation, 149*f*
Bioaccumulation of toxics, 197*f*, 336
Biocultural restoration, 178
Biodegradable pollutants, **27**
Biodegradable waste, composting of, 396
Biodiversity, **45–46**
    in aquatic (marine) life zones, 179–80
    ecosystems and preservation of (*see*
        Ecosystem approach to preserving bio-
        diversity)
    effects of climate change on, 376*f*
    effects of speciation on, 70–72
    extinction of (*see* Extinction of species)
    over geological time, 74*f*
    goals, strategies, and tactics for protect-
        ing, 156*f*
    human impacts on, 72–73, 155–56, 157*f*
    mountains as islands of, 94
    protecting with laws and treaties,
        198–201
    reasons for caring about, and protecting,
        156, 179
    of species (*see* Species; Species
        diversity)

    species approach to preserving (*see*
        Species approach to preserving biodi-
        versity)
    sustaining terrestrial, 177*f*
Biodiversity hot spots, 175, 176*f*
    in United States, 199*f*
Biofuels, **319**
Biogeochemical (nutrient cycles, **53**
Biological community, 35, **38**. See also
        Ecosystem(s)
    ecological succession in, 118, 119*f*, 120*f*
    human impacts on, 123–27
    keystone species in, 108
    population dynamics and carry capacity
        of, 120–23
    species diversity and ecological stability
        of, 110–11
    species diversity and niche structure in,
        109–10
    species interactions in, 114–18
    species types in, 111–14
    structure of, 109
Biological diversity. See Biodiversity
Biological evolution, **65**. *See also* Evolution
Biological extinction of species, 184. *See also*
        Extinction of species
Biological hazards, 328–34. *See also* Disease;
        Pathogens
    bioterrorism and, 333–34
    HIV and AIDS, 330–31
    malaria as case study, 331–32
    nontransmissible and transmissible dis-
        ease as, 328–29
    reducing incidence of infectious disease,
        332–33
    risk analysis of, 340–43
    tuberculosis as case study, 329–30
    viral disease, 330
Biological income, 6
Biological magnification, 197*f*, 336
Biological pest control, 232
Biological weathering, 274
Biomagnification of toxins, 197*f*, 336
Biomass, **47**
    principal types of fuel made from,
        319*f*
    producing energy from, 286, 319, 320*f*
    producing gaseous and liquid fuels from,
        320*f*, 321*f*
    productivity rates in conversion of, 49,
        50*f*
Biomass plantations, 319
Biome(s), **40**, 83–95
    along 39th parallel in U.S., 41*f*
    altitude, latitude, and, 84, 85*f*
    climate and, 81*f*, 83–84
    desert, 85, 86*f*, 87*b*
    earth's major, 83*f*
    forests and mountains, 89–95
    global air circulation and, 81*f*
    grasslands, 85–87, 88*f*, 89*f*
    precipitation, temperature, and, 84*f*
Biomimicry, 392, 393*f*
Biomining, 282
Biopharming, 74
Bioremediation, 404
Biosphere, **38**

Biosphere 2, failure of nutrient cycling in,
    412
Bioterrorism, infectious agents used for, 333,
    334*t*
Biotic components in ecosystems, 41, 42*f*,
    43*f*
Biotic potential of populations, **120**
Birds
    aesthetic value of, 157*f*
    effect of pesticides on embryos of, 21*b*
    evolutionary divergence of honeycreep-
        ers, 70*f*
    extinction of passenger pigeon, 183
    reconciliation ecology for protection of
        bluebirds, 204*b*
    resource partitioning among warblers,
        115*f*
    songbirds, 192*f*
    threatened species of, 192*f*
    threat of habitat loss to, 190–91, 192*f*
Birth rate (crude birth rate), **129**
    average, for select countries, 130*f*
    factors affecting fertility rates and, 132,
        133*f*
    population size and, 120
    reducing, by empowering women,
        138–39
Bitumen, 294
Black carbon aerosols, 375
Blair, Tony, 242
Bolivia, private ownership of freshwater
        resources in, 242
Boreal forests, 91, 93*f*, 94*f*
Botanical gardens, 202
Brain, human, 63
Brazil
    Curitiba, as sustainable city in, 152–53,
        429
    demographic factors in Nigeria, U.S.,
        and, 136*f*
Breeder nuclear fission reactors, **305**
Broadleaf deciduous trees, 91, 92*f*
Broadleaf evergreen plants, 90
Bronchitis, chronic, 359
Brown, Lester R., 235
Brown-air smog, 349
Brownfields, 392, 401
Buckyballs, 282
Buffer (acid neutralization), 353
Buffer zone concept, 174
Buildings
    hazardous waste storage in above-
        ground, 406
    naturally cooled, 313–14
    saving energy in, 285, 309–12
    sick-building syndrome, 357
    solar heated, 313*f*, 314*f*
Bureau of Land Management, U.S., 157
Burning
    biomass as fuel, 319
    solid and hazardous wastes, 398, 399*f*,
        404–6
Buses as transportation, 150*f*
    in Curitiba, Brazil, 152*f*
Bush, George W., climate change policies of,
    381
Bushmeat, 196

Business, reducing water use by, 250
Butterflies
    population of monarch, 35f, 37f
    protection against predation in, 116

Cahn, Robert, 413
California, geothermal energy in, 321
California Water Transfer Project, 243, 244f
Camouflage as defense against predation, 115, 116f
Canada
    oil reserves in, 291, 292f
    oil sand in, 294
Cancer
    lung, 359
    skin, 384, 385f
Cap-and-trade emissions program, 362–63
Capital, 6. See also Natural capital
Captive breeding, 202
Carbohydrates, 24
Carbon cycle, **55**, 56–57f
    effects of human activities on, 55
Carbon dioxide (CO₂) as greenhouse gas, 348, 369, 370t
    carbon cycle and, 55
    coal burning and emissions of, 26, 294f, 297f
    emissions of, per units of energy by various fuels, 294f
    global temperature and levels of, 370f
    increase in average concentrations of, 371f
    individual role in reducing, 382f
    photosynthesis and levels of, 375
    slowing global warming by removing and storing, 379, 380f
    storage of, in oceans, 374
Carbon monoxide as air pollutant, 349t
Carbon taxes, 380
Carcinogens, **335**
Carnivores, 115
Carrying capacity (K), **121**, 122f
    effects of exceeding, 121, 122f
    logistic growth of population and, 121f
Cars, hybrid and fuel-cell, 308, 309f, 310f.
    See also Motor vehicles
Carson, Rachel, work of, 229b
Cash crops, 208
Cell(s), 25f, **36**
Center-pivot low-pressure sprinkler, 248, 249f
Centers for Disease Control (CDC), 327, 407
Central Arizona Project, 244f
Ceramics, 283
Chain reaction, nuclear, 27, 28f
Channelization of streams, 253
Chateaubriand, Francois-Auguste-René de, 155
Chemical(s)
    analysis of, for finding minerals, 276
    hazardous (see Chemical hazards)
    inorganic, as water pollutant, 254t
    interactions of, 336
    ozone depleting, 383
    sensitivity to multiple, 336
    toxic (see Toxic chemicals)
Chemical bonds, 22

Chemical change in matter, **26**
Chemical components of ecosystems, 53
Chemical energy, 29
Chemical equations, balancing, S13
Chemical evolution, 64
Chemical formula, **24**
Chemical hazards, 328, 334–40. See also Hazardous wastes; Toxic chemicals
    effects on human immune, nervous, and endocrine systems, 335
    in homes, 402f
    persistent organic pollutants (POPs), 127, 340
    risk analysis of, 340–43
    terrorism and, 402
    toxic and hazardous, 334–35
    toxicology and assessment of, 335–40
Chemical nature of pollutants, 26
Chemical reaction, **26**
Chemical warfare as protection against predation, 116
Chemical weathering, 274
Chemosynthesis, **44**
Chernobyl nuclear power accident, radioactive pollutants from, 300–301, 345
Chesapeake Bay, pollution in, 261–62
Children
    fertility rates and factors related to, 132
    malnourished, 218
    protecting from lead poisoning, 407f
    protecting from toxic chemicals, 339
    saving from malnutrition and disease, 14f
China
    acid deposition in, 353
    basic demographic data for, 139f
    biogas production in, 320
    carbon dioxide emissions by, 371, 381–82
    coal reserves of, 296, 297
    oil imports by, 293
    paper production in, 168
    population, and population regulation in, 140
    water resources and pollution in, 237, 248, 256, 257
Chlorinated hydrocarbons, 24
Chlorofluorocarbons (CFCs)
    ozone depletion and role of, 383
    substitutes for, 386b
Chromosomes, **24**, 25f
Circle of poison, 231b
Cities. See Urban (metropolitan) areas
Civil disobedience, nonviolent, protection of forests through, 166b
Clarke, Arthur C., 389
Clean Air Acts, U.S., 361, 362
Clean Water Act, U.S., 264, 266
Clear-cutting of trees and forests, 161, 162f
    advantages and disadvantages of, 163f
Climate, 78–107
    aquatic life zones and, 95–106
    Aral Sea and, 245
    biomes and, 83–95 (see also Biome(s))
    change in (see Climate change; Global warming)
    defined, **79**
    factors affecting, 79, 80f

gases in atmosphere affecting, 81 (see also Greenhouse gases)
    global air circulation and, 80f
    global zones of, 80f
    land topography and local, 81–83
    ocean currents and winds affecting, 80–81
    solar energy and, 79
    weather and, 78, 79
    wind and, 79, 80f
Climate change, 367–83. See also Global warming
    as cause of species extinction, 197
    effects of, projected, 375–77
    factors affecting earth's average temperature and, 374–75
    human activities leading to, 370–74
    natural greenhouse effect and past, 368–70
    strategies for dealing with threat of, 378–83
    volcanoes and study of, 367
Climax community, 119
Clone, 74
Cloud cover, climate and changes in, 374–75
Clownfish, mutualism and, 117f
Coal, **296**–97
    advantages and disadvantages of, 297f
    carbon dioxide produced by burning of, 26, 294f, 297f
    converting, into gaseous and liquid fuel, 297, 298f
    formation of, 296f
    environmental impact of burning, 297f
    nuclear energy versus, 302f
Coal-burning facilities, reducing air pollution from, 363f
Coal gasification, **297**
Coal liquefaction, **297**
Coastal coniferous forests, 94
Coastal wetlands, **97**, 99f
Coastal zones, **97**
    coniferous forests in, 94
    effects of global warming on, 376f
    integrated coastal management, 180
    as marine life zone, 98f
    pollution in, 260–61, 262, 263f
    protecting, 262, 263f
    saltwater intrusions into freshwater sources of, 247f
    urbanization in, 145
Cockroaches
    as generalist species, 69b
    as opportunist species, 123f
Coevolution, **67**
Co-generation, **307**
Cohen, Michael J., 346
Cold desert, 85
Cold front, **S15**
Coloration, 67
Combined heat and power (CHP) systems, 307
Commensalism, **118**
Commercial energy, 286
    flow of, through U.S. economy, 306f
    industrialized agriculture and use of, 211f
    types of global, 286–87
    U.S. use of, 287f, 288f

Commercial inorganic fertilizer, **216**, 217
Common-property (free-access) resources, **10**–11
Community. *See* Biological community; Ecosystem(s)
Comparative risk analysis, 340*f*
Competition among species
  interspecific, 114
  reducing and avoiding, 114, 115*f*
Competitor (K-selected) species, 122, 123*f*
Compost and composting, **216**
  of biodegradable organic waste, 396
Composting toilet systems, 264
Compounds, **22**, 24
Comprehensive Environmental Response, Compensation, and Liability Act (Superfund program), U.S., 401
Concentration of pollutants, 26
Coniferous evergreen trees, 91
Coniferous forests, 91, 93*f*, 94*f*
  soils of, 52*f*
Consensus science, **22**
Conservation of energy, law of, 30
Conservation of matter, law of, **26**
Conservation-tillage farming, **216**
Constitutional democracy, 424
Contraception, methods of, 133*f*
Consumers, **44**
Consumption
  excessive, 14–15
  reducing, 391
Containers, refillable, 394–95
Continental drift, 72*f*
Continental shelf, 97, 98*f*
Contour farming, **216**, 217*f*
Contour strip mining, **277**
Controlled experiments, 21
Controlled nuclear fusion, 28
Convention on International Trade in Endangered Species (CITES), 198
Convergent plate boundaries, **272**, 274*f*
Cooling, natural, 313–14
Cooling services, leasing of, 393
Copenhagen Protocol, 386
Coral reefs, 96*f*, 100–101
  components and interactions in, 102*f*
  major threats to, 103*f*
Core, earth's, **270**, 271*f*
Costa Rica
  ecological restoration in, 178–79
  nature reserves in, 174
Cowie, Dean, 1
Cradle-to-grave management of hazardous wastes, 401
Crime, improving urban environments by reducing, 147*b*
Critical mass, **27**
Critical thinking, 3–4
Crop(s)
  cash, 208
  genetically modified, 220, 221*f*
  grain crops, 207, 217–18
  increasing world yields of, 209, 210*f*
  monoculture, 159, 160, 222
  multiple, 209
  pesticide use, and reduced loss of, 230*b*
  residues from, as biofuel, 319

Croplands, 207
  ecological and economic services provided by, 209*f*
Crop rotation, **216**
Crossbreeding to increase food production, 220
Crown fire in forests, 166*f*, 167
Crude birth rate. *See* Birth rate (crude birth rate)
Crude oil, **290**. *See also* Oil
  as pollutant, 262, 293*f*, 294*f*
  refining, 290*f*
Crust, earth's, **270**, 271*f*, 272*f*
Cultural eutrophication of lakes, **257***f*
Cultural hazards, 328
Culture, environment and human, 16–17
Curitiba, Brazil, as sustainable ecocity, 152–53, 429
Cuyahoga River, 255–56
Cyanide heap leaching, 280

Dams and reservoirs. *See also* Lakes
  for flood control, 254
  freshwater supplies and, 242, 243*f*
  hydropower at, 317*f*
Darwin, Charles, 64
DDT, bioaccumulation and biomagnification of, 197*f*
Death
  causes of human, 341*f*
  infectious disease as cause of, 329*f*, 332
  premature, 14
  tobacco use as cause of, in U.S., 327*f*
Death rate (crude death rate), 120, **129**, **130**
  average, for select countries, 130*f*
  factors affecting, 132–33
  population decline from rising, 137
Debt-for-nature swaps, 172
Decomposers, **44***f*
  in aquatic life zones, **95**
Deep-well disposal of hazardous wastes, 405*f*
Deforestation, 19, 161, **163**
  flooding caused by, 252, 253*f*
  global extent of, 163–64
  individual action to prevent, 166*b*
  tropical, 169–72
Degradable (nonpersistent) pollutants, **27**
  in water, 258
Democracy, **424**
  making environmental policies in, 424–27
Demographic transition, **137***f*–8
Demographic trap, 137
Denitrification, 56
Denmark, detoxification hazardous wastes in, 403–4
Department of Energy (DOE), U.S., 288
Depletion time, mineral resources, 280–81
Desalination, water, **247**–48
Desert(s), **85**
  ecology of temperate, 86*f*
  human impacts and degradation of, 87*b*
  soil of, 52*f*
Desertification, **213**, 214*f*, 215*f*, 223
Detoxification of hazardous wastes, 403–4
Detritus, **44**
Detrivores, **44***f*

Deuterium-tritium and deuterium-deuterium nuclear fusion reactions, 28*f*
Developed countries, **9**. *See also* names *of individual countries, e.g.* Canada; United States
  age structure of select, 135*f*
  connections between environmental problems and causes in, 15–16
  freshwater supplies in, 242
  global outlook, 9*f*
  projected population size in, 9*f*8
  water pollution in, 255–56
Developing countries, 9*f See also* names *of individual countries, e.g.* Brazil
  age structure of select, 135*f*
  connections between environmental problems and causes in, 15–16
  freshwater supplies in, 242
  global outlook for, 9*f*
  population growth projected for, 9*f*
  reducing poverty in, 422
  urban poverty in, 146–47
  water pollution in, 256
Diet, alternative human, 206
Differential reproduction, **67**
"Dirty bomb," 304
Discharge trading policy, 263
Disease
  bioterrorism and, 333, 334*t*
  deaths per year due to select, 329*f*
  emphysema and lung, 360*f*
  HIV and AIDS, 330–31
  infectious, 332–33, 334*t*
  malaria, 331–32
  pathogen resistance to antibiotics and, 329
  relationship of, to poverty and malnutrition, 14*f*
  tobacco, smoking, and, 327
  transmissible and nontransmissible, 328–29
  tuberculosis, 329–30
  viral, 330
Dissolved oxygen (DO) content, **43**
Divergent plate boundaries, **272**, 274*f*
DNA (deoxyribonucleic acid), 25*f*
Document services, 393
Dorr, Ann, 283
Dose, toxic chemical, **335**
  lethal, 337*t*
Dose-response curve, 337*f*
  nonthreshold dose-response model, 338*f*
  threshold dose-response model, 338*f*
Doubling time of population growth, **130**
Drainage basin, **104**, 106*f*, **238**
Drinking water. *See also* Fresh water
  arsenic in, 259
  bottled, 267
  lack of access to clean, 14
  quality of, 255, 266–68
  purification of, 264
  reducing disease by disinfecting, 264
Drip irrigation systems, 249*f*
Dubos, René, 6
Dunes, primary and secondary, 102*f*
Durant, Will, 270

Early conservation era of U.S. environmental history, 17
Earth
    atmosphere (see Atmosphere)
    chemical and biological evolution of, 64–65
    climate, local, and topography of, 81–83
    climate and rotating axis of, 79, 80f
    climate change (see Climate change; Global warming)
    crust and upper mantle of, 271f, 272f
    evolution and adaptation of life on, 65–70
    factors affecting temperature of, 374–75
    geologic processes of, 270, 271–74
    life-support systems, 38–40
    as ocean planet, 97f
    origins of life on, 64–65
    poles of, ozone depletion and, 383–86
    structure of, 270, 271f
    tectonic plates of, 272f, 273f, 274f
    water budget of, 238f
Earthquakes, S18
Earth tubes, 314
Easter Island, environmental degradation on, 19, 122
Ebola virus, 330
Ecocities, 151
    Curitiba, Brazil, as, 152–53, 429
Ecoindustrial revolution, 392, 393f
Eco-labeling programs, 420f, 421
Ecological economists, 413, 414f
Ecological efficiency, 47
Ecological extinction of species, 184
Ecological footprint
    human, on earth's surface, S2–S5
    human, in North America, S6–S7f
    natural capital use and degradation and, 10
    per capita, 11
    of U.S., Netherlands, and India, 11f
Ecological niche, 67–70
    of alligators, 108
    of bird species in wetlands, 68–69f
    fundamental, and realized, 67, 114
    generalist and specialist species and, 68–69
    of rocky and sandy shores, 100, 101f, 102f
    structure of, in biological communities, 109–10
    in tropical rain forests, 91f
Ecological restoration, 177–79
    in Costa Rica, 178–79
    of San Pedro River, Arizona, 178f
Ecological services, 6, 7f
    provided by croplands, 209f
    provided by forests, 160f
    provided by freshwater systems, 104f
    provided by marine (saltwater) systems, 97f
    provided by rivers, 243f
Ecological stability, species diversity and, 110–11. See also Sustainability
Ecological succession, 118–20
    predictability of, 119
    primary, 118, 119f
    secondary, 118, 119f
Ecologists, methods of, 60–62

Ecology, 6, 35, 36–38
    populations, communities, ecosystems, and, 37–38 (see also Biological community; Ecosystem(s); Population(s)
    research in, 60–62
    species and, 36–37 (see also Species)
Economically depleted resource, 12, 280
Economic development, 9
    demographic transition and, 137–38
    environmentally sustainable, 17, 416f
    good and bad news about (trade-offs) of, 10f
    reduced birth rates linked to, 137–38
    urbanization and, 144
Economic growth, 9
Economic incentives, improving environmental quality using, 416–21
Economic indicators, 416–17
Economic resources, 413, 414f
Economics, 413–23
    economic resources and, 413
    economic views of natural capital and sustainability, 413–15
    energy resources and, 325, 326f
    environmentally sustainable, 17–18
    environmental problems and, 9, 13–15
    forests in American and factors of global, 168–69
    globalization, 14, 168–69
    growth and development, 8–9
    improving environmental quality using, 416–21
    of mineral resource supplies, 281
    of nuclear energy, 304–5
    poverty and, 13–14, 421–23
    recycling, reuse, and factor of, 398
    reducing air pollution through, 362–63
    shift to services, 392–93
    transition to environmentally sustainable, 17
Economic services
    provided by croplands, 209f
    provided by forests, 160f
    provided by freshwater systems, 104f
    provided by saltwater (marine) systems, 97f
Economic systems, 413
Economy(ies), high-throughput (high-waste), 32, 33f. See also Economics
Ecosystem(s), 35–62. See also Biological community
    abiotic and biotic components of, 41, 42f, 43f
    adaptive management of, 175
    aquatic, 40–41 (see also Aquatic life zones)
    biodiversity and, 45–46 (see also Biodiversity)
    biological components of, 43–45
    biomes (terrestrial), 40–41 (see also Biome(s); Terrestrial ecosystems)
    defined, 38
    as earth's life-support systems, 38–40
    ecology and, 36 (see also Ecology)
    energy flow in, 46–49
    factors limiting population growth in, 43
    feeding (trophic) levels in, 46, 47f
    food chains and food webs in, 46, 47f, 48f

    human modification of natural, 123–25
    importance of soils in, 50–53
    matter cycling in, 39f, 53–60
    methods of learning about, 60–62
    populations, communities, and ecosystems, 37–38
    role of insects in, 35
    roles species play in, 36–37
Ecosystem approach to preserving biodiversity, 154–82
    aquatic systems, 179–80, 181f
    ecological restoration and, 177–79
    establishing priorities for, 181–82
    forest management and sustenance as, 159–65
    forest resources and management in United States as, 165–69
    human impacts on biodiversity and, 155–56, 157f
    national parks, 172–73
    nature reserves, 173–77
    public lands in United States as, 157–59
    reintroduced species, 154
    tropical deforestation and adverse effects on, 169–72
Egg pulling, 202
Einstein, Albert, 3, 22, 410
Electrical energy, 29
Electricity, generation of
    from burning solid biomass, 319, 320f
    decentralized system for producing of, 324f
    fuels from biomass, 320–21
    geothermal energy and, 321, 322f
    solar cells and, 315–16
    solar generation of high-temperature, 314–15
    tides, waves and dams (flowing water), 317
    wind and, 317–18
Electromagnetic radiation, 29
Electromagnetic spectrum, 29f
Electronic waste (e-waste), 390
Electrons (e), 24
Elements, 22
Elephants, reduction in ranges/habitat for, 191f
Emissions trading policy, 362–63
Endangered species, 184
    protecting with international treaties and laws, 198–201
    reintroduction of gray wolf, 154
    sanctuaries for, 202–3
    white ukari, 46f
Endangered Species Act of 1973, U.S., 154, 198–201
    accomplishments of, 201b
    arguments for strengthening, 201
    arguments for weakening, 200–201
    protection of marine animals with, 200
    scope of, 198–200
Endemic species, 190
Endocrine system, chemical hazards to, 335
Energy, 29–32. See also Energy resources
    defined, 29
    flow (see Energy flow)
    kinetic, and potential, 29

Energy (cont'd)
  laws governing changes in, 30–32
  matter and, 32–34
  one-way flow of high-quality, and life on earth, 39f
  quality of, 30
  sustainability and economies of, 32–34
Energy content, 29
Energy efficiency, **31**, 306–12
  of common devices, 306–7
  energy waste and need for, 306–7
  in homes and offices, 285, 309–12
  in industry, 307
  in transportation, 307–9
Energy flow
  in ecosystems, 46–49
  pyramid of, 47, 49f
  to and from sun, 39f, 40f, 45f
Energy productivity, **31**
Energy quality, **30**, 31f
Energy resources, 12, 276, 285–326
  best alternative, 324–25
  biomass as, 319–21
  coal as nonrenewable, 296–98
  economics and use of, 325
  efficiency as, 285, 306–12 (see also Energy efficiency)
  evaluating, 286–90
  geothermal as renewable, 321, 322f
  hydrogen as, 322–23
  hydropower as renewable, 317
  incineration of solid wastes for production of, 398, 399f
  natural gas as nonrenewable, 295–96
  net energy and, 289
  nuclear energy as nonrenewable, 298–306
  renewable, 285
  renewable, for production of heat and electricity, 312–21
  oil as nonrenewable, 290–95
  solar as renewable, 312–16
  strategies for developing sustainable, 324–26
  wind as renewable, 317–18
Energy taxes, 380
Energy waste, 306–7, 312
  reducing, 307f, 326f
England, private ownership of freshwater resources in, 242
Environment, 6
  human impact on (see Human impact on environment)
  problems in (see Environmental problems)
Environmental audits, 428
Environmental citizenship, 435
Environmental degradation, **10**. See also Environmental problems; Natural capital degradation
  on Easter Island, 19, 122
Environmental economists, 414–15
Environmental ethics, 7, **431**
  genetic engineering and, 75–76
  ownership of freshwater sources, 241–42
  sustainable living and (see Sustainable living)
Environmental groups
  on campuses, 428–29

grassroot, 410–10, 427–28
  international, 430–31
  mainline, 427–28
  successes of, 429
Environmental indicators, 191, 416
Environmentalism, **6**
Environmental leadership, 425
Environmental literacy, 433–34
Environmentally sustainable economic development, **17**–18, 416f
Environmentally sustainable economy (eco-economy), 415
  poverty and transition to, 422, 423f
  transition to, 416–21
Environmentally sustainable society, **8**
Environmental movement
  grassroots groups in, 409–10, 427–28
  mainstream groups in, 427–28
Environmental pessimists, 17
Environmental policy, **424**
  democratic political systems and making of, 424
  environmental action by students and, 428–29
  environmental groups and, 427–28
  environmental leadership and, 425
  global, 430–31
  individual's influence on, 424, 425f, 426f
  life cycle, 428f
  principles of, 424
  solutions for developing environmentally sustainable, 429–30
  in United States, 425–27 (see also United States environmental laws)
Environmental problems, 5–18
  author's study of, 1
  biodiversity loss (see Extinction of species)
  causes and connections among, 13–16
  economic growth and development, and, 8–9 (see also Economics)
  food and (see Food; Food production)
  human cultural changes and sustainability issues, 16–17
  human population growth and, 5, 8 (see Human population growth)
  key, and basic causes of, 13f, 14f
  pollution as, 12–13 (see also Air pollution; Water pollution)
  poverty as key, 13–14, 421 (see also Poverty)
  resources and, 9–12
  sustainability of present course and, 17–18
  sustainable living and, 6–8 (see also Sustainability; Sustainable living)
  waste and (see Waste)
Environmental protection
  establishing policy for, in U.S., 424–30
  U.S. history and eras of, 16, S8–S11
  U.S. legislation and (see United States environmental laws)
Environmental Protection Agency (EPA), 288, 350
Environmental quality, improving, using economic incentives, 416–21
Environmental resistance, **121**

Environmental revolution, components of, 435–36
Environmental science, **6**
  reasons for studying, 1
Environmental tax reform, 419
Environmental wisdom worldview, 432f, **433**
Environmental worldview(s), **431**–33
  frontier, 17
  human-centered, 431–32
  life-centered, 432–33
Epidemiological studies on toxic chemicals, 337–38
Epiphytes, 118
Erosion, **273**. See also Soil erosion
Estuary, 97, 98f
Ethanol as fuel, 320f
Ethics. See Environmental ethics
Euphotic zone, 95
  ocean, 98f, 103
Eutrophication, lake, **257**
Even-age forest management, **160**, 161f
Evergreen coniferous forests, 91
  ecosystem components and interactions in, 93f
  human impact and degradation of, 94f
Evolution, 63–77
  adaptation and, 65–67
  biological, 65
  coevolution, 67
  ecological niches, adaptation, and, 67–70
  future of, 74–76
  of humans, 63
  macroevolution, 65
  microevolution and, 65–67
  misconceptions about, 70
  origins of life, 64–65
  overview of biological, 66f
  of species, and biodiversity, 70–73
  theory of, 65
Evolutionary divergence of species, 68, 70f
Existence value of biodiversity, 156
Exotic species, threat of, 196–97
Experiments, scientific, **20**
Exponential growth, **5**, 121
  human population growth and, 5f, 8, 129–30
  overshoot, and population crash, 122f
External costs, 417
Extinction of species, 71, 184–88
  background, 71
  causes of, 189–97
  characteristics of being prone to, 188f
  effects of human activities on, 5, 156, 185–88
  endangered and threatened species and, 184, 186–87f
  estimating rates of, 184–85
  habitat loss, degradation, and fragmentation as cause of, 189–90
  hunting and poaching as cause of, 195–96
  killing predators and pests as cause of, 196
  laws and treaties to prevent, 198–201
  mass, 71, 72, 73f
  nonnative species as cause of, 192–95
  passenger pigeon as case study of, 183

role of climate change and pollution in, 197

role of market for exotic pets and plants in, 196–97

sanctuary approach to preventing, 202–3

species endangered by (*see* Endangered species)

species threatened with, 108, 184

three types of, 184

trade in exotic pets and plants as cause of, 196–97

Extinction spasm, 184

Extrinsic value of biodiversity, 156

Exxon Valdez oil spill, 262

Family planning, **138**

empowering women for better, 138–39

Fast-growing crops, paper production from, 168, 169*f*

Feedlots, 208

Fertility, human, **131**

factors affecting birth rates and, 132, 133*f*

reduced, and population declines, 136

replacement level, 131

total rate of (TFR), 131

in United States, 131–32

Fertilizers, organic and inorganic, 216, 217

Field research, ecological, 60, 61*f*

Financial income, 6

Fire in forests, 165–67

reducing damage of, 167

First law of thermodynamics, **30**

Fish

aquaculture of, 224, 226, 227*f*

biodiversity of, and threats to, 156

as food source, 207–8

government-subsidized fleets for harvesting, 226

harvesting of, 224, 225*f*

harvest yields, 225f

Fisheries, **224**

managing, 180, 181*f*

oceanic, 180, 181*f*, 207

overfishing and habitat degradation of, 224–25

Fish farming, **226**

Fish ranching, **226**

Flesh flower (*Rafflesia*), 46

Flooding, 251–54

in Bangladesh, 252–53

causes and effects of, 252, 253*f*

reducing risk of, 253–54

Flood irrigation, 248

Floodplains, 105, 106*f*, **251**

Bangladesh, 252–53

Food. *See also* Food production

alternative protein sources, 206

bushmeat as, 196

genetically engineered and modified, 75*f*, 76, 220, 221*f*

grain, 217–18

human nutrition, malnutrition, undernutrition, and, 218, 219

human overnutrition and, 219

important plant and animal sources of, 207–8

micronutrients in human 218–19

Food and Drug Administration (FDA), 231

Food chains, **46, 47***f*

bioaccumulation and bio-magnification of toxins in, 197*f*

Food production, 207. *See also* Food

environmental effects of, 219, 220*f*

global grain production, 217, 218*f*

green revolution and, 209, 210*f*, 221–22

human nutrition and, 217–20

increasing, 209, 220–27

industrial, and traditional forms of, 208–9

industrial, in United States, 209–11

low-input traditional, 211

methods of, 207–11

perennial polyculture and, 211, 222

protecting with pest management, 227–34

soil conservation and, 216–17

soil erosion and degradation resulting from, 211–16

sustainable, 234–35

world's principal types of, 208*f*

Food Quality Protection Act (FQPA) of 1996, 231

Food web, **47**

in Antarctic, 48*f*

industrial ecosystem modeled as, 393*f*

Forest(s), **89–95**

boreal (coniferous), 89, 91, 93*f*, 94

building roads into, 160, 161*f*, 171

deforestation (*see* Deforestation)

ecological and economic services provided by, 160*f*

effect of climate change on, 376*f*, 377*f*

fires in, 165–67

harvest methods in, 160–61, 162*f*

W. Maathai and Kenyan reforestation movement, 172*b*

management systems for, 160, 161*f*, 168

managing for sustainability, 164–65

sustainable management of, 164–65

temperate deciduous, 89, 92*f*

tropical rain forest, 89, 90*f*, 91*f* (*see also* Tropical forests; Tropical rain forests)

types of, 159–60

in United States, 156, 157, 165–69

Forestry, sustainable, 164*f*

Forest Stewardship Council (FSC), 165

Fossil fuels, 55. *See also* Coal; Natural gas; Oil

air pollution and, 347

carbon cycle and, 55

global warming and burning of, 56*f*

Fossils, **64**, 65*f*

Foundation species, **113–14**

Free-access (common-property) resources, **10–11**

French, Hilary F., 365

Fresh water

availability of, 237, 238*f*

desalination and production of, 247–48

ecological disasters involving sources of, 244–45

global uses of, 239

groundwater, 238, 239*f*, 246–47

increasing supplies of, 242–43

ownership and management of, 241–42

resources of, in U.S., 239, 240*f*

shortages of, 240–41

transfers of, advantages/disadvantages, 243–44

Freshwater life zones, 41, 42*f*, 95, **104**–6

ecological and economic services of, 104*f*

human impact on, 106

inland wetlands, 105–6

lakes, 104

streams and rivers, 104–5, 106*f*

Frogs as indicator species, 111, 112*f*

Frontier environmental worldview, **17**

Frontier era of U.S. environmental history, 17

Frontier science, **22**

Frost wedging, 273, 274

Fuel cells, 285, 309

hydrogen, 322

Fuel cell vehicles, 309

Fuels

biofuels, 319–21

carbon dioxide emissions, by select, 294*f*

coal, 296–97

ethanol and methanol as, 320*f*, 321*f*

natural gas, 295–96

oil as, 290–95

real cost of gasoline, 308*b*

synthetic, 296, 297*f*

Fuelwood shortage, 286. *See also* Biomass

Full cost pricing, 415, 417–18

Fundamental niche, **67**

Fungi

lethal effects of, on coral reefs, 78

mycorrhizae, 117*f*, 118

Gandhi, Mahatma, 435

Gangue, 278

Garbage. *See* Solid waste

Gases, atmospheric, 81. *See also* Greenhouse gases

Gasoline

real environmental cost of, 308*b*

real price of, in U.S. years 1920-2005, 308*f*

Gene(s), **24**, 25*f*

Gene banks, 202

Gene pool, **65**

Generalist species, 68–69

cockroaches as, 69*b*

General Mining Law of 1872, U.S., 269

Gene splicing, **74**

Genetically modified food, 75, 76*f*, 220, 221*f*

Genetically modified organisms (GMOs), **74**, 75*f*

Genetic diversity, 37, 38*f*. *See also* Genetic variability

Genetic engineering, 74, 75*f*, 76*f*

ethics and, 75–76

increasing crop yields with, 220–21

protecting food plants against pests with, 232*f*

Genetic information, preserving species, 188

Genetic makeup and effects of toxic chemicals, 335–36

Genetic variability, 65, 67. *See also* Genetic diversity

Genuine progress indicator (GPI), 416–17

Geographic information systems (GIS), 60, 61*f*

Geographic isolation, speciation and, **70**, 71*f*
Geology, **269**–84
    continental drift, 72*f*
    earth structure, 270, 271*f*
    geologic processes, 271–74
    mineral resources, 269, 276–83
    minerals, rocks, and rock cycle, 274–76
    soil conservation, 216–17
    soil erosion, 211–16
    soil resources, **50**–53
Geothermal energy, **321**
    advantages/disadvantages of, 322*f*
Geothermal exchange (geoexchange), 321
Giant pandas, 68
Glaciers and sea ice, global warming and, 371, 372*f*
Global cooling, 368
Globalization, 14
    effects on American forests and, 168–69
Global outlook, development, population size, and, 8*f*
Global security, environmental policy and, 430
Global sustainability movement, 428
Global warming, 368. *See also* Climate change
    approaches to dealing with problem of, 378
    current levels of, 369*f*
    current responses to, 381–82
    difficulties in responding to threat of, 378
    government role in reducing threat of, 380
    greenhouse gases and, 81, 370*t*, 371*f*
    Kyoto Protocol on greenhouse gas emissions and, 380–81
    nuclear energy as means of reducing, 304
    ocean currents and, 374*f*
    preparing for, 382–83
    projected effects of, 375, 376*f*, 377*f*
    reasons for concern about, 373–74
    removing and storing CO_2 as solution for, 379, 380*f*
    scientific consensus on, 371–73
    solutions for reducing threat of, 378–79
    tropospheric warming, 370–71
    volcanoes and study of, 367
Gorilla, hunting and poaching threat to, 196*f*
Government(s)
    agricultural policies of, 226–27
    democratic, 424
    environmental laws (*see* United States environmental laws)
    environmental policy and, 424–30
    fishing fleets subsidized by, 226
    role of, in reducing threat of climate change, 380
    subsidies from, for improving environmental quality, 418
Grain crops, 207–8
    global production of, 217, 218*f*
    increasing yields of, 220–21
Grameen Bank, Bangladesh, 423*b*
Grasshopper effect, smog formation as, 352
Grasslands, **85**–88
    arctic (polar), 87, 89*f*

components and interactions in temperate, 88*f*
    grazing and browsing animals in, 87
    human impact and degradation of, 88*f*, 89*f*
    soil of, 52*f*
Grassroots environmental groups, 409–10
Gravimeter, 276
Gravity, 39
Gray-air smog, 351
Green Belt Movement, Kenya, 172*b*
Greenhouse effect, **81**, 368–69
    global warming and, 81
    natural, 82*f*, 368–69
Greenhouse gases, **39**, 369
    Kyoto Protocol and, 380–81
    major, from human activities, 370*t*
    reducing, 381, 382*f*
Green manure, **216**
Green revolution, **209**, 210*f*
    expanding, 221–22
Green Seal eco-labeling, 420*f*
Green taxes and fees, 415, 418–19
    advantages and disadvantages of, 418*f*
Gripping mechanisms, 67
Gross domestic product (GDP), **9**
    as measure of economic and environmental health, 416, 417*f*
    net primary productivity versus, 49*f*
Gross primary productivity (GPP), **49**
Ground fires in forests, 167
Groundwater, **238**, 239*f*
    arsenic concentrations in U.S., 259
    pollution of, 258–59
    preventing and cleaning up, 259*f*
    sources of contamination in, 258*f*, 260*f*
    withdrawal of, 246*f*, 247*f*
Gulf of Mexico
    oil drilling in, 292*f*
    oxygen-depleted areas of, 261*f*
Gulf Stream, 80*f*, 81
Gully erosion, 212
Gut inhabitant mutualism, 118

Habitat, **38**, **67**
    effects of climate change on wildlife, 197
    Endangered Species Act and protection of critical, 198–99
    of fish, 224–25
    loss, degradation, and fragmentation of, as threat to species, 109, 189–92
Habitat conservation plans (HCPs), 199
Habitat islands, 190
Hansen, James, 367
Hardin, Garrett, 10
Hardrock minerals, 269
Hawken, Paul G., 6, 394, 429
Hazard(s)
    biological, 328–34
    chemical, 334–35
    disease as (*see* Disease)
    major types of, 328
    natural (earthquakes, volcanoes), S18
    risk analysis of, 340–43
    toxicology and assessment of chemical, 335–40
Hazardous air pollutants (HAPs), 361

Hazardous chemicals, **334**. *See also* Chemical hazards; Hazardous wastes
Hazardous wastes, **401**–6
    Bhopal, India, disaster involving, 403
    burning and burying, 404–6
    detoxifying, 403–4
    grassroots action on managing, 409–10
    harmful chemicals used in homes and production of, 402*f*
    individual's role in reducing and managing, 406*f*
    international treaty on transfer of, 410
    land disposal of, 406*f*
    at Love Canal, New York, 388
    phytoremediation of, 404*f*, 405*f*
    priorities for dealing with, 403*f*
    science and politics of, 401
    surface impoundments of, 405*f*, 406
    terrorism and release of, 402
    toxic metals as, 406–9
Health. *See* Human health
Healthy Forests Initiative of 2003, U.S., 167
Heat, **29**
    high temperature industrial, 289*f*
    as water pollutant, 254*t*
Heating
    energy ratios for space, 289f
    geothermal, 321
    solar, 313–14
    water, 312
Heliostats, 314
Hepatitis B virus, 330
Herbivores, 115
Heterotrophs, **44**
High (weather), **S15**
High-level radioactive wastes, 302–3
    in U.S., 303
High-quality energy, **30**, 31*f*
    one-way flow of, and life on earth, 39*f*
High-quality matter, 25*f*, **26**
High-throughput (high-waste) economy, **32**, 33*f*
High-waste society, 390
    shift from things to services in, 392–93
    waste management strategy for, 390–92
Hill, Julia Butterfly, 166
HIPPO (Habitat destruction and fragmentation, Invasive (alien) species, Population growth, Pollution, and Overharvesting), **190**
Hirschorn, John, 343
HIV (human immunodeficiency virus), 330, 331
Homes
    designing for energy efficiency, 285, 309
    hazardous chemicals in, 402*f*
    naturally cooled, 313–14
    reducing energy waste in, 310–12
    reducing water use and waste in, 250
    solar-heated, 313*f*, 314*f*
Hooker Chemicals and Plastics Company, Love Canal disaster and, 388
Hormonally-active agents (HAAs), 335
Hormone(s), controlling insect pests by disrupting, 232, 233*f*

Human(s)
cultural epochs of, and sustainability, 16–17
deaths (*see* Death)
evolution of, 63
population (*see* Human population; Human population growth)
respiratory system of, 359, 360
Human-centered environmental world-views, 431–32
Human health, 327–44. *See also* Human nutrition
biological hazards (disease) affecting, 328–34
chemical hazards affecting, 334–3
effects of acid deposition on, 353
effects of air pollution on, 347, 349*t*, 359–61
effects of climate change on, 376*f*
effects of ozone depletion and ultraviolet radiation on, 384–86
pesticides and, 232
poisons and (*see* Poison)
poverty and, 14, 15*f*, 341
risk, risk assessment, and hazards to, 328
risk analysis of hazards to, 340–43
tobacco, smoking, and damage to, 327
toxicology and assessment of chemical hazards to, 335–40
in urban areas, 144
Human immunodeficiency virus (HIV), 330
Human impact on environment, 123–27
air pollution as (*see* Air pollution)
on aquatic environments, 179–80
on biodiversity, 72–73, 155–56, 157*f*, 179–80
on carbon cycle, 55, 56*f*
climate change, global warming, and, 81, 370–74
on deserts, 87*f*
effects on ecosystems, 123–25
epochs of human cultural change and, 16
in forests, 94*f*
on freshwater systems, 106
on grasslands, 88*f*, 89*f*
on marine (saltwater) systems, 103
mining mineral resources and, 278, 279*f*, 280
in mountains, 95*f*
on nitrogen cycle, 56–57
pesticide use and, 230–31
on phosphorus cycle, 58–59
species extinction and, 5, 185–88
on sulfur cycle, 59–60
sustainability principles and, 126–27
unforeseen consequences of, 125*b*
urbanization as, 144–47
on water cycle, 54–55
Human nutrition
malnutrition, 14, 15*f*, 218, 219
micronutrients and, 218–19
overnutrition, 219
reducing malnutrition and disease in children, 219
undernutrition, 218, 219

Human population, 128–53
age structure of, 134–37, 331*f*
case studies, India and China, 139–40
dieback of, 122
distribution of, and urbanization, 140–43
effects of climate change on, 376*f*
effects in decline of, from reduced fertility, 136
effects of excessive, 128
factors affecting size of, 120, 129–34
factors affecting urbanization of, 140–43
growth of (*see* Human population growth)
improving urban livability and sustainability for, 150–53
influencing size of, 137–39
transportation, urban development, and, 148–50
urban resource and environmental problems linked to, 144–47
Human population growth, 8
birth and death rates, and, 129, 130*f*
carrying capacity and, 122
case studies on slowing, 139–40
doubling time, 130–31
effect of age structure on, 120, 135, 136*f*
environmental problems linked to, 128
exponential growth of, 5*f*, 8, 130
factors in reducing, 137–39
global, 130
poverty and, 14 (*see also* Poverty)
reasons for, 130–31
Human resources/human capital, **413**, 414*f*
Human respiratory system, 359, 360*f*
Humus, **50**
Hunting, species extinction from, 195–96
Hutchinson, G. Evelyn, 36
Hydrocarbons, 24
Hydrogen as fuel, 285, 322–23
advantages/disadvantages of, 322*f*
production of, from green algae, 323*b*
Hydrological poverty, 240
Hydrologic (water) cycle, **53**, 54*f*
availability of fresh water and, 237
effects of human activities on, 54–55
Hydropower, 317
Hydrosphere, **38**
Hydrothermal ore deposits, 283*f*

Ice, 38
Identified nonrenewable mineral resources, **276**
Igneous rock, **274**
Immigration and emigration
effects of, on population size, 120
into United States, 134*f*
Immigration Reform and Control Act of 1986, 134
Immune system, chemical hazards to human, 335
Incineration
of biomass, 319
of solid and hazardous wastes, 398, 399*f*, 404–6
Income, global distribution of, 5, 421*f*
India
basic demographic data for, 139*f*

Bhopal toxic-chemical disaster in, 403
ecological footprint of, 11*f*
industrial smog in, 351*f*
population and population regulation in, 139–40
poverty in, 421*f*
Indicator species, **111**
for air pollution, 345
for water pollution, 111, 112*f*
Individuals, influence of
on energy use and energy waste, 326*f*
environmental policy and role of, 424, 425*f*, 426*f*
preventing deforestation, 166*b*
protecting species, 204*f*
reducing CO_2 emissions, 382*f*
reducing exposure to indoor air pollution, 365*f*
reducing exposure to UV radiation, 386*f*
reducing production of hazardous wastes, 406
reducing production of solid wastes, 391*f*
reducing water use and waste, 251*f*
reusing resources, 395*f*
Indoor air pollution, 347, 357–59
individual role in reducing exposure to, 365*f*
radon gas, 358, 359*f*
strategies for reducing, 364*f*, 365*f*
types and sources of, 358*f*
Industrial ecosystem modeled as food web, 393*f*
Industrial forestry, 160, 161*f*
Industrialized (high-input) agriculture, **208**
in United States, 209–11
Industrial-medical revolution, 16
Industrial smog, 350, **351**
factors influencing formation of, 351–52
Industry
brownfields and abandoned, 392, 401
ecoindustrial revolution, 392, 393*f*
net energy ratios for high-temperature heat used in, 289*f*
reducing water waste in, 250
saving energy use and money in, 307
toxic-release disasters, 402, 403
Infant mortality rate, **133**
Infectious agents. *See* Pathogens
Infectious disease, **329**
antibiotic resistance in pathogens and effects on, 329
bioterrorism and use of, 333, 334*t*
deadliest, 329*f*
HIV and AIDS, 330–31
malaria, 331–32
reducing incidence of, 332, 333*f*
tuberculosis, 329–30
viral, 330
Infiltration, water, **51**
Influenza (flu), 330
Information-globalization revolution, 16
Inland wetlands, freshwater, **105**–6
Innovation-friendly regulation, 419
Inorganic compounds, **24**
as water pollutant, 254*t*
Inorganic fertilizers, 217
Input pollution control, **12**

Insects
    African "killer" bees, 111
    chemical pesticides for control of, 228–31
    cockroaches, 69b, 123f
    disrupting reproductive hormones of, 232, 233f
    ecological role of, 35
    as pests, 227
    sex attractants (pheromones) for, 232
    threats from introduced nonnative, 193f, 195f
Instrumental value of biodiversity, **156**, 188
Integrated pest management (IPM), **233–34**
Intercropping, **211**
Interface Corporation, 394b
Intergovernmental Panel on Climate Change (IPCC), 370, 371, 373, 377
Internal combustion engine, energy waste in, 306
Internal costs, 417
Internet, accuracy of information on, 3
Interplanting, **211**
Interspecific competition, **114**
Intertidal zones, **100**, 101f
Intrinsic rate of increase (r), population growth, **120**
Intrinsic value of biodiversity, **156**
Introduced species, 192–94. *See also* Nonnative species
    reducing threats from, 195
    role of accidentally introduced, 194–95
    role of deliberately introduced, 111, 194
Invasive species, **111**. *See also* Nonnative species
Ions, **24**
Ireland, 122
Irrigation
    increasing amount of land under, 222
    major systems of, 249f
    reducing water waste in, 248–49, 250f
    soil salinization and waterlogging caused by, 215f
Island(s), barrier, 100
Islands of biodiversity, mountains as, 94
Isotopes, **25**

Janzen, Daniel, 178
Jet streams, S15
J-shaped curve of human population growth, 5f
Junk science, **22**

Kenaf, 168, 169f
Kennedy, Donald, 368
Kenya, Green Belt Movement in, 172b
Keystone species, **112**
    alligators as, 108
Key terms, 2
Kinetic energy, **29**
K-selected species, **122**, 123f
Kudzu vine, 194f
Kyoto Protocol, 380–81

Laboratory research in ecology, 61
Lacey Act of 1900, U.S., 198

Lakes, 96f, **104**. *See also* Dams and reservoirs
    Aral Sea water ecological disaster, 244–45f
    eutrophication and cultural eutrophication in, 257–58
    water pollution in, 256–57
    zones of, 104, 105f
Land
    availability of, transportation, and urbanization, 148
    burying solid wastes in, 399–401, 406f
    impact of mining on, 278, 279f, 280
    increasing amount of cultivated, 222
    increasing amount of irrigated, 222
Land degradation, soil erosion and, **211**–12
Las Vegas, Nevada, urban sprawl of, 141f
Latitude, climate, biomes, and effects of, **84**, 85f
Law, environmental, 419
    protecting endangered species using, 198–201
    in U.S. (*see* United States environmental laws)
Law of conservation of energy, **30**
Law of conservation of matter, **26**
Law of progressive simplification, 14
Laws of thermodynamics, 30–32
Law of tolerance, **41**
Leachate, landfills and, 400
Leaching, soil, **51**
Lead
    as air pollutant, 349t
    health threat of, 406–7
    preventing poisoning of children by, 407f
Learning skills, 1–4
    improving, 1–3
Leewenhock, Antoine van, 77
Lemur, threatened species of, 192f
Leopold, Aldo, 17, 182, 184, 188, 207
Levees to control flooding, 253
Lichens, as indicators of air pollution, 345
Life
    levels of organization in, 23f
    origins of, 64–65
Life-centered environmental worldviews, 432–33
Life expectancy, **133**
Life-support systems, Earth's, 38–40
    as favorable for life, 40
    interconnected factors in, 39
    major components of, 38–39
    solar energy and, 39–40
Lighting, energy efficient, 307f
Light pollution, 146
Light-water reactors (LWR), nuclear, 298, 299f
Limiting factor, **43**
Limiting factor principle, **43**
Limnetic zones, lakes, 104, 105f
Liquified natural gas (LNG), **295**
Liquified petroleum gas (LPG), **296**
Lithosphere, 38, **271**
Littoral zones, lakes, 104, 105f
Living sustainably. *See* Sustainable living
Living systems. *See also* Systems
    precautionary principle for protecting, 126

second law of thermodynamics in, 32f
    sustainability of, 126–27
Loams, 52
Local extinction of species, 184
Logging in forests, 160–61, 162f, 163f
    in U.S., 168f
Logistic growth, population, **121**f
London, air pollution disasters in, 350b
Los Angeles, California, air pollution problems, 352
Love Canal, New York, hazardous waste dump in, 388
Lovins, Amory, 285
Low (weather), **S15**
Low-energy precision application (LEPA) sprinklers, 248, 249f
Low-quality energy, **30**, 31f
Low-quality matter, 25f, **26**
Low-throughput (low-waste) economy, **33**f, 34
Lungs, emphysema in human, 360f

Maathai, Wangari, reforestation work of, 172b
MacArthur, Robert H., 79
Macroevolution, **65**
Madagascar, 98f
Magma, 321
Magnetometer, 276
Malaria, 331–32
    global distribution, 331f
    life cycle of, 332f
    spraying pesticides to control insects carrying, 125b, 228
Maldives islands, threat of rising sea levels to, 377f
Malnutrition, **218**
    relationship of, to poverty and disease, 14, 15f, 218
Mangroves, 96f
Mantle, earth's, **270**, 271f, 272f
Manufactured resources/manufactured capital, **413**, 414f
Marine life zones, 41, 95, 96–103. *See also* Ocean(s); Saltwater (marine) life zones
    ecological and economic services provided by, 97f
    human impact on, 179–80
    sustaining biodiversity in, 180, 181f
Marine protected areas (MPAs), 180
Marine species, protection of, 200
Market-based economic system, 413
Market forces. *See also* Economics
Marsh, saltwater, 99f
Marsh, George Perkins, 109
Mass depletion, **72**
Mass extinctions, 71, 72, 73f
Mass number, **24**
Mass transit rail as transportation, 149f
Material(s)
    effects of air pollution on, 353
    substitute, for mineral resources, 283
Material efficiency, **26**
Materials-recovery facilities (MRFs), 396
Matter, **22**–28
    atoms, ions, and compounds, 24–25
    energy and, 32–34

law of conservation of, 26
levels of organization in, 23*f*
nuclear changes in, 27–28
physical and chemical changes in, 26
pollutants and characteristics of, 26–27
quality of, 25–26
types of, 22–23
Matter cycling in ecosystems, 39*f*, 45*f*, 53–60
    biogeochemical (nutrient) cycles and, 53
    carbon cycle, 55, 56–57*f*
    nitrogen cycle, 55–57, 58*f*
    phosphorus cycle, 57–58, 59*f*
    sulfur cycle, 59, 60*f*
    water (hydrologic) cycle, 53–55, 237
Matter quality, **25***f*–26
Matter-recycling-and-reuse economy, **32**, 33*f*
Mead, Margaret, 18, 387
Measure, units of, S1
Meat production, 222–23
    conversion of grain into, 224*f*
    environmental consequences of, 223
    sustainable, 223–24
Mechanical energy, 29
Mechanical weathering, 273
Megacities, 141*f*
Megalopolis, 141*f*
Megareserves, 174
Melanoma, 385*f*
Mercury
    cycling of, in aquatic environments, 408*f*
    health threat of, to humans, 113, 408
    preventing and controlling inputs of, 409*f*
Metallic mineral resources, 12, 276
Metamorphic rock, **275**
Metastasis, **335**
Methane, increase in average atmospheric concentrations of, 371*f*, 375
Methanol as fuel, 320, 321*f*
Metropolitan areas. *See* Urban (metropolitan) areas
Mexico City, environmental problems of, 147, 351*f*
Microbes, 36
Microevolution, **65**
    natural selection of, 65–67
Microirrigation, 249*f*
Microlending (microfinance), 423*b*
Microlivestock, 206
Micronutrients, human nutrition and, 218–19
Micropower systems, 324*f*
Middle East, water conflicts in, 236
Mimicry, 67
Mineral(s), **274**–76
Mineral resources, 274
    classification of nonrenewable, 275, 276*f*
    finding, removing, and processing nonrenewable, 276–77
    life cycle of, 279*f*
    limits on extraction and use of, 279–80
    ocean, 282, 283*f*
    rocks and minerals, 274
    rock types and rock cycle, 274, 275*f*
    substitutes for, 283

supplies of, 280–83
    U.S. laws on, 269
Minimum-tillage farming, 216
Mining
    environmental impact of, 279*f*
    possible limits to, 279–80
    surface, and subsurface, 276–77
    on U.S. public lands, 276
Mining Law of 1872, U.S., 159
Minnesota Mining and Manufacturing Company (3M), 392
Model(s), **20**
Molecular economy, 282
Molina, Mario, 383
Monoculture, 222
    tree plantations as, 159, 160*f*
Monomers, 24
Montague, Peter, 264
Montreal Protocol, 386
Mopani food, 206
Moths, 67
Motor vehicles
    advantages and disadvantages of, 148–49
    alternatives to, 149–50
    fuel cell, 309
    hybrid gas-electric, 308, 309*f*, 310*f*
    reducing air pollution from, 364*f*
    reducing use of, 149
    in United States, 148
Mountains, 94–95
    ecological importance of, 94
    human impact on, 95*f*
    rain shadow effect and, 82*f*
Mountaintop removal (mining), **277**, 278*f*
Mount Pinatubo, Philippines, 78, 367
Mouse (mice), genetically engineered, 76*f*
Muir, John, 106
Multiple chemical sensitivity (MCS), 336
Multiple cropping, 209
Municipal solid waste (MSW), **389**
    burning, 398, 399*f*
    land disposal of, 399, 400*f*, 401*f*
    recycling, 396–97
Mutagens, **65**, **335**
Mutations, **65**
Mutualism, **117***f*–18
Mycorrhizae fungi, 117*f*, 118
Myers, Norman, 430

Nanotechnology, producing new materials using, 282
National ambient air quality standards (NAAAQS), 361
National Forest System, U.S., 157
    logging in, 168*f*
    management of, 168
National Marine Fisheries Service (NMFS), 198
National Park Service, U.S., 157
National Park System, U.S., 157–58, 172–73
    solutions for sustaining and expanding, 173*f*
National Resource Lands, U.S., 157
National security and global environmental policy, 430

National Wilderness Preservation System, U.S., 157, 176–77
National Wildlife Refuges, U.S., 157, 202
Native species, **111**
Natural capital, 1, **6**, 7*f*, **413**
    endangered and threatened, 186–87*f*, 188*f*, 192*f*, 193*f*, 195*f*, 199*f*, 200*f*
    levels of organization of matter in nature as, 23*f*
    lost, 185*f*
    natural resources and ecological services as, 6, 7*f*
    restoration of, 154
    restoration of (*see* Natural capital restoration)
    use and degradation of, 10*f* (*see also* Natural capital degradation)
Natural capital degradation, 7*f*
    acid deposition, 354, 355*f*, 356*f*, 357*f*
    air pollution, 348*f*, 356*f*
    biodiversity, 155*f*, 156, 157*f*, 179*f*
    carbon dioxide emissions as, 294*f*
    coral reefs, 103*f*
    deforestation and forest degradation, 164*f*, 170*f*, 253*f*
    depletion and extinction of wild species as, 190*f*, 191*f*, 194*f*, 195*f*, 196*f*, 197*f*, 199*f*
    deserts, 87*f*
    on Easter Island, 19
    food production as cause of, 220*f*
    forests, 94*f*, 161*f*
    freshwater sources, 241*f*, 245*f*, 246*f*, 247*f*, 253*f*
    human impact on natural ecosystems, 125*f*
    human interference in biogeochemical cycles as, 56*f*, 58*f*
    major environmental problems and, 12*f*, 13*f*, 14*f*
    marine (saltwater) ecosystems, 103*f*, 179*f*
    mineral resources, 269*f*, 277*f*, 278*f*, 279*f*, 280*f*
    in mountains, 95
    overgrazed rangelands, 223*f*
    point-source air pollution as, 11*f*
    poverty and, 14*f*
    soil erosion and degradation, 212, 213*f*, 214*f*, 215*f*
    urban areas as unsustainable, 145*f*
    urban sprawl as, 144*f*
    of water resources, 241*f*, 245*f*, 246*f*, 247*f*, 253*f*, 257*f*, 258*f*, 260*f*, 261*f*
Natural capital restoration, gray wolf reintroduction as, 154
Natural gas, **295**
    advantages and disadvantages of, 295, 296*f*
    liquefied petroleum gas and liquefied natural, 295
    synthetic, **297**
Natural greenhouse effect, **40**
Natural radioactive decay, **27**
Natural recharge, **238**, 239*f*
Natural resources, 6, 7*f*, **413**. *See also* Natural capital; Resource(s)

Natural Resources Conservation Service, U.S., 213
Natural selection, 65
  conditions required for, 67
  microevolution by, 65–67
Natural services, 7. *See also* Ecological services; Economic services
Nature Conservancy, work of, 173
Nature reserves, 173–77
  adaptive ecosystem management of, 175
  as biodiversity hotspots, 175, 176*f*
  biosphere reserves, 174*f*
  in Costa Rica, 174
  protecting, from human exploitation, 173
  wilderness as, 175–77
Nekton, **95**
Neoclassical economists, 413
Nervous system, chemical hazards to human, 335
Net energy, **289**
Net energy ratios for select types of energy systems, 289*f*
Netherlands, ecological footprint of, 11*f*
Net primary productivity (NPP), **49**
  estimated, of select ecosystems, 50*f*
  gross primary productivity versus, 49*f*
  in oceans, 95–96
  species diversity and, 110
Neurotoxins, **335**
Neutrons (n), **24**
New urbanism tools, 151*f*
Niche, **67**. *See also* Ecological niche
Nicotine as addictive drug, 327
Nigeria, demographic factors in Brazil, U.S., and, 136*f*
Nile River, 236
Nitrate fertilizers, 58*f*
Nitrates (NO$_3$-) as water pollutant, 57
Nitrogen cycle, 55–56, 58*f*
  effects of human activities on, 56–57
Nitrogen dioxide (NO$_2$) as air pollutant, 349*t*
Nitrogen fixation, 55
Nitrogen oxide (N$_2$O) as air pollutant, 56–57
Nitrous oxide, increase in average concentrations atmospheric, 371*f*
Noise pollution, 146*f*
Nondegradable pollutants, **27**
Nonnative species, **111**, 192–95
  characteristics of invader, 195*f*
  examples of, 193*f*
  reducing threat of, 195
Nonpoint sources of pollution, **12**
  reducing, in surface water, 263
  water pollution, **255**
Nonprofit nongovernmental organizations (NGOs), 424
  making of environmental policy and role of, 427
Nonrenewable energy sources, 286–87
  coal, 296–97
  crude oil, 290–95
  natural gas, 295–96
  nuclear energy, 298–306
Nonrenewable mineral resources, 275–76
  commercial energy production dependent on, 286, 287*f*

finding, removing, and processing, 276–77
  life cycle of, 278, 279*f*
Nonrenewable resources, 10, **12**. *See also* Nonrenewable energy sources; Nonrenewable mineral resources
Nonthreshold dose-response model, 338*f*
Nontransmissible disease, 328–29
Nonuse values of biodiversity, 156
North America
  fossil fuel deposits in, 292*f*
  human ecological footprint in, S6–S7*f*
North Pole, disappearing Arctic sea ice at, 372*f*
No-till farming, 216
Nuclear change, 27–27
Nuclear energy, 29, 298–306
  advantages/disadvantages of conventional, 301*f*, 302
  brief history of, 298–300
  Chernobyl nuclear disaster, 300–301
  coal energy versus, 302*f*
  economics of, 304–5
  fuel cycle, 290, 300*f*, 301*f*
  future of, in U.S., 305–6
  net energy ratio, 290
  nuclear fusion, 305
  reactors and production of (*see* Nuclear fission reactors)
  reducing oil dependency, and global warming with, 304
  threat of "dirty" bombs and, 304
  waste produced by, 302–3
Nuclear fission, **27**, 28*f*
Nuclear fission reactors, 298, 299*f*
  breeder, 305
  decommissioning old plants, 303–4
  energy waste in, 306
  functioning of, 299*f*
  safety risk of, 341
Nuclear fuel cycle, 290, 300*f*
  advantages and disadvantages of, 301*f*
Nuclear fusion, **27**, 28*f*, 305
Nuclear Regulatory Commission (NRC), 301
Nuclear weapons, 304
Nucleic acids, 24
Nucleus (atom), **24**, 25*f*
Nutrient cycles (biogeochemical cycles), **53**
Nutrient overload causing cultural eutrophication, 257–58
Nutrition. *See* Human nutrition

Ocean(s), 41, 96*f*. *See also* Saltwater (marine) life zones
  as carbon-storage and heat reservoir, 374
  coastal zones, 97, 98*f*
  currents of, CO$_2$ and heat storage, 374*f*
  currents of, wind, and climate, 80*f*, 81
  earth's, 97*f*
  human impact on biodiversity in, 179*f*, 180
  mineral deposits in, 282, 283*f*
  net primary productivity of, 95–96
  open sea, 98*f*, 103
  reasons for caring for, 96–97

sea ice, 371, 372*f*
  sea levels, 371, 376, 377*f*
Oceanic fisheries, 180, 181*f*, 207
Ocean pollution, 259–62
  Chesapeake Bay as case study, 261–62
  coastal areas and, 260–61, 262, 263*f*
  oil as pollutant in, 262
  tolerance levels for, 259–60
Ogallala Aquifer, depletion of, 246*f*
Oil, 290–95
  advantages and disadvantages of conventional, 293*f*, 294*f*
  crude (petroleum), 290–91
  environmental impact of, 293, 294*f*
  history of Age of Oil, S20
  nuclear energy as means of reducing dependency on, 304
  oil shale and tar sands as sources of, 294–95
  as pollutant, 262, 293*f*, 294*f*
  price of, in U.S., 292*f*
  projected global reserves of, 291–93
  refining crude, 290*f*
  U.S. supplies of, 291, 292*f*
Oil and fat soluble toxins, 336
Oil sand, **294**, 295*f*
Oil shale, 294
Oil spills, 262
Old-growth forests, **159**
Oligotrophic lake, 257
Omnivores, **44**
Open dumps, **399**
Open-pit mining, 277
Open sea, **103**
  life zones of, 98*f*, 103
Opportunist (r-selected) species, 122, 123*f*
Orangutans, 157*f*
Ore, **276**
  life-cycle, 278, 279*f*
  mining lower-grade, 281–82
Organic agriculture, **234**
Organic compounds, **24**
  as water pollutant, 254*t*
Organic fertilizers, **216**
Organisms, **36**. *See also names of specific organisms, eg.* Animal(s)
  in aquatic life zones, 95
  genetically modified, 75*f*, 76, 220–21
Organization of Petroleum Exporting Countries (OPEC), 291
Orr, David W., 151
Other nonrenewable mineral resources, **276**
Outdoor air pollution
  acid deposition and regional, 352–56
  effects of, 349*t*
  global warming and, 375
  photochemical and industrial smog as, 348–52
  public health disasters involving, 350*b*
  reducing, 363–64, 365*f*
  types and sources of, 347–48, 349*t*
Output pollution control, **12**
Overburden, **277**
Overfishing, **224**, 226
Overgrazing, **223**

Overnutrition, **219**
Oxygen-demanding wastes as water pollutant, 254*t*
Oxygen-depleted coastal zones, 261*f*
Oxygen sag curve, 255, 256*f*
Ozone
 photochemical (O_3), 349*t*
 protecting stratospheric, 386–87
Ozone depletion, 383–86
 causes of, 383
 at earth's poles, 383
 effects of, 383–84
 reasons for concern about, 384
 threats of, to human health, 384–86
Ozone hole, 384

Paper, alternative plants for production of, 168, 169*f*
Paracelsus, 328
Parasitism, **117**
Parks, national, 172–73
Partial zero-emission vehicles (PZEVs), 363
Passenger pigeon, extinction of, 183
Passive solar heating system, **313**
Pathogens. *See also* Disease
 antibiotic resistance in, 329
 bioterrorism and use of, 333, 334*t*
 in water, 254*t*
Pay-as-you-throw (PAUT) system, 397, 398
Per capita ecological footprint, **11**
Per capita GDP, **9, 416**
Permafrost, 38, **87**, 371
Perpetual resource, **10**
Persistence
 of pollutants, 26, **27**
 of toxins, 336
Persistent organic pollutants (POPs), 127, 340, 410
Pest(s), 35, **227**
 alternative methods of controlling, 232, 234*f*
 chemical control of, 228–31
 ideal, 229–30
 integrated management of, 233–34
 natural control of, 227
 species extinction threats in control of, 196
Pesticides, **228**
 advantages of modern synthetic, 228–29
 alternatives to, 232, 233*f*
 bioaccumulation and biomagnification of DDT, 197*f*
 R. Carson on, 229*b*
 circle of poison and, 231*b*
 disadvantages of modern synthetic, 230–31
 effects of, on human health, 335
 evaluating effectiveness of, 230*b*
 experiment on effects of, on bird embryos, 21*b*
 ideal, 229–30
 integrated pest management as alternative to, 233–34
 regulation of, in U.S., 231
 unforeseen consequences of using, 125*b*
Petrochemicals, **291**
Petroleum, **290**. *See also* Oil

Pets, exotic, 196–97
pH
 scale, 53*f*
 of soil, 53
Pharmaceutical products from wild species, 189*f*
Phosphorus cycle, **57**–58, 59*f*
 effect of human activities on, 58–59
Photochemical smog, **348**–50
 formation of, 351–52
Photosynthesis, **43**
 in aquatic life zones, 95
 CO_2 levels and, 55, 375
Photovoltaic (PV) cells, **315***f*
Physical change in matter, **26**
Physical hazards, 328
Physical weathering, 273
Physiological adaptations, 67
Phytoplankton, 43, 95
Phytoremediation, 404*f*
 advantages and disadvantages, 405*f*
Phytostabilization, 404
Pimentel, David, 230
Planetary management worldview, **431**, 432*f*
Plankton, 113
Plant(s). *See also* Trees
 decorative, as threat to native species, 196–97
 effects of acid deposition on, 354–55, 356*f*
 genetic engineering of, 75*f*, 220–21
 most important food, 207–8
Plantation agriculture, **208**
Plant nutrients as water pollutant, 254*t*
Plasmodium, 331–32
Plastics, recycling, 397
Plate tectonics, 271*f*, 272*f*, 273*f*, 274*f*
Poaching of wildlife, 195–96
Point sources of pollution, **12**
 air pollution, 11*f*
 reducing, in water, 263–64, 266
 water pollution, **255**
Poison, **337**. *See also* Toxic chemicals
 lethal doses of, 337*t*
 as protection against predation, 116
 toxicity ratings and lethal doses for, 337*t*
Polar grasslands, 87, 89*f*
Poles, ozone thinning over earth's, 383–86
Politics, **424**–30
 developing environmentally sustainable, 429–30
 environmental action and, 428–29
 environmental groups and, 427–28, 429
 environmental laws in U.S. and, 427*f*
 environmental leadership and, 425
 environmental policy, global, 430–31
 environmental policy, U.S., 425–27
 global warming and, 380–82
 grassroots action on managing solid and hazardous wastes, 409–10
 individual's role in making environmental policy, 424–25
 knowledge about toxic chemicals and, 339
 nuclear power in U.S., 305–6
 ownership and management of freshwater resources, 241–42

 principles of making environmental policy, 424
 problems of democracy for solving environmental problems, 424
 ozone depletion and, 386–87
 recycling, reuse, and factor of, 398
 reducing water pollution, 266
 U.S. chemical plants and threat of terrorism, 402
Pollutants, chemical nature, concentration, and persistence of, 26–27
Polluter-pays principle, 401
Pollution, **12**–13
 air (*see* Air pollution)
 economic solutions for (*see* Economics)
 noise, 146
 point, and nonpoint sources of, 12 (*see also* Nonpoint sources of pollution; Point sources of pollution)
 solutions to problem of, 12–13
 tradable, 419–20
 water (*see* Water pollution)
Pollution cleanup, **12**
Pollution prevention, **12**
 integrated pest management as, 233–34
 toxic chemicals, 339–40
Polyculture, **211**, 222
Polymers, 24
Polyps, coral, 100
Polyvarietal cultivation, **211**
Population(s), **37**
 biotic potential of, 120
 dynamics of, 20–23
 growth (*see* Population growth; Human population growth)
Population, human. *See* Human population; Human population growth
Population change, **129**. *See also* Human population growth
Population distribution, human, 140–43
Population dynamics, 120–23
 carrying capacity (K) and, 121–22
 factors affecting population size, 120
 population growth, 120–21
 reproductive patterns and, 122–23
Population ecology, human. *See* Human population
Population growth, 8–9
 exponential and logistic, 5*f*, 8, 121*f*, 122*f*, 129–30
 human (*see* Human population growth)
 intrinsic rate of increase (r), 120
 limits on, 120–21
Population size. *See also* Population growth
 carrying capacity and, **121**, 122*f*
 factors affecting human, 137–39
 factors and variables affecting, 120
Potential energy, **29**–30
Poverty, **421**
 distribution of world's wealth and, 5, **421**–22
 harmful environmental effects linked to, 5, 14
 as human health risk, 14*f*, 341

Poverty (cont'd)
microlending as solution for, 423b
nutritional deficiencies caused by, 218
relationship of malnutrition, and disease
to, 14, 15f
reducing, 422
transition to eco-economy and problem
of, 422, 423f
in urban areas, 141, 146f–47
Power tower (solar energy), 314
Prairie, **85**, 87, 88f. *See also* Grasslands
Precautionary principle, **126**–27, 410
chemical hazards and, **339**–40
species extinction and, 188
Precipitation
acid deposition, 352–56
climate and average, 84f
freshwater availability and, 240f
Predation, **114**–15. *See also* Predators; Prey
defenses against, 115–16
Predators, 114–15
defenses against, 115–16
sharks as, 113
species extinction threats to, 196
strategies of, 115
Prescribed fires in forests, 167
Prey
avoidance or defense against predators
among, 115–16
relationship of, to predators, 114–15
Pricing, full-cost, 415, 417–18
Primary air pollutants, **347**
Primary (closed loop) recycling, 396
Primary sewage treatment, **264**, 265f
Primary succession, ecological, **118**, 119f
Primm, Stuart, 72, 185
Private ownership of freshwater resources,
241–42
Probability, risk and, 328
Producers, **43**f
Profundal zones, lake, 104, 105f
Protective cover, 67
Proteins, 24
bush meat as source of, 196
conversion of, grain into animal, 224f
Protons (p), **24**
Public lands, U.S., 157–58
description of, 157, 158f
management of, 159
mineral resources and mining on, 269
Pyramid of energy flow, **47**, 49f

Radiation-measuring equipment, 276
Radioactive "dirty" bomb, 304
Radioactive isotopes (radioisotopes), **27**
Radioactive materials
sites contaminated by, 303
as water pollutant, 254t
Radioactive wastes, 302–3
in U.S., 303
Radioisotopes, **27**
Rain shadow effect, 82f
Rangelands, 207
overgrazing of, 223
Range of tolerance, **41**
for temperature in aquatic ecosystem, 43f
Rapid rail as transportation, 150f

Realized niche, **67**
Reconciliation ecology, **203**–4
protecting bluebirds with, 204b
Recreational pleasure, 188–89
Recycling, **12**, 396–98
advantages and disadvantages of, 397f,
398
composting as, 396
economic and political factors affecting,
398
encouraging reuse and, 398
materials-recovery facilities, 396
plastics, 397
solid waste, 396–97
two types of, 396
Red Sea, 96f
Red tides, 78
Redwood sorrel, 118
Refinery, oil, 290
Regulation, environmental, 419. *See also*
United States environmental laws
Reliable runoff, **238**
Remote sensing, 60
Renewable energy resources, 285, 286
providing heat and electricity with,
312–21
Renewable resources, 9, **10**
energy (*see* Renewable energy
resources)
Replacement-level fertility, **131**
Reproduction, differential, **67**
Reproductive isolation, speciation and, **71**
Reproductive patterns, 122–23
competitor (K-selected), 122, 123f
opportunist (r-selected), 122f, 123f
R-selected species, **122**f
generalized characteristics of, 123f
Reserves
of nonrenewable mineral resources, **276**
of oil, 291–93
Reservoirs. *See* Dams and reservoirs
Resource(s), **9**–12
affluence and consumption of, 14–15
common-property (free-access), 10–11
human ecological footprint affecting, 11
nonrenewable, 10–12
nonrenewable mineral, 275–83
perpetual and renewable, 10
solid waste as, 390
Resource Conservation and Recovery Act
(RCRA), U.S., 401
Resource partitioning, 114, 115f
Resource productivity (material efficiency),
26
Resource-use permits, 419–20
Respiratory system, human, 359, 360f
Reuse, 12, 394–95
advantages and disadvantages of, 394
economic and political factors affecting,
398
individual role in, 395f
using refillable containers, 394–95
Revkin, Andrew C., 22
Rhinoceros, reduction in range of black,
191f
Rice, 210f
Richter sale, S18

Risk, **328**
greatest, to humans, 341f, 342f
perception of, 342–43
probability and, 328
Risk analysis, 340–43
comparative, 340f
estimating risks, 340
estimating risks from technologies,
341–42
improving, 343
perceiving risks, 342–43
Risk assessment, **328**, 340
Risk management, **328**, 340
Rivers and streams, 96f, 104–5
channelization of, 253
ecological services of, 243f
levees along, 253
pollution in, 255–56
reducing flooding of, 252f, 253–54
stresses on world, 241f
in U.S., 240f
zones in downhill flow of, 106f
Road-building in forests, 160, 161f, 171
Rock, **274**
Rock cycle, **275**f
Rocky Mountain Institute, 285
Rocky shores, 100, 101f
Rojstaczer, Stuart, 49
Roofing, energy efficient, 309
Roosevelt, Theodore, 202
Rosenblatt, Roger, 248
Rowland, Sherwood, 383
Rule of 70, **130**
Runoff
reliable, 238
surface, 54, 104, 238

Salinity, **43**
in aquatic life zones, 95
Salinization, **214**, 215f
preventing and cleanup of, 215f
Salt marsh ecosystem, 99f
Saltwater desalination, 247–48
Saltwater intrusions, 247f
Saltwater (marine) life zones, 95, 96–103
coastal zones, 97
coral reefs, 100–101, 102f, 103f
earth's major, 96f
ecological and economic services provided by, 97f
ecological importance of oceans, 96–97
ecological niches on rocky and sandy
shores, 100, 101f, 102f
estuaries, coastal wetlands, and mangrove swamps of, 97–100
food web in Antarctic, 48f
impact of human activities on, 103
open sea, biological zones in, 103
reasons for being concerned about
oceans, 96–97
salt marsh ecosystem, 99f
Sanctuary approach to protecting wild
species, 202–3
Sandy shores, 100, 102f
Sanitary landfill, **399**–400
advantages/disadvantages of, 401f
secure, for hazardous wastes, 406f

SARS (severe acute respiratory syndrome), 330
Saudi Arabia, oil reserves in, 291, 292, 293
Savanna, 87
Science, **20**–22
  consensus science versus frontier, 22
  junk, 22
  methods and process of, 20–21*f*
Scientific Certification Systems (SCS), 165
Scientific data, **20**
Scientific hypotheses, **20**, 21
Scientific (natural) law, **21**
Scientific methods, **21**–22
Scientific theory, **20**
Seal(s), protection of marine, 200*f*
Sea levels, climate change and rising, 371, 376, 377*f*
Secondary air pollutants, **347**
Secondary recycling (downcycling), 396
Secondary sewage treatment, **264**, 265*f*
Secondary succession, ecological, **118**, 120*f*
Second-growth forests, **159**
Second law of thermodynamics, **30**–32
Secure hazardous waste landfills, 406*f*
Sedimentary rock, **275**
Sediment as water pollutant, 254*t*
Seed banks, 202
Seed-tree cutting of trees, 161, 162*f*
Seismic surveys, 276
Sele, Peter, 153
Selective cutting of trees, 161, 162*f*
Sense of place, 434
Septic tank, **264***f*
Sewage, treatment of, 264, 265*f*, 266
Sewage treatment plants, 264, 265*f*
Shale oil, **294**
Sharks, 113
Shellfish, harvesting, 224
Shelterbelts, **216**
Shelterwood cutting of trees, 161, 162*f*
Shiva, Vandana, 222
Shop-'til-you drop virus, 14
Short-grass prairies, 87
Sick-building syndrome, 357
Silent Spring (Carson), 229*b*
Simple living, 434–35
Skin cancer, 384, 385*f*
Slowly degradable (persistent) pollutants, **27**
Smart growth concept, **150**, 151*f*
Smelting, **278**, 279*f*
Smog, 348–52
Smoking, health hazards of, 327
Socolow, Robert, 57
Soil(s), **50**–53
  absorption of CO_2 by, 379–80*f*
  erosion of (*see* Soil erosion)
  impact of food production on, 211–16
  layers in mature, 50–53
  pH, 53*f*
Soil conservation, **216**–17
  inorganic commercial fertilizers and, 217
  methods of, 216, 217*f*
  organic fertilizers and, 216–17
  tillage methods for, 216, 217*f*

Soil erosion, 211–16
  causes of, 211–12
  defined, **212**
  desertification and, 213–14
  extent of global, 212, 213*f*
  extent of U.S., 213
  reducing (*see* Soil conservation)
  salinization, and waterlogging as, 214–16
Soil fertility, 212
Soil horizons, 50, 51*f*
Soil profile, **50**, 51*f*
  of principal soil types, 52*f*
Soil sequestration, 379
Solar capital, **6**, 7*f*, 39*f*, 40*f*, 286. *See also* Solar energy
Solar cells, 285, 315*f*, 316
  advantages and disadvantages of, 316*f*
Solar energy, **6**, 286, 312
  climate and, 79
  flow of energy to and from earth, 39*f*, 40*f*
  generating high-temperature heat and electricity with, 314–15
  heating houses with active and passive, 313, 314*f*
  natural greenhouse effect and, 39–40
  solar cells for production of electricity, 315–16
  sustenance of life on earth and, 39
Solar thermal plant, 314, 315*f*
Solar thermal systems, 314
Solid waste, **389**
  burning and burying, 398–401, 404–6
  grassroots action on managing, 409–10
  hazardous (*see* Hazardous waste)
  as indicator of high-waste society, 390
  municipal, 389
  priorities for managing and reducing, 391
  producing less, 390–92
  production of, in U.S., 389, 390*f*
  recycling, 396–98
  reuse of, 394–95
  shift from things to services and reduction of, 392–93
Songbirds, 192*f*
Sound science, **22**
Source zone, streams, 105, 106*f*
Space heating, net energy ratios for, 289*f*
Specialist species, **68**
  resource partitioning and, 114*f*
Speciation, **70**–71
  crisis of, 187
Species, **36**–37, 111–18
  amphibians as case study of, 111–12
  categories of, 37*f*
  characteristics of vulnerable, 188*f*
  classifying and naming, S14
  commensalisms among, 118
  competition and avoiding competition, 114, 115*f*
  defense against predation, 115–16
  diversity (*see* Species diversity)
  ecological niches of, 67–69
  evolutionary divergence of, 68, 70*f*
  evolution of, 70–71 (*see also* Speciation)
  extinction of (*see* Extinction of species)
  foundation, 113–14
  generalist and specialist, 68–69

  indicator, 111
  interactions of, 114
  keystone, 108, 112
  mutualism among, 117–18
  native, and nonnative, 111
  parasitism among, 117
  predator and prey interactions, 114–15
  protecting wild, using sanctuary approach, 202–3
  reasons for preserving wild, 188–89
  reproductive patterns, 122–23
  sharks as case study of, 113
  types of, in biological communities, 111
Species approach to preserving biodiversity, 183–205
  causes of premature extinction of, 189–97
  extinction of species and, 183, 184–88
  importance of wild species, 188–89
  legal approach to protecting wild species, 198–201
  reconciliation ecology and, 203–4
  sanctuary approach to protecting wild species, 202–3
Species-area relationship, 185
Species diversity, **109**. *See also* Biodiversity goals, strategies, and tactics for preserving, 156*f*
  niche structure in biological communities and, 109–10
Spoils, mining, **277**
Squamous cell carcinoma, 384, 385*f*
S-shaped population growth curve, 121*f*
Stegner, Wallace, 176
Stewardship worldview, **432**
Stewart, Richard B., 381
Stratosphere, **38**, **346**–47
  ozone in, 383, 386–87 (*see also* Ozone depletion)
Strawbale house, 309, 311*f*
Streams. *See* Rivers and streams
Strip cropping, **216**, 217*f*
Strip cutting of trees, 161, 162*f*
Strong, Maurice, 286
Structural adaptations, 67
Study skills, improving, 1–3
Subduction zone, 272, 274*f*
Subsidence, 278, S18
Subsurface mining, 276, 277*f*
Sulfur cycle, 59, 60*f*
  effects of human activities on, 59–60
Sulfur dioxide (SO_2) as air pollutant, 59, 349*t*
Superfund Act, U.S., 401
Superinsulated houses, 309, 310*f*
Surface fire in forests, 166*f*, 167
Surface impoundments of hazardous wastes, 405*f*, 406
Surface litter layer (O horizon), soil, 50, 51*f*
Surface mining, **276**, 277*f*
Surface runoff, 54, **104**, **238**, 239*f*
Surface water, **104**, 105*f*, 238
  preventing and reducing pollution in, 263–66
  tapping reliable supply of, 238–39
Suspended particulate matter (SPM) as air pollutant, 349*t*

Sustainability, **6**–8
  ecocity concept and, 151–53
  energy resources and, 324–26
  human cultural changes and, 16–17
  low-waste society and, 409–10
  natural capital and, 6–8, 7f
  path to, 7f
  of present course, 16–17
  principles of, 126–27
  of urban areas, 151–52
  in water resources, 250f, 251f
Sustainable (low-input) agriculture,
    234
  making transition to, 234–35
Sustainable living, 6–8, 433–36
  becoming environmental citizens and,
    435
  components of environmental revolution
    and, 435–36
  environmental literacy and, 433–34
  environmental science, environmental-
    ism, and, 6
  environmental wisdom and, 432f, 433
  learning from earth and developing sense
    of place, 434
  path to, 6–8
  simple living and, 434–35
  solutions for developing, 434f
Sustainable yield, **10**
Synfuels, 297, 298f
Synthetic natural gas (SNG), **297**, 298f
Systems
  analysis of, and ecological research, 61f,
    62
  characteristics of nature, and human-
    dominated, 124f
  earth's life support, 38–40
  living, 32f, 126–27

Taigas, 91, 93f, 94f
Tall-grass prairie, 87, 88f
Tar sand, **294**, 295f
Taxes
  green, 415, 418–19
  improving environmental quality using,
    418
  reducing threat of global warming using,
    380
Taxpayer-pays programs, 401
Technological optimists, 17
Technological systems, estimating risks
    from, 341–42
Technology transfer, 380
Tectonic plates, **271**f, 272f, 273f
  boundaries between, 272–73
  divergent and convergent boundaries of,
    272, 274f
Temperate deciduous forests, 91
  ecosystem of, 92f
  soil of, 52f
Temperate desert, 85, 86f
  ecosystem components and interactions
    in, 86f
Temperate grasslands, 87, 88f
Temperate rain forest, 94
Temperature
  climate and average, 84f

of earth, and life systems, 40
  range of tolerance for, 43f
Temperature inversions, **352**
Teratogens, **335**
Terracing, **216**, 217f
Terrestrial ecosystems. *See also* Biome(s)
  components of, 41, 42f
  net primary productivity of, 50f
Terrorism
  nuclear energy and potential, 300, 301
  release of toxic chemicals as act of, 402,
    403
  use of pathogens for purposes of, 333,
    334t
Thatcher, Margaret, 242
Theory, 20, 21
Theory of evolution, **65**
Thermal pollution, 254t
Thinking skills, 3–4
Thoreau, Henry David, 18
Threatened species, 108, **184**
Threshold dose-response model, 338f
Thumb, opposable, 63, 67
Thunderheads, S15
Tiger, reduced range of Indian, 191f
Tiger salamanders, 68
Timber
  certifying sustainably-grown, 165
  harvesting, from forests, 160–61, 162f,
    168f
Tobacco, health hazards of, 327
Todd, John, 237, 265
Tool libraries, 395
Topsoil, A horizon, 50, 51f
Tornadoes, **S16**f
Total fertility rate (TFR), **131**
  decline in, for select countries, 136
  for U.S., years 1917-2005, 131f
Toxic chemicals, **334**
  acute and chronic effects of, 336
  estimating toxicity of, 337–39
  dose of, 335, 337t
  as human health hazard, 334–35
  metals as, 406–9
  persistent organic pollutants (POPs), 127,
    340
  as poisons, 337t (*see also* Poison)
  precautionary principle applied to,
    339–40
  protecting children from, 339
  release of, by industrial plants, 402, 403
  trace levels of, 336–37
  water- and oil-soluble, 336
Toxicity, **335**–36
  case reports and epidemiological studies
    to estimate, 337–38
  estimating chemical, 337–38
  laboratory experiments for estimating
    toxicity, 338–39
  lethal, 337t
  trace levels of chemicals and, 336–37
Toxicology, 335–40
  effects of trace levels, 336–37
  estimating toxicity of chemicals, 337–39
  factors affecting toxicity of chemicals,
    336
  precautionary principle and, 339–40

Toxic Release Inventory (TRI), 361
Toxin(s), **337**
Toxin-laced mining wastes, 278
Trace levels of toxic chemicals, 336–37
Tradable pollution permits, 419–20f
Trade-offs (compromises), 7f, 8
  of aquaculture, 226f
  of bicycles and mass transit rail, 149f
  of buses and rapid rail, 150f
  of clear-cutting forests, 163f
  of coal, 297f
  of coal versus nuclear, 302f
  of conventional natural gas, 296f
  of conventional nuclear fuel cycle, 301f
  of conventional oil use, 293f
  of economic development, 10f
  of environmental taxes and fees, 418f
  of ethanol fuel, 320f
  of genetically-modified foods and crops,
    221f
  of geothermal energy, 322f
  of global efforts on environmental prob-
    lems, 431f
  of heavy oils from oil shale and oil sand,
    295f
  of hydrogen fuel, 322f
  of large-scale hydropower, 317f
  of logging in U.S. national forests, 168f
  of methanol fuel, 321f
  of phytoremediation, 405f
  of recycling, 397–98
  of sanitary landfills, 401f
  of solar cells, 316f
  of solar energy for high-temperature heat
    and electricity, 315f
  of solid biomass fuels, 320f
  of synthetic fuels, 298f
  of tradable environmental permits, 420f
  of waste disposal in deep underground
    wells or surface impoundments, 405f
  of wind power, 318f
  of withdrawing groundwater, 246f
Traditional agriculture
  global outlook for, 211
  intensive, **208**–9
  subsistence, **208**
Tragedy of the commons, **11**, 19
Transform faults, **272**, 274f
Transgenic organisms, **74**
Transition zone, streams, 105, 106f
Transmissible diseases, 328–29. *See also*
    Infectious disease
Transportation, 148–50
  alternative, 149–50
  motor vehicles, 148-49
  net energy ratios for, 289f
  saving energy in, 307–10
Treaties
  climate and global-warming related,
    380–81
  protecting endangered species with inter-
    national, 198
Tree plantations (tree farms), **159**, 160f
  management of, 161f
Trees
  alternatives to, for paper production,
    168, 169f

effects of acid deposition on, 354–55, 356f

effects of climate change on, 376, 377f

fire and, 165–67

harvesting methods, 160–61, 162f, 163f, 168f

Trenches, at converging tectonic plates, 272, 274f

Tribal era of U.S. environmental history, 16

Trophic levels in ecosystems, **46**, 47f

Tropical cyclones, **S16**, S17f

Tropical desert, 85

Tropical forests, 89. *See also* Tropical rain forests

deforestation of, 169–72

ecological restoration of, 178–79

reducing deforestation and degradation of, 171f, 172

Tropical rain forests, 89, 90

components and interactions in, 90f

plants of, 90

soil of, 52f

specialized plant and animal niches in, 91f

Troposphere, **38**, **346**

greenhouse effect and warming of, 82f, 370–71, 372f, 373

possible effects of warming in, 375, 376f

Tsunamis, S18

Tuberculosis, 329–30

Tundra, 87, 89f

Turner, Ray, eco-businessman, 386b

Turtles, protection of marine, 200f

Ultraviolet (UV) radiation, ozone depletion and human health problems caused by, 384–86

Uncontrolled nuclear fusion, 28

Undernutrition, **218**, 219

Undiscovered nonrenewable mineral resources, **276**

Uneven-aged forest management, **160**

United Nations Children's Fund (UNICEF), 219

United States

agriculture and food production in, 209–11

air pollution disasters in, 350b

baby boom generation in, 135, 136f

baby bust generation in, 135

biodiversity hot spots in, 199f

biomes along 39th parallel of, 41f

birth control methods used in, effectiveness of, 133f

Chesapeake Bay, pollution in, 261–62

commercial energy flow through economy of, 306

consumption of resources in, 128

deaths from air pollution in, 361f

deaths from tobacco use in, 327f

demographic factors in Nigeria, Brazil, and, 136f

ecological footprint of, 11f

energy consumption by fuel in, 288f

energy use and energy future of, 288

energy waste in, 306–7

environmental policy in, 425–27

eras and events in environmental history of, 16, S8–S11

fertility rates in, 131–32

forests in, 156, 165–69

freshwater resources in, 239, 240f

government subsidized fishing fleets in, 226

groundwater contamination in, 259, 260f

groundwater withdrawal in, 246f, 247

hazardous waste in, 401–2

as high-waste society, 390

immigration into, 134

Kyoto Protocol and nonparticipation of, 380–81

motor vehicle use in, 148

nuclear energy in, 299, 303–6

pesticide effectiveness, 230b

pesticide use and regulation in, 230b, 231

precipitation averages in, 240f

public lands in (*see* Public lands, U.S.)

real cost of gasoline in, 308b

social and demographic changes in, years 1900-2000, 132f

solid wastes production in, 389, 390f

urbanization in, 141f, 142–43

urban problems of, 143f, 144–50

water consumption and waste in, 248

United States environmental laws

on air quality, 350b, 361–62

on forests, 167

on hazardous wastes, 401

history of, S8–S12

legislative process for passage of, 426f

major, 427f

on mining, 269

on pesticides, 231

on protecting endangered species, 154, 198–201

on water quality, 264, 266

on wilderness preservation, 175, 176–77

U.S. Fish and Wildlife Service (USFWS), U.S., 154, 157

Urban (metropolitan) areas

crime reduction and environment in, 147b

ecocity Curitiba, Brazil, 152–53

environmental and resource problems of, 144–47

global, 141f

increasing sustainability and livability of, 150–53

megacity and megalopolis, 141f

Mexico City, 147

microclimates produced by, 82–83

poverty in, 141, 146f–47

sprawl of, 143–44

U.S., 141f, 142f, 143f

water quality in, 266–67

Urban growth

patterns of, 143

problems and challenges of, 140–41

smart growth concept applied to, 150, 151f

sustainability and, 151–52

transportation and, 148–50

Urban heat island, 146

Urbanization, 140–43

advantages of, 144

environmental disadvantages of, 144–46

fertility rates and, 132f

of poverty, 141, 146f–47

sustainability and, 150–53

transportation and, 148–50

in U.S., 141f, 142–43

Urban sprawl, **143f**, 144f

Use value of biodiversity, **156**

Van Doren, Mark, 1

Vegetation patches, flooding caused by removal of, 252, 253f

Viruses, 36

HIV, and AIDS disease, 330–31

as pathogens, 330

Vitousek, Peter, 49

Volatile organic compounds (VOCs), 348

Volcanoes, 78, 274f, **S18**–19

climate change and study of, 367

Voluntary simplicity, 435

Ward, Barbara, 6

Warm front, **S15**

Warning coloration, 116f

Waste, 388–11

achieving low-waste society, 409–10

burning and burying, 398–401

electronic (e-waste), 390

of energy, 306–7

hazardous, 401–6

oxygen-demanding, as water pollutant, 254t

preconsumer, and postconsumer, 396

producing less, 390–92

radioactive, 302–3

recycling as means of reducing, 396–98

reducing water, 248–51

reuse as means of reducing, 394–95

selling services instead of things to reduce, 392–93

solid, 389–90

toxic industrial chemical, 388

toxic metals as, 406–9

Waste-to-energy incinerator, 398, 399f

Water. *See also* Water pollution; Water resources

drinking, 255

Earth's hydrosphere, 38

efficient heating of, 312

fresh (*see* Fresh water)

groundwater, 238, 239f (*see also* Groundwater)

importance and availability of fresh, 237, 238f

producing electricity from moving, 317–18

reducing waste of, 248–51

soil erosion caused by, 212

underpricing, 248

Water conservation, 248–51

benefits of reducing waste for, 248

individual contribution to, 251f

Water conservation (cont'd)
  reducing waste in industry, homes, and
    businesses, 250
  reducing waste in irrigation, 248–49,
    250*f*
  using water more sustainably, 250–51
Water (hydrologic) cycle, **53,** 54*f*, 55, 237
Waterlogging, 215*f*, **216**
Water pollutants
  arsenic as, 259
  in groundwater, 258–59, 260*f*
  major categories of, 254*t*
  oil as, 262
Water pollution, **254**–55
  in freshwater streams, lakes, and
    aquifers, 255–59
  individual role in reducing, 268*f*
  in oceans, 259–62
  point and nonpoint sources of, 255
  preventing and reducing surface,
    263–66
  reducing, 267–68
  quality of drinking water and, 255,
    266–68, 255
  types, effects, and sources of, 254–55
Water resources
  conflicts over, in Middle East, 236
  Earth's water budget, 238*f*
  effects of climate change on, 376*f*
  excessive amounts of (flooding), 251–54
  importance of, 237–40
  increasing supply of, 242–47
  ownership of, 241–42
  polluted (*see* Water pollution)
  reducing waste of, 248–51

  shortages of, 240–41
  sustainability in future use of, 251*f*
  U.S. hot spots for, 240*f*
Watershed, **104,** 106*f*, **238**
Water-soluble toxins, 336
Water table, **238**
Water transfers, 243–44
Water vapor, 38
  as greenhouse gas, 369
Wave(s), energy transmission as, 29
Wavelength, 29
Wealth, global distribution of, 421*f*
Weather, **79, S15**–S17
  climate and, 78, 79
  climate change and possible extremes of,
    376*f*
  highs and lows, S15
  tornadoes and tropical cyclones,
    S16–S17
Weathering, **273**
Weaver, Warren, 20
West Nile virus, 330
Wetlands
  coastal, 97, 99*f*
  flooding caused by draining, 252, 254
  inland, 105–6
  protection of, 254
  treating sewage using, 265–66
Wiener, Jonathan B., 381
Wilderness, 175–77
  in United States, 175, 176–77
Wilderness Act of 1964, U.S., 175, 176
Wildlife, causes of premature extinction of,
    189–97. *See also* Extinction of species
  effects of climate change on, 197

  effects of habitat loss, degradation, and
    fragmentation on, 189–90
  poaching and smuggling of, 195–96
  reasons for protecting, 188–89
Wilson, Edward O., 72, 112, 116, 155, 185
  priorities for protecting biodiversity,
    181
Wind
  climate and global patterns of, 80*f*
  ecological effects of global, 78
  generating electricity with, 317, 318*f*
  soil erosion caused by, 212
Windbreaks, **216,** 217*f*
Windows, energy loss from, 311*f*
Wind turbines, 318*f*
Wolf, reintroduced populations of gray, 154
Women, reducing birth rates by empower-
    ing, 138–39
World Health Organization (WHO), 125,
    219, 331, 332, 347, 407
World Trade Organization (WTO), 242
Wright, Frank Lloyd, 4

Xeriscaping, 250

Yellowstone National Park, reintroduction
    of gray wolves into, 154
Yucca Mountain, Nevada, radioactive waste
    disposal at, 303

Zone of saturation, **238**
Zoos, protecting endangered species in,
    202–3
Zooxanthellae, 100